SAXON MATH™
Course 3

Student Edition

Stephen Hake

A Harcourt Achieve Imprint

www.SaxonPublishers.com
1-800-531-5015

ACKNOWLEDGEMENTS

This book was made possible by the significant contributions of many individuals and the dedicated efforts of a talented team at Harcourt Achieve.

Special thanks to:

- Melody Simmons and Chris Braun for suggestions and explanations for problem solving in Courses 1–3,

- Elizabeth Rivas and Bryon Hake for their extensive contributions to lessons and practice in Course 3,

- Sue Ellen Fealko for suggested application problems in Course 3.

The long hours and technical assistance of John and James Hake on Courses 1–3, Robert Hake on Course 3, Tom Curtis on Course 3, and Roger Phan on Course 3 were invaluable in meeting publishing deadlines. The saintly patience and unwavering support of Mary is most appreciated.

– Stephen Hake

Staff Credits

Editorial: Jean Armstrong, Shelley Farrar-Coleman, Marc Connolly, Hirva Raj, Brooke Butner, Robin Adams, Roxanne Picou, Cecilia Colome, Michael Ota

Design: Alison Klassen, Joan Cunningham, Deborah Diver, Alan Klemp, Andy Hendrix, Rhonda Holcomb

Production: Mychael Ferris-Pacheco, Heather Jernt, Greg Gaspard, Donna Brawley, John-Paxton Gremillion

Manufacturing: Cathy Voltaggio

Marketing: Marilyn Trow, Kimberly Sadler

E-Learning: Layne Hedrick, Karen Stitt

ISBN-13 978-1-5914-1884-9
ISBN-10 1-5914-1884-4

5 6 7 8 9 0868 12 11 10

ABOUT THE AUTHOR

Stephen Hake has authored five books in the Saxon Math series. He writes from 17 years of classroom experience as a teacher in grades 5 through 12 and as a math specialist in El Monte, California. As a math coach, his students won honors and recognition in local, regional, and statewide competitions.

Stephen has been writing math curriculum since 1975 and for Saxon since 1985. He has also authored several math contests including Los Angeles County's first Math Field Day contest. Stephen contributed to the 1999 National Academy of Science publication on the Nature and Teaching of Algebra in the Middle Grades.

Stephen is a member of the National Council of Teachers of Mathematics and the California Mathematics Council. He earned his BA from United States International University and his MA from Chapman College.

EDUCATIONAL CONSULTANTS

Nicole Hamilton
Consultant Manager
Richardson, TX

Joquita McKibben
Consultant Manager
Pensacola, FL

John Anderson
Lowell, IN

Beckie Fulcher
Gulf Breeze, FL

Heidi Graviette
Stockton, CA

Brenda Halulka
Atlanta, GA

Marilyn Lance
East Greenbush, NY

Ann Norris
Wichita Falls, TX

Melody Simmons
Nogales, AZ

Benjamin Swagerty
Moore, OK

Kristyn Warren
Macedonia, OH

Mary Warrington
East Wenatchee, WA

CONTENTS OVERVIEW

TABLE OF CONTENTS

Integrated and Distributed Units of Instruction

Maintaining & Extending

Power Up
Facts pp. 6, 12, 19, 26, 31, 36, 41, 47, 54, 60
Mental Math Strategies
pp. 6, 12, 19, 26, 31, 36, 41, 47, 54, 60
Problem Solving Strategies
pp. 6, 12, 19, 26, 31, 36, 41, 47, 54, 60
Enrichment
Early Finishers pp. 11, 25, 40, 53

TABLE OF CONTENTS

Maintaining & Extending

Power Up
Facts pp. 139, 146, 153, 159, 163, 169, 176, 181, 186, 192

Mental Math Strategies
pp. 139, 146, 153, 159, 163, 169, 176, 181, 186, 192

Problem Solving Strategies
pp. 139, 146, 153, 159, 163, 169, 176, 181, 186, 192

Enrichment
Early Finishers pp. 158, 180, 185, 191

Section 4 *Lessons 31–40, Investigation 4*

Math Focus:
Algebra • Measurement

Distributed Strands:
Number & Operations • Algebra • Geometry • Measurement • Data Analysis & Probability
• Problem Solving

Maintaining & Extending

Power Up
Facts pp. 202, 210, 218, 223, 229, 237, 245, 250, 257, 264

Mental Math Strategies
pp. 202, 210, 218, 223, 229, 237, 245, 250, 257, 264

Problem Solving Strategies
pp. 202, 210, 218, 223, 229, 237, 245, 250, 257, 264

Enrichment
Early Finishers pp. 217, 222, 244, 263, 270

Extensions p. 275

Section 5   *Lessons 41–50, Investigation 5*

Math Focus:
Number & Operations • Algebra

Distributed Strands:
Number & Operations • Algebra • Geometry • Measurement • Problem Solving

Maintaining & Extending

Power Up
Facts pp. 277, 287, 294, 300, 308, 313, 319, 326, 330, 336

Mental Math Strategies
pp. 277, 287, 294, 300, 308, 313, 319, 326, 330, 336

Problem Solving Strategies
pp. 277, 287, 294, 300, 308, 313, 319, 326, 330, 336

Enrichment
Early Finishers pp. 307, 341

TABLE OF CONTENTS

Maintaining & Extending

Power Up
Facts pp. 346, 354, 360, 367, 375, 382, 389, 394, 400, 406

Mental Math Strategies pp. 346, 354, 360, 367, 375, 382, 389, 394, 400, 406

Problem Solving Strategies pp. 346, 354, 360, 367, 375, 382, 389, 394, 400, 406

Enrichment
Early Finishers pp. 353, 366, 374, 381, 399

Section 7   *Lessons 61–70, Investigation 7*

Math Focus:
Algebra

Distributed Strands:
Number & Operations • Algebra • Geometry • Measurement • Data Analysis & Probability
• Problem Solving

Maintaining & Extending

Power Up
Facts pp. 415, 422, 429, 435, 440, 446, 452, 457, 463, 470

Mental Math Strategies
pp. 415, 422, 429, 435, 440, 446, 452, 457, 463, 470

Problem Solving Strategies
pp. 415, 422, 429, 435, 440, 446, 452, 457, 463, 470

Enrichment
Early Finishers pp. 462, 475

TABLE OF CONTENTS

Maintaining & Extending

Power Up
Facts pp. 479, 486, 491, 496, 502, 507, 514, 519, 525, 531

Mental Math Strategies
pp. 479, 486, 491, 496, 502, 507, 514, 519, 525, 531

Problem Solving Strategies
pp. 479, 486, 491, 496, 502, 507, 514, 519, 525, 531

Enrichment
Early Finishers pp. 485, 530

Section 9 — *Lessons 81–90, Investigation 9*

Math Focus:
Algebra • Measurement

Distributed Strands:
Number & Operations • Algebra • Geometry • Measurement • Data Analysis & Probability • Problem Solving

Maintaining & Extending

Power Up
Facts pp. 545, 550, 557, 563, 568, 574, 580, 585, 593, 599

Mental Math Strategies
pp. 545, 550, 557, 563, 568, 574, 580, 585, 593, 599

Problem Solving Strategies
pp. 545, 550, 557, 563, 568, 574, 580, 585, 593, 599

Enrichment
Early Finishers pp. 567, 573, 584, 592, 598, 605

TABLE OF CONTENTS

Maintaining & Extending

Power Up
Facts pp. 610, 617, 624, 629, 634, 640, 646, 651, 658, 664
Mental Math Strategies pp. 610, 617, 624, 629, 634, 640, 646, 651, 658, 664
Problem Solving Strategies pp. 610, 617, 624, 629, 634, 640, 646, 651, 658, 664

Enrichment
Early Finishers pp. 623, 645, 657, 669

Section 11 — *Lessons 101–110, Investigation 11*

Math Focus:
Algebra • Data Analysis & Probability

Distributed Strands:
Algebra • Data Analysis & Probability • Measurement • Geometry • Number & Operations • Problem Solving

Maintaining & Extending

Power Up
Facts pp. 675, 681, 686, 691, 697, 702, 707, 712, 717, 722

Mental Math Strategies
pp. 675, 681, 686, 691, 697, 702, 707, 712, 717, 722

Problem Solving Strategies
pp. 675, 681, 686, 691, 697, 702, 707, 712, 717, 722

Enrichment
Early Finishers p. 690

TABLE OF CONTENTS

Maintaining & Extending

Power Up
Facts pp. 731, 737, 742, 748, 754, 758, 763, 768, 773, 778

Mental Math Strategies
pp. 731, 737, 742, 748, 754, 758, 763, 768, 773, 778

Problem Solving Strategies
pp. 731, 737, 742, 748, 754, 758, 763, 768, 773, 778

Enrichment
Early Finishers pp. 736, 741, 747, 781

Dear Student,

We study mathematics because of its importance to our lives. Our school schedule, our trip to the store, the preparation of our meals, and many of the games we play involve mathematics. You will find that the word problems in this book are often drawn from everyday experiences.

As you grow into adulthood, mathematics will become even more important. In fact, your future in the adult world may depend on the mathematics you have learned. This book was written to help you learn mathematics and to learn it well. For this to happen, you must use the book properly. As you work through the pages, you will see that similar problems are presented over and over again. *Solving each problem day after day is the secret to success.*

Your book is made up of daily lessons and investigations. Each lesson has three parts.

1. The first part is a Power Up that includes practice of basic facts and mental math. These exercises improve your speed, accuracy, and ability to do math "in your head." The Power Up also includes a problem-solving exercise to familiarize you with strategies for solving complicated problems.

2. The second part of the lesson is the New Concept. This section introduces a new mathematical concept and presents examples that use the concept. The Practice Set provides a chance to solve problems involving the new concept. The problems are lettered a, b, c, and so on.

3. The final part of the lesson is the Written Practice. This problem set reviews previously taught concepts and prepares you for concepts that will be taught in later lessons. Solving these problems helps you remember skills and concepts for a long time.

Investigations are variations of the daily lesson. The investigations in this book often involve activities that fill an entire class period. Investigations contain their own set of questions instead of a problem set.

Remember, solve every problem in every practice set, written practice set, and investigation. Do not skip problems. With honest effort, you will experience success and true learning that will stay with you and serve you well in the future.

Temple City, California

HOW TO USE YOUR TEXTBOOK

Saxon Math Course 3 is unlike any math book you have used! It doesn't have colorful photos to distract you from learning. The Saxon approach lets you see the beauty and structure within math itself. You will understand more mathematics, become more confident in doing math, and will be well prepared when you take high school math classes.

Power Yourself Up!

Start off each lesson by practicing your basic skills and concepts, mental math, and problem solving. Make your math brain stronger by exercising it every day. Soon you'll know these facts by memory!

Learn Something New!

Each day brings you a new concept, but you'll only have to learn a small part of it now. You'll be building on this concept throughout the year so that you understand and remember it by test time.

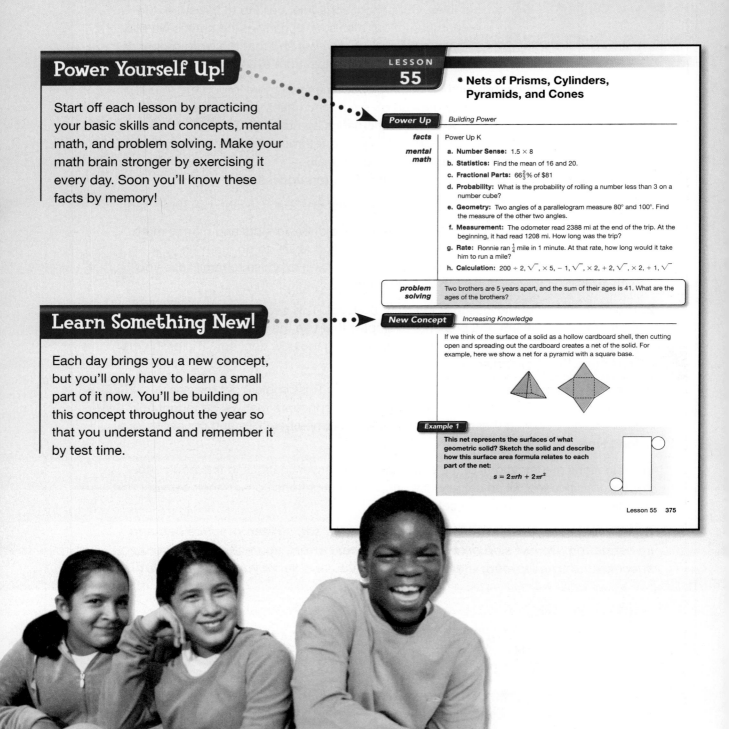

LESSON 55

• Nets of Prisms, Cylinders, Pyramids, and Cones

Power Up   *Building Power*

facts   Power Up K

mental math
- **a. Number Sense:** 1.5×8
- **b. Statistics:** Find the mean of 16 and 20.
- **c. Fractional Parts:** $66\frac{2}{3}\%$ of $81
- **d. Probability:** What is the probability of rolling a number less than 3 on a number cube?
- **e. Geometry:** Two angles of a parallelogram measure 80° and 100°. Find the measure of the other two angles.
- **f. Measurement:** The odometer read 2388 mi at the end of the trip. At the beginning, it had read 1208 mi. How long was the trip?
- **g. Rate:** Ronnie ran $\frac{1}{4}$ mile in 1 minute. At that rate, how long would it take him to run a mile?
- **h. Calculation:** $200 \div 2, \sqrt{\ }, \times 5, -1, \sqrt{\ }, \times 2, +2, \sqrt{\ }, \times 2, +1, \sqrt{\ }$

problem solving   Two brothers are 5 years apart, and the sum of their ages is 41. What are the ages of the brothers?

New Concept   *Increasing Knowledge*

If we think of the surface of a solid as a hollow cardboard shell, then cutting open and spreading out the cardboard creates a net of the solid. For example, here we show a net for a pyramid with a square base.

Example 1

This net represents the surfaces of what geometric solid? Sketch the solid and describe how this surface area formula relates to each part of the net:

$$s = 2\pi rh + 2\pi r^2$$

The front, top and right-side views are the three different rectangles that appear in the net. Each rectangle appears twice, because the front and back faces are congruent, as are the top and bottom faces and the left and right faces.

	Top	
Left	Front	Right
	Bottom	
	Back	

Activity
Net of a Cone

Materials needed: unlined paper, compass, scissors, glue or tape, ruler.

Using a compass, draw a circle with a radius of at least two inches. Cut out the circle and make one cut from the edge to the center of the circle. Form the lateral surface of a cone by overlapping the two sides of the cut. The greater the overlap, the narrower the cone. Glue or tape the overlapped paper so that the cone holds its shape.

cut

To make the circular base of the cone, measure the diameter of the open end of the cone and use a compass to draw a circle with the same diameter. (Remember, the radius is half the diameter.) Cut out the circle and tape it in place using two pieces of tape.

Now disassemble the cone to form a net. Cut open the cone by cutting the circular base free on one side. Unroll the lateral surface by making a straight cut to the point (apex) of the cone. The net of a cone has two parts, its circular base and a sector of a circle that forms the lateral surface of the cone.

Thinking Skill
Connect

The formula for the surface area of a cone is $s = \pi r l + \pi r^2$. Which parts of the formula apply to which parts of the net?

base *lateral surface*

Extend An alternate method for calculating the area of the lateral surface of a cone is to calculate the area of the portion of a circle represented by the net of the lateral surface. Use a protractor to measure the central angle of the lateral surface of the cone you created. The measure of that angle is the fraction of a 360° circle represented by the lateral surface. Use a ruler to measure the radius. Find the area of a whole circle with that find the area of the sector by multiplying the area of the whole c fraction $\frac{central\ angle}{360}$.

central angle *radius*

Get Active!

Dig into math with a hands-on activity. Explore a math concept with your friends as you work together and use manipulatives to see new connections in mathematics.

Check It Out!

The Practice Set lets you check to see if you understand today's new concept.

Practice Set

a. What is the lateral surface area of a tissue box with the dimensions shown?

4 in.
5 in.
9 in.

b. What is the total surface area of a cube with edges 2 inches long?

c. Estimate the surface area of a cube with edges 4.9 cm long.

d. **Analyze** Kwan is painting a garage. Find the lateral surface area of the building.

20 ft *8 ft*
20 ft

e. Find a box at home (such as a cereal box) and measure its dimensions. Sketch the box on your paper and record its length, width, and height. Then estimate the number of square inches of cardboard used to construct the box.

Written Practice *Strengthening Concepts*

1. Desiree drove north for 30 minutes at 50 miles per hour. Then, she drove
(7) south for 60 minutes at 20 miles per hour. How far and in what direction is Desiree from where she started?

2. In the forest, the ratio of deciduous trees to evergreens is 2 to 7. If there
(34) are 400 deciduous trees in the forest, how many trees are in the forest?

3. Reginald left his house and rode his horse east for 3 hours at 9 miles
(7) per hour. How fast and in which direction must he ride to get back to his house in an hour?

* 4. **Analyze** What is the volume of a shipping box with dimensions
(42) $1\frac{1}{2}$ inches × 11 inches × 12 inches?

5. How many square inches is the surface area of the shipping box
(43) described in problem **4?**

* 6. How many edges, faces, and vertices does a
(Inv. 4) triangular pyramid have?

* 7. Graph $y = -2x + 3$. Is (4, −11) on the line?
(41)

* 8. Find **a** the area and **b** circumference of the circle with a radius of 6 in.
(39, 40) Express your answer in terms of π.

Lesson 43 **297**

Exercise Your Mind!

When you work the Written Practice exercises, you will review both today's new concept and also math you learned in earlier lessons. Each exercise will be on a different concept — you never know what you're going to get! It's like a mystery game — unpredictable and challenging.

As you review concepts from earlier in the book, you'll be asked to use higher-order thinking skills to show what you know and why the math works.

The mixed set of Written Practice is just like the mixed format of your state test. You'll be practicing for the "big" test every day!

Become an Investigator!

Dive into math concepts and explore the depths of math connections in the Investigations.

Continue to develop your mathematical thinking through applications, activities, and extensions.

INVESTIGATION 2

Focus on
• Pythagorean Theorem

The longest side of a right triangle is called the **hypotenuse.** The other two sides are called **legs.** Notice that the legs are the sides that form the right angle. The hypotenuse is the side opposite the right angle.

Every right triangle has a property that makes right triangles very important in mathematics. **The area of a square drawn on the hypotenuse of a right triangle equals the sum of the areas of squares drawn on the legs.**

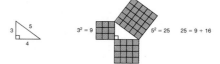

As many as 4000 years ago ancient Egyptians knew of this important and useful relationship between the sides of a right triangle. Today the relationship is named for a Greek mathematician who lived about 550 B.C. The Greek's name was Pythagoras and the relationship is called the Pythagorean Theorem.

The **Pythagorean Theorem** is an equation that relates the sides of a right triangle in this way: the sum of the squares of the legs equals the square of the hypotenuse.

$$\text{leg}^2 + \text{leg}^2 = \text{hypotenuse}^2$$

If we let the letters a and b represent the lengths of the legs and c the length of the hypotenuse, then we can express the Pythagorean Theorem with the following equation.

Pythagorean Theorem
If a triangle is a right triangle, then the sum of the squares of the legs equals the square of the hypotenuse.
$a^2 + b^2 = c^2$

• Problem Solving

As we study mathematics we learn how to use tools that help us solve problems. We face mathematical problems in our daily lives, in our careers, and in our efforts to advance our technological society. We can become powerful problem solvers by improving our ability to use the tools we store in our minds. In this book we will practice solving problems every day.

four-step problem-solving process

Solving a problem is like arriving at a destination, so the process of solving a problem is similar to the process of taking a trip. Suppose we are on the mainland and want to reach a nearby island

Problem-Solving Process	Taking a Trip
Step 1: (*Understand*) Know where you are and where you want to go.	We are on the mainland and want to go to the island.
Step 2: (*Plan*) Plan your route.	We might use the bridge, the boat, or swim.
Step 3: (*Solve*) Follow the plan.	Take the journey to the island
Step 4: (*Check*) Check that you have reached the right place.	Verify that you have reached your desired destination.

When we solve a problem, it helps to ask ourselves some questions along the way.

Follow the Process	Ask Yourself Questions
Step 1: Understand	What information am I given? What am I asked to find or do?
Step 2: Plan	How can I use the given information to solve the problem? What strategy can I use to solve the problem?
Step 3: Solve	Am I following the plan? Is my math correct?
Step 4: Check *(Look Back)*	Does my solution answer the question that was asked? Is my answer reasonable?

Below we show how we follow these steps to solve a word problem.

Example

Josh wants to buy a television. He has already saved $68.25. He earns $35 each Saturday stocking groceries in his father's store. The sizes and prices of the televisions available are shown at right. If Josh works and saves for 5 more weekends, what is the largest television he could buy?

> **Televisions**
> 15" $149.99
> 17" $199.99
> 20" $248.99

Solution

Step 1: Understand the problem. We know that Josh has $68.25 saved. We know that he earns $35.00 every weekend. We are asked to decide which television he could buy if he works for 5 more weekends.

Step 2: Make a plan. We cannot find the answer in one step. We make a plan that will lead us toward the solution. One way to solve the problem is to find out how much Josh will earn in 5 weekends, then add that amount to the money he has already saved, then determine the largest television he could buy with the total amount of money.

Step 3: Solve the problem. (Follow the Plan.) First we multiply $35 by 5 to determine how much Josh will earn in 5 weekends. We could also find 5 multiples of $35 by making a table.

Weekend	1	2	3	4	5
Amount	$35	$70	$105	$140	$175

$$\begin{array}{r} \$35 \\ \underline{\times\ 5} \\ \$175 \end{array}$$

Josh will earn $175 in 5 weekends.

Now we add $175 to $68.25 to find the total amount he will have.

$$\begin{array}{r} \$175 \\ +\quad 68.25 \\ \hline \$243.25 \end{array}$$

We find that Josh will have $243.25 after working 5 more weekends. When we compare the total to the prices, we can see that **Josh can buy the 17″ television.**

Step 4: Check your answer. (Look Back.) We read the problem again. The problem asked which television Josh could buy after working five weekends. We found that in five weekends Josh will earn $175.00, which combined with his $68.25 savings gives Josh $243.25. This is enough money to buy the 17″ television.

1. List in order the four steps in the problem-solving process.

2. What two questions do we answer to help us understand a problem?

Refer to the text below to answer problems **3–8**.

Mary wants to put square tiles on the kitchen floor. The tile she has selected is 12 inches on each side and comes in boxes of 20 tiles for $54 per box. Her kitchen is 15 feet 8 inches long and 5 feet 9 inches wide. How many boxes of tile will she need and what will be the price of the tile, not including tax?

3. What information are we given?

4. What are we asked to find?

5. Which step of the four-step problem-solving process have you completed when you have answered problems **3** and **4**?

6. Describe your plan for solving the problem. Besides the arithmetic, is there anything else you can draw or do that will help you solve the problem?

7. Solve the problem by following your plan. Show your work and any diagrams you used. Write your solution to the problem in a way someone else will understand.

8. Check your work and your answer. Look back to the problem. Be sure you used the information correctly. Be sure you found what you were asked to find. Is your answer reasonable?

As we consider how to solve a problem we choose one or more strategies that seem to be helpful. Referring to the picture at the beginning of this lesson, we might choose to swim, to take the boat, or to cross the bridge to travel from the mainland to the island. Other strategies might not be as effective for the illustrated problem. For example, choosing to walk or bike across the water are strategies that are not reasonable for this situation.

Problem-solving **strategies** are types of plans we can use to solve problems. Listed below are ten strategies we will practice in this book. You may refer to these descriptions as you solve problems throughout the year.

Act it out or make a model. Moving objects or people can help us visualize the problem and lead us to the solution.

Use logical reasoning. All problems require reasoning, but for some problems we use given information to eliminate choices so that we can close in on the solution. Usually a chart, diagram, or picture can be used to organize the given information and to make the solution more apparent.

Draw a picture or diagram. Sketching a picture or a diagram can help us understand and solve problems, especially problems about graphs or maps or shapes.

Write a number sentence or equation. We can solve many word problems by fitting the given numbers into equations or number sentences and then finding the unknown numbers.

Make it simpler. We can make some complicated problem easier by using smaller numbers or fewer items. Solving the simpler problem might help us see a pattern or method that can help us solve the complex problem.

Find a pattern. Identifying a pattern that helps you to predict what will come next as the pattern continues might lead to the solution.

Make an organized list. Making a list can help us organize our thinking about a problem.

Guess and check. Guessing a possible answer and trying the guess in the problem might start a process that leads to the answer. If the guess is not correct, use the information from the guess to make a better guess. Continue to improve your guesses until you find the answer.

Make or use a table, chart, or graph. Arranging information in a table, chart, or graph can help us organize and keep track of data. This might reveal patterns or relationships that can help us solve the problem.

Work backwards. Finding a route through a maze is often easier by beginning at the end and tracing a path back to the start. Likewise, some problems are easier to solve by working back from information that is given toward the end of the problem to information that is unknown near the beginning of the problem.

9. Name some strategies used in this lesson.

The chart below shows where each strategy is first introduced in this textbook.

Strategy	Lesson
Act It Out or Make a Model	Lesson 1
Use Logical Reasoning	Lesson 15
Draw a Picture or Diagram	Lesson 1
Write a Number Sentence or Equation	Lesson 31
Make It Simpler	Lesson 4
Find a Pattern	Lesson 19
Make an Organized List	Lesson 38
Guess and Check	Lesson 8
Make or Use a Table, Chart, or Graph	Lesson 21
Work Backwards	Lesson 2

writing and problem solving

Sometimes a problem will ask us to explain our thinking. Writing about a problem solving can help us measure our understanding of math.

For these types of problems, we use words to describe the steps we used to follow our plan.

This is a description of the way we solved the problem about tiling a kitchen.

> *First, we round the measurements of the floor to 6 ft by 16 ft. We multiply to find the approximate area: 6 × 16 = 96 sq. ft. Then we count by 20s to find a number close to but greater than 96: 20, 40, 60, 80,* ***100****. There are 20 tiles in each box, so Mary needs to buy 5 boxes of tiles. We multiply to find the total cost: $54 × 5 = $270.*

10. Write a description of how we solved the problem in the example.

Other times, we will be asked to write a problem for a given equation. Be sure to include the correct numbers and operations to represent the equation.

11. Write a word problem for the equation $b = (12 \times 10) \div 4$.

• Number Line: Comparing and Ordering Integers

facts

Power Up A

mental math

a. **Calculation:** Danielle drives 60 miles per hour for 2 hours. How far does she drive in total?

b. **Number Sense:** 100×100

c. **Estimation:** 4.034×8.1301

d. **Measurement:** How long is the key?

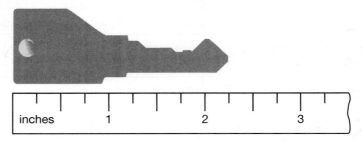

e. **Number Sense:** $98 + 37$

f. **Percent:** 50% of 50

g. **Geometry:** This figure represents a

     A line   **B** line segment   **C** ray

h[1]. **Calculation:** $7 + 5, + 3, + 1, \div 8, \times 2, \times 4$

problem solving

Some problems describe physical movement. To solve such problems, we might act out the situation to understand what is happening. We can also draw a diagram to represent the movement in the problem. We try to keep diagrams simple so they are easy to understand.

Problem: A robot is programmed to take two steps forward, then one step back. The robot will repeat this until it reaches its charger unit, which is ten steps in front of the robot. How many steps back will the robot take before it reaches the charger unit?

(**Understand**) We are told a robot takes two steps forward and then one step back. We are asked to find how many steps back the robot will take before it reaches its charger unit, which is ten steps away.

(**Plan**) We will draw a diagram to help us count the steps. Your teacher might also select a student to act out the robot's movements for the class.

---

[1] As a shorthand, we will use commas to separate operations to be performed sequentially from left to right. This is not a standard mathematical notation.

Solve We use a number line for our diagram. We count two spaces forward from zero and then one space back and write "1" above the tick mark for 1. From that point, we count two spaces forward and one space back and write "2" above the 2. We continue until we reach the tick mark for 10. The numbers we write represent the total number of backward steps by the robot. We reach 10 after having taken eight steps back. The robot reaches the charger unit after taking eight steps back.

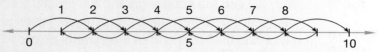

Check To check for reasonableness, we rethink the problem situation with our answer in mind. It makes sense that the robot takes eight steps back before reaching the charger unit. Two steps forward and one step back is the same as one step forward. After completing this pattern 8 times, we expect the robot to be able to take two steps forward and reach its destination.

New Concept   *Increasing Knowledge*

The numbers we use in this book can be represented by points on a number line.

Number line

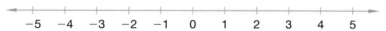

Points to the right of zero (the *origin*) represent positive numbers. Points to the left of zero represent negative numbers. Zero is neither positive nor negative.

The tick marks on the number line above indicate the locations of **integers,** which include **counting numbers** (1, 2, 3, …) and their opposites (… −3, −2, −1) as well as zero. Thus all **whole numbers** (0, 1, 2, …) are integers, but not all integers are whole numbers. We will study numbers represented by points between the tick marks later in the book.

Classify Give an example of an integer that is not a whole number.

Counting numbers, whole numbers, and integers are examples of different **sets** of numbers. A set is a collection of items, like numbers or polygons. We use braces to indicate a set of numbers, and we write the **elements** (members) of the set within the braces.

The set of integers

{…, −3, −2, −1, 0, 1, 2, 3, …}

The set of whole numbers

{0, 1, 2, 3, …}

Analyze How are the sets of whole numbers and integers different?

Reading Math

The three dots, called an *ellipsis,* mean that the sequence continues without end.

The **absolute value** of a number is the distance between the number and the origin on a number line. Absolute value is always a positive number because it represents a distance. Thus, the absolute value of −5 (negative five) is 5, because −5 is 5 units from zero. Two vertical bars indicate absolute value.

"The absolute value of negative five is five."

$$|-5| = 5$$

We can order and compare numbers by noting their position on the number line. We use the equal sign and the greater than/less than symbols (> and <, respectively) to indicate a comparison. When properly placed between two numbers, the small end of the symbol points to the lesser number.

Example 1

Arrange these integers in order from least to greatest.

−3, 3, 0, 1

Solution

We write the integers in the same order as they appear on the number line.

−3, 0, 1, 3

Example 2

Compare:

a. −3 ◯ −1   **b.** |−3| ◯ |−1|

Solution

Thinking Skill

Connect

When you move from left to right on a number line, do the numbers increase or decrease? What happens to the numbers when you move from right to left?

a. We replace the circle with the correct comparison symbol. Negative three is less than negative one.

−3 < −1

b. The absolute value of −3 is 3 and the absolute value of −1 is 1, and 3 is greater than 1.

|−3| > |−1|

We graph a point on a number line by drawing a dot at its location. Below we graph the numbers −4, −2, 0, 2, and 4.

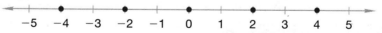

These numbers are members of the set of **even numbers.** The set of even numbers can be expressed as a **sequence,** an ordered list of numbers that follows a rule.

The set of even numbers

{... −4, −2, 0, 2, 4, ...}

Example 3

Graph the numbers in this sequence on a number line.

$$\{\ldots, -3, -1, 1, 3, \ldots\}$$

Solution

This is the sequence of **odd numbers.** We sketch a number line and draw dots at −3, −1, 1, and 3. The ellipses show that the sequence continues, so we darken the arrowheads to show that the pattern continues.

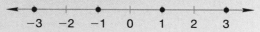

Example 4

Graph two numbers that are four units from zero.

Solution

Both **4** and **−4** are four units from zero.

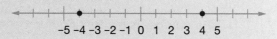

Practice Set

a. Arrange these integers in order from least to greatest.

$$-4, 3, 2, -1, 0$$

b. Which number in problem **a** is an even number but not a whole number?

c. Compare: $-2 \bigcirc -4$

d. (Model) Graph the numbers in this sequence on a number line.

$$\ldots, -4, -2, 0, 2, 4, \ldots$$

Simplify.

e. $|-3|$ **f.** $|3|$

g. (Analyze) Write two numbers that are ten units from zero.

h. Write an example of a whole number that is not a counting number.

Written Practice   *Strengthening Concepts*

* 1. Graph these numbers on a number line.
(1)

$$-5, 3, -2, 1 \qquad {}^{-5}$$

* 2. (Analyze) Arrange these numbers from least to greatest.
(1)

$$-5, 3, -2, 1$$

---

* Asterisks indicate exercises that should be completed in class with teacher support as needed.

* 3. Use braces and digits to indicate the set of whole numbers.
(1)

* 4. Use braces and digits to indicate the set of even numbers.
(1)

5. Which number in problem **1** is an even number?
(1)

6. Which whole number is *not* a counting number?
(1)

7. Write the graphed numbers as a sequence of numbers. (Assume the
(1) pattern continues.)

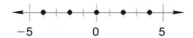

* 8. The sequence of numbers in problem 7 is part of what set of
(1) numbers?

Analyze Decide if the statements in **9–11** are true or false, and explain why.

9. All whole numbers are counting numbers.
(1)

10. All counting numbers are whole numbers.
(1)

* 11. If a number is a counting number or the opposite of a counting number,
(1) then it is an integer.

12. What is the absolute value of 21?
(1)

13. What is the absolute value of -13?
(1)

14. What is the absolute value of 0?
(1)

15. Compare: $5 \bigcirc -7$
(1)

16. Compare: $-3 \bigcirc -2$
(1)

* 17. **Evaluate** Compare: $|-3| \bigcirc |-2|$
(1)

18. Graph these numbers on a number line: $-5, 0, 5$
(1)

* 19. If $|n| = 5$, then n can be which two numbers?
(1)

20. Write two numbers that are five units from zero.
(1)

21. Graph these numbers on a number line.
(1)
$$-3, 0, 3$$

22. Write two numbers that are 3 units from 0.
(1)

23. Graph the numbers in this sequence on a number line.
(1)
$$\ldots, -15, -10, -5, 0, 5, 10, 15, \ldots$$

24. What number is the opposite of 10?
(1)

25. What is the sum when you add 10 and its opposite?
(1)

* 26. What number is the opposite of -2?
(1)

* 27. **Conclude** What is the sum when you add -2 and its opposite?
(1)

For multiple choice problems **28–30,** choose *all* correct answers from the list.

    A counting numbers (natural numbers)

    B whole numbers

    C integers

    D none of these

* 28. Negative seven is a member of which set of numbers?
(1)

29. Thirty is a member of which set of numbers?
(1)

30. One third is a member of which set of numbers?
(1)

Early Finishers
Real-World Application

A group of eight students wants to ride a roller coaster at the local fair. Each passenger must be at least 48 inches tall to ride. The list below represents the number of inches each student's height differs from 48 inches. (A negative sign indicates a height less than 48 inches.)

$$-4 \quad -3 \quad -1 \quad 5 \quad 0 \quad -2 \quad 1 \quad 4$$

  a. Write each student's height in inches, then arrange the heights from least to greatest.

  b. How many of the students represented by the list may ride the roller coaster?

• Operations of Arithmetic

facts | Power Up A

mental math

a. **Calculation:** Kyle rides his bicycle 15 miles per hour for 2 hours. How far does he travel?

b. **Number Sense:** 203 + 87

c. **Measurement:** How long would a row of 12 push pins be?

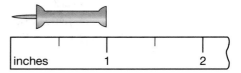

inches   1   2

d. **Percent:** 10% of 100

e. **Estimation:** 25.032 + 49.994

f. **Geometry:** This figure represents a

   A line   **B** line segment   **C** ray

g. **Number Sense:** 2 × 16

h. **Calculation:** 4 + 3, + 2, + 1, + 0

problem solving

A number of problems in this book will ask us to "fill in the missing digits" in an addition, subtraction, multiplication, or division problem. We can work backwards to solve such problems.

Problem: Copy this problem and fill in the missing digits:

$$
\begin{array}{r}
\$\_0\_8 \\
- \ \$432\_ \\
\hline
\$4\_07
\end{array}
$$

(**Understand**) We are shown a subtraction problem with missing digits. We are asked to fill in the missing digits.

(**Plan**) We will work backwards and use number sense to find the missing digits one-by-one. This is a subtraction problem, so we can look at each column and think of "adding up" to find missing digits.

(**Solve**) We look at the first column and think, "7 plus what number equals 8?" (1). We write a 1 in the blank. We move to the tens column and think, "0 plus 2 is what number?" We write a 2 in the tens place of the minuend (the top number). Then we move to the hundreds column and think, "What number plus 3 equals a number that ends in 0? We write a 7 in the difference and remember to regroup the 1 in the thousands column. We move to the thousands column and think, "4 plus 4 plus 1 (from regrouping) equals what number?" We write a 9 in the remaining blank.

Check We filled in all the missing digits. Using estimation our answer seems reasonable, since $4700 + $4300 = $9000. We can check our answer by performing the subtraction or by adding.

$$
\begin{array}{r} \$9028 \\ -\ \$4321 \\ \hline \$4707 \end{array}
\qquad
\begin{array}{r} \$4707 \\ +\ \$4321 \\ \hline \$9028 \end{array}
$$

New Concept   *Increasing Knowledge*

The fundamental operations of arithmetic are addition, subtraction, multiplication, and division. Below we review some terms and symbols for these operations.

Terminology

addend + addend = sum
minuend − subtrahend = difference
factor × factor = product
dividend ÷ divisor = quotient

Example 1

What is the quotient when the product of 3 and 6 is divided by the sum of 3 and 6?

Solution

First we find the product and sum of 3 and 6.

product of 3 and 6 = 18

sum of 3 and 6 = 9

Then we divide the product by the sum.

$18 \div 9 = \mathbf{2}$

Three numbers that form an addition fact also form a subtraction fact.

$5 + 3 = 8$

$8 - 3 = 5$

Likewise, three numbers that form a multiplication fact also form a division fact.

$4 \times 6 = 24$

$24 \div 6 = 4$

We may use these relationships to help us check arithmetic answers.

Example 2

Simplify and check:

a. 706 − 327   b. 450 ÷ 25

Solution

Thinking Skill

Represent

Form another addition fact and another subtraction fact using the numbers 706, 327, and 379.

We simplify by performing the indicated operation.

a.
```
  706
− 327
  379
```

b.
```
    18
25)450
```

We can check subtraction by adding, and we can check division by multiplying.

a.
```
  379
+ 327
  706 ✓
```

b.
```
    25
  × 18
   200
   250
   450 ✓
```

Symbols for Multiplication and Division

"three times five"	3×5, $3 \cdot 5$, $3(5)$, $(3)(5)$
"six divided by two"	$6 \div 2$, $2\overline{)6}$, $\frac{6}{2}$

Math Language

The term **real numbers** refers to the set of all numbers that can be represented by points on a number line.

The table below lists important properties of addition and multiplication. In the second column we use letters, called **variables** to show that these properties apply to all real numbers. A variable can take on different values.

Some Properties of Addition and Multiplication

Name of Property	Representation	Example
Commutative Property of Addition	$a + b = b + a$	$3 + 4 = 4 + 3$
Commutative Property of Multiplication	$a \cdot b = b \cdot a$	$3 \cdot 4 = 4 \cdot 3$
Associative Property of Addition	$(a + b) + c = a + (b + c)$	$(3 + 4) + 5 = 3 + (4 + 5)$
Associative Property of Multiplication	$(a \cdot b) \cdot c = a \cdot (b \cdot c)$	$(3 \cdot 4) \cdot 5 = 3 \cdot (4 \cdot 5)$
Identity Property of Addition	$a + 0 = a$	$3 + 0 = 3$
Identity Property of Multiplication	$a \cdot 1 = a$	$3 \cdot 1 = 3$
Zero Property of Multiplication	$a \cdot 0 = 0$	$3 \cdot 0 = 0$

We will use these properties throughout the book to simplify expressions and solve equations.

Represent Demonstrate the Associative Property of Addition using the numbers 3, 5, and 8.

Example 3

Show two ways to simplify this expression, justifying each step.

$$(25 \cdot 15) \cdot 4$$

Solution

There are three factors, but we multiply only two numbers at a time. We multiply 25 and 15 first.

Step:	Justification:
$(25 \cdot 15) \cdot 4$	Given
$375 \cdot 4$	Multiplied 25 and 15
1500	Multiplied 375 and 4

Instead of first multiplying 25 and 15, we can use properties of multiplication to rearrange and regroup the factors so that we first multiply 25 and 4. Changing the arrangement of factors can make the multiplication easier to perform.

Step:	Justification:
$(25 \cdot 15) \cdot 4$	Given
$(15 \cdot 25) \cdot 4$	Commutative Property of Multiplication
$15 \cdot (25 \cdot 4)$	Associative Property of Multiplication
$15 \cdot 100$	Multiplied 25 and 4
1500	Multiplied 15 and 100

Notice that the properties of addition and multiplication do not apply to subtraction and division. Order matters when we subtract or divide.

$$5 - 3 = 2 \quad \text{but} \quad 3 - 5 = -2$$
$$8 \div 4 = 2 \quad \text{but} \quad 4 \div 8 = \frac{1}{2}$$

Example 4

Simplify $100 - 365$.

Solution

Reversing the order of the minuend and the subtrahend reverses the sign of the difference.

$$365 - 100 = 265$$
$$\text{so } 100 - 365 = \mathbf{-265}$$

Example 5

Find the value of x and y in the equations below.

a. $6 + x = 30$ b. $6y = 30$

The letters x and y represent numbers and are unknowns in the equations.

a. We can find an unknown addend by subtracting the known addend(s) from the sum.

$$30 - 6 = 24$$

Thus, **$x = 24$.**

b. We can find an unknown factor by dividing the product by the known factor(s).

$$30 \div 6 = 5$$

Thus, **$y = 5$.**

Practice Set

Name each property illustrated in **a–d.**

a. $4 \cdot 1 = 4$

b. $4 + 5 = 5 + 4$

c. $(8 + 6) + 4 = 8 + (6 + 4)$

d. $0 \cdot 5 = 0$

e. What is the difference when the sum of 5 and 7 is subtracted from the product of 5 and 7?

f. Simplify: $36 - 87$

g. *Justify* Lee simplified the expression $5 \cdot (7 \cdot 8)$. His work is shown below. What properties of arithmetic did Lee use for steps 1 and 2 of his calculations?

$5 \cdot (7 \cdot 8)$	Given
$5 \cdot (8 \cdot 7)$	_____
$(5 \cdot 8) \cdot 7$	_____
$40 \cdot 7$	$5 \cdot 8 = 40$
280	$40 \cdot 7 = 280$

h. *Explain* Explain how to check a subtraction answer.

i. *Explain* Explain how to check a division answer.

For **j** and **k,** find the unknown.

j. $12 + m = 48$

k. $12n = 48$

Written Practice   *Strengthening Concepts*

* 1. *Analyze* The product of 20 and 5 is how much greater than the sum of 20 and 5?
(2)

* 2. *Analyze* What is the quotient when the product of 20 and 5 is divided by the sum of 20 and 5?
(2)

3. The sum of 10 and 20 is 30.
(2)
$$10 + 20 = 30$$

Write two subtraction facts using these three numbers.

4. The product of 10 and 20 is 200.
(2)
$$10 \cdot 20 = 200$$

Write two division facts using the same three numbers.

* 5. a. Using the properties of multiplication, how can we rearrange the
(2)   factors in this expression so that the multiplication is easier?

$$(25 \cdot 17) \cdot 4$$

b. What is the product? _

c. Which properties did you use?

6. Use braces and digits to indicate the set of counting numbers.
(1)

7. Use braces and digits to indicate the set of whole numbers.
(1)

8. Use braces and digits to indicate the set of integers.
(1)

Justify Decide if the statements in **9–10** are true or false, and explain why.

9. All whole numbers are counting numbers.
(1)

10. All counting numbers are integers.
(1)

11. Arrange these integers from least to greatest: 0, 1, −2, −3, 4
(1)

12. a. What is the absolute value of −12?
(1)
b. What is the absolute value of 11?

Analyze In **13–17**, name the property illustrated.

* 13. $100 \cdot 1 = 100$
(2)

* 14. $a + 0 = a$
(2)

* 15. $(5)(0) = 0$
(2)

* 16. $5 + (10 + 15) = (5 + 10) + 15$
(2)

* 17. $10 \cdot 5 = 5 \cdot 10$
(2)

18. a. Name the four operations of arithmetic identified in Lesson 2.
(2)
b. For which of the four operations of arithmetic do the Commutative
and Associative Properties not apply?

* 19. If $|n| = 10$, then n can be which two numbers?
(1)

* 20. *Analyze* Compare:
(1)
a. 0 ◯ −1   **b.** −2 ◯ −3   **c.** $|-2|$ ◯ $|-3|$

* 21. Graph the numbers in this sequence on a number line:
(1)

$$\ldots, -4, -2, 0, 2, 4, \ldots$$

22. What number is the opposite of 20?
(1)

23. Which integer is neither positive nor negative?
(1)

For multiple choice problems **24–26,** choose *all* correct answers.

* 24. One hundred is a member of which set of numbers?
(1)
  A counting numbers (natural numbers)

  B whole numbers

  C integers

  D none of these

* 25. Negative five is a member of which set of numbers?
(1)
  A counting numbers (natural numbers)

  B whole numbers

  C integers

  D none of these

* 26. One half is a member of which of these set of numbers?
(1)
  A counting numbers (natural numbers)

  B whole numbers

  C integers

  D none of these

Generalize Simplify

27. 5010 − 846
(2)

* 28. 846 − 5010
(2)

29. 780(49)
(2)

30. $\dfrac{5075}{25}$
(2)

• Addition and Subtraction Word Problems

Power Up | *Building Power*

facts | Power Up A

mental math

a. **Calculation:** Sophie drives 35 miles per hour for 3 hours. How far does she drive?

b. **Estimation:** 21 + 79 + 28

c. **Number Sense:** 1108 + 42

d. **Measurement:** If object A weighs 7 pounds, how much does object B weigh?

e. **Percent:** 50% of 100

f. **Geometry:** This figure represents a

   A line   **B** line segment   **C** ray

g. **Number Sense:** Find the two missing digits: $9 \times \square = \square 1$

h. **Calculation:** 3 dozen roses, minus 5 roses, plus a rose, minus a dozen roses, ÷ 2

problem solving | Thirty-eight cars are waiting to take a ferry across a lake. The ferry can carry six cars. If the ferry starts on the same side of the lake as the cars, how many times will the ferry cross the lake to deliver all 38 cars to the other side of the lake?

New Concept | *Increasing Knowledge*

When we read a novel or watch a movie we are aware of the characters, the setting, and the plot. The plot is the storyline. Although there are many different stories, plots are often similar.

Many word problems we solve with mathematics also have plots. Recognizing the plot helps us solve the problem. In this lesson we will consider word problems that are solved using addition and subtraction. The problems are like stories that have plots about combining, separating, and comparing.

Problems about **combining** have an addition thought pattern. We can express a combining plot with a **formula,** which is an **equation** used to calculate a desired result.

$$\text{some} + \text{more} = \text{total}$$
$$s + m = t$$

See how the numbers in this story fit the formula.

In the first half of the game, Heidi scored 12 points. In the second half, she scored 15 points. In the whole game, Heidi scored 27 points.

In this story, two numbers are combined to make a total. (In some stories, more than two numbers are combined.)

$$s + m = t$$
$$12 + 15 = 27$$

A story becomes a word problem if one of the numbers is missing, as we see in the following example.

Example 1

In the first half of the game, Heidi scored 12 points. In the whole game, she scored 27 points. How many points did Heidi score in the second half?

Solution

Heidi scored some in the first half and some more in the second half. We are given the total.

$$s + m = t$$
$$12 + m = 27$$

The "more" number is missing in the equation. Recall that we can find an unknown addend by subtracting the known addend from the sum. Since $27 - 12 = 15$, we find that Heidi scored **15 points** in the second half. The answer is reasonable because 12 points in the first half of the game plus 15 points in the second half totals 27 points.

Example 1 shows one problem that we can make from the story about Heidi's game. We can use the same formula no matter which number is missing because the missing number does not change the plot. However, the number that is missing does determine whether we add or subtract to find the missing number.

• If the sum is missing, we add the addends.

• If an addend is missing, we subtract the known addend(s) from the sum.

$$s + m = t$$

Find by subtracting. ⎯⎯⎯⎯⎯⎯⎯ Find by adding.

Now we will consider word problems about **separating.** These problems have a subtraction thought pattern. We can express a separating plot with a formula.

starting amount − some went away = what is left

$$s - a = l$$

Here is a story about separating.

Alberto went to the store with a twenty dollar bill. He bought a loaf of bread and a half-gallon of milk for a total of $4.83. The clerk gave him $15.17 in change.

In this story some of Alberto's money went away when he bought the bread and milk. The numbers fit the formula.

$$s - a = l$$
$$\$20.00 - \$4.83 = \$15.17$$

We can make a problem from the story by omitting one of the numbers, as we show in example 2. If a number in a subtraction formula is missing, we add to find the minuend. Otherwise we subtract from the minuend.

$$s - a = l$$

Find by adding. ⟶ ↑ ↑ ↑ ⟵ Find by subtracting.

Example 2

Alberto went to the store with a twenty dollar bill. He bought a loaf of bread and a half-gallon of milk. The clerk gave him $15.17 in change. How much money did Alberto spend on bread and milk? Explain why your answer is reasonable.

Solution

This story has a separating plot. We want to find how much of Alberto's money went away. We write the numbers in the formula and find the missing number.

$$s - a = l$$
$$\$20.00 - a = \$15.17$$

We find the amount Alberto spent by subtracting $15.17 from $20.00. We find that Alberto spent **$4.83** on bread and milk. Since Alberto paid $20 and got back about $15, we know that he spent about $5. Therefore $4.83 is a reasonable answer. We can check the answer to the penny by subtracting $4.83 from $20. The result is $15.17, which is the amount Alberto received in change.

Example 3

The hike to the summit of Mt. Whitney began from the upper trailhead at 8365 ft. The elevation of Mt. Whitney's summit is 14,496 ft. Which equation shows how to find the elevation gain from the trailhead to the summit?

A $8365 + g = 14{,}496$ **B** $8365 + 14{,}496 = g$

C $g - 14{,}496 = 8365$ **D** $8365 - g = 14{,}496$

Solution

A $8365 + g = 14{,}496$

One way of **comparing** numbers is by subtraction. The difference shows us how much greater or how much less one number is compared to another number. We can express a comparing plot with this formula.

$$\text{greater} - \text{lesser} = \text{difference}$$
$$g - l = d$$

We can use a similar formula for **elapsed time.**

$$later - earlier = difference$$

$$l - e = d$$

Here is a comparing story. See how the numbers fit the formula.

> *From 1990 to 2000, the population of Fort Worth increased from 447,619 to 534,694, a population increase of 87,075.*

$$greater - lesser = difference$$

$$534,694 - 447,619 = 87,075$$

In word problems a number is missing. We find the larger number (or later time) by adding. Otherwise we subtract.

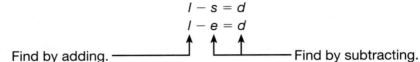

$$l - s = d$$
$$l - e = d$$

Find by adding. ⸻⸻ Find by subtracting.

Example 4

President John F. Kennedy was elected in 1960 at the age of 43. In what year was he born?

Solution

A person's age is the difference between the given date and their birth. We are missing the year of birth, which is the earlier date.

$$l - e = d$$

$$1960 - e = 43$$

We subtract and find that John F. Kennedy was born in **1917.**

Practice Set

a. Which three words identify the three types of plots described in this lesson?

b. In example 2 we solved a word problem to find how much money Alberto spent on milk and bread. Using the same information, write a word problem that asks how much money Alberto gave to the clerk.

c. ⟨Formulate⟩ Write a story problem for this equation.

$$\$20.00 - a = \$8.45$$

⟨Evaluate⟩ For problems **d–f,** identify the plot, write an equation, and solve the problem.

d. From 1990 to 2000 the population of Garland increased from 180,635 to 215,768. How many more people lived in Garland in 2000 than in 1990?

e. Binh went to the theater with $20.00 and left the theater with $10.50. How much money did Binh spend at the theater? Explain why your answer is reasonable.

f. In the three 8th-grade classrooms at Washington school, there are 29 students, 28 students, and 31 students. What is the total number of students in the three classrooms?

g. Which equation shows how to find how much change a customer should receive from $10.00 for a $6.29 purchase?

    A $10.00 + $6.29 = c   **B** $10.00 - $6.29 = c

    C $6.29 + $10.00 = c   **D** $10.00 + c = $6.29

Written Practice   *Strengthening Concepts*

Evaluate For problems **1–6,** identify the plot, write an equation, and solve the problem.

* 1. The seventh-grade winner ran the mile in 5 minutes and 8 seconds.
(3) The eighth-grade winner ran the mile in 4 minutes and 55 seconds. How many more seconds did it take for the seventh-grade winner to run a mile?

* 2. Laurie, Moesha, and Carrie went to the mall. Laurie spent $20,
(3) Moesha spent $25, and Carrie spent $30. How much did they spend in all?

* 3. Karl bought a foot-long sandwich. After he cut off a piece for his friend,
(3) his sandwich was 8 inches long. How much did he cut off for his friend?

* 4. Irina and her sister put their money together to buy a gift. Irina
(3) contributed $15. Together, they spent $32. How much did Irina's sister contribute?

* 5. Last week, Benji read 123 pages. This week Benji read 132 pages. How
(3) many more pages did Benji read this week than last week?

* 6. Ahnly's mother bought a dozen eggs. Ahnly ate two eggs on her salad.
(3) How many eggs are left?

7. Write a story for this equation: $10 - $4.05 = c.
(3)

8. Which two values of n make this equation true?
(1)
$$|n| = 3$$

9. Arrange these numbers from least to greatest: $-6, 5, -4, 3, -2$.
(1)

10. Compare:
(1)
    a. $-5 \bigcirc 1$ **b.** $-1 \bigcirc -2$

11. Compare:
(1)
    a. $-10 \bigcirc 10$ **b.** $|-10| \bigcirc |10|$

12. Compare: $|-5| \bigcirc$ (the distance from zero to -5 on a number line)
(1)

13. Refer to these expressions to answer the questions that follow:
(2)

$$6 + 4 = 10 \qquad 12 - 7 = 5$$

   a. Which number is the minuend?

   b. Which numbers are the addends?

14. Refer to these expressions to answer the questions that follow:
(2)

$$2 \cdot 3 = 6 \qquad \frac{20}{5} = 4$$

   a. Which number is a divisor?

   b. Which numbers are factors?

* 15. Consider the following expression:
(2)

$$4 \cdot (37 \cdot 25)$$

   a. How can the numbers in this expression be arranged to make the multiplication easier?

   b. Which properties did you use to rearrange the factors?

   c. What is the product?

16. Simplify, then compare:
$(1, 2)$

$$36 - 17 \bigcirc 17 - 36$$

17. Use the numbers 5 and 12 to provide an example of the Commutative
(2) Property of Addition.

18. Use the numbers 2, 3, and 5 to provide an example of the Associative
(2) Property of Addition.

19. Graph the numbers in this sequence on a number line:
(1)

$$\dots, -3, -1, 1, 3, \dots$$

20. When 15 is subtracted from 10, what is the difference?
(2)

21. If $6 + 3 = 9$, then what other addition fact and two subtraction facts are
(2) true for 6, 3, and 9?

22. If $2 \cdot 4 = 8$, then what other multiplication fact and two division facts are
(2) true for 2, 4, and 8?

23. Which whole number is not a counting number?
(1)

24. What number is the opposite of -5?
(1)

Analyze For problems **25–27,** find the value of the variable and name the property illustrated.

* 25. If $5t = 5$, then t is what number?
(1)

* 26. If $5 + u = 5$, then u is what number?
(1)

* 27. If $4x = 0$, then x is what number?
(1)

28. $100 − $90.90
(1)

29. 89 · $0.67
(1)

30. $\dfrac{\$72.18}{18}$
(1)

Early Finishers
Real-World Application

On Monday, Robert's bank account balance was $140.00. On Tuesday, Robert withdrew $56.00. On Thursday, Robert deposited money into his account. On Friday, Robert's account balance was $180.00. How much money did Robert deposit in his bank account on Thursday?

• Multiplication and Division Word Problems

facts | Power Up A

mental math |
a. **Calculation:** Ivan exercises about 40 minutes a day. About how many minutes does he exercise during the month of September?

b. **Number Sense:** 1000×1000

c. **Percent:** 50% of 30

d. **Measurement:** What mass is indicated on this scale?

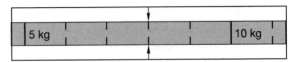

e. **Estimation:** 99×102

f. **Geometry:** A polygon with 6 sides is called a

  A pentagon.   **B** hexagon.   **C** heptagon.

g. **Number Sense:** $98 + 102$

h. **Calculation:** One score is 20. Two score is 40. Three score and 5 is 65. How many is four score and 7?

problem solving |
There are often multiple ways to solve a problem. We might find that certain solution methods are easier than others. Sometimes we can find a way to make it simpler. For example, to find the sum of 7, 8, 9, and 10, we can add 7 and 8 to get 15, then add 9 to get 24, and then add 10 to get 34. But notice how we can find the sum more simply:

$$7 + 8 + 9 + 10 \qquad 17 \times 2 = 34$$

We paired addends that have equal sums. Then we multiplied the sums by the number of pairs.

Problem: What is the sum when we add 7, 8, 9, 10, 11, 12, 13, 14, 15, and 16?

(*Understand*) We are asked to find the sum of a set of numbers.

(*Plan*) We can pair the addends at the ends of the list to get a sum 23. We can work inward to form other pairs that each have a sum of 23. Then we can count the pairs and multiply that number by 23.

Solve We count five pairs (7 and 16; 8 and 15; 9 and 14; 10 and 13; 11 and 12). Those five pairs account for all the addends in the problem. We multiply 23 by 5 and get 115.

$$7 + 8 + 9 + 10 + 11 + 12 + 13 + 14 + 15 + 16$$

Check Multiplication is a short way of adding equal groups, so it is reasonable that we can simplify an addition problem by multiplying. By making the problem simpler, we found a quick and easy solution method.

New Concept   *Increasing Knowledge*

In Lesson 3 we considered word problems that involve addition and subtraction. In this lesson we will look at problems that involve multiplication and division. Many word problems have an equal groups plot that can be expressed with this formula.

number of groups × number in group = total

$$n \times g = t$$

See how the numbers in the following story fit the formula.

In the auditorium there were 24 rows of chairs with 15 chairs in each row. There were 360 chairs in all.

The words "in each" are often used when describing the size of a group (*g*). We write the numbers in the formula.

$$n \times g = t$$
$$24 \times 15 = 360$$

An equal groups story becomes a problem if one of the numbers is missing. If the missing number is the product, we multiply. If the missing number is a factor, we divide the product by the known factor.

$$n \times g = t$$

Find by dividing. ——————⤴⤴ ⤴————— Find by multiplying.

Example 1

Analyze

What information are we given? What are we asked to find?

In the auditorium the chairs were arranged in rows with 15 chairs in each row. If there were 360 chairs, how many rows of chairs were there?

Solution

We use the formula to write an equation:

$$n \times g = t \qquad n \times 15 = 360$$

To find the missing factor in the equation, we divide 360 by 15. We find that there were **24 rows.**

Example 2

The total cost for five student tickets to the county fair was $28.75. The cost of each ticket can be found using which of these equations?

A $5t = \$28.75$

B $\$28.75t = 5$

C $\dfrac{t}{5} = \$28.75$

D $\dfrac{t}{\$28.75} = 5$

Solution

A $5t = \$28.75$

Example 3

Cory sorted 375 quarters into groups of 40 so that he could put them in rolls. How many rolls can Cory fill with the quarters? Explain why your answer is reasonable.

Solution

We are given the total number of quarters (375) and the number in each group (40). We are asked to find the number of groups.

$$n \times 40 = 375$$

We find the unknown factor by dividing 375 by 40.

$$375 \div 40 = 9 \text{ r } 15$$

The quotient means that Cory can make 9 groups of 40 quarters, and there will be 15 quarters remaining. Therefore, **Cory can fill 9 rolls.**

The answer is reasonable because 375 quarters is nearly 400 quarters, and 400 is ten groups of 40. So 375 quarters is not quite enough to fill ten rolls.

Practice Set

a. Carver gazed at the ceiling tiles. He saw 30 rows of tiles with 32 tiles in each row. How many ceiling tiles did Carver see?

b. **Analyze** Four student tickets to the amusement park cost $95.00. The cost of each ticket can be found by solving which of these equations?

A $\dfrac{4}{\$95.00} = t$

B $\dfrac{t}{4} = \$95.00$

C $4 \cdot \$95.00 = t$

D $4t = \$95.00$

c. Amanda has 632 dimes. How many rolls of 50 dimes can she fill? Explain why your answer is reasonable.

Written Practice *Strengthening Concepts*

1. The band marched in 14 rows with an equal number of members in
(4) each row. If there were 98 members in the band, how many members were in each row?

2. Olga bought five cartons of one dozen eggs. How many eggs did she
(4) buy?

3. The 5 km (5000 meters) hiking trail will have markers placed every
(4) 800 m. How many markers will be used? Explain why your answer
is reasonable.

* 4. **Represent** Fadi bought 3 pounds of apples at $1.98 per pound. The
(4) total cost of the apples can be found using which equation?

A $C = \dfrac{\$1.98}{3}$ 　　　　　　　　　**B** $C = 3 \times \$1.98$

C $\$1.98 = C \cdot 3$ 　　　　　　　　　**D** $\dfrac{3}{C} = \$1.98$

* 5. There are a total of two hundred students and chaperones going on a
(4) field trip. Each bus can hold 60 passengers. How many buses will be
used for the field trip? Explain why your answer is reasonable.

6. On Lee's first attempt at the long jump, she jumped 297 cm. On her
(3) second attempt she jumped 306 cm. How much farther was her second
mark than her first mark?

7. Zoe, Ella, and Mae ate lunch at the deli. The cost of Zoe's meal was $8,
(3) Ella's was $9, and Mae's was $7. What was the total of their bill (before
tax)?

8. Aaron was awake for 14 hours. He spent six of these hours at school.
(3) For how many of Aaron's wakeful hours was he not at school?

9. Write a word problem for this equation:
(3)
$$\$5 - p = \$1.50$$

10. Write a word problem for this equation:
(4)
$$n (\$25) = \$125$$

11. **Analyze** Arrange these numbers from least to greatest.
(1)
$$-1, 2, -3, -4, 5$$

12. Compare:
(1)
a. $-7 \bigcirc -8$ 　　　　　　　　　**b.** $5 \bigcirc -6$

13. Compare:
(1)
a. $|-7| \bigcirc |-8|$ 　　　　　　　　**b.** $|11| \bigcirc |-11|$

14. Name two numbers that are 7 units from zero on a number line.
(1)

15. Which two integer values for n make this equation true?
(1)
$$|n| = 1$$

16. Use braces and digits to indicate the set of odd numbers.
(1)

17. Graph the set of odd numbers on a number line.
(1)

18. Refer to these expressions to answer the questions that follow:
(2)

$$6 \times 5 = 30 \qquad \frac{12}{3} = 4$$

    a. Which numbers are factors?

    b. Which number is the quotient?

19. Refer to these expressions to answer the questions that follow:
(2)

$$3 + 5 = 8 \qquad 17 - 7 = 10$$

    a. Which number is the sum?

    b. Which number is the difference?

* 20. (Justify) Consider the following expression:
(2)

$$3 + (28 + 17)$$

    a. How can the numbers in the expression be arranged to make the addition easier?

    b. Which properties did you use?

* 21. (Analyze) Simplify and compare: $12 - 5 \bigcirc 5 - 12$
(1, 2)

22. If $3 \cdot 7 = 21$, what other multiplication fact and what two division facts are true for 3, 7, and 21?
(2)

23. If $2 + 4 = 6$, then what other addition fact and what two subtraction facts are true for 2, 4, and 6?
(2)

24. What number is the opposite of 10?
(1)

* 25. Which is the greater total: 4 groups of 20 or 5 groups of 15?
(4)

* 26. (Evaluate) Which change in height is greater: 54 in. to 63 in. or 48 in. to 59 in.?
(3)

* 27. (Analyze) Which total cost is less: 25 tickets for $5 each or 20 tickets for $6 each?
(4)

* 28. Use braces and digits to indicate the set of whole numbers.
(1)

* 29. (Classify) Consider the sets of integers and counting numbers. Which set includes the whole numbers?
(1)

* 30. Which two numbers satisfy this equation?
(1)

$$|n| = \frac{1}{2}$$

• Fractional Parts

Building Power

facts | Power Up A

mental math

   a. Calculation: Pedro completes six multiplication facts every ten seconds. How many does he complete in one minute?

   b. Estimation: $1.49 + $2.49 + $5.99

   c. Number Sense: Find the two missing digits: $11 \times \square = \square 7$

   d. Geometry: In this triangle, x must be greater than how many cm?

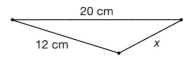

   e. Percent: 10% of 90

   f. Measurement: Sam measured from the wrong end of the ruler. How long is his pencil?

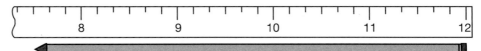

   g. Number Sense: 230 + 230

   h. Calculation: $3 + 5, -2, \times 7, -2, \div 5, \times 2, -1, \div 3$

problem solving | What are the next three numbers in this sequence?

5, 3, 8, 6, 11, …

Increasing Knowledge

We often use fractions to describe part of a group. A fraction is composed of a **numerator** and a **denominator.**

$$\begin{array}{cc} \text{numerator} & \dfrac{1}{3} \quad \text{number of parts described} \\ \text{denominator} & \phantom{\dfrac{1}{3}} \text{number of equal parts} \end{array}$$

We can find a fractional part of a group by dividing by the denominator and multiplying by the numerator.

A quart is 32 ounces. To find the number of ounces in $\frac{3}{4}$ of a quart we can divide 32 ounces by 4 ($32 \div 4 = 8$) to find the number of ounces in $\frac{1}{4}$. Then we can multiply by 3 ($3 \times 8 = 24$) to find the number of ounces in $\frac{3}{4}$.

$$\frac{3}{4} \times 32 \text{ oz} = 24 \text{ oz}$$

Example 1

There were 30 questions on the test. One third of the questions were true-false, and two fifths were multiple choice.

    a. How many questions were true-false?

    b. How many questions were multiple choice?

Solution

    a. To find $\frac{1}{3}$ of 30 we can divide 30 by the denominator 3. There were **10 true-false questions.**

    b. To find $\frac{2}{5}$ of 30 we can divide 30 by the denominator 5 and find there are 6 questions in each fifth. Then we multiply 6 by the numerator 2 to find the number of questions in $\frac{2}{5}$ of 30. There were **12 multiple choice questions.**

We will learn many strategies to compare fractions. One strategy we can use is to determine if the fraction is greater than or less than $\frac{1}{2}$. To estimate the size of a fraction, we compare the numerator to the denominator. For example, if Tyler has read $\frac{3}{10}$ of a book, then he has read less than half the book because 3 is less than half of 10. If Taylor has read $\frac{3}{4}$ of a book, then she has read more than half of the book because 3 is greater than half of 4.

Example 2

Arrange these fractions from least to greatest.

$$\frac{3}{6}, \frac{3}{5}, \frac{3}{8}$$

Solution

Thinking Skill

Connect

What do you notice about the denominators of the fractions when listed from least to greatest? The numerators? Use your observations to quickly arrange these fractions from least to greatest: $\frac{1}{10}, \frac{1}{4}, \frac{1}{8}, \frac{1}{12}$

For each fraction we can ask, "Is the fraction less than $\frac{1}{2}$, equal to $\frac{1}{2}$, or greater than $\frac{1}{2}$?" To answer the question we compare the numerator and the denominator of each fraction.

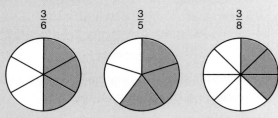

$$\frac{3}{6} \qquad \frac{3}{5} \qquad \frac{3}{8}$$

We find that $\frac{3}{8}$ is less than $\frac{1}{2}$, that $\frac{3}{6}$ is equal to $\frac{1}{2}$, and that $\frac{3}{5}$ is greater than $\frac{1}{2}$. Therefore, from least to greatest, the fractions are

$$\frac{3}{8}, \frac{3}{6}, \frac{3}{5}$$

Example 3

Compare $\frac{2}{3}$ of an hour and $\frac{2}{5}$ of an hour. Explain why your answer is reasonable.

Solution

$$\frac{2}{3} \text{ of an hour} > \frac{2}{5} \text{ of an hour}$$

The answer is reasonable because $\frac{2}{3}$ of an hour is greater than half an hour and $\frac{2}{5}$ of an hour is less than half an hour.

Practice Set

a. How many minutes is $\frac{1}{6}$ of an hour?

b. **Justify** Three fifths of the 30 questions on the test were multiple-choice. How many multiple-choice questions were there? Explain why your answer is reasonable.

c. Greta drove 288 miles and used 8 gallons of fuel. Greta's car traveled an average of how many miles per gallon of fuel?

d. Arrange these fractions from least to greatest.

$$\frac{5}{10}, \frac{5}{6}, \frac{5}{12}$$

Written Practice   *Strengthening Concepts*

* **1.** **Connect** How many yards is $\frac{1}{4}$ of a mile? (A mile is 1760 yards.)
  (5)

* **2.** Arrange these fractions from least to greatest.
  (5)
  $$\frac{2}{4}, \frac{2}{5}, \frac{2}{3}$$

* **3.** Two thirds of the game attendees were fans of the home team. If
  (5) 600 people attended the game, how many were fans of the home team?

* **4.** Risa typed 192 words in 6 minutes. She typed on average how many
  (5) words per minute?

* **5.** To prepare for the banquet, Ted bought eleven packages of rolls with
  (4) 24 rolls in each package. How many rolls did he buy?

* **6.** Patricia has a cell phone plan that includes 500 minutes of use per
  (3, 5) month. During the first week of a month, Patricia spoke on her phone for two and a half hours. How many minutes of her plan for the month remain?

* **7.** Volunteers collected non-perishable food items and packaged 14 items
  (4) per box. If the volunteers collected a total of 200 items, how many boxes could they fill?

8. Jenna budgeted $300 to redecorate a room. If she spent $54 on paint,
(3) how much money remains?

* 9. Three-fourths of the voters supported the bond measure. If there were
(5) 8000 voters, how many supported the bond measure?

* 10. **Represent** The art teacher bought 19 sketchbooks for $2.98 per book.
(4) What equation can be used to find the total cost of the sketchbooks?

A $c = 19 \cdot \$2.98$

B $\dfrac{19}{\$2.98} = c$

C $\$2.98 \cdot c = 19$

D $\dfrac{\$2.98}{19} = c$

11. The children sold lemonade for 25¢ per cup. If they sold forty cups of
(4) lemonade, how much money did they collect?

12. If the children collected $10 at their lemonade stand but had spent $2
(3) on supplies, what was their profit?

13. Write a word problem for this equation.
(4)
$$3 \times 20 \text{ sec} = t$$

14. Write a word problem for this equation.
(3)
$$10 \text{ lb} - w = 3 \text{ lb}$$

15. The water level in the beaker was 10 mL before the small toy was
(3) placed into it. After the toy was submerged in the water, the level was
23 mL. The toy has as much volume as how many mL of water?

16. Arrange these numbers from least to greatest.
(1)
$$-5, 7, 4, -3, 0$$

* 17. **Analyze** Compare: $\dfrac{1}{2} \bigcirc \dfrac{7}{15}$
(5)

18. What two numbers have an absolute value of 6?
(1)

19. Name a pair of addends with a sum of 6.
(2)

20. Name a pair of factors with a product of 6.
(2)

* 21. Use the numbers 2, 3, and 4 to provide an example of the Associative
(2) Property of Multiplication.

22. Graph the numbers in this sequence on a number line:
(1)
$$\ldots, -6, -3, 0, 3, 6, \ldots$$

23. Identify two numbers that are 50 units from zero on a number line.
(1)

24. If $10 - 6 = 4$, what other subtraction fact and what two addition facts
(2) are true for 10, 6, and 4?

25. If $6 \times 5 = 30$, what other multiplication fact and what two division facts
(2) are true for 6, 5, and 30?

26. What is the sum of 5 and its opposite?
(1)

27. Compare: $-8 \bigcirc -6$
(1)

Explain Decide if the statements in **28–29** are true or false and explain why. If the answer is false, provide a counterexample (an example that disproves the statement).

* **28.** Some integers are counting numbers.
(1)

* **29.** All fractions are integers.
(1, 5)

30. Use the Associative and Commutative Properties to make this
(1)   calculation easier. Justify the steps.

$$6 \cdot (17 \cdot 50)$$

• Converting Measures

facts | Power Up B

mental math
a. **Calculation:** The trolley rolled at a rate of 10 feet per second. How far does it go in a minute?

b. **Estimation:** 29×98

c. **Geometry:** In this triangle, *x* must be more than how many cm?

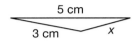

d. **Number Sense:** 25×200

e. **Measurement:** How long is this object?

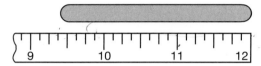

f. **Power/Roots:** $\sqrt{100} \times \sqrt{1}$

g. **Percent:** 50% of 10

h. **Calculation:** $63 \div 9, \times 3, + 1, \div 2, \times 3, - 1, \div 4, + 1, \div 3, - 4$

problem solving
Alberto, Bonnie, and Carl are sitting around a small table. They share a pencil. Alberto writes the number 1 with the pencil, and passes it to Bonnie. Bonnie writes the number 2 and passes the pencil to Carl. Carl writes the number 3 and passes the pencil to Alberto. If the pattern continues, who will write the number 11?

New Concept | *Increasing Knowledge*

Most math problems are about counts or measures. In the following tables, we show some common units from the two systems of measure used in the United States. The tables also show equivalent measures and abbreviations.

Abbreviations

Customary			
inches	in.	ounces	oz
feet	ft	pints	pt
		quarts	qt
yards	yd	gallons	gal
miles	mi	pounds	lb

Metric			
millimeters	mm	milliliters	mL
centimeters	cm	liters	L
meters	m	grams	g
kilometers	km	kilograms	kg

Equivalent Measures

Measure	U.S. Customary	Metric
Length	12 in. = 1 ft 3 ft = 1 yd 5280 ft = 1 mi	1000 mm = 1 m 100 cm = 1 m 1000 m = 1 km
	1 in. = 1 mi ≈	2.54 cm 1.6 km
Capacity	16 oz = 1 pt 2 pt = 1 qt 4 qt = 1 gal	1000 mL = 1 L
	1 qt ≈	0.95 Liters
Weight/ Mass	16 oz = 1 lb 2000 lb = 1 ton	1000 mg = 1 g 1000 g = 1 kg 1000 kg = 1 tonne
	2.2 lb ≈ 1.1 ton ≈	1 kg 1 metric tonne

Reading Math

The symbol ≈ means "approximately equal to."

Converting from one unit to another unit involves multiplication or division by a constant that defines the relationship between units.

number of first unit × constant = number of second unit

Example 1

Reggie took 36 big steps to walk the length of the hallway. If each big step was about a yard, then the hallway was about how many feet long?

Solution

The units are yards and feet. The constant is 3 ft per yd.

36 yd × 3 ft per yd = **108 ft**

The hallway is about 108 feet long. The answer is reasonable because it takes more feet than yards to measure a distance.

Example 2

A full 2-liter beverage bottle contains how many milliliters of liquid?

2-liter bottle

Solution

The units are liters (L) and milliliters (mL). The constant is 1000 mL per L.

2 L × 1000 mL per L = **2000 mL**

Example 3

The 5000 meter run is an Olympic event. How many kilometers is 5000 meters?

Solution

The units are meters (m) and kilometers (km). The constant is 1000 m per km.

$$k \cdot 1000 \text{ m per km} = 5000 \text{ m}$$

By dividing we find that 5000 m is **5 km.**

Example 4

James weighed 8 pounds 10 ounces at birth. How many ounces did James weigh?

Solution

This is a two-step problem. The first step is an equal groups problem. There are 8 groups (pounds) and 16 oz in each group.

$$n \times g = t$$
$$8 \text{ lbs} \times 16 \text{ oz per lb} = t$$

We multiply 8 pounds times 16 ounces, which is 128 ounces.

The second step of the problem is a combining problem. The expression "8 pounds 10 ounces" means 8 pounds plus 10 ounces, so we add 10 ounces to 128 ounces.

$$s + m = t$$
$$128 \text{ oz} + 10 \text{ oz} = t$$

We find that James weighed **138 ounces** at birth.

Practice Set

a. A room is 15 feet long and 12 feet wide. What are the length and width of the room in yards?

b. Nathan is 6 ft 2 in. tall. How many inches tall is Nathan?

c. Seven kilometers is how many meters?

Written Practice   *Strengthening Concepts*

* 1. (6)   *Analyze*   A movie is 180 minutes long. How long is the movie in hours?

* 2. (6)   A recipe calls for a half pint of cream. How many ounces of cream are required?

* 3. (6)   The bottle of water weighed 2 kg. What is its weight in grams?

* 4. (5)   Three quarters of the students wore the school colors. If there are 300 students, how many wore school colors?

5. Each book that Violet carried weighed 2 pounds. If she carried
(4) 12 books, how heavy was her load?

6. Jorge read 23 pages before dinner and 18 pages after dinner. How
(3) many pages did he read in all?

* 7. **Analyze** The students were divided evenly into five groups to work
(5) on a project. Two of the groups completed the project before class
ended. What fraction of the students completed the project before class
ended?

8. After the demolition, there were 20 tons of rubble to remove. Each truck
(4) could carry 7 tons of rubble. How many trucks were required to carry
the rubble away?

9. The driving distance from San Antonio to Dallas is 272 miles. Along the
(3) route is Austin, which is 79 miles from San Antonio. What is the driving
distance from Austin to Dallas?

10. A serving of three cookies of a certain kind contains 150 calories. There
(4) are how many calories in one cookie of this kind?

11. How many hours are in a week?
(4, 6)

12. Valerie rode her bicycle from home to the market in 12 minutes, spent
(3) 20 minutes at the market, then rode her bicycle home in 15 minutes.
When she arrived home, how much time had passed since Valerie left
home?

* 13. The glass is three-quarters full. If the glass can hold 12 ounces, how
(5) many ounces are in it now?

14. Arrange these numbers from least to greatest: $-2, \frac{5}{7}, 1, 0, \frac{1}{2}$
(1, 5)

15. Compare: $-5 \bigcirc 4$ **16.** Compare: $|-2| \bigcirc |-3|$
(1)   (1)

17. Compare: $5 \bigcirc |-5|$
(1)

18. Use the numbers 5 and 3 to provide an example of the Commutative
(2) Property of Addition.

19. Use braces and digits to indicate the set of even counting numbers.
(1)

20. What is the absolute value of zero?
(1)

* 21. **Model** Graph 1 on a number line and two numbers that are 3 units
(1) from 1.

22. If $12 + 3 = 15$, what other addition fact and what two subtraction facts
(2) are true for 12, 3, and 15?

23. If $40 \div 8 = 5$, what other division fact and what two multiplication facts
(2) are true for 40, 8, and 5?

24. What number is the opposite of 100?
(1)

* 25. Give an example of an integer that is not a whole number.
(1)

* 26. Give an example of an integer that is a whole number.
(1)

Explain Decide if the statements in problems **27–29** are true or false, and explain why. If the statement is false, give a counterexample (an example that disproves the statement).

* 27. Every counting number is greater than every integer.
(1)

* 28. Every positive number is greater than every negative number.
(1)

* 29. Every fraction is less than 1.
(1)

* 30. Connect Find two solutions for this equation.
(1)

$$|x| = 7$$

Early Finishers
Real-World Application

Recycling helps to preserve landfill space and save natural resources for the future. Suppose Americans recycle approximately $\frac{3}{5}$ of the 62 million newspapers bought daily. At that rate, how many newspapers are recycled in one week?

• Rates and Average
• Measures of Central Tendency

facts | Power Up B

mental math |

a. **Calculation:** How many miles per minute is 300 miles per hour?

b. **Estimation:** $2.99 + $9.99 + $6.99

c. **Number Sense:** 45 × 200

d. **Percent:** 10% of 50

e. **Measurement:** Two of these sewing needles laid end-to-end would be how long?

f. **Geometry:** In this triangle, x must be more than how many inches?

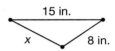

g. **Number Sense:** 1527 + 23

h. **Calculation:** 16 ÷ 2, ÷ 2, ÷ 2, ÷ 2, ÷ 2

problem solving | The sum of the odd numbers from one to seven is 16.

$$1 + 3 + 5 + 7 \qquad 8 \times 2 = 16$$

Find the sum of the odd numbers from one to fifteen.

New Concepts   *Increasing Knowledge*

rates and average | A **rate** is a division relationship between two measures. We usually express a rate as a **unit rate,** which means that the number of the second measure is 1. Here are some unit rates:

65 miles per hour (65 mph)

24 miles per gallon (24 mpg)

32 feet per second (32 ft/sec)

15 cents per ounce ($0.15/oz)

Rate problems involve the two units that form the rate.

number of **hours** × 65 **miles per hour** = number of **miles**

number of **gallons** × 24 **miles per gallon** = number of **miles** number of **seconds** × 32 **feet per second** = number of **feet**

number of **ounces** × 15 **cents per ounce** = number of **cents**

Example 1

Driving at an average speed of 55 mph, about how far will the driver travel in 6 hours?

Solution

We are given the rate and the number of hours. We will find the number of miles.

$$6 \text{ hours} \times 55 \text{ miles per hour} = n \text{ miles}$$

The driver will travel about 55 miles each hour, so in 6 hours the trucker will drive 6 × 55, or **330 miles.**

Example 2

Heather finished the 100 km bike race in four hours. What was her average speed in kilometers per hour?

Solution

We are given the two units and asked for the rate. The desired rate is kilometers per hour. This means we divide the number of kilometers by the number of hours.

$$\frac{100 \text{ km}}{4 \text{ hr}} = 25 \frac{\text{km}}{\text{hr}}$$

We divide the numbers. We can use a division bar to indicate the division of units as shown or we can use the word "per." Heather's average speed was **25 kilometers per hour.**

Example 3

Traveling at an average rate of 58 miles per hour, about how long would it take a car to travel 20 miles?

Solution

At 60 mph a car travels one mile each minute. Since 58 mph is nearly 60 mph, the trip would take **a little more than 20 minutes.**

In the examples above, we used the term *average speed.* The speed of the truck in example 1 probably fluctuated above and below its average speed. However, the total distance traveled was the same as if the truck had traveled at a steady speed—the average speed.

In common usage, the **average** of a set of numbers is a central number that is found by dividing the sum of the elements of a set by the number of elements. This number is more specifically called the **mean.**

Calculating an average often involves two steps. First we find the total by *combining,* then we make *equal groups.* As an example, consider these five stacks of coins:

There are 15 coins in all (1 + 5 + 3 + 4 + 2 = 15). Finding the average number of coins in each stack is like leveling the stacks (15 ÷ 5 = 3).

Referring to the original stacks, we say the average number of coins in each stack is three.

Example 4

In the four 8th grade classrooms at Keeler School there are 28, 29, 31, and 32 students. What is the average number of students in an 8th grade classroom at Keeler School?

Solution

Thinking Skill

Predict

Without calculating, could we predict whether 35 is a reasonable answer for this problem? Explain.

The average is a central number in a range of numbers. The average is more than the least number and less than the greatest number. We add the four numbers and then divide the sum by four.

$$\text{Average} = \frac{28 + 29 + 31 + 32}{4} = \frac{120}{4} = 30$$

The average number of students in an 8th grade classroom at Keeler School is **30 students.**

measures of central tendency

The average, or mean, is one measure of central tendency. Other measures include the **median,** which is the central number in a set of data, and the **mode,** which is the most frequently occurring number.

To find the median we arrange the numbers in order from least to greatest and select the middle number or the average of the two middle numbers. To find the mode we select the number that appears most often.

Example 5

Below we show the prices of new homes sold in a certain neighborhood. Find the mean, median, and mode of the data. Which measure should a researcher use to best represent the data?

Home Prices (in thousand $)

170	191
208	175
185	175
209	187
181	195
183	219

Prices of New Homes Sold (in thousand $)

A bar graph with vertical axis labeled 0 to 4, and horizontal axis labeled 170, 180, 190, 200, 210, 220. Bars: 170–180 = 3, 180–190 = 4, 190–200 = 2, 200–210 = 2, 210–220 = 1.

Visit www.
SaxonPublishers.
com/ActivitiesC3
*for a graphing
calculator activity.*

Solution

Mean:

$$\text{mean} = \frac{170 + 191 + 208 + 175 + 185 + 175 + 209 + 187 + 181 + 195 + 183 + 219}{12}$$

mean ≈ **190**

Median: Order the numbers to find the central value.

170 175 175 181 183 185 187 191 195 208 209 219

$$\text{median} = \frac{185 + 187}{2} = \textbf{186}$$

Mode: The most frequently occurring number is **175.**

If the researcher is reporting to someone interested in property tax revenue, the **mean** is a good statistic to report. It may be useful in predicting revenue from property taxes.

However, if the researcher is reporting to prospective buyers, the **median** is a good statistic to report. It tells the buyer that half of the new homes cost less than $186,000.

The mean may be misleading to a buyer—its value is greater than the median because of a few expensive homes. Only $\frac{1}{3}$ of the homes were sold for more than $190,000.

We may also refer to the **range** of a set of data, which is the difference between the greatest and least numbers in the set. The range of home prices in example 4 is

$$\$219,000 - \$170,000 = \textbf{\$49,000}$$

Practice Set

a. Alba ran 21 miles in three hours. What was her average speed in miles per hour?

b. How far can Freddy drive in 8 hours at an average speed of 50 miles per hour?

c. Estimate If a commuter train averages 62 miles per hour between stops that are 18 miles apart, about how many minutes does it take the train to travel the distance between the two stops?

d. Analyze If the average number of students in three classrooms is 26, and one of the classrooms has 23 students, then which of the following must be true?

A At least one classroom has fewer than 23 students.

B At least one classroom has more than 23 students and less than 26 students.

C At least one classroom has exactly 26 students.

D At least one classroom has more than 26 students.

e. What is the mean of 84, 92, 92, and 96?

f. The heights of five basketball players are 184 cm, 190 cm, 196 cm, 198 cm, and 202 cm. What is the average height of the five players?

The price per pound of apples sold at different grocery stores is reported below. Use this information to answer problems **g–i**.

$0.99	$1.99	$1.49	$1.99
$1.49	$0.99	$2.49	$1.49

g. Display the data in a line plot.

h. Compute the mean, median, mode, and range of the data.

i. (Verify) Rudy computed the average price and predicted that he would usually have to pay $1.62 per pound for apples. Evaluate Rudy's prediction based on your analysis of the data.

Written Practice   *Strengthening Concepts*

1. In the act there were 12 clowns in each little car. If there were
(4)   132 clowns, how many little cars were there?

* 2. There were 35 trees on the property. Four fifths are deciduous (they lose
(5)   their leaves in the fall and winter). How many trees on the property are deciduous?

* 3. Melody weighed 7 pounds, 7 ounces at birth. How many ounces did
(6)   Melody weigh?

* 4. The bird flew 80 km in four hours. What was the bird's average speed in
(7)   km/hr?

* 5. If the plane travels at 400 miles per hour, how far will it fly in 5 hours?
(7)

* 6. How many pints are in a gallon of milk?
(6)

* 7. The number of points scored by the starting players on the basketball
(7)   team were 8, 12, 16, 19, and 20. What was the average number of points scored per starting player?

In problems **8–11,** identify the plot, write an equation, and solve the problem.

* 8. In the package were three rows of 8 colored pencils. How many colored
(4)   pencils were there in all?

* 9. Felicia scored 15 points in the first half and 33 points in all. How many
(3)   points did she score in the second half?

* 10. There were 20 ounces of birdseed in the feeder before the birds came.
(3)   When the birds left, there were 11 ounces of birdseed in the feeder. How many ounces of birdseed did the birds remove from the feeder?

* 11. President Ronald Reagan ended his second term in office in 1989. He
(3)   had served two terms of 4 years each. What year did he begin his first term?

12. Write a word problem for this equation: 5 lbs × 9 = w
(4)

13. Write a word problem for this equation: $14 − x = $3
(3)

14. Arrange these numbers from least to greatest: $0, -1, \frac{2}{3}, 1, \frac{2}{5}$.
(1, 5)

* 15. Compare: $-11 \bigcirc -10$
(1)

* 16. **Analyze** For each letter, write a word to describe its role:
(2)

$$a + b = c \qquad\qquad d \cdot e = f$$

17. Use the numbers 3 and 8 to provide an example of the Commutative
(2) Property of Multiplication.

18. Use the numbers 1, 2, and 3 to provide an example of the Associative
(2) Property of Addition.

19. Graph the numbers in this sequence on a number line:
(1)

$$\ldots, -5, -2, 1, 4, 7, \ldots$$

20. When 20 is subtracted from zero, what is the result?
(2)

21. If $7 - 5 = 2$, what other subtraction fact and what two addition facts are
(2) true for 7, 5, and 2?

22. If $12 \div 4 = 3$, what other division fact and what two multiplication facts
(1) are true for 3, 4 and 12?

23. True or false: Every whole number is a counting number. Explain your
(1) answer.

24. What number is the absolute value of -90?
(1)

25. What number is the opposite of 6?
(1)

* 26. Find the value of the variable that makes the equation true and name
(2) the property illustrated.

    a. If $6x = 6$, then x is what number?

    b. If $7 + y = 7$, then y is what number?

Generalize Simplify.

* 27. $2020 - 10{,}101$
(2)

28. $48 \cdot \$0.79$
(2)

29. $\dfrac{\$60.60}{12}$
(2)

30. Use the Associative and Commutative Properties to make the
(2) calculations easier. Justify the steps.

$$4 \cdot (12 \cdot 75)$$

• Perimeter and Area

facts | Power Up B

mental math

a. Calculation: Dante rode his bike a mile in 5 minutes. At that rate, how many miles can he ride in an hour?

b. Number Sense: Find the two missing digits: $7 \times \square = \square 1$

c. Measurement: What is the mass of 3 pumpkins this size?

d. Powers/Roots: $\sqrt{16} + \sqrt{4}$

e. Number Sense: $651 + 49$

f. Percent: 10% of 40

g. Geometry: In this triangle, x must be more than how many inches?

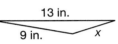

h. Calculation: 3 dozen, $\div\ 4$, $+\ 1$, $\div\ 2$, $+\ 1$, $\div$ 2, $+\ 1$, $\div\ 2$

problem solving

Guess and check is a strategy in which we try a possible answer and then check the result. If our guess is incorrect, we use the result to improve our next guess. To solve some problems, we might find it useful to simply try all the possible answers and then compare the results.

Problem:

The numbers 3, 4, and 5 are placed in the shapes above and the numbers are multiplied and added as shown. In order to make the result the least, which number should go in the triangle?

(**Understand**) Two of the three numbers are multiplied, and then the third number is added to the product.

(**Plan**) We will try each possible arrangement and find the one with the least result, then record which number goes in the triangle.

(**Solve**) There are six possible arrangements:

$3 \cdot 4 + 5 = 17$	$3 \cdot 5 + 4 = 19$	$4 \cdot 5 + 3 = 23$
$4 \cdot 3 + 5 = 17$	$5 \cdot 3 + 4 = 19$	$5 \cdot 4 + 3 = 23$

We might notice that the order of the first two numbers does not affect the result, since reversing factors does not change their product. In both arrangements with a result of 17 (the least result), we see that 5 is in the third location, which corresponds to the triangle.

Check We tried each possible arrangement and then found the answer. It makes sense that 5 must be in the triangle to produce the least result. When we multiply numbers greater than 2, we get greater results than when add the same numbers. So multiplying the lesser two of the three numbers will produce the least result.

New Concept   *Increasing Knowledge*

The floor of a room can help us understand perimeter and area. Many rooms have a baseboard trim around the edge of the floor. The baseboard represents the **perimeter** of the room, which is the distance around the room. The total length of the baseboards in a room is a measure of the perimeter of the room. (The actual perimeter of a room includes the opening for each doorway as well.)

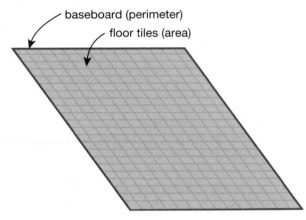

baseboard (perimeter)
floor tiles (area)

Some rooms are covered with floor tiles. The tiles cover the **area** of the floor, which is a measure of its surface. The number of floor tiles used to cover the floor is a measure of the area of the room.

Example 1

A rectangular room is five yards long and four yards wide.

a. How many feet of baseboard are needed to reach around the room?

b. How many square yards of carpet are needed to cover the floor?

c. Describe whether the tasks in a and b are similar to finding the perimeter or the area.

5 yd

4 yd

Solution

a. To find the distance around the room we add the lengths of the four sides.

$$5 \text{ yd} + 4 \text{ yd} + 5 \text{ yd} + 4 \text{ yd} = 18 \text{ yd}$$

We are asked for the number of *feet* of baseboard needed. Each yard is 3 feet so we multiply:

$$18 \text{ yd} \times 3 \text{ ft per yd} = 54 \text{ ft}$$

We find that **54 feet** of baseboard are needed. (In actual construction we would subtract from the total length the opening for each doorway.)

b. A one yard-by-one yard square is a square yard. We can find the number of square yards needed by multiplying the length in yards by the width in yards.

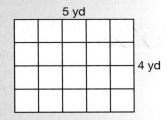

$$5 \text{ yd} \cdot 4 \text{ yd} = 20 \text{ sq. yd}$$

We find that **20 square yards** of carpet are needed to cover the floor.

c. In part a we found the distance around the room, which is its **perimeter.** In part b we found the **area** of the room.

We often refer to the length and width of a rectangle.

To find the perimeter of a rectangle, we add two lengths and two widths. To find the area of a rectangle, we multiply the length and width.

**Formulas for the Perimeter
and Area of a Rectangle**

$$P = 2l + 2w$$
$$A = lw$$

In example 2 we compare the effects of doubling the length and width of a rectangle on the perimeter and area.

Example 2

The figures below show the dimensions of room A and room B.

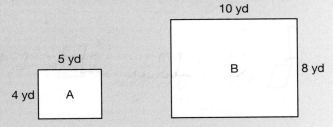

a. **Find the perimeter of each room in yards.**

b. **Find the area of each room in square yards.**

c. **How does doubling the length and width affect the perimeter of the rectangle?**

d. **How does doubling the length and width affect the area of the rectangle?**

a. The perimeter of room A is **18 yd,** and the perimeter of room B is **36 yd.**

b. The area of room A is **20 sq. yd,** and the area of room B is **80 sq. yd.**

c. **Doubling the length and width also doubles the perimeter.**

d. **Doubling the length and width results in an area four times as large.**

The floor plan of a room with a closet might look like two joined rectangles. To find the perimeter we add the lengths of each side. To find the area we add the areas of the two rectangles. In the following example we first find the lengths of the sides that are not labeled.

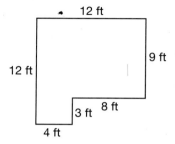

Example 3

Find the perimeter and area of this figure. All angles are right angles (square corners).

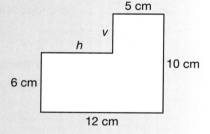

Solution

The horizontal segments labeled *h* and 5 cm must total 12 cm, so *h* must be 7 cm. The vertical segments labeled 6 cm and *v* must total 10 cm, so *v* must be 4 cm. We add the six sides to find the perimeter.

Perimeter = 5 cm + 10 cm + 12 cm + 6 cm + 7 cm + 4 cm

= **44 cm**

The area is the sum of the areas of the two rectangles.

Area = 4 cm · 5 cm + 6 cm · 12 cm

= 20 sq. cm + 72 sq. cm

= **92 sq. cm**

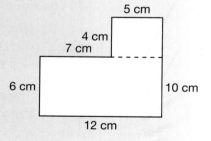

Thinking Skill

Analyze

Describe another way we could find the area of the figure in example 3 by first finding the area of two different rectangles.

Practice Set

a. **Evaluate** Jarrod is placing baseboards and one-foot-square floor tiles in a 12 ft-by-8 ft room. Ignoring doorways, how many feet of baseboard does he need? How many tiles does he need?

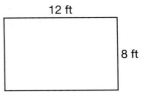
12 ft

8 ft

Find the perimeter and area of each rectangle.

b.

3 cm

4 cm

c.

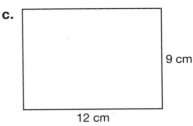

9 cm

12 cm

d. **Analyze** The length and width of the rectangle in **c** are three times the length and width of the rectangle in **b.** The perimeter and area of the rectangle in **c** are how many times those measures in **b?**

e. Placing 12 tiles side by side, Pete can make a 12-by-1 rectangle. Name two other rectangles Pete can make with the 12 tiles.

f. Find the perimeter and area of a room with these dimensions.

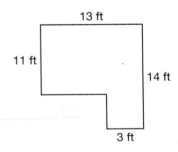
13 ft

11 ft

14 ft

3 ft

Written Practice   *Strengthening Concepts*

* **1.** **Connect** Joel will fence a rectangular yard measuring 20 yards by 15 yards. How many yards of fencing will he need? What will be the area of the fenced yard?
  (8)

* **2.** **Evaluate** Joel fences a larger yard with dimensions that are four times the dimensions in problem 1. Find the dimensions of the larger yard and then find its perimeter and area.
  (8)

* **3.** The perimeter and area of the yard in problem 2 are how many times those measured in problem 1?
  (8)

* **4.** Placing 30 tiles side by side, Patricia can make a 30-by-1 rectangle. Name three other rectangles Patricia can make with the 30 tiles.
  (8)

5. Each volume contains 12 chapters. If there are 5 volumes, how many chapters are there in all?
  (4)

* **6.** The net weight of the bag of almonds is 2 lb 8 oz. How many ounces of almonds are there?
  (6)

* 7. A video game was discounted and now costs two thirds of its original
 (5) price. If it originally cost $45, how much does it cost now?

* 8. **Evaluate** Mahesh rides a bus to work. The number of minutes his daily
 (7) bus ride took over a two-week period are recorded below:

	M	T	W	Th	F
Week 1	12	8	10	9	12
Week 2	8	7	12	9	10

   a. Find the mean, median, mode, and range of the data.

   b. Which measure would Mahesh use to describe how many minutes
   the duration of his bus ride may differ from day to day?

9. The auditorium has twelve rows of 24 seats. How many seats are there
 (4) in all?

10. What is the quotient when the sum of 15 and 12 is divided by the
 (2) difference of 15 and 12?

* 11. A mile-high city is how many feet above sea level?
 (6)

* 12. Tina drove 168 miles in 3 hours. What was her average rate in miles per
 (7) hour?

13. Write a word problem for this equation:
 (4)
$$5 \text{ mi} \times 4 = d$$

14. Write a word problem for this equation:
 (3)
$$12 \text{ in.} - x = 7 \text{ in.}$$

15. Arrange these numbers from least to greatest.
 (1, 5)
$$\frac{1}{2}, -1, \frac{5}{7}, -2, \frac{2}{6}, 1$$

16. Compare: $-100 \bigcirc 10$
 (1)

17. Compare: $-3 \bigcirc -4$
 (1)

18. Compare: $|-3| \bigcirc |-4|$
 (1)

* 19. Use the numbers 5 and 7 to provide an example of the Commutative
 (2) Property of Addition.

20. Graph the numbers in this sequence on a number line.
 (1)
$$\ldots, -7, -3, 1, 5, 9, \ldots$$

21. What is the absolute value of 15?
 (1)

22. If $3 + 4 = 7$, what other addition fact and what two subtraction facts are
 (2) true for 3, 4, and 7?

23. What is the opposite of 3?
(1)

24. What two numbers are 5 units from 0?
(1)

Explain For problems **25–27,** state whether the sentence is true or false, and explain why. If the answer is false, provide a counterexample.

* 25. Every negative number is an integer.
(1)

26. Every natural number is an integer.
(1)

27. The sum of two whole numbers is a whole number.
(1)

28. Find two solutions to this equation: $|x| = 15$.
(1)

29. Find the product of 37 and 68.
(2)

* 30. **Analyze** Three out of every four restaurant patrons did not order the
(5) special. What fraction did order the special?

Early Finishers
Real-World Application

The Tour de France is a well-known cycling competition in which cyclists from around the world are chosen to participate. The Tour's length varies from year to year but is usually between 3000 and 4000 km in length. If one mile is approximately 1.6 km, approximately how long is the Tour de France in miles?

• Prime Numbers

facts | Power Up B

mental math |

a. Calculation: Eight cups per day is how many cups per week?

b. Estimation: $99.50 × 4

c. Number Sense: 2 × 32

d. Measurement: Find the mass of 10 of these cups.

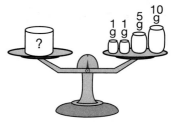

e. Geometry: In this triangle, *x* must be greater than what number?

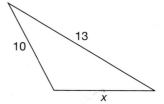

f. Number Sense: 1307 + 13

g. Percent: 50% of 40

h. Calculation: 49 ÷ 7, × 2, + 1, ÷ 3, × 9, + 5, ÷ 5, − 11

problem solving | Fraction problems can be difficult to visualize but are often easy to draw. Drawing a fraction problem can help us sort through what is actually being asked.

Problem: Given that $\frac{1}{7}$ of a number is $\frac{1}{5}$, what is $\frac{5}{7}$ of that number?

[**Understand**] We have been told that $\frac{1}{7}$ of a number is $\frac{1}{5}$ and have been asked to determine what $\frac{5}{7}$ of that number is.

[**Plan**] We know that $\frac{1}{7}$ of a number is one of seven pieces the number can be divided into. We will draw a picture of the "number" divided into its seven fractional parts.

[**Solve**] We draw a whole divided into seven equal parts. Each seventh represents one-seventh of the value of the whole, unknown number. We then label one of the sevenths as $\frac{1}{5}$, because we are told that each seventh has the value $\frac{1}{5}$:

$\frac{1}{5}$						

We want to know the value of $\frac{5}{7}$ of the number, so we label five of the sevenths as $\frac{1}{5}$:

We see that $\frac{5}{7}$ of the number is $\frac{1}{5} + \frac{1}{5} + \frac{1}{5} + \frac{1}{5} + \frac{1}{5}$, which is **1**.

(Check) It may seem unusual to consider that $\frac{1}{7}$ of a whole can have the value $\frac{1}{5}$, or that $\frac{5}{7}$ of a whole can have the value 1, but drawing a diagram showed us that it is possible.

New Concept
Increasing Knowledge

Prime numbers are counting numbers greater than 1 that have exactly two different counting number factors: the number itself and 1. For example, the numbers 2, 3, 5, 7, 11, and 13 are prime. We may use an area model to illustrate numbers that are prime and numbers that are not prime.

Suppose we have six square tiles. Fitting the tiles side by side we can form two different rectangles, a 6-by-1 rectangle and a 3-by-2 rectangle.

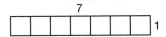

These two rectangles illustrate the two counting number factor pairs of 6, which are 6 × 1 and 3 × 2. If we have 7 square tiles, we can form only one rectangle, a 7-by-1 rectangle. This is because there is only one factor pair for 7, which is 7 × 1.

A prime number has only one factor pair, so a prime number of tiles can only form a "by 1" rectangle. The rectangles above illustrate that 7 is prime and that 6 is not prime.

Counting numbers with more than two factors are **composite numbers.** Six is a composite number. The numbers 4, 8, 9, 10, and 12 are other examples of composite numbers. Counting numbers greater than 1 are either prime or composite.

Composite numbers are so named because they are *composed* of two or more prime factors. For example, 6 is composed of the prime factors 2 and 3. We write the **prime factorization** of a composite number by writing the number as a product of prime numbers. The prime factorization of 6 is 2 · 3.

Example 1

Write the prime factorization of

   a. 36.

   b. 45.

a. A factor tree can help us find the prime factorization of a number. Here we show one possible factor tree for 36. We split 36 into the factors 6 and 6. Then we split each 6 into the factors 2 and 3. We stop when the factors are prime; we do not use the number 1 in a factor tree. We order the prime numbers at the end of the "branches" to write the prime factorization.

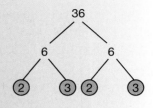

$$36 = 2 \cdot 2 \cdot 3 \cdot 3$$

b. Another method for finding the prime factorization of a number is by dividing the number by a prime number and then dividing the quotient by a prime number. We continue dividing in this way until the quotient is prime.

$$\begin{array}{r} 3 \\ 3\overline{)9} \\ 5\overline{)\;45} \end{array}$$

We use the divisors, 5 and 3, and the final quotient, 3, to write the prime factorization.

$$45 = 3 \cdot 3 \cdot 5$$

One way to test if a counting number is prime or composite is to determine if it is **divisible** (can be divided) by a counting number other than 1 and itself. For example, 2 is prime, but other multiples of 2 are composite because they are divisible by 2. Thus, even numbers greater than 2 are composite. In the table below we list some tests for divisibility.

Divisibility Tests

Condition	Number is Divisible by	Example Using 3420
the number is even (ends with 0, 2, 4, 6, or 8)	2	342<u>0</u>
the sum of the digits is divisible by 3	3	3 + 4 + 2 + 0 = <u>9</u> and 9 is divisible by 3
the number ends in 0 or 5	5	342<u>0</u>

We can combine these tests to build divisibility tests for other numbers. Here are some examples:

- A number divisible by 2 and 5 is divisible by 10 (2 × 5). Thus, 3420 is divisible by 10.

- A number divisible by 2 and 3 is divisible by 6 (2 × 3). Thus, 3420 is divisible by 6.

- A number is divisible by 9 if the sum of its digits is divisible by 9 (3 × 3). Thus, 3420 is divisible by 9 because the sum of its digits (3 + 4 + 2 + 0 = 9) is divisible by 9.

• A number is divisible by 4 if its last two digits are divisible by 4. Thus, 3420 is divisible by 4 since its last two digits (20) are divisible by 4.

Example 2

Thinking Skill

Predict

Is the number 16,443,171 prime or composite? How do you know?

Determine whether the following numbers are prime or composite and state how you know.

a. 1,237,526

b. 520,611

Solution

a. The number is **composite** because it is even and thus divisible by 2.

b. The number is **composite** because the sum of its digits is divisible by 3 (5 + 2 + 0 + 6 + 1 + 1 = 15).

Practice Set

a. **Model** Is 9 a prime or composite number? Draw rectangles using 9 squares to support your answer.

b. Write the first 10 prime numbers.

c. **Represent** Draw a factor tree for 36. Make the first two branches 4 and 9.

d. Find the prime factors of 60 by dividing by prime numbers.

Write the prime factorization of each number in **e–g.**

e. 25

f. 100

g. 16

h. Choose a number to complete the statement. Justify your choice.

A multi-digit number might be prime if its last digit is _____.

A 4   **B** 5   **C** 6   **D** 7

i. Six squares can be arranged to form two different rectangles.

The models illustrate that 6 is composite and not prime. Illustrate that 12 is composite by sketching three different rectangles with 12 squares each.

Written Practice   *Strengthening Concepts*

* **1.** Three thousand, six hundred ninety people finished the half-marathon.
  (5) If $\frac{2}{3}$ of the finishers walked a portion of the course, how many finishers walked during the race?

2. A doorway that is 2 meters high is how many centimeters high?
(6)

3. Driving at 80 km per hour, how far will Curtis drive in 2 hours, in 3 hours,
(7) and in 4 hours?

* 4. **Evaluate** Identify the plot, write an equation, and solve the problem:
(3, 7) The railroad workers laid 150 miles of track in the first month. At the end of the two months, the total was up to 320 miles.

   a. How many miles did they lay in the second month?

   b. The workers laid an average of how many miles of track each month?

5. **Explain** A traffic counter recorded the number of vehicles heading
(7) north through an intersection for a seven-day period. The following counts were recorded for the seven days.

86, 182, 205, 214, 208, 190, 126

   a. What was the median number of vehicles counted?

   b. What was the mean number of vehicles counted?

   c. Are most of the counts closer to the median or mean? Why?

6. A mile is 5280 ft. How many yards is a mile?
(6)

7. List the first eight prime numbers.
(9)

Simplify.

8. 100 cm − 68 cm
(2)

9. $2970.98 − $1429.59
(2)

10. 5280 ÷ 30
(2)

11. $123.45 ÷ 3
(2)

* 12. Find the perimeter and area.
(8)

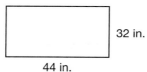

32 in.

44 in.

* 13. Find the perimeter.
(8)

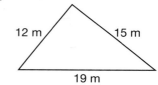

12 m   15 m

19 m

14. Find the perimeter and area of this figure.
(8)

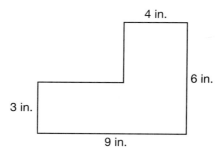

4 in.

6 in.

3 in.

9 in.

* 15. **Connect** A rectangular room is 15 feet long and 12 feet wide.
(8)

   a. How many feet of baseboard are needed to reach around the room? (Ignore door openings.)

   b. How many one-foot-square tiles are needed to cover the floor?

For problems **16–19,** find the value of the variable that makes the equation true and state the property illustrated.

16. 5 + 4 = 4 + a
(2)

17. 17 · 18 = b · 17
(2)

18. 20 · c = 20
(2)

19. 21d = 0
(2)

20. **a.** What is the absolute value of 9?
(1)

     b. Simplify and compare: $|-12|$ ◯ $|11|$

21. If 13 and 5 are the factors, what is the product?
(2)

22. If 225 is the dividend and 15 is the divisor, then what is the quotient?
(2)

For problems **23–26,** choose *all* the correct answers.

23. Twenty-seven is a member of which set(s) of numbers?
(1)
     A whole numbers   **B** counting numbers   **C** integers

24. Zero is a member of which set(s) of numbers?
(1)
     A whole numbers   **B** counting numbers   **C** integers

25. Negative two is a member of which set(s) of numbers?
(1)
     A whole numbers   **B** counting numbers   **C** integers

* 26.   Analyze   The number 5280 is divisible by which of these numbers?
(9)
     A 2   **B** 3   **C** 4   **D** 5

* 27.   List   List all the prime numbers between 30 and 50.
(9)

* 28. Use a factor tree to find the prime factors of 490. Then write the prime factorization of 490.
(9)

* 29. Use division by primes to find the prime factors of 48. Then write the prime factorization of 48.
(9)

30. Rearrange factors to make the calculation easier. Then simplify. Justify your steps.
(2)

$$40 \cdot (23 \cdot 50)$$

• Rational Numbers
• Equivalent Fractions

facts

Power Up B

mental math

a. Calculation: The windmill rotated once every 15 seconds. How many rotations per minute is this?

b. Number Sense: $324 + 16$

c. Algebra: Solve for x: $5 + x = 14$.

d. Measurement: Francisco measured the florist box with a yardstick. How long is the florist box?

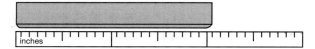

e. Geometry: In this triangle, x must be less than what number?

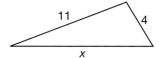

f. Number Sense: 2×64

g. Power/Roots: $\sqrt{81} - \sqrt{1}$

h. Calculation: $|-10| - 1, \sqrt{}, \times 5, + 1, \sqrt{}, \sqrt{}, - 2$

problem solving

Alan, Bianca, and Charles are counting in turn. Alan says "one," Bianca says "two," and Charles says "three." Then it's Alan's turn again and he says "four." If the pattern continues, who will say "fifty-nine"?

New Concepts · *Increasing Knowledge*

rational numbers

In the loop is the set of whole numbers.

• If we ádd any two whole numbers, is the sum a whole number?

• If we multiply any two whole numbers, is the product a whole number?

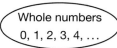

Whole numbers
0, 1, 2, 3, 4, …

We say that the set of whole numbers is **closed** under addition and multiplication because every sum or product is a whole number. Referring to our illustration we might say that we can find any sum or product of whole numbers within the whole numbers loop.

Let us consider subtraction.

• If we subtract any two whole numbers, is the result a whole number?

Certainly $3 - 1$ is a whole number, but $1 - 3$ is -2, which is not a whole number. The result is outside the loop.

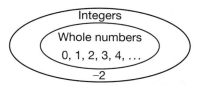

More inclusive than the set of whole numbers is the set of integers, which includes the whole numbers as well as the negatives of whole numbers. The set of integers is closed under addition, subtraction and multiplication. Every sum, difference, and product can be found inside the integers loop.

Now we will consider division.

• If we divide any two integers, is the result an integer?

The divisions $1 \div 2$, $3 \div 2$, $3 \div 4$, $4 \div 3$, and many others all have quotients that are fractions and not integers. These quotients are examples of rational numbers.

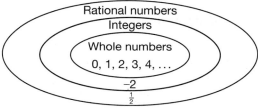

Thinking Skill

Explain

Explain why every integer is a rational number but not every rational number is an integer.

Rational numbers are numbers that can be expressed as a ratio of two integers. Any integer can be written as a fraction with a denominator of 1, so integers are rational numbers. The set of rational numbers is closed under addition, subtraction, multiplication, and division. That means any sum, difference, product, or quotient of two rational numbers is a rational number, with one exception: we guard against division by zero. That is, we do not use zero as a divisor or denominator. Reasons for this provision appear later in the book.

On the number line, rational numbers include the integers as well as many points between the integers.

Examples of Rational Numbers

Example 1

Describe each number in a–c as a whole number, an integer, or a rational number. Write every term that applies. Then sketch a number line and graph each number.

a. −3   b. 2   c. $\frac{3}{4}$

Solution

a. The number −3 is an **integer** and a **rational number.**

b. The number 2 is a **whole number,** an **integer,** and a **rational number.**

c. The number $\frac{3}{4}$ is a **rational number.**

We draw dots on the number line to graph the numbers. To graph $\frac{3}{4}$, we divide the distance between 0 and 1 into four equal parts and graph the point that is three fourths of the way to 1.

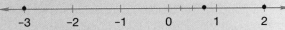

equivalent fractions

One way to express a rational number is as a fraction. Recall that a fraction has two terms, a numerator and a denominator.

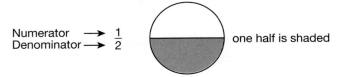

Numerator $\longrightarrow \frac{1}{2}$
Denominator $\longrightarrow$ one half is shaded

Many different fractions can name the same number. Here we show four different ways to name the same shaded portion of a circle.

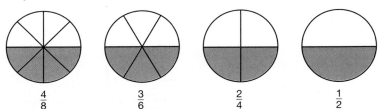

$\frac{4}{8}$ $\frac{3}{6}$ $\frac{2}{4}$ $\frac{1}{2}$

The fractions $\frac{4}{8}, \frac{3}{6}, \frac{2}{4}$, and $\frac{1}{2}$ are **equivalent fractions.** We can find equivalent fractions on an inch ruler.

Math Language

When we *reduce* a fraction we write it in its simplest form. We can also say that we *simplify* the fraction.

The segment is $\frac{1}{2}$ in., $\frac{2}{4}$ in., $\frac{4}{8}$ in., and $\frac{8}{16}$ in. long. The last three fractions **reduce** to $\frac{1}{2}$. We reduce a fraction by removing pairs of factors that the numerator and denominator have in common. For example, we can reduce $\frac{12}{18}$ by dividing the terms of the fraction by a number that is a factor of both 12 and 18, such as 2 or 3 or 6.

$$\frac{12 \div 2}{18 \div 2} = \frac{6}{9} \qquad \frac{12 \div 3}{18 \div 3} = \frac{4}{6} \qquad \frac{12 \div 6}{18 \div 6} = \frac{2}{3}$$

The last result, $\frac{2}{3}$, cannot be reduced further, so we say that it is in its "simplest form." If we begin by dividing the terms by 2 or 3, then we need to reduce twice to write the fraction in its simplest form. We only reduce once if we divide the terms by 6 because 6 is the greatest common factor of 12 and 18. The **greatest common factor** of a set of numbers is the largest whole number that is a factor of every number in the set.

In this lesson we will practice reducing fractions by first writing the prime factorization of the terms of the fraction. Then we remove pairs of like terms from the numerator and denominator and simplify.

$$\frac{4}{8} = \frac{\overset{1}{\cancel{2}} \cdot \overset{1}{\cancel{2}}}{\underset{1}{\cancel{2}} \cdot \underset{1}{\cancel{2}} \cdot 2} = \frac{1}{2} \qquad \frac{3}{6} = \frac{\overset{1}{\cancel{3}}}{2 \cdot \underset{1}{\cancel{3}}} = \frac{1}{2} \qquad \frac{2}{4} = \frac{\overset{1}{\cancel{2}}}{\underset{1}{\cancel{2}} \cdot 2} = \frac{1}{2}$$

Example 2

Using prime factorization, reduce $\frac{72}{108}$.

Solution

We write the prime factorization of the numerator and denominator.

$$\frac{72}{108} = \frac{2 \cdot 2 \cdot 2 \cdot 3 \cdot 3}{2 \cdot 2 \cdot 3 \cdot 3 \cdot 3}$$

Each identical pair of factors from the numerator and denominator reduces to 1 over 1.

$$\frac{\overset{1}{\cancel{2}} \cdot \overset{1}{\cancel{2}} \cdot 2 \cdot \overset{1}{\cancel{3}} \cdot \overset{1}{\cancel{3}}}{\underset{1}{\cancel{2}} \cdot \underset{1}{\cancel{2}} \cdot 3 \cdot \underset{1}{\cancel{3}} \cdot \underset{1}{\cancel{3}}} = \frac{2}{3}$$

We can form equivalent fractions by multiplying a fraction by a fraction equal to 1. Here we illustrate some fractions equal to 1.

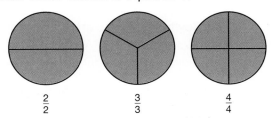

$$\frac{2}{2} \qquad\qquad \frac{3}{3} \qquad\qquad \frac{4}{4}$$

Notice that a fraction equals 1 if the numerator and denominator are equal. Multiplying by a fraction equal to 1 does not change the size of the fraction, but it changes the name of the fraction.

$$\frac{1}{2} \cdot \frac{2}{2} = \frac{2}{4} \qquad \frac{1}{2} \cdot \frac{3}{3} = \frac{3}{6} \qquad \frac{1}{2} \cdot \frac{4}{4} = \frac{4}{8}$$

Connect   What property of multiplication are we using when we multiply by a fraction equal to 1?

Example 3

Write a fraction equivalent to $\frac{1}{2}$ that has a denominator of 100.

We multiply $\frac{1}{2}$ by a fraction equal to 1 that changes the denominator from 2 to 100.

$$\frac{1}{2} \cdot \frac{50}{50} = \frac{\mathbf{50}}{\mathbf{100}}$$

Writing fractions with common denominators is one way to compare fractions.

Example 4

Compare: $\frac{2}{3} \bigcirc \frac{3}{5}$

A common denominator for the two fractions is 15.

$$\frac{2}{3} \cdot \frac{5}{5} = \frac{10}{15} \qquad \frac{3}{5} \cdot \frac{3}{3} = \frac{9}{15}$$

We see that $\frac{2}{3}$ equals $\frac{10}{15}$ and $\frac{3}{5}$ equals $\frac{9}{15}$. It is clear that $\frac{10}{15}$ is greater than $\frac{9}{15}$, therefore,

$$\frac{\mathbf{2}}{\mathbf{3}} > \frac{\mathbf{3}}{\mathbf{5}}$$

An **improper fraction** is a fraction equal to or greater than 1. These fractions are improper fractions:

$$\frac{5}{2}, \frac{4}{4}, \frac{10}{3}, \frac{12}{6}$$

The following fractions are not improper fractions.

$$\frac{2}{5}, \frac{1}{4}, \frac{3}{10}, \frac{6}{12}$$

> **Conclude** How can we tell from observation that a fraction is improper?

A **mixed number** is a whole number plus a fraction, such as $2\frac{3}{4}$. Since an improper fraction is equal to or greater than 1, an improper fraction can be expressed as an integer or mixed number. The shaded circles illustrate that $\frac{5}{2}$ equals $2\frac{1}{2}$.

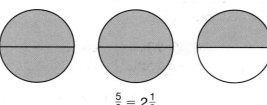

$$\frac{5}{2} = 2\frac{1}{2}$$

We see that $\frac{5}{2} = \frac{2}{2} + \frac{2}{2} + \frac{1}{2} = 2\frac{1}{2}$.

Example 5

Express each improper fraction as a whole or mixed number.

a. $\dfrac{10}{3}$

b. $\dfrac{12}{6}$

Solution

a. The thirds form three groups of $\dfrac{3}{3}$ leaving $\dfrac{1}{3}$.

$$\frac{10}{3} = \frac{3}{3} + \frac{3}{3} + \frac{3}{3} + \frac{1}{3} = 3\frac{1}{3}$$

b. Twelve sixths form two groups of $\dfrac{6}{6}$.

$$\frac{12}{6} = \frac{6}{6} + \frac{6}{6} = 2$$

We can quickly convert an improper fraction to a whole or mixed number by performing the division indicated by the fraction bar and writing any remainder as a fraction.

$$\frac{10}{3} = 3\overline{)10} \;\; \begin{array}{r} 3\frac{1}{3} \\ \hline -9 \\ \hline 1 \end{array} \qquad \frac{12}{6} = 6\overline{)12} \;\; \begin{array}{r} 2 \\ \hline \end{array}$$

Practice Set

Classify Describe each number in **a–c** as a whole number, an integer, or a rational number. Use every term that applies.

a. 5

b. −2

c. $-\dfrac{2}{5}$

Generalize Use prime factorization to reduce each fraction in **d–f.**

d. $\dfrac{20}{36}$

e. $\dfrac{36}{108}$

f. $\dfrac{75}{100}$

Complete each equivalent fraction in **g–i**. Show the multiplication.

g. $\dfrac{3}{5} = \dfrac{}{20}$

h. $\dfrac{3}{4} = \dfrac{}{20}$

i. $\dfrac{1}{4} = \dfrac{}{100}$

j. Compare: $\dfrac{3}{5} \bigcirc \dfrac{3}{4}$

k. Sketch a number line and graph the points representing $-1, \dfrac{3}{4}, 0, \dfrac{3}{2},$ and $-\dfrac{1}{2}$.

l. *Model* Which mixed number is equivalent to $\dfrac{9}{4}$? Illustrate your answer with a sketch.

m. What property of multiplication did you use to form the equivalent fractions in **g–i?**

n. *Represent* Write an example to show that the set of whole numbers is not closed under subtraction.

* **1.** **Model** Illustrate that 15 is composite by drawing two different
(9) rectangles that can be formed by 15 squares.

* **2.** **Represent** Draw a factor tree for 90, then list the prime factors of 90.
(9)

* **3.** Which of these numbers has 5 and 3 as factors?
(9)
   A 621   **B** 425   **C** 333   **D** 165

* **4.** Find the prime factors of 165 by dividing by prime numbers.
(9)

For **5–6,** use prime factorization to reduce each fraction.

* **5.** $\dfrac{22}{165}$
(10)

* **6.** $\dfrac{35}{210}$
(10)

For **7–8,** complete each equivalent fraction.

* **7.** $\dfrac{2}{3} = \dfrac{}{45}$
(10)

* **8.** $\dfrac{2}{5} = \dfrac{}{100}$
(10)

* **9.** Sketch a number line and graph the points representing $\frac{2}{3}$, $\frac{3}{2}$, 0, and 1.
(10)

10. What are the perimeter and area of a room that is 13 feet by 10 feet?
(8)

* **11.** The daily high temperatures (in degrees Fahrenheit) for a week are
(7) shown below. Find the mean, median, mode, and range of the data.

   84   85   88   89   82   78   82

* **12.** Which of the four measures from problem 11 would you report as the
(7) temperature which an equal number of data points are greater or less
than?

13. A two and a half hour performance is how many minutes?
(6)

14. Three quarters of the 84 crayons are broken. How many crayons are
(5) broken?

15. A total of 100 coins are arranged in 25 equal stacks. How many coins
(4) are in each stack?

16. Phil planned to send 178 invitations. If he has already sent 69, how
(3) many are left to be sent?

17. Write a word problem for this equation:
(3)
$$5 \text{ lb} - w = 3\frac{1}{2} \text{ lb}$$

18. Write a word problem for this equation:
(4)
$$\$2 \times n = \$18$$

* **19.** Arrange the numbers from least to greatest: $0, \frac{3}{4}, -\frac{4}{3}, 1, -1$.
(10)

20. Compare: $-7 \bigcirc -6$
(1)

21. Compare: $|-7| \bigcirc |-6|$
(1)

22. Compare: $-\frac{3}{4} \bigcirc -\frac{1}{4}$
(1, 10)

23. What is the absolute value of $-\frac{3}{4}$?
(1)

24. What two numbers are 7 units from 0?
(1)

Explain For problems **25–27,** state whether the sentence is true sometimes, always, or never. Explain your answer.

* **25.** A whole number is an integer.
(2)

* **26.** A mixed number is a whole number.
(2)

* **27.** A rational number is an integer.
(2)

28. $30.00 ÷ 8
(2)

29. $8.57 × 63
(2)

30. The product of 17 and 6 is how much greater than the sum of 17 and 6?
(2)

Focus on
• The Coordinate Plane

Two perpendicular number lines form a **coordinate plane.** The horizontal number line is called the **x-axis.** The vertical number line is called the **y-axis.** The point at which the *x*-axis and the *y*-axis **intersect** (cross) is called the **origin.**

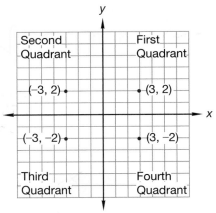

Thinking Skill

Infer

What are the signs of the *x*- and *y*-coordinates of points in each quadrant?

We can identify any point on the coordinate plane with two numbers called **coordinates** of the point. The coordinates are written as a pair of numbers in parentheses, such as (3, 2). The first number shows the horizontal (↔) direction and distance from the origin. The second number shows the vertical (↕) direction and distance from the origin. The sign of the number indicates the direction. Positive coordinates are to the right or up. Negative coordinates are to the left or down. The origin is at point (0, 0).

The two axes divide the plane into four regions called **quadrants,** which are numbered counterclockwise beginning with the upper right as first, second, third, and fourth.

Example 1

Math Language

A **vertex** (plural: vertices) of a polygon is a point where two sides meet.

Find the coordinates of the vertices of △ABC on this coordinate plane. Then count squares and half squares on the grid to find the area of the triangle in square units.

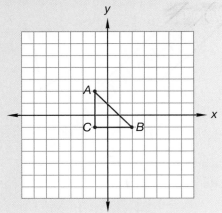

We first find the point on the *x*-axis that is directly above, below, or on the designated point. That number is the first coordinate. Then we determine how many units above or below the *x*-axis the point is. That number is the second coordinate.

Point *A* (−1, 2)

Point *B* (2, −1)

Point *C* (−1, −1)

There are 3 full squares and 3 half squares within the triangle, so the area of the triangle is **4$\frac{1}{2}$ square units.**

We can visually represent or graph pairs of related numbers on a coordinate plane. Suppose Kim is a year older than Ivan. We can let *x* represent Ivan's age and *y* represent Kim's age. Then we can record pairs of ages for Kim and Ivan in a table and graph the pairs on a coordinate plane.

x	y
1	2
2	3
3	4
4	5

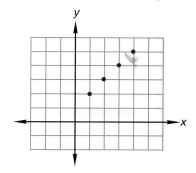

Each point represents two numbers, Kim's age and Ivan's age.

Example 2

Visit www. SaxonPublishers. com/ActivitiesC3 *for a graphing calculator activity.*

Graph the pairs of numbers from this table on a coordinate plane.

x	y
−2	−1
−1	0
0	1
1	2

Solution

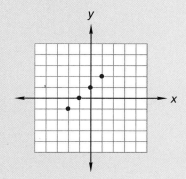

Coordinate Plane

Materials needed:

- Graph paper or copies of Lesson Activity 1
- Ruler

We suggest students work in pairs or in small groups. If using graph paper, begin by darkening two perpendicular lines on the graph paper to represent the *x*-axis and *y*-axis. For this activity we will let the distance between adjacent lines on the graph paper represent one unit. Label the tick marks on each axis from −5 to 5.

1. Graph and label the following points on a coordinate plane:

   a. (3, 4)   **b.** (2, −3)   **c.** (−1, 2)   **d.** (0, −4)

2. **Explain** Explain how you located each point in problem 1.

3. The vertices of rectangle *ABCD* are located at *A*(2, 2), *B*(2, −1), *C*(−1, −1), and *D*(−1, 2). Draw the square and find its perimeter and area.

4. Which vertex of rectangle *ABCD* in problem 3 is located in the third quadrant?

5. **Model** Graph these three points: (4, 2), (2, 0), and (−1, −3). Then draw a line that passes through these points. Name a point in the fourth quadrant that is also on the line.

6. One vertex of a square is the origin. Two other vertices are located at (−3, 0) and (0, −3). What are the coordinates of the fourth vertex?

7. **Connect** Find the perimeter and area of a rectangle whose vertices are located at (3, −2), (−2, −2), (−2, 4), and (3, 4).

8. Points (4, 4), (4, 0), and (0, 0) are the vertices of a triangle. Find the area of the triangle by counting the whole squares and the half squares.

9. The point (−3, 2) lies in which quadrant?

10. A pint is half of a quart. Letting x represent the number of pints and y the number of quarts, we can make a table of x, y pairs. Graph the pairs from this table on a coordinate plane.

x	y
0	0
2	1
4	2
6	3

11. *Represent* On a coordinate plane, graph these points and draw segments from point to point in the order given.

    1. (0, 4)   **2.** (−3, −4)

    3. (5, 1)   **4.** (−5, 1)

    5. (3, −4)   **6.** (0, 4)

12. *Model* Plan and create a straight-segment drawing on graph paper. Determine the coordinates of the vertices. Then write directions for completing the dot-to-dot drawing for other classmates to follow. Include the directions "lift pencil" between consecutive coordinates of points not to be connected.

• Percents

facts

Power Up C

mental math

a. Calculation: The rocket traveled 2000 feet per second. How many feet per minute did the rocket travel?

b. Estimation: 11×29

c. Geometry: What is the measure of ∠QSR?

d. Number Sense: $123 + 77$

e. Algebra: What are the missing digits?
$9 \times \square = \square 4$

f. Measurement: The door was 6 ft $7\frac{1}{2}$ inches high. How many inches is that?

g. Percent: 100% of 351

h. Calculation: 7×8, $+ 4$, $\div 2$, $+ 6$, $\sqrt{}$, $+ 4$, square that number

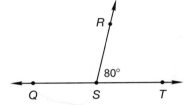

problem solving

Six friends have organized themselves by age. Molly is older than Jenna, but younger than Brett. Brett is younger than Cynthia, but older than Jenna. Jenna is older than Tina and Marcus. Marcus is younger than Cynthia and Tina. Who is the oldest and who is the youngest?

(**Understand**) Six friends are organized by age, and we are asked to determine which friend is the oldest and which is the youngest.

1. Molly is older than Jenna, but younger than Brett.

2. Brett is younger than Cynthia, but older than Jenna.

3. Jenna is older than Tina and Marcus.

4. Marcus is younger than Cynthia and Tina.

(**Plan**) We will act it out, arranging the actors from oldest to youngest, from left to right.

(**Solve**)

Step 1: We position Molly between Brett and Jenna. Brett is the oldest, so we position him on the left. Jenna is the youngest, so we position her on the right.

Step 2: Brett is in position as older than Jenna. We position Cynthia to his left.

Step 3: Marcus and Tina are younger than Jenna, so we will position them both to her right until step 4 clarifies which one of them is the youngest.

Step 4: Cynthia is already in the postion of oldest. Marcus is younger than Tina, so he is the youngest, and will be in the right-most position.

New Concept   *Increasing Knowledge*

In Lesson 10 we compared fractions with common denominators by rewriting them. One way to express rational numbers with common denominators is as percents. The word percent means *per hundred.* The denominator 100 is indicated by the word *percent* or by the symbol %. One hundred percent equals one whole.

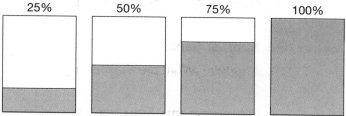

Thus, 50% means $\frac{50}{100}$, which reduces to $\frac{1}{2}$, and 5% means $\frac{5}{100}$, which equals $\frac{1}{20}$.

Example 1

Write each percent as a reduced fraction.

   a. 1%   **b. 10%**   **c. 100%**

Solution

We write each percent as a fraction with a denominator of 100. Then we reduce if possible.

   a. $1\% = \dfrac{1}{100}$ **b.** $10\% = \dfrac{10}{100} = \dfrac{1}{10}$ **c.** $100\% = \dfrac{100}{100} = 1$

One way to write a fraction as a percent is to find an equivalent fraction with a denominator of 100.

$$\frac{1}{4} = \frac{25}{100} = 25\%$$

Example 2

Write each fraction as a percent.

   a. $\dfrac{3}{4}$ **b.** $\dfrac{3}{5}$ **c.** $\dfrac{3}{10}$

Solution

We write each fraction with a denominator of 100.

   a. $\dfrac{3}{4} \cdot \dfrac{25}{25} = \dfrac{75}{100} = \mathbf{75\%}$

   b. $\dfrac{3}{5} \cdot \dfrac{20}{20} = \dfrac{60}{100} = \mathbf{60\%}$

   c. $\dfrac{3}{10} \cdot \dfrac{10}{10} = \dfrac{30}{100} = \mathbf{30\%}$

Another way to convert a fraction to a percent is to multiply the fraction by 100%. This method applies the Identity Property of Multiplication since 100% is equal to 1. For example, to convert $\frac{2}{3}$ to a percent we can multiply $\frac{2}{3}$ by 100% (or $\frac{100\%}{1}$).

$$\frac{2}{3} \cdot \frac{100\%}{1} = \frac{200\%}{3}$$

We complete the process by dividing 200% by 3 and writing any remainder as a fraction.

$$\frac{200\%}{3} = 66\frac{2}{3}\%$$

Example 3

Write each fraction as a percent.

a. $\frac{1}{3}$ b. $\frac{5}{6}$ c. $\frac{3}{8}$

Solution

Since the denominators 3, 6, and 8 are not factors of 100, we cannot easily find an equivalent fraction with a denominator of 100. Instead we multiply each fraction by 100%. We express remainders as reduced fractions.

a. $\frac{1}{3} \cdot \frac{100\%}{1} = \frac{100\%}{3} = \mathbf{33\frac{1}{3}\%}$

b. $\frac{5}{6} \cdot \frac{100\%}{1} = \frac{500\%}{6} = \mathbf{83\frac{1}{3}\%}$

c. $\frac{3}{8} \cdot \frac{100\%}{1} = \frac{300\%}{8} = \mathbf{37\frac{1}{2}\%}$

To write a percent as a fraction we replace the percent sign with a denominator of 100 and reduce if possible. If the percent is a mixed number, we rewrite it as an improper fraction before multiplying the denominator by 100 to remove the percent symbol.

Example 4

Write each percent as a reduced fraction.

a. 4%   b. 40%   c. $33\frac{1}{3}\%$

Solution

a. $4\% = \frac{4}{100} = \mathbf{\frac{1}{25}}$

b. $40\% = \frac{40}{100} = \mathbf{\frac{2}{5}}$

c. $33\frac{1}{3}\% = \frac{100}{3}\% = \frac{100}{300} = \mathbf{\frac{1}{3}}$

Notice in **c** that we converted the mixed number $33\frac{1}{3}$ to $\frac{100}{3}$. The denominator is 3. To remove the percent sign, we multiplied the denominator by 100, making the denominator 300.

Fractional part problems are often expressed as percents.

Forty percent of the 30 students rode the bus.

To perform calculations with percents we first convert the percent to a fraction or decimal. In this lesson we will convert percents to fractions. In the next lesson we will convert percents to decimals.

Example 5

Forty percent of the 30 students rode the bus. How many of the students rode the bus?

Solution

Thinking Skill

Connect

If 100% represents all of the students, what percent of the students did *not* ride the bus?

We convert 40% to a fraction and find the equivalent fractional part of 30.

$$40\% = \frac{40}{100} = \frac{2}{5}$$

$$\frac{2}{5} \text{ of } 30 = 12$$

Twelve of the students rode the bus.

Example 6

A quart is what percent of a gallon?

Solution

Since four quarts equal a gallon, a quart is $\frac{1}{4}$ of a gallon. We convert $\frac{1}{4}$ to a percent to answer the question.

$$\frac{1}{4} = \frac{25}{100} = 25\%$$

A quart is 25% of a gallon.

Practice Set

Write each fraction as a percent.

a. $\frac{4}{5}$ **b.** $\frac{7}{10}$ **c.** $\frac{1}{6}$

Write each percent as a reduced fraction.

d. 5% **e.** 50% **f.** $12\frac{1}{2}\%$

Arrange these numbers in order from least to greatest.

g. 75%, 35%, 3%, 100% **h.** $\frac{1}{2}$, $33\frac{1}{3}\%$, $\frac{1}{10}$, 65%

Solve.

i. Three of the five basketball players scored more than 10 points. What percent of the players scored more than 10 points?

j. Janice scored 30% of the team's 50 points. How many points did Janice score?

1. There are 1000 envelopes in unopened boxes in the office closet. If
(4) one box holds 250 envelopes, how many boxes of envelopes are in the
office closet?

* 2. **Analyze** A dump truck delivered four cubic yards of sand for the
(4, 6) playground. The sand weighed five tons and 800 pounds. How many
pounds of sand were delivered?

3. The heights of the rose bushes in the garden are 98 cm, 97 cm, 100 cm,
(7) 109 cm, and 101 cm. What is the average height of the rose bushes in
the garden?

4. Identify the plot, write an equation, and solve the problem: Tucker
(3) purchased comic books. He paid with a $10 bill and received $3.25 in
change. What was the cost of the comic books?

* 5. **Generalize** Using prime factorization, reduce $\frac{60}{72}$. Show your work.
(10)

* 6. Complete the equivalent fraction. Show the multiplication: $\frac{2}{3} = \frac{\square}{18}$
(10)

* 7. What property of multiplication is used to find equivalent
(10) fractions?

For problems **8** and **9** refer to this figure.

* 8. Find the perimeter of the figure.
(8)

9. What is the area of the figure?
(8)

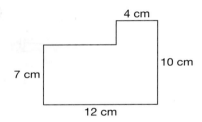

* 10. **Conclude** If each side of a square classroom is 25 feet, then what is the
(8) perimeter of the room? What is the area of the floor of the room?

11. Jim rode his bike for four hours at 16 miles per hour. How far did Jim
(7) ride in two hours, in three hours, and in four hours?

12. Steph was paid $50 for 4 hours of work. What was her rate of pay in
(7) dollars per hour?

13. Of the 150 people surveyed, $\frac{7}{10}$ said they owned at least one pet. How
(5) many of the people surveyed owned pets?

14. Find the perimeter of a triangle if each side measures 9 cm.
(8)

* 15. <u>*Analyze*</u> Compare:
(1, 10)

    a. $|-4| \bigcirc -3$ **b.** $\dfrac{4}{5} \bigcirc \dfrac{7}{10}$

For problems **16–18,** find the missing number and name the property illustrated.

16. $7 + 2 = 2 + \square$
(2)

17. $13 \cdot 200 = 200 \cdot \square$
(2)

18. $5 + (4 + 7) = (5 + 4) + \square$
(2)

19. What two equivalent fractions are illustrated by the shaded circles?
(10)

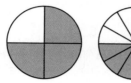

20. What is the difference between the product of 20 and 2 and the
(2) difference of 20 and 2?

21. If 3 is a factor, and 21 is the product, then what is the other factor?
(2)

* 22. <u>*Justify*</u> Rearrange factors to make the calculations easier. Then
(2) simplify. Justify your steps.

$$8 \cdot (7 \cdot 5)$$

<u>*Classify*</u> For problems **23–25,** choose *all* the correct answers from the list below.

    A whole numbers   **B** counting numbers

    C integers   **D** rational numbers

* 23. The number $\frac{22}{23}$ is a member of which sets of numbers?
(10)

* 24. The number -5 is a member of which sets of numbers?
(1)

* 25. Zero is a member of which sets of numbers?
(1)

* 26. List all the prime numbers between 50 and 70.
(9)

* 27. **a.** Use a factor tree to find the prime factors of 81. Then write the prime
(9) factorization of 81.

    b. Use division by primes to find the factors of 80. Then write the prime factorization of 80.

Simplify.

28. $\dfrac{9450}{30}$ **29.** $100.00 - \$69.86$ **30.** $\dfrac{3}{4}$ of 48
(2)   (2)   (5)

• Decimal Numbers

facts | Power Up C

mental math

a. Calculation: Samantha skates 8 miles per hour for $\frac{1}{2}$ an hour. How far does she skate?

b. Number Sense: What is the average (mean) of 30 and 40?

c. Geometry: What is the measure of $\angle B$?

d. Algebra: $\frac{20}{x} = 5$

e. Measurement: Ten yards is how many feet?

f. Power/Roots: $\sqrt{36} + \sqrt{9}$

g. Percent: 50% of 90

h. Calculation: Half a dozen $\times 4$, $+ 1$, $\sqrt{\ }$, $\times 10$, $- 1$, $\sqrt{\ }$, $+ 1$, $\div 4$

problem solving

The numbers 2, 4, and 6 are placed in the shapes. The same number is put into both triangles. What is the least possible result? What is the greatest possible result?

New Concept | Increasing Knowledge

Our number system is a base ten system. There are ten different digits, and the value of a digit depends on its place in the number. In the United States we use a decimal point to separate whole-number places from fraction places.

Decimal Place Values

1,000,000		100,000	10,000	1,000		100	10	1	.	$\frac{1}{10}$	$\frac{1}{100}$	$\frac{1}{1,000}$	$\frac{1}{10,000}$	$\frac{1}{100,000}$	$\frac{1}{1,000,000}$
millions	,	hundred thousands	ten thousands	thousands	,	hundreds	tens	ones	decimal point	tenths	hundredths	thousandths	ten-thousandths	hundred-thousandths	millionths

Discuss How do the place values in our number system relate to one another?

To read a decimal number, we read the whole number part (to the left of the decimal point), say "and" at the decimal point, and then read the fraction part (to the right of the decimal point). To read the fraction part of a decimal number we read the digits as though the digits formed a whole number, and then we name the place value of the final digit.

Example 1

Use words to write each number.

a. 12.05   b. 0.125

Solution

a. **twelve and five hundredths**

b. **one hundred twenty-five thousandths**

Example 2

Order these numbers from least to greatest.

0.5   0.41   0.05   0.405

Solution

Writing the numbers with decimal points aligned helps us compare the digits in each place.

0.5
0.41
0.05
0.405

All four numbers have a zero in the ones place. In the tenths place 0.05 has a zero, so it is least, and 0.5 has a 5 so it is greatest. Two numbers have 4 in the tenths place so we look at the hundredths place to compare. We see that 0.405 is less than 0.41. In order, the numbers are:

0.05, 0.405, 0.41, 0.5

Naming a decimal number also names a fraction. To write a decimal number as a fraction or mixed number, we write the digits to the right of the decimal point as the numerator and write the denominator indicated by the place value of the last digit. We reduce the fraction if possible.

$$12.25 = 12\frac{25}{100} = 12\frac{1}{4}$$

$$0.125 = \frac{125}{1000} = \frac{1}{8}$$

Example 3

Name each decimal number. Then write the decimal number as a reduced fraction.

a. 2.8   b. 0.75

a. two and eight tenths; $2.8 = 2\frac{8}{10} = 2\frac{4}{5}$

b. seventy-five hundredths; $0.75 = \frac{75}{100} = \frac{3}{4}$

To express a fraction as a decimal number, we perform the division indicated. For example, to convert $\frac{3}{4}$ to a decimal number, we divide 3 by 4. We divide as though we are dividing money. We write 3 as 3.00 and place the decimal point in the quotient directly above the decimal point in the dividend.

$$
\begin{array}{r}
0.75 \\
4\overline{)3.00} \\
\underline{2\,8} \\
20 \\
\underline{20} \\
0
\end{array}
$$

Notice that we used a decimal point and zeros to write 3 as 3.00 so that we could perform the division. Some fractions convert to decimal numbers that have repeating digits in the quotient (they never divide evenly). We will address decimal numbers with repeating digits in Lesson 30.

Express each fraction or mixed number as a decimal number.

a. $\frac{3}{8}$

b. $2\frac{1}{2}$

a. We divide 3 by 8. We place the decimal point on the 3 and write as many zeros as necessary to complete the division. We divide until the remainder is zero or until the digits in the quotient begin to repeat.

$$
\begin{array}{r}
0.375 \\
8\overline{)3.000} \\
\underline{2\,4} \\
60 \\
\underline{56} \\
40 \\
\underline{40} \\
0
\end{array}
$$

b. The mixed number $2\frac{1}{2}$ is a whole number plus a fraction. The whole number, 2, is written to the left of the decimal point. The fraction $\frac{1}{2}$ converts to 0.5.

$$2\frac{1}{2} = \mathbf{2.5}$$

Note that converting a rational number to decimal form has three possible outcomes.

1. The rational number is an integer.

2. The rational number is a terminating decimal number.

3. The rational number is a non-terminating decimal number with repeating digits. We will learn more about this type of rational number in Lesson 30.

We can use the Identity Property of Multiplication to convert a number to a percent. Since 100% is equivalent to 1, we can multiply a decimal by 100% to find an equivalent percent.

$$0.4 \times 100\% = 40\%$$

Converting decimals to percents shifts the decimal point two places to the right.

Example 5

Convert each number to a percent by multiplying each number by 100%

    a. 0.2   **b. 0.125**   **c. 1.5**

Solution

    a. $0.2 \times 100\% = 20\%$

    b. $0.125 \times 100\% = 12.5\%$

    c. $1.5 \times 100\% = 150\%$

Fractions and decimals greater than 1 convert to percents greater than 100%.

Converting from a percent to a decimal shifts the decimal point two places to the left.

Example 6

Convert the following percents to decimals.

    a. 5%   **b. 225%**

Solution

    a. $5\% = \dfrac{5}{100} = \mathbf{0.05}$

    b. $225\% = \dfrac{225}{100} = \mathbf{2.25}$

Example 7

Arrange in order from least to greatest: $\frac{1}{2}$, 12%, 1.2, −1.2.

Solution

Converting fractions and percents to decimals can help us compare different forms of rational numbers.

$$\frac{1}{2} = 0.5$$

$$12\% = 0.12$$

The negative number is least. Then 0.12 is less than 0.5, which is less than 1.2. The correct order is **−1.2, 12%, $\frac{1}{2}$, 1.2.**

Example 8

a. How much money is 80% of $40?

b. How much money is 75% of $40?

c. Estimate 8.25% sales tax on a $19.95 purchase.

Solution

To find a percent of a number we change the percent to a decimal or fraction and then multiply. Usually one form or the other is easier to multiply.

a. We choose to change 80% to a decimal.

$$80\% \text{ of } \$40$$

$$0.8 \times \$40 = \mathbf{\$32}$$

b. We choose to change 75% to a fraction.

$$75\% \text{ of } \$40$$

$$\frac{3}{\underset{1}{4}} \times \$\overset{10}{40} = \mathbf{\$30}$$

c. We round 8.25% to 8% and 19.95 to $20.

Step:	**Justification:**
8% of $20	Rounded
0.08 × $20	8% = 0.08
$1.60	Multiplied

Practice Set

Use words to write the decimal numbers in problems **a** and **b**.

a. 11.12   **b.** 0.375

Write each fraction in problems **c–e** as a decimal number.

c. $\frac{3}{5}$ **d.** $2\frac{1}{4}$ **e.** $\frac{1}{200}$

Write each decimal in problems **f–i** as a fraction.

f. 0.05   **g.** 0.025   **h.** 1.2   **i.** 0.001

Write each decimal in problems **j–m** as a percent.

j. 0.8   **k.** 1.3   **l.** 0.875   **m.** 0.002

Write each percent in problems **n–q** as a decimal.

n. 2%   **o.** 20%   **p.** 24%   **q.** 0.3%

r. Order from least to greatest: $-0.4, 2.3, 0.6, \frac{1}{2}$.

s. What length is 75% of 12 inches?

t. If the sales-tax rate is 7.5%, what is the sales tax on a $48.00 purchase?

u. Estimate 8.25% sales tax on a $39.79 purchase.

1. Lashonna passed back 140 papers. If these papers were in stacks of
(4) 28, how many stacks of papers did Lashonna pass back?

2. The players on the junior high basketball team had heights of 71″, 69″,
(7) 69″, 67″, and 64″. What was the average height of the players on the
team?

3. (Evaluate) Identify the plot, write an equation, and solve: This week, Kay
(3) ran a mile in 8 minutes 6 seconds. Last week, her time was 8 minutes
25 seconds. How much did her time improve?

* 4. Using prime factorization, reduce $\frac{50}{75}$.
(10)

* 5. Use words to write 4.02.
(12)

* 6. Write $\frac{5}{8}$ as a decimal number.
(12)

* 7. Convert 0.17 to a fraction.
(12)

* 8. How much money is 15% of $60?
(12)

* 9. Complete the equivalent fraction:
(10)

$$\frac{1}{7} = \frac{\square}{35}$$

* 10. Compare: $\frac{5}{5} \bigcirc \frac{7}{7}$
(10)

For problems **11** and **12**, refer to the figure at right.

11. Find the perimeter.
(8)

12. Find the area.
(8)

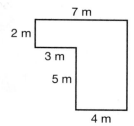

13. Brad mowed the field for 4 hours and mowed 2 acres per hour. How
(7) many acres did Brad mow?

14. Mr. Osono was paid $168 for 8 hours of work. What was his hourly rate
(7) of pay?

15. Find the perimeter of a square with side length 9 m.
(8)

(Analyze) For problems **16–18**, find the missing number and state the
property illustrated.

16. $6 \cdot 4 = 4 \cdot \square$
(2)

17. $4 + (5 + 1) = (4 + 5) + \square$
(2)

18. $6\frac{1}{2} + 1\frac{1}{8} = 1\frac{1}{8} + \square$
(2)

19. Compare:
(1, 10)

    a. $|-2| \bigcirc 2$ **b.** $\frac{3}{4} \bigcirc \frac{3}{8}$

20. What equivalent fractions are illustrated?
(10)

21. The product of two numbers is 10. Their sum is 7. Name the
(12) numbers.

22. The product of two numbers is 6. Their difference is 1. Name the
(2) numbers.

23. Arrange from least to greatest: $-0.12, -1.2, -\frac{1}{2}$.
(12)

* 24. The number -4 is a member of which sets of numbers?
(1, 10)

* 25. List all the prime numbers between zero and twenty.
(9)

* 26.   **Represent**   Write the prime factorization of 60.
(9)

27. What is $\frac{2}{3}$ of 18?
(5)

28. Suzanne is collecting money for T-shirts from the 12 members of
(4) her band section. Each shirt costs \$11.40. What is the total amount
Suzanne should collect?

29. Darren ran three miles in 27 minutes. At that rate, how long will it take
(7) him to run 5 miles?

* 30. Simplify: $18.7 + 9.04 - 1.809$
(10)

• Adding and Subtracting Fractions and Mixed Numbers

facts | Power Up C

mental math

a. **Calculation:** Miguel rode his bike at a rate of 18 miles per hour for $1\frac{1}{2}$ hours. How far did Miguel ride?

b. **Estimation:** The tires cost $78.99 each. About how much would four new tires cost?

c. **Measurement:** Two inches is about how many centimeters?

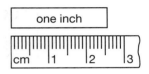

d. **Algebra:** $3x = 33$

e. **Geometry:** In this triangle, x must be between what two lengths?

f. **Fractional Parts:** $\frac{3}{4}$ of 8 muffins

g. **Number Sense:** 33×3000

h. **Calculation:** $\sqrt{81} + 1, \times 5, - 1, \sqrt{}, \times 5, + 1, \sqrt{}, - 7$

problem solving | Find the sum of the even numbers from 2 to 20.

New Concept | Increasing Knowledge

Leftover from the pizza party were $\frac{3}{8}$ of a ham and pineapple pizza and $\frac{2}{8}$ of a vegetarian pizza. What fraction of a whole pizza was left? If Mario eats two of the slices of ham and pineapple pizza, what fraction of the ham and pineapple pizza will be left?

We can add $\frac{3}{8}$ and $\frac{2}{8}$ to find the fraction of a whole pizza remaining. We can subtract $\frac{2}{8}$ from $\frac{3}{8}$ to find how much ham and pineapple pizza will be left if Mario eats two slices.

$$\frac{3}{8} + \frac{2}{8} = \frac{5}{8} \qquad\qquad \frac{3}{8} - \frac{2}{8} = \frac{1}{8}$$

Together the remaining pieces are $\frac{5}{8}$ of a whole pizza. If Mario eats two slices of the ham and pineapple pizza, $\frac{1}{8}$ of that pizza will be left.

Equal parts of a circle, like slices of pizza, is one model for fractions. Another model for fractions is a number line.

On the number lines below, we represent fraction addition and subtraction.

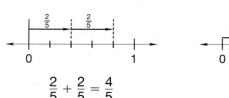 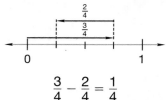

$$\frac{2}{5} + \frac{2}{5} = \frac{4}{5}$$

$$\frac{3}{4} - \frac{2}{4} = \frac{1}{4}$$

Math Language

Recall that when two fractions have the same denominator, we say that they have **"common denominators."**

Notice that the fractions in each example have common denominators. To add or subtract fractions that have common denominators, we add or subtract the numerators and leave the denominators unchanged.

Example 1

Simplify.

a. $\dfrac{3}{4} + \dfrac{3}{4}$

b. $\dfrac{3}{5} - \dfrac{3}{5}$

Solution

a. The fractions have common denominators. We add the numerators.

$$\frac{3}{4} + \frac{3}{4} = \frac{6}{4}$$

Math Language

Recall that an **improper fraction** is a fraction equal to or greater than 1. A mixed number is a whole number plus a fraction.

Fraction answers that can be reduced should be reduced. Improper fractions should be converted to whole or mixed numbers.

$$\frac{6}{4} = \frac{\overset{1}{\cancel{2}} \cdot 3}{\underset{1}{\cancel{2}} \cdot 2} = \frac{3}{2}$$

$$\frac{3}{2} = \frac{2}{2} + \frac{1}{2} = 1\frac{1}{2}$$

We illustrate this addition on a number line.

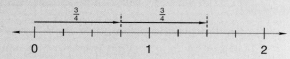

b. The fractions have common denominators. We subtract the numerators.

$$\frac{3}{5} - \frac{3}{5} = \frac{0}{5} = 0$$

We illustrate the subtraction.

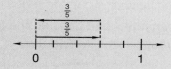

We often need to add or subtract fractions that do not have common denominators. We can do this by first renaming one or more of the fractions so that they have common denominators. Recall from Lesson 11 that we use the Identity Property of Multiplication to rename a fraction by multiplying the fraction by a fraction equal to 1.

$$\frac{1}{2} \cdot \frac{2}{2} = \frac{2}{4} \qquad \frac{1}{2} \cdot \frac{3}{3} = \frac{3}{6} \qquad \frac{1}{2} \cdot \frac{4}{4} = \frac{4}{8}$$

Forming equivalent factions enables us to add or subtract fractions that have different denominators.

Example 2

Simplify.

a. $\dfrac{2}{3} + \dfrac{3}{4}$
 b. $\dfrac{3}{4} - \dfrac{1}{6}$

Solution

a. One way to find a common denominator is to multiply the denominators of the fractions that are being added or subtracted. The denominators are 3 and 4. Since $3 \times 4 = 12$, we rename the fractions so that both denominators are 12.

Find the common denominator: Rename the fractions:

$$
\begin{array}{ll}
\dfrac{2}{3} = \dfrac{}{12} & \dfrac{2}{3} \cdot \dfrac{4}{4} = \dfrac{8}{12} \\[2mm]
+\dfrac{3}{4} = \dfrac{}{12} & +\dfrac{3}{4} \cdot \dfrac{3}{3} = \dfrac{9}{12} \\[2mm]
& \dfrac{17}{12} \text{ or } 1\dfrac{5}{12}
\end{array}
$$

b. The denominators are 4 and 6, and $4 \times 6 = 24$, so a common denominator is 24. However, 24 is not the least common denominator. Since 12 is the least common multiple of 4 and 6, the least common denominator of the two fractions is 12. We perform this subtraction twice using 24 and then 12 as the common denominator.

$$
\begin{array}{ll}
\dfrac{3}{4} \cdot \dfrac{6}{6} = \dfrac{18}{24} & \dfrac{3}{4} \cdot \dfrac{3}{3} = \dfrac{9}{12} \\[2mm]
-\dfrac{1}{6} \cdot \dfrac{4}{4} = \dfrac{4}{24} & -\dfrac{1}{6} \cdot \dfrac{2}{2} = \dfrac{2}{12} \\[2mm]
\dfrac{14}{24} = \dfrac{7}{12} & \dfrac{7}{12}
\end{array}
$$

Using the least common denominator often makes the arithmetic easier and often avoids the need to reduce the answer.

Discuss Why did we arrange the fractions vertically in example 2 but not in example 1?

When adding and subtracting mixed numbers it is sometimes necessary to regroup.

Example 3

Simplify:

a. $3\dfrac{1}{2} + 1\dfrac{3}{4}$
 b. $3\dfrac{1}{2} - 1\dfrac{3}{4}$

It is helpful to arrange the numbers vertically. We rewrite the fractions with common denominators.

a. We add the fractions, and we add the whole numbers.

$$\begin{array}{rcl} 3\frac{1}{2} & = & 3\frac{2}{4} \\ +\,1\frac{3}{4} & = & +\,1\frac{3}{4} \\ \hline & = & 4\frac{5}{4} \end{array}$$

In this case the sum includes an improper fraction. We convert the improper fraction to a mixed number and combine it with the whole number to simplify the answer.

$$4\frac{5}{4} = 4 + \frac{5}{4} = 4 + 1\frac{1}{4} = \mathbf{5\frac{1}{4}}$$

b. We rewrite the fractions with common denominators.

$$\begin{array}{rcl} 3\frac{1}{2} & = & 3\frac{2}{4} \\ -\,1\frac{3}{4} & = & -\,1\frac{3}{4} \\ \hline \end{array}$$

In this case we must rename a whole number as a fraction so that we can subtract, because $\frac{2}{4}$ is less than $\frac{3}{4}$. We sometimes call this process regrouping.

We rename $3\frac{2}{4}$ as $2\frac{6}{4}$.

$$3\frac{2}{4} = 2 + \frac{4}{4} + \frac{2}{4} = 2\frac{6}{4}$$

Now we can subtract.

$$\begin{array}{r} 2\frac{6}{4} \\ -\,1\frac{3}{4} \\ \hline \mathbf{1\frac{3}{4}} \end{array}$$

In this table we summarize the steps for adding and subtracting fractions and mixed numbers.

Adding and Subtracting Fractions and Mixed Numbers
1. Write fractions with common denominators.
2. Add or subtract numerators as indicated, regrouping if necessary.
3. Simplify the answer if possible by reducing.

Example 4

The end of a 2″ by 4″ piece of lumber is actually about $1\frac{1}{2}$-by-$3\frac{1}{2}$ in. If one 2-by-4 is nailed on top of another 2-by-4, what will be the dimensions of the combined ends?

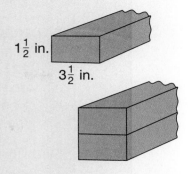

$1\frac{1}{2}$ in.

$3\frac{1}{2}$ in.

Solution

The width of the ends is not changed. We add $1\frac{1}{2}$ in. and $1\frac{1}{2}$ in. to find the boards' combined thickness.

$$1\frac{1}{2} + 1\frac{1}{2} = 2\frac{2}{2} = 3$$

The combined boards will be about **3 inches thick and $3\frac{1}{2}$ inches wide.**

Example 5

Visit www. SaxonPublishers. com/ActivitiesC3 for a graphing calculator activity.

The carpenter cut $15\frac{1}{2}$ inches from a 2-by-4 that was $92\frac{5}{8}$ inches long. How long was the resulting 2-by-4?

Solution

We subtract $15\frac{1}{2}$ inches from $92\frac{5}{8}$ after writing the fractions with common denominators.

$$92\frac{5}{8} = 92\frac{5}{8}$$
$$- 15\frac{1}{2} = 15\frac{4}{8}$$
$$\overline{\phantom{-15\frac{1}{2}=}77\frac{1}{8}}$$

The resulting 2-by-4 is **$77\frac{1}{8}$ inches long.**

Practice Set

Simplify:

a. $\frac{4}{9} + \frac{5}{9}$

b. $\frac{2}{3} - \frac{2}{3}$

c. $\frac{1}{8} + \frac{1}{4}$

d. $\frac{5}{6} - \frac{1}{2}$

e. $\frac{4}{5} + \frac{1}{2}$

f. $\frac{1}{3} - \frac{1}{4}$

g. $3\frac{1}{3} + 2\frac{1}{2}$

h. $6\frac{5}{6} - 2\frac{1}{3}$

i. $3\frac{1}{2} + 1\frac{2}{3}$

j. $7\frac{1}{3} - 1\frac{1}{2}$

k. (_Justify_) In problem **h,** did you need to rename a whole number to subtract? Why or why not?

l. Nelson wants to replace a piece of molding across the top of a door. The door frame is $32\frac{1}{4}$ inches wide and is trimmed with molding on each side that is $1\frac{5}{8}$ inches wide. How long does the molding across the top need to be?

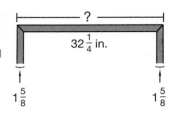

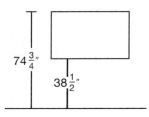

m. If the top of a bulletin board is $74\frac{3}{4}$ inches above the floor and the bottom is $38\frac{1}{2}$ inches above the floor, then what is the vertical measurement of the bulletin board?

Written Practice _Strengthening Concepts_

* 1. (_Analyze_) There were 65,000 people at the stadium. Eighty percent of
(12)   the people had been to the stadium previously. How many of the people at the stadium had been there previously?

2. A dime is about 1 mm thick. How many dimes are in a stack 1 cm
(6)   high?

3. Flying at an average speed of 525 miles per hour, how far can a
(7)   passenger fly in two hours, in three hours, and in four hours?

For problems **4** and **5,** identify the plot, write an equation, and solve the problem.

4. A whole pizza was cut into eighths. If Hector eats $\frac{3}{8}$ of the pizza and Rita
(3, 13)   eats $\frac{1}{4}$ of the pizza, then what fraction of the pizza remains?

5. The ceiling fan company sold 7400 fans this year and 6900 fans last
(3)   year. How many more fans did they sell this year compared to last year?

(_Generalize_) Simplify.

6. $\frac{5}{8} + \frac{2}{8}$
(13)

7. $\frac{5}{8} - \frac{2}{8}$
(13)

* 8. $\frac{2}{3} + \frac{1}{6}$
(13)

* 9. $\frac{2}{3} - \frac{1}{6}$
(13)

10. $1\frac{1}{4} + 2\frac{1}{4}$
(13)

11. $2\frac{1}{4} - 1\frac{1}{4}$
(13)

* 12. $3\frac{4}{5} - 2\frac{1}{10}$
(13)

* 13. $3\frac{4}{5} + 2\frac{1}{10}$
(13)

* 14. (_Justify_) Rearrange the addends to make the calculations easier.
(2, 13)   Then simplify. Justify your steps.

$$3\frac{1}{2} + \left(4\frac{9}{10} + 2\frac{1}{2}\right)$$

* 15. Write each fraction as a decimal.
(12)

a. $\frac{3}{25}$ **b.** $3\frac{2}{5}$

* **16.** Write each decimal as a reduced fraction.
  (12)
      a. 0.15   **b.** 0.015

* **17.** Choose every correct answer. The set of whole numbers is closed
  (1, 10) under
      A addition.   **B** subtraction.

      C multiplication.   **D** division.

* **18.**   *Generalize*   Use prime factorization to reduce $\frac{75}{125}$. Show your work.
  (9, 10)

* **19.**   *Connect*   Write a fraction equivalent to $\frac{3}{5}$ with a denominator of 100.
  (10)

  20. Find the perimeter of a triangle with side lengths $1\frac{1}{2}$ in., 2 in., and
  (8, 13) $2\frac{1}{2}$ in.

  21. Find the area of the ceiling of a room that is 14 feet wide and
  (8) 16 feet long.

  22. List all the prime numbers between 70 and 90.
  (9)

  23. Write the prime factorization of 120.
  (9)

  24. Find the average (mean) of 1, 3, 5, and 7.
  (7)

  25. The lengths of the tennis rackets on the shelf were 22 inches, 24 inches,
  (7) 26 inches, 28 inches, and 20 inches. What was the average length of
  the racquets?

  26. Kimberly and Kerry delivered 280 papers in 4 weeks. What was the
  (7) average number of papers delivered per week?

  27.   *Represent*   Graph the numbers $\frac{1}{2}$, $\frac{1}{3}$, and $\frac{5}{6}$ on a number line. (*Hint:* First
  (1, 10) find the common denominator of the fractions to determine the proper
  spacing on the number line.)

Simplify. Then choose one of the problems **28–30** to write a word problem
involving the given numbers and operation.

28. $64.32 ÷ 16   **29.** 2007 − 1918   **30.** $\frac{\$52.70}{17}$
(2)   (2)   (2)

• Evaluation
• Solving Equations by Inspection

facts | Power Up C

mental math

a. Calculation: Kerry reads a page every two minutes. How many pages does she read per hour?

b. Measurement: How long is the segment?

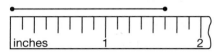

c. Power/Roots: $\sqrt{25} \times \sqrt{25}$

d. Fractional Parts: What number is $\frac{5}{8}$ of 48?

e. Number Sense: What are the missing digits? $1\square \times 9 = 1\square 3$

f. Geometry: What is the measure of $\angle S$?

g. Percent: 50% of 44

h. Calculation: $5^2 - 4$, $\div 3$, $+ 2$, $\sqrt{}$, $\times 5$, $+ 1$, $\sqrt{}$, $\times 12$, $+ 1$, $\sqrt{}$

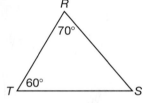

problem solving

Sometimes the given information in a problem requires us to work backwards to find an answer.

Problem: Samantha's age is half of Juan's age. Paul's age is half of Samantha's age. If Paul is eight years old, how old is Juan?

⌐Understand⌐ We are given Paul's age. We are told how Paul's age relates to Samantha's age and how Samantha's age relates to Juan's age. We are asked to find Juan's age.

⌐Plan⌐ We will work backwards from Paul's age to first find Samantha's age and then Juan's age.

⌐Solve⌐ Paul is eight years old, and we know that his age is half of Samantha's. So we multiply 8 by 2 and find that Samantha is 16 years old.

We are told that Samantha's age is half of Juan's age, so we multiply 16 by 2 to find that Juan is 32 years old.

⌐Check⌐ We worked backwards from Paul's age to find Juan's age. To find Juan's age, we first had to find Samantha's age. We check our math and find that 8 years (Paul's age) is half of 16 years (Samantha's age), which in turn is half of 32 years (Juan's age).

evaluation

Math Language
Recall that an **expression** is a mathematical phrase made up of variables, numbers, and/or symbols. To **evaluate** an expression means "to find the value of" the expression.

Sometimes we use a letter to represent an unidentified number. For example, "Four times a number s" can be expressed as $4 \cdot s$, or $4s$.

In the expression $4s$, the number 4 is a **constant,** because its value does not change, and the letter s is a **variable,** because its value can change. (If an algebraic term is composed of both constant and variable factors, we customarily write the constant factor first.)

Recall that a formula is an expression or equation used to calculate a desired result. For example, the formula for the perimeter (P) of a square guides us to multiply the length of a side (s) by 4.

$$P = 4s$$

We **evaluate** an expression by substituting numbers in place of variables and calculating the result.

For example, in the expression $4s$, if we substitute 6 for s we get $4(6)$, which equals 24.

Example 1

A formula for the perimeter of a rectangle is $P = 2l + 2w$. Find P when l is 15 cm and w is 12 cm.

Solution

We substitute 15 cm for l and 12 cm for w in the formula. Then we perform the calculations.

$$P = 2l + 2w$$
$$P = 2(15 \text{ cm}) + 2(12 \text{ cm})$$
$$P = 30 \text{ cm} + 24 \text{ cm}$$
$$P = \textbf{54 cm}$$

Explain How can we use the formula $A = lw$ to find the area of the same rectangle?

solving equations by inspection

An **equation** is a mathematical sentence stating that two quantities are equal. An equation can have one or more variables. A **solution** to an equation with one variable is a number that satisfies the equation (makes the equation true). In the equation $4x = 20$, the solution is 5 because $4(5) = 20$.

Formulate Write and solve an equation that has one operation and one variable. Use any operation and variable you wish.

In this book we will learn and practice many strategies for solving equations, but first we will practice solving equations by inspection. That is, we will study equations and mentally determine the solutions.

Example 2

If a taxi company charges a $2 flat fee plus $3 per mile (*m*), then the cost (*c*) in dollars of a taxi ride is shown by this formula:

$$3m + 2 = c$$

Suppose that the cost of a taxi ride is $20. Find the length of the taxi ride by solving the equation $3m + 2 = 20$.

Solution

This equation means that the number of miles is multiplied by 3, then 2 is added making the total 20.

$$3m + 2 = 20$$

That means that $3m$ equals 18. Since $3m$ equals 18, m must equal 6.

$$m = 6$$

So the taxi ride was **6 miles.**

Discuss Why does $3m + 2 = 20$ mean that $3m = 18$?

Practice Set

a. A formula for the area (*A*) of a parallelogram is $A = bh$. Find the area of a parallelogram with a base (*b*) of 12 in. and a height (*h*) of 4 in.

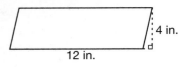

b. In the expression $4ac$, what is the constant?

c. Evaluate $4ac$ for $a = 1$ and $c = 12$.

Analyze Solve each equation by inspection.

d. $w + 5 = 16$

e. $m - 6 = 18$

f. $25 - n = 11$

g. $5x = 30$

h. $\dfrac{d}{4} = 8$

i. $\dfrac{12}{z} = 6$

j. $2a - 1 = 9$

k. $20 = 3f - 1$

l. Admission to the county fair is $3. Each ride costs $2. Nate has $15. Solve the following equation to find how many rides he can take.

$$2r + 3 = 15$$

Written Practice *Strengthening Concepts*

1. There were 256 farmers in the county 30 years ago. Five eighths of
(5) those farmers have changed occupations. How many of the 256 farmers changed occupations?

2. Averaging 35 pages per hour, how many pages can Karina read in
(7) 2 hours, in 3 hours, and in 4 hours?

3. Each package of light bulbs contains 4 bulbs. If an employee at a
$_{(4)}$ construction office needs to order 480 bulbs, how many packages
should she order?

For problem **4,** identify the plot, write an equation, and solve the problem.

4. When Betsy Ross made the American flag, it had 13 stars. On July 4,
$_{(3)}$ 1960, the 50th star was added to the flag for Hawaii. How many more
stars were on the flag by July 4, 1960, than were on the Betsy Ross
flag?

Analyze Solve by inspection.

5. $15 = m - 4$
$_{(14)}$

6. $15 - m = 4$
$_{(14)}$

7. $3w = 36$
$_{(14)}$

* 8. $5x + 5 = 20$
$_{(14)}$

* 9. $\dfrac{x}{3} = 4$
$_{(14)}$

* 10. $\dfrac{x}{2} + 3 = 12$
$_{(14)}$

Generalize Simplify.

* 11. $\dfrac{3}{8} + \dfrac{1}{3}$
$_{(13)}$

* 12. $1\dfrac{1}{4} - \dfrac{3}{16}$
$_{(13)}$

* 13. **Justify** Rearrange the addends to make the calculations easier. Then
$_{(2,\ 13)}$ simplify. Justify your steps.

$$\frac{3}{8} + \left(1\frac{1}{2} + 2\frac{5}{8}\right)$$

14. Arrange in order from least to greatest.
$_{(1,\ 12)}$

$$\frac{1}{2},\ -1,\ 0,\ \frac{3}{8},\ 0.8,\ 1,\ -0.5$$

Evaluate For problems **15–18,** use the given information to find the value of
each expression.

* 15. Find ac when $a = 2$ and $c = 9$.
$_{(14)}$

* 16. The formula for the perimeter of a square is $P = 4s$. Find P when
$_{(14)}$ $s = 12$.

* 17. For $y = x - 9$, find y when $x = 22$.
$_{(14)}$

* 18. For $d = rt$, find d when $r = 60$ and $t = 4$.
$_{(14)}$

19. Arrange these fractions in order from least to greatest.
$_{(10)}$

$$\frac{2}{3},\ \frac{3}{4},\ \frac{5}{12}$$

20. Write the prime factorization of 990.
$_{(9)}$

21. What is the difference between the product of 6 and 8 and the sum of 8
$_{(2)}$ and 12?

22. If 4 is the divisor, and 8 is the quotient, what is the dividend?
$_{(2)}$

* 23. If each side of a square is $2\frac{1}{2}$ in., then what is the perimeter of the
(8, 13)   square?

24. Compare:
(1)
  a. $|-5| \bigcirc |-6|$ **b.** $-5 \bigcirc -6$

25. The data below show the number of students in seven classrooms.
(7)   What is the median number of students?

$$27, 30, 29, 31, 30, 32, 29$$

26. Estimate 8.25% sales tax on a $39.89 purchase.
(12)

27. Express each percent as a decimal.
(12)
  a. 125%   **b.** 8.25%

Find each perimeter.

28.
(8)

* 29.
(8)

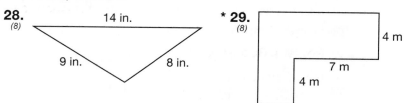

* 30. (Generalize) Leah wants to carpet a room that is 12 feet wide and 15 feet
(6, 8)   long. Carpeting is sold by the square yard. Find the length and width
of the room in yards. Then find the number of square yards needed to
carpet the room.

• Powers and Roots

Power Up | Building Power
_____

facts | Power Up C

mental math

  a. Calculation: Wendy walked 8 miles in 2 hours. What was her average rate in miles per hour?

  b. Estimation: $19.95 × 4

  c. Fractional Parts: $\frac{1}{4}$ of $84

  d. Measurement: Jeremiah's forearm (elbow to wrist) is 11 inches long. His hand is 6 inches long. He measures a cabinet by placing his hand and forearm against it. He has to do this twice to reach the end. About how long is the cabinet?

  e. Powers/Roots: $\sqrt{100} - \sqrt{25}$

  f. Geometry: In this triangle, *x* must be between what two numbers?

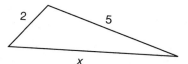

  g. Percent: 10% of $6.50

  h. Calculation: $200 - 50$, $\div 3$, $- 1$, $\sqrt{}$, $- 5$, $\times 50$, $\sqrt{}$, $+ 3$, minus a dozen

problem solving

In some problem situations, we are given enough information to use logical reasoning to find other information.

Problem: Dora, Eric, and Francis went to the restaurant and each ordered a different sandwich. Dora ordered a tuna sandwich, and Eric did not order a ham sandwich. If their order was for tuna, grilled cheese, and ham sandwiches, what did each person order?

(**Understand**) In the problem, three people order three sandwiches. We know what one person ordered, and we know what one other person did **not** order. We are asked to find each person's sandwich order.

(**Plan**) We will make a table to keep track of the information. We will place checkmarks to represent sandwich orders that we know are true and X's to represent orders that we know are false.

(**Solve**) We draw a table with a row for each person and a column for each sandwich. We use abbreviations for the names. We start by placing a checkmark in the box that corresponds to "Dora orders tuna sandwich." Then we place an X in the box that corresponds to "Eric orders ham sandwich."

	T	H	GC
D	✓		
E		X	
F			

Now we use logical reasoning to fill the table. Dora ordered the tuna sandwich, so it is implied that neither Eric nor Francis ordered tuna. We write X's in the boxes below Dora's checkmark. By ordering tuna, it is also implied that Dora did not order the ham or grilled cheese. So we write X's in the boxes to the right of Dora's checkmark.

	T	H	GC
D	✓	X	X
E	X	X	
F	X		

Each person's order was different, so only one checkmark should appear in each row and column. Since there are two X's in the column for "ham sandwich," we conclude that Francis ordered the ham sandwich. Now we know Francis did not order the grilled cheese, so we place an X in the lower-right box. The only possible order for Eric is grilled cheese.

	T	H	GC
D	✓	X	X
E	X	X	✓
F	X	✓	X

Check We created a table and filled in the information we knew. Then we used logical reasoning to find each person's order (Dora: tuna; Eric: grilled cheese; Francis: ham).

New Concept   *Increasing Knowledge*

To show repeated multiplication we can use an exponent.

$$\text{base} \longrightarrow 5^2 \longleftarrow \text{exponent}$$

If the exponent is a counting number, it indicates how many times the base is used as a factor.

$$5^3 = 5 \cdot 5 \cdot 5 = 125$$

We read 5^2 as "five squared." We read 5^3 as "five cubed." We say "to the *n*th power" if the exponent *n* is greater than 3. For example, 5^4 is read as "five to the fourth power" or just "five to the fourth."

Example 1

Simplify:

a. 5^4

b. 10^4

Solution

Thinking Skill

Explain

Explain the difference between 5^4 and $5 \cdot 4$.

The exponent, 4, indicates that the base is a factor four times.

a. $5 \cdot 5 \cdot 5 \cdot 5 = \mathbf{625}$

b. $10 \cdot 10 \cdot 10 \cdot 10 = \mathbf{10,000}$

Notice that the number of zeros in 10,000 matches the exponent of 10.

Example 2

Write the prime factorization of 72 using exponents.

Solution

The prime factorization of 72 is $2 \cdot 2 \cdot 2 \cdot 3 \cdot 3$. We can group like factors with an exponent.

$$72 = \mathbf{2^3 \cdot 3^2}$$

We can use exponents with units of length to indicate units of area. We show this in example 3 with the area of a square. The formula for the area of a square is $A = s^2$. In this formula, A represents area and s represents the length of the side.

Example 3

The figure shows a square floor tile that is one foot on each side. Find the area covered by the tile in square inches using the formula $A = s^2$.

12 in.

Solution

Squaring a length results in square units.

Reading Math

We read in.^{2} as "square inch."

12 in.

12 in.

144 square inches

Step:	Justification:
$A = s^2$	Formula for area of a square
$= (12 \text{ in.})^2$	Substituted 12 in. for s and squared
$= \mathbf{144 \text{ in.}^2}$	Notice that one square foot equals 144 square inches.

Exponents can be applied to variables. If the same variable is a factor in an expression a number of times, we can simplify the expression by writing the variable with an exponent.

Example 4

Express with exponents.

$$2xxyyyz$$

Solution

Since x is a factor twice, we use the exponent 2, and since y is a factor 3 times, we use the exponent 3.

$$\mathbf{2x^2y^3z}$$

The inverse of raising a number to a power is taking a root of a number. We may use a **radical** sign, $\sqrt{}$, to indicate a root of a number.

$$\sqrt{25} = 5 \qquad \text{"The \textbf{square root} of 25 is 5."}$$
$$\sqrt[3]{125} = 5 \qquad \text{"The \textbf{cube root} of 125 is 5."}$$

The number under the radical sign is the **radicand**.

The small 3 in $\sqrt[3]{125}$ is the **index** of the root. If the index is 4 or more, we say **"the *n*th root."**

$$\sqrt[5]{32} \text{ is read "the fifth root of 32"}$$
$$\sqrt[5]{32} = 2 \text{ because } 2 \cdot 2 \cdot 2 \cdot 2 \cdot 2 = 32$$

Example 5

Visit www. SaxonPublishers. com/ActivitiesC3 for a graphing calculator activity.

Simplify:

a. $\sqrt{144}$ **b.** $\sqrt[3]{27}$

Solution

a. We find the number which, when multiplied by itself, equals 144. Since $12 \cdot 12 = 144$, $\sqrt{144}$ equals **12.**

b. We find the number which, when used as a factor three times, equals 27. Since $3 \cdot 3 \cdot 3 = 27$, $\sqrt[3]{27}$ equals **3.**

A number that is a square of a counting number is a **perfect square.** For example, 25 is a perfect square because $5^2 = 25$.

Analyze The number 64 is both a perfect square and a perfect cube. Is $\sqrt{64}$ less than or greater than $\sqrt[3]{64}$?

Example 6

The floor of a square room is covered with square-foot floor tiles. If 100 tiles cover the floor, how long is each side of the room?

100 ft²

Solution

The side length of a square is the square root of the area of the square. Therefore, the length of each side is $\sqrt{100 \text{ ft}^2}$, which is **10 ft.**

Example 7

Name the first three counting numbers that are perfect squares. Then find their positive square roots.

Solution

Here we show several counting numbers in order.

$$1, 2, 3, 4, 5, 6, 7, 8, 9, 10, 11, 12, \ldots$$

The numbers 1 (1×1) and 4 (2×2) and 9 (3×3) are perfect squares. We find the square roots.

$$\sqrt{1} = 1 \qquad \sqrt{4} = 2 \qquad \sqrt{9} = 3$$

Practice Set

Simplify:

a. 10^5

b. 3^4

c. $(15 \text{ cm})^2$

d. $\sqrt{121}$

e. $\sqrt[3]{8}$

f. $\sqrt{5^2}$

Express with exponents.

g. $5xyyyzz$

h. $aaab$

i. $3xyxyx$

j. Mr. Chin wants to cover the floor of his garage with a non-slip coating. His garage measures 20 feet on each side. Use the formula $A = s^2$ to find the number of square feet of floor that need to be coated.

20 ft

Written Practice *Strengthening Concepts*

1. At the baseball game, 40% of the fans joined in singing the national anthem, while the rest listened quietly. If there were 25,000 fans in the stadium, how many joined in singing?
(12)

2. A gallon of milk contains 128 oz. How many 8-oz cups can be filled with a gallon of milk?
(4)

3. Jackie played the drums in the band. If he beat the drum at a rate of 110 beats per minute, how many beats would he play in $4\frac{1}{2}$ minutes?
(7)

Generalize Simplify.

* 4. $(4 \text{ in.})^3$
(15)

* 5. $\sqrt[3]{1}$
(15)

* 6. $\sqrt{8^2}$
(15)

7. Express with exponents: $2xxxmrr$
(15)

Analyze Solve by inspection.

* 8. $\frac{x}{4} = 12$
(14)

* 9. $2x + 5 = 5$
(14)

* 10. $9 - m = 7$
(14)

Simplify.

* 11. $\frac{4}{5} - \frac{1}{2}$
(13)

* 12. $1\frac{1}{4} + 1\frac{1}{3}$
(13)

* 13. **Justify** Compare and justify your answer with a property of addition.
(2, 13)
$$\left(\frac{1}{3} + \frac{1}{2} + \frac{2}{3}\right) \bigcirc \left(\frac{1}{3} + \frac{2}{3} + \frac{1}{2}\right)$$

14. Arrange in order from least to greatest: $0, -2, \frac{1}{3}, 0.3, 3, -1.5$.
(1, 12)

* 15. Find $4ac$ when $a = 1$ and $c = 9$.
(14)

* 16. What is the difference between the sum of $\frac{1}{2}$ and 10 and the product of $\frac{1}{2}$ and 10?
(2)

* 17. For $y = \frac{1}{3}x + 2$, find y when $x = 6$.
(14)

* 18. How much money is 20% of $65?
(11, 12)

* 19. **Analyze** Arrange these fractions in order from least to greatest.
(10)

$$\frac{3}{8}, \frac{2}{5}, \frac{1}{4}$$

* 20. Write the prime factorization of 630 using exponents.
(9, 15)

21. A formula for the area of a square is $A = s^2$. Find A when $s = 15$ cm.
(14)

22. If the subtrahend is 12 and the difference is 12, what is the
(2) minuend?

23. If each side of a square is $7\frac{1}{2}$ in., what is the perimeter of the
(8, 13) square?

24. Compare: $0.625 \bigcirc \frac{3}{5}$
(12)

25. What is the median number of bears in the zoos in the state if the zoos
(7) contain 11 bears, 3 bears, 18 bears, 2 bears, and 5 bears?

26. Write 0.07 as a fraction. Then name both numbers.
(12)

27. Explain why the following statement is false: "Every whole number is
(1) positive."

Find the perimeters.

28.
(8)

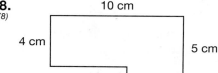

29.
(8)

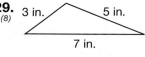

* 30. A 15-foot square room is going to be carpeted. Use $A = s^2$ to find the
(15) number of square feet of carpet that is needed.

•Irrational Numbers

facts | Power Up D

mental math

a. Calculation: Chrissy drove 100 miles in 2 hours. What was her average speed in miles per hour?

b. Geometry: What is the measure of ∠*PMR*?

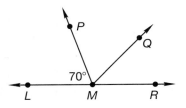

c. Fractional Parts: $\frac{1}{5}$ of 35

d. Measurement: Half of a kilogram is how many grams?

e. Algebra: What is the mass of 8 of the objects in the left tray?

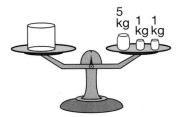

f. Calculation: 18, ÷ 3, ÷ 3, × 3, × 3, + 2, + 1, ÷ 3, × 2, + 1, ÷ 3, square it, $\sqrt{}$

problem solving

Find the next four numbers in the sequence:

$$\frac{1}{24}, \frac{1}{12}, \frac{1}{8}, \frac{1}{6}, \frac{5}{24}, \text{---}, \text{---}, \text{---}, \text{---}$$

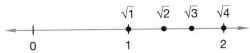

New Concept | *Increasing Knowledge*

Recall that the set of rational numbers is closed under addition, subtraction, multiplication, and division. However, the square roots of many rational numbers, such as $\sqrt{2}$ and $\sqrt{3}$, are not rational numbers. Since $\sqrt{2}$ and $\sqrt{3}$ are greater than $\sqrt{1}$ and less than $\sqrt{4}$, their values are more than 1 and less than 2.

To find the approximate value of $\sqrt{2}$ and $\sqrt{3}$, we can use a calculator.

Most calculators have a square root key, $\boxed{\sqrt{}}$. If we enter a number and then press $\boxed{\sqrt{}}$, the display shows the square root of the entered number.

• Use a calculator to find $\sqrt{1}$ and $\sqrt{4}$.

• Use a calculator to find $\sqrt{2}$ and $\sqrt{3}$.

How do the displays for $\sqrt{2}$ and $\sqrt{3}$ differ from the displays for $\sqrt{1}$ and $\sqrt{4}$?

No matter how many digits a calculator displays, the displays for $\sqrt{2}$ and $\sqrt{3}$ will be filled with digits without a repeating pattern. The display fills because there is no rational number—no fraction or repeating decimal—that equals $\sqrt{2}$ or $\sqrt{3}$.

The numbers $\sqrt{2}$ and $\sqrt{3}$ are examples of **irrational numbers.** The prefix *ir-* means *not,* so "irrational numbers" means "non-rational numbers." The square root of any counting number that is not a perfect square is an irrational number. Later we will consider other examples of irrational numbers.

Example 1

Which number is irrational?

A $\sqrt{0}$ **B** $\sqrt{4}$ **C** $\sqrt{8}$ **D** $\sqrt{16}$

Solution

The numbers 0, 4, and 16 are perfect squares.

$$\sqrt{0} = 0 \qquad \sqrt{4} = 2 \qquad \sqrt{16} = 4$$

Since 8 is not a perfect square, $\sqrt{8}$ is an irrational number. (Since $\sqrt{8}$ is between $\sqrt{4}$ and $\sqrt{9}$, its value is greater than 2 and less than 3.)

Thinking Skill

Explain

How can you determine which two counting numbers the square root of 108 falls between?

Rational numbers and irrational numbers together form the set of **real numbers.**

Real Numbers

Rational Numbers	Irrational Numbers

All of the real numbers can be represented by points on a number line, and all of the points on a number line represent real numbers.

Examples of Real Numbers

Example 2

Arrange these real numbers in order from least to greatest.

$$2, \sqrt{5}, \frac{5}{9}, 0, 3, 1.2, 1$$

Solution

The irrational number $\sqrt{5}$ is greater than $\sqrt{4}$, which equals 2, and less than $\sqrt{9}$, which equals 3.

$$0, \frac{5}{9}, 1, 1.\,2, 2, \sqrt{5}, 3$$

Example 3

The area of this square is 2 cm². What is the length of each side?

Solution

The side length of a square is the square root of its area. Ther
length of each side is $\sqrt{2}$cm², which is $\sqrt{2}$ **cm.**

Notice that $\sqrt{2}$ cm is an actual length that is approximately ec

Practice Set

a. If a real number is not rational, then it is what type of numb

b. Which of the following numbers is irrational?

    A -3 **B** $\sqrt{9}$ **C** $\dfrac{9}{2}$ **D** $\sqrt{.}$

c. **Explain** Which of these numbers is between 2 and 3 on the num.
line? How do you know?

    A $\sqrt{2}$ **B** $\sqrt{3}$ **C** $\sqrt{4}$ **D** $\sqrt{5}$

d. Arrange these real numbers in order from least to greatest.

$$0, 1, \sqrt{2}, \frac{3}{4}, 0.5, -0.6$$

Find the length of each side of these squares.

e.
```
400 mm²
```

f.
```
3 cm²
```

Use a calculator to find the square roots of these numbers. Round to the nearest hundredth.

    g. $\sqrt{10}$ h. $\sqrt{20}$ i. $\sqrt{40}$

j. Refer to your answers to problems **g, h,** and **i** to answer this question:
Which number is twice as much as the square root of ten, $\sqrt{20}$ or $\sqrt{40}$?

Written Practice   *Strengthening Concepts*

1. Bonnie drove 108 miles in 2 hours. What was her average speed?
(7)

2. Before the gift was revealed, only $\frac{2}{7}$ of the crowd guessed correctly. If
(5) there were 294 people in the crowd, how many guessed correctly before it was revealed?

For problems **3** and **4,** identify the plot, write an equation, and solve the problem.

3. Roger completed 28 of 41 math exercises during school. How many
(3) exercises are left to complete after school?

4. In Major League Baseball in 1960 there were two leagues of eight teams
(3) each. By 1998 the total number of teams had expanded to 30. How many teams were added from 1960 to 1998?

Generalize Simplify.

* 5. $\sqrt[3]{64}$
(15)

* 6. 10^3
(15)

* 7. $\dfrac{6}{7} + \dfrac{1}{14}$
(13)

* 8. $2\dfrac{5}{8} - 1\dfrac{1}{4}$
(13)

9. Find the perimeter and area of this
(8) figure.

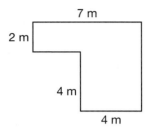

10. **Represent** Express with exponents:
(15)

    a. 4xyzxy

    b. 3aabbbc

11. **Analyze** Compare:
(15)

    a. $\sqrt{16} \bigcirc 7$

    b. $\sqrt{81} \bigcirc 3^2$

Analyze Solve problems **12–15** by inspection.

* 12. $4z - 2 = 30$
(14)

* 13. $200 - a = 140$
(14)

* 14. $50 = 5m + 10$
(14)

* 15. $7 = \dfrac{x}{3}$
(14)

16. Compute 75% of $800.
(11)

17. Find 4ac when $a = 3$ and $c = 39$.
(14)

18. Rearrange the factors to make the calculation easier. Then simplify,
(2) justifying your steps.

$$8 \cdot (49 \cdot 25)$$

* 19. **Conclude** Momentum is the product of mass and velocity, $p = mv$. Find
(14) p when $m = 40$ and $v = 11$.

20. What is the mean and the median of 2, 4, 6, and 8?
(7)

21. Write the prime factorization of 400. Use exponents for repeated
(9, 15) factors.

22. What is the product of the sum of 10 and 5 and the difference of 10
(2) and 5?

23. Express $\frac{1}{8}$ as a decimal and as a percent.
(12)

24. ⬤ Conclude ⬤ The relationship between rate (r), time (t), and distance (d) is
(14) expressed in the formula $d = rt$. Find d when $r = 55$ miles per hour and
$t = 2$ hours.

25. List the even prime numbers.
(1, 9)

Simplify.

26. $123 - 321$
(2)

27. $\dfrac{2001 \text{ hours}}{3}$
(2)

⬤ Classify ⬤   For problems **28–30,** choose from this list *all* correct answers.

    A integers   **B** whole numbers   **C** natural numbers

    D rational numbers   **E** irrational numbers **F** real numbers

* 28. The number -2 is a member of which set(s) of numbers?
(1, 16)

* 29. The number $\sqrt{2}$ is a member of which set(s) of numbers?
(1, 16)

* 30. The number $\frac{1}{2}$ is a member of which set(s) of numbers?
(1, 16)

Early Finishers
*Real-World
Application*

Uranus is the seventh planet from the Sun and is composed primarily
of rocks and ice. Uranus is turquoise in appearance due to its
atmosphere, which is made up of approximately 82% hydrogen, 15%
helium, and 3% methane. Express each percent as a fraction and as
a decimal.

• Rounding and Estimating

facts | Power Up D

mental math

a. Calculation: Tammy types 20 words per minute. How many words can she type in $2\frac{1}{2}$ minutes?

b. Powers/Roots: 25×10^2

c. Fractional Parts: If the folded piece of wire is unfolded, about how long will it be?

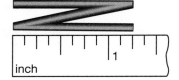

d. Algebra: Solve for x: $7x - 1 = 20$.

e. Number Sense: 14×5

f. Measurement: If Roger walks one kilometer in 8 minutes, how long does it take him to walk 500 meters?

g. Percent: 25% of 100

h. Calculation: $444 \div 2$, $\div 2$, $- 11$, $\div 4$, $\sqrt{}$, $\times 7$, $\times 2$, $+ 20$, $+ 9$, $\div 3$, $\div 3$, $+ 100$

problem solving | Daria, Evan, and Fernando went to the cafeteria and each ordered a drink. Daria did not order lemonade. Evan ordered cranberry juice. If their order was for lemonade, cranberry juice, and orange juice, what did each person order?

Parade officials reported that one million people lined the route of the Rose Parade.

Thinking Skill

Connect

What are some everyday activities that require estimation?

The number used in the statement, one million, is a round number and is not exact. The number was obtained not by counting every person along the route but by estimating the size of the crowd. We often use rounding and estimating for situations in which an exact or nearly exact number is not necessary.

We **round** whole numbers by finding a nearby number that ends with one or more zeros. For example, we might round 188 to 190 or to 200 depending on whether we are rounding to the nearest ten or nearest hundred.

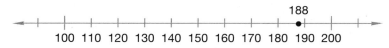

The mathematical procedure for rounding involves place value. We round a number to a certain place value by inspecting the digit that follows that place and deciding whether to round up or down.

Example 1

Round 1,481,362 to the nearest hundred thousand.

Solution

The number is between the round numbers 1,400,000 and 1,500,000.
To help us decide which round number is nearer, we inspect the digit in the
ten-thousands place. If the digit is 5 or greater we round up. If the digit is
less than 5 we round down.

$$1,4\textcircled{8}1,362$$

The digit in the ten-thousands place is 8, so we round up to **1,500,000.**

We round decimal numbers by finding a decimal number with fewer decimal
places. The procedure is similar to the one used for whole numbers except
that we do not use trailing zeros.

Example 2

Round 3.14159 to two decimal places.

Solution

"Two decimal places" means two places to the right of the decimal point,
so we are rounding the number to the nearest hundredth. Thus our choice
is between 3.14 and 3.15. We inspect the digit in the thousandths place to
decide which number is closer.

$$3.14\textcircled{1}59$$

The digit in the thousandths place is 1, which is less than 5, so we round
down to **3.14.**

To round a mixed number to the nearest whole number, we decide if the
fraction is more than or less than $\frac{1}{2}$. If the numerator is less than half the
denominator, then the fraction is less than $\frac{1}{2}$, and we round down. If the
numerator is half or greater than half of the denominator, we round up.

Example 3

Round each number to the nearest whole number.

a. $13\frac{5}{12}$ b. **13.512**

Solution

a. The mixed number is between 13 and 14. The fraction $\frac{5}{12}$ is less than $\frac{1}{2}$
because 5 is less than half of 12. We round down to **13.**

b. The decimal point separates the whole number from the fraction. Again
we choose between 13 and 14. The digit in the first decimal place is 5,
so we round up to **14.**

To **estimate** is to determine an approximate value. We make estimates in our activities throughout the day. In a baseball game, for example, the batter estimates the speed and location of the pitch and swings the bat to arrive at the expected location at just the right time. The awaiting fielder hears the crack of the bat and—in a split second—estimates the ball's trajectory and begins moving to the point where the ball is expected to land. Both players adjust their estimates as they visually gather more information about the path of the baseball.

The estimates we make in athletics are extremely complex, yet with practice we perform them with ease. We can also become proficient at estimating measures and calculations as we practice these skills.

Example 4

Estimate the length of one wall of your classroom in feet.

Solution

There are a variety of ways to estimate a length of several feet. Some people walk a distance using big steps counting three feet (one yard) to a step. Other people will use a reference, such as a door which is about seven feet high, to visually form an estimate. Make a thoughtful estimate, and then use a tape measure or yardstick to find the length of the wall, rounded to the nearest foot.

Discuss Explain how you might estimate the perimeter of the classroom.

Estimating calculations helps us decide if an answer is **reasonable.** For example, suppose we are calculating the cost of 48 square yards of carpeting priced at $28.95 per square yard. We could round the quantity to 50 square yards and the price to $30 per square yard. Multiplying these numbers mentally we get $1500. If the calculated product is very different, then the results are not reasonable and we will review our calculations.

Example 5

Ivan wants to tile the floor of a kitchen that is 12 feet long and 8 feet wide. The installed tile is $7.89 per square foot. Ivan calculates that the cost to tile the floor is $315.60. Is Ivan's calculation reasonable?

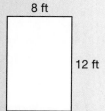

Solution

The cost is based on the number of square feet. The area of the kitchen is 8 feet × 12 feet, or 96 square feet. We can estimate the cost of tiling the floor by multiplying 100 square feet by $8 per square foot. A reasonable estimate of the cost is $800, which is very different from Ivan's calculation. Since Ivan's calculation is **not reasonable,** he should check his calculations.

Example 6

Richard is paid $11.85 per hour. About how much money does he earn in a year (52 weeks) if he works 40 hours per week?

Solution

We use round numbers and multiply the three values to estimate how much Richard earns in a year.

Total Pay ≈ $12 · 40 · 50   Rounded $11.85 and 52

≈ $12 · 2000   Multiplied 50 · 40

≈ $24,000   Multiplied

Richard earns about **$24,000** per year.

Practice Set

a. In 2000 the population of Dallas was 1,188,580. Round that number to the nearest hundred thousand.

b. Round 3.14159 to four decimal places.

c. The price of regular unleaded gasoline was 2.19\frac{9}{10}$. Round that price to the nearest cent.

d. _Estimate_ Estimate the height of the ceiling in your classroom in meters and in feet.

e. Estimate the sum of 3879 and 5276.

f. Estimate the product of $12\frac{3}{8}$ and 9.75 by rounding each number to the nearest whole number before multiplying.

g. _Evaluate_ Gus works 39 hours for $9.85 per hour. He calculates that his weekly paycheck should be $38,415. What is a reasonable estimate of what Gus earns? How do you suppose Gus arrived at his calculation?

h. A rectangular room is $11\frac{3}{4}$ feet long and $10\frac{1}{2}$ feet wide. Grace calculated the area to be about 124 square feet. Explain why her calculation is or is not reasonable.

i. It is 6:05 p.m. and Eduardo is driving to Dallas. If he is 157 miles away and driving at an average speed of 62 miles per hour, is it reasonable that he can reach Dallas by 9:00 p.m.? Explain your answer.

Written Practice   _Strengthening Concepts_

1. In the orchard there were 10 rows with 12 trees in each row. How many
(4)   trees were in the orchard?

2. At 8 a.m. it was 57°F, and at 1 p.m. it was 72°F. By how many degrees
(3)   did the temperature change between 8 a.m. and 1 p.m.?

3. Write a sequence of numbers with 1 as the first term and a constant
(2)   difference of 2 between terms.

* 4. *(6)* **Explain** Brandon said that 48 yards is equal in length to 4 feet. Explain why his statement is unreasonable. Forty-eight yards is equal to how many feet?

Generalize Simplify.

* 5. *(15)* 11^2

* 6. *(15)* $\sqrt{121}$

* 7. *(13)* $\dfrac{1}{2} + \dfrac{3}{4}$

* 8. *(13)* $1\dfrac{2}{3} + \dfrac{5}{6}$

9. *(8)* Justine will fence a rectangular piece of land that measures 15 yards by 18 yards, then lay sod within the fencing. Find the length of fencing and the number of square yards of sod that she will need.

10. *(15)* **Represent** Express with exponents:

  a. 5*xxy*   **b.** 6*xyyxy*

11. *(15)* Compare:

  a. $\sqrt{100} \bigcirc -10$ **b.** $\sqrt{25} \bigcirc \sqrt{36}$

Solve by inspection.

12. *(14)* $5x = 45$

13. *(14)* $\dfrac{x}{12} = 3$

14. *(14)* $x + 7 = 15$

15. *(14)* $18 - x = 10$

16. *(2, 3)* Select an equation from problems **12–15** and write a story for it.

17. *(14)* Evaluate $P = 2(l + w)$ when $l = 7$ in. and $w = 5$ in.

18. *(9)* Write the prime factorization of 90. Use exponents for repeated factors.

19. *(9)* Patrice has 18 square tiles that she will lay side-by-side to form a rectangle. Name the dimensions of three different rectangles she can form.

Use the following information to answer problems **20** and **21**.

For two weeks, Destiny recorded the number of red lights at which she had to stop between home and school. These numbers are listed below.

$$3, 4, 3, 2, 5,$$
$$3, 4, 6, 3, 2$$

20. *(7)* Find the mean, median, mode, and range of the data.

21. *(7)* Which measure would you use to report the number of stops that Destiny most frequently made?

22. *(2)* Rearrange the numbers to make the calculations easier. What properties did you use? What is the sum?

$$17 + 28 + 3$$

23. *(Inv. 1)* Graph these points on the coordinate plane: $(4, 0)$, $(-2, 3)$, $(0, -\frac{7}{2})$. Which appears to be farthest from the origin?

24. Find the area and perimeter of a rectangle with vertices at $(-1, 2)$, $(3, 2)$,
<small>(8, Inv. 1)</small> $(3, -5)$, and $(-1, -5)$.

25. Write these numbers from least to greatest:
<small>(10)</small>

$$1.3, 0, 1, \frac{3}{4}, \frac{4}{3}, 1\frac{1}{12}$$

26. Which is a better discount?
<small>(5, 11)</small>

    A Price of $20 is reduced by 30%.

    B Price of $20 reduced by $\frac{1}{4}$.

    C Price of $20 is reduced by $4.

* 27. <u>**Estimate**</u> Chris budgets $5000 to purchase 6 computers at $789 each.
<small>(17)</small> Is the budget reasonable? Estimate the cost of the computers.

* 28. Round to the nearest whole number.
<small>(17)</small>

    a. $4\frac{5}{8}$ **b.** 3.199

29. Estimate 8.25% sales tax on a $789 computer.
<small>(12)</small>

* 30. Write 0.24 as a reduced fraction and as a percent.
<small>(12)</small>

• Lines and Angles

Power Up   *Building Power*

facts   Power Up D

mental math

a. Calculation: Dash galloped 20 miles per hour for $\frac{1}{4}$ hour. How far did Dash gallop?

b. Measurement: How many pints are shown?

c. Number Sense: 88×5

d. Proportions: If 3 cans cost $6, how much would 4 cans cost?

e. Fractional Parts: $\frac{5}{8}$ of 88

f. Percent: 25% of 40

g. Power/Roots: $\sqrt{10}$ is between which two consecutive integers?

h. Calculation: $24 \div 6$, $+ 3$, $\times 9$, $+ 1$, $\sqrt{}$, $+ 1$, $\sqrt{}$, $- 5$, $+ 2$, $\times 17$, $+ 5 =$

problem solving

The Mystery Man gave interesting responses. When Erica said 5, the Mystery Man said 9. When Joseph said 8, the Mystery Man said 15. When Humberto said 12, the Mystery Man said 23. When Wally whispered, the Mystery Man said 19. What did Wally whisper?

New Concept   *Increasing Knowledge*

A **line** is a straight path that extends without end in both directions. A **ray** is a part of a line that begins at a point and extends without end in one direction. A **segment** is part of a line with two endpoints.

A B	A B	A B
line AB ($\overleftrightarrow{AB}$)	ray AB ($\overrightarrow{AB}$)	segment AB ($\overline{AB}$)
line BA ($\overleftrightarrow{BA}$)		segment BA ($\overline{BA}$)

Justify   Can we change the order of the letters we use to name rays? Why or why not?

A **plane** is a flat surface that extends without end. Two lines on the same plane either intersect or do not intersect. If the lines do not intersect, then they are **parallel** and remain the same distance apart. Lines that intersect and form square corners are **perpendicular.**

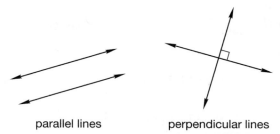

parallel lines   perpendicular lines

Lines that are in different planes that do not intersect are **skew lines.** For example, a wall and the ceiling represent different planes. A line on a wall and a line on the ceiling that do not intersect are skew lines.

Intersecting lines form angles. An **angle** is two rays with the same endpoint. The endpoint is the vertex and the rays are sides of the angle.

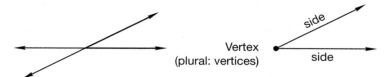

We can name angles in a variety of ways. If we use three letters, the middle letter is the vertex. We can name an angle by a single letter at the vertex if there is no chance of confusion. Sometimes we use a number to name an angle.

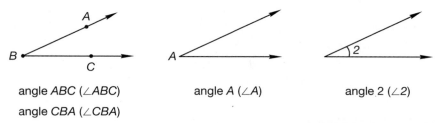

angle *ABC* (∠*ABC*)   angle *A* (∠*A*)   angle 2 (∠2)

angle *CBA* (∠*CBA*)

We can measure angles in **degrees.** A full turn is 360 degrees (360°), so a quarter turn is 90°.

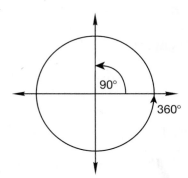

Angles are commonly described as **acute, right, obtuse,** or **straight** depending on their measure. Right angles are sometimes indicated by a small square.

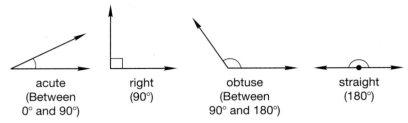

acute
(Between
0° and 90°)

right
(90°)

obtuse
(Between
90° and 180°)

straight
(180°)

If two angles together form a straight angle, the angles are called a **linear pair** and the sum of their measures is 180°. In the figure below, ∠QSR and ∠RST are a linear pair.

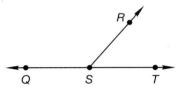

A **protractor** is a tool used to measure angles. To measure an angle we center the protractor on the vertex of the angle and position one of the zero marks on one side of the angle. We find the measure of the angle where the other side of the angle passes through the scale.

Example

Find m∠*RST.*

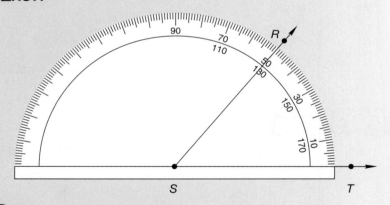

Solution

The abbreviation m∠*RST* means "measure of angle *RST.*" Protractors have two scales, one starting from the left side and one starting from the right side. Since side *ST* of this angle is on the right-side zero mark, we read the scale that counts up from zero from the right side. The angle measures **50°.** Another way to decide which scale to use is to determine if we are measuring an acute angle or an obtuse angle. Since this angle is an acute angle, we use the scale with numbers less than 90.

Practice Set

Name each figure in problems **a–c**.

a. P Q (line with arrows both ends, points P and Q)

b. P Q (ray, points P and Q)

c. P Q (segment, points P and Q)

Describe each pair of lines in problems **d** and **e**.

d. (two parallel lines with arrows)

e.

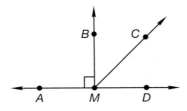

Classify Describe each angle in problems **f–i** as acute, obtuse, right, or straight.

f. ∠CMD

g. ∠CMA

h. ∠AMB

i. ∠AMD

j. If the measure of ∠DMC is 45°, then what is the measure of ∠CMA? Explain how you found your answer.

Written Practice   *Strengthening Concepts*

1. **Estimate** Michaela estimated that there were nine tables in the room
(4) with about six people seated at each table. About how many people were seated at the tables?

2. **Analyze** The temperature at 8 a.m. was 65°F. By 11 a.m. the
(3) temperature had increased 17°. What was the temperature at 11 a.m.?

3. Of the fifty customers, 35 bought the advertised item. What fraction of
(5, 11) the customers bought the advertised item? What percent?

4. How many cups are in 3 gallons?
(6)

5. Cleo walked 3 miles in 45 minutes. What was his average rate in
(7) minutes per mile? In miles per hour?

Generalize Simplify.

6. 12^2
(15)

7. $\sqrt{144}$
(15)

* 8. $\dfrac{5}{8} - \dfrac{2}{6}$
(13)

* 9. $3\dfrac{1}{4} + 1\dfrac{11}{12}$
(13)

10. Compare:
(15, 16)

a. $|-10| \bigcirc \sqrt{100}$

b. $\sqrt{5} \bigcirc 2$

Solve by inspection.

11. $x - 13 = 20$
(14)

12. $5 + x = 13$
(14)

13. $\dfrac{2x}{5} = 10$
(14)

14. $8x = 64$
(14)

15. Write a sequence of four numbers with the first term 5 and a constant
(2) difference of 5 between terms.

16. Jenna plans to paint the ceiling of a 15 foot-by-10 foot room, then
(8) mount crown molding along the upper edge of the walls. Jenna should buy enough paint to cover what area? What length of crown molding will she need?

17. Write the prime factorization of 128. Use exponents for repeated
(9, 15) factors.

18. How many one-digit counting numbers are prime?
(9)

19. Write these numbers in order from least to greatest.
(10)

$$0, 1, -1, -\dfrac{4}{5}, \dfrac{5}{4}$$

20. Compare:
(10, 12)

a. $\left|-\dfrac{4}{5}\right| \bigcirc \left|\dfrac{5}{4}\right|$

b. $0.8 \bigcirc \dfrac{7}{10}$

* 21. **Connect** Write 0.125 as a fraction and as a percent.
(12)

*22. Compute a 15% tip for a dinner bill of $40.
(11)

* 23. **Estimate** The sales-tax rate is 8.25%. What is the approximate sales
(12, 17) tax for a $104 purchase?

* 24. **Analyze** Plot these numbers on a number line.
(8, 16)

$$\sqrt{4}, \sqrt{5}, \sqrt{9}, \sqrt{10}, \sqrt{15}, \sqrt{16}$$

* 25. **Estimate** What is the side length of a square with an area of 40 cm²?
(8, 16) Measured with a ruler, this side length is between what two consecutive whole centimeters?

Use this diagram to complete problems **26** and **27**.

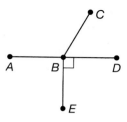

* 26. (18) **Classify** Describe each angle by type.

    a. ∠ABC   **b.** ∠ABD

    c. ∠CBD   **d.** ∠DBE

* 27. (18) Suppose m∠CBD = 60°. Find the following angle measures.

    a. m∠ABC   **b.** m∠ABE

* 28. (7, 17) Kiersti drove 58 mph for 1 hour and 52 minutes. Estimate the distance she traveled.

For problems **29** and **30**, choose from this list *all* correct answers.

    A integers   **B** whole numbers   **C** natural numbers

    D rational numbers   **E** irrational numbers   **F** real numbers

* 29. (1, 16) The number 0 is a member of which set(s) of numbers?

* 30. (1, 16) The number $\sqrt{7}$ is a member of which set(s) of numbers?

Early Finishers
Real-World Application

Earth has only one natural satellite, the Moon. The Moon has many effects on Earth. For example, the Moon's gravity affects the tides within the oceans, even though the Moon is 384,403 kilometers away from Earth. If 1 kilometer is 0.62 miles, estimate the distance from Earth to the Moon in miles (round to the nearest thousand).

• Polygons

facts

Power Up D

mental math

a. **Calculation:** Erica rides a snowmobile 20 miles per hour for $2\frac{1}{2}$ hours. How far does Erica ride?

b. **Powers/Roots:** 15×10^2

c. **Fractional Parts:** Three fourths of a pound is how many ounces?

d. **Geometry:** What is the measure of $\angle ACB$?

e. **Proportions:** If 2 tickets cost $10, then how much would four tickets cost?

f. **Algebra:** $3x = 300$

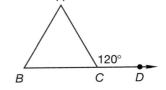

g. **Probability:** A number cube is rolled once. What is the probability of rolling a number greater than or equal to 4?

h. **Calculation:** $12 \times 2, \div 6, \sqrt{}, \times 10, + 5, \sqrt{}, \times 4, \times 5, + 1$

problem solving

What is the last number in the 20th row?

Row 1			2	
Row 2		4	6	
Row 3	8	10	12	
Row 4	14	16	18	20
Row 5	22	24	26	28	30

(Understand) We have been given a visual pattern of the even numbers arranged in a triangle, and are asked to find the last number in the 20th row.

(Plan) We have been asked to find a specific term, so we can *make the problem simpler* by only concerning ourselves with the row numbers and the final numbers in each row. We will *find a pattern* and *write an equation* to determine the final term in the 20th row.

(Solve)

Row Number:	1	2	3	4	5
Final Number:	2	6	12	20	30

We notice that the final number in each row equals the row number times the next row number. For example, the final number in the 4th row equals 4 times 5. We can write an equation for this pattern. The equation for finding the final number, F, in the nth row is $F = n(n + 1)$.

The final number in the 20th row will be $20(21) = $ **420.**

Check We could have completed the triangle to determine the number, but it would have been time consuming.

Polygons are closed plane figures with straight sides. They have the same number of sides as angles.

Example 1

Which of these figures is a polygon?

A 　　B 　　C 　　D

Solution

Figure A is closed but it has a curved side, so it is not a polygon.

Figure B is not closed, so it is not a polygon.

Figure C represents a cube, which is not a plane figure, so it is not a polygon.

Figure D is a plane, closed figure with straight sides, so it is a polygon.

A polygon is **regular** if all its sides are the same length and all its angles are the same size.

Example 2

Which polygon is not regular?

A 　　B 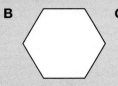　　C 　　D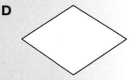

Solution

The sides of each figure are equal in length. The angles in figures A, B, and C are of equal size. However, **D** has two acute angles and two obtuse angles, so the figure is not regular.

Polygons are named by the number of their sides and angles.

Names of Polygons

Name of Polygon	Number of Sides	Name of Polygon	Number of Sides
Triangle	3	Octagon	8
Quadrilateral	4	Nonagon	9
Pentagon	5	Decagon	10
Hexagon	6	Hendecagon	11
Heptagon	7	Dodecagon	12

A polygon with more than 12 sides may be referred to as an *n*-gon, with *n* being the number of sides. Thus, a polygon with 15 sides is a 15-gon.

The point where two sides of a polygon meet is called a **vertex** (plural: **vertices**). A particular polygon can be identified by naming the letters of its vertices in order. We may name any letter first. The rest of the letters can be named clockwise or counterclockwise.

Math Language

Recall that moving **clockwise** means moving in the same direction as the hands of a clock. Moving **counterclockwise** means moving in the direction opposite to the hands of a clock.

Discuss One name for the polygon below is *USTV.* What are the other seven names?

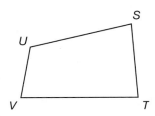

Figures are **congruent** if they are the same shape and size. Figures A and D below are congruent, even though they do not have the same **orientation.** We could slide and turn figure A to perfectly align with figure D.

Figures A and C and figures C and D are not congruent, because they are not the same size. They are **similar** because they have the same shape. Figure C appears to be a **dilation** (enlargement) of figures A and D. Using a magnifying glass, we could enlarge the image of figure A or figure D to match figure C. Figure B is not similar to the other figures because its shape is different.

We use the symbol ~ to indicate similarity.

Figure A ~ Figure C

"Figure A is similar to figure C."

Since congruent figures are both similar in shape and equal in measure, we use the symbol ≅ to indicate congruence.

Figure A ≅ Figure D

"Figure A is congruent to figure D."

Example 3

a. Which triangles below appear to be similar?

b. Which triangles appear to be congruent?

A   B   C   D

Solution

a. Triangles A, B, and C appear to be similar.

b. Triangles A and C appear to be congruent.

Practice Set

a. Name this polygon. Is it regular or irregular?

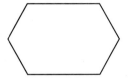

b. Which of the following is not a way to name this rectangle?

   A ☐ *ABCD*   **B** ☐ *ADCB*

   C ☐ *BACD*   **D** ☐ *BADC*

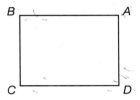

c. What is the name for the shape of a stop sign?

d. The block H shown is a polygon. Count the sides and name the polygon.

e. Represent   Use four letters to correctly name the figure that looks like a square. Start the name with the letter *A*.

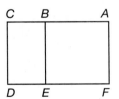

f. In example 3, which triangle appears to be a dilation of triangle **A**?

g. Which figures below appear to be similar and which appear to be congruent?

    A   **B**   **C**   **D**

1. One pack of trading cards sells for $1.12. For how much would seven
(4) packs sell?

2. In the 1968 Olympics in Mexico City, Bob Beamon broke Ralph
(3) Boston's long jump world record by leaping 890 cm. The former record
was 835 cm. How much greater was Beamon's record than Boston's
record?

3. Kerry had 17 bottles in her collection. Some friends gave her more. Now
(3) Kerry has 33 bottles. How many bottles did her friends give her?

4. Kim swam one kilometer in 40 minutes. The length of the pool was
(4) 25 meters. About how long did it take Kim to swim each length?

5. What is the mean of the prime numbers between 10 and 20?
(7, 9)

6. Sketch a number line and plot the approximate location of the following
(10, 15) points.

$$-2, |-3|, \frac{1}{2}, \sqrt{2}$$

7. Using prime factorization, reduce $\frac{375}{1000}$.
(10)

* 8. (Analyze) Compare:
(10, 15)
    a. $4 \bigcirc \dfrac{7}{2}$ **b.** $7 \bigcirc \sqrt{49}$

(Generalize) Simplify.

* 9. $\dfrac{1}{5} + \dfrac{4}{9}$ *** 10.** $\dfrac{1}{2} - \dfrac{1}{10}$
(13)   (13)

11. Write a fraction and a decimal equivalent to 40%.
(11, 12)

12. Compute 150% of $70.
(11)

13. Express with exponents:
(15)
    a. $y \cdot y \cdot y \cdot y$ **b.** $3xxyxy$

* 14. Three coordinates of a rectangle are (0, 4), (2, 4), and (2, −3). Find the
(8, Inv. 1) fourth coordinate and the area and perimeter of the rectangle.

Simplify.

15. 2^5 **16.** 10^6 **17.** $\sqrt{144}$
(15)   (15)   (15)

* 18. $(12 \text{ in.})^2$ *** 19.** 8^3 *** 20.** $\sqrt[3]{8}$
(15)   (15)   (15)

* 21. (Analyze) What is the quotient when 30^2 is divided by the product of 5
(1, 15) and 6?

An area of the playground will be seeded with grass and fenced until the grass grows in. Use this information and the figure to answer problems **22** and **23**.

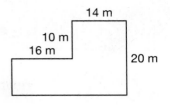

22. The grass seed needs to cover how many square meters?
(8)

23. How many meters of fencing are needed to surround the area?
(8)

For problems **24–27,** solve by inspection. Select one equation and write a word problem to match.

* **24.** $72 = 36 + m$
(14)

25. $w - 11 = 39$
(14)

* **26.** $7c = 56$
(14)

27. $8 = \dfrac{72}{x}$
(14)

28. Estimate the distance covered by a car traveling at 62 mph for 3 hours and 50 minutes.
(7, 17)

* **29.** All rectangles are also parallelograms. The formula for the area of a parallelogram is $A = bh$. Find A when $b = 7$ mm and $h = 9$ mm.
(14)

30. The number 0 is a member of which sets of numbers?
(1)

    A integers   **B** whole numbers   **C** counting numbers

    D rational numbers   **E** irrational numbers   **F** real numbers

---

Early Finishers
Real-World Application

Pete, Jim, and John want to change the oil and oil filter in each of their cars. For each oil change, Pete's car needs $4\frac{3}{4}$ quarts of oil, Jim's needs $4\frac{1}{2}$ quarts, and John's needs 5 quarts. Each boy also wants to keep one full quart of oil in his trunk for emergencies. Given that the boys already have one case of oil containing 12 quarts, how many more quarts of oil do they need to buy at the store?

• Triangles

facts | Power Up D

mental math

a. Calculation: Glenda takes 20 breaths per minute. How many breaths does she take in $4\frac{1}{2}$ minutes?

b. Estimation: $35.8 \div 3.9$

c. Measurement: If the loop of string is cut and laid out lengthwise, about how long will it be?

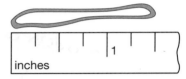

d. Number Sense: 1.5×2 inches

e. Geometry: $\angle A$ corresponds to $\angle X$. $\angle B$ corresponds to which angle?

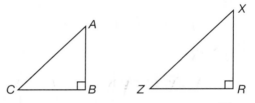

f. Calculation: $7 - 8, + 2, + 10, \times 7, - 2, \div 3, \sqrt{\ }, \times 7, + 7, \div 7, - 7$

problem solving | Find the next three numbers in the sequence:

$$\frac{1}{16}, \frac{1}{8}, \frac{3}{16}, \frac{1}{4}, \frac{5}{16}, \underline{\quad}, \underline{\quad}, \underline{\quad}$$

New Concept | Increasing Knowledge

Recall that polygons are closed, plane figures with straight sides. A **triangle** is a polygon with three sides. We classify triangles based on their angle measures and based on the relative lengths of their sides.

Triangles are classified as acute, right, or obtuse based on the measure of their *largest* angle. **The measures of the three angles of a triangle total 180°,** so at least two angles of a triangle are acute.

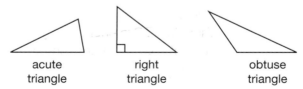

acute triangle   right triangle   obtuse triangle

Triangles are also classified by relative side lengths.

- **Equilateral** triangles have three equal-length sides (and three equal angles).

- **Isosceles** triangles have at least two equal-length sides (and at least two equal angles).

- **Scalene** triangles have sides of different lengths (and angles of different measures).

We can use tick marks and arcs to indicate sides and angles that have equal measures.

equilateral   isosceles   scalene

Any triangle can be classified both by angles and by sides.

Conclude   Can an equilateral triangle be an obtuse triangle? Why or why not?

Example 1

Classify triangle *ABC* by angles and by sides.

Solution

The triangle is a **right triangle** and a **scalene triangle.**

Example 2

The roof of the building in the drawing slopes 35°. What is the measure of the angle at the peak of the roof?

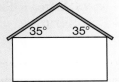

Solution

The roof structure forms an isosceles triangle. The measures of the acute base angles total 70°. All three angles total 180°, so the angle at the peak of the roof is **110°.**

$$35° + 35° + 110° = \mathbf{180°}$$

Now we will consider how to find the area of a triangle. Triangle *ABC* is enclosed by a rectangle. What fraction of the area of the rectangle is the area of triangle *ABC*?

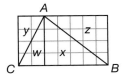

Math Language

Recall that *congruent* means "having the same shape and size."

The letters *w, x, y,* and *z* refer to four regions of the rectangle. Regions *w* and *x* are within the triangle. Regions *y* and *z* are outside the triangle. Notice that regions *w* and *y* are the same size. Also notice that regions *x* and *z* are the same size. These pairs of congruent regions show us that triangle *ABC* occupies half of the rectangle. We will use this fact to help us understand the formula for the area of a triangle.

We start with a rectangle. If we multiply two perpendicular dimensions, the product is the area of a rectangle.

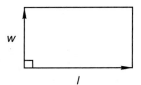

The perpendicular dimensions of a *triangle* are called the **base** and **height.**

Multiplying the base and height of a triangle gives us the area of a rectangle with the same base and height as the triangle. Since the triangle occupies half the area of the rectangle, we can find the area of the triangle by finding half the area of the rectangle.

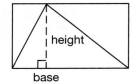

Area of a Triangle

Area of triangle $= \frac{1}{2}$ base $\cdot$ height

$$A = \frac{1}{2}bh$$

Example 3

Find the area of each triangle.

a. 6 in. 10 in. 8 in.

b. 10 cm 6 cm 6 cm

c. 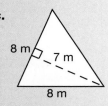 8 m 7 m 8 m

Solution

a. One angle is a right angle, so two sides are perpendicular. We multiply the perpendicular sides and find half the product.

$$A = \frac{1}{2}bh$$

$$A = \frac{1}{2}(8 \text{ in.})(6 \text{ in.})$$

$$A = \textbf{24 in.}^2$$

b. We multiply the perpendicular dimensions. As the triangle is oriented the base is 6 cm and the height is 6 cm. We find half the product of the base and height.

$$A = \frac{1}{2}bh$$

$$A = \frac{1}{2}(6\ cm)(6\ cm)$$

$$A = \textbf{18 cm}^2$$

c. The base does not need to be horizontal and the height vertical. The base is 8 m and the height is 7 m. (Turn the book to help you see the relationship.)

$$A = \frac{1}{2}bh$$

$$A = \frac{1}{2}(8\ m)(7\ m)$$

$$A = \textbf{28 m}^2$$

Practice Set

Classify Classify each triangle in problems **a–c** by angles and by sides.

a.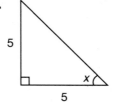

b.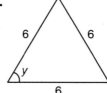

c.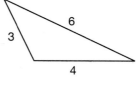

d. In problems **a** and **b,** the letters x and y represent how many degrees?

Find the area of each in problems **e–g.**

e.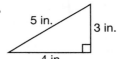

f.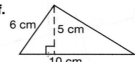

g.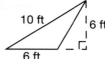

h. Mr. Torres arranges square floor tiles diagonally in a room. Near each wall he needs to cut the tiles to fit. Sketch the shape of the triangle pieces he needs to cut. Classify the triangle by angle and by sides. Write the measure of each angle.

Written Practice   *Strengthening Concepts*

1. The tram driver collected $50.40 from the 45 passengers. On average, how much did the driver collect per passenger?
(4)

2. Mary could carry 12 liters, while Martha could carry 19 liters. How many liters could they carry together?
(3)

3. Practicing quick shots, Roberto made three baskets in ten seconds. At this rate, how many baskets could he make in one minute?
(4)

4. Conrad brought home three gallons of milk. How many quarts is that?
(4)

5. Write the prime factorization of 60. Use exponents for repeated factors.
(9)

* 6. Plot the following points on a coordinate plane: $(-1, 1)$, $(-1, -6)$, (5, 1) and (5, −6). Then draw segments connecting the points to make a rectangle. What is the perimeter of the rectangle?
(4, Inv. 1)

Evaluate Use the following information for problems **7** and **8**.

Sergio priced a popular game at several game stores. The prices are recorded below.

$$\$52, \$49, \$48, \$50, \$49, \$50, \$51$$

* 7. a. Find the mean, median, mode, and range of the data.
(7)

    b. Which measure would you use to report the spread in the prices?

8. If Sergio found the game on sale at a different store for $30, which measure of central tendency (mean, median, or mode) would be affected most?
(7)

9. Compare: **a.** $\frac{2}{3} \bigcirc \frac{3}{2}$ **b.** $|{-7}| \bigcirc |7|$
(10)

10. Write a fraction equivalent to $\frac{3}{25}$ that has a denominator of 100. What decimal is equal to $\frac{3}{25}$?
(10)

* 11. Plot the approximate location of $\sqrt{5}$ on a number line.
(16)

* 12. **Classify** Classify each triangle by angles and by sides.
(24)

    a.   **b.**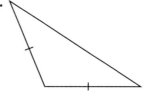

* 13. Refer to the figure to answer **a–c.**
(18)

    a. Which angle appears to measure 60°?

    b. Which angle appears to measure 120°?

    c. Which angle appears to measure 180°?

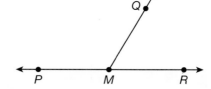

* 14. Find the area of this triangle:
(20)

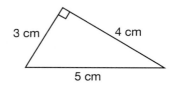

Simplify.

* 15. 9^3
(15)

16. 13^2
(15)

17. $\sqrt{64}$
(15)

* 18. $\sqrt[3]{216}$
(15)

19. $\dfrac{1125}{15}$
(2)

20. $\dfrac{3}{4} - \dfrac{5}{12}$
(13)

* 21. Write in exponential form:
(15)

    a. $a \cdot a \cdot a \cdot a \cdot b \cdot b$ **b.** $2xyxyxyx$

* 22. **Analyze** The coordinates of three vertices of a square are $(-1, -1)$,
(Inv. 1) $(3, -1)$, and $(3, 3)$. What are the coordinates of the fourth vertex?

For problems **23–26,** solve by inspection.

* 23. $m + 90 = 110$
(14)

24. $90 - 3x = 60$
(14)

* 25. $38 = 2x - 2$
(14)

26. $\dfrac{x}{100} = 1$
(14)

27. Find $4ac$ when $a = 3$ and $c = 12$.
(14)

28. For $A = s^2$, find A when $s = 25$ ft.
(14)

* 29. Rearrange the addends to make the calculation easier. Then simplify,
(13)   justifying each step.

$$3\frac{1}{6} + \left(4\frac{1}{6} + 1\frac{2}{3}\right)$$

30. The number -9 is a member of which set(s) of numbers?
(10)

    A whole numbers   **B** counting numbers   **C** integers

    D irrational numbers   **E** rational numbers   **F** real numbers

Focus on

• Pythagorean Theorem

The longest side of a right triangle is called the **hypotenuse.** The other two sides are called **legs.** Notice that the legs are the sides that form the right angle. The hypotenuse is the side opposite the right angle.

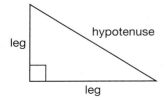

Every right triangle has a property that makes right triangles very important in mathematics. **The area of a square drawn on the hypotenuse of a right triangle equals the sum of the areas of squares drawn on the legs.**

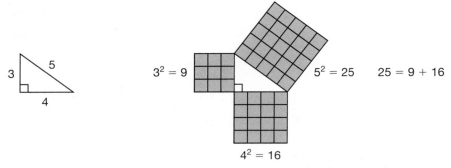

As many as 4000 years ago ancient Egyptians knew of this important and useful relationship between the sides of a right triangle. Today the relationship is named for a Greek mathematician who lived about 550 B.C. The Greek's name was Pythagoras and the relationship is called the Pythagorean Theorem.

The **Pythagorean Theorem** is an equation that relates the sides of a right triangle in this way: the sum of the squares of the legs equals the square of the hypotenuse.

$$\text{leg}^2 + \text{leg}^2 = \text{hypotenuse}^2$$

If we let the letters a and b represent the lengths of the legs and c the length of the hypotenuse, then we can express the Pythagorean Theorem with the following equation.

Pythagorean Theorem
If a triangle is a right triangle, then the sum of the squares of the legs equals the square of the hypotenuse. $$a^2 + b^2 = c^2$$

It is helpful to use an area model to solve problems that involve the Pythagorean Theorem. Work through the following examples on your own paper.

Example 1

Copy this triangle. Draw a square on each side. Find the area of each square. Then find *c*.

Solution

We copy the triangle and draw a square on each side of the triangle as shown.

We were given the lengths of the legs. The areas of the squares on the legs are 36 cm^2 and 64 cm^2. The Pythagorean Theorem says that the sum of the smaller squares equals the area of the largest square.

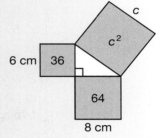

$$36 \text{ cm}^2 + 64 \text{ cm}^2 = 100 \text{ cm}^2$$

This means that each side of the largest square must be 10 cm long because $(10 \text{ cm})^2$ equals 100 cm^2. Therefore, *c* = **10 cm.**

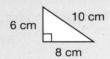

Example 2

Copy this triangle and draw a square on each side. Find the area of each square. Then find *a*.

Solution

We copy the triangle and draw a square on each side. The area of the largest square is 169 in.^{2}. The areas of the smaller squares are 144 in.^{2} and a^2. By the Pythagorean Theorem a^2 plus 144 must equal 169.

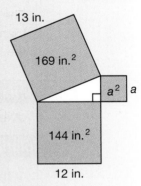

$$a^2 + 144 = 169$$

By inspection we see that

$$a^2 = 25$$

Thus *a* equals **5 in.**, because 5^2 is 25.

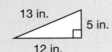

1. Write the name of the property of right triangles illustrated in examples 1 and 2.

2. What is the name for the two sides of a right triangle that form the right angle?

3. What is the name for the side of a right triangle that is opposite the right angle (across the triangle from the right angle)?

4. **Represent** Using the letters a and b for the shorter sides of a right triangle and c for the longest side of a right triangle, write the Pythagorean Theorem as an equation.

5. **Evaluate** Copy the triangle and draw squares on all three sides. Find the area of each square. Then find the length of the longest side of the triangle.

6. Use the Pythagorean Theorem to find a.

The triangles we have considered so far have sides that are whole numbers. Three whole numbers that can be the side lengths of a right triangle are called a **Pythagorean triple.**

Some Pythagorean Triples

3, 4, 5

5, 12, 13

8, 15, 17

Thinking Skill

Verify

Verify that this multiple of the 5-12-13 triple is a Pythagorean triple: 25-60-65.

Multiples of Pythagorean triples are also Pythagorean triples. If we multiply 3, 4, and 5 by two, the result is 6, 8, and 10, which is a Pythagorean triple. As we see in this test:

$$a^2 + b^2 = c^2$$
$$6^2 + 8^2 = 10^2$$
$$36 + 64 = 100$$
$$100 = 100$$

Example 3

Are the numbers 2, 3, and 4 a Pythagorean triple?

Solution

If three numbers are a Pythagorean triple, then the sum of the squares of the two lesser numbers equals the square of the greatest number. Since 2^2 plus 3^2 does not equal 4^2, the three numbers **2, 3, and 4 are not a Pythagorean triple.**

a^2	b^2	c^2
2^2	3^2	4^2
4	9	16

The Pythagorean Theorem applies to all right triangles and only to right triangles. If a triangle is a right triangle, then the sides are related by the Pythagorean Theorem. It is also true if the sides of a triangle are related by the Pythagorean Theorem, then the triangle is a right triangle. The latter statement is the **converse** of the Pythagorean Theorem.

Converse of Pythagorean Theorem
If the sum of the squares of two sides of a triangle equals the square of the third side, then the triangle is a right triangle.

We use the converse of the Pythagorean Theorem to determine if a triangle is a right triangle.

Example 4

Cassidy sketched a triangle and carefully measured the sides. Is her triangle a right triangle?

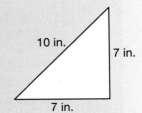

Solution

The triangle might look like a right triangle, but to be a right triangle the side lengths must satisfy the Pythagorean Theorem.

a^2	b^2	c^2
7^2	7^2	10^2
49	49	100

The sum of 49 and 49 is 98, which is nearly 100, but not quite. Therefore, **the triangle is not a right triangle.** (The largest angle of a triangle with these side lengths is about 91°.)

7. Write a multiple of 5, 12, 13, that is a Pythagorean triple.

8. Are the numbers 8, 15, and 17 a Pythagorean triple? Explain your answer.

9. Does a right triangle with the dimensions 8 cm, 9 cm, and 12 cm exist? Explain your answer.

Pythagorean triples are used in the building trades to assure that corners being constructed are right angles.

Before pouring a concrete foundation for a building, construction workers build wooden forms to hold the concrete. Then a worker or building inspector can use a Pythagorean triple to check that the forms make a right angle. First the perpendicular sides are marked at selected lengths.

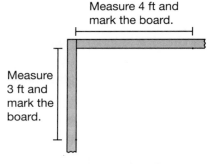

Measure 4 ft and mark the board.

Measure 3 ft and mark the board.

Then the distance between the marks is checked to be sure the three measures are a Pythagorean triple.

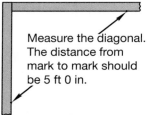

Measure the diagonal. The distance from mark to mark should be 5 ft 0 in.

If the three measures are a Pythagorean triple, the worker can be confident that the corner forms a 90° angle.

You can test your ability to draw a right angle with a pencil and ruler. Draw an angle with sides at least four inches long and mark one side at 3 in. and the other at 4 in. from the vertex.

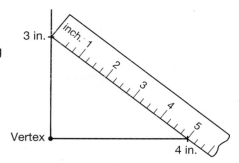

3 in.

Vertex

4 in.

Then measure from mark to mark to see if the distance is 5 inches. If the distance is less than 5 inches, the angle is less than 90°. You can use a protractor to confirm the size of the angle.

10. Britney is framing a wall. She wants the angle at *C* to be a right angle. She marks point *A* 8 feet from *C* and point *B* 6 feet from *C*. She measures from *A* to *B* and finds the distance is 10 ft $1\frac{1}{2}$ in. To make the angle at *C* a right angle, should Britney push *B* toward *A* or pull *B* away from *A*? Why?

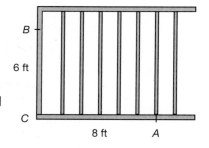

B

6 ft

C

8 ft   *A*

As we have seen, some right triangles have three sides with lengths that can be expressed as whole numbers. However, most right triangles have at least one side that cannot be expressed as a whole number.

For example, consider the length of the diagonal of a square. A diagonal of a square is the hypotenuse of two isosceles right triangles. Using the Pythagorean Theorem we find that the length of the hypotenuse is an irrational number.

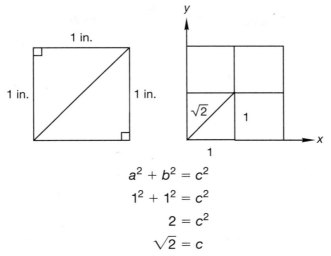

$$a^2 + b^2 = c^2$$
$$1^2 + 1^2 = c^2$$
$$2 = c^2$$
$$\sqrt{2} = c$$

Since the square of the hypotenuse is 2 square inches, the hypotenuse is $\sqrt{2}$ inches. Use an inch ruler to find the length of the diagonal of the square.

From the figure above, we see that on a coordinate plane the diagonal of each square is $\sqrt{2}$ units.

Analyze Find the irrational numbers that represent the lengths of the unmeasured side of each triangle.

11.

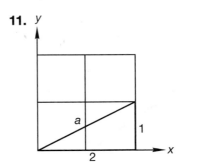

12.

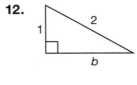

If the side length is an irrational number we can use a calculator to find the approximate length of the side. We will practice doing so in later lessons.

Activity

Pythagorean Puzzle

Materials needed:

- Photocopies of Investigation Activity 3 (1 each per student)
- Scissors
- Envelopes or plastic locking bags (optional)

Cut out and rearrange the pieces of the squares drawn on the legs of the right triangle *ABC* to form a square on the hypotenuse of the triangle.

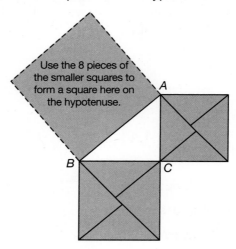

Use the 8 pieces of the smaller squares to form a square here on the hypotenuse.

• Distributive Property
• Order of Operations

facts | Power Up E

mental math

a. Calculation: Damian rides his bike 10 miles per hour for $2\frac{1}{2}$ hours. How far does he ride?

b. Estimation: $3.49 + $1.49 + $6.49 + $3.49

c. Algebra: $34 - x = 25$

d. Power/Roots: $\sqrt{144}$

e. Number Sense: 16×5

f. Fractional Parts: Three fourths of a dollar is how many cents?

g. Calculation: 12×3, $\sqrt{}$, $\times 7$, $- 2$, $\div 5$, $+ 1$, $\sqrt{}$, $\div 2$

problem solving

Cameron and Ann are the two finalists in a chess tournament. The first player to win either two consecutive games, or a total of three games, wins the match and the tournament. How many different ways can their match be played out?

(**Understand**) Two players are in a chess tournament, and the winner of the match will be the player who either wins two games in a row, or is the first player to win a total of three games.

(**Plan**) We will make a table of the possible outcomes of the match. We will include in our list the number of games it is necessary to play, who won each game, and who the won the match.

(**Solve**)

# of Games Played	Who Won Each Game	Match Winner
2 games	CC	Cameron
	AA	Ann
3 games	CAA	Ann
	ACC	Cameron
4 games	CACC	Cameron
	ACAA	Ann
5 games	CACAC	Cameron
	ACACA	Ann
	CACAA	Ann
	ACACC	Cameron

Check Tournaments are a part of almost any competitive activity, and you may be familiar with other formats for determining a winner. Our organized list shows that there are ten different ways the chess match can be played out.

New Concepts — *Increasing Knowledge*

distributive property

These two formulas can be used to find the perimeter of a rectangle.

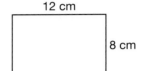

1. $P = 2(l + w)$

2. $P = 2l + 2w$

Using the first formula we add the length and width (12 cm + 8 cm = 20 cm). Then we double this sum (2 × 20 cm = 40 cm).

Using the second formula we double the length (2 × 12 cm = 24 cm) and double the width (2 × 8 cm = 16 cm). Then we add these products (24 cm + 16 cm = 40 cm).

Both formulas produce the same result because the formulas are equivalent.

$$2(l + w) = 2l + 2w$$

These two formulas illustrate the **Distributive Property** (sometimes called the *Distributive Property of Multiplication over Addition*). The Distributive Property gives us a choice when dealing with expressions such as this one.

$$2(12 + 8)$$

One choice is to add before multiplying, 2(20). The other choice is to multiply each addend before adding.

$$2(12 + 8) = 2 \cdot 12 + 2 \cdot 8$$

The second choice distributes the multiplication over the addition. Here we show in symbols the Distributive Property over addition and subtraction.

Distributive Property

$$a(b + c) = a \cdot b + a \cdot c$$
$$a(b + c + d) = a \cdot b + a \cdot c + a \cdot d$$
$$a(b - c) = a \cdot b - a \cdot c$$

The Distributive Property works in two ways. We can **expand** and we can **factor.**

- We expand $2(a + b)$ and get $2a + 2b$.
- We factor $2a + 2b$ and get $2(a + b)$.

Notice that we used the word *factor* as a verb. We factor an expression by writing the expression as a product of two or more numbers or expressions.

Explain How can we factor the expression 15 + 10? Explain and rewrite the expression.

Example 1

a. Expand: $3(w + m)$

b. Factor: $ax + ay$

Solution

a. To expand, we distribute the multiplication $3(w + m)$.

$$3(w + m) = \mathbf{3w + 3m}$$

b. To factor, we divide both terms by a.

$$ax + ay = \mathbf{a(x + y)}$$

Example 2

a. Expand: $3(x + 2)$

b. Factor: $6x + 9$

Solution

a. We multiply each term within the parentheses by 3.

$$3(x + 2) = \mathbf{3x + 6}$$

b. We see that 3 is a factor of both terms. We remove by division a factor of 3 from each term.

$$6x + 9 = \mathbf{3(2x + 3)}$$

When factoring, we may prefer to write each term as a product of factors before applying the distributive property.

$6x + 9$	Given expression
$2 \cdot 3 \cdot x + 3 \cdot 3$	Factored each term
$\mathbf{3(2x + 3)}$	Factored 3 from each term

order of operations

If there is more than one operation in an expression, we follow this order.

1. Simplify within **parentheses.**

2. Simplify **exponent** expressions.

3. **Multiply** and **divide** in order from left to right.

4. **Add** and **subtract** in order from left to right.

The first letters of the boldfaced words are also the first letters of the words in the following sentence, which can help us remember this order.

Please **e**xcuse **m**y **d**ear **A**unt **S**ally.

If there are operations within the parentheses we follow the order of operations to simplify within the parentheses.

Example 3

Simplify: $20 - 2 \cdot 3^2 + (7 + 8) \div 5$

Solution

Step:	Justification:
$20 - 2 \cdot 3^2 + 15 \div 5$	Simplified within parentheses
$20 - 2 \cdot 9 + 15 \div 5$	Applied exponent
$20 - 18 + 3$	Multiplied and divided
5	Added and subtracted

Brackets [ ] and braces { } are used as grouping symbols like parentheses. We use brackets to enclose parentheses and braces to enclose brackets.

$$10 - \{8 - [6 - (5 - 3)]\}$$

To simplify expressions with multiple grouping symbols, we begin from the inner most symbols.

Example 4

Simplify: $10 - \{8 - [6 - (5 - 3)]\}$

Solution

We simplify within the parentheses, then the brackets, then the braces.

Step:	Justification:
$10 - \{8 - [6 - (5 - 3)]\}$	Given expression
$10 - \{8 - [6 - 2]\}$	Simplified in parentheses
$10 - \{8 - 4\}$	Simplified in brackets
$10 - 4$	Simplified in braces
6	Subtracted

Absolute value symbols, division bars, and radicals sometimes group multiple terms as well.

Example 5

Simplify:

a. $|2 - 7|$ b. $\sqrt{9 + 16}$ c. $\dfrac{12 \times 12}{3 + 3}$

Solution

a. We simplify within the absolute value symbols.

$$|2 - 7| = |-5| = \mathbf{5}$$

b. We simplify under the radical.

$$\sqrt{9 + 16} = \sqrt{25} = \mathbf{5}$$

c. We simplify above and below the division bar first.

$$\frac{12 \times 12}{3 + 3} = \frac{144}{6} = 24$$

Example 6

Consider this expression: $3 + 3 \times 3 - 3 \div 3$

a. As written, this expression equals what number?

b. Copy the expression and place one pair of parentheses so that the expression equals 3. Justify your answer.

Solution

a. 11

b. $3 + 3 \times (3 - 3) \div 3$

$3 + 3 \times 0 \div 3$	Simplified within parentheses
$3 + 0$	Multiplied and divided left to right
3	Added

Practice Set

Perform the indicated multiplication.

a. $2(3 + w)$ **b.** $5(x - 3)$

Factor each expression.

c. $9x + 6$ **d.** $8w - 10$

e. (Analyze) Which property is illustrated by the following equation?

$$x(y + z) = xy + xz$$

f. (Explain) Why might you use the Distributive Property to simplify $3(30 - 2)$?

Simplify.

g. $2 + 5 \cdot 2 - 1$ **h.** $(2 + 5) \cdot (2 - 1)$ **i.** $4 + (11 - 3^2) \cdot 2^3$

j. $10 - [8 - (6 - 3)]$ **k.** $\sqrt{36 + 64}$ **l.** $|8 - 12|$

m. The expression $2^2 + 2 \times 2 - 2 \div 2$ equals 7. Copy the expression and insert one pair of parentheses to make the expression equal 11.

* 1. Bob drove from his home near Los Angeles to his destination in
(3)   Sacramento in 7 hours. The next time he made his trip, he flew.
He drove $\frac{1}{2}$ hour to the airport, spent $1\frac{1}{2}$ hours at the airport, had
a 1 hour flight, and arrived at his destination 1 hour after landing.
Considering Bob's total travel time, how much time did it save Bob to
fly rather than drive?

2. George chopped 8 times in the first 30 seconds. Then he chopped
(3, 4)   5 times in the next 20 seconds. Finally he chopped 2 times in the last
10 seconds. How many times did he chop? How long did it take him?
When was he chopping the most rapidly?

3. John read 66 books in one year. At this rate, how many would he read in
(7)   four months?

* 4.   **a.** How can these numbers be rearranged to make the addition easier?
(2)
$$7 + (18 + 13)$$

b. What two properties did you use?

c. What is the sum?

5. Use the Associative and Commutative Properties of Multiplication to
(2)   find the product of these numbers mentally.

$$12 \cdot 13 \cdot 5 \cdot 10$$

Write the order in which you multiplied the numbers. What is the
product?

* 6.   **Formulate**   Linda reserved two hotel rooms for $75 each room plus a tax
(21)   of $7.50 each. Illustrate the Distributive Property by writing two ways to
compute the total. What is the total?

7. Estimate the area of this figure.
(8, 17)

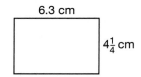

6.3 cm

$4\frac{1}{4}$ cm

* 8.   **a.** Expand: $2(x + 7)$
(21)
  b. Factor: $9x + 12$

* 9. Classify this triangle by sides and by angles and state the measure of
(20)   each angle.

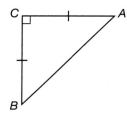

10. The numbers of students playing soccer after school Monday through
(7) Friday were 20, 22, 19, 25, and 24. What was the mean number of students playing soccer that week?

Generalize Simplify.

* 11. $1 + 2 + 3 \times 4 + 5$
(21)

* 12. $\dfrac{2 + 5}{1 + 20}$
(10, 21)

13. $\dfrac{20}{21} - \dfrac{2}{3}$
(13)

14. $1\dfrac{3}{5} + 1\dfrac{1}{10}$
(13)

15. $\dfrac{3660}{12}$
(2)

* 16. $\sqrt{169}$
(15)

* 17. $\sqrt[3]{27}$
(15)

* 18. 3^4
(15)

19. Plot the following points, connect the points to make a rectangle, and
(Inv. 1, 8) find the area and perimeter of the rectangle. $(-5, 7)$, $(3, 7)$, $(3, -2)$, $(-5, -2)$

* 20. *Analyze* Compare. $20 - 33 \bigcirc |20 - 33|$
(1, 21)

21. Find the two values for r which make this statement true: $|r| = 2$.
(1)

* 22. Write in exponential form: $\dfrac{bbrr}{yyyy}$
(15)

* 23. Find the perimeter and area of this triangle.
(20, Inv. 2)

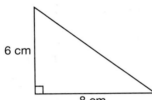

6 cm

8 cm

For problems **24–27,** solve by inspection.

24. $27 + w = 35$
(14)

25. $\dfrac{32}{y} = 2$
(14)

26. $8n + 2 = 90$
(14)

27. $2b - 3 = 5$
(14)

* 28. *Evaluate* Find $b^2 - 4ac$ when $a = 2$, $b = 3$, $c = 1$
(14, 15)

29. How many square yards of carpet are needed to cover the floor of a
(6, 8) classroom that is 30 ft long and 30 ft wide?

* 30. Name four sets of numbers of which zero is a member.
(10, 16)

• Multiplying and Dividing Fractions

facts | Power Up E

mental math

a. Calculation: Madrid read 20 pages per hour. How many pages can she read in $1\frac{1}{4}$ hours?

b. Geometry: When we write a congruence statement for a polygon, we write the corresponding vertices in the same order for both polygons. Copy and complete this statement: $\triangle ABC \cong \triangle X\_\_$

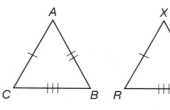

c. Number Sense: $14 \div 4$

d. Fractional Parts: $\frac{1}{4}$ of 120

e. Proportions: If four of the boxes weigh 20 pounds, how much would six of the boxes weigh?

f. Power/Roots: $\sqrt{16} \cdot \sqrt{16}$

g. Percent: 25% of 400

h. Calculation: $1000 \div 2, \div 2, \div 2, -25, \div 2, \div 2, -25, \div 2, -25$

problem solving

These two congruent squares of note paper can be overlapped to form polygons with as few as 4 sides or as many as 16 sides. Are there polygons with other numbers of sides that can be formed by overlapping congruent squares?

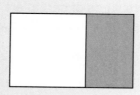
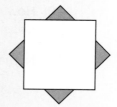

(**Understand**) We know that a polygon is a closed shape. The polygons given in the problem are both concave, because it is possible to connect two points located inside the polygon with a segment that goes outside the polygon.

(**Plan**) This is a good problem to solve by using a model and acting it out, so we will need to locate or create two congruent squares.

Solve We will leave one of our squares stationary, and rotate and slide the second square:

  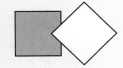

7 sides   8 sides   9 sides   13 sides

Look Back We have constructed polygons with 7, 8, 9, and 13 sides by sliding the two squares. We may be able to form different polygons with other numbers of sides. A natural extension of this problem would be to work with a variety of shapes, such as two congruent rectangles, or a square and a rectangle

---

New Concept — *Increasing Knowledge*

Pictured below are containers for a gallon, a quart, and a pint. We can use these measures to illustrate the multiplication of fractions. A quart is $\frac{1}{4}$ of a gallon, and a pint is $\frac{1}{2}$ of a quart. What fraction of a gallon is a pint?

1 gallon   1 quart   1 pint

A pint is $\frac{1}{2}$ of $\frac{1}{4}$ of a gallon. By multiplying $\frac{1}{2}$ and $\frac{1}{4}$, we find that a pint is $\frac{1}{8}$ of a gallon.

$$\frac{1}{2} \cdot \frac{1}{4} = \frac{1}{8}$$

Notice that the numerator of $\frac{1}{8}$ is the product of the numerators, and the denominator is the product of the denominators. Also notice that $\frac{1}{8}$ is less than both factors.

An area model can also help us visualize the result of multiplying fractions. Here we show a 1 inch-by-1 inch square. Within the square is a small, shaded square that is $\frac{1}{2}$ inch-by-$\frac{1}{2}$ inch. Notice the shaded square is $\frac{1}{4}$ the area of the larger square. This model illustrates the multiplication of $\frac{1}{2}$ and $\frac{1}{2}$.

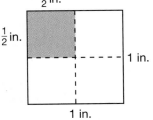

$$\frac{1}{2} \cdot \frac{1}{2} = \frac{1}{4}$$

To multiply fractions we multiply the numerators to find the numerator of the product, and we multiply the denominators to find the denominator of the product.

Example 1

Multiply $\frac{1}{3}$ and $\frac{2}{3}$. Use an area model to illustrate the multiplication.

Multiply the numerators. Multiply the denominators

$$\frac{1}{3} \cdot \frac{2}{3} = \frac{2}{9}$$

We draw a square and divide two adjacent sides into thirds, extending the divisions across the square. Then we shade an area with sides $\frac{1}{3}$ and $\frac{2}{3}$ long. We see that $\frac{2}{9}$ of the square is shaded.

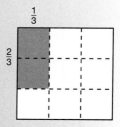

Example 2

What number is $\frac{3}{4}$ of $\frac{8}{9}$?

Solution

To find $\frac{3}{4}$ of $\frac{8}{9}$ we multiply the fractions. We may reduce before we multiply (cancel).

$$\frac{\overset{1}{\cancel{3}}}{\underset{1}{\cancel{4}}} \cdot \frac{\overset{2}{\cancel{8}}}{\underset{3}{\cancel{9}}} = \frac{2}{3}$$

Discuss   Why might we want to reduce before we multiply?

If the product of two fractions is 1, the fractions are **reciprocals.**

$$\underbrace{\frac{2}{3} \cdot \frac{3}{2}}_{\text{reciprocals}} = \frac{6}{6} = 1$$

We can form the reciprocal of a fraction by reversing the numbers of the numerator and denominator of the fraction.

$$\frac{3}{4} \bigtimes \frac{4}{3}$$

Some say we *invert* a fraction to form its reciprocal. This expression comes from another name for reciprocal, **multiplicative inverse.**

Example 3

a. What is the reciprocal of $\frac{5}{6}$?

b. What is the multiplicative inverse of 3?

c. How many $\frac{1}{4}$s are in 1?

d. How many $\frac{2}{3}$s are in 1?

Solution

a. The reciprocal of $\frac{5}{6}$ is $\frac{6}{5}$.

b. The multiplicative inverse of 3 (or $\frac{3}{1}$) is $\frac{1}{3}$.

c. The number of $\frac{1}{4}$s in 1 is **4** (or $\frac{4}{1}$), which is the reciprocal of $\frac{1}{4}$.

d. The number of $\frac{2}{3}$s in 1 is the reciprocal of $\frac{2}{3}$, which is $\frac{3}{2}$.

We use reciprocals to help us divide fractions. If we want to know the number of $\frac{3}{4}$s in 2, we divide 2 by $\frac{3}{4}$.

$$2 \div \frac{3}{4}$$

We perform the division in two steps. First we find the number of $\frac{3}{4}$s in 1, which is the reciprocal of $\frac{3}{4}$.

$$1 \div \frac{3}{4} = \frac{4}{3}$$

There are twice as many $\frac{3}{4}$s in 2 as there are in 1, so we multiply $\frac{4}{3}$ by 2.

$$2 \times \frac{4}{3} = \frac{8}{3} = \mathbf{2\frac{2}{3}}$$

We can combine the two steps and multiply by the reciprocal of the divisor.

$$2 \div \frac{3}{4}$$

$$2 \times \frac{4}{3} = \frac{8}{3} = \mathbf{2\frac{2}{3}}$$

Example 4

Thinking Skill

Connect

There are 2 cups in a pint. How many cups are in a gallon? What fraction of a gallon is one cup? How are the two numbers related?

A quart is $\frac{1}{4}$ of a gallon. A pint is $\frac{1}{8}$ of a gallon. How many $\frac{1}{8}$s are in $\frac{1}{4}$?

Solution

The question "How many are in...?" is a division question. We divide $\frac{1}{4}$ by $\frac{1}{8}$.

$$\frac{1}{4} \div \frac{1}{8}$$

$$\frac{1}{4} \times \frac{8}{1} = \frac{8}{4} = \mathbf{2}$$

The answer is reasonable. Two $\frac{1}{8}$s equal $\frac{1}{4}$, and two pints equal a quart.

Example 5

Simplify: $\frac{3}{4} \div \frac{1}{3}$

Solution

We show two methods.

First method: Find the number of $\frac{1}{3}$s in 1.

$$1 \div \frac{1}{3} = 3$$

Then find $\frac{3}{4}$ of this number.

$$\frac{3}{4} \cdot 3 = \frac{9}{4} = \mathbf{2\frac{1}{4}}$$

Second method: Rewrite the division problem as a multiplication problem and multiply by the reciprocal of the divisor.

$$\frac{3}{4} \div \frac{1}{3}$$

$$\frac{3}{4} \cdot \frac{3}{1} = \frac{9}{4} = 2\frac{1}{4}$$

Practice Set

a. **Model** Multiply $\frac{1}{4}$ and $\frac{1}{2}$. Use an area model to illustrate the multiplication.

b. One half of the students in the class are boys. One third of the boys walk to school. Boys who walk to school are what fraction of the students in the class?

Generalize Simplify.

c. $\frac{1}{2} \times \frac{3}{4}$

d. $\frac{5}{6} \cdot \frac{2}{3}$

e. What number is $\frac{1}{2}$ of $\frac{3}{4}$?

f. What number is $\frac{2}{3}$ of $\frac{3}{4}$?

For **g** and **h** write the multiplicative inverse (reciprocal) of the number.

g. $\frac{3}{8}$

h. 4

i. How many $\frac{2}{3}$ s are in 1?

j. $1 \div \frac{3}{5}$

k. $1 \div \frac{1}{6}$

l. How many $\frac{1}{2}$ s are in $\frac{3}{4}$? $(\frac{3}{4} \div \frac{1}{2})$

Simplify.

m. $\frac{2}{3} \div \frac{3}{4}$

n. $\frac{3}{4} \div \frac{2}{3}$

Written Practice Strengthening Concepts

1. Mariya ran 14 miles in one week. At that rate, how many miles would she run in 5 days?
(7)

* 2. **Analyze** The triathlete swam 2 miles in 56 minutes. Then, he rode his bicycle 100 miles in 5 hours. Finally, he ran 20 miles in 160 minutes. How many minutes did it take him to complete the course? During which portion of the race did he have the greatest average speed?
(3, 7)

* 3. Expand: $3(x + 7)$
(21)

* 4. Factor: $5x + 35$
(21)

* 5. $\frac{6 + 4}{7 - 2}$
(21)

* 6. $6 + 5 \times 4 - 3 + 2 \div 1$
(21)

* 7. Compare: $|6 - 19| \bigcirc (6 - 19)$
(1, 21)

* 8. $\frac{3}{4} \cdot \frac{1}{3}$
(22)

* 9. What number is $\frac{1}{3}$ of $\frac{3}{5}$?
(22)

* 10. **Generalize** What is the reciprocal of $\frac{1}{7}$?
(22)

* 11. What is the multiplicative inverse of $\frac{2}{3}$?
(22)

12. Find $x^2 + 2yz$ when $x = 4$, $y = 5$, and $z = 7$.
(14)

* 13. $\frac{3}{4} \div \frac{1}{3}$
(22)

* 14. **Generalize** Find the perimeter and area of this triangle.
(8, 20)

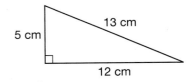

* 15. **Classify** Classify this triangle by angles and sides, and state the measure of each angle.
(20)

16. Find the area of a square with a side length of 17 in.
(8)

17. **Explain** Is this figure a polygon? Explain why or why not.
(19)

18. Can an obtuse triangle be regular? Explain why or why not.
(20)

19. Choose the appropriate symbol to complete this statement:
(19) $\triangle EFG \bigcirc \triangle XYZ$

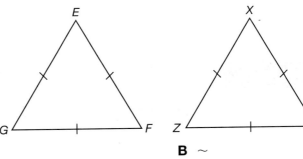

A $\cong$ **B** $\sim$

C $=$ **D** $<$

20. $\angle WYX$ and $\angle XYZ$ form a linear pair. If $m\angle XYZ$ is 60°, find $m\angle WYX$.
(18)

21. Name this figure
(18)

A $\overrightarrow{EF}$ **B** $\overrightarrow{FE}$ **C** either **A** or **B**

22. A freeway passes under a street. If we think of the freeway and the street as lines, we would say the lines are:
(18)

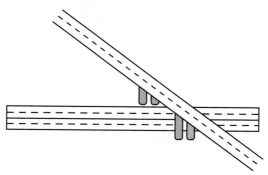

    A intersecting   **B** parallel   **C** skew

23. Round $403,500,000 to the nearest million.
(17)

24. Estimate the product $(17.89)(9\frac{7}{8})$.
(17)

25. Name a set of numbers that includes 1, but not zero.
(1)

26. $\sqrt[3]{64}$ **27.** $1\frac{1}{3} + 1\frac{1}{6}$
(15)   (13)

28. Write $\frac{7}{8}$ as a decimal number.
(12)

29. What is $\frac{3}{5}$ of 35?
(5)

* 30.   *Generalize*   A home in the mountains is going
(20)   to be built with a steep roof to keep heavy
snow from accumulating on the roof. If the
roof is at a 70° angle as shown, what is the
angle at the peak?

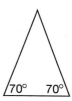

• Multiplying and Dividing Mixed Numbers

facts | Power Up E

mental math

a. Calculation: Robbie ran ten miles per hour. How many minutes would it take Robbie to run 2 miles?

b. Geometry: Two angles of a triangle are 90° and 35°. What is the measure of the third angle?

c. Number Sense: $18 ÷ 4

d. Proportions: If 2 pounds cost $5, how much would 3 pounds cost?

e. Percent: 25% of 104

f. Algebra: $\frac{54}{x} = 0.54$

g. Fractional Parts: $\frac{1}{3}$ of 93

h. Calculation: $7 + 3$, $× 10$, $÷ 2$, $÷ 10$, $+ 2$, $× 9$, $+ 1$, $\sqrt{\ }$, $+ 1$, $\sqrt{\ }$, $+ 1$, $\sqrt{\ }$, $- 1$, $\sqrt{\ }$, $- 1$

problem solving

We can use tree diagrams to help us identify possible combinations or possible outcomes of probability experiments.

Problem: The security guard can choose from three different shirts and two different pairs of pants for his uniform. How many different combinations does he have to choose from?

(Understand) We are told that the guard has three different shirts and two different pants. We are asked to find the numbers of different uniform combinations he has to choose from.

(Plan) We will make a tree diagram to show the possible combinations. We will use the first set of branches for the three shirts (*A, B, C*) and the second set of branches for pants (1, 2).

(Solve)

Combination Tree

Shirt-Pants Combinations

Shirt A	Pants 1	A 1
	Pants 2	A 2
Shirt B	Pants 1	B 1
	Pants 2	B 2
Shirt C	Pants 1	C 1
	Pants 2	C 2

We identify **6 different possible combinations.**

New Concept   *Increasing Knowledge*

Thinking Skill

Explain

How do we write whole numbers as improper fractions?

To multiply or divide mixed numbers we first write each mixed number (or whole number) as a fraction. A mixed number can be converted into a fraction as we illustrate.

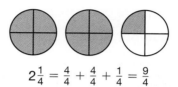

$$2\frac{1}{4} = \frac{4}{4} + \frac{4}{4} + \frac{1}{4} = \frac{9}{4}$$

Another way to perform the conversion is to multiply the whole number by the denominator and add the numerator to find the numerator of the new fraction.

$$4 \times 2 + 1 = 9$$
$$2\frac{1}{4} = \frac{9}{4}$$

If the answer is an improper fraction we might choose to convert the improper fraction to a mixed number. We do so by dividing the numerator by the denominator.

$$\frac{9}{4} = 4\overline{)9} \;\; \begin{array}{r} 2 \\ -8 \\ \hline 1 \end{array} = 2\frac{1}{4}$$

The remainder becomes the numerator of the fraction part of the mixed number.

Visit www. SaxonPublishers. com/ActivitiesC3 for a graphing calculator activity.

Example 1

Simplify:

a. $2\frac{1}{3} \times 1\frac{1}{2}$

b. $1\frac{2}{3} \div 2\frac{1}{2}$

Solution

a. First we rewrite each mixed number as an improper fraction. Then we multiply.

$$2\frac{1}{3} \times 1\frac{1}{2}$$

$$\frac{7}{3} \times \frac{3}{2} = \frac{21}{6} = 3\frac{3}{6} = 3\frac{1}{2}$$

The product is an improper fraction which we can convert to a mixed number. We reduce the fraction.

b. We write the mixed numbers as improper fractions. Instead of dividing, we multiply by the reciprocal of the divisor.

$$1\frac{2}{3} \div 2\frac{1}{2}$$

$$\frac{5}{3} \div \frac{5}{2}$$

$$\frac{5}{3} \times \frac{2}{5} = \frac{10}{15} = \frac{2}{3}$$

a and **b** could be reduced before multiplying. We may reduce (or cancel) to eliminate the need to reduce after multiplying.

a. $\dfrac{7}{\overset{}{\underset{1}{\cancel{3}}}} \times \dfrac{\overset{1}{\cancel{3}}}{2} = \dfrac{7}{2} = 3\dfrac{1}{2}$
　　　　　　　　b. $\dfrac{\overset{1}{\cancel{5}}}{3} \times \dfrac{2}{\underset{1}{\cancel{5}}} = \dfrac{2}{3}$

Example 2

A carpenter is nailing horizontal boards to the outside surface of a building. The exposed portion of each board is $6\frac{3}{4}$ in. How many rows of boards are needed to reach to the top of an 8 ft wall?

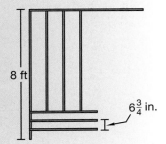

8 ft

$6\frac{3}{4}$ in.

Solution

We begin by converting 8 ft to 96 in. Then we divide 96 in. by $6\frac{3}{4}$ in.

$$96 \div 6\frac{3}{4}$$

$$\frac{96}{1} \div \frac{27}{4}$$

$$\frac{\overset{32}{\cancel{96}}}{1} \cdot \frac{4}{\underset{9}{\cancel{27}}} = \frac{128}{9} = 14\frac{2}{9}$$

Fourteen boards is not enough. To reach to the top of the wall the carpenter will need **15 rows of boards.**

The following table shows the steps we take to multiply and divide fractions and mixed numbers.

**Multiplying and Dividing
Fractions and Mixed Numbers**

1. Write whole numbers or mixed numbers as fractions.	
2. **Multiplying** Multiply numerators. Multiply denominators.	**Dividing** Instead of dividing, multiply by the reciprocal of the divisor.
3. Simplify the answer by reducing if possible. Improper fractions may be expressed as whole or mixed numbers.	

Practice Set

a. What is the area of a rectangle $2\frac{1}{2}$ inches long and $1\frac{1}{2}$ inches wide?

b. A recipe calls for $1\frac{3}{4}$ cups of milk. How many cups of milk are needed if the recipe is doubled?

c. A tile setter is covering a 5 ft by 5 ft square shower wall. Each tile covers a $4\frac{5}{8}$ in. by $4\frac{5}{8}$ in. square. How many rows of tile are needed to reach 5 feet? How many tiles are needed to cover the 5 ft by 5 ft square?

5 ft

5 ft

Write the multiplicative inverse (reciprocal) of each number.

d. $2\frac{2}{3}$

e. $\frac{5}{6}$

f. 5

Simplify.

g. $1\frac{1}{2} \times 1\frac{2}{3}$

h. $2\frac{1}{2} \cdot 2\frac{1}{2}$

i. $1\frac{1}{2} \div 2\frac{2}{3}$

j. $2\frac{2}{3} \div 1\frac{1}{2}$

Written Practice   *Strengthening Concepts*

1. *(3)* **Analyze** At the wedding the guests sat on either side of the banquet hall. There were 109 guests seated on the left side. If there were 217 guests in all, how many sat on the right side of the aisle?

2. *(5)* Visitors to the school's website can follow a link to the school's chorale pictures. If the website had 250 visitors in one week and $\frac{3}{10}$ of them visited the chorale pictures, how many visited the chorale pictures?

3. *(3)* Martin Luther was born in 1483, which was twenty-seven years after the invention of the printing press. In what year was the printing press invented?

* **4.** *(13)* $\frac{5}{6} + \frac{1}{2}$

* **5.** *(13)* $\frac{5}{6} - \frac{1}{2}$

* **6.** *(22)* $\frac{5}{6} \cdot \frac{1}{2}$

* **7.** *(22)* $\frac{5}{6} \div \frac{1}{2}$

* 8. $\frac{5}{7} \cdot \frac{7}{10}$
(22)

* 9. $\frac{5}{7} \div \frac{7}{10}$
(22)

* 10. $\frac{3}{7} \cdot 2$
(22)

* 11. $\frac{3}{7} \div 2$
(22)

* 12. $2\frac{1}{2} \cdot 1\frac{2}{3}$
(23)

* 13. $2\frac{1}{2} \div 1\frac{2}{3}$
(23)

Use order of operations to simplify.

* 14. $16 - [8 - 4(2 - 1)]$
(21)

* 15. $6 + 7 \times (3 - 3)$
(21)

16. A rectangle has vertices at: (6, 2), (6, −6,), (−1, 2). Write the
(Inv. 1, 8) coordinates of the fourth vertex and find the area and perimeter of the
rectangle.

* 17. **Generalize** Compare:
(15, 21)

    a. $(5 - 7) \bigcirc |5 - 7|$ **b.** $\sqrt{64} \bigcirc \sqrt[3]{64}$

18. Find the two values that make this statement true: $|x| = 12$
(1)

19. Express with exponents:
(15)

    a. $yyyy$ **b.** $2ababa$

* 20. What is the area of a square with sides $2\frac{1}{2}$ in. long?
(8, 23)

21. Name the properties illustrated below:
(2)

    a. $a + 5 = 5 + a$ **b.** $a \cdot 0 = 0$

* 22. Find the perimeter and area of this triangle. (Dimensions are in meters.)
(8, 20)

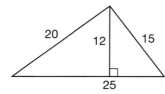

For problems **23–28,** solve by inspection.

23. $5 + x = 22$
(14)

24. $4x + 13 = 25$
(14)

* 25. $\frac{1}{2} \cdot p = \frac{1}{6}$
(14)

26. $\frac{x}{30} = \frac{2}{3}$
(14)

27. Evaluate $b^2 - 4ac$ when $a = 1$, $b = 6$, and $c = 8$.
(14)

* 28. For $A = \frac{1}{2}bh,$ find A when b is 6 cm and h is 4 cm.
(14)

* 29. (Analyze) **a.** Expand: $5(x + 3)$
(21)

       b. Factor: $6x - 10y$

* 30. A steel cable providing tension on a power
(Inv. 2) pole is anchored 16 ft from the pole and
reaches 12 feet up the pole. How long is the
cable?

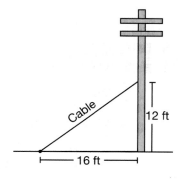

Cable

12 ft

16 ft

Early Finishers
Real-World
Application

To complete the bookcase he is building for his mother, Tucker needs
10 pieces of wood each measuring $2\frac{3}{4}$ ft long. Each piece must be whole; he
cannot splice together two smaller pieces to create a piece $2\frac{3}{4}$ ft long. One
8-foot length of lumber costs $14.95 and one 10-foot length costs $16.95.

   a. If Tucker only purchased 8-foot lengths of wood, how much money
would he spend? How much wood would he waste?

   b. If Tucker only purchased 10-foot sections of wood, how much money
would he spend? How much wood would he waste?

   c. What is the most economical combination of 8-foot and 10-foot lengths
Tucker could buy? What would be the cost and how much wood would
be wasted?

LESSON
24

• Adding and Subtracting Decimal Numbers

Power Up   *Building Power*

facts   Power Up E

mental math

   a. Calculation: Maria reads 40 pages per hour. How many pages does she read in 2 hours and 15 minutes?

   b. Measurement: What is the diameter of the ring?

   c. Number Sense: $46 \div 4$

   d. Geometry: Which angle in $\triangle XYZ$ corresponds to $\angle B$?

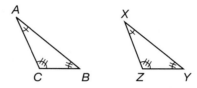

   e. Proportions: If Darrel jogs 2 miles in 16 minutes, how long will it take Darrel to jog 3 miles at that pace?

   f. Power/Roots: $\sqrt{400}$

   g. Fractions: $\frac{3}{5} = \frac{x}{25}$

   h. Calculation: $- 5 + 10, + 4, \sqrt{}, + 1, \sqrt{}, \times 8, \sqrt{}, + 1, \times 20, \sqrt{}$

problem solving

Franklin, Grover, and Herbert went to a restaurant and each ordered a soup and a salad. Franklin ordered chicken noodle soup and did not order a Caesar salad. Herbert did not order French onion soup but did order a house salad. If their order was for chicken noodle, French onion, and vegetable soups, and their salads were Caesar, house, and fruit, what did each person order?

New Concept   *Increasing Knowledge*

Performing arithmetic with decimal numbers is similar to arithmetic with whole numbers. We follow a few more rules to keep track of the decimal point. When adding or subtracting decimal numbers, we first align the decimal points. By lining up the decimal points we assure that we are adding or subtracting digits with the same place value.

Example 1

Simplify:

a. 12.5 + 3.75 + 2

b. 5.2 − 2.88

Solution

Thinking Skill

Explain

Why can we attach zeros to the end of a decimal number?

a. We align the decimal points. Since 2 is a whole number we place the decimal point to the right of the 2. We may use zeros to write all the decimal numbers with two decimal places. We add in columns and place the decimal point in the sum aligned with the other decimal points.

```
  12.50
   3.75
+  2.00
─────────
  18.25
```

b. We subtract 2.88 from 5.2. We align the decimal points and fill in with zero so that we can subtract.

```
   5.20
−  2.88
─────────
   2.32
```

Example 2

Gregory's temperature was 100.2°F. The thermometer marked 98.6°F as normal body temperature. How many degrees above normal was Gregory's temperature?

Solution

We are given two temperatures and are asked to find the difference. Therefore we subtract the lower temperature from the higher temperature.

$$100.2 - 98.6 = 1.6$$

Gregory's temperature is **1.6 degrees Fahrenheit** above normal.

Practice Set

a. Arrange these numbers from least to greatest.

$$\frac{1}{2}, 0, -1, 2, 0.3, 1.75$$

b. Compare 0.036 ◯ 0.0354

Simplify:

c. 7.5 + 12.75

d. 4.2 + 12

e. 0.3 + 0.8

f. 11.46 − 3.6

g. 5.2 − 4.87

h. 3 − 2.94

i. *Explain* How can you check your answers in problems **f–h?**

j. The weather report stated that the recent storm dropped 1.50 inches of rain raising the seasonal total to 26.42 inches. What was the seasonal total prior to the recent storm? Explain how you found your answer.

1. Irina ran $\frac{1}{2}$ mile in $3\frac{1}{2}$ minutes. At that rate, how long would it take her to
(7) run 2 miles?

2. There were 180 people watching the football game. If $\frac{7}{10}$ of the people
(5) watching the game were sitting on the home team bleachers, how many
people were sitting on the home team bleachers?

3. The Pacific team defeated the Atlantic team by a score of 5 runs to 3.
(5) What fraction of the runs were scored by the Pacific team?

* 4. $1.23 + 12.3 + 123$
(24)

* 5.   *Analyze*   The average low temperature for November 1 in a certain town
(24) is 41.5°F. The low on November 1 last year was 38.6°F. How much lower
was that temperature than the average?

Generalize   Solve.

* 6. $\frac{4}{5} \cdot 2$ *** 7.** $\frac{4}{5} \div 2$
(23)   (23)

* 8. $1\frac{1}{3} \cdot \frac{3}{4}$ *** 9.** $1\frac{1}{3} \div \frac{3}{4}$
(23)   (23)

* 10. A square has sides $1\frac{2}{3}$ in. long. What is the area of the square?
(23)

* 11. $\frac{4}{9} \cdot \frac{2}{3}$ *** 12.** $\frac{4}{9} \div \frac{2}{3}$
(22)   (22)

* 13.   *Generalize* $4 + 3 \times (5 - 1)$
(21)

* 14. Compare: $|16 - 20|$ ◯ $(16 - 20)$
(21)

15. Find the perimeter and area of this triangle.
(8, 20)

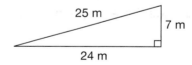

16. Classify this triangle by sides and angles.
(20)

* 17.   *Connect*   Terrell built a scale model of his house. The front door of his
(19) model is _____ to the front door of his house.

     A congruent   **B** equal   **C** similar

18. Which of these is regular?
(19)

     A   **B**   **C**   **D** all of the above

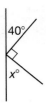
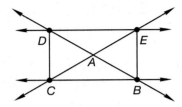

20. Does $\overrightarrow{BA}$ intersect $\overline{CD}$? If so, what is the
(18) point of intersection?

21. Round 6.029 to the nearest tenth.
(17)

22. Estimate: $\dfrac{24.42}{8\frac{1}{8}}$
(17)

23. Name a set of numbers that includes zero, but not negative one.
(1)

24. $\sqrt[3]{125}$
(15)

25. $4^2 + 3^2$
(15)

26. $1\frac{1}{2} + 1\frac{1}{5}$
(13)

27. $\dfrac{1}{3} - \dfrac{1}{4}$
(13)

28. Find the prime factorization of 84.
(9)

29. The capacity of a small glass is 120 mL. How many small glasses can
(6) be completely filled from a 2L pitcher of water?

30. There is a 25% discount on bongo drums at the music shop. The
(12) regular price is $56. How much money will be saved?

• Multiplying and Dividing Decimal Numbers

Building Power

facts | Power Up E

mental math

a. Calculation: Luis walked 3 miles per hour on his way home from school, a distance of $1\frac{1}{2}$ miles. How many minutes did it take for Luis to walk home?

b. Estimation: $24.97 + $1.97 + $9.99

c. Measurement:

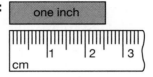

one inch

cm   1   2   3

About how many cm are four inches? (Round to the nearest cm.)

d. Number Sense: $22 divided equally among 4 people

e. Measurement: If four sugarless gum sticks were laid end to end, they would be how long?

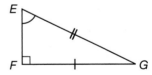

inch   1   2

f. Geometry: Rewrite and complete this congruence statement:
$\triangle CBD \cong \triangle G\_\_$

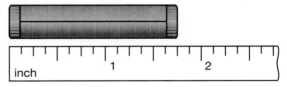

g. Proportions: If four pencils cost $1.00, how much would eight pencils cost?

h. Fractional Parts: $\frac{3}{4}$ of $84

i. Calculation: $17 + 8$, $\sqrt{}$, $+ 1$, square it, $\div 9$, $\sqrt{}$, $- 3$, $\times 5$, $+ 10$, $+ 2$

problem solving

A trivia game gives 5 points for each correct answer, 0 points for an answer left blank, and −1 points for each incorrect answer. The game has 20 questions. For each of these scores {95, 96, 99, 94}, determine how many answers were correct, how many were left blank, and how many were incorrect. If the score is not possible, say so.

To multiply decimal numbers with pencil and paper we do not need to align the decimal points. We instead multiply to find the product and then count all the decimal places in the factors to locate the proper position for the decimal point in the product. (Decimal places are places to the right of the decimal point.) In part **a** of the example below, we show the same multiplication in decimal form and in fraction form.

Example 1

Simplify:

a. 0.3 × 0.12   b. (1.5)²   c. 4.25 × 10

Solution

a. Three tenths times twelve hundredths equals 36 thousandths.

$$\frac{3}{10} \cdot \frac{12}{100} = \frac{36}{1000}$$

$$\begin{array}{r} 0.12 \\ \times\ \ 0.3 \\ \hline .036 \end{array}$$ 3 places in factors

.036 3 places in product

We count over three places from the end of the product, filling the empty place with zero. The product, **0.036,** is equivalent to $\frac{36}{1000}$.

b. To simplify (1.5)² we multiply 1.5 by 1.5.

$$\begin{array}{r} 1.5 \\ \times\ 1.5 \\ \hline 75 \\ 15\ \ \\ \hline 2.25 \end{array}$$ 2 places

2 places

Note that the result also means that $\sqrt{2.25}$ is 1.5.

c. We multiply, then count.

$$\begin{array}{r} 4.25 \\ \times\ \ \ 10 \\ \hline 42.5 \end{array}$$

Multiplying a decimal number by 10 simply shifts the decimal point one place to the right.

Generalize What is 4.25 × 10,000? How many places is the decimal shifted to the right?

Example 2

What is the cost of 0.86 pounds of cheese at $3.90 per pound? Explain why your answer is reasonable.

We multiply the quantity in pounds by the price per pound.

$$\$3.90 \times 0.86 = \$3.3540$$

We round the answer to the nearest cent, **$3.35.** The answer is reasonable because 0.86 pounds is a little less than one pound and $3.35 is a little less than $3.90, which is the price per pound.

We distinguish between dividing by a whole number and dividing by a decimal number.

Whole number divisor	Decimal number divisor
$0.42 \div \mathbf{6}$	$0.42 \div \mathbf{0.6}$

When dividing by a whole number the decimal point in the quotient is placed above the decimal point of the dividend. In this example the quotient and dividend have the same number of decimal places. We fill empty places with zero if necessary.

$$\begin{array}{r} 0.07 \\ 6\overline{)0.42} \end{array}$$

Example 3

Simplify:

   a. 0.48 ÷ 8   **b. 4.8 ÷ 5**   **c. 4.8 ÷ 10²**

Solution

a. We place the decimal point in the quotient above the decimal point in the dividend. Then we divide like we divide whole numbers. We use zero as a place holder, so **0.06** is the quotient.

$$\begin{array}{r} 0.06 \\ 8\overline{)0.48} \\ \underline{48} \\ 0 \end{array}$$

b. We place the decimal point in the quotient, and then we divide. We do not write remainders with decimal numbers. Instead we affix zero(s) and continue dividing (or we round). We find that **0.96** is the quotient.

$$\begin{array}{r} 0.96 \\ 5\overline{)4.80} \\ \underline{45} \\ 30 \\ \underline{30} \\ 0 \end{array}$$

c. Since 10^2 equals 100, we divide 4.8 by 100. Notice that the quotient has the same digits as the dividend with the decimal points shifted two places.

$$\begin{array}{r} \mathbf{0.048} \\ 100\overline{)4.800} \end{array}$$

When dividing by a decimal number we take an extra step to make an equivalent problem in which the divisor is a whole number. Multiplying the dividend and the divisor by 10 shifts the decimal points one place without changing the quotient. Instead of showing the multiplication we can shift both decimal points the same number of places before we divide.

$$\frac{0.42}{0.6} \cdot \frac{10}{10} = \frac{4.2}{6}$$

$$06.\overline{)04.2}$$

Example 4

Simplify:

a. 3.6 ÷ 0.6   **b. 3 ÷ 0.06**

Solution

Thinking Skill

Connect

A fraction bar and a division sign both indicate division. What fraction name for 1 do we multiply 3.6 ÷ 0.6 by to make the equivalent division problem 36 ÷ 6?

a. We rewrite the problem, shifting both decimal points one place. Then we divide. The quotient, **6,** is a whole number, so the decimal point does not need to be shown.

$$\begin{array}{r} 6. \\ 06.\overline{)36.} \\ 36 \\ \hline 0 \end{array}$$

b. The dividend is the whole number 3. To make 0.06 a whole number we shift the decimal point two places. The decimal point to the right of 3 shifts two places, which are filled with zeros. The quotient, **50,** is a whole number and can be written without a decimal point.

$$\begin{array}{r} 50. \\ 006.\overline{)300.} \end{array}$$

Example 5

Chloe drove her car 236.4 miles on 9.6 gallons of gas. Find the number of miles per gallon the car averaged to the nearest mile per gallon.

Solution

To find miles per gallon we divide miles by gallons. Since the divisor is a decimal number we take an extra step to make the divisor a whole number.

$$\frac{mi}{gal} \; \frac{236.4}{9.6} \cdot \frac{10}{10} = \frac{2364}{96}$$

$$\begin{array}{r} 24.625 \\ 96\overline{)2364.000} \end{array}$$

Rounding to the nearest whole number, we find Chloe's car averaged **25 miles per gallon.** The answer is reasonable because 9.6 gallons is nearly 10 gallons and dividing 236 miles by 10 gallons is 23.6 miles per gallon, and this estimate is close to our calculated answer.

Decimal Arithmetic Reminders

Operation	+ or −	×	÷ by whole (*W*)	÷ by decimal (*D*)
Memory cue	line up $\pm\ \overset{.}{\underset{.}{.}}$	×; then count $\times\ \overset{._-}{._=}$ $._{--}$	up $W\overline{)\ .}^{\,.}$	over, over, up $D.\overline{)\underset{\smile\smile}{.}}^{\,.}$

You may need to …
• Place a decimal point to the right of a whole number.
• Fill empty places with zeros.

Practice Set

Simplify.

a. 0.4×0.12

b. $(0.3)^2$

c. 6.75×10^3

d. $0.144 \div 6$

e. $1.2 \div 0.06$

f. $2.4 \div 0.5$

g. Use the answer to **b** to find $\sqrt{0.09}$.

h. What is the cost of 10.2 gallons of gas at $2.249 per gallon?

i. If 1.6 pounds of peaches cost $2.00, then what is the cost per pound?

Written Practice   *Strengthening Concepts*

* 1. Mr. Villescas typed 16 names in a minute. At that rate, how many names
(7)   could he type in one and a half minutes?

2. The number of students in four classrooms is 27, 29, 30, and 30. What
(7)   is the average (mean) number of students in the four classrooms?

* 3. *Analyze* Al Oerter won the gold medal in the discus in four consecutive
(3, 24)   Olympics, establishing a new Olympic record each time. He won his
first gold medal in 1956 in Melbourne with a throw of 56.36 meters. He
won his fourth gold medal in Mexico City in 1968, throwing 64.78 m. His
1968 winning mark was how much farther than his 1956 mark?

Simplify.

4. $\dfrac{1}{2} + \dfrac{2}{3}$
(13)

* 5. $\dfrac{1}{2} \cdot \dfrac{2}{3}$
(22)

* 6. $\dfrac{1}{2} \div \dfrac{2}{3}$
(22)

7. $\dfrac{2}{3} - \dfrac{1}{2}$
(13)

* 8. $0.9(0.11)$
(25)

* 9. $(3.1)^2$
(15, 25)

* 10. $4.36 + 0.4$
(24)

* 11. $4.2 - 0.42$
(24)

* 12. $\dfrac{0.144}{4}$
(25)

* 13. $\dfrac{4.36}{0.4}$
(25)

14. Arrange these numbers in order from least to greatest.
(12)

$$0.25 \qquad 0.249 \qquad 0.251$$

15. **a.** Name a pair of angles in this figure that form a linear pair.

(18)

b. If ∠AGB measures 120°, and if ∠BGD is a right angle, then what is the measure of ∠BGC and ∠CGD?

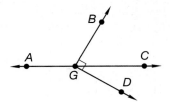

16. Find the area and perimeter of a rectangle with vertices at $(-7, -1)$, $(-7, -10)$, $(2, -10)$, and $(2, -1)$.

(8, Inv. 1)

17. **a.** Expand: $3(t + 4)$

(21)

b. Factor: $10x - 15$

18. Simplify and compare:

(15, 21)

a. $|11 - 7| \bigcirc |7 - 11|$ **b.** $\sqrt{121} \bigcirc \sqrt[3]{1000}$

19. Find the two values for z that make this statement true: $|z| = 7$

(1)

20. Write this expression in exponential form.

(15)

$$\frac{k \cdot k \cdot u \cdot u \cdot u}{r \cdot r \cdot r \cdot r \cdot t}$$

21. What is the area of a square window that is $3\frac{1}{2}$ feet on each side?

(8, 23)

22. Name the properties illustrated.

(2)

a. $4 + (3 + 1) = (4 + 3) + 1$ **b.** $0.5 \times 1 = 0.5$

* 23. Find the perimeter and area of this triangle:

(20, Inv. 2)

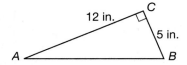

* 24. Classify the triangle in problem **23** by angles and sides. If the measure of ∠A is $22\frac{1}{2}°$, what is the measure of ∠B?

(20)

For problems **25–28,** solve by inspection.

25. $70 = 2r + 10$ **26.** $1 - x = \dfrac{2}{3}$

(14)   (13, 14)

27. $\dfrac{1}{4}p = \dfrac{1}{8}$ **28.** $\dfrac{100}{m} = \dfrac{1}{2}$

(14, 22)   (10, 14)

Evaluate.

29. Find $\frac{1}{2}bh$ when $b = 3$ and $h = 6$.

(14)

30. **Evaluate** Find $b^2 - (4ac)$ when $a = 1$, $b = 4$, and $c = 4$.

(14)

• Transformations

facts | Power Up F

mental math

a. Calculation: The laundromat washes about 90 loads of laundry per day. About how many loads of laundry does it wash per week?

b. Geometry: Two angles of a triangle are 90° and 10°. What is the third angle?

c. Number Sense: $1\frac{1}{2} \times 12$

d. Power/Roots: What whole number is closest to $\sqrt{37}$?

e. Proportions: On the long downhill slope the elevation dropped 100 feet every quarter mile. How much did the elevation drop in one mile?

f. Percent: 25% of $36

g. Fractional Parts: $\frac{2}{3}$ of the 36 students

h. Calculation: $100 - 50, + 14, \sqrt{\ }, + 1, \sqrt{\ }, + 1, \sqrt{\ }, - 1, \sqrt{\ }, - 1, \sqrt{\ }, - 1, \times 9$

problem solving

Determine the number of small triangles in the first three figures. Draw the fourth figure, and determine the number of small triangles in it. How are the areas of the four figures related?

Transformations are operations on a geometric figure that alter its position or form. Transformations are used for computer graphics and in animation. In this lesson we will consider four different transformations. Below we illustrate three transformations that move $\triangle ABC$ to its image $\triangle A'B'C'$(we read $\triangle A'B'C'$ as "A prime B prime C prime"). The three transformations are **reflection** (flip), **rotation** (turn), and **translation** (slide). Notice that the size of the figure is not changed in these transformations.

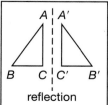 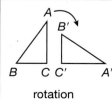 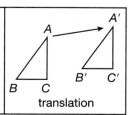

reflection   rotation   translation

A reflection occurs across a line. Here we show the reflection of △ABC across the y-axis. Each point in the reflection is the same distance from the y-axis as the corresponding point in the original figure. A segment between corresponding points is perpendicular to the line of reflection. Notice that if we were to fold this page along the line of reflection, the figures would coincide exactly.

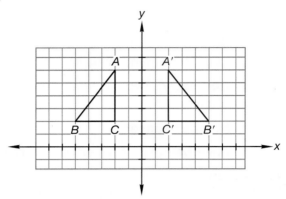

A positive rotation turns a figure counter-clockwise about (around) a point. Here we show a 90° rotation about point B. Point B is the point of rotation —its location is fixed, and the figure spins around it. If we trace the path of any other point during this rotation, we find that it sweeps out an arc of 90°.

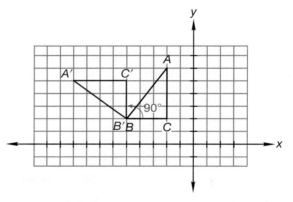

A translation slides a figure a distance and direction without flipping or turning. Here we show a translation of (6, 1), which is 6 units to the right and 1 unit up. For any translation (a, b), a describes the horizontal shift, and b describes the vertical shift for each point of the figure.

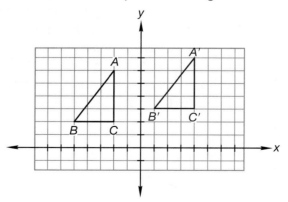

Example 1

Describe the transformations that move △ABC to the location of
△A‴B‴C‴.

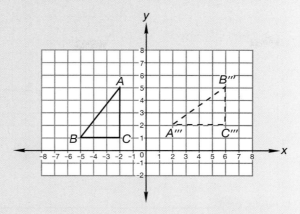

We will use all three transformations. We begin by **reflecting** △ABC across
line AC.

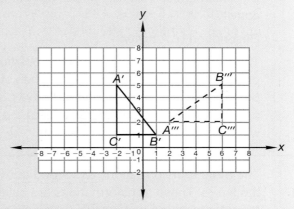

Then we **rotate** △A′B′C′ **90° about point C.**

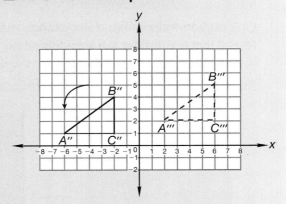

We finish by **translating △A″B″C″ 8 units to the right and 1 unit up.**

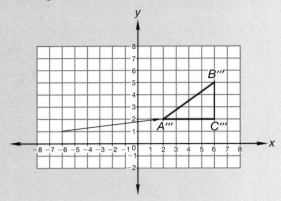

Activity

Describing Transformations

Materials needed:

Copies of Lesson Activity 4. Cut out the two triangles before beginning the activity.

Description of Activity:

The teacher or a student volunteer positions the two figures on the same plane but in different orientations. Then students direct the volunteer to perform the transformations that move one of the figures to the location and orientation of the other figure. Repeat the activity with other volunteers and with other positions of the polygons.

Math Language
The word "isometries" comes from the Greek words "iso" (meaning "equal") and "meter" (meaning "measure").

The three transformations described above do not change the size of the figure. These transformations are called **isometries** or **congruence transformations** because the original figure and its image (result of the transformation) are congruent.

One transformation that does change the size of a figure is a dilation. Recall that a **dilation** is a transformation in which the figure grows larger. (A **contraction** is a transformation in which the figure grows smaller.)

An example of a dilation is a photographic enlargement.

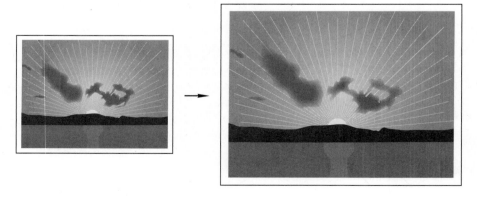

Although a dilation changes the dimensions of a figure, it does not change its shape. The original figure and its image are similar and corresponding lengths are proportional. Thus, dilations and contractions are **similarity transformations.**

Dilations of geometric figures occur away from a fixed point that is the center of the dilation.

Dilation of Rectangle ABCD

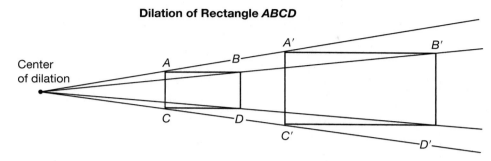

Note that corresponding vertices are on the same rays from the center of dilation and that the corresponding segments of △ABC and its image △A′B′C′ are parallel.

Example 2

Measure the length and width of rectangle *ABCD* above and its dilated image rectangle *A′B′C′D′*.

   a. What is the scale factor of the dilation?

   b. The perimeter of the image is what multiple of the perimeter of rectangle *ABCD*?

   c. The area of the image is what multiple of the area of rectangle *ABCD*?

Solution

Rectangle *ABCD* is 2 cm by 1 cm. Rectangle *A′B′C′D′* is 4 cm by 2 cm.

   a. The scale factor is **2.**

   b. The perimeter of the image (2 × 2 cm + 2 × 4 cm = 12 cm) is **2 times** the perimeter of rectangle *ABCD* (2 × 1 cm + 2 × 2 cm = 6 cm).

   c. The area of the image (2 cm × 4 cm = 8 cm^2) is **4 times** the area of rectangle *ABCD* (1 cm × 2 cm = 2 cm^2).

Discuss How is the scale factor of the dilation related to the relative perimeters and areas?

Practice Set

Classify For **a–d** write the meaning of each transformation by selecting a word from this list: turn, flip, enlargement, slide.

   a. Translation   **b.** Rotation

   c. Reflection   **d.** Dilation

Rectangle *PQRS* is dilated with a scale factor of 3 forming the rectangle *P′Q′R′S′*. Use this information for **e–g.**

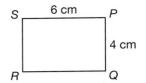

e. Sketch rectangle *P′Q′R′S′* and label its length and width.

f. The perimeter of the dilated image is how many times the perimeter of rectangle *PQRS*?

g. The area of the image is how many times the area of rectangle *PQRS*?

Written Practice   *Strengthening Concepts*

1. Marco Polo, the Italian trader, journeyed from Italy to China with his
(3) father and uncle. They returned to Italy in 1295, which was twenty-four years after they had left Italy. In what year did they leave Italy for China?

2. Of the fifty states in the United States, $\frac{4}{25}$ of them begin with the letter
(5) "M." How many of the fifty states begin with the letter "M"?

3. On the safari, the average speed traveled by the vehicle was 25 miles
(7) per hour. If the vehicle traveled for 6 hours, how far did it travel?

Generalize   Simplify.

4. $1\frac{7}{8} + 1\frac{1}{6}$
(13)

5. $\frac{7}{8} - \frac{1}{6}$
(13)

* 6. $\frac{7}{8} \cdot \frac{1}{6}$
(22)

* 7. $\frac{7}{8} \div \frac{1}{6}$
(23)

* 8. $0.9 + 0.85$
(24)

* 9. $0.9 - 0.85$
(24)

* 10. $(0.9)(0.85)$
(25)

* 11. $\frac{0.828}{0.9}$
(25)

12. Write the expressions in exponential from:
(15)
   a. *mmnmn*   **b.** *2xyyxy*

Use order of operations to simplify.

* 13. $5 \times 4 - [3 \times (2 - 1)]$
(21)

14. $|25 - 16| - \sqrt{25 - 16}$
(15, 21)

15. Find the area and perimeter of a rectangle with vertices at $(-3, -3)$,
(8, Inv. 1) $(-3, 6)$, $(0, -3)$, and $(0, 6)$.

* 16. **a.** Expand: $5(4 - p)$
(21)
   b. Factor: $8y - 12$

17. Compare:
(15, 21)
   a. $|1 - 9| \bigcirc (9 - 1)$ **b.** $\sqrt{9} + \sqrt{16} \bigcirc \sqrt{9 + 16}$

18. Find two values for *x* that make this statement true: $|x| = \frac{1}{2}$
(1)

Estimate For problems **19** and **20,** estimate each answer.

* 19. $14.3 \div 1.9$
₍₁₇₎

* 20. 2983×7.02
₍₁₇₎

21. Arrange these numbers in order from least to greatest.
₍₁₂₎
1.2, 0, 0.12, 1, 2

22. Which property is illustrated?
_(2, 21)
    a. $3(x - 9) = 3x - 27$ **b.** $3(4 \cdot 2) = (3 \cdot 4)2$

* 23. _Analyze_ Find the perimeter and area of this
_(8, 20) triangle. Dimensions are in mm.

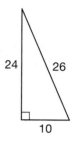

Solve by inspection. Choose one equation and write a story
for it.

24. $\dfrac{5 + m}{2} = 10$
₍₁₄₎

25. $6m + 2 = 50$
₍₁₄₎

26. $2m = 104$
₍₁₄₎

27. $m + \dfrac{1}{4} = \dfrac{3}{4}$
_(13, 14)

Evaluate.

28. Find $b^2 - 4ac$ when $a = 2$, $b = 3$, and $c = 1$.
_(14, 15)

29. Write the prime factorization of 200 using exponents.
_(9, 15)

* 30. _Evaluate_ Banks made a kite with paper
_(8, Inv. 2) and string and perpendicular wooden
cross pieces with the dimensions shown (in
inches).

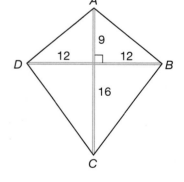

    a. What is the distance from A to B?

    b. What is the distance from B to C?

    c. If there is a loop of string around the kite,
       how long is the loop?

• Laws of Exponents

Power Up Building Power

facts Power Up F

mental math

a. Calculation: Roper does 20 pushups per minute. If he can do 50 pushups at this rate, how long will it take?

b. Number Sense: $2\frac{1}{2} \times 12$

c. Geometry: Find x.

d. Algebra: $4x - 3 = 25$

e. Fractional Parts: $\frac{2}{5}$ of $55

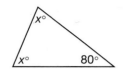

f. Proportions: Walking up 3 stair steps, Vera rose 2 feet. How many feet did Vera rise walking up 12 stair steps?

g. Percent: 25% of 444

h. Calculation: $\sqrt{100}$, -1, $\sqrt{}$, $\times 12$, $\sqrt{}$, $\times 4$, $+1$, $\sqrt{}$, $\times 10$, -1, $\sqrt{}$

problem solving

Sharon took Susan to a playground. Use the graph to write a brief story about what Susan was doing on the playground.

Susan's Trip to the Playground

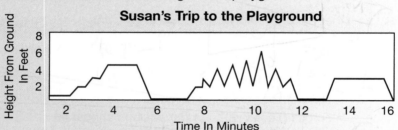

(**Understand**) We have been given a graph that shows how many feet Susan was above the ground over a span of 17 minutes of play on a playground. We are asked to write a brief story of what she was doing while on the playground.

(**Plan**) The motion graph that we have been provided shows three distinct events in terms of height. The first event shows a gradual rise up, and then a sudden drop down. The second event shows a consistent up-and-down. The third event shows an extended amount of time at approximately 4 feet.

(**Solve**) Susan arrived at the play area and immediately ran to the slide. She waited her turn to climb up the rungs to the top, where she hesitated for a while before finally sliding down. She then walked over to the swings where she began to swing, gradually gaining altitude, and then slowing down so she could get off. Finally, Susan decided to play in the "Tubes and Tunnels." She climbed up the steps, and then spent several minutes crawling through the tunnels before exiting.

(**Check**) We were able to write a story because Susan's movement on the playground had been graphed. Other stories may differ from ours.

Math Language

The **exponent,** or *n,* shows the number of times the **base,** or *x,* is used as a factor.

Recall that a counting-number exponent indicates how many times a base is a factor.

$$x^n = x \cdot x \cdot x \ldots x \qquad (n \text{ factors of } x)$$

Applying this knowledge we can begin to develop the **laws of exponents** that describe relationships between exponents for certain operations.

Example 1

Find the missing exponent: $x^5 \cdot x^3 = x^{\square}$

Solution

We apply the meaning of the exponents.

$$x^5 \cdot x^3 = (x \cdot x \cdot x \cdot x \cdot x)(x \cdot x \cdot x) = x^{\boxed{8}}$$

We see that multiplying x^5 and x^3 means that x is a factor eight times. Notice that the exponent 8 is the sum of the exponents 5 and 3.

Example 2

Find the missing exponent: $\dfrac{3^5}{3^3} = 3^{\square}$

Solution

We apply the meaning of the exponents and reduce the fraction.

$$\frac{3^5}{3^3} = \frac{3 \cdot \overset{1}{\cancel{3}} \cdot \overset{1}{\cancel{3}} \cdot \overset{1}{\cancel{3}} \cdot 3}{\underset{1}{\cancel{3}} \cdot \underset{1}{\cancel{3}} \cdot \underset{1}{\cancel{3}}} = 3^{\boxed{2}}$$

Notice that the exponent 2 is the difference of the exponents 5 and 3.

Example 3

Find the missing exponent: $(x^3)^2 = x^{\square}$

Solution

The exponent 2 means that the base x^3 is a factor twice.

$$(x^3)^2 = x^3 \cdot x^3 = xxx \cdot xxx = x^{\boxed{6}}$$

Notice that the exponent 6 is the product of the exponents 2 and 3.

Examples 1, 2, and 3 illustrate the following three laws of exponents.

Thinking Skill

Connect

Why does the table specify that for division, $x \neq 0$?

Laws of Exponents for Multiplication and Division

$x^a \cdot x^b = x^{a+b}$
$\dfrac{x^a}{x^b} = x^{a-b}$ for $x \neq 0$
$(x^a)^b = x^{a \cdot b}$

Summarize State the laws of exponents as shown in the table in your own words.

We will expand on these laws in later lessons.

Example 4

Write each expression as a power of ten.

a. $10^8 \cdot 10^6$ b. $\dfrac{10^8}{10^6}$

Solution

The bases are the same, so we can apply the laws of exponents.

a. We add the exponents.

$$10^8 \cdot 10^6 = 10^{14}$$

b. We subtract the exponent of the divisor from the exponent of the dividend.

$$\dfrac{10^8}{10^6} = 10^2$$

Practice Set

Find the missing exponents.

a. $x^5 \cdot x^2 = x^{\square}$ b. $\dfrac{x^5}{x^2} = x^{\square}$ c. $(x^2)^3 = x^{\square}$

Simplify.

d. $2^3 \cdot 2^2$ e. $\dfrac{2^5}{2^2}$ f. $(2^5)^2$

Write each expression as a power of 10.

g. $10^6 \cdot 10^3$ h. $\dfrac{10^6}{10^2}$

Written Practice *Strengthening Concepts*

1. Cynthia planted 12 potted plants in the last 3 weeks. At that rate, how many would she plant in 5 weeks?
(7)

2. The volleyball team has 5 wins and 7 losses. What fraction of their games will they have won if they win their next 3 games?
(5)

3. Of the 2,800 visitors to the fair, only $\frac{3}{70}$ won a prize. How many visitors won prizes?
(5)

* 4. *(26)* **Analyze** Reflect the point (1, 4) across the *y*-axis. What is the image?

* 5. *(26)* **Classify** What transformation is shown at right?

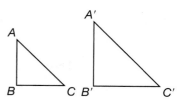

* 6. *(26)* **Model** If this figure is rotated 90° clockwise, draw the image.

Generalize Simplify.

* 7. *(27)* $3^2 + 3^1 + 3^0$

* 8. *(25)* $2.1 \div 0.7$

* 9. *(25, 27)* 6.02×10^6

10. *(24)* $4.21 + 42.1 - 0.421$

* 11. *(23)* $1\frac{1}{2} \cdot \frac{4}{5}$

* 12. *(23)* $1\frac{1}{2} \div \frac{4}{5}$

13. *(22)* $\frac{2}{5} \div \frac{4}{5}$

14. *(22)* $\frac{4}{5} \div \frac{2}{5}$

15. *(21)* $4^2 - 4(2)(2)$

16. *(15)* $\sqrt[3]{8} + \sqrt{9}$

17. *(13)* $1\frac{1}{10} + \frac{1}{2}$

18. *(21)* Compare: $|16 - 20| \bigcirc (16 - 20)$

19. *(20)* Classify this triangle by sides and angles.

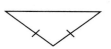

20. *(8, 20)* Find the perimeter and area of this triangle.

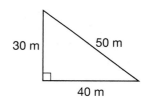

30 m   50 m   40 m

21. *(19)* A regular triangle has a perimeter of 51 m. How long is each side?

* 22. *(19)* **Verify** Give a counterexample to disprove the statement "all squares are congruent."

23. *(18)* Choose the best word to complete the following sentence. Two lines that are in different planes and do not intersect are

  A parallel.

  B skew.

  C perpendicular.

  D diagonal.

24. *(18)* These lines are perpendicular.

$m \angle 1 + m \angle 3 = $ _____.

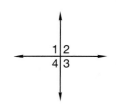

25. Round 7,654,321 to the nearest hundred thousand.
(17)

26. Estimate the difference: $15.81 - 6\frac{1}{4}$
(17)

27. Name a set of numbers that includes -2, but does not include -2.3.
(10)

28. Anthony is building a rectangular pen 3.54 m long and 2.8 m wide for his dog. How many meters of fence will Anthony need?
(8, 24)

29. Write the prime factorization of 144 using exponents.
(9)

30. Solve by inspection.
(14)
    a. $35 - d = 17$ **b.** $8n = 72$

Early Finishers

Real-World Application

Each of the 50 states on a US flag are represented by a 5-point star. There is a story that Betsy Ross showed George Washington how to make a 5-point star with one clip of her scissors. To make a 5-point star, fold and cut an $8\frac{1}{2}$ by 10 sheet of unlined paper according to the following directions.

1. left corner

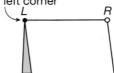

Fold the paper in half to create a left (L) and a right (R) corner.

2. midlines

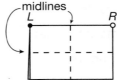

Crease the folded paper at midlines. Then unfold.

3.

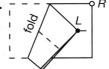

Fold down left corner from vertical midline to horizontal midline.

4.

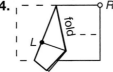

Fold left corner back to fold shown in step 3.

5.

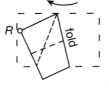

Fold right corner to left across fold shown in step 4.

6.

Fold right corner back to fold shown in step 5.

7.

Use scissors to make one cut at about the angle shown.

8. Unfold star. To create a three-dimensional look make ridges of the folds that radiate to the points. Then make valleys of the folds between the points.

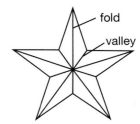

• Scientific Notation for Large Numbers

facts | Power Up F

mental math

a. Calculation: The odometer read 235 miles. Seth drove north until the odometer read 245 miles. Then Seth drove south until the odometer read 254 miles. How far is Seth from where he started?

b. Geometry: If a spoke is 12 inches long, then the circumference of the wheel is a little more than how many feet?

12 in. radius

c. Number Sense: $10\frac{1}{2} \times 4$

d. Proportions: If 3 gallons cost $7.50, how much would 4 gallons cost?

e. Power/Roots: $\sqrt{900}$

f. Fractional Parts: $\frac{5}{6}$ of 48

g. Fractions: $\frac{1}{7} = \frac{3}{x}$

h. Calculation: $180 - 80$, $\div 2$, $\div 2$, $\sqrt{}$, $\times 9$, $+ 4$, $\sqrt{}$, $- 6$, $\times 17$, $- 1$, $\sqrt{}$, $\sqrt{}$

problem solving

Aoki, Breann, Chuck, and Dan line up from left to right to take a picture. How many different arrangements are there?

New Concept *Increasing Knowledge*

Scientific notation is a method of writing a number as a decimal number times a power of 10. Scientific notation is a useful way to express and manipulate very large or very small numbers that take many places to express in standard form. For example, the speed of light is about 300,000,000 meters per second. In scientific notation, we write this number as follows:

$$3.0 \times 10^8$$

"Three point zero times ten to the eighth."

The first number (3.0) is the **coefficient.** The coefficient is written with one non-zero digit to the left of the decimal point. The power of 10 is the multiplier that turns 3.0 into 300,000,000. Notice that the exponent, 8, corresponds to the number of places the decimal point shifts when the number is written in standard form.

In other words, powers of 10 can be used to indicate place value.

Place Value Table

Trillions			Billions			Millions			Thousands			Units (ones)		
hundreds	tens	ones	hundreds	tens	ones	hundreds	tens	ones	hundreds	tens	ones	hundreds	tens	ones
10^{14}	10^{13}	10^{12}	10^{11}	10^{10}	10^{9}	10^{8}	10^{7}	10^{6}	10^{5}	10^{4}	10^{3}	10^{2}	10^{1}	10^{0}

Example 1

Earth's average distance from the sun is 93,000,000 miles. Write 93,000,000 in scientific notation.

Solution

The coefficient is written with one digit to the left of the decimal point, so we place the decimal point between 9 and 3. This is seven decimal places from the actual location of the decimal point, so we use the exponent 7. We drop trailing zeros. The answer is read "nine point three times ten to the seventh."

$$9.3 \times 10^7$$

Example 2

Visit www. SaxonPublishers. com/ActivitiesC3 for a graphing calculator activity.

When Brigita finished the 1500 meter run she announced that she had run "one point five times ten to the sixth millimeters." Write that number a in scientific notation and b in standard form.

Solution

a. 1.5×10^6

b. The power of 10 indicates that the actual location of the decimal point is six places to the right.

$$1\underset{\smile\smile\smile\smile\smile\smile}{500000}.$$

Brigita ran **1,500,000** mm.

Scientific calculators will display the results of an operation in scientific notation if the number would otherwise exceed the display capabilities of the calculator. For example, to multiply one million by one million, we would enter

1 0 0 0 0 0 0 ×
1 0 0 0 0 0 0 =

The answer, one trillion, contains more digits than can be displayed by most calculators. Instead of displaying one trillion in standard form, the calculator displays one trillion in some modified form of scientific notation such as

or perhaps

Connect On a calculator multiply the speed of light (300,000,000 m/sec), times the approximate number of seconds in a year (31,500,000) to find a light year in meters.

Connect Use your calculator to multiply the measure of a light year in meters by 10. Use Laws of Exponents to explain the result.

Practice Set

a. Write "two point five times ten to the sixth" with digits.

b. **Represent** Use words to write 1.8×10^8.

Write each number in standard form.

c. 2.0×10^5 　　　　　　　　**d.** 7.5×10^8

e. 1.609×10^3 　　　　　　　**f.** 3.05×10^4

Write each number in scientific notation.

g. 365,000 　　　　　　　　**h.** 295,000,000

i. 70,500 　　　　　　　　　**j.** 25 million

Written Practice 　　　*Strengthening Concepts*

* 1.
(3)
The lowest point on dry land is the shore of the Dead Sea, which is about 1292 feet below sea level. Less than 20 miles from the Dead Sea is the Mount of Olives, which is 2680 feet above sea level. What is the elevation difference between the shore of the Dead Sea and the Mount of Olives?

* 2.
(11)
Infer About $\frac{7}{10}$ of Greece's mainland is mountainous. About what percent of Greece's mainland is not mountainous?

3.
(4)
The Great Wall of China is about 1500 miles long. If there are about 8 towers distributed along each mile, about how many towers are on the Great Wall of China?

4.
(20)
Classify this triangle by angles and sides and find its area.

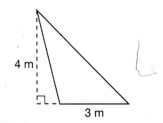

5.
(15)
If the largest angle of the triangle in problem **4** is 104°, then the sum of the measures of the two acute angles is how many degrees?

* 6.
(6, 25)
On earth a kilogram mass weighs about 2.2 pounds. A 16-pound bowling ball has a mass of about how many kilograms? Express your answer to the nearest whole kilogram.

Simplify using rules of exponents.

* **7.** **a.** $m^3 \cdot m^4$ **b.** $\dfrac{m^6}{m^2}$
(27)

Simplify. Remember to follow the order of operations.

* **8.** $\dfrac{1}{3} + 1\dfrac{1}{3} \cdot 2\dfrac{1}{4}$ * **9.** $\dfrac{3}{4} - \dfrac{1}{5} \div \dfrac{4}{5}$
(21, 22)   (21, 22)

* **10.** $\dfrac{2 + 1.2}{2 - 1.2}$ * **11.** $(1.5)^2 - 1.5$
(21, 25)   (21, 25)

* **12.** Find the measure of $\angle WQY$.
(18)

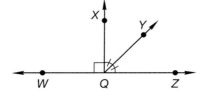

* **13.** Evaluate: $x^2 - 2x$ when $x = 5$
(14, 15)

* **14.** Round each number to the nearest whole number.
(17)
     a. $3\dfrac{3}{4}$ **b.** $3\dfrac{3}{10}$ **c.** 3.10

* **15.** Plot the following points which are vertices of a rectangle. Locate the
(8, fourth vertex, and find the area and perimeter of the rectangle. $(-1, 5)$,
Inv. 1) $(-1, -2)$, $(-6, 5)$

16. **a.** Expand: $3(x + r + 5)$ **b.** Factor: $4x + 6y$
(21)

* **17.** **a.** Write 2.97×10^5 in standard notation.
(28)

     b. Write 4,030,000 in scientific notation.

* **18.** Simplify: $1^3 + 2^3 + 3^1 + 4^0$
(27)

* **19.** Estimate: $\dfrac{6870}{6.9}$
(17)

20. Compare: $\sqrt{25} \bigcirc \sqrt[3]{27}$
(17)

* **21.** What is the multiplicative inverse of $2\dfrac{1}{2}$?
(23)

22. Which property is illustrated?
(2)
     a. $xy = yx$ **b.** $1 \cdot x = x$

23. If a triangle is equilateral, then each angle measures how many
(20) degrees?

Solve by inspection:

24. $10 \cdot 5 = 10 + h$ **25.** $1 - f = \dfrac{9}{10}$
(14)   (13, 14)

* **26.** $\dfrac{10^6}{10^2} = 10^x$ **27.** $\dfrac{18}{x} = \dfrac{9}{10}$
(27)   (10, 14)

28. Simplify.
(21)
     a. $2 + 3 \cdot 10 \div 2$ **b.** $2 \times (3 + 12) + 10$

29. *Analyze* Arrange these numbers in order from least to greatest.
(12)

$$2, 0.2, -2, 0, 0.02$$

* 30. What is the length of the hypotenuse of this
(15,
Inv. 2) isosceles right triangle?

1 cm

1 cm

Early Finishers

Real-World
Application

Flies belong to a group of insects called *Diptera.* The *Diptera* order is one of the four largest groups of living organisms. There are over 120,000 different kinds of flies known in the world today. If a particular species of fly weighs approximately 0.0044 of a pound, how many grams would 12 flies weigh (1 pound = 453.5923 g)? Round your answer to the nearest ten-thousandth.

• **Ratio**

facts | Power Up F

mental math

a. Calculation: Frank flipped 2 hotcakes every 3 seconds. At that rate, how many can he flip in a minute?

b. Estimation: $449.8 \div 49.7$

c. Geometry: What is the value of x?

d. Number Sense: 7.5×4

e. Scientific Notation: Write 7.5×10^6 in standard notation.

f. Fractional Parts: $\frac{3}{8}$ of $32

g. Percent: $33\frac{1}{3}\%$ of $90

h. Calculation: $11, \times 9, + 1, \sqrt{}, + 6, \sqrt{}, + 5, \sqrt{}, + 7 =$

problem solving | Mr. Roberts faced west. He walked ten steps forward. Then, without turning, he took five sideways steps to his right. Then, he took seven steps backward. Then, he turned left and walked six steps forward. Then, he turned right and took three steps backward. Which direction is Mr. Roberts facing, and how far is he from where he started?

New Concept Increasing Knowledge

A **ratio** is a comparison of two numbers by division. For example, if there are 12 girls and 16 boys in a class, then we can express the ratio of girls to boys as a reduced fraction.

$$\frac{girls}{boys} \quad \frac{12}{16} = \frac{3}{4}$$

We read this ratio as "3 to 4." Here we show four ways to write a ratio:

with the word **to**	3 to 4
as a fraction	$\frac{3}{4}$
as a decimal	0.75
with a colon	3:4

Example 1

There are 9 girls and 15 boys in a class. What is the ratio of boys to girls in the class?

Solution

We write the ratio in the order stated—boys to girls.

$$\frac{boys}{girls} \qquad \frac{15}{9} = \frac{5}{3}$$

The ratio of boys to girls is 5 to 3. Notice that we reduce the ratio, but we do not write the ratio as a mixed number.

Example 2

A 30 ft tall tree cast a shadow 18 ft long. What was the ratio of the height of the tree to the length of its shadow?

Solution

We write the ratio in the order specified: height to length.

$$\frac{\text{height of tree}}{\text{length of shadow}} \qquad \frac{30 \; \cancel{ft}}{18 \; \cancel{ft}} = \frac{5}{3}$$

Thinking Skill

Represent

Use words to state the exact ratio of advancers to decliners.

In some situations we round numbers to express a ratio. For example, in a day of stock-market trading perhaps 789 stocks gained value and 597 lost value. The ratio of advancers to decliners is estimated as

$$\frac{\text{advancers}}{\text{decliners}} \qquad \frac{789 \; \cancel{stocks}}{597 \; \cancel{stocks}} \approx \frac{800}{600}$$

The rounded ratio is reduced, so we might hear a business report stating that advancers led decliners 4 to 3.

Example 3

At the high school football game 1217 fans sat on the home-team side and 896 fans sat on the visiting-team side. What was the approximate ratio of home-team fans to visiting-team fans?

Solution

We arrange the numbers in the stated order, and then we round and reduce the ratio.

$$\frac{\text{home-team fans}}{\text{visiting-team fans}} \qquad \frac{1217 \; \cancel{fans}}{897 \; \cancel{fans}} \approx \frac{1200}{900}$$

The ratio reduces to 12 to 9 which further reduces to $\frac{4}{3}$.

Recall that a rate is a ratio of two measures with different units. The units do not cancel but instead remain part of the rate.

Example 4

Rosa kept a record of the number of pages and number of minutes she read each night. Find Rosa's approximate reading rate for the book in pages per minute.

Reading Record

Day	Pages	Minutes
M	15	30
T	11	20
W	14	30
Th	20	40
F	27	50

Solution

We are asked for the rate in pages per minute, so the units are pages and minutes. To find Rosa's reading rate we have some choices. We can find the rate for each day to see if the rate is consistent, or we can find the total number of pages and minutes and calculate an average rate. We show both methods in the table. We see that each ratio is equal or very close to 1 page every 2 minutes or about $\frac{1}{2}$ **page per minute.**

Reading Record

Day	Pages	Minutes	p/m
M	15	30	$\frac{1}{2}$
T	11	20	$\frac{11}{20}$
W	14	30	$\frac{7}{15}$
Th	20	40	$\frac{1}{2}$
F	27	50	$\frac{27}{50}$
5 days	87	170	$\frac{87}{170}$

Practice Set

Connect Refer to this information to answer questions **a** and **b**.

In a bag are 14 marbles. Six marbles are red and eight are blue.

a. What is the ratio of red marbles to blue marbles?

b. What is the ratio of blue marbles to red marbles?

Refer to this statement to answer questions **c** and **d**.

A 20 ft flagpole casts a shadow 24 feet long.

c. What is the ratio of the height of the flagpole to the length of the shadow?

d. What is the ratio of the length of the shadow to the height of the flagpole?

e. If 479 stocks advanced in price and 326 declined in price, then what was the approximate ratio of advancers to decliners?

f. Using the table in example 4, find Rosa's approximate reading rate in minutes per page.

1. One thousand pounds of grain were packaged in five-pound bags. How
(4) many bags were there?

2. The minute hand of a clock changes position by 30° in five minutes.
(10) What fraction of a full 360° turn is this?

3. Compute an 18% gratuity on a $3000 banquet fee.
(11)

4. If the base and height of a triangle are the lengths of two sides of the
(20) triangle, then the triangle is

   A acute   **B** right

   C obtuse   **D** cannot be determined

5. Val planted 36 pumpkin seeds in the community garden. In two weeks,
(11) 27 of the seeds had germinated. What percent of the seeds had
germinated?

Simplify.

* 6. $5^2 \cdot 5^1 \cdot 5^0$
(27)

* 7. $10^2 - 10^1 - 10^0$
(15, 27)

8. $xxwxw$
(15)

9. $(xy)(xy)$
(15)

10. $5\frac{1}{5} \cdot 2\frac{1}{2}$
(23)

11. $\frac{1}{2} + \frac{2}{3} \div \frac{3}{4}$
(13, 22)

12. $\dfrac{1.35 + 0.07}{0.2}$
(24, 25)

13. $0.01(101 + 10.1)$
(24, 25)

14. Cruz filled his car with 10.2 gallons of gas at $2.89 per gallon. If he
(17) fills his car with about this much gas at about this price once a week,
estimate how much he will have spent on gas after four weeks.

* 15. *Analyze* Rectangles *ABCD* and *MNOP* are similar. What scale factor is
(26) used to dilate rectangle *ABCD* to rectangle *MNOP*? Round your answer
to the thousandth.

16. A triangle has vertices $(-2, 1)$, $(3, 1)$, and $(1, 4)$. What is the area of the
(Inv. 1, 20) triangle?

* 17. *Represent* The product 53(120) is equal to 53(100 + 20). Which of the
(21) following shows how to rewrite 53(100 + 20)?

   A $53 + 100 \times 53 + 20$ **B** $53 \times 100 + 20$ **C** $53 \times 100 + 53 \times 20$

18. Explain how the answer to problem **17** can help us calculate 53(120).
(21)

19. Arrange in order from least to greatest.
(1, 12)

$$\frac{2}{3}, -1, 0.5, 0, \frac{3}{7}$$

20. Chef Paul forgot to put pepper in the gumbo. If each bowl of gumbo
(23) needs $1\frac{1}{3}$ teaspoons of pepper, how much pepper will Chef Paul need for 6 bowls of gumbo?

21. Solve by inspection.
(14)
    a. $22 - n = 10$ **b.** $\frac{d}{4} = -8$

22. What is the median for this set of data?
(7)

Days	Number of Cars Sold This Week
Monday	13
Tuesday	9
Wednesday	11
Thursday	5
Friday	3
Saturday	16

23. Tedd drove 294 miles in 7 hours. What was his average speed in miles
(7) per hour?

Simplify.

24. Expand: $3(2x - y)$
(21)

25. Factor: $18x - 27$
(21)

* 26. **Formulate** The orbit of the moon is elliptical. Its distance from earth
(28) ranges from about 226,000 miles to 252,000 miles.

    a. Write the greater distance in scientific notation.

    b. The greater distance is about what fraction of a million miles?

27. Classify this triangle by angles and sides.
(20)

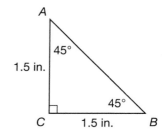

* 28. **Analyze** Marion, a store owner, uses the formula $c \times 1.75 = p$ to price
(14, 25) items (c is how much she pays for the item and p is selling price). How much will she charge for music boxes that she purchased for $16 each?

29. Use this figure to find *a* and *b*.
(18, 20)

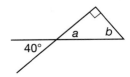

30. Miles is building a right triangle shaped pen for his chickens. He has a
(Inv. 2) 45 foot roll of chicken wire and plans to make the two shorter sides 12 ft and 16 ft. Will he have enough chicken wire to go all the way around?

Early Finishers
Real-World Application

Computers are able to perform complex and repetitive procedures quickly and reliably by breaking the procedures down into a series of simple instructions. The clock rate of a Central Processing Unit (CPU) measures how many instructions per second a computer can process. Northwood Mid-High is upgrading their office's computer system from a CPU that runs at a clock speed of 1,300,000 cycles/second to a CPU that runs at 2,000,000 cycles/second. What is the difference in clock speed between the two processors? Write your answer in standard form and in scientific notation.

• Repeating Decimals

Power Up   Building Power

facts   Power Up F

mental math

a. **Calculation:** Ronald ran 7 laps of the park in an hour. Enrique ran 4 laps in half an hour. Whose average speed was faster?

b. **Geometry:** Rewrite and complete this congruence statement: $\triangle ABC \cong$

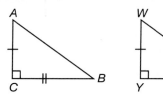

c. **Fractional Parts:** Jackie rode his bike 20 miles per hour for $1\frac{1}{2}$ hours. How far did he ride?

d. **Number Sense:** 4.5×4

e. **Scientific Notation:** Write 750,000 in scientific notation.

f. **Proportions:** Gina earned $24 in 3 hours. At that rate, how much would she earn in 5 hours?

g. **Power/Roots:** What whole number is closest to $\sqrt{101}$?

h. **Calculation:** $28, \div 4, + 2, \times 7, + 1, \sqrt{\ }, + 5, \times 2, - 1, \sqrt{\ }, - 3, \times 50$

problem solving   Copy the problem and fill in the missing digits:

$$8\overline{)\begin{array}{r} \_\_ \\ \_\_\_ \end{array}}$$
$$-\ \underline{\_\_}$$
$$-\ \underline{48}$$
$$0$$

New Concept   Increasing Knowledge

Recall that to express a ratio as a decimal number, we perform the division indicated. For example, to convert $\frac{3}{4}$ to a decimal number, we divide 3 by 4.

$$\begin{array}{r} 0.75 \\ 4\overline{)3.00} \end{array}$$

Math Language

Recall that the three dots (…) is called an *ellipsis*. It indicates that a sequence continues without end.

Notice that we used a decimal point and zeros to write 3 as 3.00 so that we could perform the division. Some fractions convert to decimal numbers that have repeating digits such as 0.272727…. The repeating digits are called the **repetend.** We can indicate repeating digits with a bar over the repetend. We write 0.272727… as $0.\overline{27}$.

Example 1

Express each fraction or mixed number as a decimal number.

a. $\frac{1}{6}$

b. $2\frac{1}{3}$

Solution

a. We divide 1 by 6 and find a repeating pattern. The division does not terminate. Instead the digit 6 repeats in the quotient. The repeating digit or set of digits is called the repetend. We place a bar over the first 6 to show that every digit that follows is a 6. Thus, **0.1$\overline{6}$** is the decimal equivalent of $\frac{1}{6}$.

$$\begin{array}{r} 0.166 \\ 6\overline{)1.000} \\ \underline{6} \\ 40 \\ \underline{36} \\ 40 \\ \underline{36} \\ 4 \end{array}$$

b. The mixed number $2\frac{1}{3}$ is a whole number plus a fraction. The whole number, 2, is written to the left of the decimal point. The fraction $\frac{1}{3}$ converts to 0.333..., which we write as .$\overline{3}$.

$$2\frac{1}{3} = \mathbf{2.\overline{3}}$$

Recall that converting a rational number to decimal form has three possible outcomes.

1. The rational number is an integer.

2. The rational number is a terminating decimal number.

3. The rational number is a non-terminating decimal with repeating digits.

Non-terminating, non-repeating decimal numbers are *irrational* and do not result from converting a rational number to a decimal.

To perform calculations with repeating decimals we first round the number to a suitable number of decimal places.

Example 2

Find $\frac{5}{6}$ of $12.47.

Solution

Before multiplying, we choose to convert $\frac{5}{6}$ to the decimal 0.8$\overline{3}$, rounding to 0.83.

$$0.83 \times 12.47 = 10.3501$$

We round to the nearest cent, **$10.35.**

Explain Marsha multiplied $\frac{5}{6}$ and $12.47 and got $10.39. Which answer is more accurate? Why is there a difference in the two answers?

Example 3

Arrange in order from least to greatest.

$$0.3 \quad 0.\overline{3} \quad 0.33$$

Solution

We will write each decimal number with three places to make the comparison easier.

$$0.3 = 0.300$$
$$0.\overline{3} = 0.33\overline{3}$$
$$0.33 = 0.330$$

Now we arrange the given numbers in order.

$$0.3, \quad 0.33, \quad 0.\overline{3}$$

Many calculators do not have functions that enable the user to perform arithmetic with fractions. To convert a fraction to a decimal, we enter the numerator, press the ÷ key, then key in the denominator and press =. Convert these fractions to decimals.

a. $\frac{10}{11}$ **b.** $\frac{11}{12}$ **c.** $\frac{17}{80}$

Thinking Skill

Analyze

Write $\frac{1}{6}$ as a decimal rounded to two decimal places. Double the result. Add $\frac{1}{6}$ and $\frac{1}{6}$ as fractions and write the sum as a decimal. Which decimal sum is more accurate?

To add fractions it is not necessary to enter long repeating decimals and it is not necessary to round the decimal equivalents. For example, to add $\frac{5}{6}$ and $\frac{8}{3}$ on a calculator that has algebraic logic (follows the order of operations), we can use this key stroke sequence:

[5] [÷] [6] [+] [8] [÷] [3] [=]

If your calculator has algebraic logic, the final display should read 3.5, which equals $3\frac{1}{2}$. Try these operations and write the decimal result.

d. $\frac{11}{12} + \frac{11}{6}$ **e.** $\frac{17}{24} - \frac{1}{2}$

Many calculators do not have a percent key. To find a given percent of a number with a calculator, we mentally shift the decimal point two places to the left before we enter the number.

f. Estimate 6.25% of $72.59. Then use a calculator to perform the actual calculation and round the answer to the nearest cent. What decimal number did you enter for 6.25%?

Practice Set

Write each fraction in **a–d** as a decimal number.

a. $\frac{3}{11}$ **b.** $\frac{2}{9}$ **c.** $2\frac{2}{3}$ **d.** $\frac{1}{30}$

Round each decimal number to the nearest thousandth.

e. $0.\overline{3}$ **f.** $0.\overline{6}$ **g.** $0.\overline{36}$ **h.** $1.8\overline{6}$

Arrange in order from least to greatest.

i. $0.6, 0.\overline{6}, 0.66$ **j.** $\frac{1}{2}, 0.\overline{5}, \frac{1}{3}, 0.3, 0.0\overline{6}$

1. There are 20 students in the algebra class. One-half of the students
(5) were boys. One-half of the boys wore T-shirts. How many boys wore
T-shirts to the algebra class? Boys wearing T-shirts were what fraction
of the students in the class?

2. The big room measured 19 ft 10 in. by 22 ft 1 in. About how many square
(8, 17) feet of carpet are needed to re-carpet the big room?

* 3. *Evaluate* The four pencils in Julio's desk have these lengths: $7\frac{1}{2}$ in., $5\frac{3}{4}$ in.,
(7, 13) 5 in. $2\frac{3}{4}$ in. What is the average (mean) length of the four pencils?

4. *Analyze* Find the area of the triangle. (Dimensions are in mm.)
(20)

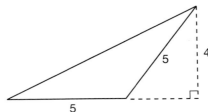

* 5. *Classify* Classify the triangle in problem **4** by angles and sides. If each
(20) acute angle measures $26\frac{1}{2}°$, then what is the measure of the obtuse
angle?

6. The $80.00 DVD player is on sale for 25% off. Find 25% of $80.
(12)

* 7. What is the ratio of a side length to the perimeter of a square?
(29)

Simplify.

8. $\dfrac{1}{4} + \dfrac{2}{3} \cdot \dfrac{1}{2}$
(13, 22)

9. $\dfrac{7}{8} - \dfrac{1}{8} \div \dfrac{1}{4}$
(13, 22)

* 10. $\dfrac{1.23 + 0.4 + 5}{0.3}$
(24, 25)

11. $4^0 + 4 \cdot 5 \cdot 2 - 1^3$
(21, 27)

* 12. $0.8 - \{0.6 - [0.4 - (0.2)^2]\}$
(21, 24)

* 13. **a.** Write $\dfrac{2}{9}$ as a decimal and a percent.
(30)
    b. Which of the three forms is the most convenient for finding $\dfrac{2}{9}$ of $27?

* 14. Write 70% as a decimal and a fraction.
(11, 12)

* 15. Plot these points and find the area and perimeter of the rectangle with
(8, Inv. 1) these vertices: (5, 11), (5, 1), (−2, 1), (−2, 11).

16. **a.** Expand: $5(x + y + z)$
(14)
    b. Factor: $wx + wy$

* 17. Simplify: $\dfrac{x^5 \cdot x^2}{x}$
(15)

* 18. Write 28,000,000 in scientific notation.
(28)

19. Write the prime factorization of 1,000,000 using exponents.
(9, 15)

* 20. Simplify and compare: $\sqrt{169} \bigcirc (3.7)^2$
(15)

21. Which property is illustrated below?
(1)

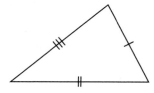

a. $\dfrac{1}{2} \cdot \dfrac{1}{4} = \dfrac{1}{4} \cdot \dfrac{1}{2}$ **b.** $\dfrac{1}{3} + \left(\dfrac{2}{3} + \dfrac{1}{2} \right) = \left(\dfrac{1}{3} + \dfrac{2}{3} \right) + \dfrac{1}{2}$

22. Arrange in order from least to greatest.
(12)
$$0.3, \dfrac{2}{5}, 35\%, 0, 1$$

23. Classify this triangle by side length and angle measure:
(20)

Solve by inspection.

* 24. $2.2 = 2m + 0.2$
(14, 24)

25. $\dfrac{4}{5} = 1 - r$
(10, 14)

26. $\dfrac{7}{20} = \dfrac{7}{10}h$
(10, 14)

27. $\dfrac{t + 3}{4} = 5$
(10, 14)

Evaluate.

28. Find $b^2 - 4ac$ when $a = 4, b = 9, c = 5$.
(14, 20)

29. **Connections** Suspended objects such as apples in a tree or hailstones
(14, 25) in a cloud have potential energy equal to their mass (m) times the acceleration of gravity (g) times their height (h). For the equation $E = mgh$, find E when $m = 2$, $g = 9.8$, and $h = 5$.

* 30. **Analyze** In the figure, $\overline{AC}$ is 1 unit and $\overline{BC}$
(Inv. 1, is 2 units. Use the Pythagorean Theorem to
Inv. 2) find $\overline{AB}$.

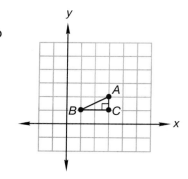

Focus on

• Classifying Quadrilaterals

Recall that a four-sided polygon is called a quadrilateral. Refer to the quadrilaterals shown below to answer the problems that follow.

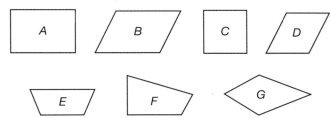

1. Which figures have two pairs of parallel sides?

2. Which figure has just one pair of parallel sides?

3. Which figures have no pairs of parallel sides?

One way to classify quadrilaterals is by the number of pairs of parallel sides. A quadrilateral with two pairs of parallel sides is called a **parallelogram.** Here we show four parallelograms. Notice that the sides that are parallel are the same length.

4. Which of the figures *A–G* are parallelograms?

A quadrilateral with one pair of parallel sides is a **trapezoid.** The figures shown below are trapezoids. Identify each pair of parallel sides. Notice that the parallel sides of a trapezoid are not the same length.

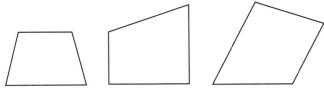

Thinking Skill

Explain

Does the length of the sides affect whether or not the sides are parallel? Explain.

5. Which of the figures *A–G* is a trapezoid?

If the non-parallel sides of a trapezoid are equal in length, then the trapezoid is called an **isosceles trapezoid.**

A quadrilateral with no pairs of parallel sides is called a **trapezium.** Below we show two examples. The figure on the right is called a **kite.** A kite has two pairs of adjacent sides of equal length.

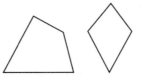

6. Which of the figures *A–G* are trapeziums? Which is a kite?

We have sorted figures *A–G* into three categories of quadrilaterals.

Quadrilaterals

Parallelograms (A, B, C, D)	Trapezoids (E)	Trapeziums (F, G)

We can further classify some types of parallelograms.

7. Which of the figures *A–G* have four right angles?

A quadrilateral with four right angles is a special kind of parallelogram called a **rectangle.**

8. Which of the figures *A–G* have four sides of equal length?

A quadrilateral with four equal-length sides is a specific kind of parallelogram called a **rhombus.**

9. Which figure in *A–G* is both a rectangle and a rhombus?

A figure that is both a rectangle and a rhombus is a **square.** This Venn Diagram illustrates the relationship among parallelograms. A **Venn diagram** is a diagram that uses overlapping circles to represent the relationships among sets.

Thinking Skill

Classify

Nita's polygon had four sides of equal length and no right angles. Sketch a quadrilateral Nita might have drawn. Use the Venn diagram to classify it.

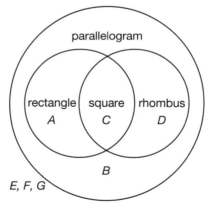

Refer to the Venn diagram and to figures *A–G* to answer questions **10–14.**

10. Which figure is a parallelogram but not a rectangle or a rhombus?

11. Which figure is a parallelogram and a rhombus but not a rectangle?

12. *Conclude* Which figure is a parallelogram and a rectangle but not a rhombus?

13. Which figure is a parallelogram, a rectangle, and a rhombus?

14. *Explain* Which figures on the Venn diagram are not parallelograms? Explain why the figures are not parallelograms.

One special type of rectangle identified by ancient Greeks and found frequently in art and architecture is the "Golden Rectangle." The ratio of the sides of a golden rectangle is not a rational number. If the width is 1, then the length is $1 + \frac{1}{2}(\sqrt{5} - 1)$. Here is an example of a golden rectangle:

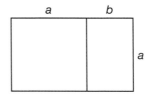

If we draw a segment through a golden rectangle to form a square in the rectangle, we also form a smaller golden rectangle.

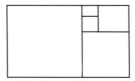

That is, the original rectangle and the smaller rectangle are similar. Therefore, the process of dividing can continue indefinitely.

All the rectangles that are formed are similar.

15. With a calculator, find the length of the largest golden rectangle shown above to the nearest tenth.

16. What are some whole number pairs of side lengths that form rectangles that approximate a golden rectangle?

A student made a model of a rectangle out of straws and pipe cleaners (Figure J). Then the student shifted the sides so that two angles became obtuse and two angles became acute (Figure K).

 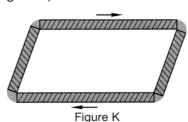

Figure J   Figure K

Refer to figures J and K from the previous page to answer problems **17–27.**

17. Is figure K a rectangle? Is figure K a parallelogram?

18. Did the lengths of the sides change when the angles changed?

19. Does the perimeter of figure K equal the perimeter of figure J?

20. Does the area of figure K equal the area of figure J?

21. As the sides of figure K are shifted and the obtuse angles become large and the acute angles smaller, in what way does the area change?

22. The four angles of figure J total how many degrees? How do you know?

23. What is the sum of the measures of the four angles of figure K?

24. How many degrees is a full turn?

25. If you trace with your eraser figures J or K, you will find your eraser heading in a direction opposite to the direction it started after tracing two angles. This means your eraser has finished half of a turn. How many degrees is half a turn?

26. In figure K, the measure of an obtuse angle and an acute angle together total how many degrees?

27. If an obtuse angle of figure K measures 100°, what does an acute angle measure?

28. Figure *ABCD* is a parallelogram. Angle *A* measures 70°. What does ∠*B* measure?

29. In parallelogram *ABCD,* ∠*A* measures 70°, what does ∠*D* measure?

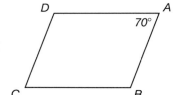

30. Use the information you found in questions 28 and 29 to find the measure of ∠*C*.

A figure has **reflective symmetry** if it can be divided in half by a line so that the halves are mirror images of each other. The triangle in the illustration below has reflective symmetry.

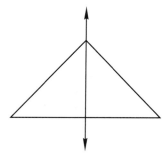

The line that divides the triangle into mirror images is a **line of symmetry.**

Some quadrilaterals have one or more lines of symmetry. Trace figures *A, B, C,* and *D,* from the beginning of the lesson and sketch their lines of symmetry.

31. Figure *A* has how many lines of symmetry?

32. Figure *B* has how many lines of symmetry?

33. Figure *C* has how many lines of symmetry?

34. Figure *D* has how many lines of symmetry?

35. Does a line of symmetry for figure *A* pass through sides or angles?

36. Does a line of symmetry for figure *D* pass through sides or angles?

37. Does a line of symmetry for figure *C* pass through sides or angles?

A figure has **rotational symmetry** if, as the figure turns, its original image reappears in less than a full turn. For example, as we rotate a square, its original image reappears every quarter (90°) turn.

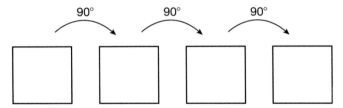

Since the original image appears four times in a full turn we say that a square has rotational symmetry of order four.

Predict Refer to figures *A–G* and state whether the following figures have rotational symmetry. If a figure has rotational symmetry, state its order.

38. Figure *A*   **39.** Figure *B*   **40.** Figure *D*

41. Figure *E*   **42.** Figure *F*   **43.** Figure *G*

- **Adding Integers**
- **Collecting Like Terms**

Power Up | *Building Power*

facts | Power Up G

mental math | **a. Number Sense:** Seventeen people have $200 each. How much money do they have altogether?

b. Fractions: $\frac{5}{4} = \frac{25}{x}$

c. Fractional Parts: 75% of $200

d. Measurement: How long is the paperclip? If 10 paperclips are placed end to end, how far would they reach?

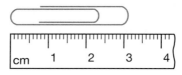

e. Rate: If the space shuttle orbits earth 16 times a day, how many times does it orbit in $2\frac{1}{2}$ days?

f. Geometry: Find *x*

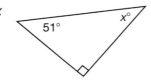

51°

x°

g. Powers/Roots: Estimate $\sqrt{99} - \sqrt{80}$.

h. Calculation: $45 - 40$, $\times 2$, $+ 1$, $\times 6$, $- 2$, $\sqrt{\ }$, $\times 3$, $+ 1$, $\sqrt{\ }$, $- 10$

problem solving | A rectangle has a length of 10 meters and a width of 8 meters. A second rectangle has a length of 6 meters and a width of 4 meters, and overlaps the first rectangle as shown. What is the difference between the areas of the two non-overlapping regions of the two rectangles?

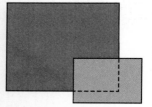

(**Understand**) We have been given the dimensions of two rectangles that overlap. We have not been given the dimensions of the overlapping area, but have been asked to find the difference between the two **non-overlapping** regions of the two rectangles.

(*Plan*) We will use the equation for the area of a rectangle ($A = l \times w$). We will need to subtract the overlapping area from the area of each rectangle, and then subtract the remaining areas of each. We will make it simpler by drawing diagrams of the extremes to see if there is any sort of pattern.

Solve The two extremes of this problem are:

1. None of the smaller rectangle overlaps the larger rectangle.
2. All of the smaller rectangle overlaps the larger rectangle.

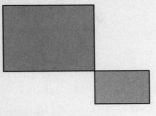

$$(10 \times 8) - (6 \times 4) =$$
$$80 - 24 = 56 \text{ sq. m}$$

$$[(10 \times 8) - (6 \times 4)] - [0] =$$
$$[80 - 24] - [0] =$$
$$56 - 0 = 56 \text{ sq. m}$$

In both cases, the difference of the areas of the non-overlapping regions is 56 sq. m.

Check To verify that the solution is always going to be 56 sq. m, we can create a simple example of an area of overlap:

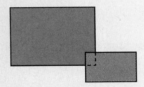

$$[(10 \times 8) - (1 \times 1)] - [(6 \times 4) - (1 \times 1)] =$$
$$[80 - 1] - [24 - 1] =$$
$$79 - 23 = 56 \text{ sq. m}$$

As the smaller rectangle slides into the larger rectangle, the same amount of area is subtracted from both the larger and the smaller rectangles, resulting in no difference of their non-overlapping areas.

New Concepts   *Increasing Knowledge*

adding integers

Recall that integers include the counting numbers (1, 2, 3, ...), the opposites of the counting numbers (..., −3, −2, −1), and zero. In this lesson we show examples of adding integers.

We see integers on thermometers. On the Celsius scale, 0° marks the freezing temperature of water. Negative numbers indicate temperatures below freezing. Positive numbers indicate temperatures above freezing.

Example 1

If the temperature in the morning is −4 degrees but increases 10 degrees by noon, then what is the noontime temperature?

Solution

Counting up 10 degrees from −4, we find that the noontime temperature is **6 degrees above zero.**

$$-4 + 10 = 6$$

Another application of positive and negative numbers is debt and credit. A debt is recorded as a negative number and a credit as a positive number.

Example 2

Sam has no money and owes a friend $5. If Sam earns $15 washing cars, then he can repay his friend and have money left over. Write an equation for this situation and find how much money Sam has after he repays his friend.

Solution

We show Sam's debt as −5 and Sam's earnings as +15.

$$-5 + 15 = 10$$

Sam has **$10** after he pays his friend.

We can show the addition of integers on a number line. We start at zero. Positive integers are indicated by arrows to the right and negative integers by arrows to the left. Here we show three examples.

A. $(-3) + (+5) = (+2)$

B. $(+3) + (-5) = (-2)$

C. $(-5) + (-3) = -8$

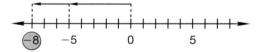

Look at number sentences **A** and **B** above and answer these questions:

- Do the addends have the same sign or opposite signs?
- How can we determine whether the sign of the sum will be positive or negative?
- How can we determine the number of the sum?

Now look at number sentence **C** and answer the same three questions. Then try completing the exercises in example 3 before reading the solution.

Example 3

Find each sum.

    a. $(-6) + (+2)$ b. $(+6) + (-2)$

    c. $(-6) + (-2)$ d. $(-6) + (+6)$

Solution

Math Language

Recall that the **absolute value** of a number is its distance from 0 on the number line.

a. The signs of the addends are not the same. The number with the greater absolute value is negative, so the sum is negative. The difference of the absolute values is 4, so the sum is **−4.**

b. The signs are different. The number with the greater absolute value is positive. The difference of the absolute values is 4. The sum is positive **4.**

c. The signs are the same and negative, so the sum is **−8.**

d. The numbers are opposites. The sum of opposites is **0.**

Math Language

Recall that the **subtrahend** is the number being subtracted.

Recall that the opposite of a number is also called the **additive inverse.** We can change any subtraction problem into an addition problem by adding the opposite of the subtrahend. For example, instead of subtracting 2 from 6, we can add −2 to 6.

$$6 - 2 = 4 \qquad \text{Subtract 2 from 6.}$$

$$6 + (-2) = 4 \qquad \text{Or add } -2 \text{ to 6.}$$

We use this concept to change subtraction of integers to addition of integers.

Addition with Two Integers

> To find the sum of addends with different signs:
>
> **1.** Subtract the absolute values of the addends.
>
> **2.** Take the sign of the addend with the greater absolute value.
>
> To find the sum of addends with the same sign:
>
> **1.** Add the absolute values of the addends.
>
> **2.** Take the sign of the addends.

collecting like terms

An algebraic **term** is an expression that includes a positive or negative number and may include one or more variables.

Here are some examples of terms:

$$3y, \ x, \ -w^2, \ 3x^2y, \ -2$$

The number part of a term is called its **numerical coefficient** or just coefficient for short. Each of the first four terms has a coefficient that is a positive or negative number.

        The coefficient of 3y is +3.

        The numerical coefficient of x is +1.

        The numerical coefficient of $-w^2$ is −1.

        The coefficient of $3x^2y$ is +3.

Notice that if the coefficient is 1, it is not written. It is understood to be 1. A term without a variable, such as −2, is called a constant.

Example 4

Identify the numerical coefficient of each of these terms.

    a. a^2b **b.** $3y$ **c.** $-x^2$

Solution

    a. $+1$

    b. $+3$

    c. -1

We can identify like terms by looking at the variables. **Like terms** have identical variable parts, including exponents. The variables may appear in any order.

Like terms: $-5xy$ and yx

Unlike terms: $2xy$ and $-5x^2y$

Discuss Why are $-5xy$ and yx like terms? Why are $2xy$ and $-5x^2y$ unlike terms?

We combine or "collect" like terms by adding their numerical coefficients. The variables do not change.

Step:	Justification:
$-5xy + yx$	Given expression
$-5xy + 1xy$	Express $+yx$ as $+1xy$
$-4xy$	Combined terms $(-5 + 1 = -4)$

Example 5

Simplify this expression by collecting like terms.

$3x + 2 - x + 3$

Solution

Thinking Skill

Justify

The Commutative Property of Addition does not apply to subtraction. Why can we use the Commutative Property of Addition to rearrange the terms in this expression that includes subtraction?

Using the commutative property, we rearrange the terms. (The sign moves with the term.)

Step:	Justification:
$3x + 2 - x + 3$	Given expression
$3x - x + 2 + 3$	Commutative Property of Addition

We combine the terms $3x - x$ (which is $3x - 1x$). We also combine the $+2$ and $+3$.

$2x + 5$	Added

Example 6

What is the perimeter of this rectangle?

3

2x

Solution

We add the lengths of the four sides.

Step:	Justification:
$P = 2x + 3 + 2x + 3$	Added four sides
$= 2x + 2x + 3 + 3$	Commutative Property of Addition
$= 4x + 6$	Collected like terms

Practice Set

Simplify:

a. $(-12) + (-3)$

b. $(-12) + (+3)$

c. $(+12) + (-3)$

d. $(-3) + (+12)$

e. *Model* Choose any of the problems **a–d.** Show the operation on a number line.

f. At 6:00 a.m. the temperature was $-12°C$. By noon the temperature increased 8 degrees. Write an equation for this situation and find the temperature at noon.

g. The hikers started on the desert floor, 182 feet below sea level. After an hour of hiking they had climbed 1,018 ft. Write an equation for the situation and find the hikers' elevation after an hour of hiking.

Collect like terms:

h. $3x + 2xy + xy - x$

i. $6x^2 - x + 2x - 1$

j. $2a^3 + 3b - a^3 - 4b$

k. $x + y - 1 - x + y + 1$

l. $P = L + W + L + W$

m. $P = s + s + s + s$

n. *Connect* What is the perimeter of this rectangle?

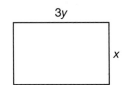

3y

x

Analyze   Solve problems **1–3**.

1.
(25)
The Acme Scrap Metal Company pays 45¢ per pound for aluminum cans. Troy wants to sell 54.2 lbs of cans. How much money should he receive?

* 2.
(23)
A recipe for chili calls for $4\frac{1}{4}$ lbs of beans. Azza wants to make $2\frac{1}{2}$ times the recipe. How many pounds of beans will she need?

* 3.
(24)
The atmosphere that we live in is 78.08 percent nitrogen and 20.95 percent oxygen. What percentage of the atmosphere is made up of nitrogen and oxygen?

* 4.
(26)
Marcia developed a roll of film and received 4-inch by 6-inch prints. She wants to have one picture enlarged to an 8-inch by 12-inch print. What kind of transformation is this? What is the scale factor?

* 5.
(28)
In 1875, an estimated twelve-and-a-half trillion locusts appeared in Nebraska. Write the number of locusts in scientific notation.

6.
(29)
Patrick has 7 tetras, 4 goldfish, and 22 snails in his fish tank. What is the ratio of snails to goldfish in Patrick's fish tank?

Simplify.

7.
(21)
$4 + 10 \div 5 \times 2$

8.
(15, 21)
$\dfrac{5 + 3 \cdot 10}{5^2 - 10}$

9.
(13, 22)
$\dfrac{3}{4} - \dfrac{1}{6} \div \dfrac{1}{3}$

10.
(13, 22)
$\dfrac{2}{3} - \dfrac{1}{12} \cdot \dfrac{1}{2}$

11.
(15)
Simplify and compare: $\sqrt{9} + \sqrt{16} \bigcirc \sqrt{9 + 16}$

12.
(21)
a. Expand: $4(x + 2y + 3)$

b. Factor: $6x + 3y$

13.
(2)
Rearrange the numbers to make multiplication easier. Then simplify. What properties did you use?

$$2 \times (5 \times 27)$$

* 14.
(31)
Conclude   The addition of the absolute values of two negative integers will always be

A a negative integer.   **B** a positive integer.

C zero.   **D** any positive number.

Evaluate:

15.
(14, 31)
Find $a + b + c$ when $a = 5$, $b = -2$, and $c = -4$.

* 16.
(14, 25)
Evaluate   The weight of an object on Earth is equal to its mass times the acceleration of gravity. For the equation $W = mg$, find W when $m = 5$ and $g = 9.8$.

17. **Classify** Classify this triangle by angles
(20) and by sides. Then find the area of the
triangle.

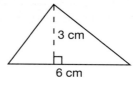
3 cm
6 cm

18. Collect like terms: $3x^2 + 2x + 3x - 1$
(31)

Simplify:

19. $47 - 79$
(2)

20. $\dfrac{x^6 \cdot x^2}{x \cdot x^3}$
(27)

21. $6\dfrac{1}{4} - 3\dfrac{3}{8}$
(13)

22. $6 - 3.67$
(24)

* 23. **a.** $(-3) + (-12)$ **b.** $(-12) + (-3)$
(14)

24. It costs $10 plus $8 per hour to rent a canoe. Brandon and Erin have
(14) $50. Solve the following equation to find how many hours they can rent
a canoe.

$$8h + 10 = 50$$

* 25. Express $\dfrac{3}{11}$ as a decimal number.
(30)

26. Use prime factorization to reduce $\dfrac{27}{45}$.
(10)

* 27. Julie wrote a 500-word essay. Following the suggestions made by her
(31) English teacher, Julie deleted a 25-word long paragraph and added
a new paragraph 17 words long. How many words long is Julie's final
essay?

28. Name each property illustrated.
(2)
    a. $62 \times 0 = 0$ **b.** $27 \times 1 = 27$

* 29. **a.** Find the missing exponent: $x^8 \cdot x^2 \cdot x^{\square} = x^{15}$
(27)
    b. Simplify: $2^3 + 2^2 + 2^1 + 2^0$

30. Loose birdseed is priced at $4.29 per pound. Raoul scoops out a small
(25) bag for his birdfeeder. It weighs 1.3 pounds. How much will he pay for
the birdseed (rounded to the nearest cent)?

• Probability

facts | Power Up G

mental math

a. **Number Sense:** 110×200

b. **Algebra:** $15 + x = 60$

c. **Measurement:** How long is this rectangle?

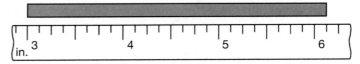

d. **Rate:** At a rate of 40 words per minute, how many words can Valerie type in 20 minutes?

e. **Scientific Notation:** Write 280,000,000 km in scientific notation.

f. **Geometry:** In this triangle, x must be between what two numbers?

g. **Powers/Roots:** List the perfect squares from 49 to 100.

h. **Calculation:** $25 \times 4, \div 2, - 1, \sqrt{}, \times 7, + 1, \times 2, - 1, \div 11, \sqrt{}, \times 7$

problem solving | The diameter of a penny is $\frac{3}{4}$ in. How many pennies placed side by side would it take to make a row of pennies $\frac{3}{4}$ ft long? How many would it take to measure $\frac{3}{4}$ of a yard?

Probability is the likelihood that a particular event will occur. To represent the probability of event A we use the notation $P(A)$. We express probability as a number ranging from zero to one.

Range of Probability

$$0 \qquad \frac{1}{2} \qquad 1$$

Impossible   Unlikely   Likely   Certain

Math Language
Recall that a **ratio** is a comparison of two numbers by division.

Probability is a ratio of favorable outcomes to possible outcomes.

$$P(\text{Event}) = \frac{\text{number of favorable outcomes}}{\text{number of possible outcomes}}$$

Example 1

The spinner shown at right is spun once. What is the probability the spinner will stop

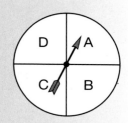

    a. in sector *A*?

    b. in sector *A* or *B*?

    c. in sector *A*, *B*, *C*, or *D*?

Solution

The face of the spinner is divided into fourths. There are four equally likely outcomes.

    a. One outcome, *A*, is favorable.

$$P(A) = \frac{1}{4}$$

    b. To find the probability of *A* or *B* we add the probabilities of each.

$$P(A \text{ or } B) = P(A) + P(B)$$

$$= \frac{1}{4} + \frac{1}{4}$$

$$= \frac{1}{2}$$

    c. The probability of *A*, *B*, *C*, or *D* is **1** because every spin results in one of the listed outcomes.

$$P(A, B, C, \text{ or } D) = \frac{4}{4} = 1$$

Probability may be expressed as a fraction, as a decimal, or as a percent. We often use the word **chance** when expressing probability in percent form. In example 1 *P(A)* is 0.25, which means that the spinner has a 25% chance of stopping in sector *A*.

Odds is the ratio of the number of favorable to unfavorable outcomes and is often expressed with a colon. We will calculate the odds of the spinner stopping in sector *A* in example 2.

Example 2

If the spinner in example 1 is spun once, what are the odds the spinner will stop in sector *A*?

Solution

Of the four equally likely outcomes, one is *A* and three are not *A*. So the odds the spinner will stop in sector *A* are **1:3.**

The **sample space** of an experiment is the collection of all possible outcomes. We can record the sample space in a variety of ways, including a list or a table. For example, if a coin is tossed twice, there are four possible outcomes.

In the list and table at right we use H for "heads" and T for "tails."

Sample space = {HH, HT, TH, TT}

Knowing the sample space for an experiment helps us calculate probabilities. Referring to the sample space above, can you find the probability of getting heads at least once in two tosses of a coin?

Sample Space

		First Toss	
		H	**T**
Second Toss	**H**	HH	TH
	T	HT	TT

Formulate If Javier spins the spinner in example 1 twice, what is the sample space for the experiment?

The **complement** of an event is the set of outcomes in the sample space that are not included in the event. The complement of getting heads at least once in two tosses is not getting heads at least once. The probability of an event and the probability of its complement total one, because it is certain that an event either will occur or will not occur. If we know the probability of an event, we can find the probability of its complement by subtracting the probability from 1.

Example 3

The spinner in example 1 is spun twice.

  a. **Predict: Is the probability of spinning A more likely with one spin or with two spins?**

  b. **Find the probability of getting A at least once.**

  c. **Find the probability of not getting A at least once.**

Solution

  a. We guess that we will be **more likely to get A with two spins than with one spin.**

  b. To find the probability of A we first determine the sample space. We can make a table or a tree diagram. For a two-part experiment, a table is convenient.

		1st Spin			
		A	*B*	*C*	*D*
2nd Spin	*A*	AA	BA	CA	DA
	B	AB	BB	CB	DB
	C	AC	BC	CC	DC
	D	AD	BD	CD	DD

Seven of the 16 outcomes result in A at least once, so $P(A) = \frac{7}{16}$. Notice that $\frac{7}{16}$ is greater than $P(A)$ with one spin, which is $\frac{1}{4}$.

  c. The probability of not A is the complement of the probability of A. One way to find $P(\text{not } A)$ is by subtracting $P(A)$ from 1.

$$P(not\ A) = 1 - P(A)$$

$$P(not\ A) = 1 - \frac{7}{16}$$

$$P(not\ A) = \frac{9}{16}$$

The answer is reasonable because 9 of the 16 possible outcomes in the table do not include A.

Example 4

Nathan flips a coin three times.

  a. **Predict: Which is more likely, that he will get heads at least once or that he will not get heads at least once.**

  b. **Find the probability of getting heads at least once.**

  c. **Find the probability of not getting heads at least once.**

Solution

  a. The probability of getting heads in one toss is $\frac{1}{2}$. Since the opportunity to get heads at least once increases with more tosses, **we predict that getting heads at least once is more likely than not getting heads.**

  b. We use a tree diagram to find the sample space. We find eight equally likely possible outcomes.

First Toss	Second Toss	Third Toss	Outcome
H	H	H	HHH
		T	HHT
	T	H	HTH
		T	HTT
T	H	H	THH
		T	THT
	T	H	TTH
		T	TTT

Since 7 of the 8 outcomes have heads at least once, $P(\text{H}) = \frac{7}{8}$.

  c. The probability of not getting heads is $\frac{1}{8}$.

We distinguish between the **theoretical probability,** which is found by analyzing a situation, and **experimental probability,** which is determined statistically. Experimental probability is the ratio of the number of times an event occurs to the number of trials. For example, if a basketball player makes 80 free throws in 100 attempts, then the experimental probability of the player making a free throw is $\frac{80}{100}$ or $\frac{4}{5}$. The player might be described as an 80% free-throw shooter.

In baseball and softball, a player's batting average is the experimental probability, expressed as a decimal, of the player getting a hit.

Example 5

Emily is a softball player who has 21 hits in 60 at-bats. Express the probability that Emily will get a hit in her next at-bat as a decimal number with three decimal places. Emily's coach needs to decide whether to have Emily bat or to use a substitute hitter whose batting average is .300 (batting averages are usually expressed without a 0 in the ones place). How would you advise the coach?

Solution

The ratio of hits to at-bats is 21 to 60. To express the ratio as a decimal we divide 21 by 60.

$$21 \div 60 = 0.35$$

We write the probability (batting average) with three decimal places: **.350.** Since Emily's batting average is greater than the batting average of the substitute hitter, **we advise the coach to have Emily bat.**

Practice Set

a. A number cube is rolled once. What is the probability of rolling an even number? Express the probability as a fraction and as a decimal.

b. A number cube is rolled once. What are the odds of rolling a 6?

c. **Justify** If the chance of rain tomorrow is 20%, then is it more likely to rain or not rain? What is the chance it will not rain tomorrow? Explain your reasoning.

d. A coin is flipped three times. What is the sample space of the experiment?

e. Referring to the experiment in **d,** what is the probability of getting heads exactly twice? What is the probability of not getting heads exactly twice?

f. Quinn runs a sandwich shop. Since she added turkey melt to the menu, 36 out of 120 customers have ordered the new sandwich. What is the probability that the next customer will order a turkey melt? If Quinn has 50 customers for lunch, about how many are likely to order a turkey melt?

g. State whether the probabilities found in problems **e** and **f** are examples of theoretical probability or experimental probability.

Written Practice   *Strengthening Concepts*

1. If postage for an envelope costs 39¢ for the first ounce and 24¢
(3, 4) for each additional ounce, then what is the postage for a 12 ounce envelope?

2. An author writes an average of eight pages per day. At this rate, how
(4) long would it take to write a 384 page book?

3. Pete saw that he was 200 yards from shore when he dove out of the
(4) boat. It took him 49 seconds to swim the first 50 yards. At that rate, how much longer will it take to reach the shore?

* 4. **Evaluate** Alonzo chose a name from a box, read the name, then
(32) replaced it and mixed the names. He repeated this experiment several times. Out of the 20 names he chose, 12 were eighth-graders.

    a. What is the experimental probability of choosing the name of an eighth-grader based on Alonzo's experiments?

    b. If half of the names in the box are eighth-graders, what is the theoretical probability of choosing the name of an eighth-grader?

* 5. **Analyze** There are 18 marbles in Molly's bag. Two of the marbles are
(32) yellow. If Molly chooses one marble at random, find the theoretical probability that she will choose a yellow marble. What is the probability that she will not choose a yellow marble?

* 6. **Connect** Use the figure at right to answer **a**
(8) and **b.**

    a. What is the area of this triangle?

    b. What is the perimeter of this triangle?

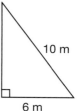

10 m

6 m

7. Estimate: $\dfrac{610 + 195}{398}$
(17)

Simplify.

8. $\dfrac{1}{5} - \dfrac{1}{3} \cdot \dfrac{4}{5}$
(13, 23)

9. $\dfrac{7}{9} - \dfrac{1}{3} \div \dfrac{3}{4}$
(13, 23)

10. $2.05 - (3.1)(0.1)$
(24, 25)

11. $7^2 - 6^1 + 5^0$
(15, 27)

12 Simplify and compare:
(15, 21)

    a. $\sqrt{3^2 + 4^2} \bigcirc \sqrt{3^2} + \sqrt{4^2}$

    b. $|-1 + 5| \bigcirc |-1| + |5|$

* 13. **Represent** Write $\dfrac{5}{9}$ as a decimal and a percent.
(30)

* 14. **a.** Write 60% as a decimal and a reduced fraction.
(11, 16)

    b. Which form would you find most convenient to use to find 60% of $15? Explain.

15. These points are vertices of a rectangle: $(3, -5)$, $(7, -5)$, $(7, 2)$. Plot
(8, Inv. 1) these points and locate the fourth vertex. Then find the area and perimeter of the rectangle.

16. **a.** Expand: $3(x + 12)$
(21)

    b. Factor: $x^2 + 6x$

* 17. Write 6.02×10^{10} in standard notation.
(28)

Generalize Simplify.

* 18. **a.** $(m^4)^3$ **b.** $\dfrac{r^2 r^5}{r^3}$
(27)

19. Predict **a.** On a number line, graph the points 2 and 6. Then graph the
(1, 3) mean of 2 and 6.

b. On a number line, graph the points -5 and 1. Use the number line to predict the mean of -5 and 1.

20. What numbers are members of both the set of whole numbers and the
(1) set of integers?

21. Rearrange the addends in this addition problem to make the calculation
(2) easier. Then simplify. What properties did you use?
$$23 + (7 + 18)$$

22. Find the factors of 2100 using a factor tree. Then write the prime
(9, 15) factorization of 2100 using exponents.

23. The spinner shown is spun twice. What is the
(32) sample space of the experiment? What is the
probability of spinning A at least once?

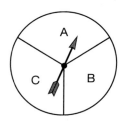

24. Jon and Meg play a game with the spinner from problem **23.** They take
(32) turns spinning the spinner twice. Jon earns a point if a player spins at
least one A in a turn. Meg earns a point if a player spins "doubles" (AA,
BB, or CC). Predict who you think will win more often if they play several
times, and justify your prediction.

25. Order these numbers from least to greatest.
(12)
$$0, 2.3, -3.2, 4.5, -5.4$$

26. When a page of print is viewed in a mirror the words appear to be
(26) printed backwards. The image in the mirror is an example of which type
of transformation?

27. Solve by inspection: $3x + 20 = 80$
(14)

Evaluate.

28. Find $\sqrt{b^2 - 4ac}$ when $a = 2$, $b = 5$, $c = 2$.
(14, 15)

* 29. Evaluate The kinetic energy of an object is given by the equation
(14, 15) $E = \frac{1}{2}mv^2$. Find E when $m = 2$, $v = 3$.

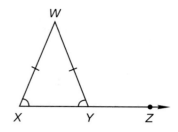

* 30. (18, 20) **Analyze** Triangle ∠WXY is isosceles. If
m∠W is 40°, then what is the measure of the
following angles?

   a. ∠X

   b. ∠WYX

   c. ∠WYZ

Early Finishers

Real-World Application

To award door prizes at the winter carnival, the Anderson Intermediate
School principal drew the names of five students from a hat. Only seventh
and eighth graders go to Anderson Intermediate School.

   a. One possible outcome is the principal drawing five seventh-grade
students. What are the other possible outcomes? (Order does not
matter, so 7-7-7-8-7 is the same as 8-7-7-7-7.)

   b. If sixty percent of the names drawn were eighth graders, how many
names drawn were seventh graders?

• Subtracting Integers

facts | Power Up G

mental math

a. **Number Sense:** 21×400

b. **Fractional Parts:** $33\frac{1}{3}\%$ of $390

c. **Measurement:** How long is this strip of paper?

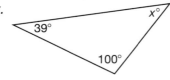

d. **Rate:** Albert threw only 64 pitches in 7 innings. About how many pitches per inning did Albert throw?

e. **Scientific Notation:** Write twelve million in scientific notation.

f. **Geometry:** Find x.

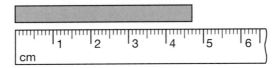

g. **Estimation:** Sam bought 7 at $2.97 each plus 4 at $9.97 each. About how much did he spend?

h. **Calculation:** $32 \div 4$, $+ 1$, $\sqrt{\ }$, $+ 5$, square it, $+ 1$, $+ 10$, $\div 3$, $\sqrt{\ }$, $\times 7$

problem solving | Andrew, Amelia, Beth, and Freddy decided to take pictures of two people at a time. How many different pairs of people can be made from these four people?

Recall that we use a method called algebraic addition to subtract integers. Instead of subtracting a number, we add its opposite. Consider the following examples.

Visit www. SaxonPublishers. com/ActivitiesC3 *for a graphing calculator activity.*

Example 1

Change each subtraction to addition and find the sum.

a. $(-3) - (+2)$

b. $(-3) - (-2)$

Solution

We change the subtraction sign to an addition sign and reverse the sign of the subtrahend (the number being subtracted). Then we add.

Expression a.	Justification	Expression b.
$(-3) - (+2)$	Given	$(-3) - (-2)$
$(-3) + (-2)$	Added opposite	$(-3) + (+2)$
−5	Simplified	**−1**

Thinking Skill

Verify

Use a number line to prove that $(-3) - (+2)$ and $(-3) + (-2)$ are equal.

Example 2

Jocelyn has a checking balance of $1286. In the mail she receives a rebate check for $25 and a utility bill for $128. She deposits the rebate and writes a check for the bill. Write an equation with integers for the situation and find her checking balance after the transactions.

Solution

To find the balance b we add the rebate to and subtract the utility bill from her current balance.

$$b = 1286 + 25 - 128$$

Instead of subtracting 128, we may add negative 128.

$$b = 1286 + 25 + (-128)$$

Both methods result in a balance of **$1183.**

Addition and Subtraction with Two Integers

Operation	Rule
+	To find the sum of addends with different signs: **1.** Subtract the absolute values of the addends. **2.** Take the sign of the addend with the greater absolute value. To find the sum of addends with the same sign: **1.** Add the absolute values of the addends. **2.** Take the sign of the addends.
−	Instead of subtracting a number, add its opposite.

Practice Set

Simplify:

a. $(-12) - (-3)$ 　　　　　　　**b.** $(-12) - (+3)$

c. $(-3) - (-12)$ 　　　　　　　**d.** $(-3) - (+12)$

e. *Model* Choose any of the problems **a–d.** Show the operation on a number line.

Lesson 33   219

f. Victor owed $386. His creditor forgave (subtracted) $100 of the debt. Write an equation for the situation and find out how much Victor still owes.

Written Practice | *Strengthening Concepts*

* 1. (31) **Analyze** At midnight, the temperature was −5°F. By 10:00 a.m. the next morning it had increased by 17°. What was the temperature at 10:00 a.m.?

A 22°F   **B** 12°F   **C** 10°F   **D** 8°F

* 2. (28) Write the approximate distance from Earth to Venus, 25,000,000 miles, in scientific notation.

3. (25) Enrique bought 6 bags of grass seed to reseed his lawn. Each bag cost $13.89. How much did Enrique spend on grass seed?

4. (11) In the city of Austin, Texas, the state sales tax is 6.25%, the city sales tax is 1%, and the transportation board sales tax is 1%. What percent do the people of Austin pay in sales taxes? Ken, who lives in Austin, purchased a bicycle for $189. How much sales tax did he pay?

5. (19, Inv. 1) What kind of figure is created when these points are connected by line segments in the given order?

$$(-2, 2), (1, 5), (5, 5), (8, 2), (-2, 2)$$

* 6. (31) **Evaluate** In a whale ecology study, gray whales were tagged in order to study their feeding habits. One day, a biologist recorded that a gray whale feeding at a depth of 10 meters (−10 m) suddenly descended 32 meters. What is the new depth of the gray whale?

Generalize Simplify.

* 7. (33) **a.** $8 - (-1)$ **b.** $-1 - (-2)$

* 8. (31) **a.** $-6 - 5$ **b.** $-8 + (-12)$

* 9. (31) **Analyze** Collect like terms: $3a + 2a^2 + a - a^2$

* 10. (30) **Formulate** Write $\frac{7}{9}$ as a decimal and a percent.

11. (22) A spool contains 9 yards of ribbon. How many quarter-yard pieces can be cut from the spool of ribbon?

12. (Inv. 2) What is the length of side *BC*?

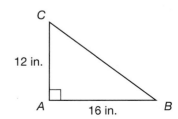

* 13. (22, 15) **Generalize** Simplify: $\sqrt{\left(\frac{1}{3}\right)^2 \cdot \left(\frac{1}{4}\right)^2}$

14. Solve by inspection. $5x - 3 = 27$
(14)

* 15. **Estimate** Estimate and then calculate. 29×0.3
(25)

16. Add:
(31, 12)
    a. $-17 + (-5)$ **b.** $16 + -4$ **c.** $-12 + 27$ **d.** $-\dfrac{1}{4} + 0.75$

17. **a.** Write 20% as decimal and a reduced fraction.
(11)
    b. Which form would you find most convenient to use to calculate a tip of 20% on a meal that cost $45.00?

Generalize Simplify.

* 18. $9^2 - 7^2 + 6^0$
(15, 27)

19. $1.09 + 0.055 + 3.2$
(24)

20. $3.14 - (0.2)(1.5)$
(21, 24)

21. $0.5 \div 0.025$
(25)

* 22. In a shipping carton of juice there are 8 boxes. In each box there are
(15) 8 containers. In each container there are 8 ounces of juice. How many ounces of juice are in 8 shipping cartons of juice? Use exponents to find the answer.

23. Order these numbers from least to greatest.
(10, 12)

$$\frac{1}{8}, \ 0.12, \ -1.8, \ 0.22, \ -\frac{1}{10}$$

* 24. A triangle has vertices $(-2, 1)$, $(2, 4)$ and $(6, 1)$. Sketch this triangle and
(20, Inv. 2) refer to your sketch to answer questions **a–c.**
    a. Classified by sides, what type of triangle is it?
    b. What is the area of the triangle?
    c. What is the perimeter of the triangle?

25. If the triangle in problem **24** is transformed so that the vertices of
(26) the image are at $(-2, 1)$, $(1, -3)$, and $(-2, -7)$, then what type of transformation was performed?

* 26. Tanya made a scale drawing of a bookcase she is going to build. The
(29) height of the bookcase is going to be 60 inches. Her drawing shows the height as 4 inches. What is the ratio of the scale drawing height to the actual height of the bookcase?

27. Chef Lorelle has a 79.5 ounce block of cheese. She asks her assistant
(25) to slice the cheese into 0.75 ounce slices. What is the maximum number of slices they can get from the block of cheese?

28. Keisha is doing a science experiment. She is measuring the growth
(24) differences that result from using different amounts of water and
sunlight on bean seedlings. In one of the pots, a seedling is 6.7 cm
tall. In another, the seedling is 5.8 cm tall. How much taller is the first
seedling than the second?

* 29. **Formulate** Use the Distributive Property to find $\frac{1}{2}$ of $6\frac{3}{4}$ in the following
(21) expression: $\frac{1}{2}(6 + \frac{3}{4})$.

30. Ryan had an $8\frac{1}{2}$ in.-by-11 in. piece of paper. He cut it into quarters as
(8, 23) shown. What is the length and width of each quarter? What is the area
of each quarter?

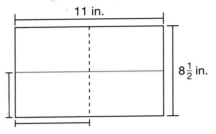

"Sea level" refers to the level of the ocean's surface and is used as a
benchmark to measure land elevation. Death Valley, California, the lowest
point in the United States, is approximately 282 feet below sea level. Denver,
the largest city and capital of the state of Colorado, is 5431 feet above sea
level. What is the difference in elevation between these two places?

• Proportions
• Ratio Word Problems

facts | Power Up G

mental math

a. **Number Sense:** 6.50×40

b. **Fractional Parts:** $66\frac{2}{3}\%$ equals $\frac{2}{3}$. Find $66\frac{2}{3}\%$ of $45.

c. **Measurement:** Find the temperature indicated on this thermometer.

d. **Proportions:** Caesar finished 12 problems in 20 minutes. At that rate, how many problems can he do in 60 minutes?

e. **Percent:** 20% of 90

f. **Geometry:** Find the area of this rectangle.

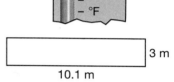

3 m

10.1 m

g. **Powers/Roots:** List all the perfect squares from 16 to 64, then approximate $\sqrt{26} \times 4.9$.

h. **Calculation:** $10 - 9$, $\times 8$, $+ 7$, $\div 5$, $\times 6$, $+ 4$, $- 3$, $+ 2$, $\div 1$, $\div 3$, $\times 2$, $- 4$, $\times 5$, $+ 6$, $\div 7$

problem solving | Every block in this cube is labeled with a number and a letter. The block labeled 9A has two blocks behind it (9B and 9C). Which block is not visible from any angle?

proportions | A **proportion** is a statement that two ratios are equal. The ratios $\frac{2}{4}$ and $\frac{6}{12}$ are equal and form a proportion.

$$\frac{2}{4} = \frac{6}{12}$$

Equal ratios reduce to the same ratio. Both $\frac{2}{4}$ and $\frac{6}{12}$ reduce to $\frac{1}{2}$. Notice these multiplication relationships between the numbers in the proportion.

$$2 \times 3 = 6$$
$$\frac{2}{4} = \frac{6}{12}$$
$$4 \times 3 = 12$$

$$2 \times 2 = 4 \quad \downarrow \frac{2}{4} = \frac{6}{12} \downarrow \quad 6 \times 2 = 12$$

Thinking Skill

Connect

What mathematical property is illustrated by the equation $\frac{2}{4} \cdot \frac{3}{3} = \frac{6}{12}$?

One ratio in a proportion can be expressed as the other ratio by multiplying the terms by a constant factor.

$$\frac{2}{4} \cdot \frac{3}{3} = \frac{6}{12}$$

We can use this method to test whether ratios form a proportion. For example, to park for 2 hours, a lot charges $3. To park for 3 hours, the lot charges $4.

Time (hr) $\frac{2}{3} \frac{3}{4}$
Charge ($)

Justify   Is the time parked and the fee charged by the lot a proportional relationship? Why or why not?

Example 1

Nora is paid $12 an hour. Is her pay proportional to the number of hours she works?

Solution

We can make a table of her pay for various hours of work. We see that the ratio of pay to hours worked is constant. The ratio does not change. Any two ratios form a proportion.

$$\frac{\text{Pay (\$)}}{\text{Time (hr)}} \quad \frac{12}{1} = \frac{36}{3}$$

Therefore, Nora's pay **is proportional** to the number of hours she works.

Nora's Pay

Hours	Pay	$\dfrac{\text{Pay}}{\text{Hours}}$
1	12	$\dfrac{12}{1}$
2	24	$\dfrac{24}{2} = \dfrac{12}{1}$
3	36	$\dfrac{36}{3} = \dfrac{12}{1}$
4	48	$\dfrac{48}{4} = \dfrac{12}{1}$

Example 2

Nelson has a paper route. If he works by himself the job takes 60 minutes. If he splits the route with a friend, it takes 30 minutes. If two friends help, the job takes 20 minutes. Is the amount of time it takes to complete the route proportional to the number of people working?

Solution

The table shows the number of workers and the number of minutes needed to complete the job. The ratio of time to workers changes.

$$\frac{60}{1} \neq \frac{15}{1}$$

There is a relationship between the number of workers and the time to complete a job, but the relationship **is not proportional.**

Time for Paper Route

Number Working	Time (min.)	Time / Workers
1	60	$\frac{60}{1}$
2	30	$\frac{30}{2} = \frac{15}{1}$
3	20	$\frac{20}{3}$

We can use proportions to solve problems where one of the numbers in the proportion is missing. A variable represents the missing number in the proportion.

$$\frac{2}{8} = \frac{6}{x}$$

One way to find the missing number in a proportion is to use the multiple between the terms of the ratios. This method is like finding equivalent fractions.

$$2 \times 3 = 6$$
$$\frac{2}{8} = \frac{6}{x}$$
$$8 \times 3 = 24$$

By multiplying $\frac{2}{8}$ by $\frac{3}{3}$, we find that the missing term is 24. Below we show another relationship we can use to find a missing number in a proportion.

$$2 \times 4 = 8 \quad \downarrow\frac{2}{8} = \frac{6}{x}\downarrow \quad 6 \times 4 = 24$$

Again we find that the missing number is 24.

Example 3

Solve: $\frac{24}{m} = \frac{8}{5}$

Solution

We solve the proportion by finding the missing number. We inspect the proportion to find a relationship that is convenient to use in order to calculate the missing number. This time we multiply both terms from right to left.

$$8 \times 3 = 24$$
$$\frac{24}{m} = \frac{8}{5}$$
$$5 \times 3 = 15$$

We find that **$m = 15$.**

Ratio word problems can include several numbers, so we will practice using a table with two columns to sort the numbers. In one column we write the ratio numbers. In the other column we write the actual counts. We can use a ratio table to help us solve a wide variety of problems.

Example 4

The ratio of boys to girls in the class is 3 to 4. If there are 12 girls, how many boys are there?

Solution

We see ratio numbers and one actual count. We record the numbers in a ratio table with two columns and two rows. We let b stand for the actual number of boys. Now we use two rows from the table to write a proportion.

	Ratio	Actual Count
Boys	3	b
Girls	4	12

$$\rightarrow \frac{3}{4} = \frac{b}{12}$$

By multiplying each term of the first ratio by 3, we find that there are **9 boys.**

$$\frac{3}{4} \cdot \frac{3}{3} = \frac{9}{12}$$

Practice Set

a. Which pair of ratios forms a proportion?

A $\frac{3}{6}, \frac{6}{9}$ **B** $\frac{3}{6}, \frac{6}{12}$ **C** $\frac{3}{6}, \frac{6}{3}$

b. *Justify* How do you know your choice is correct?

Solve each proportion.

c. $\frac{4}{6} = \frac{c}{18}$ **d.** $\frac{9}{3} = \frac{18}{d}$

e. $\frac{e}{9} = \frac{5}{15}$ **f.** $\frac{4}{f} = \frac{3}{12}$

g. A wholesaler offers discounts for large purchases. The table shows the price of an item for various quantities. Is the price proportional to the quantity? How do you know?

Quantity	Price
10	$30
100	$200
1000	$1000

h. A retailer sells an item for $5. Brenda needs several of the items. Is the total price proportional to the quantity purchased? How do you know?

Analyze Solve problems **i** and **j** by creating a ratio table and writing a proportion.

i. The ratio of the length to width of a rectangular room is 5 to 4. If the room is 20 feet long, how wide is the room?

j. The teacher-student ratio in the primary grades is 1 to 20. If there are 100 students in the primary grades, how many teachers are there?

* 1. *(Analyze)* Marc's cell phone company charges users $33\frac{1}{3}$ cents for every
(23)   minute they go over their plan. If Marc goes 27 minutes over his plan,
      how much money will his cell phone company charge him?

2. Juan ran the 100-meter dash in 11.91 seconds. Steve finished behind
(24)   Juan with a time of 12.43 seconds. How many seconds slower than
      Juan was Steve?

* 3. The ratio of the base of a certain triangle to its height is 3 to 2. If the
(34)   base of the triangle is 24 inches, what is the height of the triangle?

* 4. A sculpture in the center of the bank lobby casts a triangular shadow
(26)   on the west floor when sunlight shines through the east window in the
      morning. In the afternoon, the sun shines through the west window
      casting the sculpture's shadow on the east floor. The change in the
      position of the shadow from morning to afternoon corresponds to what
      kind of geometric transformation?

* 5. Simplify: $(-3) + (-4) - (-5) - (+6)$
(31, 33)

* 6. *(Explain)* True or false: Some rectangles are rhombuses. Explain.
(Inv. 3)

Simplify.

7. $\dfrac{7^7 \cdot 7^0}{7^5 \cdot 7^2}$
(27)

8. $\dfrac{3 + 2.7}{3 - 2.7}$
(24, 25)

9. $3^3 + 4[21 - (7 + 2 \cdot 3)]$
(15, 21)

10. $\dfrac{9}{5} \cdot \dfrac{1}{3} - \dfrac{1}{10}$
(13, 23)

11. $\sqrt{1^3 + 2^3 + 3^3}$
(15, 21)

* 12. $-2 - (-13 + 20)$
(33)

* 13. *(Evaluate)* The spinner at right is spun
(32)   once. What is the probability that the
      spinner will stop

   a. in sector A?

   b. in sector B?

   c. in sector A or C?

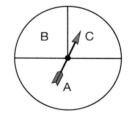

* 14. Collect like terms to simplify.
(31)
   a. $7xy - 3y + 4y - 18yx$ **b.** $13b^2 + 8ac - ac + 4$

15. On one shipping pallet of water there are 30 boxes. In each box are
(3, 24)   10 cases. In each case are 24 bottles. In each bottle are 20 fluid ounces
      of water. How many ounces are in 2 shipping pallets?

* 16. **Analyze** Parallelograms *ABCD* and
   WXYZ are similar.
   (19, Inv. 3)

   a. What is the measure of ∠*Y*?

   b. What is the length of side *WX*?

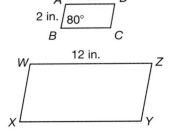

For problems **17–20**, find all values of *b* that make the equation true.

17. $3|b| = 21$
(14)

18. $b + 2b = 15$
(14, 31)

19. $\dfrac{81}{b} = 9$
(14)

20. $|b| + |b| = 12$
(14)

* 21. Write $\frac{1}{3}$ as a decimal and a percent.
(11, 30)

22. Alfred was studying the mass of several planets. He found that Pluto's
(28)   mass is 4.960×10^{27} kg, Neptune's is 1.028×10^{27} kg, Jupiter's is
      1.894×10^{30} kg, and Earth's is 5.976×10^{27} kg. Order the planets from
      greatest to least mass.

23. **Model** On a number line, graph -3 and two numbers that are 4 units
(1)   away.

24. Brittanie's gasoline credit card bill was $70.92 in September, $38.15 in
(32)   October, and $52.83 in November. Altogether, what were her gasoline
      charges for these months?

   A $109.07   **B** $150.90   **C** $161.90   **D** $207.70

25. **Connect** The hockey puck slid across the ice from point A to point B.
(26)   This movement corresponds to what geometric transformation?

Solve for *y*.

* 26. $\dfrac{10}{y} = \dfrac{2}{7}$
(34)

* 27. $\dfrac{4}{5} = \dfrac{16}{y}$
(34)

* 28. **Evaluate** Vostok, Antarctica holds the world's record for the coldest
(33)   temperature, which is $-129°F$. Death Valley, California holds the record
      for the hottest recorded temperature, which is $134°F$. Write an equation
      with integers for finding the difference between these two temperatures
      and then solve it.

29. Evaluate $a^2 + b^2 - 2ab$ for $a = 4$ and $b = 2$.
(14)

30. In which quadrant is the point $(2, -3)$?
(Inv. 1)

• Similar and Congruent Polygons

facts | Power Up G

mental math

   a. **Number Sense:** $22 \div 4$

   b. **Algebra:** $x - 12 = 48$

   c. **Measurement:** Find the temperature indicated on this thermometer.

   d. **Rate:** Mei Ling walked 100 yards in 1 minute. How many feet per minute is that?

   e. **Scientific Notation:** Write five billion in scientific notation.

   f. **Geometry:** Two sides of a triangle are 7 and 9. The third side must be between what two numbers?

   g. **Estimation:** 3 at $19.98 plus 2 at $3.48

   h. **Calculation:** $25 \times 3, + 6, \sqrt{}, \sqrt{}, \times 9, - 2, \sqrt{}, \times 9, + 4, \sqrt{}$

problem solving

Find each missing digit. (Hint: If the remainder is 8, then the divisor must be greater than 8.)

```
       __ r 8
   _)1__
     ==
     --
     =
     -
```

Recall that two figures are **similar** if they have the same shape even though they may vary in size. In the illustration below, triangles A, B and C are similar. To see this, we can imagine dilating △B as though we were looking through a magnifying glass. By enlarging △B, we could make it the same size as △A or △C. Likewise, we could reduce △C to the same size as △A or △B.

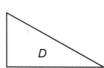

△D is not similar to the other three triangles, because its shape is different. Reducing or enlarging △D will change its size, but not its shape.

Also recall that figures that are the same shape and size are not only similar, they are also **congruent**. All three of the triangles below are similar, but only triangles △ABC and △DEF are congruent. Figures may be reflected (flipped) or rotated (turned) without affecting their similarity or congruence.

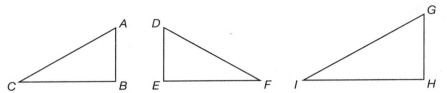

When inspecting polygons to determine whether they are similar or congruent, we compare their **corresponding parts**. Referring back to the illustration of △ABC and △DEF, we identify the following corresponding parts:

Corresponding Angles	**Corresponding Sides**
∠A corresponds to ∠D	$\overline{AB}$ corresponds to $\overline{DE}$
∠B corresponds to ∠E	$\overline{BC}$ corresponds to $\overline{EF}$
∠C corresponds to ∠F	$\overline{CA}$ corresponds to $\overline{FD}$

We use tick marks on corresponding side lengths that have equal length and arcs on corresponding angles that have the same measure.

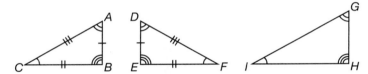

The symbol ~ represents similarity. Since △ABC is similar to △GHI, we can write △ABC ~ △GHI.

The symbol ≅ represents congruence. Since △ABC and △DEF are congruent, we can write △ABC ≅ △DEF.

Notice that when we write a statement about the similarity or congruence of two polygons, we name the letters of the corresponding vertices in the same order. Below we define similar and congruent polygons in a way that enables us to draw conclusions about their side lengths and angle measures.

Similar Polygons

> **Similar polygons have corresponding angles which are the same measure and corresponding sides which are proportional in length.**

Congruent Polygons

> **Congruent polygons have corresponding angles which are the same measure and corresponding sides which are the same length.**

Example 1

Which two quadrilaterals are similar?

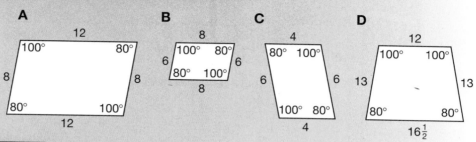

Solution

Polygons are similar if:

1. The corresponding angles are congruent.
2. The corresponding sides are proportional.

All four quadrilaterals have two 100° angles and two 80° angles. Figures A and B have the same orientations so it is clear that their corresponding angles are congruent. We could rotate (turn) figure C so that it has the same orientation as figures A and B and see that its corresponding angles are also congruent. The sequence of angles in figure D does not match the corresponding angles of the other figures, so we remove figure D from consideration for similarity.

Now we consider the side lengths to see if corresponding sides have proportional lengths.

	A	B	C
shorter sides / longer sides	$\frac{8}{12}$	$\frac{6}{8}$	$\frac{4}{6}$

We see that $\frac{8}{12}$ and $\frac{4}{6}$ are equal ratios, as both reduce to $\frac{2}{3}$. Thus the side lengths of figures A and C are proportional, and **figures A and C are similar.**

The ratio of side lengths of figure B does not reduce to $\frac{2}{3}$, so it is not similar to figures A and C.

The relationship between the sides and angles of a triangle is such that the lengths of the sides determine the size and position of the angles. For example, with three straws of different lengths, we can form one and only one shape of triangle. Three other straws of twice the length would form a similar triangle with angles of the same measure. Knowing that two triangles have proportional corresponding side lengths is enough to determine that they are similar.

Side-Side-Side Triangle Similarity

> **If two triangles have proportional corresponding side lengths, then the triangles are similar.**

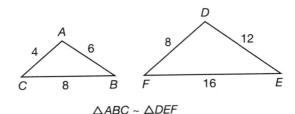

$$\triangle ABC \sim \triangle DEF$$

Consider the three triangles below. They have corresponding angles with equal measures.

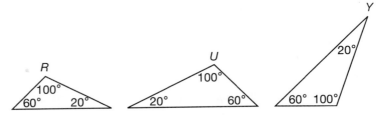

Analyze How can you determine whether or not these triangles are similar?

We might make the conjecture that these triangles are similar and that any triangle with the same angle measures would be similar to these. We will accept this basic assumption about similar triangles: all triangles with the same set of angle measures are similar. Thus, knowing that two triangles have congruent corresponding angles is sufficient information to conclude that the triangles are similar.

Angle-Angle-Angle Similarity

> If the angles of one triangle are congruent to the angles of another triangle, then the triangles are similar and their corresponding side lengths are proportional.

Example 2

Thinking Skill

Analyze

Is the perimeter of the triangle on the right three times the perimeter of the triangle on the left? Is the area three times as great?

The triangles are similar. Find x.

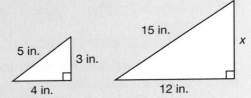

Solution

We pick two sides from the right triangle to write a ratio and include *x*. We pick the two corresponding sides from the left triangle to write a ratio.

Left Triangle	Right Triangle
4	12
3	x

Since the triangles are similar, the ratios form a proportion. We solve the proportion for *x*.

$$4 \times 3 = 12$$

$$\frac{4}{3} = \frac{12}{x}$$

$$3 \times 3 = 9$$

The constant factor that relates the ratios of similar polygons is called the **scale factor**. In this case the scale factor is 3. Since 3 × 3 is 9, we find that the side of the triangle is **9 inches.**

Example 3

An architect makes a scale drawing of a building.

   a. If one inch on the drawing represents 4 feet, then what is the scale factor from the drawing to the actual building?

   b. What is the length and width of a room that is 5 in. by 4 in. on the drawing?

Solution

A scale drawing is similar to the view of the object it represents.

   a. If 1 inch represents 4 feet, then 1 inch represents 48 inches. Therefore, the scale is 1:48 and the scale factor is **48.**

   b. Each inch represents 4 ft, so 5 in. represents 5 × 4 = 20 ft, and 4 in. represents 4 × 4 = 16 ft. The room is **20 ft long and 16 ft wide.**

Practice Set

Refer to the following figures for problems **a–e.**

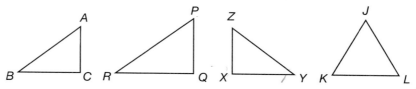

   a. Which triangle is congruent to △*ABC*?

   b. Which triangle is similar but not congruent to △*XYZ*?

   c. Which triangle is not similar to △*PQR*?

   d. Which angle in △*ZYX* corresponds to ∠*A* in △*ABC*?

   e. Which side of △*PRQ* corresponds to side *XY* in △*ZYX*?

   f. *Justify* Are these two triangles similar? How do you know?

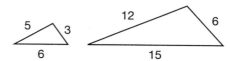

g. These quadrilaterals are similar. Find the missing side length. What is the scale factor from the smaller figure to the larger? (Units are inches.)

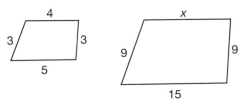

For **h–j,** draw two rectangles, one 2 cm-by-1 cm and the other 4 cm-by-2 cm. Then answer the following questions.

h. What is the scale factor from the smaller to larger rectangle?

i. The perimeter of the larger rectangle is how many times the perimeter of the smaller rectangle?

j. The area of the larger rectangle is how many times the area of the smaller rectangle?

k. On a scale drawing of a house, one inch represents 8 feet. How wide is the garage if it is $2\frac{1}{2}$ inches wide on the drawing?

Written Practice   *Strengthening Concepts*

* 1. Mr. Connors drives his car 11 miles each day to and from work. Is the
(34)   distance Mr. Connors drives commuting to work proportional to the number of days he works? How do you know?

* 2. **Analyze** If a number cube is rolled, what is the probability of rolling a
(9, 32)   prime number?

3. Adam worked 8 hours per week before his workload was increased.
(3, 4)   After the increase, he worked 4 hours per day for six days each week. How many more hours per week did he work after the increase than he worked before the increase?

* 4. One hundred residents of the city were randomly selected for a survey
(32)   regarding the new gymnasium design. Thirty-two residents approved of Design A and 85 residents approved of Design B.

    a. **Evaluate** Use these experimental results to calculate the probability that any city resident selected at random would approve of Design A. Then calculate the probability for Design B.

    b. Based on the results of the survey, which design would you recommend?

* 5. Expand then collect like terms: $3(x + 2) + x + 2$
(21, 31)

6. Plot these points on a coordinate plane, then draw segments from
(20,   point to point to form a triangle: (0, 0), (4, 0), (0, 3). Use the Pythagorean
Inv. 2)   Theorem to find the length of the hypotenuse. What is **a** the perimeter and **b** the area of the triangle?

7. Find the
(Inv. 2)
    a. area and

    b. perimeter of the triangle.

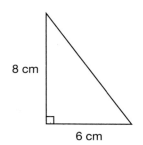

8 cm

6 cm

Generalize Simplify.

* 8. **a.** $(-3) + (-15)$ **b.** $(-3) - (-15)$
(31, 33)

  9. **a.** $(-15) + (-3)$ **b.** $(-15) - (-3)$
(31, 33)

* 10. $1\frac{2}{3} + 2\frac{3}{4}$ **11.** $\frac{1}{8} + \frac{7}{8} \div \frac{7}{8}$
(13)   (13, 22)

12. $\frac{9}{10} - \frac{2}{3} \cdot \frac{3}{5}$ **13.** $\frac{(2.54) + 1.21}{0.03}$
(13, 22)   (24, 25)

* 14. $7^2 - 4(7 + 3) + 1^{20} - \sqrt{64}$ **15.** $(0.3^2 - 0.2^2) - (0.3 - 0.2)^2$
(15, 21)   (24, 25)

16. Find each missing exponent.
(27)
    a. $10^5 \cdot 10^6 = 10^{\square}$ **b.** $\frac{10^5}{10^2} = 10^{\square}$

17. Triangles *ABE* and *ACD* are similar.
(35)
    a. Which side of $\triangle ABE$ corresponds to
    side *CD?*

    b. Compare: m$\angle AEB \bigcirc$ m$\angle D$

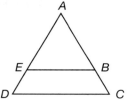

For problems **18** and **19** refer to this spinner. It is
divided into five congruent sectors.

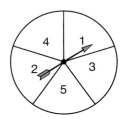

18. **a.** What is the ratio of even numbers to odd
(32)   numbers on the spinner?

    b. What is the probability of spinning an
    even number?

    c. What is the probability of not spinning an even number?

19. **a.** If a coin is tossed once and the spinner is spun once, what is the
(32)   sample space of the experiment?

    b. What is the probability of getting heads and an even number?

* 20. **a.** **Formulate** Write $\frac{1}{8}$ as a decimal and a percent.
(12)
    b. Which of the three forms would be most convenient to use to find a
    part of 4 inches?

For problems **21–24**, use inspection to find all values of x which make the equations true.

21. $|x| = 4$
(1, 14)

* 22. $2x + 5 = 29$
(14)

* 23. $\dfrac{3}{x} = \dfrac{12}{20}$
(14, 34)

* 24. $\dfrac{7}{28} = \dfrac{9}{x}$
(14, 34)

25. Use prime factorization to reduce $\dfrac{144}{360}$.
(9)

26. On a number line, graph the negative integers.
(1)

27. Estimate: $\dfrac{(39)(1.9) + 1.1}{8.89}$
(17)

28. **Evaluate** The floor of Memorial Arena measures 100 ft wide by 200 ft
(35) long. The facility is used for arena football games, which are played on a similar rectangular field that is 150 ft long.

a. What is the width of the football field?

b. What is the perimeter of the football field?

29. One sheet of paper in a notebook is $8\frac{1}{2}$ in. × 11 in. A smaller notebook
(35) has a similar shape, and its pages are $5\frac{1}{2}$ in. long. What is the width of the paper in the smaller notebook?

30. **Analyze** Triangle *PQR* is isosceles and $\angle PRS$ measures 130°. Find:
(20)
    a. $m\angle PRQ$

    b. $m\angle Q$

    c. $m\angle P$

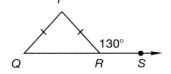

• Multiplying and Dividing Integers
• Multiplying and Dividing Terms

Power Up | *Building Power*

facts | Power Up H

mental math

   a. Number Sense: $280 ÷ 40

   b. Fractional Parts: $66\frac{2}{3}$% of $66

   c. Measurement: Approximate the measure of this angle.

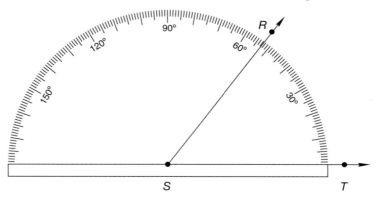

   d. Rate: Kerry wrote 12 pages in 2 hours. What was her rate in pages per hour?

   e. Proportions: The ratio of sheep to goats in a petting zoo is 3 to 1. There are 12 sheep. How many goats are there?

   f. Geometry: Two angles of a triangle measure 30° and 60°. Find the measure of the third angle.

   g. Powers/Roots: List all the perfect squares from 1 to 100. Then, approximate $\sqrt{101} - \sqrt{38}$.

   h. Calculation: $8 + 8$, $\sqrt{}$, $\sqrt{}$, $\times 10$, $+ 1$, $\div 3$, $\times 7$, $\sqrt{}$, $- 8$

problem solving | The item cost $4.59. Abraham paid with a $5 bill. The clerk gave Abraham 4 coins. What were they?

New Concepts | *Increasing Knowledge*

multiplying and dividing integers | On this number line we illustrate multiplication and division of integers. The arrows show that 3 times −2 is −6.

$$3(-2) = -6$$

$$-2 \quad -2 \quad -2$$

$$-6 \quad\quad -4 \quad\quad -2 \quad\quad 0$$

The illustration also shows that if we divide -6 into 3 parts, each part is -2.

$$\frac{-6}{3} = -2$$

Likewise, the illustration shows that the number of -2s in -6 is 3.

$$\frac{-6}{-2} = 3$$

From these illustrations notice these relationships:

$$3(-2) = -6$$

different signs → negative product

$$\frac{-6}{3} = -2$$

different signs → negative quotient

$$\frac{-6}{-2} = 3$$

same signs → positive quotient

When we multiply two negative numbers like $(-3)(-2)$, we can think about the multiplication as "the opposite of 3 times -2." Since $3(-2)$ is -6, the opposite of $3(-2)$ is the opposite of -6, which is positive 6.

$$(-3)(-2) = 6$$

same signs → positive product

Example 1

Visit www. SaxonPublishers. com/ActivitiesC3 for a graphing calculator activity.

Simplify:

a. $(-10)(-5)$

b. $(+10)(-5)$

c. $\dfrac{-10}{-5}$

d. $\dfrac{10}{-5}$

Solution

We multiply or divide as indicated. If the two numbers have the same sign, the answer is positive. If the two numbers have different signs, the answer is negative.

a. $(-10)(-5) = \mathbf{50}$

b. $(10)(-5) = \mathbf{-50}$

c. $\dfrac{-10}{-5} = \mathbf{2}$

d. $\dfrac{10}{-5} = \mathbf{-2}$

Example 2

Every month $25 is automatically deducted from Newton's checking account to pay for Internet service. Write an expression using integers that shows the effect on Newton's account of six months of charges.

Each month's charge subtracts $25 from Newton's checking account. We can represent each change as -25. Therefore, six months of charges is **6(-25)**, or -150. Six months of charges would deduct $150 from Newton's account.

Since the product of two negative numbers is a positive number, squaring a negative number results in a positive number.

$$(-5)^2 = (-5)(-5) = 25$$

However, cubing a negative number results in a negative number.

$$(-2)^3 = (-2)(-2)(-2)$$

The product of -2 and -2 is 4.

$$[(-2)(-2)](-2) = (4)(-2)$$

The product of 4 and -2 is -8.

$$(4)(-2) = (-8)$$

Example 3

Simplify.

    a. $(-1)^5$ **b. $(-1)^6$**

Solution

Every pair of negative factors has a positive product, so an even number of negative factors is positive and an odd number is negative.

    a. $(-1)^5 = -1$ **b. $(-1)^6 = 1$**

Example 4

    a. For $y = x^2$, find y if x is -2.

    b. If $x^2 = 9$, then x can be which two numbers?

Solution

    a. We are asked to find the value of x^2 when x is -2. Since x^2 means $x \cdot x$, we multiply $(-2)(-2)$. The product is **4.**

    b. Both **3** and **-3** make the equation true.

$$3^2 = 9 \qquad (-3)^2 = 9$$

We could have solved example **4b** by finding the square root of x^2 and 9.

We distinguish between the words "square root" and the ($\sqrt{}$). The radical indicates the **principal square root**, which is the positive square root of a positive real number.

- The square roots of 9 are 3 and -3.
- $\sqrt{9} = 3$

Thinking Skill

Connect

Use digits and symbols to show both square roots of 41.

Below we summarize the rules for arithmetic with two integers.

Arithmetic with Two Integers

Operation	Rule
+	To find the sum of addends with different signs: **1.** Subtract the absolute values of the addends. **2.** Take the sign of the addend with the greater absolute value. To find the sum of addends with the same sign: **1.** Add the absolute values of the addends. **2.** Take the sign of the addends.
−	Instead of subtracting a number, add its opposite.
× ÷	**1.** Multiply or divide as indicated. **2.** If the two numbers have the same sign, the answer is positive. If the two numbers have different signs, the answer is negative.

Activity

Reversing Signs on a Calculator

Most calculators have a [+/−] key that reverses the sign of the displayed number. To key in a negative number, such as −5, we first key in its absolute value, in this case 5, and then press the [+/−] key to make the entry negative. Here is the keystroke sequence for adding −5 and −3 on most calculators:

The display should be −8. Try the following exercises:

   a. $(-5) + (-3)$ **b.** $(-15) - (-3)$

   c. $(-12)(-8)$ **d.** $\dfrac{144}{-8}$

multiplying and dividing terms

In Lesson 31 we added and subtracted terms. In this lesson we will multiply and divide terms.

Example 5

Simplify.

   a. $(6x^2y)(-2xyz)$ **b.** $\dfrac{6x^2y}{-2xyz}$

Solution

   a. We can use the commutative and associative properties of multiplication to rearrange and regroup the factors.

Step:	Justification:
$(6x^2y)(-2xyz)$	Given expression
$(6)(-2)x^2 \cdot x \cdot y \cdot y \cdot z$	Commutative and associative properties
$-12x^3y^2z$	Multiplied and grouped with exponents

b. We can factor the dividend and divisor and then reduce. The quotient assumes that no variables are zero.

Step:	Justification:
$\dfrac{6x^2y}{-2xyz}$	Given expression
$\dfrac{2 \cdot 3 \cdot x \cdot x \cdot y}{-2 \cdot x \cdot y \cdot z}$	Factors
$\dfrac{\overset{1}{2} \cdot 3 \cdot \overset{1}{\cancel{x}} \cdot x \cdot \overset{1}{\cancel{y}}}{-\underset{1}{\cancel{2}} \cdot \underset{1}{\cancel{x}} \cdot \underset{1}{\cancel{y}} \cdot z}$	Reduced
$-\dfrac{3x}{z}$	Simplified

Notice that in example 5, part **b,** we expressed the quotient as a negative fraction. The following three forms are equivalent; however, we avoid expressing a denominator as a negative number.

$$\frac{3x}{-z} = \underbrace{\frac{-3x}{z}}_{\text{Preferred}} = -\frac{3x}{z}$$

Example 6

Expand.

a. $3x(2x - 4)$
b. $-3x(2x - 4)$

Solution

a. We apply the distributive property and multiply the terms within the parenthesis by $3x$.

$$3x(2x - 4) = 6x^2 - 12x$$

b. We multiply by $-3x$. Notice that $-3x$ times -4 is $+12x$.

$$-3x(2x - 4) = -6x^2 + 12x$$

Practice Set

Simplify:

a. $(-12)(-3)$

b. $(-12)(+3)$

c. $\dfrac{-12}{-3}$

d. $\dfrac{-12}{+3}$

e. $(-6)^2$

f. $(-1)^2$

g. $(-3)^3$

h. $(-2)^4$

i. Solve: $x^2 = 100$ (Write two possible solutions.)

j. What are the square roots of 100? What is $\sqrt{100}$?

k. ⟨**Represent**⟩ Drilling for core samples in the sea floor, the driller used 12 drill rods 8 meters long to reach from the surface of the ocean to the lowest core sample. Write an equation using integers that shows the depth of the core sample with respect to sea level.

Simplify:

l. $(-12xy^2z)(3xy)$

m. $\dfrac{-12xy^2z}{3xy}$

Expand:

n. $2x(2x - 5)$

o. $-2x(2x - 5)$

Written Practice   *Strengthening Concepts*

1. A water company charges customers $1.081 for each CCF
(4, 25) (hundred cubic feet) of water used. If one customer uses 70 CCF of water, how much will the water company charge the customer?

2. Each box of cards contains forty-five packs. Each pack contains
(4) eight cards. How many cards are in two boxes?

3. Doug read three books that averaged 156 pages each. He also read
(3, 4) 12 short stories that were each six pages long. How many pages did Doug read altogether?

* 4. ⟨**Analyze**⟩ A bank prefers to have a customer-to-teller ratio of 3 to 1.
(29) If there are 9 customers in the bank, how many tellers should be available?

⟨**Justify**⟩ Solve for x. Check your solution.

* 5. $\dfrac{x}{12} = \dfrac{2}{3}$
(34)

* 6. $\dfrac{1}{5} = \dfrac{4}{x}$
(34)

* 7. ⟨**Evaluate**⟩ Find the **a** area and **b** perimeter of the triangle with vertices
(Inv. 1, 20) (2, 3), (5, 3), and (5, −1).

* 8. Quadrilaterals *ABCD* and *PQRS* are similar.
(19)

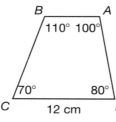 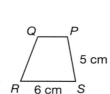

a. What is the measure of ∠*S*?

b. What is the length of side *AD*?

9. a. Expand: $-2(3x - 4)$
(21, 36)

b. Factor: $7x - 21$

* 10. **Analyze** What is the probability of rolling an odd number on a number
(17) cube? How does that compare to the probability of rolling an even
number?

Simplify problems **11–18.**

* 11. $(-4)\left(\dfrac{1}{2}\right)(-3)$
(22, 36)

* 12. $\dfrac{-72xy}{6}$
(36)

13. $\dfrac{2}{3} + \dfrac{3}{4} \div \dfrac{1}{2}$
(21, 22)

14. $\dfrac{1}{2} - \dfrac{2}{3} \cdot \dfrac{3}{4}$
(21, 22)

15. $(-2)^2 \cdot (-2)^3$
(36)

16. $\dfrac{10^3 \cdot 10^2}{10^4}$
(27)

17. $1\dfrac{2}{3} + 3\dfrac{3}{4} \div 1\dfrac{1}{2}$
(21, 23)

18. $\dfrac{64}{0.16}$
(25)

* 19. **a.** Write $\dfrac{1}{25}$ as a decimal and as a percent.
(12, 25)

    b. Instead of dividing a number by 25, we can multiply by any form
of $\dfrac{1}{25}$. Show how to use multiplication and one of the forms of the
reciprocal of 25 to find $80 \div 25$.

20. Arrange these numbers in order from least to greatest.
(12, 30)

$$12.5\%, \frac{1}{6}, 0.15$$

* 21. Draw quadrilateral WXYZ so that sides WX and YZ are parallel and sides
(18, 19) XY and YZ are perpendicular. Make side YZ longer than side WX.

For problems **22–25,** find all values of x which make the equation true.

22. $-3|x| = -6$.
(1, 14)

23. $x + x = 26$
(14)

* 24. $2x - 1 = 19$
(14)

* 25. $x^2 - 1 = 24$
(14, 36)

26. **Model** On a number line, graph the whole numbers less than 5.
(1)

27. Fritz recorded the daily high temperature in degrees Fahrenheit for a
(7) week.

S	M	T	W	T	F	S
81°	78°	79°	74°	70°	77°	80°

Find the mean high temperature for the week.

Evaluate.

* 28. Find $\sqrt{b^2 - 4ac}$ when $a = 2$, $b = -5$, $c = -3$.
(21, 36)

* 29. **Evaluate** Both spinners are spun once.
(32)

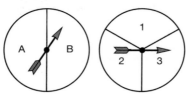

    a. What is the sample space of the experiment?

    b. What is the probability that the spinners stop on a vowel and an odd number?

30. **Classify** In the figure, right triangle *ABC* is
(20) divided into two smaller triangles by segment *CD.* Classify the three triangles by sides.

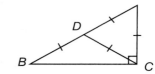

    a. △*ABC*

    b. △*BCD*

    c. △*ADC*

Early Finishers
Real-World Application

To make a small fruit smoothie, Prianka used juice, ice, 5 strawberries and 15 blueberries. She is planning a party this weekend and would like to make small smoothies for all her guests.

    a. If Prianka used 270 blueberries to make all the smoothies for the party, how many strawberries did she use?

    b. How many guests attended Prianka's pool party?

• Areas of Combined Polygons

facts   Power Up H

mental math

a. **Number Sense:** $180 \div 40

b. **Measurement:** Find the temperature indicated on this thermometer.

c. **Proportions:** $\frac{4}{16} = \frac{2}{x}$

d. **Rate:** The walkers maintained a pace of 4 miles per hour for $3\frac{1}{2}$ hours. To complete the 15-mile course, how much farther do they need to walk?

e. **Scientific Notation:** Write 125,000 in scientific notation.

f. **Geometry:** Two angles of a triangle measure 30° and 50°. Find the measure of the third angle.

g. **Estimation:** 7 at $4.97 each plus 2 at $34.97 each

h. **Calculation:** $77 \div 7, -2, \times 11, +1, \sqrt{}, -1, \sqrt{}, -10$

problem solving   A newspaper is made of folded sheets of paper. The outside sheet contains pages 1, 2, 35, and 36. The next sheet has pages 3, 4, 33, and 34. The inside sheet contains what pages?

The area of some polygons can be found by dividing the polygon into smaller parts and finding the area of each part.

Example 1

Find the area of this figure.

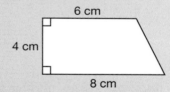

One way to find the area of this figure is to divide the polygon into a rectangle and a triangle. Then we find the area of each part and add the areas.

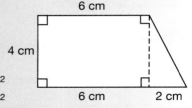

6 cm

4 cm

6 cm   2 cm

$$\begin{array}{llll} \text{Area of rectangle} = & 6 \text{ cm} \cdot 4 \text{ cm} = & 24 \text{ cm}^2 \\ + \text{ Area of triangle} & = \dfrac{1}{2} \cdot 4 \text{ cm} \cdot 2 \text{ cm} = & 4 \text{ cm}^2 \\ \hline \text{Area of polygon} & & = \mathbf{28 \text{ cm}^2} \end{array}$$

Evaluate There is another way to find the area of the polygon by dividing it into parts. Describe this way and use it to find the area of the polygon. Compare the result to the answer above.

Example 2

Use a calculator to estimate the area of this lot.

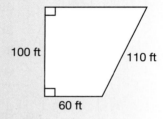

100 ft

110 ft

60 ft

Solution

We divide the figure into a rectangle and a triangle. We use the Pythagorean Theorem to find the unknown dimension of the right triangle.

$$x^2 + 100^2 = 110^2$$
$$x^2 + 10{,}000 = 12{,}100$$
$$x^2 = 2100$$
$$x \approx 46$$

The base and height of the triangle are 100 ft and about 46 ft. We add the area of the triangle and the area of the rectangle.

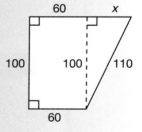

60   x

100   100   110

60

$$\begin{array}{lll} \text{Area of triangle} & \approx \dfrac{1}{2}(100 \text{ ft})(46 \text{ ft}) \approx & 2300 \text{ ft}^2 \\ + \text{ Area of rectangle} = & (100 \text{ ft})(60 \text{ ft}) = & 6000 \text{ ft}^2 \\ \hline \text{Area of lot} & \approx & \mathbf{8300 \text{ ft}^2} \end{array}$$

Example 3

In the figure, *AD* is 40 cm and *DC* is 30 cm. Point *B* is the midpoint of $\overline{AC}$ and Point *E* is the midpoint of $\overline{AD}$. Find the perimeter and area of quadrilateral *BCDE*.

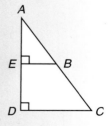

Solution

Since triangle *ACD* is a right triangle we can find that *AC* is 50 cm. A midpoint of a segment divides a segment in half. We sketch the figure again recording the dimensions (in cm) we are given and have calculated.

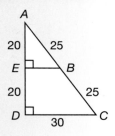

To find the length of segment *EB* we can use the Pythagorean Theorem or we can solve a proportion since the smaller and larger triangles are similar.

	△ACD	△ABE
Longer Leg	40	20
Shorter Leg	30	*x*

$$\frac{40}{30} = \frac{20}{x}$$

Thinking Skill

Analyze

By what number did we multiply 30 to solve the proportion for *x*?

Since the dimensions of the smaller triangle are half the dimensions of the larger triangle, segment *EB* is 15 cm long.

We find the perimeter of *BCDE* by adding the lengths of the four sides.

Perimeter of *BCDE* = 25 cm + 30 cm + 20 cm + 15 cm

= **90 cm**

We can find the area of figure *BCDE* by subtracting the area of the smaller triangle from the area of the larger triangle. The dimensions of the larger triangle are 2 times the dimensions of the smaller triangle, so the area is 2^2 times the area of the smaller triangle.

Area of △ACD	$(\frac{1}{2})$(40 cm)(30 cm) = 600 cm²
− Area of △ABE	$(\frac{1}{2})$(20 cm)(15 cm) = 150 cm²
Area of figure *BCDE*	= **450 cm²**

Practice Set

Find the area of each polygon by dividing the polygons into rectangles or triangles. (Angles that look like right angles are right angles).

a.

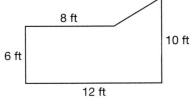

b.

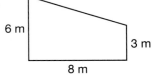

c. Find the area of the shaded region by subtracting the area of the smaller triangle from the area of the larger triangle. Dimensions are in cm.

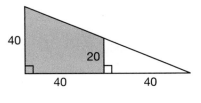

d. Use a calculator to help estimate the area of a lot with the given dimensions. Dimensions are in feet.

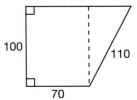

Lesson 37   247

1. Of the 1800 people at the zoo, three-fifths were children. How many
(4)   children were at the zoo?

* 2. **Analyze** At the zoo, the ratio of spotted cats to striped cats was 2 to 5.
(34)   If there were 12 spotted cats, how many striped cats were there?

* 3. In the reptile section, the ratio of lizards to snakes was 2 to 9. If there
(34)   were 36 snakes, how many lizards were there?

* 4. If the space shuttle travels completely around the world in 90 minutes,
(4)   how many times will it circle the globe in a full week?

* 5. **Generalize** The perimeter of a square is four
(8)   times the length of one of its sides. In this
table we show the perimeter of a square
with a side length of 3 units. Copy this table
on your paper. Imagine two more squares
with different side lengths. Record the side
lengths and perimeters of those squares in
your table.

$P = 4s$

Side Length	Perimeter
3	12

6. **Connect** A stop sign is the shape of a regular octagon. Many stop signs
(6, 8)   have sides 12 inches long.

   a. What is the perimeter of a regular octagon with sides 12 inches
   long?

   b. Convert your answer in part **a** to feet.

* 7. Triangles *ABD* and *ECD* are similar.
(18, 20)
   a. If $m\angle A = 60°$, then what is the measure
   of $\angle CED$?

   b. What is the measure of $\angle D$?

   c. Which side of $\triangle ECD$ corresponds to
   side *AD* of $\triangle ABD$?

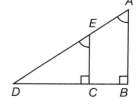

8. Solve $\dfrac{6}{10} = \dfrac{x}{15}$
(34)

9. Solve $\dfrac{x}{5} = \dfrac{4}{20}$
(34)

10. Find the **a** area and **b** perimeter of the triangle with the vertices (0, 0),
(8, Inv. 1)   (−5, 0) and (0, −12).

11. Find the **a** area and **b** perimeter of this
(8, 37)   pentagon. Dimensions are in ft.

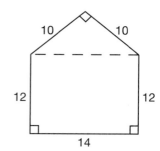

* 12. (21, 36) **Analyze** **a.** Expand: $-3(x - y + 5)$ **b.** Factor: $4x - 2$

13. (12) Write each of the following as a reduced fraction and as a decimal:

a. 99%   **b.** $\dfrac{180}{360}$

Generalize Simplify.

* 14. (36) $\dfrac{35x}{-7}$ **15.** (27) $\dfrac{x^3}{x^4 x}$

16. (15, 21) $10 + 10(10^2 - 10 \cdot 3^2)$ *** 17.** (31, 36) $(-2)^3 + (-3)^2$

18. (13, 22) $\dfrac{9}{10} - \dfrac{3}{4} \cdot \dfrac{2}{5}$ **19.** (13, 22) $\dfrac{6}{7} \div \dfrac{2}{7} + \dfrac{1}{2}$

Collect like terms.

20. (36) $x^2 + x + 1 + 3x + x^2 + 2$

21. (36) $5x - x + 2y - y$

Find all values for x which make the equation true.

22. (7, 8) $x^2 = 1$ **23.** (1, 7) $|x| + 1 = 3$

24. (31, 36) Simplify: $(-2) + (-3) - (-4) + (-5)(-2)$

25. (7) Chandra bowled five games. Her scores were 89, 97, 112, 104, and 113. What is her mean score for the five games?

26. (10) The number $-\dfrac{1}{2}$ is a member of which sets of numbers?

A real numbers   **B** integers

C rational numbers   **D** irrational numbers

27. (9, 15) Write the prime factorization of 77,000 using exponents.

28. (14) Find $n(n - 1)$ when n is 10.

* 29. (32) **Evaluate** The teacher asked for two volunteers. Andy, Benito, Chen, and Devina raised their hands. If the teacher picks two of the four volunteers,

a. what is the sample space?

b. what is the probability Benito will be picked?

30. (18, 20) The measure of $\angle ABD$ is 145°. Find the measure of

a. $\angle CBD$

b. $\angle CDB$

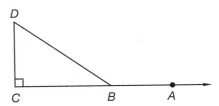

• Using Properties of Equality to Solve Equations

facts | Power Up H

mental math

a. **Number Sense:** 55×4

b. **Fractional Parts:** $\frac{7}{8}$ of $72

c. **Measurement:** Find the diameter of the quarter in millimeters.

d. **Percent:** 30% of $60

e. **Proportion:** Daniel finished 4 problems in two minutes. At that rate, how many problems can he finish in 20 minutes?

f. **Geometry:** Two sides of a triangle measure 7 inches and 12 inches. The length of the third side must be between what two lengths?

g. **Powers/Roots:** Approximate $\sqrt{26} + \sqrt{80} - \sqrt{5}$.

h. **Calculation:** $\sqrt{1} + \sqrt{4} + \sqrt{9} + \sqrt{16} + \sqrt{25} + \sqrt{36} + \sqrt{49} - \sqrt{64} - \sqrt{81} - \sqrt{100}$

problem solving

Problem solving usually requires us to use multiple strategies.

Problem: Sergio used 20 yards of fence to enclose a rectangular garden on the side of his house. He used the wall of his house for one side of the garden. He wanted the rectangle to have the greatest possible area. Find the dimensions and the area of Sergio's garden.

(**Understand**) Twenty yards of fencing material is used to enclose three sides of a garden. We are asked to determine the dimensions that will result in the greatest possible area for the garden.

(**Plan**) We will begin by drawing a diagram of the house and fenced garden. We will then write an equation for the perimeter of the three fenced sides of the garden, and use the formula for the area of a rectangle when we make an organized list of possible garden sizes.

Solve Our diagram of the yard shows a garden that is adjacent to the house and that is fenced on only three sides.

The formula for the perimeter of a rectangle is $2l + 2w = P$. Since one side of the garden is enclosed by the wall, we will use the formula $l + 2w = 20$.

The possible dimensions and the resulting area for the garden are:

Length	Width	Resulting Area
18 yards	1 yard	18 sq. yards
16 yards	2 yards	32 sq. yards
14 yards	3 yards	42 sq. yards
12 yards	4 yards	48 sq. yards
10 yards	5 yards	50 sq. yards
8 yards	6 yards	48 sq. yards
6 yards	7 yards	42 sq. yards
4 yards	8 yards	32 sq. yards
2 yards	9 yards	18 sq. yards

A garden measuring 10 yd by 5 yd will enclose the maximum area using 20 yd of fencing.

Check We found the dimensions for the garden that result in the maximum garden area. A garden that is 10 yd × 5 yd will use 20 yards of fencing ($10 + 2(5) = 20$), and will have 50 sq. yards of area in which to plant.

New Concept   *Increasing Knowledge*

We have practiced solving equations by inspection using logical reasoning. In this lesson we will use inverse operations to solve equations. Inverse operations "undo" each other.

- Addition and subtraction are inverse operations. ($n + 5 - 5 = n$)

- Multiplication and division are inverse operations. ($n \times 5 \div 5 = n$)

Any number may be used for n in the equations above to perform the operations. The result in every case is n.

We use inverse operations to isolate the variable in an equation. We can illustrate the process with a balance-scale model:

The balanced scale shows that the weight on the left side equals the weight on the right side. We can represent the relationship with this equation:

$$x + 5 = 12$$

To isolate x we need to remove the 5. Since 5 is added to x, we remove the 5 by subtracting (which is the inverse operation of addition). However, if we subtract 5 from the left side, we must also subtract 5 from the right side to keep the scale balanced.

$$x + 5 - 5 = 12 - 5$$
$$x = 7$$

We find that x equals 7.

Isolating a variable means that the variable, like x, is "alone" on one side of the equals sign.

Notice these two aspects of isolating a variable.

1. We choose the operation that isolates the variable.

2. We perform the operation on both sides of the equals sign to keep the equation balanced.

Performing the same operation on both sides of an equation preserves the equality of the equation. Below are four **properties of equality** we can use to manipulate equations.

Operation Properties of Equality		
Addition:	If $a = b$, then	$a + c = b + c$.
Subtraction:	If $a = b$, then	$a - c = b - c$.
Multiplication:	If $a = b$, then	$ac = bc$.
Division:	If $a = b$, then	$\frac{a}{c} = \frac{b}{c}$ if c is not 0.

Formulate   Use numbers to give an example of each property.

Example 1

After spending $2.30, Robert had $5.70. How much money did he have before he spent $2.30? Solve this equation to find the answer.

$$x - 2.30 = 5.70$$

Solution

We see that 2.30 is subtracted from x. To isolate x we add 2.30 to the left side to undo the subtraction, and we add 2.30 to the right side to keep the equation balanced. The result is a simpler equivalent equation.

Step:	Justification:
$x - 2.30 = 5.70$	Given equation
$x - 2.30 + 2.30 = 5.70 + 2.30$	Added 2.30 to both sides
$x = 8.00$	Simplified

We find that Robert had **$8.00.**

Example 2

If a 12-month subscription to a magazine costs $8.40, what is the average cost per month? Solve this equation to find the answer.

$$12m = 8.40$$

Solution

The variable is multiplied by 12. To isolate the variable we divide the left side by 12. To keep the equation balanced we divide the right side by 12.

Step:	Justification:
$12m = 8.40$	Given equation
$\dfrac{12m}{12} = \dfrac{8.4}{12}$	Divided both sides by 12
$1m = 0.70$	Simplified
$m = 0.70$	Simplified

Explain   In the context of the problem, what does $m = 0.70$ mean?

Example 3

Solve: $\dfrac{2}{3}w = \dfrac{3}{4}$

Solution

Thinking Skill

Connect

When the variable is multiplied by a fraction, why do we multiply by the reciprocal to help us isolate the variable?

The variable is multiplied by $\frac{2}{3}$. Instead of dividing by $\frac{2}{3}$, we multiply both sides by the multiplicative inverse (reciprocal) of $\frac{2}{3}$.

Step:	Justification:
$\dfrac{2}{3}w = \dfrac{3}{4}$	Given equation
$\dfrac{3}{2} \cdot \dfrac{2}{3}w = \dfrac{3}{2} \cdot \dfrac{3}{4}$	Multiplied both sides by $\dfrac{3}{2}$
$1w = \dfrac{9}{8}$	Simplified
$w = \dfrac{9}{8}$	Simplified

It is customary to express a solution with the variable on the left side of the equal sign. Since the quantities on either side of an equal sign are equal, we can reverse everything on one side of an equation with everything on the other side of the equation. This property of equality is called the **symmetric property.**

Symmetric Property of Equality

If $a = b$, then $b = a$.

Example 4

Solve: $1.5 = x - 2.3$

Solution

We may apply the Symmetric Property at any step.

Step:	Justification:
$1.5 = x - 2.3$	Given equation
$1.5 + 2.3 = x - 2.3 + 2.3$	Added 2.3 to both sides
$3.8 = x$	Simplified
$x = 3.8$	Symmetric Property

Practice Set

a. What does "isolate the variable" mean?

b. Which operation is the inverse of subtraction?

c. Which operation is the inverse of multiplication?

Justify Solve each equation. Justify the steps.

d. $x + \dfrac{1}{2} = \dfrac{3}{4}$ **e.** $x - 1.3 = 4.2$

f. $1.2 = \dfrac{x}{4}$ **g.** $\dfrac{3}{5}x = \dfrac{1}{4}$

h. Peggy bought 5 pounds of oranges for $4.50. Solve the equation $5x = \$4.50$ to find the price per pound.

i. From 8 a.m. to noon, the temperature rose 5 degrees to 2°C. Solve the equation $t + 5 = 2$ to find the temperature at 8 a.m.

Written Practice   *Strengthening Concepts*

1. It is 4:35 p.m. and dinner is at 5:00 p.m. If each task takes five minutes
(3, 4)   to complete, how many tasks can be completed before dinner?

* 2. The ratio of whole milk bottles to skim milk bottles at the dairy was 3 to
(34)   7. If there were 210 bottles of skim milk, how many bottles of whole milk were there?

* 3. *Justify* Printed on the label of the one-gallon milk bottle was 3.78 L.
(6, 25)

    a. How many milliliters is 3.78 liters?

    b. Which is more, a liter of milk or a quart of milk? How do you know?

4. *Analyze* Bob drives 28 miles every half hour. If he drives for $2\frac{1}{2}$ hours,
(7)
how far does he drive?

Analyze Solve problems **5–8** using inverse operations.

* 5. $x - 2.9 = 4.21$ *** 6.** $\frac{2}{3} + x = \frac{5}{6}$
(38)   (38)

* 7. $\frac{3}{5}x = \frac{6}{7}$ *** 8.** $\frac{x}{2.2} = 11$
(38)   (38)

* 9. Solve $\frac{7}{x} = \frac{21}{30}$
(34)

10. Find the **a** area and **b** perimeter of the triangle with vertices $(-4, 1)$,
(Inv. 1, 20) $(2, 1)$, and $(2, -7)$.

* 11. *Evaluate* Find the **a** area and **b** perimeter of the figure below.
(8, 20) (Dimensions are in ft.)

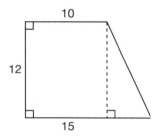

12. **a.** Expand: $-4x(2x - 9)$
(21, 36)

    b. Factor: $3x + 3$

13. Write 32% as a decimal and as a reduced fraction.
(11, 12)

Simplify.

14. $26 + 3(8^2 - 4 \cdot 4^2)$ **15.** $21 - 20(0.5)^2$
(8, 11)   (15, 25)

16. $\dfrac{-48mx}{-8x}$ **17.** $\dfrac{2}{5} - \dfrac{4}{7} \div 1\dfrac{5}{7}$
(36)   (13, 23)

18. $\dfrac{1}{9} + \dfrac{4}{6} \cdot \dfrac{4}{12}$ **19.** $\dfrac{x^5}{x^2 x} \cdot \dfrac{y^4}{y}$
(13, 22)   (27)

Generalize Combine like terms to simplify.

* 20. $xyz + yxz - zyx - x^2yz$
(31)

* 21. $gh - 4gh + 7g - 8h + h$
(31)

22. The triangles are similar. Find x and y.
(19, 20)

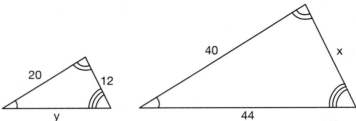

For problems **23** and **24,** find all values of x which make the equations true.

23. $x^2 = 100$
(14, 15)

24. $3 + |x| = 9$
(1, 14)

25. There are twelve sheep in the field. Eduardo notices that there are
(29) more goats in the field than there are sheep. Which of the following is a possible ratio of sheep to goats?

A $\dfrac{12}{5}$ **B** $\dfrac{12}{12}$ **C** $\dfrac{17}{12}$ **D** $\dfrac{12}{17}$

* 26. *Evaluate* A strain of bacteria doubles every minute. Beginning with one
(15) bacteria, how many would there be after 12 minutes? Write an equation using exponents and powers of 2. You may use a calculator to find the number.

27. *Model* On a number line, graph the integers that are greater than -3
(1) and less than 3.

28. Dharma has a bag that contains six pens; three are black, two are blue
(32) and one is red. If Dharma takes one pen out of the bag without looking,

    a. what is the probability of selecting a blue pen?

    b. what is the complement of selecting a blue pen?

29. Use 3.14 for π to find $2\pi r$ when $r = 5$.
(14, 25)

30. Refer to this figure to answer the following
(18) questions.

    a. Which angle is acute?

    b. Which angle is obtuse?

    c. Which angle is straight?

    d. Which two angles form a linear pair?

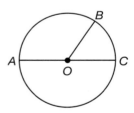

• Circumference of a Circle

facts | Power Up H

mental math

a. Number Sense: $4\frac{1}{2}$ inches $+ 13\frac{1}{2}$ inches

b. Algebra: $\frac{1}{x} = \frac{1}{2}$

c. Measurement: Find the length of this object in inches.

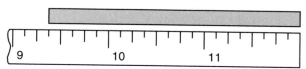

d. Rate: Peter picked 6 peppers per minute. What was his rate in peppers per hour?

e. Proportions: There are 10 computers to every 30 students in the school. What is the ratio of computers to students? If there are 600 students in the school, how many computers does the school own?

f. Scientific Notation: Write 2×10^6 in standard notation.

g. Geometry: Approximate the area of this rectangle.

<table>
<tr><td></td><td>4.9 cm</td></tr>
</table>

20.2 cm

h. Calculation: $48 \div 4, -3, \sqrt{}, -4, \times 2, +3, \times 9, \times 4, \sqrt{}, -6$

problem solving

A robot is programmed to follow this sequence of commands: forward 3, rotate 90° clockwise, backward 1, rotate 90° counterclockwise, forward 2. If the robot begins at space 7 D facing the top of the page, where does it end, and which direction is it facing?

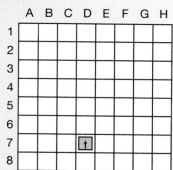

To draw a circle we can use a compass, a tool with a pivot point and a drawing point. As we swing the compass around its pivot point the drawing point traces a circle. Every point on the circle is the same distance from the center. The distance from the center to a point on the circumference of the circle is its **radius** (plural: **radii**).

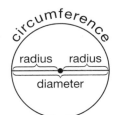

Thinking Skill

Generalize

How does the diameter of a circle relate to its radius? How does the radius of a circle relate to its diameter?

The distance across a circle measured through its center is its **diameter.**

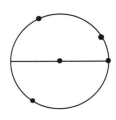

The distance around a circle is its **circumference.** Circumference is the specific name for the perimeter of a circle. The distance around a circle is related to the distance across the circle. If we take lengths equal to the diameter and wrap them around a circle, we would find that it takes a little more than three diameter lengths to complete the circumference.

The circumference of a circle is about 3.14 or $3\frac{1}{7}$ diameters, but these numbers are approximations of a special constant. A circumference is exactly π **(pi)** diameters. Thus the ratio of the circumference to the diameter of a circle is π.

$$\frac{\text{circumference}}{\text{diameter}} = \pi$$

These are two commonly used formulas for the circumference of a circle.

Circumference of a Circle

$$c = \pi d \qquad c = 2\pi r$$

To find a circle's circumference using the formula on the left, we multiply its diameter by π.

Explain How would you use the formula on the right to find circumference?

Multiplying by π involves approximation, because π is an irrational number. There is no decimal or fraction that exactly equals π. This book will use two commonly used approximations for π.

Reading Math

Recall that when numbers are approximate, we use $\approx$ in place of $=$.

$$\pi \approx 3.14 \qquad \pi \approx \frac{22}{7}$$

Applications that require greater precision may use closer approximations for π, such as 3.14159. For rough mental estimations or to check the reasonableness of a calculation we may use 3 for π.

Some calculators have a key or keystroke sequence for π.

a. If your calculator displays π, write its value to the number of places shown.

b. Arrange these numbers in order from least to greatest.

$$3.14, \frac{22}{7}, \pi$$

Example 1

The diameter of the truck wheel is 35 inches. About how far does the wheel roll in one turn? (Use $\frac{22}{7}$ for π.)

Solution

The distance the wheel rolls in one turn equals the circumference of the wheel. To find the circumference we multiply the diameter by the given approximation for π.

$$c = \pi d$$

$$c \approx \frac{22}{7} \cdot 35 \text{ inches}$$

$$c \approx 110 \text{ inches}$$

We calculate that the wheel rolls about **110 inches** in one turn. The answer seems reasonable because π is a little more than 3, and 3 × 35 in. is 105 in.

Discuss Why might you choose to use $\frac{22}{7}$ for π instead of 3.14?

Example 2

The fan belt on Fernando's car wraps around two circular pulleys. One pulley has a diameter of 8 cm. The other has a diameter of 12 cm. What is the ratio of the pulleys' circumferences?

Solution

The circumferences are 8π and 12π, so the ratio of the circumferences is

$$\frac{8\pi}{12\pi} = \frac{8}{12} = \frac{2}{3}$$

All circles are similar, so the radii, diameters, and circumferences of any two circles are proportional. The scale factor from the large to small pulley is $\frac{2}{3}$. From the smaller to the larger pulley, the scale factor is $\frac{3}{2}$.

Example 3

What effect does doubling the diameter of a circle have on the circumference?

The circumference is a length. Applying a scale factor to a figure applies the scale factor to all lengths. Therefore, **doubling the diameter doubles the circumference.** For example, if the diameter of a 6 inch circle is doubled to 12 inches, the circumference doubles from 6π inches to 12π inches.

Example 4

George Ferris designed a Ferris Wheel for the 1893 World Columbian Exposition in Chicago. The wheel was 75 meters in diameter. Calculate the circumference of the wheel to the nearest meter. (Use a calculator if available.)

Solution

Using a calculator we can multiply 75 m by 3.14 or by using the π function. Here we show sample displays.

Using 3.14 = 235.5

Using π key = 235.619449

Rounding to the nearest meter we find the circumference of the wheel was about **236 m.**

Practice Set

a. The diameter of a circle on the playground is 12 ft. What is the radius of the circle?

b. A spoke on the bike wheel is about 12 inches long. The diameter of the wheel is about how many inches?

c. What is the name for the perimeter of a circle?

d. How many diameters equal a circumference?

e. Write two formulas for the circumference of a circle.

f. Which of these terms does not apply to π?

A rational number   **B** irrational number   **C** real number

g. Justify   Explain your choice for problem **f.**

h. Write a decimal number and a fraction commonly used as approximations for π.

i. What is incorrect about the following equation?

$$\pi = 3.14$$

j. To roughly estimate calculations involving π or to check if a calculation is reasonable, we can use what number for π?

k. Estimate   A bicycle tire with a diameter of 30 inches rolls about how many inches in one turn?

l. *Justify* The diameter of a circle painted on the playground is 7 yards. The distance around the circle is about how many yards? What number did you use for π and why did you use that number?

m. The diameter of the earth at the equator is about 7927 miles. About how many miles is the circumference of the earth at the equator? Use a calculator with a π function if available. Round to the nearest hundred miles.

Written Practice   *Strengthening Concepts*

1. The dump truck traveled north 20 miles to pick up a load of rocks. Then
(3) it traveled south 7 miles to dump the load. How far is the truck from where it started? (It might be helpful to draw a picture.)

2. A puzzle has 330 pieces. How long will it take to complete the puzzle if
(4) 60 pieces are put together each hour?

3. Sarah laughed once the first hour, twice the second hour, three times
(3) the third hour, etc. If this pattern continued nine hours, how many times did Sarah laugh in all?

* 4. Leah will win the board game on her next turn if she rolls a number
(30, 32) greater than 2 with one number cube. Find her chances of winning on her next turn. Express your answer as a ratio in simplest form and as a percent rounded to the nearest whole percent.

* 5. *Analyze* In a coin collection, the ratio of buffalo nickels to non-buffalo
(34) nickels is 2 to 8. If there are 96 non-buffalo nickels, how many buffalo nickels are in the collection?

Simplify.

* 6. $(-3)(-5)(-2)$
(36)

* 7. $\dfrac{(-5)(-12)}{-10}$
(36)

* 8. $\dfrac{-6}{-3} + \dfrac{-10}{5}$
(36)

* 9. $\dfrac{-12x}{-2}$
(36)

* 10. $1\dfrac{1}{4} + 2\dfrac{1}{2} \div 1\dfrac{1}{3}$
(13, 23)

11. $\dfrac{1}{5} - \dfrac{1}{5} \cdot \dfrac{1}{2}$
(21, 22)

12. $\dfrac{4.28 - 1}{(2)(0.2)}$
(24, 25)

13. $\sqrt{5^2 + 12^2}$
(15, 21)

* 14. $\dfrac{x^0 \cdot x^5}{x^2}$
(27)

*15. $2^6 \cdot 3 \cdot 5^6$ (*Hint:* pair 2's and 5's)
(15, 27)

16. One side of a rectangle has vertices $(-2, -5)$ and $(-9, -5)$. What is the
(8,
Inv. 1) length of an adjacent side if the area of the rectangle is 21 square units? Find two possible locations for the other pair of vertices.

17. A corner was cut from a 5-inch square. Find
(8, 20) the **a** area of the square, **b** the area of the triangle that was cut from the original square, and **c** the area of the remaining shape. (Hint: subtract)

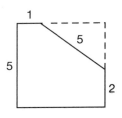

* 18. ⬛ Justify **a.** Write 25% as a reduced fraction and a decimal.
(12)

b. Which form would you find most convenient to compute 25% of $24? Explain.

* 19. ⬛ Analyze Consider these upper-case letters:
(26,
Inv. 3)
$$X \; Y \; Z$$

a. Which letter does not have reflective symmetry?

b. Which letter does not have rotational symmetry?

20. Estimate by rounding first: $\dfrac{2.8^2 + 1.1^2}{1.9}$
(17)

* 21. A coin is tossed and a number cube is rolled. What is the sample
(32) space for the experiment? What is the probability of tossing heads and rolling 1?

22. How would you rearrange these numbers so they are easier to multiply?
(1, 22) Name the properties you use.
$$3\left(2 \cdot \frac{1}{3}\right)$$

Solve by inspection.

* 23. Find two values of x that make this equation true.
(15, 36)
$$x^2 = 16$$

⬛ Analyze Solve using inverse operations.

* 24. $0.9 + x = 2.7$ *** 25.** $0.9x = 2.7$
(38)   (38)

26. The moon is about a quarter of a million miles from earth.
(5, 28)
a. The moon is about how many miles from earth?

b. Express your answer to part **a** in scientific notation.

27. Refer to the diagram.
(18)
a. Find m∠ *a.*

b. Find m∠ *b.*

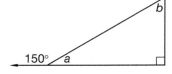

Evaluate.

* 28. Sketch a circle with a diameter of 2 inches. What is the circumference of
(39) a circle with a diameter of 2 inches? (Use 3.14 for π.)

* 29. Rafters are lengths of wood, often 2 by 6s or
(Inv. 2)   2 by 8s, that support the roof. Find the length
of the rafters for the building shown.

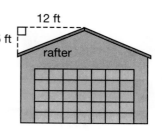

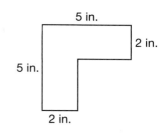

* 30. A smaller square is cut from a larger square.
(37)   What is the area of the remaining figure?

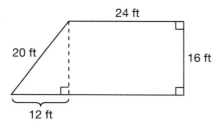

Early Finishers
*Real-World
Application*

Luz wants to place a fence around the edge of her garden and fertilize the
soil inside. A scale drawing of her garden is shown below. She plans to put
support posts 4 feet apart around the edge of the garden and to string wire
fencing between the posts. Each post costs $5.00 and the wire fencing costs
$1.50 per foot. One box of fertilizer costs $4.00 and covers 500 sq. ft. How
much will it cost Luz to fence and fertilize her garden?

• Area of a Circle

facts   Power Up H

mental math

a. **Number Sense:** 2×128

b. **Probability:** The probability of rolling a 3 on a number cube is $\frac{1}{6}$, because there is only one "3" out of six possible outcomes. What is the probability of rolling a 6?

c. **Fractional Parts:** $\frac{3}{8}$ of $48

d. **Measurement:** The odometer read 0.8 miles when she left. When she returned, it read 2.2 miles. How many miles did she travel?

e. **Rate:** Danielle drinks 8 cups of fluids a day. How long will it take her to drink 56 cups of fluids?

f. **Geometry:** Find the area of the rectangle, and then find the area of the shaded triangle.

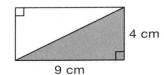

4 cm

9 cm

g. **Estimation:** $6 \times \$2.98 + 2 \times \$7.48 + \$39.98$

h. **Calculation:** $200 - 50, \div 3, -1, \sqrt{}, +2, \sqrt{}, \times 20, +4, \sqrt{}, +4, \div 4$, square it

problem solving   The mystery man listened and replied. When Bobby said 7, the mystery man said 16. When Jack said 11, the mystery man said 24. When Bao said 99, the mystery man said 200. Wally whispered. The mystery man said 40. What did Wally whispered?

The area of a circle is related to the area of a square on the radius. The area of a circle is π times the area of a square on the radius. A formula for the area of a circle is

$$A = \pi r^2$$

The area of a circle is related to the area of a square on its radius. The area is greater than the area of three of these squares but less than the area of four such squares. The area of the circle is π times the area of the square of the radius as we see in the formula above.

area of circle $> 3r^2$ area of circle $< 4r^2$

Example 1

A circular table in the library has a diameter of 6 feet. If four students sit around the table, then each student has about how many square feet of work area?

Solution

First we find the area of the tabletop. The diameter is 6 feet, so the radius is 3 feet, and the radius squared is 9 square feet. We multiply 9 square feet by π, using 3.14.

$$A = \pi r^2$$
$$A \approx 3.14 \, (9 \text{ ft}^2)$$
$$A \approx 28.26 \text{ ft}^2$$

The area of the tabletop is about 28 square feet. If four students equally share the area, then each student has about **7 square feet** of work area.

Example 2

Express the answers to a and b in terms of π.

    a. What is the circumference of this circle?

    b. What is the area of this circle?

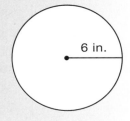

6 in.

Solution

Instead of multiplying by an approximation for π, we leave π in the answer as a factor. The parts of the answer are written in this order: rational number, irrational number, units.

    a. The circumference is π times the diameter. The diameter is 12 inches. Thus the circumference is **12π in.**

    b. The area of a circle is π times the radius squared. The square of the radius is 36 in.^{2}. Thus the area of the circle is **36π in.^{2}.**

A **sector** of a circle is a portion of the interior of a circle enclosed by two radii and an **arc** which is part of the circle.

The angle formed by the radii, called a **central angle,** determines the fraction of the area of the circle the sector occupies.

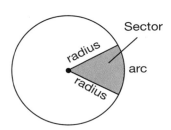

Sector

radius

arc

radius

Example 3

A lawn sprinkler waters a circular portion of a yard. The radius of the circular portion is 7 ft. Calculate the area of the yard watered by the sprinkler to the nearest square foot.

Solution

Thinking Skill

Discuss

Why might we choose to use $\frac{22}{7}$ for π in this instance?

We choose to use $\frac{22}{7}$ for π.

$$A = \pi r^2$$

$$\approx \frac{22}{7}(7 \text{ ft})^2$$

$$\approx \frac{22}{7}(\overset{7}{\cancel{49}} \text{ ft}^2)$$

$$\approx 154 \text{ ft}^2$$

The sprinkler system waters about **154 ft²**.

Example 4

What effect does doubling the diameter of a circle have on the area of the circle?

Solution

Doubling the diameter doubles the radius. The radius is squared to calculate the area, so doubling the radius results in a circle with an area **four times** (2^2) as large. For example, if the diameter is doubled from 6 in. to 12 in. the area of the circle increases fourfold from 9π in.² to 36π in.².

Example 5

Find the area of each sector. Express in terms of π.

a.

b.

16 cm

Solution

a. Sector *RMS* is bounded by angle *RMS* and arc *RS*. The 45° central angle indicates that the sector occupies $\frac{1}{8}$ of the area of the circle.

$$\frac{45°}{360°} = \frac{1}{8}$$

The area of a circle with a radius of 4 cm is 16π cm².

Step:	Justification:
$A = \pi r^2$	Area formula
$= \pi(4 \text{ cm})^2$	Substituted 4 cm for r
$= 16\pi \text{ cm}^2$	Simplified

The area of the sector is $\frac{1}{8}$ of the area of the circle.

$$\text{Area of sector} = \frac{1}{8} \cdot 16\pi \text{ cm}^2$$

$$= 2\pi \text{ cm}^2$$

b. A sector that is half a circle is called a **semicircle.** We first find the area of the whole circle. The diameter is 16 cm, so the radius is 8 cm.

Step:	Justification:
$A = \pi r^2$	Area formula
$= \pi(8 \text{ cm})^2$	Substituted 8 cm for r
$= 64\pi \text{ cm}^2$	Simplified

The area of the semicircle is half of the area of the circle.

$$\text{Area of sector} = \frac{1}{2}\left(64\pi \text{ cm}^2\right)$$

$$= 32\pi \text{ cm}^2$$

Practice Set

For **a** and **b**, find the area of each circle.

a.

5 cm

(Use $\pi = 3.14$)

b.

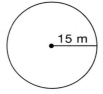

15 m

(Express in terms of π)

c. What is the scale factor from the circle in **a** to the circle in **b?** The area of the circle in **b** is how many times the area of the circle in **a?**

d. A 140 ft diameter helicopter landing pad covers an area of about how many square feet? What value did you use for π and why did you choose it?

e. A semicircular window with a radius of 36 inches allows light through an area of how many square feet? You may use a calculator if one is available. Round your answer to the nearest square foot.

In the figure, segment AC is a diameter of circle O and measures 12 cm. Central angle BOC measures 60°. Find the area of the sectors named in **f–h** below. Express each area in terms of π.

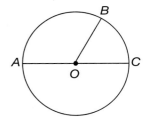

f. Sector BOC

g. Sector BOA

h. Sector AOC

1. How many hours and minutes is it from 3:45 p.m. on Tuesday to
(3) 6:30 a.m. on Wednesday?

2. The ratio of heroes to evil characters in the trilogy is 7 to 8. How many
(34) evil characters were there if there were 84 heroes?

3. The four books Sergio read during the summer contained 186 pages,
(7) 210 pages, 246 pages, and 206 pages. What was the mean number of
pages in the four books?

* 4. Find the area and circumference of the circle with diameter 6 m. Express
(40) answers in terms of π.

* 5. Find the **a** area and **b** circumference of the circle with radius 10 m. Use
(40) 3.14 for π.

* 6. The rule for this table is $y = \frac{1}{2}x$. Graph the
(Inv. 1) (x, y) pairs on a coordinate plane.

x	y
−4	−2
−2	−1
0	0
2	1
4	2

* 7. The triangles are similar.
(19, 20)
   a. Find x.

   b. Find y.

   c. What is the scale factor from the smaller
     to larger triangle?

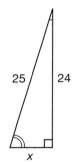

* 8. Arrange the numbers in order from least to greatest:
(30)
  A $\frac{2}{5}$ **B** $0.8\overline{8}$ **C** 0.8 **D** $0.4\overline{4}$

Analyze Solve.

* 9. $\frac{1}{4}x = 5$ *** 10.** $2.9 + x = 3.83$
(38)   (24, 38)

* 11. **Evaluate** Find the **a** area and **b** perimeter
(Inv. 2, of the figure. (Dimensions are in ft.)
37)

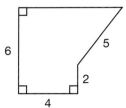

* 12. Find the **a** area and **b** perimeter of a triangle with vertices located at
(Inv. 1, 20) (2, 1), (2, −5), (−6, −5).

13. Expand and collect like terms: $x(2x − 3) − x^2 − x$
(21, 36)

Simplify.

14. $\dfrac{27x^2y^2}{3xy}$
(27)

15. $(7^2 − 2^2) − (4^2 − 1^2)$
(15, 21)

16. $\left(\dfrac{1}{2}\right)^2 − \left(\dfrac{1}{2}\right)^3$
(15, 22)

17. $5 + 5(10)^2$
(15, 21)

18. $\dfrac{1}{2} − \dfrac{3}{4} \cdot 1\dfrac{1}{2}$
(13, 23)

19. $\dfrac{2}{7} + \dfrac{5}{14} \div \dfrac{1}{2}$
(13, 22)

20. $(−2) − (−6) + (−4)$
(31, 33)

21. $\dfrac{(−2)(−6)}{(−4)}$
(36)

For problems **22** and **23** find all values of x which make the equations true.

22. $x^2 + 1 = 26$
(14, 15)

23. $|x| − 16 = 2$
(1, 14)

* 24. *Classify* Which terms below describe
(Inv. 3) this figure? Choose all terms that
apply.

  A rectangle

  B parallelogram

  C trapezoid

  D quadrilateral

* 25. *Analyze* Which quadrilateral named below does not have rotational
(Inv. 3) symmetry?

  A rectangle   **B** rhombus   **C** parallelogram   **D** trapezoid

26. Write 4.93×10^8 in standard form.
(28)

* 27. Segment *AC* is a diameter of circle *D* and
(18, 40) measures 8 cm. Central angle $\angle ADB$
measures 45°.

  a. Find the measure of $\angle BDC$.

  b. Find the area of sector *ADB* expressed in
     terms of π.

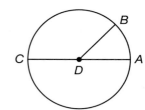

28. In baseball, the pitcher's mound has a diameter of 18 feet. Which is the
(39) best estimate of the circumference of the pitcher's mound?

  A 48 ft   **B** 50 ft   **C** 52 ft   **D** 54 ft

29. Evaluate: Find *Fd* when $F = 2.2$. and $d = 10$.
(14, 25)

30. **Evaluate** A coin is flipped and the spinner
(32) is spun. 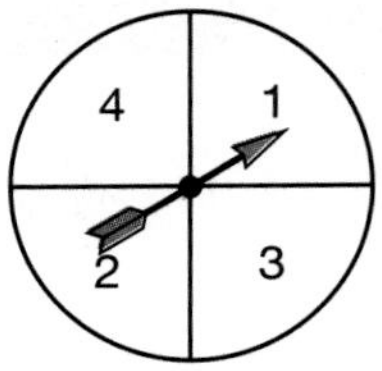
  a. Write the sample space for the
      experiment.

  b. What is the probability of getting tails and
      an odd number?

Early Finishers
Real-World
Application

The Reliant Astrodome is located in Houston, Texas. It was the first sports stadium in the nation to install Astroturf, a type of artificial grass that requires minimal maintenance. Suppose that to improve the Astrodome's outward appearance, the management wants to install new lights at 10-foot intervals around the structure of the stadium where the dome meets the vertical walls. The diameter of the Astrodome is 710 feet. How many lights would the management need to buy to implement their plan? Use 3.14 for π and round your answer to the nearest whole number.

Focus on

• Drawing Geometric Solids

Closed figures that we draw on paper, like polygons and circles, are two-dimensional. They have length and width, but they are flat, lacking depth. Physical objects in the world around us are three-dimensional; they have length and width and depth.

Two-dimensional figures are called plane figures because they are contained in a plane. Three-dimensional figures are sometimes called space figures, because they occupy space and cannot be contained in a plane. Space figures are also called **geometric solids.**

The terms **face, edge,** and **vertex** (pl. vertices) refer to specific features of solids. The face of a polyhedron is a polygon. Two faces meet at an edge and edges intersect at a vertex.

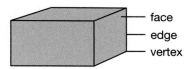

In the table below we name, illustrate, and describe some geometric solids.

Geometric Solids

Polyhedron (pl. polyhedra)		A general term that identifies a solid with faces that are polygons. A polyhedron has no curved surfaces or edges.
Prism		A type of polyhedron with parallel congruent bases.
Pyramid		A type of polyhedron with lateral surfaces that narrow to a point (apex).
Cylinder		In this book we will use the term **cylinder** to refer to a right circular cylinder as illustrated.
Cone		In this book we will use the term **cone** to refer to a right circular cone as illustrated.
Sphere		A smooth curved solid every point of which is the same distance from its center.

Prisms can be further classified by the shapes of their bases.

Triangular Prism

Rectangular Prism

Hexagonal Prism

One type of prism is a **rectangular prism** in which all faces are rectangles. One type of rectangular prism is a **cube** in which all faces are congruent squares.

Example

The pyramids of Egypt have square bases. A pyramid with a square base has how many faces, edges, and vertices?

Solution

A pyramid with a square base has **5 faces** (one square and four triangles), **8 edges**, and **5 vertices.**

Conclude How do the number of faces of a pyramid relate to the shape of its base? How do the number of faces of a prism relate to the shape of its base?

1. A triangular prism has how many faces, edges, and vertices?

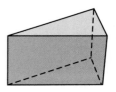

2. Name each figure. Which is not a polyhedron? Explain your choice.

A

B
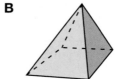

C
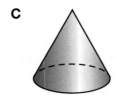

D
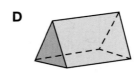

Geometric solids are three-dimensional figures. They have length, width, and depth (or height); they take up space. Flat surfaces like pages in a book, billboards, and artists' canvases are two-dimensional and lack depth. Thus, to represent a three-dimensional figure on a two-dimensional surface we need to create the illusion of depth. In this investigation we will practice ways to draw and represent three-dimensional figures in two dimensions.

Parallel projection is one way to sketch prisms and cylinders. Recall that prisms and cylinders have two congruent bases, one at each end of the figure. Using parallel projection, we draw the base twice, one offset from the other. We draw each base the same size.

Then we draw parallel segments between the corresponding vertices (or the corresponding opposite sides of the circles).

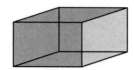
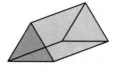

The resulting figures show all the edges as though the prisms were made with wire. If we were drawing a picture of a box or tube, the hidden edges would be erased.

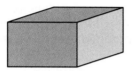

In geometric sketches we often show hidden edges with dashes.

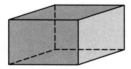
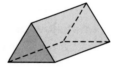

Activity 1

Sketching Prisms and Cylinders Using Parallel Projection

Materials needed:

- unlined paper
- pencil

Description of activity:

Sketch a rectangular prism, a triangular prism, and a cylinder using parallel projection. Make the bases of each figure congruent. Show hidden edges with dashes. To sketch the figures from different viewpoints, shift the position of the more distant base.

To draw pyramids and cones we can begin by sketching the base. To represent a square base of a pyramid we can draw a parallelogram. To represent the circular base of a cone we can draw an **ellipse,** which looks like a stretched or tilted circle.

Centered above the base we draw a dot for the **apex,** or peak, of the pyramid or cone. Then we draw segments to the vertices of the base of the pyramid (or opposite points of the circular base of the cone).

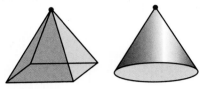

We erase hidden edges or represent them with dashes.

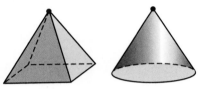

Activity 2

Sketching Pyramids and Cones

Sketch a few pyramids and cones from different viewpoints.

Model Sketch a pyramid that has a triangular base.

Discuss What steps will you follow to draw a pyramid with a triangular base?

Multiview projection displays three-dimensional objects from three perpendicular viewpoints, **top, front,** and **side** (usually the right side). An **isometric** view is often included, which is an angled view of the object, as shown in the projection below.

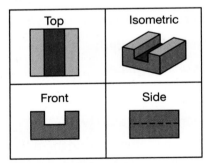

Notice that hidden edges do not appear in the isometric drawing but do appear in the other views as dashes. Notice that the width of the figure is the same in the top and front views. The height is the same in the front and side views, and the length is the same in the top and side views.

Activity 3

Create a Multiview Drawing

Materials needed:

- unlined paper
- pencil
- ruler

Description of activity:

Divide your paper into four equal sections. In the isometric section copy the sketch below. Then draw the top, front, and side views.

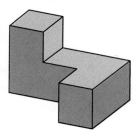

extensions

1. Build a figure from unit cubes and create a multiview projection for the figure.

2. Create a multiview projection for a building on your school's campus or in your school's neighborhood.

Parallel projection and isometric drawings do not provide the same perspective of depth our eyes perceive in the three-dimensional world. Objects appear to diminish in size the farther they are from us. For example, if we look down a straight road it may appear to nearly vanish at the horizon.

Using a **vanishing point** can provide a **perspective** to a figure as its more distant edges are shorter than corresponding nearer edges. Below we show a triangular prism drawn toward a vanishing point.

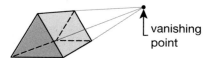

To create a one-point perspective drawing of a prism, draw a polygon in the foreground and pick a location for the vanishing point.

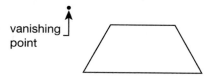

Then lightly draw segments from the vertices of the polygon to the vanishing point.

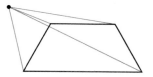

Determine the depth you want to portray and draw corresponding segments *parallel* to the sides of the polygon in the foreground. Darken the edges between the two bases. Use dashes to show hidden edges.

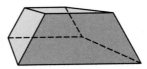

Activity 4

One-Point Perspective Drawing

Materials needed:

- unlined paper
- pencil
- ruler

Description of activity:

Create one-point perspective drawings of some prisms. Place the vanishing point in different locations to change the point of view. Try different polygons for the bases of the prisms.

• Functions

facts | Power Up I

mental math

   a. Number Sense: 19×20

   b. Probability: The probability of rolling a number less than 3 on a number cube is $\frac{2}{6}$, because there are 2 numbers less than 3 (1 and 2) out of 6 possible outcomes. Find the probability of rolling a number less than 6.

   c. Fractional Parts: 75% of $36

   d. Measurement: Find the length of this piece of cable.

   e. Scientific Notation: Write 4.06×10^6 in standard notation.

   f. Rate: The town's garbage trucks dump 7.7 tons of garbage per week. On average, how many tons per day is that?

   g. Geometry: To find volume of a rectangular prism, we multiply $l \times w \times h$. Approximate the volume of this rectangular prism.

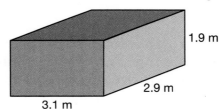

1.9 m

2.9 m

3.1 m

   h. Calculation: $41 + 14$, $\div 11$, $\times 7$, $+ 1$, $\sqrt{\ }$, $+ 3$, square it, $+ 19$, $\sqrt{\ }$, $+ 3$

problem solving | Find the sum of the whole numbers from 1 to 15.

The formula for the perimeter of a square, $P = 4s$, identifies the relationship between two variables, the length of a side of a square (s) and its perimeter (P). This formula is an example of a function.

A **function** is a mathematical rule that identifies a relationship between two sets of numbers. A function's rule is applied to an input number to generate an output number. **For each input number there is one and only one output number.** In the formula $P = 4s$, the input is the side length and the output is the perimeter.

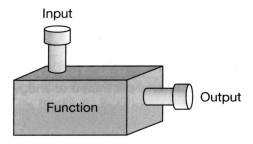

Input

Function

Output

A function may be described with words or expressed with an equation, as in the formula for the perimeter of a square. We may illustrate functional relationships in tables or in graphs.

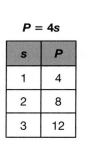

P = 4s

s	P
1	4
2	8
3	12

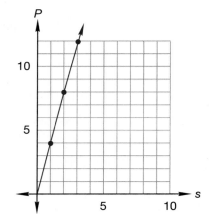

Here is a table that shows the perimeter of a square for some given side lengths, and a graph that shows other pairs of numbers that represent side lengths and perimeters of squares.

In this lesson we will consider function tables and graph some functions.

Example 1

The perimeter of an equilateral triangle is three times the length of one of its sides.

> **a. Write an equation that expresses this relationship. Use *P* for perimeter and *s* for side length.**

> **b. Choose four different side lengths. Make a function table that shows the side lengths and the perimeters of the equilateral triangles.**

> **c. Graph the pairs of numbers from the table on a coordinate plane. Let the horizontal axis represent side length (*s*) and the vertical axis represent perimeter (*P*).**

a. To find the perimeter (*P*) of an equilateral triangle we multiply the length of a side (*s*) by 3.

$$P = 3s$$

b. We can use the equation to help us fill in the table. The input number is on the left; the output number is on the right. We choose four side lengths such as 1, 2, 3, and 4. We apply the rule (multiply the side length by 3) to find each perimeter.

P = 3s

s	P
1	3
2	6
3	9
4	12

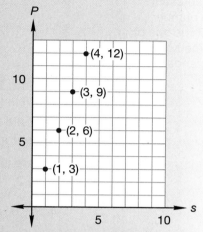

c. We graph the (*s*, *P*) pairs just as we graph (*x*, *y*) pairs. Since lengths are only positive, all possible pairs are in the first quadrant. Notice that the points are aligned.

Example 2

Xavier and Yolanda are playing a game. When Xavier says a number, Yolanda performs an operation with Xavier's number and says the result. The table shows four numbers Xavier said and Yolanda's replies. Describe the rule Yolanda uses, and write the rule as an equation. Then graph the (*x*, *y*) pairs from the table and all other pairs of numbers Xavier and Yolanda could say.

x	y
2	4
3	5
4	6
5	7

Solution

Yolanda adds 2 to the number Xavier says. The equation is

$$y = x + 2.$$

We graph the *x*, *y* pairs from the table and see that the points are aligned. We suspect other points along this line also meet the conditions of the problem. We draw a line through and beyond the graphed points and test some *x*, *y* pairs on the line to see if they fit the equation $y = x + 2$. Any point we choose on the line satisfies the equation. We conclude that the graph of all possible pairs of numbers Xavier and Yolanda could say using the rule are represented by points on the graphed line.

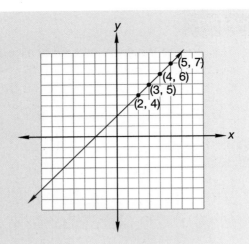

Examples 1 and 2 illustrate special classes of functions. We noted that the points plotted in both examples are aligned. If all the input-output pairs of a function fall on a line, then the function is **linear.** The functions in examples 1 and 2 are linear functions.

However, there is an important difference between the functions in example 1 and example 2. Notice that the ratio of output to input in example 2 is not constant.

Example 1
$P = 3s$

s	P	$\frac{P}{s}$
1	3	$\frac{3}{1}$
2	6	$\frac{6}{2} = \frac{3}{1}$
3	9	$\frac{9}{3} = \frac{3}{1}$
4	12	$\frac{12}{4} = \frac{3}{1}$

Proportional

Example 2
$y = x + 2$

x	y	$\frac{y}{x}$
1	3	$\frac{3}{1}$
2	4	$\frac{4}{2} = \frac{2}{1}$
3	5	$\frac{5}{3}$
4	6	$\frac{6}{4} = \frac{3}{2}$

Not Proportional

This means the relationship of perimeter to side length is **proportional,** but the relationship of Yolanda's number to Xavier's number is **not proportional.** Graphs of proportional functions have these two characteristics.

1. The function is linear.

2. The points are aligned with the origin (0, 0).

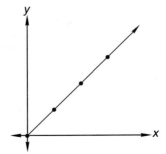

Example 3

The area of a square is a function of its side length. The graph of an area function is different from the graph of a perimeter function. For the function $A = s^2$, describe the rule with words and make a table that shows at least four pairs of numbers that satisfy the function. Then graph the (s, A) pairs and predict the graph of other points for the function.

Solution

The rule is, **"To find A, square s."** We may choose any four numbers for s to put into the function. To keep our calculations simple we choose 0, 1, 2, and 3. Then we find A for each value of s.

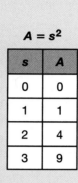

$A = s^2$

s	A
0	0
1	1
2	4
3	9

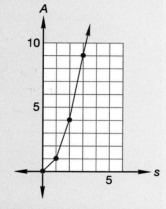

We graph the pairs of numbers from the table with s on the horizontal axis and A on the vertical axis. We see that the points are not aligned. Instead, the points seem to be on a curve that becomes steeper and steeper as the side length increases.

Discuss Is the function of side length to area linear? Is the function proportional?

Example 4

Yanos is playing a numbers game. When Xena says a number, Yanos says a number that is twice Xena's number.

    a. Write an equation for the game using x for the number Xena says and y for the number Yanos says.

    b. Make a table that shows some pairs of numbers for the game.

    c. Graph all the possible pairs of numbers for the game.

a. $y = 2x$

b. We write some numbers Xena could say in the *x*-column. Then we find each number Yanos would say for each of Xena's numbers and write them in the *y*-column.

c. The numbers in the function table form ordered pairs of numbers we can graph on a coordinate plane. We graph the ordered pairs and draw a line through the graphed points. Every point on the line represents an (x, y) pair of numbers that meets the conditions of the game.

y = 2x

x	y
0	0
1	2
2	4
−2	−4

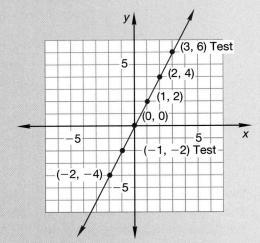

To check our work we can test some (x, y) pairs on the line to see if the pairs meet the conditions of the problem. We choose $(3, 6)$ and $(-1, -2)$. If Xena says 3, Yanos doubles the number and says 6. If Xena says -1, Yanos doubles the number and says -2. These pairs of numbers satisfy the problem, which verifies our work.

Example 5

Graph the function $y = \frac{1}{2}x - 2$.

Solution

In this lesson we will graph functions by first making a function table of (x, y) pairs that satisfy the equation. We choose values for *x* and calculate each corresponding value for *y*. Since *x* is multiplied by $\frac{1}{2}$, we choose even numbers for *x* so that *y* is a whole number. Then we plot the (x, y) ordered pairs and draw a line through them to complete graphing the equation.

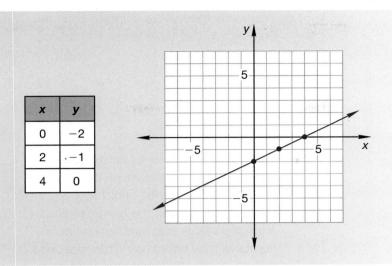

x	y
0	−2
2	−1
4	0

In this lesson, we have described functions with words and expressed functions with equations. We have listed pairs of numbers in tables that satisfy selected functions and represented these functions in graphs.

Discuss Describe some of the benefits and limitations of each of these four ways of expressing a function.

Practice Set

a. *Connect* The relationship between feet and inches is a function. Write an equation that shows how to find the number of inches (*n*, output) if you know the number feet (*i*, input). Then make a function table that shows the number of inches in 1, 2, 3, and 4 feet. Is the relationship linear? Is the relationship proportional?

b. Yolanda played a new number game with Xavier. She used this equation to generate her response to each number Xavier said:

$$y = x - 2$$

Describe with words the rule Yolanda uses. Then make a function table that shows the numbers Yolanda says for four numbers Xavier might say. Use 0 for one of Xavier's numbers and three more numbers of your choosing. Is the function linear? Is it proportional?

c. *Formulate* This table shows the capacity in ounces of a given number of pint containers. Describe with words the rule of the function and write an equation that relates pints (*p*) to ounces (*z*). Then find the number of ounces in 5 pints.

Pints	Ounces
1	16
2	32
3	48
4	64

d. State why the relationship between the numbers in this table is not a function.

x	y
0	0
1	1
1	−1
2	2

e. Yanos and Xena played a numbers game. This graph shows all the possible pairs of numbers they could say following the rule of the game. Make a function table that lists four pairs of numbers they could say. Then describe the rule Yanos followed and write an equation for the rule.

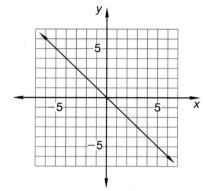

f. **Formulate** Make up a numbers game Yanos and Xena could play. Describe the rule and write an equation for the rule. Make a function table that has four pairs of numbers they could say. Graph all possible pairs of numbers for the game you create.

g. **Model** For the function $y = \frac{1}{2}x + 1$, make a function table and use it to graph the equation.

Written Practice   *Strengthening Concepts*

1. Cathy makes $18 teaching a half-hour piano lesson. She makes $32 an
(3, 4)   hour as a chef at a restaurant. How much does she make in a week if she works as a chef for 2 hours every Friday and teaches piano lessons for one hour on Mondays and 2 hours on Tuesdays?

* 2. **Analyze** The cafeteria offered six choices of vegetables and eight
(29)   choices of fruits. What was the ratio of fruit to vegetable choices?

3. Socks are sold four pairs per pack. Bobby put three complete packages
(3, 4)   of socks in the laundry. He later noticed that he had 21 socks. How many socks were misplaced in the laundry?

* 4. **Verify** Shown is a graph of the function
(41) $y = \frac{1}{2}x + 3$. Does $x = 2$ and $y = 4$ satisfy the function? Demonstrate your answer by substituting the values and simplifying.

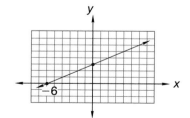

* 5. **Model** Sketch a rectangle and draw its lines of symmetry.
(Inv. 3)

* 6. **Conclude** Which pair of words correctly completes the following
(35) sentence? "Similar polygons have corresponding angles that are _____ and corresponding sides lengths that are _____."

   A similar, congruent   **B** congruent, proportional

   C acute, straight   **D** vertexes, lines

* 7. **Generalize** Use $\frac{22}{7}$ for π to find the **a** area and **b** circumference of a
(39, 40) circle with a diameter of 14 meters.

* 8. Find the **a** area and **b** circumference of the circle with radius 1 cm.
(39, 40) Express answers in terms of π.

9. At 7 p.m. the wind-chill temperature was 4°F. By 10 p.m., it had fallen
(33) 11 degrees. Write an equation to show the wind-chill temperature at
10 p.m., then solve the equation.

Solve.

* 10. $-3m = 4.2$ *** 11.** $30.7 - x = 20$
(25, 38)   (24, 38)

Analyze For problems **12** and **13**, refer to the
figure at right. (Dimensions are in meters.)

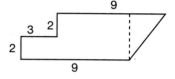

12. Find the area of the figure.
(37)

13. Find the perimeter of the figure.
(8,
Inv. 2)

14. **a.** Expand: $-4(2m - 7x + 9)$
(11, 18)

    b. Factor: $3x^2 + 3x + 3$

Simplify.

15. $\dfrac{32xmz}{-4mz}$
(27)

16. $\dfrac{2}{5} + 4\dfrac{2}{7} \div 1\dfrac{11}{49}$
(13, 23)

17. $\dfrac{2}{3} \cdot 2\dfrac{1}{4} \cdot 1\dfrac{4}{5}$
(23)

18. $36 \div [2(1 + 2)^2]$
(21)

19. $\dfrac{(-3) + (-5)}{-4}$
(31, 36)

20. $\left(\dfrac{1}{3}\right)^2 - \left(\dfrac{1}{3}\right)^3$
(6, 12)

21. $(-3)^2 + (-3)^1 + (-3)^0$
(27, 36)

22. $(2x^2y)^3$
(27, 36)

23. Write $\frac{11}{20}$ as a **a** decimal and **b** percent.
(12)

24. Three alphabet tiles are face down. The letters on the tiles are S, A,
(32) and M. Danielle picks one tile, keeps it, and picks up another.

    a. What is the sample space of the experiment?

    b. What is the probability that one of the tiles she picks up will be
an A?

Combine like terms to simplify.

25. $2xy - xy - y + 7y - xy$
(31)

26. $9x^2 - 4x + 7x - 6x^2 + 3$
(31)

Evaluate.

27. For $E = \frac{1}{2}mv^2$, find E when $m = 5$ and $v = 4$.
(14, 15)

28. The radius of the sector shown is 3 ft. State the area of
(11) the sector in terms of π.

120°

29. Florence bought 50 cm of thin leather cord to make jewelry. She
(31) used 16 cm of it for a bracelet and 34 cm for a necklace. How many
centimeters of leather cord were left?

* 30. **Evaluate** Xena and Yanos played a number
(41) game. In the table are four numbers Xena
said and the numbers Yanos said in reply.

a. Use words to describe the rule Yanos
followed.

b. Write an equation that shows the rule.

c. Sketch a graph that shows all the pairs of
numbers that fit the rule. Then draw a line
through the points to show other *x, y* pairs
of numbers that fit the rule.

X	Y
3	0
0	-3
2	-1
5	2

• Volume

Building Power

facts

mental math

Power Up I

a. **Number Sense:** 2×256

b. **Algebra:** $\frac{x}{4} = 6$

c. **Measurement:** Find the length of the piece of cable.

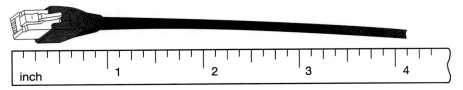

d. **Percent:** 75% of 40

e. **Scientific Notation:** Write 6.05×10^5 in standard notation.

f. **Rate:** Robert biked 12 miles in an hour. On average, how many minutes did it take him to ride each mile?

g. **Geometry:** Approximate the volume:

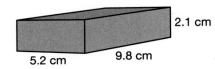

2.1 cm

5.2 cm   9.8 cm

h. **Calculation:** $\sqrt{100}$, $\times 6$, $+ 4$, $\sqrt{}$, $\times 6$, $+ 1$, $\sqrt{}$, $\times 9$, $+ 1$, $\sqrt{}$, $+ 1$, $\sqrt{}$

problem solving

In each bowling lane, 10 pins are arranged in four rows forming a triangle as shown:

If the pins from 12 lanes are combined to make one big triangle of pins, how many rows will it have?

New Concept   Increasing Knowledge

In this lesson we will find the volumes of rectangular prisms. In later lessons we will find the volumes and surface areas of other geometric solids.

The **volume** of a solid is the total amount of space occupied or enclosed by the solid. Volume is measured in cubic units such as cubic centimeters (cm^3), cubic inches (in.^{3}), and cubic feet (ft^3). The following diagram illustrates volume.

Thinking Skill

Analyze

Why are the formulas $V = Bh$ and $V = lwh$ equivalent formulas for the volume of a rectangular prism?

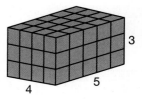

The bottom layer of this solid has 4 rows of cubes with 5 cubes in each row. Therefore, there are 20 cubes on the bottom layer. There are three layers of cubes, so there are 3×20, or 60 cubes in all.

Notice that we found the volume (V) by multiplying the area of the base (B) by the height (h). We can use this formula for any prism.

Volume of a Prism

$$V = Bh$$

The area of the base of a rectangular prism equals the length (l) times the width (w). Thus, the specific formula for a rectangular prism is:

Volume of a Rectangular Prism

$$V = lwh$$

Example 1

To calculate the heating and cooling requirement for a room, one factor architects consider is the room's volume. A classroom that is 30 feet long, 30 feet wide, and 10 feet high contains how many cubic feet?

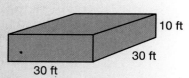

Solution

We can find the volume by multiplying the room's length, width, and height.

$$V = lwh$$
$$V = (30 \text{ ft})(30 \text{ ft})(10 \text{ ft})$$
$$\mathbf{V = 9000 \text{ ft}^3}$$

Example 2

A sculptor carved a 2-foot cube of ice into a swan. At 57 pounds per cubic foot, what did the 2-foot cube weigh before it was carved?

Solution

A 2-foot cube is a cube with edges 2 feet long. The specific formula for the volume of a cube is $V = s^3$, where s is the edge length. We will use this formula to find the volume of the cube.

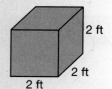

Step:	Justification:
$V = s^3$	Volume of a cube
$V = (2 \text{ ft})^3$	Substituted
$V = 8 \text{ ft}^3$	Simplified

The volume of the ice before it was carved was 8 ft³. At 57 pounds per ft³, the weight of the 2-ft cube was 8×57, or **456 pounds.**

Analyze Why are the formulas $V = Bh$ and $V = s^3$ equivalent formulas for the volume of a cube?

Activity

Volume of a Box

Find a box in the classroom such as a tissue box. Measure the length, width, and height of the box and estimate its volume after rounding the dimensions to the nearest inch.

Example 3

Many buildings (and other objects) are shaped like combinations of geometric solids. Find the volume of a building with this shape. All angles are right angles.

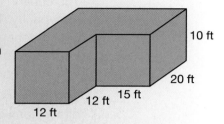

10 ft

20 ft

12 ft   15 ft

12 ft

Solution

First we find the area of the building's base. We sketch the base and divide it into two rectangles. We add the area of the two rectangles to find the area of the base (684 ft²). Now we find the volume.

$$V = Bh$$
$$= (684 \text{ ft}^2)(10 \text{ ft})$$
$$= \textbf{6840 ft}^3$$

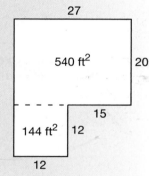

27

540 ft²   20

15

144 ft²   12

12

Example 4

Find the volume of this cube.

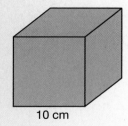

10 cm

We can choose from three formulas to calculate the volume of a cube. A cube is a prism, so we may use the formula $V = Bh$. A cube is a rectangular prism, so we may use the formula $V = lwh$. Also, since every edge (s) of a cube is the same length, we may use the formula $V = s^3$.

$$V = s^3$$
$$V = (10 \text{ cm})^3$$
$$\textbf{V = 1000 cm}^3$$

The cube in example 4 is a special cube. The metric system relates volume to capacity. The capacity of this cube (1000 cm³) is equal to one liter. That means a container of this size and shape will hold one liter of liquid. Furthermore, if the liter of liquid is water, then in standard conditions its mass would be one kilogram.

Predict How would the volume of a cube change if you doubled its dimensions?

Practice Set

a. What is the volume of a tissue box with the dimensions shown?

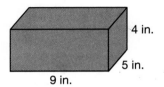

4 in.

5 in.

9 in.

b. What is the volume of a cube with edges 2 inches long?

c. Find the volume of a cube with edges 6 inches long.

d. How many times greater are the dimensions of the cube in problem **c** than the cube in problem **b?** How many times greater is the volume?

e. What is the volume of this block T? (All angles are right angles).

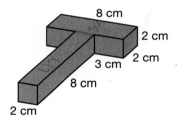

8 cm

2 cm

2 cm

3 cm

8 cm

2 cm

f. **Estimate** Find a box at home that is the shape of a rectangular prism (such as a cereal box) and measure its dimensions. Then sketch the box on your paper and record its length, width, and height. Estimate the capacity of the box in cubic inches by rounding the measurements to the nearest inch before using the volume formula.

Written Practice   *Strengthening Concepts*

1. **Evaluate** Jason has two part-time jobs. He earns $15 an hour working
(3, 4) at the help desk at a local library. He also mows lawns earning $25 per lawn. How much does he earn in a week if he works at the help desk for 2 hours each on Monday and Wednesday and for 4 hours on Friday and he mows 4 lawns on the weekend?

2. There were 35 plastic cups and 15 one-liter bottles of water on a table.
(29) What was the ratio of plastic cups to bottles of water?

3. The dimensions of a rectangular Olympic-sized swimming pool are
(42) 2 meters deep, 25 meters wide, and 50 meters long. What is the volume of the pool? If 1 m³ is equivalent to about 264 gallons, how many gallons of water does the pool hold?

4. *Justify* Shown is a graph of the function
(41) $y = \frac{1}{4}x + 1$. Does $x = 8$ and $y = 3$
satisfy the function? Demonstrate your answer by substituting the values and simplifying.

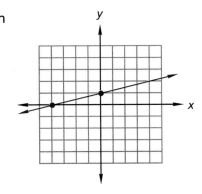

5. a. Using the figure and vanishing point shown below, draw a
(Inv. 4) polyhedron.

b. What is the name of the polyhedron you drew?

vanishing point
●

6. Lars stitched three right triangular sails for
(35) sailboats as shown. Which statement about the sails is true?

A △A and △B are congruent.

B △A and △C are congruent.

C △C and △B are similar, but not congruent.

D △B and △C are congruent and both are similar to △A

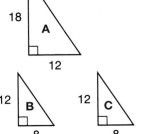

7. Use a calculator to find the **a** area and **b** circumference of a circle with a
(4) diameter of 28 feet. Use $\frac{22}{7}$ for π.

8. *Evaluate* On Saturday, Margo bought six packages of rawhide treats
(3, 4) for her dog. Each package contained 8 treats. By the following Saturday there were 44 treats left. If Margo continues to give her dog treats at this rate, how many weeks will the treats she bought last?

9. Use this table to answer **a–c.**
(18, 41)

x	y
−2	−6
0	−2
2	2
4	6

    a. Use words to describe the rule shown in the table.

    b. Write an equation that shows the rule.

    c. Sketch a graph that shows the pairs of numbers in the table. Then draw a line through the points to show other (x, y) pairs of numbers that fit the rule.

Simplify.

10. $3 \times 2(6 - 4)^3$
(15, 21)

11. $\dfrac{5}{8} + 3\dfrac{2}{3} \div 1\dfrac{1}{21}$
(13, 22)

12. $\dfrac{4}{9} \cdot \dfrac{10}{9} \cdot \dfrac{3}{8}$
(13, 22)

13. $48 \div [3(4 - 2)^2]$
(15, 21)

14. $\left(\dfrac{1}{2}\right)^3 - \left(\dfrac{1}{2}\right)^4$
(13, 22)

15. $\dfrac{105xyz}{-5yz}$
(15, 27)

Solve.

16. $-5s = 21.5$
(13, 22)

17. $45.6 - m = 25$
(13, 22)

18. If the thickness of an average sheet of 8.5-inch by 11-inch copy paper is 0.004 inches (or about 0.1 mm) then what is the volume of the sheet of paper in cubic inches?
(42)

19. **a.** Expand: $-3(4n - 6w + 5)$
(18, 21)

    b. Factor: $8y^2 + 8y + 8$

20. The spinner is spun once. What is the probability the spinner will stop
(32)

    a. in sector A?

    b. in sector A, B or C?

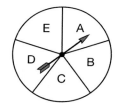

21. Find the **a** circumference and **b** area of a circle with a diameter of 8 meters. Express the answers in terms of π.
(40)

22. In a parking lot, the ratio of cars to trucks is 5 to 1. If there are 20 trucks in the parking lot, how many cars are there? Write a proportion and use it to solve the problem.
(34)

For problems **23–26**, find all values of x, if any, that make the equations true.

23. $x^2 + 3 = 7$
(14, 15)

24. $-5|x| = 25$
(14)

25. $3x^2 = 27$
(14, 15)

26. $|x| - 4 = 12$
(14)

27. **Formulate** A submarine was at a depth of 255 meters below sea
(33)   level. The submarine rose 65 m. Then it dove 25 m. Write an equation
with integers to express the situation. Then find the new depth of the
submarine.

Combine like terms to simplify.

28. $3xy - xy - 2y + 5y$
(21)

29. $5x^2 - 3x + 5x - 4x^2 + 6$
(21)

* 30. Recall that the graph of a proportional relationship has these
(41)   characteristics:

   1. The graph is a line or aligned points.

   2. The graph is aligned with the origin.

   Which graph below indicates a proportional relationship? Explain why
   the other graphs do not show a proportional relationship.

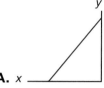

A. x

B. x

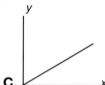

C. x

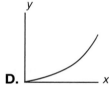

D. x

• Surface Area

facts Power Up I

mental math

a. **Number Sense:** $700 \div 200$

b. **Fractional Parts:** $\frac{1}{3}$ of $243

c. **Scientific Notation:** Write 38,100 in scientific notation.

d. **Proportions:** If Darla can read 6 pages in 10 minutes, how many pages can she read in half an hour?

e. **Geometry:** Find the area of the rectangle. Then find the area of the shaded triangle.

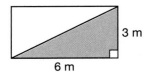

3 m

6 m

f. **Proportions:** $\frac{7}{3} = \frac{x}{9}$

g. **Powers/Roots:** Estimate $\sqrt{120} - \sqrt{99} + \sqrt{80}$

h. **Calculation:** $77 - 7, \div 7, - 7, \times 7, + 7, \div 7 + 7, \times 7$

problem solving

Rolling a number cube has 6 different possible outcomes (from 1 to 6). Rolling two number cubes has 11 different possible outcomes (from 2 to 12). How many different possible outcomes does rolling 5 number cubes have?

New Concept Increasing Knowledge

The surface area of a solid is the combined area of the surfaces of the solid. We distinguish between total surface area and lateral surface area, which is the combined area of surfaces on the sides of a solid and does not include the area(s) of the base(s). We can find the lateral surface area of prisms and cylinders by multiplying the perimeter or circumference of the base by the height.

Example 1

The figure represents a cube with edges 10 cm long.

 a. Find the total surface area of the cube.

 b. Find the lateral surface area of the cube.

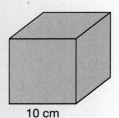

10 cm

a. All edges of a cube are the same length. So the length, width, and height of the cube are each 10 cm. Thus, the shape of each face is a square with an area of 100 cm². Since the cube has six congruent faces, the total surface area is

$$6 \times 100 \text{ cm}^2 = \mathbf{600 \text{ cm}^2}$$

b. The lateral surface area of the cube is the combined area of the four side faces, not including the area of the top or the bottom of the cube. One way to find the lateral surface area of a cube is to multiply the area of one face of the cube by 4.

Lateral surface area of the cube = $4 \times 100 \text{ cm}^2$

$$= \mathbf{400 \text{ cm}^2}$$

Another way to find the lateral surface area is to multiply the perimeter of the base by the height of the cube.

Lateral surface area = perimeter of base · height

$$= 40 \text{ cm} \cdot 10 \text{ cm}$$

$$= \mathbf{400 \text{ cm}^2}$$

Example 2

Before buying paint for the sides of this building, Malia calculated its lateral surface area. Find its lateral surface area.

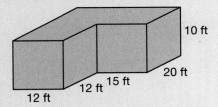

Solution

We can find the lateral surface area by multiplying the perimeter of the base by the height of the building.

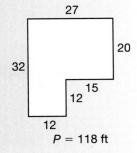

Lateral surface area = perimeter of base · height

$$= 118 \text{ ft} \cdot 10 \text{ ft}$$

$$= \mathbf{1180 \text{ ft}^2}$$

A net is a two-dimensional image of the surfaces of a solid. A net can help us visualize a solid's surfaces. Here is a net of a cube. We see the six square faces of the cube.

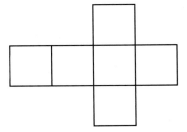

Surface Area of a Box

Find a box in the classroom that has six complete faces.

1. Identify the faces with terms such as front, back, top, bottom, left side, right side.

2. Which faces are congruent?

3. Sketch a net of the box predicting its appearance if some edges were cut and the box were unfolded.

4. Unfold the box, cutting and taping as necessary, to reveal a net of six rectangles. Find the surface area by adding the areas of the six rectangles.

Example 3

Thinking Skill

Connect

How could we use the lateral surface area of the box to find the total surface area?

The figure shows the dimensions of a cereal box. Estimate the surface area of the box.

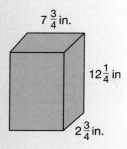

$7\frac{3}{4}$ in.

$12\frac{1}{4}$ in

$2\frac{3}{4}$ in.

Solution

We round to the nearest whole number before performing the calculations.

To estimate the surface area we estimate the area of the six surfaces.

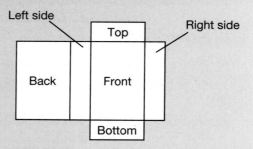

Left side

Top

Right side

Back

Front

Bottom

Area of front	8 in. × 12 in.	= 96 in.²
Area of back	8 in. × 12 in.	= 96 in.²
Area of top	8 in. × 3 in.	= 24 in.²
Area of bottom	8 in. × 3 in.	= 24 in.²
Area of left side	12 in. × 3 in.	= 36 in.²
Area of right side	12 in. × 3 in.	= 36 in.²
Total Surface Area		= **312 in.²**

The surface area of the box is **about 312 in.²**

Practice Set

a. What is the lateral surface area of a tissue box with the dimensions shown?

b. What is the total surface area of a cube with edges 2 inches long?

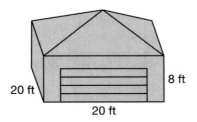
4 in.
5 in.
9 in.

c. Estimate the surface area of a cube with edges 4.9 cm long.

d. **Analyze** Kwan is painting a garage. Find the lateral surface area of the building.

20 ft
8 ft
20 ft

e. Find a box at home (such as a cereal box) and measure its dimensions. Sketch the box on your paper and record its length, width, and height. Then estimate the number of square inches of cardboard used to construct the box.

Written Practice   *Strengthening Concepts*

1. Desiree drove north for 30 minutes at 50 miles per hour. Then, she drove
(7) south for 60 minutes at 20 miles per hour. How far and in what direction is Desiree from where she started?

2. In the forest, the ratio of deciduous trees to evergreens is 2 to 7. If there
(34) are 400 deciduous trees in the forest, how many trees are in the forest?

3. Reginald left his house and rode his horse east for 3 hours at 9 miles
(7) per hour. How fast and in which direction must he ride to get back to his house in an hour?

* 4. **Analyze** What is the volume of a shipping box with dimensions
(42) $1\frac{1}{2}$ inches $\times$ 11 inches $\times$ 12 inches?

5. How many square inches is the surface area of the shipping box
(43) described in problem **4?**

* 6. How many edges, faces, and vertices does a
(Inv. 4) triangular pyramid have?

* 7. Graph $y = -2x + 3$. Is $(4, -11)$ on the line?
(41)

* 8. Find **a** the area and **b** circumference of the circle with a radius of 6 in.
(39, 40) Express your answer in terms of π.

* 9. **Formulate** A hotel shuttles 50 people per
(41) bus to an amusement park. The table at the
right charts the number of busses x and the
corresponding number of people y shuttled
to the park. Write an equation that shows the
relationship in this table.

x	y
1	50
2	100
3	150
4	200

For problems **10** and **11**, solve using inverse operations:

10. $\dfrac{x}{1.1} = 11$
(25, 38)

11. $x + 3.2 = 5.14$
(24, 38)

At right is a diagram of a grassy schoolyard.
Refer to this diagram to answer problems **12**
and **13**. Dimensions are in yards.

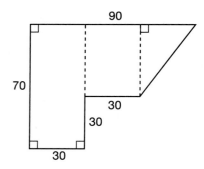

* 12. **Evaluate** How many square yards are there to mow?
(37)

13. The athletes do cardiovascular conditioning by running laps around the
(8, perimeter of the entire schoolyard. How far do they run in one lap?
Inv. 2)

* 14. **a.** Expand: $-5(x^2 - x + 4)$
(21, 36)

 b. Factor: $7x + 7$

Simplify.

15. $\dfrac{20\,wx^2}{5\,x^2}$
(27)

16. $\left(\dfrac{1}{4}\right)^2 - \left(\dfrac{1}{2}\right)^4$
(13, 27)

17. $\dfrac{7}{8} - \dfrac{12}{13} \div \dfrac{16}{13}$
(13, 22)

18. $\dfrac{1}{2} \cdot \dfrac{2}{3} \cdot \dfrac{3}{4}$
(22)

19. $\dfrac{1 + 3^2}{(1 + 3)^2}$
(15, 21)

20. $\dfrac{(-3)(-4)}{(-3) - (-4)}$
(33, 36)

For problems **21–22**, find all values of x which make the equations true.

21. $\dfrac{x^2}{2} = 18$
(14, 15)

22. $\dfrac{|x|}{5} = 10$
(1, 14)

23. Write 0.005 **a** as a percent and **b** as a reduced fraction.
(11, 12)

24. Express $\frac{5}{6}$ **a** as a decimal number, and **b** as a percent.
(30)

25. Write the prime factorization of 9000 with exponents.
(9)

26. Todd wondered if students were likely to favor certain numbers when
(32) choosing a number between 1 and 10, so he conducted a survey.
He asked 60 students to pick a number from 1 to 10 and tallied the
choices.

1	2	3	4	5	6	7	8	9	10
|||	ⵏⵏ	ⵏⵏ		|||	ⵏⵏ ⵏⵏ		ⵏⵏ	ⵏⵏ ⵏⵏ ⵏⵏ	||||	|||	ⵏⵏ

Before the survey, Todd guessed that all numbers were equally likely to
be chosen. Based on his hypothesis,

a. What is the theoretical probability that a student would choose the
number 7?

b. Based on Todd's survey, what is the experimental probability that a
student will choose the number 7?

c. Was Todd's hypothesis confirmed in the survey?

27. Combine like terms to simplify: $x^2 + 4x + 4 + 2x^2 - 3x - 4$
(31)

28. Find mgh when $m = 3$, $g = 9.8$, and $h = 10$.
(14)

29. Segment *PR* is a diameter of circle *M*
(18, 40) and measures 20 in. Central angle *QMR*
measures 60°.

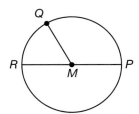

a. Find m ∠*QMP*.

b. Find the area of the semicircle *RMP* (Use
$\pi \approx 3.14$).

* 30. **Model** For **a–c**, refer to this figure
(Inv. 4, constructed of 1-cm cubes.
42, 43)

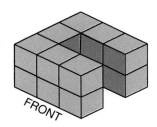

a. On grid paper draw the front view, top
view, and right-side view of this figure.

b. What is the total surface area of this
figure?

c. What is the volume of this figure?

• Solving Proportions Using Cross Products
• Slope of a Line

facts | Power Up I

mental math

a. **Number Sense:** $6\frac{1}{2} \times 20$

b. **Fractional Parts:** David tipped $\frac{1}{5}$ of the $35 meal bill. What was his tip?

c. **Measurement:** Find the temperature indicated on this thermometer.

d. **Rate:** Charlene drove 55 miles per hour for 4 hours. Ralph drove 50 miles per hour for 5 hours. How far did each drive?

e. **Geometry:** Two sides of a triangle are 16 m and 20 m. The third side is between what two lengths?

f. **Scientific Notation:** Write 2.38×10^7 in standard notation.

g. **Estimation:** Approximate the total for this shopping bill: 3 items at $1.99 each, 4 items at $2.49 each, and 2 items at $6.99 each.

h. **Calculation:** $5 + 2, \times 6, \div 7, \times 8, + 1, \div 7, \div 7, \times 12, \div 3, \times 5, + 1, \div 7$

problem solving

a. Find the square root of the sum of the first 2 positive odd numbers.

b. Find the square root of the sum of the first 3 positive odd numbers.

c. Find the square root of the sum of the first 4 positive odd numbers.

d. Describe the pattern.

e. Using the pattern you noticed, what do you think is the square root of the sum of the first 17 positive odd numbers?

solving proportions using cross products

We have solved proportions by finding equivalent ratios just as we found equivalent fractions. To solve some proportions it is helpful to use another method. In this lesson we will practice using cross products to solve proportions.

Math Language

Recall that a **proportion** shows that two ratios are equal.

A **cross product** is the result of multiplying the denominator of one fraction and the numerator of another fraction. A characteristic of equal ratios is that their cross products are equal.

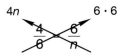

$$8 \cdot 3 = 24 \qquad\qquad 4 \cdot 6 = 24$$

If we know that two ratios are equal, we can use cross products to help us find an unknown term in one of the ratios.

$$4n \qquad\qquad 6 \cdot 6$$

We do not know n, but we know that the cross product $4n$ equals the cross product $6 \cdot 6$.

$$4n = 6 \cdot 6$$

We can solve this equation to find n.

Step:	Justification:
$4n = 6 \cdot 6$	Equal ratios have equal cross products
$\dfrac{4n}{4} = \dfrac{6 \cdot 6}{4}$	Divide both sides by 4.
$n = 9$	Simplified

Example 1

Solve using cross products: $\dfrac{m}{10} = \dfrac{15}{25}$

Solution

The ratios are equal. We find the cross products and solve for m.

Step:	Justification:
$\dfrac{m}{10} = \dfrac{15}{25}$	Given proportion
$25m = 10 \cdot 15$	Equal ratios have equal cross products.
$\dfrac{25m}{25} = \dfrac{10 \cdot 15}{25}$	Divide both sides by 25.
$m = 6$	Simplified

Justify Explain how we can check the solution.

slope of a line

In Lesson 41 we graphed lines to illustrate the relationship between two variables. Lines on a coordinate plane may be horizontal, vertical, or slanted one way or the other. We use a number called the *slope* to indicate the steepness of the line and whether the line is slanted "uphill" (positive) or "downhill" (negative).

Slope

The **slope** of a line is the ratio of the **rise** to the **run** between any two points on the line.

$$\text{slope} = \frac{\text{rise}}{\text{run}}$$

These two graphs illustrate the rise and run of two lines.

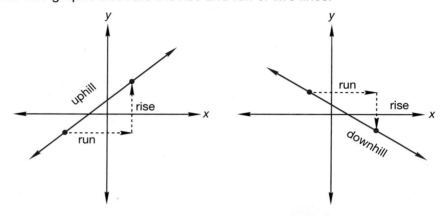

The *run* starts from a point on the line and moves over to the right.

The *rise* starts at the end of the run and moves up or down to meet the line.

Train your eyes to read lines on graphs from left to right like you read words.

"Uphill" lines have a positive slope.

"Downhill" lines have a negative slope.

Infer Which of the graphs above has a positive slope? Which has a negative slope?

Here are four examples of calculating slope.

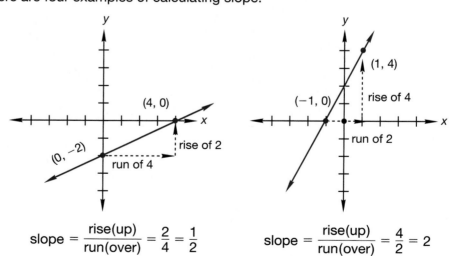

$$\text{slope} = \frac{\text{rise(up)}}{\text{run(over)}} = \frac{2}{4} = \frac{1}{2}$$

$$\text{slope} = \frac{\text{rise(up)}}{\text{run(over)}} = \frac{4}{2} = 2$$

Infer Which of the two graphs on the previous page has the steeper slope? Support your answer.

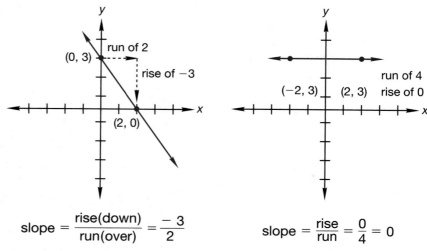

$$\text{slope} = \frac{\text{rise(down)}}{\text{run(over)}} = \frac{-3}{2}$$

$$\text{slope} = \frac{\text{rise}}{\text{run}} = \frac{0}{4} = 0$$

Thinking Skill

Explain

Can we use the words *uphill* and *downhill* to describe a line with a slope of zero? Explain.

Note that **the slope of a horizontal line is zero.** We do not assign a slope to a vertical line. Instead, we say that **the slope of a vertical line is undefined.**

Discuss Why is the slope of a vertical line undefined? (Hint: Think about the run of a vertical line.)

Two points on a graph represent an interval. The slope between the two points is the average **rate of change** of one variable relative to the other in that interval.

Example 2

Derek drives the highway at a steady rate, noting the time and distance he has traveled. This graph describes the distance Derek travels when driving at a constant rate of 50 mph. Find the average rate of change in miles per hour Derek drives over these three intervals: zero and 1 hour, 1 hour and 3 hours, and 3 hours and 4 hours.

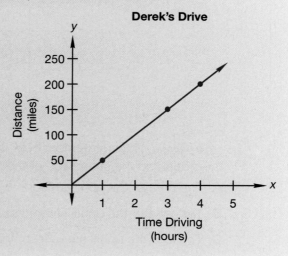

Solution

We find that the average rate of change is **50 miles per hour** for each of the three intervals.

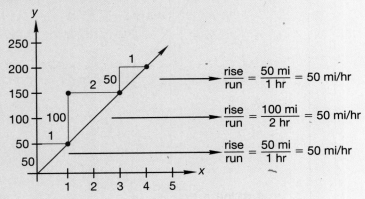

$$\frac{\text{rise}}{\text{run}} = \frac{50 \text{ mi}}{1 \text{ hr}} = 50 \text{ mi/hr}$$

$$\frac{\text{rise}}{\text{run}} = \frac{100 \text{ mi}}{2 \text{ hr}} = 50 \text{ mi/hr}$$

$$\frac{\text{rise}}{\text{run}} = \frac{50 \text{ mi}}{1 \text{ hr}} = 50 \text{ mi/hr}$$

Notice that if a function is linear (the points are aligned as on the graph above), then the slope between any two points is constant.

Practice Set

Analyze Use cross products to solve the proportions in **a–d**.

a. $\dfrac{8}{12} = \dfrac{12}{w}$

b. $\dfrac{3}{12} = \dfrac{x}{1.6}$

c. $\dfrac{y}{18} = \dfrac{16}{24}$

d. $\dfrac{0.8}{z} = \dfrac{5}{1.5}$

Find the slope of lines **e–g**.

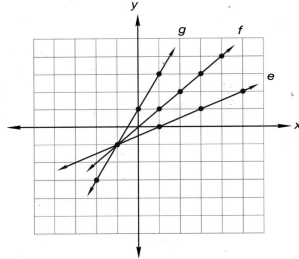

h. *Generalize* Which statement best describes the relationship between the slope and the steepness of the line?

   A The greater the uphill steepness, the less the slope.

   B The greater the uphill steepness, the greater the slope.

   C The less the uphill steepness, the greater the slope.

i. *Represent* Graph the line that passes through the origin and $(2, -4)$. Then find the slope of the line.

j. What is the slope of a line passing through $(1, 1)$ and $(5, -1)$?

1. A marine biologist reported the results of a survey that found 3 out of
(34) 5 lobsters are undersized and if caught, must be put back in the water.
If a lobsterman traps 260 lobsters, about how many should he expect
to toss back? Write a proportion and solve.

2. During their third season, a baseball team played 36 games and won $\frac{5}{9}$
(22) of them. How many games did they win?

A 20   **B** 25   **C** 30   **D** 45

* 3. **Connect** Tomás plans to build a circular pen for his dog. The radius of
(39) the pen will be 17.5 feet. Approximately, how many feet of fencing will
Tomás use for the pen?

* 4. **Analyze** The dimensions of a pasta box are 14 in. by 1.5 in. by 3 in.
(43) What is the surface area of the box?

* 5. Geri wants to fill a rectangular planter box with soil. If the dimensions
(42) of the box are $2\frac{1}{2}$ ft by $1\frac{3}{4}$ ft by 4 ft long. How many cubic feet of soil will
Geri need?

* 6. Anna makes leather coin purses and donates them to the local
(41) children's hospital fundraiser fair. The coin purses sell for $4 each.
Make a function table that relates the number of coin purses sold to
the amount of money collected. Write an equation for the rule. Is the
function linear? Is it proportional?

* 7. An artist makes rugs by painting intricate designs on canvas. He has
(40) been commissioned by a local store to create a 12-foot diameter
circular rug. He prices his rugs by their area. About what is the area of
this 12-foot rug?

* 8. **Justify** Consider the function $y = \frac{2}{3}x - 1$.
(41)
    a. Does $(9, -5)$ satisfy this function? Why or why not?

    b. If the value of x is 15, what is y? Write your answer as an ordered
    pair.

Simplify.

9. $6\frac{3}{4} - \frac{3}{5} \div \frac{1}{5}$
(13, 22)

10. $(2.5)^2 - \left(\frac{3}{2}\right)^2$
(15, 25)

11. $(-10) - (-4)$
(33)

12. $\dfrac{(-4)(-8) \div (-2)}{-2}$
(36)

* 13. $(3x^2y)(-2xy)^3$
(27, 36)

*14. $5(40) \div [-5(1 - 3)^3]$
(21, 36)

15. $\left[\frac{3}{5} - \left(-\frac{1}{2}\right)^2 + \frac{9}{10}\right]^0$
(13, 27)

16. $\dfrac{10^8}{10^3}$
(27)

17. Francisco and Rafael are baseball players. Francisco has had 31 hits in
(32) 75 at-bats, while Rafael has had 40 hits in 83 at-bats. Between the two,
who is more likely to get a hit in his next at-bat? Explain your answer.

18. A 1-inch garden hose had a small leak. Ali patched the leak by wrapping
(39) plastic tape around the hose 6 times. About how many inches of tape
did Ali use?

A 6 in. **B** 12 in. **C** 18 in. **D** 21 in.

19. There was very little rain in Austin one summer. In June, the depth
(31) of Lake Travis fell 3 inches from its usual level. In July, it went down
another 5 inches. In August it finally rained, adding 2 inches to the lake's
depth. How many inches below the usual level was the lake at that
point?

20. It is estimated that at the turn of the 20th century, one farmer in the U.S.
(34) could feed 25 people, whereas today, that ratio is about 1 to 130. How
many farmers would it take today to feed 2600 people?

21. Pedro collects baseball caps. Three-fourths of the caps are red, and he
(38) has 12 red caps. How many caps does Pedro have? Write an equation
and solve.

22. The larger triangle is a dilation of the smaller
(35) triangle. What is the scale factor?

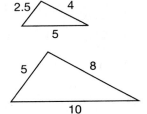

23. Sergio's front yard has the shape shown. He
(37) is going to sod his lawn. Before he buys the
sod, Sergio needs to know the area. What is
the area of the front yard?

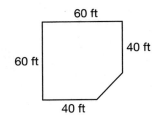

24. **a.** Write 4% as a reduced fraction and as a decimal.
(11)
b. Which form would you find most convenient to answer the following
question? "If 4% of a class of 25 students were absent, how many
were absent?"

25. Solve by inspection. $5x + 3 = 28$
(14)

26. Find the mean time of Bert's last five downhill skiing runs: 98 sec,
(7) 90 sec, 102 sec, 97 sec, and 113 sec.

27. Find the area and perimeter of the quadrilateral with vertices $(-4, -1)$,
(8, Inv. 3) $(-4, 7)$, $(4, 7)$ and $(4, -1)$. What kind of quadrilateral is it?

28. The science club charges $1.00 to join and 50¢ dues per week so that
(14) they can go on a field trip at the end of the year. The equation for dues
(d) collected and the number of weeks (w) is $d = 0.50w + 1$. Katja has
been a member for 18 weeks. How much has she paid?

29. A line passes through points (4, 2) and (0, −6). Sketch the line and find
(44) its slope.

* 30. **Evaluate** At Hilbert's hat store, Hilbert has a strategy for pricing hats.
(41) The rates are shown on the graph below.

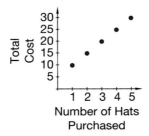

Explain Hilbert's pricing strategy. Is the relation proportional? Why or why
not? If it is a proportional relationship, write an equation that describes
the graph and state the constant of proportionality.

Early Finishers
Real-World
Application

A local rental company rents cars for $45.50 per day plus $0.10 per mile.
Azam rented a car for one day to take a drive along the coast. Write a
function that represents the cost of his trip (use c to represent cost and m to
represent distance in miles) and create a function table showing the cost of a
50-mile trip, a 125-mile trip, and a 160-mile trip.

• Ratio Problems Involving Totals

facts | Power Up I

mental math

a. **Number Sense:** $8\frac{1}{2} \times 4$

b. **Probability:** What is the probability of rolling a number less than 1 on a number cube?

c. **Algebra:** $12x = 6$

d. **Measurement:** Find the length of the object in inches:

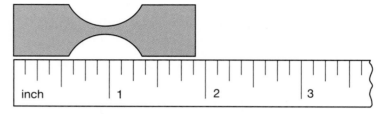

e. **Scientific Notation:** Write 40,800 in scientific notation.

f. **Rate:** Nathaniel drives north at a rate of 60 miles per hour. Sally drives south at a rate of 60 miles per hour. If they started at the same place and time, how far apart are they after one hour?

g. **Geometry:** Approximate the volume of this box:

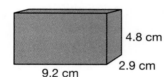

4.8 cm
2.9 cm
9.2 cm

h. **Calculation:** $10 \times 10, + 44, \sqrt{}, - 3, \sqrt{}, - 3, \sqrt{}$

problem solving | Three friends ordered 3 different sandwiches and 3 different drinks. Annie ordered tuna, but didn't order water. Bernice ordered lemonade. Calvin did not order grilled cheese. Who ordered the cranberry juice? Who ordered the chicken sandwich?

New Concept | Increasing Knowledge

Some ratio problems require us to consider the total to solve the problem. For these problems we add a third row for the total to our ratio table.

Example 1

Acrobats and clowns converged on the center ring in the ratio of 3 to 5. If a total of 24 acrobats and clowns performed in the center ring, how many were clowns?

We are given the ratio of acrobats to clowns. We can add the ratio numbers to find a ratio number for the total. We are given the actual total.

	Ratio	Actual Count
Acrobats	3	a
Clowns	5	c
Total	8	24

There are three rows in the table. We can use the numbers in two rows to write a proportion. We use the row for the number we want to find (clowns), and we use the row in which we know both numbers (total).

	Ratio	Actual Count
Acrobats	3	a
Clowns	5	c
Total	8	24

$$\rightarrow \frac{5}{8} = \frac{c}{24}$$
$$c = 15$$

We solve the proportion and find that there were **15 clowns**.

Thinking Skill

Formulate

Write and solve a proportion to find the total number of acrobats.

Connect What is the relationship between the ratios and the actual counts in example 1? How could you use this information to solve the proportion a different way?

Example 2

A bus company has small and large buses in the ratio of 2 to 7. If the company has 84 large buses, how many buses does it have?

Solution

The question involves the total number of buses, so we use a three-row table. To write the proportion, we use numbers from the row with two known numbers and from the row with the unknown we want to find.

	Ratio	Actual Count
Small Buses	2	s
Large Buses	7	84
Total	9	t

$$\rightarrow \frac{7}{9} = \frac{84}{t}$$
$$7t = 9 \cdot 84$$
$$t = 108$$

The answer is reasonable because the ratio of 2 to 7 means that there are fewer small buses than large buses. Therefore, the total number of buses is just a little more than the number of large buses.

Practice Set

a. The ratio of boys to girls at the assembly was 5 to 4. If there were 180 students at the assembly, how many girls were there? Explain why your answer is reasonable.

b. The coin jar was filled with pennies and nickels in the ratio of 7 to 2. If there were 28 nickels in the jar, how many coins were there?

c. The ratio of football players to soccer players at the park was 5 to 7. If the total number of players was 48, how many were football players?

Written Practice   *Strengthening Concepts*

* 1. *(45)* **Analyze** The ratio of house finches to goldfinches is 7 to 3. If there are 80 in all, how many goldfinches are there?

2. *(24)* If it was 113.5 degrees in Amarillo and 95.7 degrees in Phoenix, how much hotter was it in Amarillo than it was in Phoenix?

* 3. *(45)* The ratio of hours that the power is on to the hours that the power is off is 5 to 7. If there are 720 hours in the month, how many hours is the power on?

* 4. *(45)* Two guests out of every 25 guests at an amusement park bought popcorn. If there were 5175 guests at the amusement park, how many bought popcorn?

5. *(Inv. 4)* **a.** A pentagonal prism has how many faces, edges, and vertices?

b. Sketch a net of the prism.

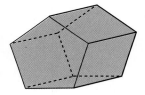

* 6. *(42)* Find the volume of a speaker cabinet with dimensions 8 in. by 8 in. by 11 in.

* 7. *(41, 44)* **Model** Graph the equation $y = -x$. Is (5, 5) on the line? What is the slope of the line?

* 8. **Evaluate** Radius *DB* measures 6,
(40) $m\angle ADB = 120°$.

    a. Find the area of the circle. (Leave in terms of π.)

    b. What fraction of the area of the circle does sector *ADB* cover?

    c. Find the area of sector *ADB*. (Leave in terms of π.)

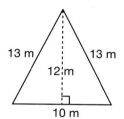

* 9. Write an equation for the function shown in
(41) the table:

x	y
−1	.5
0	1.5
1	2.5
2	3.5

10. **Classify** **a.** Classify this triangle by
(20) sides.

    b. What is its area?

    c. What is its perimeter?

Solve for *x*.

11. $\frac{2}{3}x = 10$
(13, 22)

12. $x - \frac{2}{3} = \frac{1}{9}$
(13)

For problems **13–14**, find all values of *x* which make the equations true.

13. $x^2 + 1 = 145$
(38)

14. $7 - |x| = 1$
(38)

Simplify.

15. $\frac{mn}{2m^2}$
(15)

16. $\frac{4}{5} - \frac{1}{5} \div \frac{1}{4}$
(13, 22)

17. $\frac{2}{7} \cdot \frac{3}{4} + \frac{11}{14}$
(13, 22)

18. $\left(\frac{3}{5}\right)^2 + \frac{3}{5}$
(13, 15)

19. Simplify and compare: $\dfrac{-5 + \sqrt{25-16}}{2} \bigcirc \dfrac{-5 - \sqrt{25-16}}{2}$
(21, 31)

* 20. Simplify and compare: 0.5% of 1000 $\bigcirc$ 101% of 5
(11, 25)

21. **a.** Write 45% as a decimal and as a reduced fraction.
(11)

    b. Which of the three forms would be convenient for computing a 45% discount of a $40 shirt?

22. A chalk artist draws the earth as a circle with a diameter of
(89, 40) 14 m. Find the **a** area and **b** circumference of the circle. Express the measures in terms of π.

23. Consider the drawing from problem **22.**
(39, 40)

    a. The number of square meters of a drawing gives the artist an indication of how much chalk is needed. Estimate the number of square meters the drawing will cover.

    b. The artist will rope off the drawing to protect the work in progress. Estimate the length of rope needed to rope off the art.

24. For $E = \frac{1}{2}mv^2$, find E when $m = 6$ and $v = 2$.
(14)

25. Which shows how to rewrite $7 \cdot 3 + 7 \cdot 5$?
(21)

    A $7 + 3 \times 7 + 5$ **B** $7 \times 7 + 3 \times 5$

    C $7(3 + 5)$ **D** $8(7 + 7)$

26. Frank earned $80.00 this week working at a supermarket. He owed his father $12.00 for a book and $6.00 for drawing paper that he bought for school last week. How much money does Frank still have from his weekly pay?
(31)

* 27. *Analyze* Maria wants to make a cover for the cushion on her chair. She needs to know the total surface area before she buys material. Find the total surface area of the cushion.
(43)

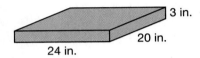

3 in.

20 in.

24 in.

* 28. Venus and Earth are often called twin planets because they are almost the same size. Earth has a circumference of 40,070 km. Venus has a diameter of 12,100 km. Which planet is larger? Hint: find the circumference of Venus.
(39)

* 29. Lisa stood next to a Ponderosa pine tree that is 5 times her height. If the tree is 6.85 meters high, how tall is Lisa? Write an equation and solve.
(38)

* 30. Leora's lasagna recipe required 3 cups of spaghetti sauce to make 4 servings. Leora is serving lasagna to a large group of people. Leora makes a table that shows the number of cups of sauce she needs for multiples of four servings. Copy and complete the table. What is the constant of proportionality? Use the information in the table to find the minimum number of cups of sauce she will need to serve 18 people.
(34)

Number (n) of Servings	Number of Cups of Sauce (s)	Ratio $\frac{s}{n}$
4	3	$\frac{3}{4}$
8	6	
12		
16		
20		

• Solving Problems Using Scientific Notation

facts | Power Up J

mental math

a. **Number Sense:** $10\frac{1}{2} \times 16$

b. **Probability:** What is the probability of rolling an odd number with one roll of a number cube?

c. **Fractional Parts:** $\frac{2}{5}$ of the 30 students were girls. How many girls were there?

d. **Ratio:** The ratio of the length to width of a rectangular field is 4 to 2. If the field is 100 yards long, how wide is the field?

e. **Measurement:** Find the length of this object:

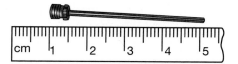

f. **Geometry:** Find the area of the rectangle. Then find the area of the shaded triangle.

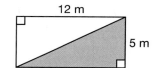

g. **Powers/Roots:** $\sqrt{10 \cdot 10}$

h. **Calculation:** $7 \times 6, + 3, \div 5, \sqrt{}, \times 8, + 1, \sqrt{}, \times 10, - 1, \sqrt{}$

problem solving | Find the sum of all the multiples of 3 from 30 to 81. (30 + 33 + 36 + ... + 75 + 78 + 81).

New Concept Increasing Knowledge

To multiply numbers written in scientific notation, we multiply the coefficients to find the coefficient of the product. Then we multiply the powers of 10 by adding the exponents.

Steps:	Justification:
$(1.2 \times 10^6)(4 \times 10^8)$	Given
$(1.2 \times 4)(10^6 \times 10^8)$	Assoc. and Comm. Properties
4.8×10^{14}	Simplified

If the product is not in the proper form of scientific notation, we revise the product so that there is one digit to the left of the decimal point.

Steps:	Justification:
$(7.5 \times 10^4)(2 \times 10^3)$	Given
$(7.5 \times 2)(10^4 \times 10^3)$	Assoc. and Comm. Properties
15.0×10^7	Simplified (but incorrect form)
$1.5 \times 10^1 \times 10^7$	$15.0 = 1.5 \times 10^1$
1.5×10^8	Proper form ($10^1 \times 10^7 = 10^8$)

Look carefully at the last three steps. The coefficient 15.0 is not the correct form. We write 15.0 in scientific notation (1.5×10^1) and simplify again.

Generalize Write a rule that explains how moving the decimal point of the coefficient changes the exponent.

Example 1

Find each product

a. $(1.2 \times 10^5)(3.0 \times 10^5)$ **b. $(4.0 \times 10^4)(5.0 \times 10^5)$**

Solution

a. We multiply the coefficients and add the exponents.

$$(1.2 \times 10^5)(3.0 \times 10^5) = 3.6 \times 10^{10}$$

b. We multiply the coefficients and add the exponents.

$$(4.0 \times 10^4)(5.0 \times 10^5) = 20.0 \times 10^9$$

The answer is not in proper form because the coefficient has two digits left of the decimal point. We reposition the decimal point.

$$20.0 \times 10^9 = \mathbf{2.0 \times 10^{10}}$$

To divide numbers in scientific notation we divide the coefficients, and we divide the powers of 10 by subtracting the exponents.

$$\frac{4.8 \times 10^6}{4.0 \times 10^3} = \frac{4.8}{4.0} \times \frac{10^6}{10^3} = 1.2 \times 10^3$$

Thinking Skill

Justify

Write 1.2×10^3 in standard form. Explain how you found your answer.

If the quotient is not in the proper form, we reposition the decimal point and change the exponent.

$$\frac{1.0 \times 10^8}{4.0 \times 10^3} = 0.25 \times 10^5 = 2.5 \times 10^4$$

When changing the exponent of a number written in scientific notation, first decide whether the shifting decimal point makes the coefficient larger or smaller, then compensate by changing the exponent.

$$0.25 \times 10^5 \longrightarrow 2.5 \times 10^4$$

The coefficient becomes 10 times larger, so we remove one power of ten.

Conclude Can the coefficient for a number written in scientific notation ever be greater than or equal to 10? Explain your answer.

Example 2

Light travels at a speed of about 300,000 kilometers per second. An hour is 3600 seconds. Write both numbers in scientific notation. Then estimate the distance light travels in an hour using scientific notation.

Solution

The rate 300,000 km/sec is equal to 3.0×10^5 km/sec.

The measure 3600 sec is equal to 3.6×10^3 sec.

The product of 3.0×10^5 km/sec and 3.6×10^3 sec is 10.8×10^8 km. The product is not in proper form for scientific notation, so we adjust the answer. The distance light travels in one hour is about 1.08×10^9 km, which is more than 1 billion kilometers.

Example 3

Find each quotient

a. $\dfrac{1.44 \times 10^{12}}{1.2 \times 10^8}$ 　　 b. $\dfrac{3.0 \times 10^8}{4.0 \times 10^4}$ 　　 c. $\dfrac{7.5 \times 10^6}{2.5 \times 10^6}$

Solution

We divide the coefficients and subtract the exponents.

a. $\dfrac{1.44 \times 10^{12}}{1.2 \times 10^8} = 1.2 \times 10^4$

b. $\dfrac{3.0 \times 10^8}{4.0 \times 10^4} = 0.75 \times 10^4$

The quotient is not in proper form. We move the decimal point one place to the right and subtract 1 from the exponent.

$$0.75 \times 10^4 = 7.5 \times 10^3$$

c. $\dfrac{7.5 \times 10^6}{2.5 \times 10^6} = 3 \times 10^0$

Recall that 10^0 equals 1. Thus the quotient is simply **3.**

Example 4

Earth's average distance from the sun is about 150 million kilometers. Light travels about 300 thousand kilometers per second. Express both numbers in scientific notation and estimate how long it takes the sun's light to reach Earth.

Solution

We rewrite 150 million km as 1.5×10^8 km.

We rewrite 300 thousand km/sec as 3.0×10^5 km/sec.

Dividing 1.5×10^8 km by 3.0×10^5 km/sec equals 0.5×10^3 sec. We rewrite the answer in proper form. To reach Earth, light from the sun travels about 5×10^2 seconds, which is 500 seconds.

Practice Set | Find each product in **a–d**.

a. $(2.4 \times 10^6)(2.0 \times 10^5)$ **b.** $(1.25 \times 10^4)(2 \times 10^6)$

c. $(4.0 \times 10^5)(4.0 \times 10^5)$ **d.** $(2.5 \times 10^6)(5 \times 10^7)$

e. One day is 86,400 seconds. Express this number in scientific notation and estimate how far the sun's light travels in a day.

f. (Analyze) What operation did you use for the exponents in problems **a–e?**

Find each quotient in **g–j**.

g. $\dfrac{3.6 \times 10^{10}}{3.0 \times 10^6}$ **h.** $\dfrac{6.0 \times 10^8}{4.0 \times 10^8}$

i. $\dfrac{1.2 \times 10^7}{3.0 \times 10^3}$ **j.** $\dfrac{3 \times 10^9}{8 \times 10^4}$

k. If Mars is 225 million kilometers from the sun, how long does it take light from the sun to reach Mars?

l. (Analyze) What operation did you apply to the exponents in problems **g–k?**

Written Practice   *Strengthening Concepts*

* 1. The ratio of the cat's weight to the rabbit's weight is 7 to 4. Together,
(45) they weigh 22 pounds. How much does the rabbit weigh?

* 2. The ratio of the weight of the Guinea pig to the weight of the Goliath
(45) beetle is 8 to 1. Together, they weigh 27 ounces. How much does the Goliath beetle weigh?

3. The mass of a flea is about 0.005 grams. How many fleas does it take to
(6, 25) make 1 gram?

* 4. Simplify: $\dfrac{3.36 \times 10^7}{1.6 \times 10^3}$ *** 5.** Simplify: $(9.0 \times 10^2)(1.1 \times 10^7)$
(46)   (46)

* 6. (Model) Graph the equation $y = \frac{3}{2}x - 2$. (Hint: For the table select even
(41, 44) numbers for x.) What is the slope of the line?

7. Find the **a** area and **b** the circumference of a circle that has a diameter
(39, 40) of 100. (Use a decimal approximation for π.)

* 8. (Analyze) Find the volume of a plastic CD case with dimensions 12.0 cm
(42) by 12.5 cm by 0.3 cm.

9. It will cost a manufacturer $45 to make
(41) machinery modifications in order to produce
a new product. Each item will cost $10 to
manufacture. To the right is a chart of the
number of items (x) that are manufactured
and the total cost (y). Write an equation that
shows the relationship in the table.

x	y
0	45
1	55
2	65
3	75

10. Find the **a** area and **b** perimeter of this figure.
(8, 37)

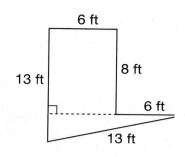

Solve for *x*.

11. $\frac{1}{2}x = \frac{3}{7}$
(22, 38)

12. $x + \frac{1}{12} = \frac{1}{2}$
(13, 38)

13. $x - 1.2 = 1.95$
(24, 38)

14. $\frac{x}{16} = \frac{30}{20}$
(36)

15. **a.** Expand: $-6(x^2 - 18x + 81)$ **b.** Factor: $9x - 3$
(21, 36)

16. **a.** Write $\frac{11}{20}$ as a decimal and percent.
(12)

    b. Which of the three forms is most appropriate for a store owner to describe a discount of $11 from a $20 item?

Simplify:

17. $\frac{mr}{3mr}$
(27)

18. $\frac{(-12) - (-2)(3)}{(-2)(-3)}$
(33, 36)

19. $\frac{1}{2} \div \frac{3}{4} - \frac{1}{3}$
(13, 22)

20. $(0.15)^2$
(25)

Combine like terms to simplify:

21. $x^3 + 2x^2y - 3xy^2 - x^2y$
(31)

* 22. **Connect** Simplify $5(x - 3) + 2(3 - x)$ (Hint: expand first using the Distributive Property.)
(31, 36)

23. Find all values of *x* which make the equation true: $x^2 - 9 = 0$
(14, 36)

24. Claire flips a coin twice.
(32)

    a. Use the sample space to find the probability of Claire getting tails at least once.

    b. Find the probability of Claire getting heads twice.

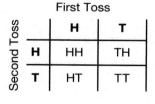

25. Consider the similar triangles.
(35)

    a. What is the scale factor from the smaller to the larger triangle?

    b. Find *x*.

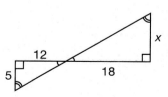

* 26. **Evaluate** Richard drew the floor plan of his dream house. On his scale
(35) drawing of a kitchen, he drew a 6 in. segment to represent a distance
of 30 ft.

    a. Each inch in the drawing represents what distance in the house?

    b. If Rick's kitchen is to measure 10 ft by 20 ft, what are the dimensions
of the floor plan of the kitchen?

* 27. **Analyze** We can use this formula to find the
(21) area of a trapezoid.

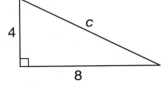

$$A = \frac{b_1 + b_2}{2} \cdot h$$

Solve for A if $b_1 = 5$, $b_2 = 7$, and $h = 3$.
(Units are in centimeters.)

28. What is the length of the hypotenuse?
(Inv. 2)

    A $\sqrt{80}$ **B** $\sqrt{81}$

    C 9.1   **D** 9.3

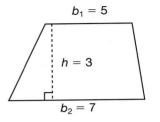

* 29. Miguel is going to build a storage cube for his sports equipment. How
(43) many square feet of wood will he need to make a three-foot cube?

* 30. A bakery produces 144 loaves (*l*) of bread per hour (*h*) every day. Which
(41) graph represents the relationship of time and the number of loaves
baked? Is this a proportional relationship? How do we know? Write
an equation that describes the relationship. What is the constant of
proportionality?

A

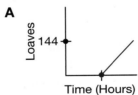

B

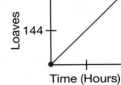

C

D

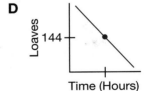

• Graphing Functions

facts | Power Up J

**mental
math**

a. **Number Sense:** $8\frac{1}{2} \times 10$

b. **Statistics:** Find the mean of 5, 6, and 10.

c. **Fractional Parts:** Calculate a 20% tip on a bill of $30.

d. **Measurement:** The odometer read 5306.5 miles at the end of her trip. When she started, the odometer read 5103.5. How long was her trip?

e. **Proportions:** If Zollie can write 40 multiplication facts in 2 minutes, how many can she write in 5 minutes?

f. **Geometry:** Two angles of a triangle measure 80° and 20°. Is the triangle a right triangle, an acute triangle, or an obtuse triangle? Explain.

g. **Estimation:** Approximate the total: $2.47 plus $3.47 plus 5 at $6.99

h. **Calculation:** Square 10, × 5, ÷ 10, × 2, $\sqrt{}$, × 4, − 4, $\sqrt{}$

**problem
solving** | Three painters paint a house in 3 days. How many houses could 6 painters paint in 6 days working at the same rate?

New Concept | *Increasing Knowledge*

Recall that a **function** is a rule that pairs one output number with each input number. The following picture of a function machine and two buckets illustrates the relationship between the **domain** (set of inputs) and **range** (set of outputs) of a function.

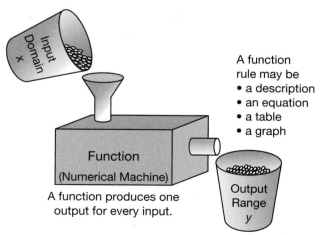

A function
rule may be
• a description
• an equation
• a table
• a graph

Input
Domain
x

Function
(Numerical Machine)

A function produces one
output for every input.

Output
Range
y

A function machine takes each element (member) from the domain "bucket" (set) and produces one element for the range "bucket."

A checkout scanner at a store is a real-world example of a function machine. The input is the product code the scanner reads and the domain is the set of codes programmed for the scanner. The output is the price the register records for that product code and the range is the set of prices of the coded items. The function pairs a code with a price.

For each coded item there is only one price. Different coded items may have the same price, but each coded item has only one price.

Functions may be expressed with descriptions, equations, tables, or graphs. The sign in example 1 is an example of a description for a function that pairs the number of hours parked and the number of dollars charged for parking.

Example 1

The sign at right describes a function—a rule that relates hours parked to dollars charged.

 a. What is the domain of the function?

 b. What is the range of the function?

 c. Graph the function.

> ## Parking
> **$1.00 for first hour**
> **$2.00 for each additional hour or part thereof**
> **$15.00 maximum**
> **Lot open 6 a.m. to 10 p.m.**
> **No overnight parking**

Solution

 a. The domain of the function is the number of hours a car can be parked in the lot. Since the lot is open from 6:00 a.m. to 10:00 p.m., the domain is from a few moments in the lot up to a maximum of 16 hours, so the domain is the set of real numbers from **0–16.**

 b. The range is the number of dollars charged for parking. For up to one hour the charge is $1.00. Any time during the second hour the charge is $3.00 ($1.00 + $2.00). During the third hour the charge is $5.00. The pattern of odd dollar charges continues to $15.00: **$1, $3, $5, ... $13, $15.**

 c. We use the horizontal axis for the domain and the vertical axis for the range.

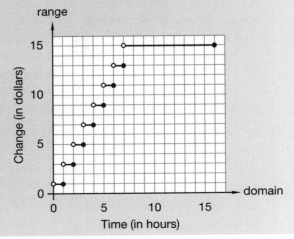

The parking charge during an hour is flat, but as the hour passes the charge increases by two dollars until the maximum charge is reached. Because of the stair step pattern, this type of function is sometimes called a **step function.** The small filled and empty circles on each step of the graph mean that at the exact hour the charge is the lower amount, not the higher amount. The next minute the charge jumps to the next level.

A formula can be an example of a function expressed as an equation. Consider the formula that computes the perimeter of a square given a side length.

$$P = 4s$$

output ⟍ output ⟋ input

Example 2

Consider the function $P = 4s$.
  a. **What is the domain?**
  b. **What is the range?**
  c. **Create a table for some function pairs.**
  d. **Graph the function.**

Solution

  a. The domain is any number that can represent a side length. Lengths are positive, so any positive number on the number line can represent a length. Thus the domain is **all positive real numbers.**

  b. The range is any number that can represent the perimeter. A perimeter is a length, so any positive number on the number line can represent the perimeter of a square. The range is **all positive real numbers.**

  c. To create a table we choose input numbers (side length) and use the function rule to find the output (perimeter). Although we may choose any positive real number for s, we decide to select numbers that are easy to compute and graph.

$P = 4s$

s	P
1	4
2	8
3	12

  d. Since lengths are positive, the graph of the function is in the first quadrant only. We graph the pairs from the table and notice that the points are aligned. We could have chosen and graphed (4, 16), (0.5, 2), (1.5, 6), (0.1, 0.4), or countless other pairs. All such pairs would be aligned with these points. By drawing a line from the origin through the points from the table, the line becomes a representation of all possible side-length and perimeter pairs for the function.

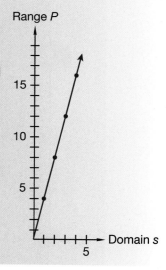

Notice how the graph in example 2 is different from the graph in example 1. The graph in example 2 is not interrupted with gaps. This means that the function $P = 4s$ is a **continuous function.**

Notice two more important features of the graph in example 2.

1. The graph begins at the origin.

2. All points in the graph are aligned.

Together, these two features indicate that the function is a **direct proportion.** We will be studying direct proportions in many lessons in this book. Any two pairs from a direct proportion form a proportion. Take (2, 8) and (3, 12) for example.

$$\frac{2}{8} = \frac{3}{12}$$

Verify How can we verify that these two ratios form a proportion?

Visit www. SaxonPublishers. com/ActivitiesC3 *for a graphing calculator activity.*

Example 3

An online auction service charges a 30¢ listing fee plus 6% of the selling price for items up to $25. The total service charge (*c*) is a function of the selling price (*p*) of an item.

$$c = 0.30 + 0.06p$$

a. Use a calculator to generate a table of pairs of values for the selling price and service charge.

b. Graph the function on axes like the ones below.

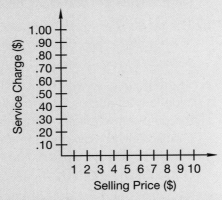

c. Is the relationship a proportion? How do you know?

Solution

a. We choose input values (selling prices) that are easy to calculate.

p	c
1	0.36
2	0.42
3	0.48
4	0.54
5	0.60

b. We graph the points from our table on the given axes and draw a line connecting the points.

c. The relationship is not proportional. The graph does not intercept zero and the price/charge ratio is not constant.

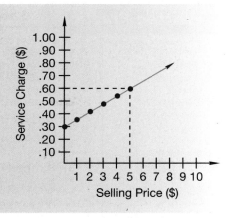

Practice Set

a. *Represent* The directions on the can of paint described a function: One gallon covers 400 sq. ft of a sealed, non-textured surface. On graph paper draw and extend axes like the ones illustrated. Then create a function table relating quantity of paint in gallons (g) to area (A) covered in square feet. Use numbers from the table to graph the function on your paper.

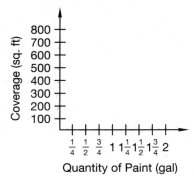

b. Referring to problem **a,** which equation below describes the relationship between the number of gallons (g) of paint and the area (A) covered in square feet?

A $A = 400 + g$ **B** $A = 400g$ **C** $A = \dfrac{400}{g}$ **D** $A = \dfrac{g}{400}$

c. Referring to problem **a,** painting 700 square feet of wall would require about how many *quarts* of paint?

d. Janine works in a high-rise building. When she wants to go to upper floors she enters an elevator and pushes a button (input), and the elevator takes her a distance above street level (output). Assuming that the building has 12 floors and that floors are 10 feet apart, graph the relationship between the floor Janine is on and her distance (elevation) above street level. Let the tick marks on the vertical axis (output) increase by tens.

e. In problem **d,** what is the domain of the function?

f. Is the relationship described in problem **d** proportional? Why or why not?

* 1.   *Analyze*   The ratio of buses to cars entering the lot was 2 to 11.
(45)   If 650 vehicles entered the lot all together, how many were cars?

2.   A pack of 8 collectable cards contains 1 rare card, 3 uncommon cards,
(45)   and 4 common cards. If Javier has 45 packs of cards, how many more
uncommon cards does he have than rare cards?

3.   Kimberly used coupons worth $1.00 off and $1.25 off and two coupons
(3, 4)   for $1.50 off. If the discount of each of these coupons were doubled,
how much would Kimberly save?

* 4.   *Generalize*   Simplify: $(4.0 \times 10^3)(3.0 \times 10^4)$
(28, 46)

* 5.   *Model*   Graph the equation $y = x - 1$. Is the point (3, 1) on the line?
(41)

6.   Find the **a** area and **b** circumference of a circle with a radius of 5 in.
(40)   Express the measures in terms of π.

7.   Find the volume of an aquarium with dimensions 1 ft $\times$ 2 ft $\times$ 1.5 ft.
(42)

8.   Polly's pie shop is famous for its 12-inch blueberry pies. Is the
(39)   circumference of Polly's pie greater or less than 1 yard?

For problems **9** and **10,** refer to the following
information. Salvador has a yard of this shape.
(Dimensions are in meters.) He would like to lay
sod and fence his yard.

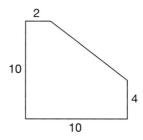

9.   How many square meters of sod will he need?
(37)

10.   What length of fencing will he need if he fences all but one 10-meter
(8,   edge of his yard?
Inv. 2)

Solve.

11. $\frac{4}{3}x = 12$
(22, 38)

12. $x - \frac{2}{5} = 12$
(13, 38)

13. $4x^2 = 64$
(36, 38)

14. $\frac{24}{x} = \frac{15}{20}$
(44)

15.   **a.** Expand: $-5(4r - 2d - 1)$ **b.** Factor: $4x - 28$
(21, 36)

* 16.   **a.** Write 0.005 as a percent and as a reduced fraction.
(12)

     b. Which of the three forms would be most convenient for finding
0.5% of $1000?

Simplify.

17. $\dfrac{24wx^2y}{-12xy}$
(27, 36)

18. $\dfrac{3}{8} - \dfrac{1}{2} \cdot \dfrac{2}{3} \cdot \dfrac{3}{4}$
(13, 22)

19. $1\frac{2}{3} \div 3\frac{1}{3} + 1\frac{5}{6}$
(13, 23)

20. $\dfrac{(-12)-(-2)(-3)}{-(-2)(-3)}$
(33, 36)

21. $\dfrac{1.2-0.12}{0.012}$
(24, 25)

22. $(-1)^2 + (-1)^3$
(36)

23. Combine like terms: $2(x + 1) + 3(x + 2)$ (Hint: Expand first using the distributive property.)
(31, 36)

* 24. *Analyze* Find the slope of the line that passes through points (0, 0) and (2, −4).
(44)

* 25. The radius of this circle is 10 ft.
(40)

    a. Using 3.14 for π, what is the area of the circle?

    b. What fraction of the circle is shaded?

    c. What is the combined area of the shaded regions? (Round to the nearest square foot.)

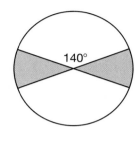

* 26. For **a–c,** refer to these triangles.
(35)

    a. Redraw $\triangle ABE$ and $\triangle ACD$ as two separate triangles.

    b. Explain how you know that $\triangle ABE \sim \triangle ACD$.

    c. What is the scale factor from $\triangle ABE$ to $\triangle ACD$?

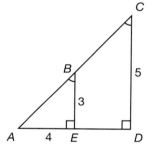

27. For the equation $E = mc^2$, find E when $m = 0.001$ and $c = 3.0 \times 10^8$.
(28)

28. Express $\frac{1}{12}$ **a** as a decimal number and **b** as a percent.
(30)

* 29. Amelia is making a scratching post for her cat. She has a rectangular prism post that is 4 inches wide by 4 inches deep by 18 inches tall. How much carpet does Amelia need to go around the post without covering the top or bottom of the post?
(43)

30. Two business partners split their profits evenly. The table shows possible x- and y-values with x representing their profit and y representing the amount they each receive.
(41)

    a. Write an equation for the function shown in the table.

    b. Graph the function.

    c. Is the relationship proportional? Explain your answer.

x	y
0	0
2	1
4	2
6	3
8	4

• Percent of a Whole

facts | Power Up J

mental math |

a. **Number Sense:** 3.25×4

b. **Measurement:** How many degrees below "freezing" is the temperature shown on the thermometer? (Water freezes at 32°F.)

c. **Percent:** How much is 75% of $1,000?

d. **Scientific Notation:** Write 63,400,000 in scientific notation.

e. **Rate:** Maria walks 4 miles per hour south, while Reginald strolls 3 miles per hour north. If they start at the same place and same time, how long will it take them to be 14 miles apart?

f. **Algebra:** $x + 3 = 5.8$

g. **Geometry:** Approximate the volume of this prism.

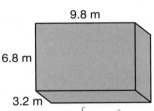

9.8 m

6.8 m

3.2 m

h. **Calculation:** $12 \div 4$, $\times 3$, $\times 4$, $\sqrt{}$, $\times 8$, $+ 1$, $\sqrt{}$, $\times 9$, $+ 1$, $\sqrt{}$

problem solving | The mystery man listened and replied. Juan said 1, the mystery man said 13. Samantha said 11, the mystery man said 113. Francisco said 5, the mystery man said 53. Wally whispered. The mystery man said 5283. What did Wally whisper?

New Concept | *Increasing Knowledge*

Percents are commonly used to describe part of a whole.

• Shanna answered 92% of the questions correctly.

• A basketball player has made 80% of her free throws.

• The governor was elected with 56% of the vote.

In this lesson we will use a three-row ratio table to help us solve percent problems about parts of a whole. A percent is a ratio, with the given percent representing the part and 100% representing the total.

Example 1

Thirty percent of the students ride the bus. If 210 do not ride the bus, how many students are there in all?

Solution

We draw a three-row ratio table and fill it in as completely as possible.

If 30% ride the bus, then 70% do not. To write a proportion we use the row with two known numbers and the row with the number we want to find.

	Percent	Count
Bus	30	a
Not Bus	(70)	210
Total	100	t

$$\frac{70}{100} = \frac{210}{t}$$

$$70t = 21{,}000$$

$$t = 300$$

There were **300 students** in all.

Example 2

Thinking Skill

Analyze

How could we use the answer to find what percent of the questions Shanna answered incorrectly?

Shanna correctly answered 17 of the 20 questions on the game show. What percent of the questions did she answer correctly?

Solution

In an earlier lesson we found the fraction of correct answers and then found the equivalent percent. Here we use a proportion to find the percent Shanna answered correctly. We draw a three-row ratio table and fill it in as completely as possible. We want to find the percent for the number correct, so we use the numbers in the first and third rows to write the proportion.

	Percent	Count
Correct	c	17
Incorrect	n	3
Total	100	20

$$\frac{c}{100} = \frac{17}{20}$$

$$20c = 17 \cdot 100$$

$$\frac{20c}{20} = \frac{17 \cdot 100}{20}$$

$$c = 85$$

We find that **85%** of Shanna's answers were correct.

Practice Set

Formulate Solve each of these problems using a proportion. Begin by making a ratio table.

a. Mariah has read 135 of the 180 pages in the book. What percent of the book has she read?

b. McGregor is growing alfalfa on 180 acres, which is 30% of his farmland. McGregor has how many acres of farmland?

c. The frequency of the letter *e* in written English is about 13%. On a page of a novel that has about 2000 letters per page, about how many occurrences of the letter *e* can we expect to find?

d. The Springfield Sluggers won 64% of their games and lost 9 games. How many games did the Sluggers win?

Written Practice   *Strengthening Concepts*

* 1. *Analyze* The ratio of people watching whales to people fishing was 5 to
(45)   7. If there were 72 in all, how many people were fishing?

2. Sergio purchased 4 items for $1.50 each. Sales tax was 8.25%. What
(4, 11)   was the total price?

3. At back-to-school night, $\frac{1}{2}$ of the algebra students came, and $\frac{1}{3}$ of
(5)   the pre-algebra students came. If there are 62 algebra students and 66 pre-algebra students, how many of the students came to back-to-school night?

4. Betsy correctly answered 19 of the 25 trivia questions. What percent of
(48)   the questions did she answer correctly?

* 5. *Evaluate* Find the slope of the line passing through points (–2, 2) and
(44)   (2, 0).

Simplify.

6. $\dfrac{a^2b^2c}{12a^2c^2}$
(27)

7. $\dfrac{4}{9} \div \dfrac{2}{3} - \dfrac{1}{2}$
(13, 22)

8. $3^3 \div 3 + 3 - 3(3)$
(15, 21)

9. $\left(\dfrac{2}{5}\right)^2 - \left(\dfrac{1}{5}\right)^2$
(13, 15)

* 10. $\dfrac{5.2 \times 10^7}{1.3 \times 10^4}$
(46)

* 11. $(5.0 \times 10^7)(3.0 \times 10^4)$
(46)

12. $(12.4 + 2)(1 - 0.998)$
(24, 25)

* 13. Graph $y = x - 3$. Is the point (3, 0) on the line?
(41)

* 14. Find the **a** area and **b** the circumference of a circle that has a diameter
(39, 40)   of 40 cm. Use 3.14 for π.

* 15. Find the volume of a drawer with dimensions 10" $\times$ 15" $\times$ 4".
(42)

16. For **a–c,** refer to this triangle.
(20,
Inv. 2)   **a.** Classify the triangle by angles.

b. Find the area of the triangle.

c. Find the perimeter of the triangle.

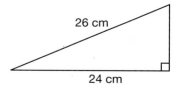

26 cm

24 cm

Solve.

17. $x - 3\dfrac{1}{3} = 2\dfrac{1}{2}$
(13, 38)

18. $x - \dfrac{2}{3} = \dfrac{5}{6}$
(13, 38)

19. $4x^2 = 400$
(36, 38)

* 20. $\frac{1.2}{3} = \frac{x}{2}$
(44)

21. **a.** Write 250% as a decimal and reduced fraction.
(11, 12)

b. Which of the three forms would you choose to find the price of a home that is now worth 250% of its original price of $160,000?

Combine like terms to simplify.

22. $abc - cab + bac - 2b^2$
(31)

23. $5x^2y - 4yx^2 + 2x - x$
(31)

* 24. At a high-rise hotel, the ground floor is 3 feet above street level. Each successive floor is 10 feet above the one below it. At the right is a chart of the number of floors (x) a person is above the ground floor and how high they are in feet (y) above street level. Write an equation for the function shown in the table. Is the function a proportional relationship? How do you know?
(41)

x	y
1	13
2	23
3	33
4	43

* 25. A circle with a radius of 4 in. is divided into congruent sectors, each with a central angle of 30°.
(40)

a. How many sectors cover the circle?

b. What is the area of one sector? (Round to the nearest square inch.)

26. A rate formula is $r = \frac{d}{t}$. Find r when $d = 26.2$ miles and $t = 2$ hours.
(14, 25)

27. An entomologist recorded the lengths of two newly discovered species of beetles. The first beetle was 2.76 cm and the second was 3.14 cm long. What is the difference in their lengths?
(24)

28. When Jon began eating lunch the clock read 12:30. Fifteen minutes later the clock read 12:45. Describe the transformation of the minute hand on the clock between 12:30 and 12:45.
(26)

* 29. **Evaluate** Rivera is creating a package for a new product. The package is a box that is 5 in. by 9 in. by 14 in. He needs to let the manufacturer know the amount of surface area for each box. What is the surface area?
(43)

* 30. The graph shows the relationship between the number of servings of soup (s) and the number of teaspoons (t) of salt in the soup. Is the relationship proportional? If the relationship is proportional, write an equation for the relationship and state the constant of proportionality, k.
(41)

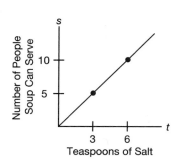

Lesson 48   **329**

• Solving Rate Problems with Proportions and Equations

Power Up | *Building Power*

facts | Power Up J

mental math

a. **Number Sense:** 10.5×4

b. **Statistics:** Find the mean of 13, 3, and 2.

c. **Fractional Parts:** Because of the sale, Liz only had to pay 75% of the original price of $28. How much did she pay? What was the percent of discount?

d. **Measurement:** Find the length of this section of cable.

e. **Geometry:** Two angles of a triangle measure 50° and 30°. Is the triangle acute, obtuse, or right?

f. **Proportions:** $\frac{6}{x} = \frac{12}{2}$

g. **Powers/Roots:** $\sqrt{3 \cdot 3} + \sqrt{2 \cdot 2}$

h. **Calculation:** Start with a dozen, $\times 4$, $+ 1$, $\sqrt{}$, $\times 5$, $+ 1$, $\sqrt{}$, $- 2$, $\sqrt{}$, $- 2$, $\sqrt{}$

problem solving | Five coins total $1. What are the coins?

New Concept | *Increasing Knowledge*

Math Language
Recall that a **rate** is a ratio that compares two different units. A **unit rate** is a rate with a denominator of 1 unit.

Recall that a rate is a ratio of two measures. For example, speed is a rate that is a ratio of distance to time. If an object's speed is constant, then the ratio of its distance traveled to time of travel is constant. A car traveling at an average speed of 50 miles per hour travels the following distances in the given time periods.

$\dfrac{50 \text{ miles}}{1 \text{ hour}}$	$\dfrac{100 \text{ miles}}{2 \text{ hours}}$	$\dfrac{150 \text{ miles}}{3 \text{ hours}}$	$\dfrac{200 \text{ miles}}{4 \text{ hours}}$

These ratios are equivalent, and each reduces to 50 miles/1 hour, which is the unit rate. Therefore, the relationship is proportional, and any two of these ratios form a proportion. Thus we may solve rate problems two ways: with proportions or by multiplying by the unit rate.

These two methods are based on two forms of the rate equation.

1. The ratio is constant: $\frac{y}{x} = k$

2. The unit rate is constant: $y = kx$

All proportional relationships can be expressed with these two equations. Examples 1 and 2 show how we can use a proportion or multiplying by the unit rate to solve the same problem.

Example 1

If 6 books weigh 15 pounds, how much would 20 books weigh?

Solution

We do not need to find the weight per book. Instead we can use the given information to write a proportion. We are given the ratio of books to weight in one case and we are asked to complete the ratio in another case. We record the information in a table with the headings Case 1 and Case 2.

Thinking Skill

Connect

What method did we use to find the weight of 20 books? Why is the relationship proportional?

	Case 1	Case 2
Books	6	20
Weight (lbs)	15	p

$$\frac{6}{15} = \frac{20}{p}$$

$$6p = 15 \cdot 20$$

$$p = \frac{15 \cdot 20}{6}$$

$$p = 50$$

We find that 20 books would weigh **50 pounds.**

Example 2

If 6 books weigh 15 pounds, what is the unit rate? What is the weight of 20 books? Write an equation that shows how to find the weight (*w*) of books knowing the number (*n*) of books and the unit rate.

Solution

To find the unit rate (weight of one book), we divide:

$$\frac{\text{weight } (w)}{\text{number } (n)} = \frac{15 \text{ pounds}}{6 \text{ books}} = \textbf{2.5 pounds per book}$$

To find the weight of a number of books, we multiply the number by the unit rate:

$$w = 2.5n$$

To find the weight of 20 books, we multiply the unit rate by 20:

$$w = 2.5 \cdot 20 = \textbf{50 lbs}$$

Example 3

Julio is riding in a 50 km bike race. He passed the 15 km mark in 27 minutes. If Julio continues to ride at the same rate, what will be his time for the whole race?

Solution

At the 15 km mark, Julio was less than one-third of the way through the race and he had been riding for nearly half an hour.

Predict Will it take Julio more or less than one hour to complete the whole race?

To solve the problem we can write a proportion. We use the given distance/time ratio for one point in the race and the incomplete ratio for the end of the race.

	One Point in the Race	End of the Race
Distance (km)	15	50
Time (min)	27	m

$$\frac{15}{27} = \frac{50}{m}$$

$$15m = 50 \cdot 27$$

$$m = \frac{50 \cdot 27}{15}$$

$$m = 90$$

If Julio continues riding at the same rate, his time for the whole race will be **90 minutes,** which is $1\frac{1}{2}$ hours.

Example 4

In example 3, what is Julio's rate in km/min? Write an equation that relates Julio's distance to time riding.

Solution

The ratio of distance to time is:

$$\frac{d}{t} = \frac{15 \text{ km}}{27 \text{ min}}$$

$$\frac{d}{t} = \frac{5 \text{ km}}{9 \text{ min}}$$

This can also be expressed:

$$d = \frac{5}{9} t$$

Example 5

Three tickets cost $20.25. How much would 7 tickets cost? What is the unit cost?

Solution

Since 7 is more than double 3, we expect 7 tickets to cost more than twice as much as 3 tickets, that is, more than $40.

We record the information in a ratio table. Then we write and solve a proportion.

	Actual	Estimate
Tickets	3	7
Price ($)	20.25	p

$$\frac{3}{20.25} = \frac{7}{p}$$

$$3p = 7 \cdot 20.5$$

$$p = \frac{7 \cdot 20.25}{3}$$

$$p = 47.25$$

Seven tickets would cost **$47.25,** which agrees with our original estimate. We find the unit cost by dividing $20.25 by 3. The unit cost is **$6.75.**

Practice Set

For problems **a–c** below record the information in a ratio table. Estimate an answer and then solve the problem by writing and solving a proportion.

a. If 5 pounds of seedless grapes cost $3.80, how much would 9 pounds cost?

b. If 8 cows eat 200 pounds of hay a day, how many pounds of hay would 20 cows eat in a day?

c. Darcie can type 135 words in 3 minutes. At that rate, how many minutes would it take her to type 450 words?

d. *Analyze* Find the price per pound of grapes in **a,** the pounds of hay per cow in **b,** and the words per minute in **c.**

e. Solve problems **a** and **b** again using a rate equation instead of a proportion.

Written Practice   *Strengthening Concepts*

Evaluate For problems **1–2,** record the information in a ratio table. Estimate and then solve by writing and solving a proportion.

* 1. Marcus can run 3 miles in 18 minutes. At that rate, how long will it take
(49)   him to run 7 miles?

* 2. Felicia's pets eat thirty pounds of food in two weeks. How many pounds
(49)   of food would they eat in five weeks?

* 3. **Analyze** In a bag of 60 marbles, 18 are green. If one marble is drawn
(32)   from the bag, what is the probability the marble is

     a. green?   **b.** not green?

* 4. **Evaluate** At the flag shop, for every 13 stripes there are 50 stars.
(45)   If there are 10,000 stars, then how many more stars are there than
     stripes?

5. The shipping charges for the store include a $3.50 base fee plus
(3, 4) $0.65 per pound. How much would it cost to ship a 10 pound
     order?

Solve.

6. $1.4x = 84$ **7.** $x + 2.6 = 4$
(25, 38)   (24, 38)

8. $\dfrac{x}{3} = 1.2$ *** 9.** $x - 7 = -2$
(25, 38)   (31, 38)

* 10. Austin has answered 21 of the 25 questions. What percent of the
(48)   questions has he answered?

* 11. **Analyze** If 70% of the members remembered the password but 21
(48)   members forgot, then how many members were there in all?

* 12. Refer to the graph of lines *a* and *b* to
(Inv. 1,   answer the following questions.
44)

     a. What is the slope of line *a*?

     b. Which line is perpendicular to the
          x-axis?

     c. Which line intersects the *y*-axis at
          positive 4?

     d. In which quadrant do lines *a* and *b*
          intersect?

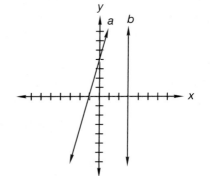

* 13. Graph $y = 3$. Is the point (6, 3) on the line?
(41)

14. At the park is a circular wading pool 20 feet in diameter. If parents
(39)   are seated on the edge of the pool about every three feet, how many
     parents are sitting on the edge of the pool?

15. Find the volume of an oven with inside dimensions 3 ft by 1.5 ft
(42)   by 3 ft.

16. How many edges, vertices, and faces does a number cube have?
(Inv. 4)

17. Which of the following is not a parallelogram?
(Inv. 3)
     A square   **B** rectangle   **C** trapezoid   **D** rhombus

18. Find *x* and *y.*
(18, 20)

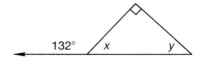

19. About twenty-one percent of the earth's atmosphere is oxygen. Write
$(11, 12)$ 21% as a decimal and a fraction.

20. Find $\frac{1}{2}mv^2$ when $m = 8$ and $v = 2$.
$(14, 15)$

* 21. Sketch a triangular prism that has an equilateral base.
$(Inv. 4)$

22. Yanos and Xena play a numbers game. The table shows some numbers
(41) Xena says and the numbers Yanos says in response.

X	6	2	1	−2
Y	3	1	$\frac{1}{2}$	−1

a. Describe the rule Yanos uses.

b. Write the rule as an equation beginning with $y =$.

c. Sketch a graph that shows all the pairs of numbers Xena and Yanos could say using the rule.

Combine like terms to simplify.

23. $5(x + 3) - 2(x + 4)$
$(31, 36)$

24. $-7x + 2(x^2 + 4x - 1)$
$(31, 36)$

Simplify.

* 25. $\dfrac{-(-3)}{-6}$
(36)

* 26. $-(-4) - 3$
$(31, 33)$

* 27. $\left(1\dfrac{1}{2}\right)^2 - 1\dfrac{1}{2}$
$(23, 13)$

* 28. $\dfrac{(5.2 \times 10^9)}{(4 \times 10^7)}$
(46)

29. Find all values of x: $3|x| = 6$
$(1, 14)$

* 30. **Evaluate** What relationship does
(41) the graph show? Is the relationship proportional? How do you know? Using the information given, calculate the number of revolutions per minute at one mile per hour. Then write an equation for the relationship. If it is proportional, state the constant of proportionality, k.

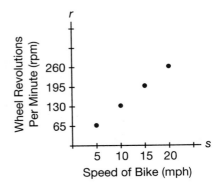

• Solving Multi-Step Equations

Power Up | Building Power

facts | Power Up J

mental math

a. **Number Sense:** 6.4×4

b. **Statistics:** Find the mean of 5, 6, and 1.

c. **Fractional Parts:** Three-eighths of the class wore jackets. There were 32 students in the class. How many wore jackets?

d. **Scientific Notation:** Write 4.05×10^5 in standard notation.

e. **Rate:** Robert rode 8 miles per hour north. Stephen rode 7 miles per hour north. If they start at the same place and time, how far apart are they after 1 hour? After 2 hours?

f. **Geometry:** Find the area of the rectangle, then the area of the shaded triangle.

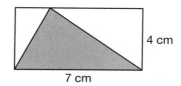

7 cm

4 cm

g. **Estimation:** Approximate the total for this shopping trip: $3.47 + 2 \times $5.99 + $39.97 + $1.47

h. **Calculation:** $8 + 7, \div 5, \times 9, - 2, \sqrt{}, \times 11, + 1, \div 7, + 1, \sqrt{}, \times 1000$

problem solving | $\sqrt{900} = 30$; $\sqrt{1600} = 40$; $\sqrt{2500} = 50$; Find $\sqrt{2025}$

New Concept | Increasing Knowledge

We have used inverse operations to isolate a variable in an equation. The equations we will solve in this lesson require two or more steps to isolate the variable. Recall that isolating a variable, such as x, actually means that $1x + 0$ is on one side of the equal sign.

Example 1

Thinking Skill

Connect

How are the steps to solving and checking an equation related to the order of operations?

Suppose a taxi charges $1.20 to start a ride plus $3 per mile. If a ride costs $5.40, how far was the ride? Let x represent the distance traveled in miles. Solve this equation to find the length of the ride.

$$3x + 1.20 = 5.40$$

We see that x is multiplied by 3 and that 1.20 is added to that product. We undo these operations in reverse order. First we subtract 1.20 from both sides of the equation, which leaves $3x$ on the left side. Then we divide both sides by 3 to isolate x. We find that x equals 1.4.

Step:	Justification:
$3x + 1.20 = 5.40$	Given
$3x + 1.20 - 1.20 = 5.40 - 1.20$	Subtracted 1.2 from both sides
$3x = 4.20$	Simplified
$\dfrac{3x}{3} = \dfrac{4.20}{3}$	Divided both sides by 3
$x = 1.4$	Simplified

The solution means that the taxi ride was **1.4 miles.**

We check the solution by substituting 1.4 for x in the original equation. Then we simplify.

$3(1.4) + 1.20 = 5.40$	Substituted
$4.20 + 1.20 = 5.40$	Simplified $3(1.4)$
$5.40 = 5.40$	Simplified

Example 2

Solve: $-2x - 5 = 9$

Solution

We see that x is multiplied by -2 and 5 is subtracted from that product. First we add 5 to both sides of the equation. Then we divide both sides by -2.

Step:	Justification:
$-2x - 5 = 9$	Given
$-2x - 5 - 5 = 9 + 5$	Added 5 to both sides
$-2x = 14$	Simplified
$\dfrac{-2x}{-2} = \dfrac{14}{-2}$	Divided both sides by -2
$x = -7$	Simplified

We check the solution in the original equation.

$-2(-7) - 5 = 9$	Substituted -7 for x
$14 - 5 = 9$	Simplified $-2(-7)$
$9 = 9$	Simplified $14 - 5$

Every step in solving an equation forms a new equation with the same solution. Step by step we form simpler equations until the final equation states the value(s) of the variable.

Example 3

Solve: $3x + 4 - x = 28$

Solution

We write a sequence of simpler but equivalent equations until the variable is isolated.

In this equation the variable appears twice. We collect like terms and then proceed to isolate the variable.

Step:	Justification:
$3x + 4 - x = 28$	Given equation
$(3x - x) + 4 = 28$	Commutative and Associative Properties
$2x + 4 = 28$	Added $3x$ and $-x$
$2x = 24$	Subtracted 4 from both sides
$x = 12$	Divided both sides by 12

We check the solution by replacing each occurrence of x with 12 in the original equation.

Step:	Justification:
$3x + 4 - x = 28$	Given equation
$3(12) + 4 - (12) = 28$	Substituted 12 for x
$36 + 4 - 12 = 28$	Simplified
$28 = 28 \checkmark$	Simplified

Practice Set

Justify Solve. Check your work.

a. $3x + 5 = 50$

b. $4x - 12 = 60$

c. $30n + 22 = 292$

d. $\dfrac{x}{5} + 4 = 13$

e. $-2x + 17 = 3$

f. $3m - 1.5 = 4.2$

g. $4x + 10 + x = 100$

h. $7x - 12 - x = 24$

i. A computer repair shop charges $40 per hour plus the cost of parts. If a repair bill of $125 includes $35 in parts, then how long did the repair shop work on the computer? Solve this equation to find the answer. (Express your answer in both decimal form and in hours and minutes.)

$$40x + 35 = 125$$

Written Practice *Strengthening Concepts*

* 1.
(45) **Analyze** The ratio of in-state to out-of-state visitors at the zoo is approximately 7 to 3. If 2000 people visited the zoo, about how many people were from out of state?

2.
(3, 4) Arnold is planning a party for his friends. Renting the facilities will cost $500. Feeding each friend will cost $15. What will the total cost of the party be if Arnold invites 50 friends?

* 3. *Generalize* Consider the quadrilaterals below.
(35)

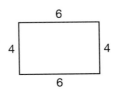 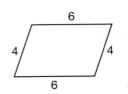 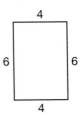

Choose the correct word to complete the conjecture: Quadrilaterals with corresponding sides of equal length are _____ (sometimes/always/never) congruent."

Analyze Solve.

* 4. $5x + 25 = 100$
(50)

* 5. $\frac{x}{2} + 8 = 16$
(50)

* 6. $2x - 1.2 = 3$
(50)

* 7. $-4m + 5.5 = 9.5$
(50)

* 8. $-2w + 22 = 30$
(50)

* 9. $\frac{1}{2}x - \frac{1}{3} = \frac{2}{3}$
(50)

Generalize Simplify.

* 10. $-3^2 + (-3)^2$
(36)

11. $\frac{1}{3} + \frac{5}{6} \cdot \frac{4}{5}$
(13, 22)

12. $\frac{mc^2xc}{mx^2}$
(27)

13. $1\frac{1}{2} \cdot 2\frac{2}{3} - 3\frac{3}{4}$
(13, 23)

* 14. $\frac{2.7 \times 10^8}{9 \times 10^3}$
(28, 46)

15. $\frac{0.24 - 0.024}{0.02}$
(24, 25)

16. Find the **a** area and **b** circumference of the circle with radius 1 meter. Express the measures in terms of π.
(39, 40)

17. Estimate the volume in cubic feet of a cabinet with height 4 feet 2 inches, depth 13 inches, and width 2 feet 11 inches.
(42)

18. Find the **a** area and **b** perimeter of this figure.
(8)

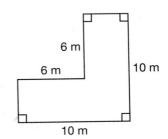

19. The company will charge $40 for each item it sells.
(41)

   a. Make a function table of possible *x*- and *y*-values with *x* representing the number of items sold and *y* representing the revenue from the sales.

   b. Write an equation for the function table:

20. Write 7.5% **a** as a decimal and **b** as a reduced fraction.
(11, 12)

* 21. Consider the triangles.
(35)

    a. Explain how you know that the triangles are similar.

    b. What is the scale factor from $\triangle ABC$ to $\triangle QRS$?

    c. Find x.

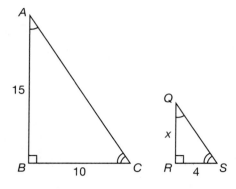

22. Find $\dfrac{r^2 - 4r}{t}$ when $r = 3$ and $t = -1$.
(14, 15)

23. Carlos was making necklaces to sell at the school carnival. He sold
(41) each necklace for $3. Make a function table that relates the number of necklaces sold to the amount of money collected. Write an equation for the rule. Is the function linear? Is it proportional?

* 24. *Analyze*   Geraldo is going to paint the interior walls of the lobby of the
(43) local history museum. The room is 40 ft by 60 ft. Its height is 16 ft. In order to purchase the right amount of paint, Geraldo must calculate the lateral surface area of the room. What is the lateral surface area?

* 25. There are approximately 2.25×10^{12} grains of sand on a stretch of
(46) beach. If a bucket holds 1.25×10^9 grains of sand, how many buckets would it take to remove all of the sand from the beach?

* 26. Last month, Andrea missed two days of school because she had a
(48) cold. What percent of school days did Andrea miss? Hint: Assume that there are 20 school days in a month.

* 27. *Evaluate*   If the bowling alley charges $3.50 for a pair of shoes and
(50) $5.50 per game, how many games will Rita be able to play with a $20 dollar bill? Write and solve an equation and check the solution.

28. Write the mixed number $1\frac{1}{9}$ as a decimal.
(30)

* 29. *(41, 44)* **Represent** Graph $y = -2x$. Is the point $(-1, -2)$ on the line? What is the slope of the line?

30. *(41)* For a promotion, the clothes shop is giving every customer 2 free T-shirts. If the customers want more T-shirts, they must buy them for $5 each. Which graph represents the relationship between the number of shirts a customer receives and the amount of money the customer pays for the shirts? Is the relationship proportional? Why or why not?

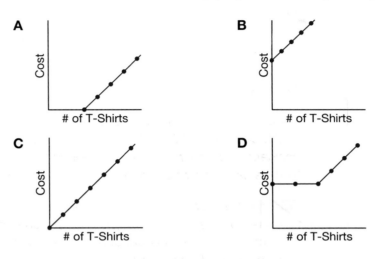

Early Finishers
Real-World Application

Ferdinand is planting four rose bushes in the city park. He is planning to place the bushes into four rectangular planters that are 3 feet long, 1 foot wide and 1.5 feet deep. How many bags containing 1.25 feet of cubic soil will he need to fill all four planters two-thirds full?

Focus on
• Graphing Transformations

Recall that **transformations** are operations on a geometric figure that alter its position or size. In this investigation we will graph different transformations on the coordinate plane.

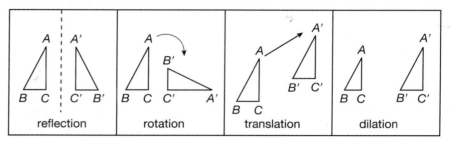

A **reflection** occurs across a line. Below we show the reflection of △ABC across the y-axis. If we positioned a vertical mirror on the line, the reflection of △ABC would appear to be on the other side of the y-axis and the same distance from the y-axis as △A'B'C' (read "A prime, B prime, C prime"). A segment between corresponding points of a figure and its reflection is perpendicular to the line of reflection. If we were to fold a graph along the line of reflection, the figures would align exactly.

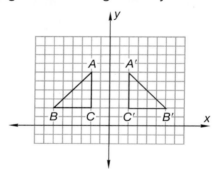

1. On a coordinate plane draw triangle ABC with A at (−2, 6), B at (−5, 2), and C at (−2, 2). Then draw the reflection of △ABC across the x-axis. Label the coordinates of the vertices.

2. Describe how the coordinates of the reflection across the x-axis compare with the coordinates of the original triangle.

3. On the same coordinate plane as problem 1, draw a vertical line through the x-axis at x = 1. (The line should be parallel to the y-axis.) Then graph the reflection of △A'B'C' across the line x = 1. Name the reflection △A"B"C" (read "A double prime, B double prime, C double prime") and label the coordinates of the vertices.

Math Language

Recall that **counter-clockwise** means "in the opposite direction as the movement of the hands of a clock," or "to the left."

A positive **rotation** turns a figure counterclockwise about (around) a point. Here we show a 90° rotation about point *B*. Point *B* is the point of rotation; its location is fixed, and the figure spins around it. If we trace the path of any other point during this rotation, we find that it sweeps out an arc of 90°.

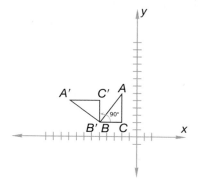

To graph a rotation it is helpful to have a piece of transparency film and a transparency marker. Otherwise use a piece of paper thin enough to use for tracing.

4. On a coordinate plane graph △*ABC* as illustrated above. We will rotate △*ABC* 90° about the origin. Place a piece of transparency film or tracing paper on the coordinate plane so that it covers △*ABC* and the origin. Trace △*ABC* and mark an inch or so of the *x* and *y*-axes through the origin as illustrated below.

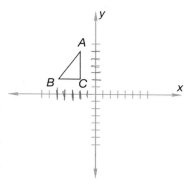

Press a pencil point on the tracing sheet at the origin and rotate the tracing sheet counterclockwise 90° so that the *x*-axis on the tracing paper aligns with the *y*-axis on the original coordinate plane below, and the *y*-axis on the tracing paper aligns with the *x*-axis below.

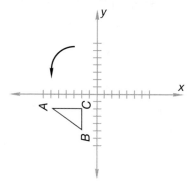

Use the position of the traced image to help you draw the rotated image on the coordinate plane. Name the rotated image △*A'B'C'* and label the coordinates of the vertices.

5. On the same coordinate plane rotate △*ABC* 180°. Name the image △*A"B"C"* and label the coordinates of the vertices.

A **translation** slides a figure a distance and direction without flipping or turning. Here we show a translation of (6, 1). For any translation (*a, b*), ***a*** describes the horizontal shift and ***b*** describes the vertical shift for each point of the figure.

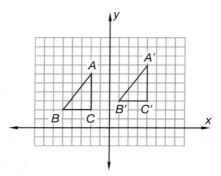

6. Draw triangle *ABC* as illustrated on a coordinate plane. Then draw its image △*A'B'C'* translated 7 units to the right and 8 units down, or (7, −8). Label the coordinates of △*A'B'C'*.

7. If △*ABC* is translated so that the image of point *A* at *A"* is located at the origin, then what would be the coordinates of points *B"* and *C"*? Describe the translation.

Visit www. SaxonPublishers. com/ActivitiesC3 for a graphing calculator activity.

Recall that reflections, rotations, and translations are called **isometries** (meaning "same measures"), or **congruence transformations,** because the original figure and its image are congruent. A **dilation** is a transformation in which the figure grows larger, while a **contraction** is a transformation in which the figure grows smaller. Dilations and contractions are **similarity transformations,** because the original figure and its image are similar and corresponding lengths are proportional.

Dilations of geometric figures occur away from a fixed point called the center of the dilation. On a coordinate plane the center of a dilation may be any point. In this book the center of dilation will be the origin.

Dilation of Triangle *ABC*

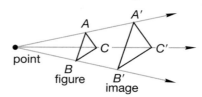

Note that corresponding vertices are on the same rays from the center of dilation and that the corresponding segments of △*ABC* and its image △*A'B'C'* are parallel.

Recall that to graph a dilation on a coordinate plane we prescribe a scale factor, such as 2. If the center of dilation is the origin, we may multiply the coordinates of the vertices of the figure by the scale factor. For example, using a scale factor of 3, a vertex at (1, 3) in the original figure would have a corresponding vertex at (3, 9) in its image after dilation.

8. On a coordinate plane draw rectangle ABCD with A at (6, 4), B at (6, −2), C at (−2, −2), and D at (−2, 4). Then draw its image □A'B'C'D' using scale factor $\frac{1}{2}$, with the center of the contraction at the origin. What are the coordinates of the vertices of □A'B'C'D'?

9. What fraction of the dimensions of □ABCD are the dimensions of □A'B'C'D'? What fraction of the perimeter of □ABCD is the perimeter of □A'B'C'D'?

10. What fraction of the area of □ABCD is the area of □A'B'C'D'?

In the following table we summarize the transformations graphed in this investigation.

Transformations

Preserving Congruence	Preserving Similarity
Translation (slide)	Dilation (scale increased)
Rotation (turn)	Contraction (scale reduced)
Reflection (flip)	

For review, graph the following transformations on a coordinate plane.

11. Draw △XYZ with X at (−2, 5), Y at (−5, 2), and Z at (−2, 2). Then draw its reflection across the x-axis. Correctly label the corresponding vertices of △X'Y'Z'.

12. Draw △PQR with P at (3, 0), Q at (3, −5), and R at (6, 0). Then draw its image △P'Q'R' after a 90° counterclockwise rotation about point P.

13. Draw △DEF with D at (−2, −2), E at (2, −2), and F at (0, 2). Then draw its image △D'E'F' after a translation of (5, 3).

14. Draw △JKL with J at (1, 2), K at (−1, −2), and L at (1, −2). Then draw its image △J'K'L' after a dilation of scale factor 2.

• Negative Exponents
• Scientific Notation for Small Numbers

Power Up | **Building Power**

facts | Power Up K

mental math |
a. **Number Sense:** 3.3×4

b. **Measurement:** What temperature is indicated on this thermometer?

c. **Algebra:** $25x = 250$

d. **Proportions:** There are 2 oranges to every 3 apples in a fruit basket. If there are 8 oranges, how many apples are in the basket?

e. **Percent:** How much is a 10% down payment on a $20,000 car?

f. **Scientific Notation:** Write 250 in scientific notation.

g. **Geometry:** Approximate the volume of this object.

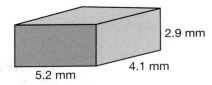

2.9 mm

4.1 mm

5.2 mm

h. **Calculation:** $\sqrt{9} + \sqrt{16} - \sqrt{25}$

problem solving | The first four numbers Hexa said were 16, 32, 48, and 64. If she keeps counting this way, what is the 99th number Hexa will say?

New Concepts | *Increasing Knowledge*

negative exponents | Recall from Lesson 27 that we subtract the exponents when dividing exponential expressions that have the same base.

$$x^5 \div x^3 = x^2$$

We can understand this law by applying what we know about exponents and division.

$$\frac{x^5}{x^3} = \frac{\overset{1}{\cancel{x}} \cdot \overset{1}{\cancel{x}} \cdot \overset{1}{\cancel{x}} \cdot x \cdot x}{\underset{1}{\cancel{x}} \cdot \underset{1}{\cancel{x}} \cdot \underset{1}{\cancel{x}}} = x^2$$

Now consider the result when we reverse the numbers in the division. Following the Laws of Exponents, the exponent of the quotient is negative.

$$x^3 \div x^5 = x^{-2} \text{ (because } 3 - 5 = -2\text{)}$$

We will apply what we know about exponents and division to understand the meaning of x^{-2}.

$$\frac{x^3}{x^5} = \frac{\overset{1}{\cancel{x}} \cdot \overset{1}{\cancel{x}} \cdot \overset{1}{\cancel{x}}}{\underset{1}{\cancel{x}} \cdot \underset{1}{\cancel{x}} \cdot \underset{1}{\cancel{x}} \cdot x \cdot x} = \frac{1}{x^2}$$

By performing the division we find that x^{-2} means $\frac{1}{x^2}$. We see that x^{-2} is the reciprocal of x^2. This fact is another law of exponents.

Math Language

Recall that the product of a number and its **reciprocal** is 1.

Law of Exponents for Negative Exponents

$$x^{-n} = \frac{1}{x^n}$$

Applying this law to powers of 10, we see the following pattern. Note that $10^0 = 1$.

$$10^2 = 100$$
$$10^1 = 10$$
$$10^0 = 1$$
$$10^{-1} = \frac{1}{10} \text{ or } 0.1$$
$$10^{-2} = \frac{1}{100} \text{ or } 0.01$$

Evaluate Use the Laws of Exponents to find the product of 10^2 and 10^{-2}.

Very small numbers may exceed the display capabilities of a calculator. One millionth of one millionth is more than zero, but it is a very small number. On a calculator we enter

```
•  0  0  0  0  0  1  ×
•  0  0  0  0  0  1  =
```

The product, one trillionth, contains more digits than can be displayed by many calculators. Instead of displaying one trillionth in standard form, the calculator displays the number in a modified form of scientific notation:

$$\boxed{1.^{-12}} \text{ or perhaps } \boxed{1. \times 10^{-12}}$$

Example 1

Which of the following does not equal 10^{-3}?

A $\frac{1}{10^3}$ B $\frac{1}{1000}$ C 0.001 D -1000

Solution

An exponential expression with a negative exponent is the reciprocal of the expression with the opposite exponent, as shown in A. Since 10^3 equals 1000, B is just an alternate form of A. Choice C is the decimal equivalent of B. The only number that does not equal 10^{-3} is choice **D, −1000.** A negative exponent does not imply a negative number. Actually, 10^{-3} is a positive number.

Example 2

Find each missing exponent.

a. $10^{-2} \cdot 10^{-4} = 10^{\square}$

b. $\dfrac{10^{-2}}{10^{-4}} = 10^{\square}$

Solution

a. To multiply exponential expressions with the same base we can add the exponents. We add -2 and -4.

$$(-2) + (-4) = -6$$

The missing exponent is **−6.**

b. To divide exponential expressions with the same base we can subtract the exponents. In this case we subtract -4 from -2. Instead of subtracting a negative, we add its opposite.

$$(-2) - (-4)$$
$$(-2) + (+4) = 2$$

The missing exponent is **2.**

Verify Find the answers to **a** and **b** by simplifying each exponential expression and performing the multiplication and division.

Example 3

Simplify:

a. 2^{-3} 　　　　　　　　　　　　　b. $(-2)^3$

Solution

We distinguish between negative powers and powers of negative numbers.

a. To simplify 2^{-3} we rewrite the expression with a positive exponent.

$$2^{-3} = \frac{1}{2^3}$$

Then we apply the positive exponent to the base.

$$\frac{1}{2^3} = \frac{1}{2 \cdot 2 \cdot 2} = \frac{1}{8}$$

b. The expression $(-2)^3$ is written with a positive exponent. Three negative factors produce a negative product.

$$(-2)^3 = (-2)(-2)(-2) = -8$$

Example 4

Simplify: $3^2 \cdot 3^{-2}$

To simplify $3^2 \cdot 3^{-2}$ we add the exponents. Adding the exponents 2 and −2 results in the exponent zero.

$$3^2 \cdot 3^{-2} = 3^0$$

Recall that $x^0 = 1$ if x is not zero. Therefore $3^0 = 1$. We can confirm this by applying what we have learned about negative exponents. We will rewrite the multiplication by expressing 3^{-2} with a positive exponent.

Step:	Justification:
$3^2 \cdot 3^{-2}$	Given expression
$3^2 \cdot \dfrac{1}{3^2}$	Wrote 3^{-2} with positive exponent
$9 \cdot \dfrac{1}{9}$	Applied exponents
1	Simplified

We confirm that the expression 3^0 equals **1**.

Example 5

Express with positive exponents and simplify: $2x^{-1}yx^2y^{-2}z$

A negative exponent indicates a reciprocal, so x^{-1} is $\frac{1}{x}$ and y^{-2} is $\frac{1}{y^2}$. A quick way to change the sign of the exponent is to draw a division bar and shift the exponential expression to the opposite side of the division bar.

$$\frac{2x^{-1}yx^2y^{-2}z}{1} = \frac{2yx^2z}{xy^2}$$

All exponents are positive. Now we reduce.

$$\frac{2yx^2z}{xy^2} = \frac{2\cancel{y}\cancel{x}xz}{\cancel{x}\cancel{y}y} = \frac{2xz}{y}$$

Connect How can we apply the Multiplication Law of Exponents to simplify the original expression?

scientific notation for small numbers

We use negative powers of 10 to write small numbers in scientific notation. By small numbers we mean numbers between 0 and 1.

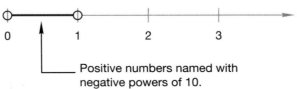

Positive numbers named with negative powers of 10.

Visit www.
SaxonPublishers.
com/ActivitiesC3
for a graphing
calculator activity.

Example 6

Write this number in standard form.

$$1.5 \times 10^{-3}$$

Solution

The power 10^{-3} equals 0.001. If we perform the multiplication we find the product is 0.0015.

$$1.5 \times 10^{-3}$$
$$1.5 \times 0.001 = \mathbf{0.0015}$$

Notice that we can find the product simply by shifting the decimal point in the coefficient three places to the left. Thus the power of ten indicates both the direction and the number of places the decimal shifts.

Visit www.
SaxonPublishers.
com/ActivitiesC3
for a graphing
calculator activity.

Example 7

The diameter of a red blood cell is about 0.000007 meters. Write that number in scientific notation.

Solution

The coefficient is 7. In standard form the decimal point is six places to the left, so the power of 10 is -6.

$$0.000007 = \mathbf{7 \times 10^{-6}}$$

Example 8

Compare: $1 \times 10^{-6} \bigcirc -10$

Solution

The expression 1×10^{-6} is a positive number, though a small one (0.000001). Any positive number is greater than a negative number.

$$1 \times 10^{-6} > -10$$

Practice Set

Simplify:

a. 3^{-2} **b.** 2^{-3} **c.** 5^0 **d.** $2^{-3} \cdot 2^3$

e. Analyze Arrange in order from least to greatest:

$$\frac{1}{2}, 0, 1, 2^{-2}, -1, 0.1$$

Evaluate Find the missing exponent in problems **f–g**. Check your answers by substituting the exponent answer into each equation and solving.

f. $10^{-3} \cdot 10^{-4} = 10^{\square}$ **g.** $\dfrac{1}{1} = 10^{\square}$

Simplify:

h. $x^{-3}y^2xy^{-1}$

i. $\dfrac{6x^{-2}y^3z^{-1}}{2xy}$

j. Write 10^{-4} as a decimal number.

k. Write 5×10^{-5} in standard form.

l. Write 2.5×10^{-2} in standard form.

m. Write 0.008 in scientific notation.

n. Write 0.000125 in scientific notation.

o. If lightning strikes a mile away the sound reaches us in about 5 seconds, but its light reaches us in about 5 millionths of a second. Write 5 millionths in scientific notation.

p. A nanometer is 10^{-9} meters. Write that number in standard form.

Written Practice   *Strengthening Concepts*

Evaluate For problems **1–2**, record the information in a ratio table. Estimate an answer and then solve by writing and solving a proportion.

* 1. Ruben wrote 2 pages in 3 hours. At that rate, how long will it take him to write 9 pages?
(49)

* 2. If 9 gallons of gas cost $20.25, how much would it cost to fill a 25 gallon tank?
(49)

* 3. On a piano, the ratio of black keys to white keys is 9 to 13. If there are 88 keys on a piano, how many are black?
(45)

* 4. Kerry mixes apple slices and raisins at a 3 to 1 ratio to make her fruit salad. If she wants 16 cups of fruit salad, how many cups of apple slices will she need?
(45)

5. The shipping charges totaled $9.26. Included in the charge was a $3.50 flat fee. The rest of the cost was for weight, at $0.64 per pound. How much did the package weigh?
(3, 4)

Solve:

* 6. $12x - 3 = 69$
(50)

* 7. $\dfrac{x}{4} + 1 = 12$
(50)

* 8. $\dfrac{x}{3} - 4 = 5$
(50)

* 9. $-x + 1 = 6$
(50)

* 10. $1 - m = -1$
(50)

* 11. $3x + 2x - 1 = 99$
(50)

* 12. Refer to the graph of lines a and b to answer the following questions.
(Inv. 1)

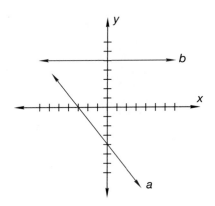

    a. Which line intersects the y-axis at -4?

    b. Which line is perpendicular to the y-axis?

    c. What is the slope of line a?

    d. In which quadrant do lines a and b intersect?

* 13. Graph $y = 4x + 1$. Is (3, 13) on the line?
(47)

* 14. The figure shows three concentric (meaning "same center") circles. The diameters of the three circles are 4 cm, 8 cm, and 12 cm. Find the area of the shaded region. Use 3.14 for π and round the answer to the nearest square centimeter.
(40)

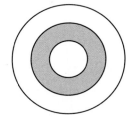

* 15. Sketch a model of a book that measures 2 in. by 10 in. by 8 in. What is the volume of the book?
(Inv. 4, 42)

* 16. Write 3.4×10^{-5} in decimal notation.
(51)

Combine like terms to simplify:

17. $9(x - 3) + 5(x + 5)$
(31, 36)

18. $3x^2 - 3x + x - 1$
(31)

Simplify:

* 19. $\dfrac{(2xy)(3x^2y)}{6xy^2}$
(36)

* 20. $-6 - (-5)$
(33)

21. $2\dfrac{1}{2} \cdot 1\dfrac{3}{5} - 2\dfrac{3}{8}$
(13, 23)

* 22. $(2.3 \times 10^4)(1.5 \times 10^3)$
(46)

23. Compare $0.62 \bigcirc \dfrac{5}{8}$
(12)

24. Write $\dfrac{7}{10}$ **a** as a decimal and **b** as a percent.
(12)

25. **a.** Redraw the triangles separately and find a and b.
(35)

    b. Explain how you know that the two triangles are similar.

    c. Find the scale factor from the smaller to the larger triangle, then find x.

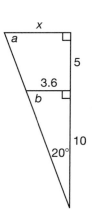

26. Find $\sqrt{b^2 - 4ac}$ when $a = 6$, $b = 5$, and $c = -1$.
(14, 36)

* 27. James is building a 10-inch by 12-inch rectangular picture frame. To
(Inv. 2) assure the frame has right angles, James measures the two diagonals
to see if they are equal. When the diagonals are equal, how long
is each diagonal? (Express your answer to the nearest tenth of an
inch.)

28. Arnold flipped a coin twice and it landed heads up both times. If he flips
(32) the coin again, what is the probability the coin will land heads up?

29. Find all possible values of x for the equation $x^2 - 16 = 0$.
(14, 36)

* 30. **a.** Which figure—A, B, or C—is a translation of figure $MNOPQR$?
(Inv. 5)

b. For the figure you chose, give the coordinates that locate M', N', O',
P', Q', and R'.

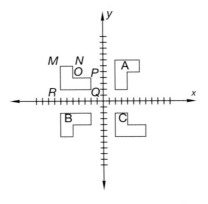

Early Finishers
Real-World
Application

In a vacuum, the speed of light is constant at approximately 300,000,000
meters/second. If the distance from the Earth to the Sun is about
149,000,000,000 meters, about how many seconds does it take light to reach
the Earth? Express your answer in scientific notation.

• Using Unit Multipliers to Convert Measures
• Converting Mixed-Unit to Single-Unit Measures

Power Up   *Building Power*

facts | Power Up K

mental math

   a. **Number Sense:** 7.5×20

   b. **Statistics:** The mean of two numbers is the number halfway between. What is the mean of 10 and 20?

   c. **Fractional Parts:** How much do you save if a $90 item is 10% off?

   d. **Scientific Notation:** Write 0.0051 in scientific notation.

   e. **Rate:** Jackie finishes 50 problems per hour. Ray finishes 40 problems per hour. How many more problems will Jackie finish than Ray if they work for 30 minutes?

   f. **Geometry:** Two angles of a triangle are 60° and 60°. Find the measure of the third angle. What type of triangle is the triangle?

   g. **Powers/Roots:** List the perfect squares from 1 to 169.

   h. **Calculation:** $22, + 2, \times 2, + 2, \times 2, \sqrt{}$

problem solving | A number is divided by 2 and then added to 11. The result is 19. What is the number? Explain your thinking.

New Concepts   *Increasing Knowledge*

using unit multipliers to convert measures

A **unit multiplier** is a ratio in which the numerator and denominator are equivalent measures but different units. Since 12 inches equal one foot, we can write these two unit multipliers for that relationship.

$$\frac{12 \text{ in.}}{1 \text{ ft}} \qquad \frac{1 \text{ ft}}{12 \text{ in.}}$$

Each of these ratios equals 1 because the measures above and below the division bar are equivalent. Since multiplying by 1 (or a form of 1) does not change a quantity, we can multiply by a unit multiplier to convert a measure from one unit to another unit.

Example 1

The sapling apple tree is 64 inches tall. Convert 64 inches to feet by multiplying 64 inches by $\frac{1 \text{ ft}}{12 \text{ in.}}$.

Solution

The unit multiplier cancels inches and leaves feet as the unit.

$$64 \text{ in.} \times \frac{1 \text{ ft}}{12 \text{ in.}} = \frac{64}{12} \text{ft} = 5\frac{1}{3} \text{ ft}$$

Example 2

A certain double feature at a theater is 270 minutes long. Use a unit multiplier to convert 270 minutes to hours.

Solution

Since 60 minutes equal one hour, we have a choice of two unit multipliers.

$$\frac{60 \text{ min}}{1 \text{ hr}} \qquad \frac{1 \text{ hr}}{60 \text{ min}}$$

Thinking Skill

Explain

Is canceling units of measure like canceling numbers? Explain.

To cancel minutes we choose the unit multiplier with minutes below the division bar.

$$270 \text{ min} \times \frac{1 \text{ hr}}{60 \text{ min}} = \frac{270}{60} \text{ hr} = 4\frac{1}{2} \text{ hr}$$

Connect In addition to canceling units in examples 1 and 2, could we also cancel numbers? Explain.

We can use unit multipliers to convert measures to desired or appropriate units. Appropriate units are correct and meaningful to the intended reader.

Example 3

Daniel is buying refreshments for a meeting. He wonders if four quarts of juice is enough for 30 students. Is four quarts a reasonable amount of juice for the meeting? If not, suggest a more reasonable quantity.

Solution

If each student has the same amount of juice, then one serving is 4 quarts divided by 30.

$$\frac{4 \text{ quarts}}{30 \text{ students}} = \frac{4}{30} \text{ quarts per student}$$

The answer is not very meaningful. A more appropriate unit is ounces. We convert 4 quarts to ounces and then we divide by 30.

$$4 \text{ qt} \cdot \frac{32 \text{ oz}}{1 \text{ qt}} = 128 \text{ oz}$$

$$\frac{128 \text{ oz}}{30 \text{ students}} = 4 \text{ ounces per student}$$

Since four ounces is half a cup, **four quarts is probably not enough juice.** If Daniel buys **eight quarts,** then each student could have a full cup (8 ounces) of juice.

Some units like feet and inches are used together to express a measure.

Mr. Seymour is 6 ft 3 in. tall.

To express Mr. Seymour's height with a single unit we could give his height in inches (75 in.) or in feet. Below we show his height in feet. Three inches is $\frac{3}{12}$ of a foot, which equals $\frac{1}{4}$ ft or 0.25 ft. We can combine 6 ft with $\frac{1}{4}$ ft or 0.25 ft.

$$6 \text{ ft } 3 \text{ in. } = 6\frac{1}{4} \text{ ft } = 6.25 \text{ ft}$$

Example 4

Samantha drove 100 miles in an hour and 45 minutes. How many hours did it take for Samantha to drive 100 miles?

Solution

We express 45 minutes as $\frac{45}{60}$ of an hour.

$$45 \text{ min} = \frac{45}{60} \text{ hr} = \frac{3}{4} \text{ hr} = 0.75 \text{ hr}$$

Samantha drove for 1 hr 45 min, which is **$1\frac{3}{4}$ hr or 1.75 hr.**

Example 5

Noel ran one mile in 7 minutes and 30 seconds. Find Noel's time in minutes. Express the result as a mixed number and as a decimal number.

Solution

We convert 30 sec to $\frac{30}{60}$ minutes which is $\frac{1}{2}$ or 0.5 min.

$$7 \text{ min } 30 \text{ sec} = 7\frac{1}{2} \text{ min} = 7.5 \text{ min}$$

Example 6

To measure a room for carpeting, Juan converts the length and width of the room from feet and inches to yards. A room that is 13 ft 6 in. long is how many yards long?

Solution

First we convert 13 ft 6 in. to feet.

$$13 \text{ ft } 6 \text{ in. } = 13\frac{6}{12} \text{ ft} = 13.5 \text{ ft}$$

Then we convert feet to yards.

$$13.5 \text{ ft} \cdot \frac{1 \text{ yd}}{3 \text{ ft}} = \frac{13.5 \text{ yd}}{3} = 4.5 \text{ yd}$$

The room is **4.5 yd** long.

Practice Set

a. _Justify_ One day is 24 hours. Write two unit multipliers that have days and hours as the units. Identify the unit multiplier you would use to convert days to hours. Explain your choice.

b. _Represent_ Write two unit multipliers for this equivalence:

$$16 \text{ oz} = 1 \text{ pt}$$

c. A gallon is 128 oz. Convert 128 oz. to pints using a unit multiplier.

d. An inch is 2.54 cm. A bookcase that is 50 inches high is how many centimeters high? Use a unit multiplier to perform the conversion.

e. Convert 24 quarts to gallons using a unit multiplier. (1 gal = 4 qt)

f. Carter claims he can run 10,000 centimeters in 14 seconds. Express his claim in more appropriate terms.

g. The newborn child was 21 inches long and weighed 8 pounds 4 ounces. Convert these measures to feet and to pounds respectively, expressed as mixed numbers and as decimal numbers.

h. Marsha swam 400 meters in 6 minutes and 12 seconds. Convert that time to minutes.

i. A room is 11 ft 6 in. long and 11 ft 3 in. wide. Find the perimeter of the room in feet.

j. Convert 11 ft 3 in. to feet. Then use a unit multiplier to convert that measure to yards.

Written Practice   _Strengthening Concepts_

1. The hen to rooster ratio in the barnyard is 11 to 2. If there are 26 hens
(45) and roosters in all, how many are roosters?

2. Sally sells popcorn for $1 and beverages for $2. What will be
(3, 4) her total sales if 19 people each buy one bag of popcorn and
one beverage?

3. For a collect call, a phone company charges a $1 connection fee, then
(3, 4) 4 cents per minute for the duration of a call. How much will it cost for an
11 minute call?

For problems **4–6,** record the information in a ratio table. Estimate an answer and then solve by writing and solving a proportion.

* 4. Cynthia sews four dresses in 3 hours. How many dresses could she sew
(49) in 9 hours?

* 5. Sergio estimates that his trip will take 3 hours and 30 minutes. Express
(52) Sergio's estimate as a mixed number of hours and as a decimal number
of hours.

* 6. A light is left on for 40 days. Convert 40 days to hours using a unit
(52) multiplier. If the light bulb has an estimated life of 1400 hours, is the light
likely to still be on after 40 days?

Analyze Solve.

* 7. $20x + 50 = 250$
(50)

* 8. $\dfrac{x}{3} + 5 = 7$
(50)

* 9. $x + 5 + x = 25$
(50)

* 10. $14 - m = 24$
(50)

11. Victoria is playing a board game with two number cubes. She rolls both
(32) cubes once.

    a. List the possible totals (sample space).

    b. Predict: Which totals do you predict are least likely and which totals
    are most likely. Justify your prediction.

* 12. Analyze Refer to the graph of lines
(Inv. 1, 44) a and b to answer the following
questions.

    a. Which line intersects the y-axis at
    5?

    b. What is the slope of
    line b?

    c. In which quadrant do lines a and b
    intersect?

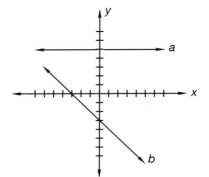

13. Triangle ABC with vertices at (2, 1), (1, 4), and (2, 2) is reflected in the
(Inv. 5) x-axis. What are the coordinates of the vertices of $\triangle A'B'C'$?

14. A 14-inch diameter circular porthole (window) is framed in a brass ring
(39, 40) and mounted in the side of a ship.

    a. Find the area of the glass. Round your answer to the nearest
    square inch.

    b. What is the inner circumference of the brass mounting ring?
    (Use $\frac{22}{7}$ for π.)

15. Find the volume of a box with dimensions 5 in. by 5 in. by 8 in. Sketch
(42) the box resting on a square face and indicate its dimensions.

16. Find the surface area of the box in exercise **15.**
(43)

Generalize Simplify:

17. $4(x - 4) - 2(x - 6)$
(31, 36)

18. $b + h + b + h$
(31)

* 19. 5^{-2}
(51)

* 20. $(-3) - (-7)$
(33)

21. $\dfrac{(4.2 \times 10^7)}{(1.4 \times 10^2)}$
(46)

22. $\left(\dfrac{2}{3}\right)^2 + \dfrac{2}{3}$
(13, 22)

* 23. *Connect* **a.** Write $\frac{19}{20}$ as a decimal and as a percent.
(12)

   b. How would you report your score on a test for which you answered 19 out of 20 questions correctly?

24. The pentagons are similar.
(35)
   a. Find a.

   b. What is the scale factor from the smaller to the larger pentagon?

   c. Find x.

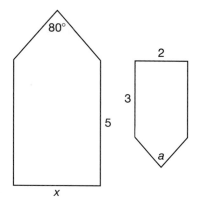

25. Simplify and express with positive exponents.
(51)

$$\frac{x^{-1}y^2}{xy}$$

* 26. Chester needs a length of 2-by-4 to make a diagonal brace for a wooden gate that has the dimensions shown. The 2-by-4 must be at least how many inches long?
(Inv. 2)

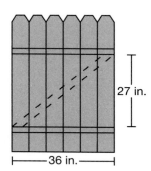

* 27. An ant weighs about 10^{-5} kg. Write this as a decimal number.
(51)

28. Sketch a quadrilateral with just one pair of parallel sides. What type of quadrilateral did you sketch?
(Inv. 3)

Find all values of x.

29. a. $7x^2 = 7$ **b.** $|x| - 15 = 5$
(14, 36)

30. Sketch the front, right side, and top views of this figure.
(Inv. 4)

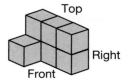

• Solving Problems Using Measures of Central Tendency

facts   Power Up K

mental math

a. **Number Sense:** 7.5×20

b. **Proportions:** $\frac{25}{100} = \frac{4}{x}$

c. **Probability:** A spinner has 8 equal sections numbered 1–8. What is the probability that the spinner will stop on a number greater than 5?

d. **Measurement:** The odometer read 2901 mi at the beginning of the trip. At the end of the trip, it read 3247 mi. How long was the trip?

e. **Scientific Notation:** Write 2.08×10^7 in standard notation.

f. **Geometry:** Find the area of the rectangle, then find the area of the shaded triangle.

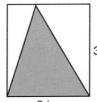

3 in.

2 in.

g. **Estimation:** Estimate the total cost of items priced at: $2.45 plus $7.45 plus 3 items at $9.95.

h. **Calculation:** $\sqrt{121}$, $\times 9$, $+ 1$, $\sqrt{}$, $- 1$, $\times 7$, $+ 1$, $\sqrt{}$, $\times 6$, $+ 1$, $\sqrt{}$

problem solving

About how many sheets of paper the thickness of this page would make a stack 1 cm high? About how many centimeters thick is this page? What should be the product of your two answers?

To summarize data, we often report an average of some kind, like the **mean, median,** or **mode.**

Suppose a popular television show received these ratings over its season:

16.0	15.2	15.3	15.5
15.7	15.1	14.8	15.6
15.0	14.6	14.4	15.8
15.3	14.7	14.2	15.7
14.9	15.4	14.9	16.1

```
                X   X   X
        X  X  XXXXXXXXXXXXX  XX
    <---+------------+-------+-------+--->
        14           15      16      17
```

The data above are displayed on a **line plot.** Each data point is represented with an X above its value on a number line.

The **mean** TV rating is:

$$\text{mean} = \frac{16.0 + 15.7 + \ldots + 16.1}{20}$$

$$\text{mean} = 15.21$$

Recall that the **median** of an ordered list of numbers is the middle number or the mean of the two central numbers. There are an even number of data, so the numbers 15.2 and 15.3 share the central location. We compute the mean of these two numbers to find the median.

$$\text{median} = \frac{15.2 + 15.3}{2} = 15.25$$

The **mode** is the most frequently occurring number in a set. In this example, there are three modes: 14.9, 15.3, and 15.7.

The **range** of this data is the difference between the highest and lowest ratings.

$$\text{range} = 16.1 - 14.2$$

$$\text{range} = 1.9$$

We may also consider separating the data into intervals such as: 14.0–14.4, 14.5–14.9, 15.0–15.4, 15.5–15.9, 16.0–16.4. Data organized in this way can be displayed in a **histogram,** as shown below.

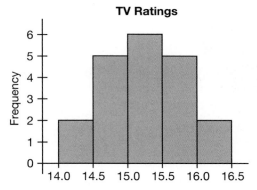

TV Ratings

Data which fall within specified ranges are counted, and the tallies are represented with a bar. For example, the tallest bar tells us that 6 data points fall within the range 15.0–15.4.

The highest peak of a histogram will always indicate the mode of the data ranges.

At the end of a season, Neilsen Ratings reports the mean season rating as a measure of a show's success. For various types of data, the mean is a frequently reported statistic. It is commonly referred to as the "average," even though the median and mode are other averages.

Example 1

The owner of a cafe was interested in the age of her customers in order to plan marketing. One day she collected these data on customers' ages.

28	33	31	41	42
27	75	73	38	35

a. Make a line plot of the data, then find the mean, median, mode, and range.

b. Which measure (mean, median, mode, or range) best represents the typical age of the customers?

Solution

a.
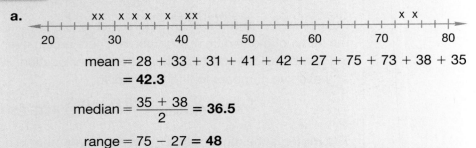

$$\text{mean} = 28 + 33 + 31 + 41 + 42 + 27 + 75 + 73 + 38 + 35$$
$$= \textbf{42.3}$$

$$\text{median} = \frac{35 + 38}{2} = \textbf{36.5}$$

$$\text{range} = 75 - 27 = \textbf{48}$$

There is no number that occurs more frequently than the others, so the mode is **not reported.**

b. The **median** gives the best description of the typical age of the customers. Half of the customers surveyed were younger than 36.5, and half were older. The mean, on the other hand, is much higher than the median because of a few customers that were older than the rest.

It is common to see medians used to report median age of residents by state, or median home prices by state.

Example 2

Suppose the median age of residents of a certain state was 30. Twenty years later it was 37. What changes in the population may have occurred in the twenty years?

Solution

When the median age was 30, half of the residents were younger than 30 and half were older. Twenty years later the median age had raised to 37. **This might happen if fewer children are born, if older people live longer, or if people younger than 37 move out of state or people older than 37 move into the state.** Any of these factors would contribute to an "older" population.

Example 3

Thinking Skill

Analyze

Which measure of central tendency would most accurately convey the salary a prospective employee could expect from the company? Why?

Consider the salaries of the employees of a small business. In an interview, a prospective employee was told that the average salary of employees is $50,000. How might this information be misleading?

Yearly Salary (in thousands $)

32	43
34	45
32	67
30	85
37	95

mean: 50

median: 40

mode: 32

range: 65

Solution

The interviewer reported the mean salary, which is much higher than the median because of a few salaries which are greater than most of the others. A prospective employee who heard that the average salary is $50,000 would be disappointed if his or her starting salary is near $30,000.

Example 4

Josiah sells three different hats to test their popularity. His sales are reported in a bar graph below. The height of each bar corresponds to the number of hats of that type Josiah sold.

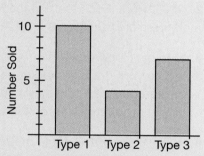

Find the mode of the data. Why are we not interested in the mean or median?

Solution

The mode of the types of hats is Type 1, the most popular choice. **In this case, the data Josiah collected was qualitative data (hat type), not quantitative data (numbers),** so mean and median are not important.

Practice Set

The amount of electricity (in kWh) used by one household each month of a year is listed below.

420 450 480 440 420 490 580 590 510 450 430 480

a. Make a line plot of the data.

b. *Analyze* Find the mean, median, mode, and range of the data.

c. Which measure would you report to describe the difference between the most electricity used in a month and the least?

d. Which measure would you report to someone who wanted to compute the total amount of electricity used in the year? Explain your answer.

e. If the greatest data point (590) were changed to 500, which measure of central tendency (mean, median, or mode) would change the most?

f. *Evaluate* Make a list of data values that fits the statement "Half of the days of February were colder than 30° F." Find the mean, median, mode and range of the data.

Written Practice   *Strengthening Concepts*

1. The ratio of adults to children attending the concert was 2 to 3. If there were 54 children, how many adults were there?
(45)

* 2. *Evaluate* Each team in the league has five starters and two alternates. If there are 30 starters in all, how many alternates are there?
(49)

3. It takes Robert one minute to gather his materials to begin his homework, then an additional minute for each problem on his assignment. If it takes him 20 minutes to complete the assignment, how many problems were on his assignment?
(3)

* 4. *Conclude* Each angle of an equilateral triangle measures how many degrees?
(20)

* 5. *Connect* The pole vaulter cleared the bar at 490 cm. Convert 490 cm to meters using a unit multiplier.
(52)

* 6. **Analyze** Find the mean, median, and
(53) mode of the five temperature readings
at each hour from 10 a.m. to 2 p.m.
According to the graph, for half the time
between 10 a.m. and 2 p.m., it was
warmer than what temperature?

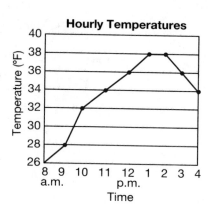

Analyze Solve.

* 7. $-2x = -16$
(50)

8. $\frac{w}{2} = 1.5$
(38)

* 9. $-z + 3 = 7$
(50)

* 10. $23 - p = 73$
(50)

11. A meter is about 3.3 ft, so 4.9 m is about
(52)

 A 2 ft   **B** 8 ft   **C** 16 ft   **D** 20 ft

12. Refer to the graphed lines to answer the
(Inv. 1, following questions.
44)

 a. What is the slope of line *b*?

 b. Line *b* intersects the *y*-axis at what
  point?

 c. In which quadrant do lines *a* and *b*
  intersect?

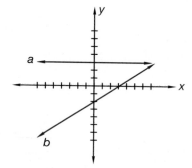

13. Graph $y = -x$ and the point $(4, -5)$. Is the point on the line?
(41)

14. Find the **a** area and **b** circumference of a circle with a diameter of
(39, 40) 18 inches. Express your answer in terms of π.

15. Find the volume of a crate with dimensions 4 ft by 3 ft by 2 ft.
(42)

16. A type of spider weighs about 1×10^{-4} lb. Write this as a decimal
(51) number.

Simplify.

17. $3(x - y) - 2(x + y)$
(31, 36)

* 18. $5x^2 - 3x - 2x^2 + 4x$
(31)

* 19. $\dfrac{2^3 \cdot 2^0}{2^1 \cdot 2^2}$
(27)

20. $\dfrac{x^3 y^3 z}{xy^3 z^3}$
(27)

* 21. $-5 - (-7)$
(31)

22. $(1.5 \times 10^3)(1.5 \times 10^{-2})$
(46)

23. **a.** Write 98% as a decimal and reduced fraction.
(12)

 b. For a $50,000 fundraising campaign, 98% of the funds have been
  raised. Which of the three forms would you choose to mentally
  calculate this amount?

24. Find $\frac{x}{2m}$ when $x = 5$ and $m = -5$.
(14, 36)

* 25. Express with all positive exponents: $6x^{-1}yyz^{-2}$
(51)

* 26. Sergio claims he can lift 10,000 grams over his head. Is his claim
(52) reasonable? Justify your answer.

* 27. A coin is flipped and the spinner is spun.
(32) Which sample space is most useful? Why?

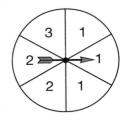

    A Sample Space {H1, H2, H3, T1, T2, T3}

    B Sample Space {H1, H1, H1, H2, H2, H3,
       T1, T1, T1, T2, T2, T3}

    Find P(H and 1).

28. Kellie high jumped 5 ft 3 in. Express that height as a mixed number of
(52) feet and as a decimal number of feet.

29. Find all values of x which are solutions.
(14, 36)
    a. $x^2 + 2 = 27$ **b.** $-5|x| = -15$

30. Estimate the **a** volume and **b** surface area
(42, 43) of this tissue box.

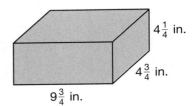

Early Finishers
Real-World
Application

While traveling on vacation, Mr. and Mrs. Rodriguez decide to spend an afternoon biking. There are two bike-rental companies nearby to choose from. Buck's Bikes charges $10 per hour per bike. Cycles Galore charges $4 per hour per bike plus an initial fixed fee of $15 per bike. Let x represent the time (in hours) that the bikes are rented and y represent the total cost.

    a. Write an equation to represent the total amount it would cost Mr. and
        Mrs. Rodriguez to rent two bikes from Buck's Bikes.

    b. Write an equation to represent the total amount it would cost Mr. and
        Mrs. Rodriguez to rent two bikes from Cycles Galore.

    c. Graph both equations on the same coordinate plane.

    d. Use your graph to determine which company would be cheaper if Mr.
        and Mrs. Rodriguez want to ride for three hours.

• Angle Relationships

facts   Power Up K

mental math

a. Number Sense: 3.5×8

b. Fractional Parts: 20% is $\frac{1}{5}$. How much is a 20% tip on a bill of $45?

c. Algebra: $x^2 = 81$

d. Scientific Notation: Write 5 trillion in scientific notation.

e. Rate: Vu and Tim began running at the same time and from the same place. Vu ran west at 8 miles per hour. Tim ran east at 9 miles per hour. After 30 minutes, how far had each run and how far were they from each other?

f. Geometry: Estimate the volume of this box.

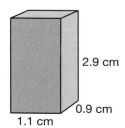

2.9 cm

0.9 cm

1.1 cm

g. Proportions: A survey of 50 students at a college found that 30 could name the Vice President. If the survey was representative of all 5000 students, then about how many of all the students could name the Vice President?

h. Calculation: $\sqrt{49}$, $\times 5$, $+ 1$, $\sqrt{}$, $\times 4$, $+ 1$, $\sqrt{}$, $\times 3$, $+ 1$, $\sqrt{}$, $\times 2$, $+ 1$, $\sqrt{}$

problem solving

At first, one-third of the class were boys. Then another boy joined the class. Now $\frac{3}{8}$ of the class are boys. How many students are in the class now? (Hint: There are fewer than 40 students in the class.)

Intersecting lines form pairs of adjacent angles and pairs of opposite angles.

Adjacent angles:

∠1 and ∠3

∠3 and ∠4

∠4 and ∠2

∠2 and ∠1

Opposite angles:

∠1 and ∠4

∠2 and ∠3

Adjacent angles share a common vertex and a common side but do not overlap.

Opposite angles are formed by two intersecting lines and share the same vertex but do not share a side. Opposite angles are also called **vertical angles.** Vertical angles are congruent.

Angles formed by two intersecting lines are related. If we know the measure of one of the angles, then we can find the measure of the other angles.

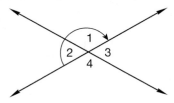

Reading Math

We can use symbols to abbreviate an angle's measure. We read the expression "$m\angle 1 = 120°$" as "the measure of angle 1 is 120 degrees."

Together $\angle 1$ and $\angle 2$ form a straight angle measuring 180°. If $\angle 1$ measures 120°, then $\angle 2$ measures 60°. Likewise, $\angle 1$ and $\angle 3$ form a straight angle, so $\angle 3$ also measures 60° and vertical angles 2 and 3 both measure 60°.

Since $\angle 3$ and $\angle 4$ together form a straight angle (as do $\angle 2$ and $\angle 4$), we find that $m\angle 4 = 120°$. Thus, vertical angles 1 and 4 both measure 120°.

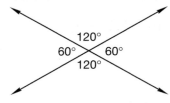

Two angles whose measures total 180° are called **supplementary angles.** We say that $\angle 2$ is the supplement of $\angle 1$ and that $\angle 1$ is the supplement of $\angle 2$. Supplementary angles may be adjacent angles, like $\angle 1$ and $\angle 2$, but it is not necessary that they be adjacent angles.

Angle Pairs Formed by Two Intersecting Lines

Adjacent angles are supplementary.
Opposite angles (vertical angles) are congruent.

Example 1

Refer to this figure to find the measures of angles 1, 2, 3, and 4.

Solution

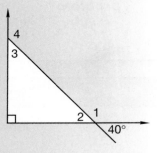

- Angle 1 is the supplement of a 40° angle, so **$m\angle 1 = 140°$.**

- Angle 2 is the supplement of $\angle 1$. Since $\angle 1$ measures 140°, **$m\angle 2 = 40°$.**

- Angle 3 and $\angle 2$ are the acute angles of a right triangle. The sum of the angle measures of a triangle is 180°, so **$m\angle 3 = 50°$.**

- Angle 4 is the supplement of $\angle 3$, so **$m\angle 4 = 130°$.**

Analyze When you know the measure of one of the acute angles of a right triangle, how do you find the measure of the other acute angle?

Notice that the measures of angles 2 and 3 in example 1 total 90°. Two angles whose measures total 90° are **complementary angles,** so ∠3 is the complement of ∠2, and ∠2 is the complement of ∠3.

Angles Paired by Combined Measures

Supplementary: Two angles totaling 180°
Complementary: Two angles totaling 90°

In the following figures we name several pairs of angles formed by parallel lines cut by a third line called a **transversal.** If the transversal were perpendicular to the parallel lines, then all angles formed would be right angles. The transversal below is not perpendicular, so there are four obtuse angles that are the same measure, and four acute angles that are the same measure. We will provide justification for these conclusions in a later lesson.

Corresponding angles are on the same side of the transversal and on the same side of each of the parallel lines. Corresponding angles of parallel lines are congruent. One pair of corresponding angles is ∠1 and ∠5. Name three more pairs of corresponding angles.

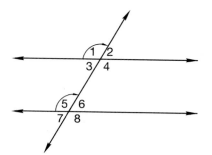

Alternate Sides of Transversal

Alternate interior angles are on opposite sides of the transversal and between the parallel lines. Alternate interior angles of parallel lines are congruent. One pair of alternate interior angles is ∠3 and ∠6. Name another pair of alternate interior angles.

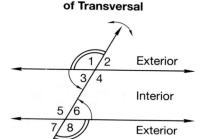

Alternate exterior angles are on opposite sides of the transversal and outside the parallel lines. Alternate exterior angles of parallel lines are congruent. One pair of alternate exterior angles is ∠1 and ∠8. Name another pair of alternate exterior angles.

Congruent Angle Pairs Formed by a Transversal Cutting Two Parallel Lines

Corresponding angles: Same side of transversal, same side of lines
Alternate interior angles: Opposite sides of transversal, between lines
Alternate exterior angles: Opposite sides of transversal, outside of lines

In summary, if parallel lines are cut by a non-perpendicular transversal, the following relationships exist.

• All the obtuse angles that are formed are congruent.

• All the acute angles that are formed are congruent.

• Any acute angle formed is supplementary to any obtuse angle formed.

Example 2

In the figure parallel lines *p* and *q* are cut by transversal *t*. The measure of ∠8 is 55°. What are the measures of angles 1 through 7?

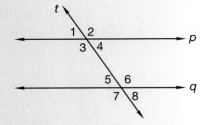

Solution

We are told that *p* and *q* are parallel. Angle 1 and ∠8 are alternate exterior angles and are congruent.

Statement:	Justification:
m∠1 = 55°	Alternate exterior angle to ∠8
m∠2 = 125°	∠2 and ∠1 are supplementary
m∠3 = 125°	∠3 and ∠1 are supplementary
m∠4 = 55°	∠4 and ∠1 are vertical and congruent
m∠5 = 55°	∠5 and ∠8 are vertical and congruent
m∠6 = 125°	∠6 and ∠8 are supplementary
m∠7 = 125°	∠7 and ∠8 are supplementary

Example 3

Structural engineers design triangles into buildings, bridges, and towers because triangles are rigid. The sides of triangles do not shift when force is applied like the sides of quadrilaterals do.

A railroad bridge built with a steel truss is strengthened by triangles that keep the bridge straight under the weight of a train.

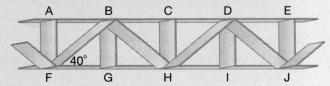

Knowing one acute angle in this truss is sufficient to find the measures of all the angles. Find the measures of angles *ABF, FBC,* and *FGB.* Segments that look parallel are parallel, and segments that look perpendicular are perpendicular.

Solution

Angle *ABF* and the 40° angle are alternate interior angles between parallel lines. So *m∠ABF* = **40°**. Together angles *ABF* and *FBC* form a straight angle (180°), so *m∠FBC* = **140°**. Angle *FGB* is a right angle, so *m∠FGB* = **90°**.

Practice Set

Conclude Refer to Figure 1 for exercises **a–c.**

a. Name two pairs of vertical angles.

b. Name four pairs of supplementary angles.

c. If $m\angle a$ is 110°, then what are the measures of angles, *b*, *c*, and *d*?

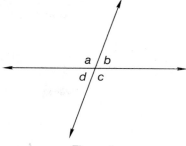

Figure 1

Refer to Figure 2 for exercises **d–f.**

d. Which angle is the complement of $\angle g$?

e. Which angle is the supplement of $\angle g$?

f. If $m\angle h$ is 130°, then what are the measures of $\angle g$ and $\angle f$?

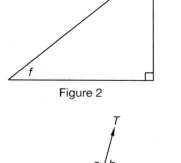

Figure 2

In Figure 3 parallel lines *R* and *S* are cut by transversal *T*. Refer to this figure for exercises **g–j.**

g. Which angle corresponds to $\angle f$?

h. Name two pairs of alternate interior angles.

i. Name two pairs of alternate exterior angles.

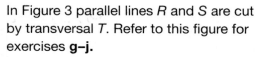

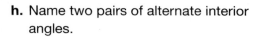

Figure 3

j. If $m\angle a$ is 105°, what is the measure of $\angle f$?

Written Practice   *Strengthening Concepts*

1. The ratio of stars to stripes is 50 to 13. If there were 400 stars in the parade, how many stripes were there?
(45)

* 2. *Connect* From the front to the back of the property Wally stepped off 40 paces. He estimated that each pace was a yard. Convert 40 yards to feet using a unit multiplier to estimate the depth of the property in feet.
(52)

3. Solve by writing and solving a proportion: The recipe calls for 2 cups of
(49) flour and 7 cranberries. Cathy is multiplying the recipe so that it feeds
more people. If the new recipe calls for 49 cranberries, how many cups
of flour are needed?

* 4. **Evaluate** For 5 minutes every day at 10 a.m., James counted the
(53) number of birds that came to the bird feeder outside his window. Over
10 days, he collected this data: 3, 5, 4, 5, 5, 4, 6, 5, 6, 6.

   a. Display the data with a line plot.

   b. Find the range and the mean, median, and mode of the
   data.

   c. James said, "Most often, there were 5 birds that came to the bird
   feeder." Which measure of central tendency did he use?

For problems **5** and **6,** use the figure of the parallel lines cut by a transversal.

* 5. Find $m\angle 1$. Justify your answer.
(54)

* 6. Find $m\angle 2$. Justify your answer.
(54)

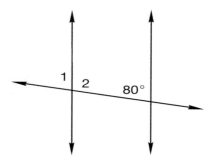

Analyze Solve. Then select one equation and write a story for it.

* 7. $3 = 2x + 3$
(50)

* 8. $3x + 7 - x = 21$
(50)

* 9. $20 - x = 1$
(50)

* 10. $-2w = -3$
(36)

11. Simplify and express with all positive exponents.
(51)
$$xyx^0y^{-2}x^{-1}$$

12. Refer to the graphed lines to answer the
(Inv. 1, following questions.
44)

   a. Which line intersects the y-axis
   at -1?

   b. What is the slope of line b?

   c. Which line is horizontal?

   d. In which quadrant do the lines
   intersect?

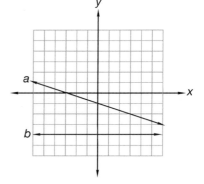

13. Graph $y = x - 2$. Is the point (9, 7) on the line?
(41)

* 14. *Generalize* Naomi cuts circles from squares with sides 20 cm long.
(40)

    a. What is the diameter of the largest circle she can cut from the square?

    b. What is the area of the largest circle? (Use 3.14 for π.)

    c. What is the area of the waste to the nearest sq. cm?

15. Find the **a** area and **b** perimeter of the figure
(37,
Inv. 2) to the right. (Units are inches.)

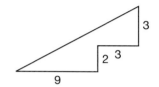

Simplify:

16. $-(x - y) - (x + y)$
(31, 36)

17. $2x^2 - 3x - x^2 + 5$
(31)

18. $\dfrac{2.9 \times 10^{12}}{2.9 \times 10^3}$
(46)

19. $-2 - (-2)$
(33)

20. $\dfrac{x^9 y}{x^8 y}$
(27)

21. $\left(\dfrac{-1}{2}\right)^3 + \dfrac{7}{8}$
(22, 36)

22. The desktop measured 76.2 cm wide. Convert this measure to meters
(51) and write in scientific notation.

23. Write 100% **a** as a decimal and **b** as a reduced fraction.
(12)

* 24. *Analyze* Raul planned a painting of his mountain cabin. The cabin
(35) stands 20 ft tall and the trees around it are 30 ft tall. The height of the cabin in his painting will be 6 inches.

    a. What is the scale from the actual cabin to the painting?

    b. How tall should Raul paint the trees?

* 25. Kyla constructed a simple tent with stakes,
(Inv. 2) ropes, and a length of plastic sheeting. She
used the plastic sheeting for the roof and
floor of the tent, leaving the ends open. The
tent stands 1 m tall and 2 m wide at the
base. What is the minimum length of plastic
used? (Round up to the next whole meter.)

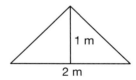

Find all solutions.

26. $\dfrac{x}{12} = \dfrac{12}{9}$
(34)

27. $\dfrac{x^2}{2} = 18$
(14, 36)

28. $-3|x| = -3$
(1, 14)

29. $\dfrac{m}{5} = 0.2$
(38)

30. Use the views shown to draw a three-dimensional view of the figure.
(Inv. 4)

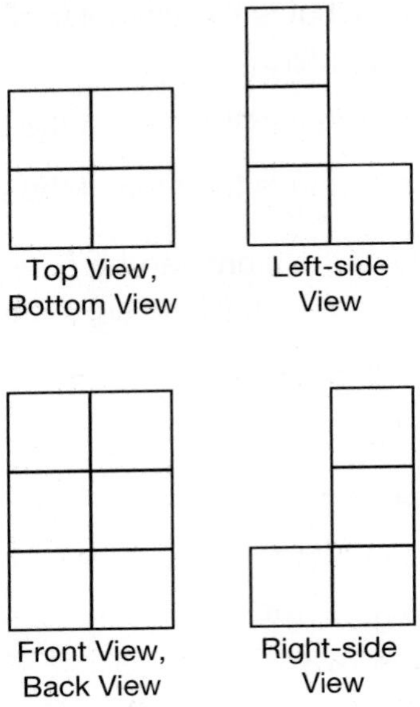

Top View,
Bottom View

Left-side
View

Front View,
Back View

Right-side
View

Early Finishers
Real-World
Application

Mike can swim the 100-yard freestyle in two minutes. If he swims at the same rate, how many seconds will it take him to swim 50 meters (1 meter ≈ 1.09 yard)?

• Nets of Prisms, Cylinders, Pyramids, and Cones

Power Up | Building Power

facts | Power Up K

mental math

a. **Number Sense:** 1.5×8

b. **Statistics:** Find the mean of 16 and 20.

c. **Fractional Parts:** $66\frac{2}{3}\%$ of $81

d. **Probability:** What is the probability of rolling a number less than 3 on a number cube?

e. **Geometry:** Two angles of a parallelogram measure 80° and 100°. Find the measure of the other two angles.

f. **Measurement:** The odometer read 2388 mi at the end of the trip. At the beginning, it had read 1208 mi. How long was the trip?

g. **Rate:** Ronnie ran $\frac{1}{4}$ mile in 1 minute. At that rate, how long would it take him to run a mile?

h. **Calculation:** $200 \div 2$, $\sqrt{}$, $\times\, 5$, $-\, 1$, $\sqrt{}$, $\times\, 2$, $+\, 2$, $\sqrt{}$, $\times\, 2$, $+\, 1$, $\sqrt{}$

problem solving | Two brothers are 5 years apart, and the sum of their ages is 41. What are the ages of the brothers?

New Concept | Increasing Knowledge

If we think of the surface of a solid as a hollow cardboard shell, then cutting open and spreading out the cardboard creates a net of the solid. For example, here we show a net for a pyramid with a square base.

Example 1

This net represents the surfaces of what geometric solid? Sketch the solid and describe how this surface area formula relates to each part of the net:

$$s = 2\pi rh + 2\pi r^2$$

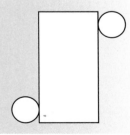

Solution

A geometric solid with two circular bases and a lateral surface that unwraps to form a rectangle is a **cylinder.** The first part of the formula, $2\pi rh$, applies to the lateral surface area (circumference, $2\pi r$, times height, h) and the second part, $2\pi r^2$, applies to the two circular bases.

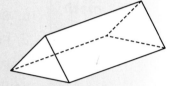

Example 2

Sketch a net for this triangular prism.

Solution

The prism has two congruent triangular bases, and three rectangular lateral faces.

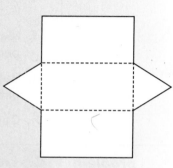

Example 3

Sketch the front, top, and right-side views of this figure. Then sketch a net for this figure. Describe how the two sets of sketches are related.

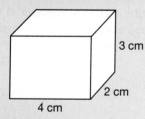

3 cm

2 cm

4 cm

Solution

Front

Top

Side

One example of a net:

The front, top and right-side views are the three different rectangles that appear in the net. Each rectangle appears twice, because the front and back faces are congruent, as are the top and bottom faces and the left and right faces.

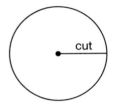

	Top	
Left	Front	Right
	Bottom	
	Back	

Activity

Net of a Cone

Materials needed: unlined paper, compass, scissors, glue or tape, ruler.

Using a compass, draw a circle with a radius of at least two inches. Cut out the circle and make one cut from the edge to the center of the circle. Form the lateral surface of a cone by overlapping the two sides of the cut. The greater the overlap, the narrower the cone. Glue or tape the overlapped paper so that the cone holds its shape.

To make the circular base of the cone, measure the diameter of the open end of the cone and use a compass to draw a circle with the same diameter. (Remember, the radius is half the diameter.) Cut out the circle and tape it in place using two pieces of tape.

Now disassemble the cone to form a net. Cut open the cone by cutting the circular base free on one side. Unroll the lateral surface by making a straight cut to the point (apex) of the cone. The net of a cone has two parts, its circular base and a sector of a circle that forms the lateral surface of the cone.

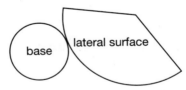

base · lateral surface

Thinking Skill

Connect

The formula for the surface area of a cone is $s = \pi r l + \pi r^2$. Which parts of the formula apply to which parts of the net?

Extend An alternate method for calculating the area of the lateral surface of a cone is to calculate the area of the portion of a circle represented by the net of the lateral surface. Use a protractor to measure the central angle of the lateral surface of the cone you created. The measure of that angle is the fraction of a 360° circle represented by the lateral surface. Use a ruler to measure the radius. Find the area of a whole circle with that radius. Then find the area of the sector by multiplying the area of the whole circle by the fraction $\frac{\text{central angle}}{360}$.

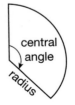

central angle · radius

Practice Set

a. The net created in this lesson's activity represents the surfaces of what geometric solid? Sketch the solid.

b. Sketch a net for this cube.

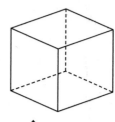

c. This pyramid is called a tetrahedron. All of its faces are congruent equilateral triangles. Draw a net of this pyramid.

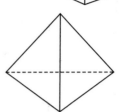

d. Sketch the back, top, and right side views of this triangular prism. Then sketch a net for the figure and label the dimensions.

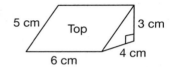

e. **Model** Build a model of the figure in exercise **c** by cutting, folding, and taping the net.

Written Practice   *Strengthening Concepts*

1. A manufacturing company has a debt to equity ratio of 3 to 2. If the company has a debt of $12 million, how much does it have in equity?
(34)

2. It takes Jack 5 minutes to drive from home to the nearest ATM machine. Then it takes 2 minutes in line at the ATM for each customer ahead of him. How long will it be before Jack can use the ATM if he leaves from home and then waits behind 3 customers?
(3, 4)

* 3. **Justify** Yueling read 5 books of these lengths: 105, 97, 96, 99, 103. Write a statement using a measure of central tendency to communicate how many pages Yueling read. Explain your choice.
(53)

4. $(-2)^2 + (-2)^3$
(31, 36)

5. $\dfrac{4}{3} = \dfrac{x}{1.5}$
(44)

*6. **Conclude** A transversal cuts parallel lines. Find x.
(54)

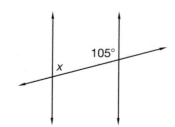

Solve. Then select one of the equations and write a word problem for it.

7. $4 = -m + 11$
(50)

8. $9x + 9 = 90$
(50)

9. $-5 - x = -9$
(50)

10. $3y - y - 1 = 9$
(36)

11. Describe each of the following views of this figure as top, right, or front.
(Inv. 4)

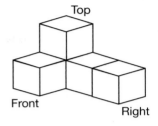
Top
Front
Right

a.

b.

c.
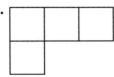

12. If the spinner is spun once, Roger says the sample space is {A, A, A, B, B, C}. Simon says the sample space is {A, B, C}. Who is correct? Explain your answer.
(32)

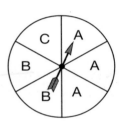

13. Garrett does an Internet search for the nearest location of a particular store. He requests a search of a location within a 60 mi radius of his home. About how many square miles does the search cover?
(40)

14. Graph $y = \frac{4}{5}x - 2$. Is the point (5, 2) a solution?
(47)

15. Find the perimeter and area of this figure. (Units are in ft.)
(37)

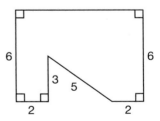

* 16. **Analyze** Moving at a uniform rate the train traveled 200 miles in five hours.
(49)

    a. Express the average speed of the train as a unit rate.

    b. How long did it take the train to travel 100 miles?

* 17. (52) **Analyze** Convert 440 ft to yards. Use a unit multiplier.

18. (42) The dimensions of an air mattress are 80 inches by 33 inches by 5 inches. What is the volume of the air mattress? If the average set of human lungs can hold about 244 in.^{3}, about how many breaths will it take to inflate the mattress?

Generalize Simplify.

* 19. (27, 51) $9^{-2}x^{-1}y^0x$

20. (27) $\dfrac{x^5 m^2}{mx}$

21. (22) $\left(\dfrac{1}{3}\right)^2 \div \dfrac{2}{3}$

22. (25) $1.2 \div 0.05$

23. (12) **a.** Write $\dfrac{1}{100}$ as a decimal and percent.

  b. How does $\dfrac{1}{100}$ compare to 0.009?

24. (21) Factor:

  a. $6x - 15$

  b. $x^2 - x$

25. (35) The triangles are similar.

  a. What is the scale factor from the small to large triangle?

  b. Find x.

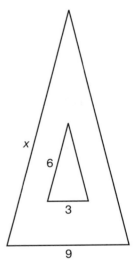

26. (52) Brooke finished the facts practice test in 1 minute and 18 seconds. Express her time as a mixed number of minutes and as a decimal number of minutes.

27. (36) Rahm is playing his favorite video game. Every time his player grabs a wrong object, Rahm loses 5 points. If Rahm accidentally grabs a wrong object 23 times, what is the effect on his score?

28. (51) A human hair measures 50 millionths (50×10^{-6}) of a meter. Which is thicker, 100 human hairs or a nickel that is $\dfrac{2}{10^3}$ meter thick?

* 29. Combine like terms to simplify.
(31, 36)

    a. $2(x + b) - (-x - b)$ **b.** $2x + 1 - 3x + 4$

30. As humans get older, their maximum heart rate decreases. Which of
(41) the following graphs illustrates this relationship? Is it a proportional
relationship?

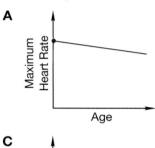

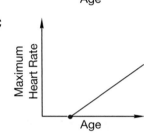

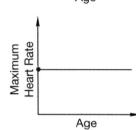

Early Finishers

Real-World
Application

A manager at the local movie theater wants to increase popcorn profits.
The cost of a bucket of popcorn is $0.05 for popcorn kernels, $0.02 for
butter, and $0.25 for the bucket. If the manager wants to sell 115 buckets
of popcorn a night and make a profit of $365.00, how much should the
manager charge for each bucket of popcorn? Note: The profit is the amount
of money made after subtracting the cost.

• The Slope-Intercept Equation of a Line

Power Up | *Building Power*

facts | Power Up L

mental math

a. **Number Sense:** 6×3.1

b. **Powers/Roots:** $\sqrt{2 \cdot 2} \cdot \sqrt{3 \cdot 3}$

c. **Scientific Notation:** Write 4.013×10^{-4} in standard form.

d. **Percent:** 75% of 84

e. **Geometry:** Can the sides of a triangle measure 7 ft, 3 ft, and 3 ft?

f. **Fractional Parts:** The gas tank holds 20 gallons when full. About how much gas is in the tank now?

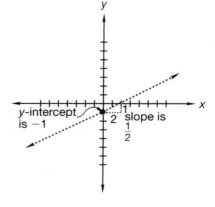

g. **Rate:** Tommy rode his bike 20 miles per hour north. Christina rode 15 miles per hour south. If they started at the same place and time, how far apart are they after 1 hour? After 2 hours?

h. **Calculation:** $6 \times 7, -2, \div 4$, square it, $-1, \div 9, +5, \div 2, \div 2, \div 2$

problem solving

A hiker estimates his hiking time by using the following rule: $\frac{1}{2}$ hour for every 1 mile, plus $\frac{1}{2}$ hour for each 1000 foot rise in elevation. A pair of hikers is planning a 12-mile hike to the summit of a mountain with a 5000 ft rise in elevation. They want to reach the summit by 3:00 p.m. At what time should they begin hiking?

New Concept | *Increasing Knowledge*

If the equation of a line is written in slope-intercept form, we can read the slope and y-intercept directly from the equation.

$$y = (\text{slope})x + (y\text{-intercept})$$

The slope-intercept equation of the line graphed on the previous page is $y = \frac{1}{2}x - 1$.

$$y = \frac{1}{2}x - 1$$

The slope is $\frac{1}{2}$. ↑   ↑ The y-intercept is -1.

Many books show the slope-intercept form this way:

Slope-intercept equation

$$y = mx + b$$

The number for **m** is the slope.

The number for **b** is the y-intercept.

Consider the following equations and their graphs.

Thinking Skill

Analyze

What is the value of b in the following equation: $y = 3x$.

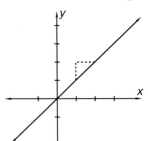

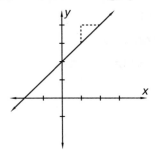

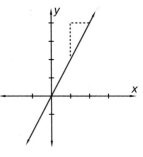

- Slope is 1
- y-intercept is zero

$y = x$

- Slope is 1
- y-intercept is $+2$

$y = x + 2$

- Slope is 2
- y-intercept is zero

$y = 2x$

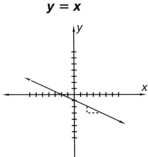

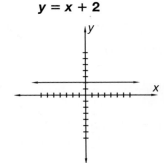

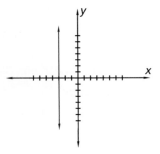

- Slope is $-\frac{1}{2}$
- y-intercept is -1

$$y = -\frac{1}{2}x - 1$$

- Slope is zero
- Intersects y-axis at $+2$

$$y = 2$$

- Slope is undefined
- Every point has x-coordinate -3

$$x = -3$$

Notice that a horizontal line has zero slope and can be expressed with slope-intercept form, $y = 0x + 2$, which simplifies to $y = 2$. A vertical line cannot be expressed in slope-intercept form.

Generalize Can the slope of a vertical line be determined? Why or why not?

Example 1

Refer to this equation to answer the questions that follow.

$$y = \frac{2}{3}x - 4$$

a. Where does the graph of the equation cross the y-axis?

b. Does the line rise to the right or fall to the right?

Solution

a. The graph of the equation crosses (intercepts) the y-axis at **−4,** which is 4 units down from the origin.

b. The line **rises to the right** because the slope is $\frac{2}{3}$ which is positive.

Example 2

Write the equations of lines *a* and *b* in slope-intercept form. At what point do lines *a* and *b* intersect?

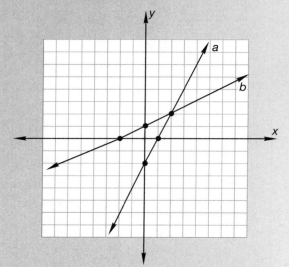

Solution

Line *a* has a slope of 2 and intercepts the y-axis at −2.

$$y = 2x - 2$$

Line *b* has a slope of $\frac{1}{2}$ and intercepts the y-axis at +1.

$$y = \frac{1}{2}x + 1$$

Lines *a* and *b* intersect at **(2, 2).**

Justify Prove that these lines intersect at (2, 2) by substituting for *x* in both equations and solving for *y*.

Example 3

Graph each equation using the given slope and y-intercept.

a. $y = 2x - 3$ b. $y = -\frac{1}{2}x + 2$

Solution

a. We study the equation to understand what the numbers mean.

$$y = 2x - 3$$

This number means
the slope is positive 2
(over 1, up 2).

This means the line
intersects the y-axis
at −3.

We start by graphing the point of the y-intercept. From there we graph additional points by going over 1 and up 2. The direction of "over" is to the right. Then we draw a line through the points.

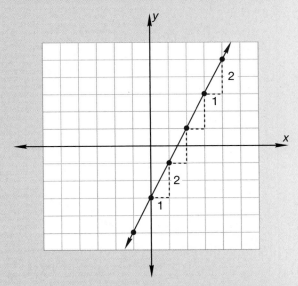

b. If the slope is a fraction, the denominator represents the "over" number and the numerator represents the "up or down" number.

$$y = -\frac{1}{2}x + 2$$

The slope is $-\frac{1}{2}$
(over 2, down 1).

y-intercept is
at +2.

We start at +2 on the y-axis and count over 2, then down 1.

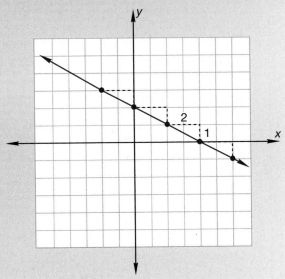

Practice Set

a. *Analyze* Which of the following equations is written in slope intercept form?

A $x = 2y + 3$ **B** $y + 2x = 3$ **C** $y = 2x + 3$

b. What is the slope and y-intercept of the graph of this equation?

$$y = -\frac{1}{3}x + 2$$

Represent Write equations for lines **c–f** in slope-intercept form.

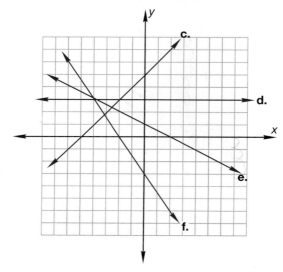

Graph the following equations using the given slope and y-intercept.

g. $y = x - 2$ **h.** $y = -2x + 4$ **i.** $y = \frac{1}{2}x - 2$

Written Practice   *Strengthening Concepts*

* 1. *(45)* *Analyze* The ratio of arable land (land that can be used for growing crops) to non-arable land in a certain county is 3 to 7. If the county has an area of 21,000 sq km, what area of the land is arable?

2. *(3, 4)* A lamp costs $10 and a package of 4 light bulbs costs $3. How much would it cost to buy 8 lamps and one light bulb for each lamp?

* 3. *(53)* *Represent* Mary measures the heights of the young plants in her garden to the nearest cm and collects this data: 4 cm, 5 cm, 6 cm, 6 cm, 7 cm, 8 cm, 6 cm, 7 cm, 5 cm

a. Display the data with a line plot.

b. Find the range and the mean, median, and mode of the data.

* 4. *(54)* *Conclude* Two of the lines are parallel. Find x.

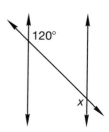

5. Use the slope-intercept method to graph the equation $y = 2x - 4$. Is
(56) (3, 2) a solution?

6. Find the **a** area and **b** circumference of the circle with radius 1. Use
(39, 40) 3.14159 for π.

* 7. The Mid-Atlantic Ridge spreads along the ocean floor an average of
(52) about 10 cm every 4 years.

    a. Express this rate as a unit rate.

    b. ⟨Connect⟩ This rate is equal to what distance of spread every one
    million years? (Express in km.)

* 8. Convert 1600 m to km. Use a unit multiplier.
(52)

Solve.

    9. $\dfrac{0.6}{x} = \dfrac{0.12}{5}$
    (44)

    * 10. $2x - x = 1.5$
    (50)

* 11. $0.6x + 1.2 = 3$
(50)

* 12. $7m - 9m = -12$
(50)

* 13. a. Write $\frac{2}{5}$ as a decimal and as a percent.
(12)

    b. Select any one of these three forms to represent the fact that
    10 students out of 25 had visited the national park.

* 14. Factor:
(21)

    a. $2x^2 + 14x$
                  b. $15x - 20$

* 15. A square with vertices at (2, 2), (−2, 2), (−2, −2), and (2, −2) is dilated
(Inv. 5) by a scale factor of 2. The area of the original square is what fraction of
its dilated image?

* 16. The figure is a net for a geometric solid.
(55) Sketch and name the solid.

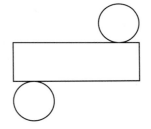

⟨Generalize⟩ Simplify:

    17. $-5(x + 2) - 2x + 9$
    (31, 36)

    * 18. $(-4)^2 - (-4)^3$
    (33, 36)

* 19. $\dfrac{(-15)(-12)}{(-15) - (-12)}$
(33, 36)

    20. $\dfrac{wr^2 d}{r^3 d}$
    (27)

21. $2 \cdot 1\frac{1}{2} - \left(1\frac{1}{2}\right)^2$
(13, 23)

    22. $\dfrac{6}{8} = \dfrac{9}{x}$
    (44)

23. $\dfrac{4.8 \times 10^7}{1.6 \times 10^4}$
(46)

    24. $\sqrt{10^2 - 8^2}$
    (15, 21)

25. Simplify and express using only positive exponents.
(27, 51)

$$a^0 b^1 a b^{-1} c^{-1}$$

26. Arrange in order from least to greatest:
(12, 30)

$$0.3, \frac{1}{3}, 0.33, 3\%$$

27. Represent Sketch a quadrilateral similar
(35) to the one shown enlarged by a scale
factor of 1.5. Label the side lengths.

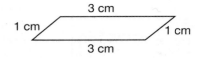

28. Find the measures of angles *x*, *y*, and *z*.
(18, 20)

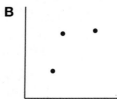

29. Find the slope of the line passing through the points (3, 4) and
(44) (−4, −3).

30. Which graph represents a proportion? Explain how you know.
(41)

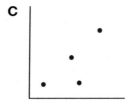

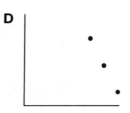

• Operations with Small Numbers in Scientific Notation

facts | Power Up L

mental math

   a. Number Sense: 25×36

   b. Statistics: Find the mean of 30 and 40.

   c. Fractional Parts: $66\frac{2}{3}\%$ of $300

   d. Probability: What is the probability of rolling a number greater than 1 with one roll of a number cube?

   e. Proportions: $\frac{x}{24} = \frac{3}{8}$

   f. Geometry: Two angles of a parallelogram measure 80° and 80°. What are the measures of the other two angles?

   g. Measurement: Find the length of this nail.

   h. Calculation: 10×8, $+ 1$, $\sqrt{\ }$, $\sqrt{\ }$, $\times 10$, $+ 6$, $\sqrt{\ }$, $- 7$, square it.

problem solving | A plane leaves Minneapolis, Minnesota, at 1 p.m. and flies non-stop to Honolulu in 8 hours. If Hawaii is four time zones earlier than Minnesota, what time does the plane arrive? If the return flight also takes 8 hours, what time does the plane arrive back in Minneapolis if it leaves Honolulu at 8 p.m.?

New Concept   *Increasing Knowledge*

In Lesson 51 we practiced writing small numbers between 0 and 1 in scientific notation. In this lesson we will multiply and divide those numbers.

Recall that to multiply powers of 10 we add the exponents and to divide we subtract.

$$10^5 \cdot 10^3 = 10^8 \qquad \frac{10^5}{10^3} = 10^2$$

The Laws of Exponents implied on the previous page apply to negative exponents as well. We exercise care adding and subtracting the exponents. Here we show some examples.

$$10^{-5} \cdot 10^{-3} = 10^{-8} \qquad 10^5 \cdot 10^{-3} = 10^2$$

$$\frac{10^{-5}}{10^{-3}} = 10^{-2} \qquad \frac{10^5}{10^{-3}} = 10^8$$

Example 1

If a sheet of notebook paper is 0.01 cm thick, how tall is a stack of 2500 sheets of notebook paper? Express each number in scientific notation and perform the calculation in scientific notation.

Solution

We multiply 1×10^{-2} and 2.5×10^3.

$$(1 \times 10^{-2})(2.5 \times 10^3) = 2.5 \times 10^1$$

The stack of paper is 2.5×10^1 cm which is **25 cm.**

Example 2

Perform each indicated calculation and express the result in scientific notation.

a. $(2.4 \times 10^{-6})(2 \times 10^{-2})$

b. $\dfrac{2.4 \times 10^{-8}}{2 \times 10^{-2}}$

c. $(4 \times 10^{-6})(5 \times 10^{-4})$

d. $\dfrac{4 \times 10^{-8}}{5 \times 10^{-2}}$

Solution

Thinking Skill

Explain

Describe how the power of 10 for each product or quotient in **a–d** is found.

a. $(2.4 \times 10^{-6})(2 \times 10^{-2}) = \mathbf{4.8 \times 10^{-8}}$

b. $\dfrac{2.4 \times 10^{-8}}{2 \times 10^{-2}} = \mathbf{1.2 \times 10^{-6}}$

c. $(4 \times 10^{-6})(5 \times 10^{-4}) = 20 \times 10^{-10}$ (improper form)

Adjustment: $20 \times 10^{-10} = 2 \times 10^1 \times 10^{-10} = \mathbf{2 \times 10^{-9}}$

d. $\dfrac{4 \times 10^{-8}}{5 \times 10^{-2}}$ (improper form)

Adjustment: $0.8 \times 10^{-6} = 8 \times 10^{-1} \times 10^{-6} = \mathbf{8 \times 10^{-7}}$

Practice Set

Find each product or quotient.

a. $(4 \times 10^{10})(2 \times 10^{-6})$

b. $(1.2 \times 10^{-6})(3 \times 10^3)$

c. $(1.5 \times 10^{-5})(3 \times 10^{-2})$

d. $(7.5 \times 10^{-3})(2 \times 10^{-4})$

e. $\dfrac{7.5 \times 10^5}{3 \times 10^{-2}}$

f. $\dfrac{4.8 \times 10^{-3}}{3 \times 10^2}$

g. $\dfrac{8.1 \times 10^{-4}}{3 \times 10^{-7}}$

h. $\dfrac{1.2 \times 10^{-8}}{3 \times 10^{-4}}$

i. A dollar bill weighs about 0.001 kg. What is the weight of 1,000,000 dollar bills? Express each number in scientific notation and perform the calculation in scientific notation.

Written Practice   *Strengthening Concepts*

1. In a certain state, the ratio of weddings held in spring or summer to
(45) weddings held in autumn or winter is about 3 to 2. If there were 46,000 weddings in a year, how many were in spring or summer?

* 2. Sixty percent of the days in the desert were hot and dry. Out of the
(48) 365 days, how many were hot and dry?

* 3. **Analyze** Martin completed 60% of his math assignment during class.
(48) If his math assignment consists of 30 problems, how many problems does he have left to do?

* 4. Allison collected these donations for a charity:
(53)

$50, $75, $100, $50,

$60, $75, $80, $50, $75

a. Display the data with a line plot.

b. Find the range and the mean, median, and mode.

c. Write a description of the data using the mode.

5. $\dfrac{6}{7} = \dfrac{x}{10.5}$
(44)

* 6. Two of the lines are parallel. Find x.
(54)

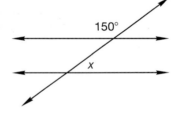

7. The spinner is spun twice.
(32)
   a. Write the sample space of 16 equally-likely outcomes.

   b. What is the probability the spinner will stop on A at least once in two spins?

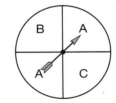

8. Use the slope-intercept method to graph $y = x - 1$. Then graph the
(56) point $(-2, -3)$. Is $x = -2, y = -3$ a solution?

Solve.

9. $0.3m - 0.3 = 0.3$
(50)

10. $4x + 7x = 99$
(50)

11. $\dfrac{2}{3}x + \dfrac{1}{2} = \dfrac{2}{3}$
(50)

12. $7 = -2p - 5p$
(50)

Refer to the following information to solve problems **13** and **14**.

A circular tetherball court 6 meters in diameter is painted on the playground. A stripe divides the court into two equal parts.

13. What is the circumference of the tetherball court to the nearest
(39) meter?

14. What is the area of each semicircle to the nearest square meter?
(40)

15. Collect like terms: $x(x + 2) + 2(x + 2)$
(31, 36)

* **16.** Simplify and compare: $\left(\dfrac{1}{2}\right)^2 \bigcirc \left(-\dfrac{1}{2}\right)^2$
(22)

* **17.** **Represent** Maggie drove 195 miles in 3 hours.
(49)
    a. At that rate, how many miles will she drive in 5 hours?

    b. Express 195 miles in 3 hours as a unit rate.

* **18.** A mile is 5280 ft. Use a unit multiplier to convert 5280 ft to yards.
(52)

Evaluate For **19** and **20** refer to this description.

Diego is building an architectural rendering of a house. For a portion of the roof Diego cuts and folds a net (pattern) for a triangular prism. The base of the prism is an isosceles triangle. The triangle has a base of 24 in. and a height of 5 in. The distance between the triangular bases is 20 in.

* **19.** **a.** Make an isometric sketch of the folded prism.
(Inv. 4,
55)
    b. Make a sketch of the net of the prism.

* **20.** **a.** Find the volume of the prism.
(42, 43)
    b. Find the surface area of the prism.

21. **Generalize** Simplify: $\dfrac{b^3 r^4}{m b^3 r^2}$
(27)

22. Simplify, then compare: $\sqrt{\dfrac{4}{9}} \bigcirc \dfrac{\sqrt{4}}{\sqrt{9}}$
(15)

23. Write $\dfrac{11}{12}$ **a** as a percent and **b** as a decimal.
(30)

* **24.** **Classify** The fare charged for a taxi ride was $1.50 plus 40¢ for each
(47) $\dfrac{1}{5}$ of a mile. What term best describes the relationship between the
distance traveled and the fare charged?

    A continuous function   **B** step function   **C** direct proportion

25. The spinner is spun twice.
(32)
    a. What is the sample space of the experiment?

    b. What is the probability the spinner stops on A at least once?

    c. What is the probability the spinner does not stop on A in two spins?

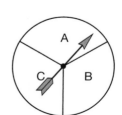

26. Evaluate the expression $-b + \sqrt{b^2 - 4ac}$ when $a = 3$, $b = 4$, and
(14, 15) $c = 1$.

27. Using the slope-intercept form,
(56) write the equation of each line at the right.

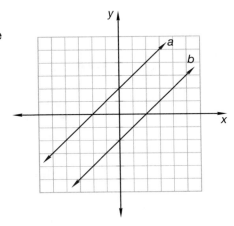

28. If a line has a slope of $\frac{1}{2}$ and if it intersects the y-axis at -1, then what is
(56) the equation of the line in slope-intercept form?

29. Find all solutions: $9x^2 = 36$
(14, 15)

30. Use this figure to answer the following
(Inv. 4) questions.

  a. Which of the following is the front view
  of this figure?

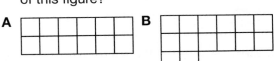

  b. Which of these is the right-side view?

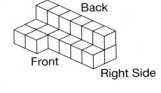

• Solving Percent Problems with Equations

Power Up | *Building Power*

facts | Power Up L

mental math

a. **Number Sense:** Can we produce a 2-digit whole number by multiplying two 2-digit whole numbers?

b. **Scientific Notation:** Write 0.0019 in scientific notation.

c. **Geometry:** Three angles of a trapezoid measure 50°, 90° and 100°. Find the measure of the fourth angle.

d. **Fractional Parts:** If coats are on sale for 40% **off** the regular price, then the sale price is what percent **of** the regular price?

e. **Algebra:** $12x + 1 = 145$

f. **Measurement:** The bicycle odometer read 24.2 miles when she left. It read 29.5 miles when she returned. How far did she ride?

g. **Rate:** Kenneth read 35 pages per hour. Shana read 40 pages per hour. After 2 hours, how many more pages has Shana read than Kenneth?

h. **Calculation:** $\sqrt{9}$, × 8, × 2, + 1, $\sqrt{}$, × 8, − 1, ÷ 5, × 3, − 1, ÷ 4

problem solving

A store sells three sizes of the same candle—a "regular" size with radius *r* and height *h*. Candle B is twice as tall as Candle A but has the same radius. Candle C is just the reverse, with twice the radius of Candle A but the same height. If both large Candles B and C are made of the same material, which of them is heavier? Why?

New Concept | *Increasing Knowledge*

We may solve percent problems using proportions (as we did in Lesson 48) or by solving a percent equation.

A **percent** of a **whole** is a **part.**

$\% \times W = P$

The factors and product in the equation above are the percent, the whole, and the part. If two of the three numbers are known, we can write and solve an equation to find the unknown number. If we are given the percent, we convert the percent to a fraction or decimal before performing the calculation.

For any percent problem we may choose to express the percent as a decimal or as a fraction. We might make the choice based on which form seems easier to calculate. For some problems the calculations are tedious, so we might choose to express the percent as a decimal and use a calculator. For other problems, expressing the percent as a fraction is the best choice.

Example 1

Thirty-two ounces is 25% of a gallon. How many ounces is a gallon?

We are given the percent and the part. We are asked for the whole. We write the percent as a decimal or as a fraction and solve for the unknown.

As a decimal:

32 oz is 25% of a gallon

$$32 = 0.25g$$

$$\frac{32}{0.25} = g$$

$$g = 128$$

As a fraction:

32 oz is 25% of a gallon

$$32 = \frac{1}{4}g$$

$$\frac{4}{1} \cdot 32 = \frac{4}{1} \cdot \frac{1}{4}g$$

$$g = 128$$

A gallon is **128 ounces.**

Example 2

Mr. Villescas bought a used car for $8,500. The sales-tax rate was 7%. How much did Mr. Villescas pay in tax?

The sales tax will be added to the price, but the amount of tax is based on the price. In this case the tax is 7% of the price. We translate the sentence to an equation using = for "is" and × for "of." We substitute the known numbers and solve for the unknown. We may write the percent as a decimal or a fraction before performing the calculation.

As a decimal:

Tax is 7% of the price

$$t = 0.07 \times \$8500$$

$$t = \$595$$

As a fraction:

Tax is 7% of the price

$$t = \frac{7}{\overset{1}{\cancel{100}}} \times \overset{\$85}{\cancel{\$8500}}$$

$$t = \$595$$

The sales tax on the car was **$595.** Notice that the amount of tax on large-dollar purchases might seem high. However, it is correct. Seven percent of the price is less than 10% but more than 5%. Since 10% of $8500 is $850, and 5% is half of that ($425), our answer of $595 is reasonable.

Example 3

For the following percent problems, decide whether it is better to express the percent as a decimal or as a fraction.

   a. The sales-tax rate is 8.25%. What is the sales tax on a $18.97 purchase?

   b. Shirts are on sale for $33\frac{1}{3}$% off the regular price. How much is saved on a $24 shirt?

Thinking skill

Connect

What is the decimal form of $33\frac{1}{3}$%? Why would we need to round the decimal to perform the calculation?

a. Whether we change the percent to a fraction or decimal, we must multiply a 4-digit number by a 3-digit number. Converting the percent to a fraction further complicates the arithmetic. To calculate an exact answer we would **convert the percent to a decimal** and use a calculator if one was readily available.

b. To write the percent as a decimal we would need to round. Expressing the percent as the fraction $\frac{1}{3}$ is more accurate, and we can perform the calculation mentally. We would **express the percent as a fraction.**

Example 4

Blanca correctly answered 23 of the 25 questions. What percent of the questions did she answer correctly?

Solution

We are given the whole and the part. We are asked for the percent. After solving the equation, we convert the fraction or decimal solution to a percent.

What percent of 25 is 23?

$$P \cdot 25 = 23$$

$$\frac{P \cdot 25}{25} = \frac{23}{25}$$

$$P = \frac{23}{25} \text{ or } 0.92$$

We convert 0.92 to **92%.**

Practice Set

Solve by writing and solving equations. In **a–d** choose whether to express the percent as a fraction or as a decimal.

a. *Justify* Six percent of $4500 is how much money? Explain why your answer is reasonable.

b. Twenty percent of what number is 40?

c. How much is a 15% tip on a $13.25 meal? Round your answer to the nearest dime.

d. How much money is $16\frac{2}{3}$% of $1200?

e. What percent of 50 is 32?

f. Dixon made a $2,000 down payment on an $8,000 car. The down payment was what percent of the price?

g. Kimo paid $24 for the shirt which was 75% of the regular price. What was the regular price?

* 1. *(45)* **Analyze** In water, the ratio of hydrogen atoms to oxygen atoms is 2 to 1. If there are 3×10^{23} atoms in a sample of water, how many are hydrogen atoms?

* 2. *(48, 58)* Sixty percent of the voters favored the initiative, and 12,000 did not favor it. How many voters favored the initiative?

3. *(3, 4)* A bank charges customers $10 per month for a checking account, plus 10 cents for each check that is written. If Jessie uses 13 checks in a month, what will this bank charge for the month?

* 4. *(53)* **Evaluate** The sales revenue of a retail store for one week was (in thousands of dollars):

$$2.9, \ 1.6, \ 1.4, \ 1.3, \ 1.5, \ 2.5, \ 2.8$$

a. Display the data in a line plot.

b. Find the mean, median, mode, and range of the data.

c. Which measure would you choose to report the sales most favorably?

5. *(32)* Brad is trying to remember the combination to a lock. He knows the three numbers are 15, 27, and 18, but he cannot remember the order.

a. What are the possible orders of the numbers?

b. What is the probability that Brad finds the correct order on his first try?

Solve for *x*.

6. *(44)* Solve for *x*. $\dfrac{x}{5} = \dfrac{14}{20}$

* 7. *(54)* These two lines are parallel. Find *x*.

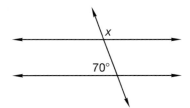

8. *(56)* Use the slope-intercept method to graph $y = \frac{1}{2}x - 2$.

9. *(39)* The circumference of the earth at the equator is about 25,000 miles. Estimate the diameter of the earth to the nearest thousand miles. Explain how you found your answer.

Analyze Solve.

* 10. *(50)* $3m - 6 + 2m = 4$ *** 11.** *(50)* $x - 3x = 18$

* 12. $\frac{2}{3}x + \frac{1}{2} = \frac{5}{6}$
(50)

13. $\frac{y}{8} = 0.375$
(38)

14. Isabel wonders whether a beach ball with a 50 inch circumference will
(39) pass through a basketball hoop with an 18 inch diameter. What do you predict? Justify your answer.

15. One corner is cut from a small rectangle of
(37) note paper. Find the **a** area and **b** perimeter of the resulting piece of paper.

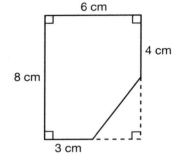

16. Sarah just got her hair cut. If Sarah's hair
(23) grows $1\frac{1}{4}$ inches every month, how much longer will it be in $4\frac{1}{2}$ months if she does not cut it?

17. Use a unit multiplier to convert 2.5 ft to inches.
(52)

Simplify.

18. $\left(\frac{2}{5}\right)^2$
(22)

19. $\frac{ssr^5}{s^2r^4}$
(27)

20. $\frac{2}{15} + \frac{2}{5} \cdot \frac{1}{6}$
(13, 22)

21. $x^2 + 2x + x + 2$
(31)

22. **a.** Write 55% as a decimal and reduced fraction.
(12, 48)

    b. Choose one form and use it to find 55% of $1200.

23. Express $\frac{1}{7}$ as a decimal rounded to three decimal places.
(30)

24. Find $\frac{n + m}{m}$ when $n = 100$ and $m = -10$.
(31, 36)

25. **a.** Explain why
(35)   the triangles are similar.

    b. Find x.

    c. Find y.

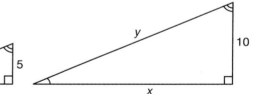

26. An ant weighs as little as 0.00001 kilograms. Write the minimum weight
(51) of an ant using an exponent.

27. Last month, Andrea missed two days of school because she had a cold.
(48) What percent of school days did Andrea miss that month. (Hint: Assume that there are 20 school days in a month.)

28. The Giant Water Lily of the Amazon River is the world's largest lily. Its
(40) pads are almost perfectly round. One lily pad usually measures about 6 feet in diameter. What is the approximate surface area of one lily pad?

29. On the equator, the earth rotates about its axis about 25,000 miles in
$^{(7)}$ 24 hours. Express this rate in miles per hour. (Round to the nearest
hundred.)

30. What set of coordinates represents the
$_{(Inv. 5)}$ translation of rectangle *ABCD* 3 units to
the right and 2 units down? Sketch the
rectangle on graph paper to illustrate
the transformation.

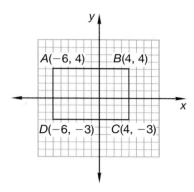

Early Finishers
Real-World
Application

Below are the weight in grams (g) of a neutron, a proton and an electron.
First write each particle's weight in standard form. Then list the particles in
order from lightest to heaviest.

neutron 1.6750×10^{-24}

proton 1.6726×10^{-24}

electron 9.1083×10^{-28}

• Experimental Probability

facts | Power Up L

mental math
a. **Number Sense:** Can we produce a 4-digit number by multiplying two 2-digit numbers?

b. **Statistics:** The mean of equally spaced numbers equals the median of the numbers. Find the mean of 30, 40, and 50.

c. **Fractional Parts:** A tip of 15% to 20% is customary. Sam tipped $\frac{1}{6}$ of the bill on a bill of $24. What was the tip?

d. **Ratio:** There are 16 pens in a drawer. Six of the pens are blue and the other 10 are black. What is the ratio of black pens to blue pens?

e. **Probability:** What is the probability of rolling an even number with one roll of a number cube?

f. **Geometry:** Two sides of a triangle are 5 inches and 13 inches. Can the third side be 12 inches long? 7 inches? 19 inches?

g. **Measurement:** What temperature is indicated on this thermometer?

h. **Calculation:** Start with a dollar, add 50%, give away half, subtract 25¢, add a dime, give away half. How much is left?

problem solving
A square is divided as shown. Regions B and C combine to equal what fraction of the area of the square?

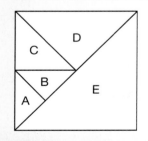

When discussing probability, we distinguish between the **theoretical probability,** which is found by analyzing a situation, and the **experimental probability,** which is determined statistically. Experimental probability is the ratio of the number of times an event occurs to the number of trials.

$$\text{experimental probability} = \frac{\text{number of times an event occurs}}{\text{number of trials}}$$

If we want to experimentally determine the probability that a flipped coin will land heads up, we would flip the coin a number of times, counting the number of trials and the number of heads.

Trial	1st	2nd	3rd	4th	5th	6th	7th	8th	9th	10th
Outcome	H	H	T	H	T	H	T	T	T	H
Probability $H \div t$	1.00	1.00	0.67	0.75	0.60	0.67	0.57	0.50	0.44	0.50

As the number of trials grows, the experimental probability tends to approach the theoretical probability of an experiment. Therefore, we want a large number of trials when conducting probability experiments.

Experimental probability is used widely in sports. If a basketball player makes 80 free throws in 100 attempts, then the probability of the player making a free throw is $\frac{80}{100}$ or $\frac{4}{5}$. The player might be described as an 80% free throw shooter. In baseball and softball, a player's batting average is the probability, expressed as a decimal, of the player getting a hit.

Example 1

A softball player has 21 hits in 60 at-bats. Express the probability the player will get a hit in her next at-bat as a decimal number with three decimal places.

Solution

The ratio of hits to at-bats is 21 to 60. To express the ratio as a decimal we divide 21 by 60.

$$21 \div 60 = 0.35$$

It is customary to write the probability (batting average) with three decimal places and without a zero in the ones place: **.350**

Thinking Skill

Analyze

Marcia has a batting average of .330. In one game she has 2 hits in 5 at bats. Will this raise or lower her batting average?

Experimental probabilities are used not only in sports but in many fields such as business, weather forecasting, insurance, and banking. Theoretical probability is commonly applied to designing games of chance.

Example 2

A salesperson distributed free samples at the mall. Fifty shoppers accepted the samples and thirty did not. What is the probability that a shopper at the mall will accept the salesperson's samples?

Solution

Fifty out of eighty shoppers accepted the sample, so the probability is $\frac{5}{8}$.

Predict If the salesperson from example 2 expects to see 400 customers at the mall during the next shift, how many samples should the salesperson have ready? Explain your reasoning.

Sometimes experiments are impractical to conduct, so to find an experimental probability, we **simulate** the event using models such as spinners, number cubes, coins, or marbles.

Example 3

The commuter plane flight arrives on schedule $\frac{2}{3}$ of the time on any given weekday. Heather wonders what the probability is that it will arrive on schedule on a Tuesday, Wednesday, and Thursday in the same week.

a. How could the experimental probability be found? Why is this impractical?

b. How could the spinner at right be used to simulate the experiment?

Solution

a. **Heather could check the flight arrival time on Tuesdays, Wednesdays, and Thursdays for several weeks.** The experimental probability is the number of times the flight arrives on schedule all 3 days in a row divided by the total number of selected 3-day periods. **This experiment would take a very long time to conduct.**

b. Each spin can represent one flight. The probability of an on-time arrival is $\frac{2}{3}$. **Heather can spin the spinner three times in a row and repeat this for many trials to find the experimental probability.**

Explain How might a number cube or marbles be used to simulate this experiment?

Represent Select one of the models described in the lesson and use it to simulate Heather's experiment. Conduct 12 trials, which is 36 spins, rolls, or draws. What experimental probability do you find?

Practice Set

a. *Explain* When a member of the opposing team fouls a basketball player who is shooting, the player shoots two free throws. Near the end of a close game a player has made 6 out of 20 free throws, and he has made 60% of his 2-point shots. When the player has the ball in 2-point range, should the opposing team foul him or risk the shot? Explain your answer in terms of probability.

b. Quinn runs a sandwich shop. Since she added a turkey melt to the menu, 36 out of 120 customers have ordered the new sandwich. What is the probability that the next customer will order a turkey melt?

c. *Predict* Quinn ordinarily has 200 customers on a busy afternoon, about how many turkey melts should she expect to sell?

d. To prepare premium rates (the amounts customers pay) for an insurance plan, an insurance company conducts an extensive risk study to determine the probability that a member of a certain group may require a pay-off. Based on these probabilities, the expected amount of payoff is charged to the customer as part of the premium. Explain why car insurance companies charge a higher premium for teenage drivers with a clean driving record than for adult drivers with a clean driving record.

e. Meghan is a 50% free throw shooter. Select a model to simulate a game in which Meghan shoots 10 free throws.

Written Practice   *Strengthening Concepts*

* 1. *(45)* *Analyze* In carbon dioxide, the ratio of carbon atoms to oxygen atoms is 1 to 2. If there are 12×10^{23} oxygen atoms in a sample of carbon dioxide, how many atoms are there in all?

* 2. *(53)* In the triple-jump competition, Lee jumped 28 ft, 32 ft, 29 ft, 30 ft, and 33 ft.

    a. Find the mean, median, mode, and range of Lee's marks.

    b. Which measure gives the difference between her shortest and longest jumps?

* 3. *(48, 58)* *Formulate* Twelve of the 30 students in the class brought their lunches from home. What percent of the students brought their lunches from home?

* 4. *(59)* If a student from the class in problem **3** were selected at random, what is the probability that the student would be one who brought his or her lunch from home?

5. *(35)* These triangles are similar. Find the scale factor from the larger triangle to the smaller triangle and solve for *x*.

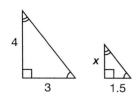

* 6. Two of the lines are parallel. Solve for *x*.
(54)

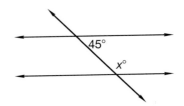

7. Use slope-intercept to graph $y = -2x + 4$. Is the point $(0, -2)$ a
(56) solution?

Analyze Solve.

* 8. $3(x - 4) = 15$
(14, 21)

* 9. $2x - x - 2 = 12$
(14, 31)

10. $\dfrac{x}{3} = 0.7$
(25, 38)

11. $\dfrac{1}{2}x = \dfrac{1}{3}$
(22, 38)

12. Find **a** the area and **b** the circumference of a circle that has a radius
(39, 40) of 13 units. Express your answer in terms of π.

* 13. **Evaluate** Compare the figure to a 4 m by
(37) 8 m rectangle.

    a. Is the area of the figure greater than or
less than the area of the rectangle?

    b. Is the perimeter of the figure greater than
or less than the perimeter of the rectangle?

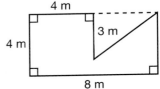

14. Express as a unit rate: 2400 miles in 48 hours.
(7)

* 15. Use a unit multiplier to convert 3.5 minutes to seconds.
(52)

16. The door is 6 ft 9 in. high. Express the height of the door as a mixed
(52) number of feet and as a decimal number of feet.

Simplify.

17. $3^{-2} \cdot 2^{-2}$
(51)

18. $\dfrac{h^3 p^2}{ph}$
(27)

19. $\dfrac{5}{18} - \dfrac{5}{18} \cdot \dfrac{1}{5}$
(13, 22)

20. $0.3 - 0.2(0.1)$
(24, 25)

21. **a.** Write 0.95 as a percent and reduced fraction.
(12)

    b. Order 0.95, $\dfrac{39}{40}$, and $\dfrac{9}{10}$ from least to greatest.

* 22. **Evaluate** Van selected two unmarked paint cans from a shelf with
(32) three unmarked cans. Inside two cans is white paint and inside one can
is yellow paint. Van will open the cans as he looks for the yellow paint.

    a. In a list, record the sample space for the experiment. (Hint: Use the
abbreviations W_1, W_2, Y.)

    b. What is the probability that one of the two cans Van selected
contains yellow paint?

23. The Wilsons traveled the 2400 miles of Route 66 from Chicago, IL to
(7) Santa Monica, CA. If their total driving time was 50 hours, what was their average rate?

* 24. What is the amount of sales tax for $315.90 if the tax rate is 7%?
(58) (Round to the nearest cent.)

* 25. If Dori answers 23 of 25 questions correctly, what percent of the
(58) questions did she answer correctly?

26. Combine like terms: $x(x + 1) - 1(x + 1)$
(31, 36)

27. Estimate the volume and surface area of a
(42, 43) cereal box that has these dimensions.

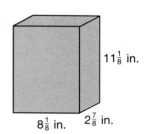

$11\frac{1}{8}$ in.

$8\frac{1}{8}$ in. $2\frac{7}{8}$ in.

28. Some lichens grow at a rate of 0.0000000000625 miles per hour. What
(51) is the same number expressed in scientific notation?

29. If the ink in the period at the end of this sentence weighs 1.3×10^{-8} lb,
(57) and if there are 30,000 periods printed in this book, then how much does this ink weigh altogether? Express each number in scientific notation and perform the calculation in scientific notation.

30. The coordinates of the vertices of
(Inv. 5) $\triangle ABC$ are $A(1, 3)$, $B(3, 6)$ and $C(4, 3)$. A
translation of $\triangle ABC$ is used to produce
$\triangle A'B'C'$. Vertex B' is located at $(5, 7)$.
Choose the coordinates of the vertices
A' and C'.

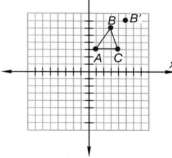

A $A'(3, 5)$ and $C'(6, 5)$

B $A'(2, 4)$ and $C'(4, 7)$

C $A'(5, 3)$ and $C'(4, 7)$

D $A'(3, 4)$ and $C'(6, 4)$

• Area of a Parallelogram

facts | Power Up L

mental math

a. **Number Sense:** Can we produce a 5-digit whole number by multiplying two 2-digit whole numbers?

b. **Algebra:** If $x^3 = 8$, then x equals what number or numbers?

c. **Geometry:** Two sides of a triangle measure 17 inches and 12 inches. Can the third side measure 4 inches? 19 inches? 29 inches?

d. **Fractional Parts:** $\frac{5}{9}$ of the 72 athletes were on the home team. How many were on the home team?

e. **Scientific Notation:** Write 0.000507 in scientific notation.

f. **Rate:** Brad ran north up the trail at 8 miles per hour. Skip walked the trail at 4 miles per hour north. After 30 minutes, how far apart are they if they started at the same place and time?

g. **Proportions:** $\frac{18}{6} = \frac{12}{x}$

h. **Calculation:** $6 \times 7, -2, \div 2, +1, \div 3, +2, \sqrt{}, \times 12, \sqrt{}, -\frac{1}{2}$

problem solving | Victor makes three different purchases at three different stores one afternoon. Each time, he spends $\frac{1}{3}$ of the money he is carrying. If he ends the day with $16, how much money did he have at the start of the day?

New Concept   *Increasing Knowledge*

Recall that to find the area of a rectangle we multiply the length and width, which are the perpendicular dimensions of a rectangle.

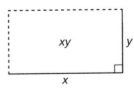

Thinking Skill

Explain

Why does the formula $A = \frac{1}{2}bh$ work for finding the area of a triangle?

Also recall that to find the area of a triangle, we multiply the perpendicular base and height as the first step to calculating the area. Then we find half of that product.

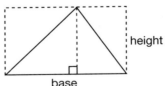

Math Language

Recall that a **parallelogram** is a quadrilateral with two pairs of parallel sides. Rectangles, squares, and rhombuses are special parallelograms.

Likewise, to find the area of a parallelogram we multiply the perpendicular base and height. Again, the product is the area of a rectangle, but the area of the rectangle is equal to the area of the parallelogram. In the figure below, notice that the part of the parallelogram outside the rectangle matches the missing portion of the rectangle.

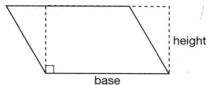

The formula for the area of a parallelogram is

$$A = bh$$

in which A represents area, b represents the length of the base, and h represents the height. Two parallel sides are the bases of a parallelogram. The perpendicular distance between the bases is the height.

Activity

Area of a Parallelogram

Materials needed: $\frac{1}{2}$ sheet of graph paper per student, scissors, ruler, pencil

On graph paper, draw a parallelogram that is not a rectangle. Trace over lines on the background grid to make one pair of parallel sides. Make sure that the vertices of the parallelogram are at intersections of the background grid and then cut out your parallelogram.

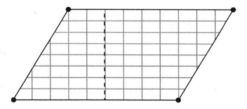

Next, on a grid line (such as the dashed line in the example), cut the parallelogram into two parts. Then, rearrange the two pieces to form a rectangle and find the area of the rectangle. Does the area of the rectangle equal the product of the base and height of the parallelogram?

Connect Repeat this activity by drawing a parallelogram and one of its diagonals. How does this area of each triangle you formed compare to the area of the parallelogram? How does the formula for the area of a triangle compare to the formula for the area of a parallelogram?

Example 1

Find the perimeter and area of this
parallelogram.

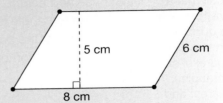

5 cm   6 cm

8 cm

Solution

We find the perimeter of the parallelogram by adding the lengths of the four
sides.

$$8 \text{ cm} + 6 \text{ cm} + 8 \text{ cm} + 6 \text{ cm} = \mathbf{28 \text{ cm}}$$

We find the area of the parallelogram by multiplying the perpendicular
dimensions of the base and the height.

Step:	Justification:
$A = bh$	Formula
$A = 8 \text{ cm} \cdot 5 \text{ cm}$	Substituted
$A = \mathbf{40 \text{ cm}^2}$	Simplified

Model Draw a parallelogram with an area of 24 in.^{2} Label its dimensions.

Example 2

a. On a coordinate plane sketch and find the area of parallelogram
   ABCD with vertices *A* (2, 1), *B* (1, −1), *C* (−2, −1) and *D* (−1, 1).

b. Graph the dilation of parallelogram *ABCD* with scale factor 3. What
   is the area of the image?

c. The area of the image is how many times the area of parallelogram
   ABCD? Why?

Solution

a. The area of parallelogram
   ABCD is **6 sq. units.**

b. The area of the dilated
   image is **54 sq. units.**

c. The area of the image
   is **9 times** the area of
   parallelogram *ABCD*
   because the scale factor
   tripled the base and tripled
   the height. Therefore, when
   we multiply the base and
   height, the product (the
   area) is 9 times as great.

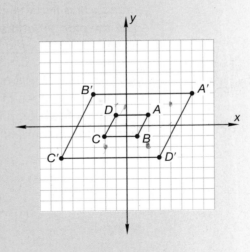

Practice Set

a. The base and the height of a parallelogram are always

    A perpendicular   **B** parallel

    C sides   **D** congruent

b. What is the area of this parallelogram?

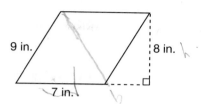

c. The sides of a square and another rhombus are 3 cm long. Find the perimeter and area of each parallelogram.

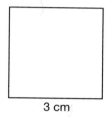

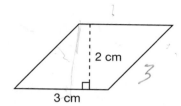

d. Find the area of a parallelogram with vertices at (4, 2), (2, −2), (−2, −2) and (0, 2). What would be the area of a dilation of this parallelogram with a scale factor of 3?

e. *Evaluate* A brass sculpture of the letter N is formed from one long metal parallelogram folded into three congruent parallelograms. What is the area of one of the smaller parallelograms? What is the total area of the long metal parallelogram?

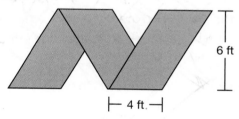

f. *Justify* How do you know your answer to problem **e** is correct?

Written Practice   *Strengthening Concepts*

---

* 1. Gina has a coupon for 15% off at the video store. If she purchases a
 (58)   video marked $16.80, how much will she pay using the coupon?

* 2. Scientists know that the diameter of a normal cell is 0.000025 cm. How
 (51)   is that number written in scientific notation?

3. A triangle has vertices at (0, 5), (−10, 1), and (10, 1). What is the area of
 (Inv. 1, the triangle?
 20)

* 4. The bill for a family of four at a restaurant was $62. Simon paid $72
(58) for the meal and tip. How much was the tip? What percent was the
tip?

5. Marshall knows that 0.01 is equal to 10^{-2}. How would he represent this
(51) using factors?

A 8 **B** $\dfrac{1}{10} \times \dfrac{1}{10}$ **C** -20 **D** $10 \times (-2)$

* 6. *Analyze* Some stores have a layaway plan for customers. This means
(58) that the customer will pay for the item in installments and will be able to
take the item home when it is fully paid for. One store requires that the
person make a minimum down payment of 25%. Kayla wants to put a
$60 item on layaway at that store. What is the minimum down payment
she must make?

Generalize Simplify.

7. $-13 + (-4)(-2)$ *** 8.** $256 \div [2(6 - (-2))^2]$
(31, 36)   (21)

9. $(1.6 \times 10^4)(2.0 \times 10^5)$ *** 10.** $(1.5)^2 (-2)^4$
(46)   (25, 36)

* 11. $\dfrac{(5^3)^2}{5^3}$
(27)

Solve.

12. $\dfrac{1}{3}x - 1 = 4$ **13.** $-5k - 11 = 14$
(50)   (50)

14. $\dfrac{x}{7} = -2$ **15.** $5(2t + 4) = 140$
(50)   (50)

16. $-4m + 1.8 = -4.2$ **17.** $27 = 8b - 5$
(50)   (50)

* 18. *Evaluate* Find the area of a parallelogram with vertices at $(-2, 0)$, $(6, 0)$,
(Inv. 5, $(1, 3)$, and $(9, 3)$. What is the area of a dilation of this parallelogram with
60) a scale factor of 2?

* 19. The Great Wall of China extends approximately 4163 miles long. A mile
(25, 52) is approximately 1.61 kilometers. Find the length of the Great Wall to the
nearest kilometer.

20. Jodi is joining a fitness center. The application fee is $75 and the
(50) monthly rate is $25. She wants to pay $125 now. Solve this equation to
find out how many months (m) she will she be able to go to the fitness
center for this amount of money.

$$25m + 75 = 125$$

* 21. *Analyze* Which of these nets is not a net of a cube?
(55)

A   **B**

C   **D**

22. In parallelogram *ABCD*, m∠B is 128°.
(54) What is the measure of ∠C?

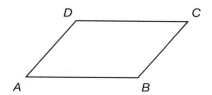

23. What is the median of this data?
(53)

24. The pet store sells guppies for 8 for
(49) $1.50. How much would 36 guppies
cost? Write a proportion and solve.

25. At Washington Middle School, 64% of
(48) the students play a sport. If 192 play
a sport, how many students attend
Washington Middle School?

26. Will a 25½-inch umbrella fit into a gift
(Inv. 2) box that is 10 inches wide by
24 inches long? (Hint: Will it fit if it
is placed diagonally? Use the
Pythagorean Theorem.)

**Average Annual Precipitation
for Selected Texas Cities**

City	Inches
Brownsville	26
Dallas-Fort Worth	32
El Paso	9
Houston	47
Midland-Odessa	14
Port Arthur	55

27. A bakery sells two kinds of bread, whole wheat and white, in a ratio of
(45) 3 whole wheat to 2 white. If the bakery made 60 loaves of bread, how
many were whole wheat?

28. Lou makes pancakes at a pancake house. He makes 12 pancakes
(44) in 4 minutes. At that rate, how long would it take him to make
54 pancakes?

29. Wallpaper comes in rolls that cover 80 square feet. Vicki wants to
(43) wallpaper the walls and ceiling of her living room. The room is 15 feet
wide by 18 feet long by 9 feet high. How many rolls of wallpaper will she
need?

* 30. *Conclude* Tran earns $7.50 per hour at
(41) his part-time job. Copy and complete the
function table for 1, 2, 3, and 4 hours of
work. Then write an equation for
the function. Start the equation with
P =. Is the relationship between the
number of hours worked and amount
of pay proportional?

Number of Hours (*h*)	Amount of Pay (*P*)
1	
2	
3	
4	

Focus on
• Collect, Display, and Interpret Data

Statistics is the science of collecting data and interpreting the data in order to draw conclusions and make predictions.

Suppose the managers of a theme park plan to build a new restaurant on its grounds, and the managers want to know the preference of the park attendees. The managers can collect this information by conducting a survey.

The group of people that researchers want to study is called the **population** of the study. In the case of the theme park, the population would be all park visitors. Often it is not practical to survey an entire population, so researchers select a smaller **sample** of the population to survey. Researchers use sampling methods to ensure that the sample is representative of the larger population and to avoid **bias** or slant toward a particular point of view. For example, the managers of the theme park might ask every 100th person who walks through the gates to participate in the survey.

1. Consider the survey that the theme park managers wish to conduct. Discuss the difficulties of surveying the entire population, that is, all park visitors.

2. If the survey were conducted only among people in one of the other restaurants at the park, would the sample be representative of the entire population?

3. Conduct a **closed-option** survey (multiple choice, not free response) among your classmates. As a class, choose three restaurant options for the theme park. Ask each student to write their preference on a piece of paper, then collect and tally the results. Display the data in a **bar graph.**

4. *Evaluate* Which type of restaurant is most popular among your classmates? Do you think the survey results would differ if it were conducted among a sample of park attendees?

In the survey, we collected **qualitative data** (data that falls into categories), in this case, types of restaurants. Data that is numerical is called **quantitative data.**

Suppose we want to know about how much time the students in your grade spend doing homework each day. We may select a sample of students, then ask them to record how much time they spent doing homework that day. Sample data for nine responses are shown below.

Time Spent on Homework

1 hr, 10 min	2 hr, 5 min	20 min
55 min	30 min	1 hr, 40 min
1 hr, 30 min	1 hr, 45 min	2 hr

We could then display this data in a **histogram.** A histogram displays quantitative data in adjacent intervals, represented by bars that touch. We show an example for the homework sample data. We sort the data into 30 minute intervals.

Thinking Skill

Discuss

Why would a bar graph *not* represent the data well?

Time	Tally	Frequency
:00–:29	I	1
:30–:59	II	2
1:00–1:29	I	1
1:30–1:59	III	3
2:00–2:29	II	2

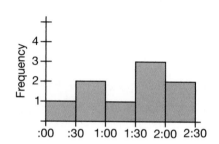

5. A movie theater collected data on the ages of customers attending a new movie. The ages are listed below. Create a frequency table for the data by setting intervals and tallying the numbers for each interval. Then create a histogram to display the data.

7	8	8	9	9	9	10	10	11	12
12	12	13	14	15	16	16	17	18	20
23	28	32	33	34	34	35	37	40	41
48	51	53	57	58	62	68	70	72	75

Discuss How does changing the interval affect the display?

Another kind of survey that is often conducted is an **opinion survey.** Sample results of an opinion survey are shown below.

"Do You Favor or Oppose Ballot Measure A?"

Response	Number	%
Favor	546	42
Oppose	520	40
Undecided	234	18

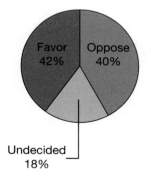

The data are summarized and displayed in a **circle graph.** A circle graph is often used to display qualitative data and is particularly useful to illustrate the relative sizes of parts of a whole. The central angle of a sector is computed by multiplying the fraction or percent of responders by 360°.

6. Conduct a closed-option opinion survey in your class, tally the results, and display the data in a circle graph.

Visit www. SaxonPublishers. com/ActivitiesC3 for a graphing calculator activity.

7. One hundred eighth graders were surveyed and asked what extra-curricular activities they planned to be involved with in high school. The students were given several choices and could select as many as they wanted. The results are shown below.

music	卌 卌 卌 卌 卌 Ⅱ
sports	卌 卌 卌 卌 Ⅰ
service clubs	卌 卌 卌 卌 卌 卌 Ⅲ
student government	卌 卌 卌
other	卌 卌 Ⅲ

Choose a type of graph (bar graph, histogram, or circle graph) for representing these data, and explain your choice. Then graph the data.

• **Sequences**

facts

Power Up M

mental math

a. **Percent:** 10% of $90

b. **Algebra:** Evaluate: $a^2 - b^2$ when a is 5 and b is 4.

c. **Estimation:** $5 \times \$0.97 + \$9.97$

d. **Fractional Parts:** $\frac{3}{4}$ of 240

e. **Rate:** Thirty minutes per day is how many hours per week?

f. **Measurement:** What temperature is indicated on this thermometer?

g. **Select a Method:** To find the number of minutes in 3 hours, which would you use?
   A mental math
   B pencil and paper
   C calculator

h. **Calculation:** $400 \div 2, \div 2, \div 2, \div 2, \div 5, \div 5, \div 5$

problem solving

Which of the following is the greatest?

A $(-95) + (-35) + (-5) + (5) + (35) + (95)$

B $(-95)^2 + (-35)^2 + (-5)^2 + (5)^2 + (35)^2 + (95)^2$

C $(-95)^3 + (-35)^3 + (-5)^3 + (5)^3 + (35)^3 + (95)^3$

A **sequence** is an ordered list of numbers, called **terms,** that follow a certain pattern or rule.

$$3, 6, 9, 12, \ldots$$

The first four terms are given. To find more terms in the sequence, we study the sequence to understand its pattern or rule, then we apply the rule to find additional terms.

Reading Math

Recall that the three dots is called an *ellipsis.* The ellipsis indicates that the sequence continues without end.

Example 1

Describe the pattern in this sequence, and then find the next three terms.

$$3, 6, 9, 12, \ldots$$

We see that the first term is 3 and each succeeding term is 3 more than the previous term.

$$\overset{+3}{\frown}\ \overset{+3}{\frown}\ \overset{+3}{\frown}\ \overset{+3}{\frown}\ \overset{+3}{\frown}\ \overset{+3}{\frown}$$
$$3,\ 6,\ 9,\ 12,\ \mathbf{15},\ \mathbf{18},\ \mathbf{21}$$

This sequence is an example of an **arithmetic sequence.** An arithmetic sequence has a **constant difference** between terms. By subtracting consecutive terms we find the constant difference.

$$\begin{array}{ccc} 6 & 9 & 12 \\ -3 & -6 & -9 \\ \hline 3 & 3 & 3 \end{array} \quad \text{and so on}$$

Example 2

Describe the pattern in this sequence, and then find the next three terms.

$$1, 3, 9, 27, \ldots$$

Each term is three times the preceding term. To find the next term we multiply 27 by 3, then that term by 3, and so on.

$$\overset{\times 3}{\frown}\ \overset{\times 3}{\frown}\ \overset{\times 3}{\frown}\ \overset{\times 3}{\frown}\ \overset{\times 3}{\frown}\ \overset{\times 3}{\frown}$$
$$1,\ 3,\ 9,\ 27,\ \mathbf{81},\ \mathbf{243},\ \mathbf{729}$$

The sequence in example 2 is a **geometric sequence.** A geometric sequence has a **constant ratio** between terms. By dividing consecutive terms we find the constant ratio.

$$\frac{3}{1} = 3 \qquad \frac{9}{3} = 3 \qquad \frac{27}{9} = 3 \qquad \text{and so on}$$

The terms of a numerical sequence have a **position** and a **value.** The position of the term, such as 1st, 2nd, 3rd, etc., is often called the number of the term. The number of the term is found by counting and may be abbreviated with the letter n. The value of the term, which is the actual number that appears in the sequence, may be abbreviated with the letter a.

Number of Term (n)	1	2	3	4	5
Value of Term (a)	3	6	9	12	15

If we know the numbers and values of several terms of a sequence, we can visually represent the sequence as a graph.

Arithmetic Sequence

n	1	2	3	4
a	3	6	9	12

Geometric Sequence

n	1	2	3	4
a	1	3	9	27

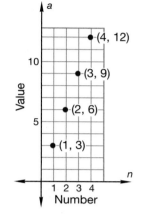

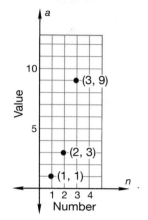

Notice that the graphed points are aligned for the arithmetic sequence. Visually we could easily predict the value of the next term or terms. The graphed points are not aligned for the geometric sequence. Instead the distance from the *x*-axis to the point triples from term to term. We could not plot the 4th term, 27, on the graph because it far exceeded the scale of the graph.

We may describe the rule of a function using words or using a formula. A formula relates the value of a term (*a*) with its number (*n*). Studying a sequence table can help us find a formula.

Example 3

Write a formula that describes this sequence.

$$3, 6, 9, 12, \ldots$$

Solution

First we make a table that shows the number and the value of the terms.

Number (*n*)	1	2	3	4
Value (*a*)	3	6	9	12

Studying the table we see that the value (*a*) is three times the number (*n*). We write this formula.

$$a_n = 3n$$

The little *n* after the *a* is called a **subscript**. The expression a_n means *the value of the nth term*. (It does not mean that *a* and *n* are multiplied.) On the right side of the formula we see that the constant difference, 3, is multiplied by n. The formula tells us that to find the nth term of the sequence, multiply n by 3. For example, the tenth term of the sequence is $3 \cdot 10$, which is 30. Using the formula, it looks like this:

$$a_{10} = 3(10)$$
$$a_{10} = 30$$

Example 4

Write a formula that describes this sequence.

1, 3, 9, 27, . . .

Solution

First we make a table.

n	1	2	3	4
a	1	3	9	27

The rule of this sequence involves a constant ratio, 3, as the base, and an exponent related to the number of the term. We try $a_n = 3^n$

$$3^1 = 3 \qquad 3^2 = 9 \qquad 3^3 = 27$$

We see that these terms are one place out of position, so we modify the formula and try again.

$$a_n = 3^{(n-1)}$$
$$a_1 = 3^{1-1} = 3^0 = 1$$
$$a_2 = 3^{2-1} = 3^1 = 3$$
$$a_3 = 3^{3-1} = 3^2 = 9$$
$$a_4 = 3^{4-1} = 3^3 = 27$$

Practice Set

Describe each sequence as arithmetic, geometric, or neither.

a. 1, 2, 4, 8, 16, . . .

b. 2, 4, 6, 8, 10, . . .

c. 1, 4, 9, 16, 25, . . .

d. What is the constant difference in this sequence?

1, 5, 9, 13, . . .

e. What is the constant ratio in this sequence?

5, 25, 125, 625, . . .

Write the first four terms of the sequence described by each formula.

f. $a_n = 2n$

g. $a_n = 2^n$

1. A shoe company makes 7 pairs of white shoes for every 4 pairs of
(45) black shoes. If it makes 121,000 pairs of white shoes and black shoes
combined, how many pairs are white?

2. When Karen serves jury duty, she will receive $15 each day for the first
(3, 4) three days she serves. After three days, she will receive $24 each day
she serves. How much will Karen receive if she serves for 10 days?

3. A truck rental costs Arturo $89 for the weekend, plus $1.30 per mile it is
(3, 4) driven. How much would Arturo have to pay for the rental if he drives
60 miles?

4. Find the mean, median, and mode of these temperatures: 56, 50, 56,
(7) 63, 75.

5. Transversal t intersects parallel lines r and s.
(54) Find m $\angle x$.

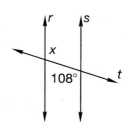

* 6. *Explain* How can you decide if the point $(-4, -1)$ is on the graph of the
(41) equation $y = \frac{3}{4}x + 2$ without graphing the equation?

7. Kataya wants to paint the outside of her house. The perimeter of the
(43) house is 160 feet and the height is nine feet. To know how much paint
to buy she needs to calculate the surface area. What is the actual
surface area of the house?

8. Kurt traces a 6-inch salad plate onto his paper. Find the **a** area and
(39, 40) **b** circumference of Kurt's circle. Use $\pi = 3.14$.

9. These triangles are similar. Find the scale
(35) factor from the smaller to larger triangle and
solve for m.

Analyze Solve.

* 10. $0.4x + 1.3 = 1.5$
(50)

* 11. $0.002x + 0.03 = 0.92$
(50)

* 12. $3x + 8 - x = 20$
(50)

* 13. $2(x + 3) = 16$
(50)

14. $\frac{2}{5}x = \frac{2}{3}$
(38)

15. $\frac{x}{4} = 1.25$
(38)

* 16. *Formulate* Suppose that a mountain grows at a rate of 10 inches in
(7) 4 years. Express the growth as a unit rate.

* 17. A window that is 54 inches wide is how many feet wide? Use a unit
(52)   multiplier.

18. Combine like terms: $x + 1 + x^2 + 4x + 4$
(31)

19. a. Write $\frac{17}{100}$ as a decimal and as a percent.
(12)

   b. Select one of these three forms and use it in the following sentence
   replacing the underlined words. <u>Seventeen out of 100</u> customers
   registered for the discount card.

* 20. **_Represent_**   Write the equation of each line in slope-intercept form.
(56)

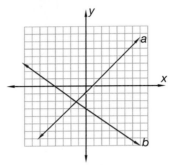

21. The advertised size of a television screen is the length of its diagonal
(Inv. 2)   measure. For example, a "27-inch" screen might be only 22 inches
   wide. If so, what would be its height to the nearest inch?

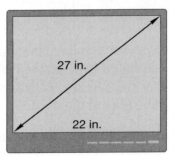

Simplify.

22. $(-3) - |-3| - 3^{-1}$
(33, 51)

23. $\dfrac{t^2 u^2 r}{r u t^3}$
(27)

24. $\dfrac{3.6 - 0.36}{0.03}$
(24, 25)

25. The favorite professional sports of 25 students are shown in the graph below.

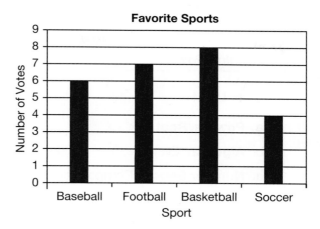

Which set of data matches the information in the graph?

A

Sport	Number of Votes
Baseball	卌
Football	卌 II
Basketball	卌 IIII
Soccer	卌

B

Sport	Number of Votes
Baseball	IIII
Football	卌 IIII
Basketball	卌 I
Soccer	卌 III

C

Sport	Number of Votes
Baseball	卌 I
Football	卌 III
Basketball	IIII IIII
Soccer	卌

D

Sport	Number of Votes
Baseball	卌 I
Football	卌 II
Basketball	卌 III
Soccer	IIII

• Graphing Solutions to Inequalities on a Number Line

facts   Power Up M

mental math
a. **Powers/Roots:** $x^4 \cdot x^3$

b. **Algebra:** Evaluate: $5^2 - 4(a)(2)$ when $a = 2$

c. **Number Sense:** Compare $\frac{1}{4} + \frac{3}{4}$ ◯ $\frac{1}{4} \times \frac{3}{4}$

d. **Proportions:** The ratio of boys to girls is 10 to 11. If there are 55 girls, how many boys are there?

e. **Probability:** If a number cube is rolled, what is the probability of rolling a prime number?

f. **Geometry:** What is the value of y?

g. **Measurement:** How many inches are in one yard? Two yards?

h. **Calculation:** $13 \times 2, + 2, \div 2, \div 2, + 2, \div 3, \div 3, \div 3$

problem solving   Jaime knows that 30^2 is 900. He wonders if knowing 30^2 can help him find 29×31. Raquel knows what 25^2 equals. How can she use that knowledge to find 24×26?

Tom is thinking of a number between 1 and 10. Of what possible numbers could he be thinking?

We might assume that Tom is thinking of a counting number greater than 1 and less than 10, such as 3, 6 or 7. However, Tom could be thinking of $2\frac{1}{2}$ or π or $\sqrt{20}$, because each of these numbers is also between 1 and 10.

We can write an **inequality** to indicate the range of possible numbers. Letting x represent a possible number, we write

$$1 < x < 10$$

We read this inequality in both directions from x, "x is greater than 1 and less than 10." Notice that both the description and the inequality exclude 1 and 10.

We can use a number line to graph the possible numbers of which Tom could be thinking.

The shaded portion of the number line represents the range of choices for Tom's number. The word *between* suggests that Tom's number is greater than 1 and less than 10, so the small circles at 1 and 10 indicate that those numbers are excluded from the choices. To show that those numbers are included we make a shaded dot.

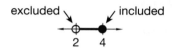

"Greater than 2 and less than or equal to 4"

Example 1

Christina is thinking of a number. If she doubles the number, the product is greater than 4. Write an inequality for this description. Graph on a number line all the possible choices for Christina's number.

Solution

If we let x represent Christina's number, then the description of the number can be represented by this inequality.

$$2x > 4$$

"Twice the number is greater than four."

We can solve this inequality like we solve an equation.

Step:	Justification:	Visual Representation:
$2x > 4$	Given inequality	$\boxed{x}\,\boxed{x} > \square\square\square\square$
$\dfrac{2x}{2} > \dfrac{4}{2}$	Divided both sides by 2	$\boxed{x}\quad\boxed{x} > \square\square\ \ \square\square$
$x > 2$	Simplified	$\boxed{x} > \square\square$

Christina's number is greater than 2. We are asked to graph all possible choices on a number line.

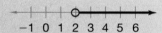

The circle at 2 indicates that the possible choices do not include 2. The shaded number line is a graph of the possible choices. The shaded arrowhead shows that the graphed choices continue without end.

Christina's possible choices are called the solution set. We can test our answer by selecting a number from the solution set and another number not in the solution set and decide if they satisfy the conditions of the problem. We pick 3, which is on the shaded part of the line, and 1, which is on the unshaded part.

• If Christina doubles 3, the answer is 6, which is greater than 4, so 3 is a number in the solution set.

• If Christina doubles 1, the answer is 2, which is not greater than 4, so 1 is not a number in the solution set.

These two tests support our answer.

Example 2

Solve and graph the solution to this inequality: $3x + 1 \leq 7$

Solution

We may solve inequalities as we solve equations. (We will learn an exception in a later lesson.)

Thinking Skill

Explain

Why is the first step to subtract 1 from each side of the inequality?

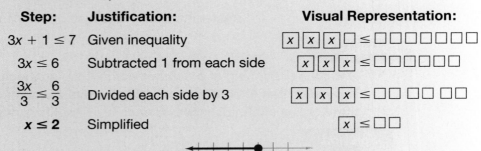

Step:	Justification:
$3x + 1 \leq 7$	Given inequality
$3x \leq 6$	Subtracted 1 from each side
$\dfrac{3x}{3} \leq \dfrac{6}{3}$	Divided each side by 3
$x \leq 2$	Simplified

We test selected numbers in the original inequality. We select 1, which is in the solution set, and 3, which is not.

Step: (try $x = 1$)	**Justification:**	**Step:** (try $x = 3$)
$3(1) + 1 \leq 7$	Substituted	$3(3) + 1 \leq 7$
$4 \leq 7$ (true)	Simplified	$10 \leq 7$ (false)
1 is a solution		3 is not a solution

Explain Why do we choose a number that is in the solution set and one that is not when testing our solution?

When applying inequalities to situations in the real world, we might be limited to positive numbers or whole numbers, as we see in the following example.

Example 3

Vincent goes to the store with $20 to buy a notebook for $4. He notices that packages of notebook paper are on sale for $2. Write and solve an inequality to show the number of packages of notebook paper he could buy if he also buys the notebook (ignoring sales tax). Then graph the solution.

Solution

The number of packages of paper (p) Vincent could buy if he buys the notebooks is

$$2p + 4 \leq 20$$

We solve the inequality.

Step:	Justification:
$2p \leq 16$	Subtracted 4 from both sides
$p \leq 8$	Divided both sides by 2.

If we were to graph this inequality for real numbers, the graph would be

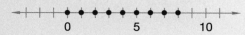

However, Vincent cannot buy a fraction of a package or a negative number of packages. He can buy none, or he can buy 1, 2, 3 or up to 8 packages of paper.

Example 4

Solve and graph the solution: $3 \leq x - 2$

Solution

We isolate x by adding 2 to both sides.

Step:	Justification:
$3 \leq x - 2$	Given inequality
$5 \leq x$	Added 2 to both sides

We may read the inequality from right to left, "x is greater than or equal to 5." Alternately, we may reverse the position of x and 5 provided we reverse the direction of the inequality symbol to $x \geq 5$. (If 5 is less than or equal to x, then x is greater than or equal to 5.)

We test selected numbers in the original inequality. We select 6, which is a solution, and 4, which is not.

Step (try $x = 6$):	Justification:	Step (try $x = 4$).
$3 \leq (6) - 2$	Substituted	$3 \leq (4) - 2$
$3 \leq 4$ (true)	Simplified	$3 \leq 2$ (false)
6 is a solution		4 is not a solution

Justify Is 5 part of the solution? How do you know?

Practice Set

a. **Formulate** Alisha has a number in mind. If she adds three to her number the result is less than five. Use this information to write and solve an inequality about Alisha's number. Then graph the solution set.

Solve these inequalities and graph their solutions on number lines.

b. $4x > -12$ **c.** $6x + 1 \geq -5$

d. $2x - 5 \leq 3$ **e.** $7 \leq x - 8$

f. Jan is participating in a 6 mile run/walk for charity. If she completes the course and runs at least half the distance, then what inequality shows how far she might walk? Graph the inequality.

Written Practice   *Strengthening Concepts*

1. The ratio of the length of the input lever arm to the length of the output
(45) lever arm was 6 to 5. If the entire lever is 132 in. long, how long is the input lever arm?

output lever arm   input lever arm

* 2. ⬛ Analyze   Eighty percent of the castles had been preserved. If there
(48) were 135 castles in all, how many had not been preserved?

3. **a.** Sketch a cube that has a volume of 1 cm³.
(Inv. 4, 42)

b. Sketch a cube with edges 2 cm long.

c. Find the volume of the larger cube. How many of the smaller cubes would fit in the larger?

4. Find the mean, median, and mode(s) of the distances in feet from home
(7) plate to the left-field fence at various ball parks: 303, 279, 297, 321, 310, 310, 301.

5. Find x.
(54)

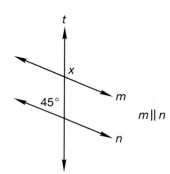

6. Graph $y = 4x - 4$. Is point $(-2, -4)$ on the line?
(41)

7. A belt is wrapped around two wheels. As the larger wheel turns it moves
(39) the belt which moves the smaller wheel. The diameter of the larger wheel
is 6 in. and the diameter of the smaller wheel is 3 in. When the larger
wheel turns once, how many times does the smaller wheel turn?

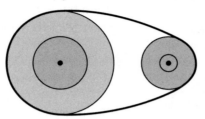

* 8. [Classify] **a.** Find the area and perimeter of the
(Inv. 2, figure. (Dimensions are in yds.)
37)
    b. What type of quadrilateral is the
figure?

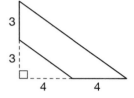

9. World class male sprinters run 100 meters in about 10 seconds. Express
(7) this as a unit rate.

10. Kilani bought a cell phone service plan that provides 300 minutes of
(52) phone use each month. Use a unit multiplier to convert 300 minutes to
hours.

11. This spinner is spun twice.
(32)
    a. Find the sample space of the experiment.

    b. What is the probability that the same
color will be spun twice?

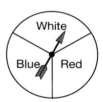

12. Write 125% **a** as a decimal and **b** as a reduced fraction.
(12)

13. The walls of buildings are usually rectangles, but rectangular structures
(Inv. 2) are not stable. They shift easily.

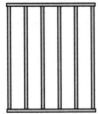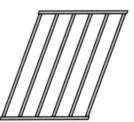

Triangles, however, are stable. So engineers require triangles in buildings to increase the stability of the building. Just one diagonal brace greatly increases the stability of a wall. Conrad will use a board to brace the wall of a shed as shown. Is a 10-foot long board long enough? Justify your answer.

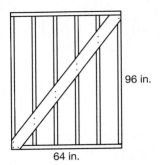

96 in.

64 in.

Generalize Simplify:

* 14. $(-1)^2 + (-1)^1 + (-1)^0 + (-1)^{-1}$
(31, 51)

15. $\dfrac{race}{car} + e$
(27)

16. $\left(\dfrac{2}{3}\right)^2 - \dfrac{1}{3} \div \dfrac{3}{4}$
(13, 22)

17. $x(x + 2) - 2(x + 2)$
(31, 36)

Analyze Solve.

* 18. $1.3x - 0.32 = 0.98$
(50)

* 19. $0.001x + 0.09 = 1.1$
(50)

* 20. $-4(2 + x) - 2(x + 3) = 4$
(50)

21. $\dfrac{2}{7}x = \dfrac{3}{14}$
(38)

22. Using slope-intercept form, write the equation of each line.
(56)

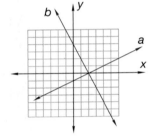

Graph each solution on a number line.

* 23. $x + 2 < 6$
(62)

* 24. $x + 5 \geq 3$
(62)

25. **a.** Square *ABCD* has these vertices: $A(-2, 2)$, $B(2, 2)$, $C(2, -2)$ and $D(-2, -2)$. Use graph paper to draw square *ABCD* and similar figure $A'B'C'D'$ using a scale factor of 4. (Assume that the origin is the center of both figures.)
(Inv. 5)

b. What are the locations of the vertices of $A'B'C'D'$?

• Rational Numbers, Non-Terminating Decimals, and Percents
• Fractions with Negative Exponents

facts | Power Up M

mental math

a. **Percent:** 30% of $80

b. **Statistics:** Find the mean of these numbers: 20, 40, 50, 30, 10

c. **Sequences:** Find the next three terms. 1, 2, 4, 8, ____, ____, ____, ...

d. **Fractional Parts:** $\frac{1}{9}$ of 7200

e. **Measurement:** The odometer read 438.1 miles before the trip. Now it reads 449.9 miles. How long was the trip?

f. **Proportion:** $\frac{2}{3} = \frac{8}{x}$

g. **Scientific Notation:** $(2.0 \times 10^5)(3.0 \times 10^6)$

h. **Calculation:** Square 10, − 1, ÷ 11, $\sqrt{}$, × 5, + 1, $\sqrt{}$, $\sqrt{}$

problem solving | Greg rode his bike up the 30 mile hill at 10 miles per hour. Then, he rode back down the hill at 30 miles per hour. How far did he go altogether? How long did it take him? What was his average speed?

New Concepts | Increasing Knowledge

rational numbers, non-terminating decimals, and percents

Reading Math

Recall that the three dots (...) is called an **ellipsis.** It indicates that the pattern of repetition continues. An ellipsis may be used in place of a bar.

Recall that every rational number can be expressed as a ratio of two integers. Converting the ratio to a decimal number results in a decimal that terminates (ends) or does not terminate but has repeating digits.

Terminating: $\frac{1}{8} = 0.125$

Non-terminating but repeating: $\frac{1}{11} = 0.0909090909...$

Also recall that we can indicate repeating digits by writing a bar over the repetend.

$$\frac{1}{11} = 0.\overline{09}$$

We do not perform computations with a decimal number that has a bar over the repetend. Instead we use the fraction equivalent, or we round the decimal number to an appropriate number of decimal places.

Analyze Which produces a more exact product for multiplying by $0.\overline{09}$, multiplying by $\frac{1}{11}$ or by 0.0909?

Example 1

Express $\frac{2}{3}$ as an equivalent decimal and percent. Then express $\frac{2}{3}$ as a decimal rounded to the thousandths place.

Solution

To express $\frac{2}{3}$ as a decimal we divide 2 by 3.

$$3\overline{)2.000} \quad 0.666\ldots$$

The decimal equivalent to $\frac{2}{3}$ is **$0.\overline{6}$.**

To express $\frac{2}{3}$ as a percent, we multiply $\frac{2}{3}$ by 100%.

$$\frac{2}{3} \cdot 100\% = \frac{200\%}{3} = \mathbf{66\frac{2}{3}\%}$$

To express $\frac{2}{3}$ as a decimal rounded to the thousandths place, we write the number with four places and then round back to three places.

$$0.\overline{6} \approx 0.6666$$
$$\approx \mathbf{0.667}$$

Example 2

Express $16\frac{2}{3}\%$ as a reduced fraction and as a decimal.

Solution

A percent is the numerator of a fraction with a denominator of 100.

$$16\frac{2}{3}\% = \frac{16\frac{2}{3}}{100}$$

We perform the division that is indicated.

$$16\frac{2}{3} \div 100$$

$$\frac{50}{3} \times \frac{1}{100} = \frac{1}{6}$$

One way to convert $16\frac{2}{3}\%$ to a decimal is to convert the equivalent fraction $\frac{1}{6}$ to a decimal.

$$16\frac{2}{3}\% = \frac{1}{6} = \mathbf{0.1\overline{6}}$$

The result is reasonable because removing the percent sign shifts the decimal point two places.

$$16\frac{2}{3}\% = 0.16\frac{2}{3}$$

Later in this book we will learn a method for converting decimals with a repetend to fractions and percents.

Example 3

Arrange these numbers in order from least to greatest.

$$\frac{4}{5}, \ 83\frac{1}{3}\%, \ 0.83$$

Solution

To help us compare we will express each number as a decimal rounded to three decimal places.

$$\frac{4}{5} = 0.800$$

$$83\frac{1}{3}\% \approx 0.833$$

$$0.83 = 0.830$$

Now we write the numbers in order.

$$\frac{4}{5}, \ 0.83, \ 83\frac{1}{3}\%$$

Example 4

The incumbent received $41\frac{2}{3}\%$ of the 1260 votes cast in the town council election. Describe how to estimate the number of votes the incumbent received. Then calculate the exact number of votes.

Solution

As an approximation we see that the incumbent received somewhat less than half the votes cast, so the number of votes is probably fewer than 600.

For a closer estimate we could find 40% of 1260. Since 10% of 1260 is 126, and since 40% is 4 × 10%, we could multiply 4 × 125 to find that the total number of votes should be a little more than 500.

Since the percent includes a non-terminating fraction we choose to convert the percent to a fraction to perform the calculation.

$$41\frac{2}{3}\% = \frac{125}{300} = \frac{5}{12}$$

We translate the problem into an equation using $\frac{5}{12}$ in place of the percent. The incumbent's votes (v) were $41\frac{2}{3}\%$ of the 1260 votes.

$$v = \frac{5}{\cancelto{1}{12}} \times \cancelto{105}{1260}$$

$$v = 525$$

The incumbent received **525 votes.**

fractions with negative exponents

Recall that a negative exponent indicates the reciprocal of the positive power of the base.

$$x^{-n} = \frac{1}{x^n} \qquad 3^{-2} = \frac{1}{3^2}$$

Also recall that with fractions we reverse the position of the numerator and the denominator to form the reciprocal.

The reciprocal of $\frac{x}{y}$ is $\frac{y}{x}$.

If the base of a negative exponent is a fraction, we make the exponent positive by replacing the fraction with its reciprocal.

$$\left(\frac{x}{y}\right)^{-n} = \left(\frac{y}{x}\right)^{n}$$

Example 5

Simplify: $\left(\frac{1}{3}\right)^{-2}$

Solution

We replace $\frac{1}{3}$ with its reciprocal, 3, and make the exponent positive. Then we apply the positive exponent.

$$\left(\frac{1}{3}\right)^{-2} = 3^2 = \mathbf{9}$$

Practice Set

Convert each fraction to a decimal and a percent. Then write the decimal rounded to the nearest thousandth.

a. $\frac{1}{3}$ **b.** $\frac{5}{6}$ **c.** $\frac{2}{11}$

Convert each percent to a fraction and a decimal. Then write the decimal rounded to the nearest hundredth.

d. $66\frac{2}{3}\%$ **e.** $8\frac{1}{3}\%$ **f.** $22\frac{2}{9}\%$

g. Arrange in order from least to greatest.

$$16\%, \frac{1}{6}, 0.165$$

h. The bill for the meal was $24.00. Dario left a tip of $16\frac{2}{3}\%$. How much was the tip?

Simplify.

i. $\left(\frac{1}{2}\right)^{-3}$ **j.** $\left(\frac{1}{10}\right)^{-2}$ **k.** $\left(\frac{3}{2}\right)^{-1}$

Written Practice   *Strengthening Concepts*

1. For the centerpieces at the party, four stems of flowers were to be
(4) placed in every vase. If there were 72 stems, how many vases could be filled?

2. Sixty-five percent of the 1600 students wore school colors on school
(48) spirit day. How many students did not wear school colors?

3. Duke carried only dimes. Diana carried just nine nickels. Together they
(3, 4) had $1.95. How many dimes did Duke carry?

4. The weights of the fish caught by a fisherman were in pounds:
(53)

$$1.7, 1.8, 3.0, 1.8, 3.5, 3.1, 3.3$$

a. Find the mean, median, mode, and range of the data.

b. Which measure of central tendency would the fisherman use to represent his fishing most favorably?

5. Which angle corresponds to $\angle 3$?
(54)

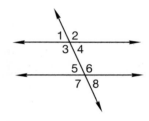

6. Graph $y = -\frac{2}{3}x + 1$. Is point $(-6, 4)$ a solution?
(41)

* 7. **Formulate** **a.** Write $83\frac{1}{3}\%$ as a fraction and as a decimal.
(63)

b. Write the decimal rounded to the nearest hundredth.

* 8. **a.** Write $\frac{4}{9}$ as a decimal and as a percent.
(63)

b. Then write $\frac{4}{9}$ as a decimal rounded to the nearest thousandth.

9. Two red marbles (R_1 and R_2), one white marble (W), and one blue marble (B) are in a bag. Two marbles are selected.
(32)

a. Find the sample space of the experiment.

b. What is the probability of selecting at least one red marble?

c. State the complement of this event and find its probability.

* 10. **Connect** If a rectangular computer screen is 17 inches wide and 12 inches high, then what is its diagonal measure to the nearest inch?
(Inv. 2)

11. Joyce's grandmother has a 4-inch cube with photos on each face. What is the surface area of the cube?
(43)

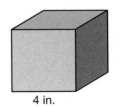

4 in.

* 12. **Analyze** Jessica built a patio shaped like the figure at right. Find the **a** area and **b** perimeter of the patio. Use $\pi = 3.14$. (Units are in feet.)
(39, 40)

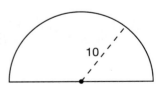

10

13. Express as a unit rate: 30 math problems in 30 minutes.
(7)

14. Use a unit multiplier to convert 2460 inches to feet.
(52)

* 15. Factor:
(21)

a. $7m^2 - 49$

b. $x^4 - 4x^3$

Simplify.

16. $(-3)^2 - 3^2 - \left(\dfrac{1}{3}\right)^{-1}$
(51)

17. $\dfrac{8cabs^2}{2bssa}$
(27)

18. $2\dfrac{1}{2} - 1\dfrac{1}{3} \cdot 1\dfrac{1}{4}$
(13, 23)

19. $\dfrac{(0.6)(0.3)}{9}$
(25)

Solve.

* 20. $0.02x + 0.1 = 1$
(50)

* 21. $1.2 + 0.2x = 0.8$
(50)

22. $\dfrac{3}{5}x = 6$
(38)

23. $3|x| = 12$
(1, 14)

24. Solve and graph the solution on a number line: $x - 11 > -4$
(62)

25. Identify the graph that best represents the relationship between the
(41) amount of Euros Simon received at an exchange rate of $0.87 Euros for every U.S. dollar.

A

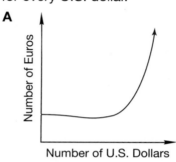

B

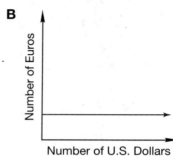

C

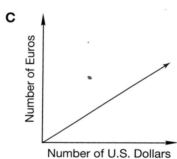

D

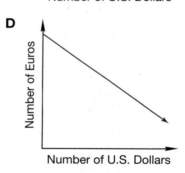

• Using a Unit Multiplier to Convert a Rate

facts | Power Up M

mental math

a. Algebra: Evaluate: $7^2 - 4(1)(c)$ when $c = 12$

b. Sequences: Find the next three terms: 20, 15, 10, 5, ...

c. Number Sense: Compare $\frac{1}{2} + \frac{1}{2} \bigcirc \frac{1}{2} \times \frac{1}{2}$

d. Powers/Roots: $x \cdot x^4$

e. Rate: Rosie is reading an edition of *The Lord of the Rings* that is 1000 pages long. If she reads 50 pages per day, and Sam reads 40 pages per day, how much longer will it take Sam to read than Rosie?

f. Geometry: Find x.

g. Measurement: How many square feet is one square yard?

h. Calculation: $2 \times 2, \times 2, \times 2, \times 2, \sqrt{}$

problem solving | If the water balloons are divided into 7 equal piles, there is one left over. If they are divided into 5 equal piles, there are two left over. If they are divided into 4 equal piles, there are two left over. If they are divided into 6 equal piles there are four left over. There are fewer than 70 water balloons. How many are there, exactly?

Math Language
Recall that a **rate** is a ratio of two measures.

A rate, such as 60 miles per hour, can be written in fraction form.

$$\frac{60 \text{ miles}}{1 \text{ hour}}$$

We can use a unit multiplier to convert a rate to a different unit such as miles per minute. In this case, we would convert hours to minutes. Since 60 minutes equal 1 hour, we have these two unit multipliers available.

$$\frac{60 \text{ min}}{1 \text{ hour}} \qquad \frac{1 \text{ hour}}{60 \text{ min}}$$

We choose the unit multiplier that cancels hours.

$$\underbrace{\frac{60 \text{ miles}}{\cancel{1 \text{ hour}}}}_{\text{Rate}} \cdot \underbrace{\frac{\cancel{1 \text{ hour}}}{60 \text{ min}}}_{\substack{\text{Unit} \\ \text{Multiplier}}} = \frac{60 \text{ miles}}{60 \text{ min}} = \frac{1 \text{ mile}}{1 \text{ min}}$$

So 60 miles per hour equals 1 mile per minute.

To convert rates to alternative units, we create a unit multiplier using the appropriate units. Then we multiply the rate by the appropriate unit multiplier to cancel the unit we want to eliminate and include the unit we want in the answer.

Example 1

Yasmine ran a mile in 6 minutes. Use a unit multiplier to find her average rate in miles per hour (mph).

Solution

We can express Yasmine's run as a rate.

$$\frac{1 \text{ mile}}{6 \text{ min}}$$

We are given miles over minutes; we are asked for miles over hours.

$$\frac{mi}{min} \times \left(\genfrac{}{}{0pt}{}{\text{Unit}}{\text{Multiplier}}\right) = \frac{mi}{hr}$$

We need to convert from minutes below the division bar to hours. Sixty minutes equals one hour, so we have a choice of two unit multipliers.

$$\frac{60 \text{ min}}{1 \text{ hr}} \qquad \frac{1 \text{ hr}}{60 \text{ min}}$$

We choose the unit multiplier with minutes above the division bar so we can cancel minutes and replace minutes with hours.

$$\underbrace{\frac{1 \text{ mile}}{6 \text{ min}}}_{\text{Rate}} \cdot \underbrace{\frac{60 \text{ min}}{1 \text{ hr}}}_{\substack{\text{Unit} \\ \text{Multiplier}}} = \frac{60 \text{ mi}}{6 \text{ hr}} = \frac{10 \text{ mi}}{1 \text{ hr}}$$

We find that Yasmine's average rate was **10 mph.**

Thinking Skill

Conclude

What unit multiplier would we use if we wanted to convert miles per hour to miles per second?

Evaluate Determine the unit multipliers that would allow us to convert miles per hour to feet per hour. Then find Yasmine's average rate in feet per hour.

Example 2

Driving at 60 miles per hour, a car travels about how many kilometers in two hours? (1 mile is about 1.6 kilometers.)

Solution

One way to solve the problem is first to convert 60 miles per hour to kilometers per hour.

$$\frac{60 \text{ mi}}{1 \text{ hr}} \cdot \frac{1.6 \text{ km}}{1 \text{ mi}} = \frac{96 \text{ km}}{1 \text{ hr}}$$

We find the distance traveled in 2 hours by multiplying rate times time.

$$d = rt$$

$$d = \frac{96 \text{ km}}{1 \text{ hr}} \cdot 2 \text{ hr}$$

$$= 192 \text{ km}$$

The car travels **192 km.**

Practice Set

Analyze Use a unit multiplier to perform the following rate conversions.

a. 3 miles per minute to miles per hour

b. 880 yards in 2 minutes to feet per minute

c. 440 yards per minute to yards per second

d. 24 miles per gallon to miles per quart

e. Peter packed 32 pints of pickles in a minute. Find Peter's pickle packing rate in ounces per minute.

f. Shannon rode her bike 6 miles in 24 minutes. Find her average rate in miles per hour.

g. Find the average riding rate from **f** in kilometers per hour. Then find the number of kilometers Shannon could ride in 30 minutes at that rate.

Written Practice   *Strengthening Concepts*

1. Each tetherball game lasted about 3 minutes and then a new pair of
(4)   students would start to play. If there are 45 min in recess, how many students will get to play ?

* 2. **Analyze** Sixty percent of the students voted in the student elections. If
(48)   there are 300 students, how many did not vote?

3. Kerry carried quarters. Doug carried nine dimes. Together they had
(3, 4) $2.65. How many quarters did Kerry carry?

4. The weights of the puppies at birth were:
(53)

16 oz, 18 oz, 17 oz, 10 oz, 15 oz, 13 oz, 16 oz

a. Find the mean, median, mode, and range of this data.

b. Which measure would you use to report the spread of the birth weights?

5. Two of the lines are parallel. Find *m*.
(54)

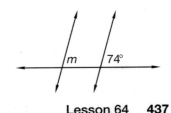

* 6. The figure is a net for what geometric solid?
(Inv. 4)   Name and sketch the solid.

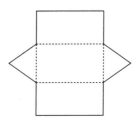

7. Use the slope-intercept method to graph $y = 3x - 5$. Is $(3, -1)$ a
(41)   solution?

* 8. **Evaluate** A coin is tossed and a number cube is rolled.
(32)
   a. What is the sample space of the experiment?

   b. What is the probability of heads and a number greater than three?

* 9. Use a unit multiplier to convert 1200 feet per minute to feet per second.
(64)

* 10. Use a unit multiplier to convert 0.3 tons per day to pounds per day.
(64)

* 11. **Represent** **a.** Sketch a rectangular prism with dimensions 2 cm by 3 cm
(42, 43)   by 4 cm.

   b. What is the volume?   **c.** What is the surface area?

* 12. **a.** Write $77\frac{7}{9}\%$ as a reduced fraction and decimal.
(63)
   b. Then round the decimal to the nearest thousandth.

* 13. **Generalize** Simplify: $\left(\frac{1}{2}\right)^2 + \left(\frac{1}{2}\right)^1 + \left(\frac{1}{2}\right)^0 + \left(\frac{1}{2}\right)^{-1}$
(27, 63)

* 14. **Analyze** The diameter of a cylindrical soup can is 7 cm. Find **a** the
(39, 40)   circumference and **b** the area of the top of the can. Use $\frac{22}{7}$ for π.

15. Find the **a** area and **b** perimeter of the figure at
(Inv. 2,   right. Dimensions are in feet.
37)

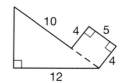

16. Factor:
(21)
   a. $-3x + 27$ **b.** $10x - 100$

17. Sketch the triangles separately.
(35)
   a. State how you know they are
       similar.

   b. Find the scale factor from the small to large
       triangle.

   c. Find x.

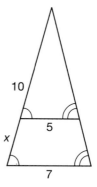

Simplify.

18. $-(-2)^2 - 2^2 - (\sqrt{2})^2$
(15, 33)

19. $\dfrac{bake^2}{acake}$
(27)

Solve.

* 20. $2.1 + 0.7x = 9.8$
(50)

21. $4(x - 4) - 3x = 10$
(50)

22. $\dfrac{4}{5}x = 16$
(38)

23. $\dfrac{12}{40} = \dfrac{30}{x}$
(44)

* 24. **Model** Solve and graph the solution on a number line: $x + 3 < 2$
(62)

* 25. The graph shows the amount of rainfall in Glen County for the months of
(Inv. 6) March through August.

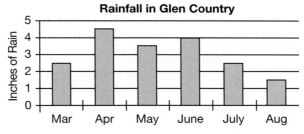

a. About how many inches of rain fell during the months of March through August?

b. Which month received about as much rain as the months of July and August combined?

Early Finishers
Real-World Application

At a certain point on their orbits, the distance from Mars to Earth is about 35,000,000 miles. If light travels through space at approximately 186,000 miles per second, how many seconds does it take for light to travel from Mars to Earth? Write each number in scientific notation before performing the calculation.

• Applications Using Similar Triangles

facts | Power Up M

mental math

a. Algebra: Evaluate: $\frac{1}{2}m^2$ when $m = 6$

b. Fractional Parts: $\frac{7}{8}$ of 4800

c. Probability: Twelve of the 28 students are boys. If the teacher calls on one student at random, what is the probability a girl will be called?

d. Proportions: The ratio of cars to trucks in the lot was 5 to 2. If there were 70 cars and trucks all together, how many were trucks?

e. Percent: 5% of $70

f. Measurement: Find the length of this rectangle.

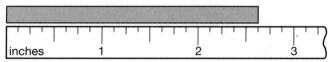

g. Select a Method: To find the square root of 200, which would you use?

A mental math
B pencil and paper
C calculator

h. Calculation: $55 \div 11$, $\times 9$, $+ 4$, $\sqrt{}$, $+ 2$, $\sqrt{} + 1$

problem solving

The price of lumber is calculated by the board foot. One board foot is the amount of lumber in a board 12″ long, 12″ wide, and 1″ thick, or the equivalent. Below we show other dimensions equal to one board foot.

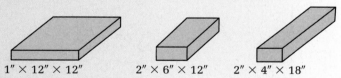

$1″ \times 12″ \times 12″$ $2″ \times 6″ \times 12″$ $2″ \times 4″ \times 18″$

The lumber yard sells one type of lumber with the end dimensions of 1″ by 6″. What length of 1″ by 6″ lumber is equal to one board foot?

Math Language

Recall that **similar figures** have the same shape but not necessarily the same size. Corresponding sides of similar figures are proportional. Corresponding angles of similar figures are congruent.

We can find the measure of some objects that are difficult to measure by using similar triangles. Recall that the side lengths of similar triangles are proportional.

Example 1

Maricruz wants to know the height of a flag pole. She stands a meter stick in a vertical position and finds that the length of its shadow is 120 cm. At the same time the shadow of the flag pole is 30 m long. How tall is the flag pole?

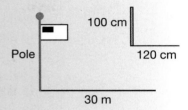

Solution

We may think of the flagpole and its shadow as one triangle and the meter stick and its shadow as a smaller but similar triangle.

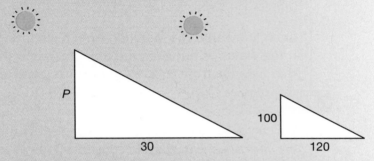

The triangles are similar because the flagpole and meter stick are vertical, and the angle of the sun is the same for both objects. Therefore, the corresponding angles are congruent, so the triangles are similar and the corresponding sides are proportional.

	Smaller	Larger
Height	p	100
Shadow	30	120

$$\frac{p}{30} = \frac{100}{120}$$

$$120p = 30 \cdot 100$$

$$p = \frac{30 \cdot 100}{120}$$

$$p = 25$$

The shadow of the pole was measured in meters, so the height of the pole is **25 meters**.

Thinking Skill

Explain

Why might Maricruz use indirect measure to measure the flagpole?

In example 1, Maricruz used **indirect measure** to find the height of the flagpole. Instead of placing a measuring tool against the flagpole, she measured the shadow of the flagpole and used proportional relationships to find the height of the flagpole. Here is another example of indirect measure.

Example 2

Josh stood facing a pond. From his viewpoint he could see two points on opposite sides of the pond: P_1 and P_2. He wanted to know the distance between P_1 and P_2 but he did not have a way to measure the distance directly.

Josh had an idea: he could indirectly measure the distance. He marked his position with a large stone, walked 10 yards towards P_1, and drove a stake into the ground. He tied a string around the bottom of the stake and stretched it out in a line parallel to the line from P_1 to P_2. Josh then returned to his starting position and walked in a straight line towards P_2. When Josh reached the string, he drove another stake into the ground and tied off the string.

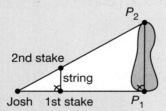

Josh knew that the distance from his starting position to P_1 is 45 yards. If the distance between the first stake and the second stake is 8 yards, the distance from P_1 to P_2 across the pond is how many yards?

Solution

We can divide the map into two triangles. The triangles are similar because the view angle was the same for each triangle and because two other corresponding angles of the triangles are also corresponding angles of parallel lines.

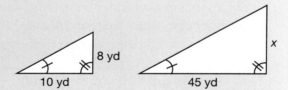

The sides of similar triangles are proportional, so we may write a proportion or use the scale factor to find the distance across the pond. This time we will use the scale factor.

$$\text{Scale factor} = \frac{45 \text{ yd}}{10 \text{ yd}} = 4.5$$

Now we multiply the length of the string (8 yards) by the scale factor to find the distance across the pond from P_1 to P_2.

$$8 \text{ yd} \cdot 4.5 = 36 \text{ yd}$$

The distance across the pond from P_1 to P_2 is about **36 yards**.

Discuss Discuss other situations in which you might use indirect measure.

Practice Set

a. **Evaluate** Trevor took big steps to find the length of the shadow of the big pine tree in front of the school. He estimated that the shadow was 16 yards long, or about 48 feet. He also stepped off the shadow cast by the two-story school building and estimated that its shadow was 18 feet long. Trevor thinks the building is 24 feet tall. About how many feet tall is the big pine tree?

b. Holding a ruler upright at arm's distance (24 in.), Ronnie aligned the bottom of the ruler with a mark on the utility pole that was about 5 feet above the ground. He saw that the top of the pole aligned with the 6-inch mark on the ruler. Then he took 40 long strides to reach the pole. If each stride was about one yard (3 feet), then the top of the pole is about how many feet high?

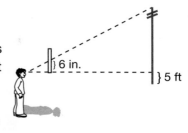

c. Using a ruler or tape measure, perform a height-and-shadow activity at school or at home to estimate the height of a building, tree, or pole you cannot measure directly. Find the shadow length of the object you cannot measure directly on level ground. Then find the shadow length and height of an object that you can measure such as an upright ruler, meter stick of fence post. Create and solve a proportion relating the heights and shadow lengths of the two objects to estimate the height of the object.

Written Practice   *Strengthening Concepts*

1. A recipe calls for 5 cans of stew and 2 cans of chili. To feed the group it (45) will take 210 cans total. How many of the 210 cans should be chili?

2. Eighty-four percent of the audience dozed during the 4-hour film. (48) If there were 75 people in the audience, how many did not doze?

3. The beagle buried 5 big bones and 3 small ones. For how long must the (3, 4) beagle search for the bones if it takes 4 minutes to find each big bone and 10 minutes to find each small one?

4. Five students measured the mass of an object. Their measurements (7) were 7.02 g, 6.98 g, 6.98 g, 6.99 g, and 6.98 g. Find the mean, median, and mode of the measurements.

5. Lines *a* and *b* are parallel. Find $m\angle x$. (54)

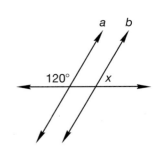

* 6. **Justify** **a.** Sketch a rectangular prism with a 10 mm square base and a
(43, Inv. 4) height of 20 mm.

   b. Find the surface area of the prism in **a.**

   c. Sketch a pyramid with a 10 mm square base and a height of 20 mm.
   Which has the greater volume, the prism or the pyramid? How do
   you know?

7. Use the slope-intercept method to graph $y = -x + 3$. Is $(3, -6)$ a
(56) solution?

8. The Big Prize was behind curtain 1, 2, or 3. Dexter gets two choices.
(32) **a.** What is the sample space of the experiment?

   b. What is the probability he will choose the Big Prize?

* 9. **Connect** A vertical meter stick casts a
(65) shadow 40 cm long at the same time a
flagpole casts a shadow 10 feet long. How
tall is the flagpole?

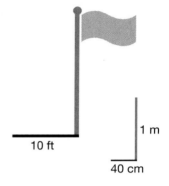

10. **Analyze** Find the perimeter of the figure
(Inv. 2, 39) to the right. Dimensions are in centimeters.
(Use 3.14 for π.)

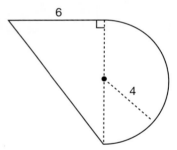

* 11. Use a unit multiplier to convert 90 meters per minute to meters per
(64) second.

* 12. Use a unit multiplier to convert 1.2 feet per second to inches per
(64) second.

13. Factor:
(21)
   a. $4y - 32$ **b.** $2x^2 - 16x$

14. **a.** Write $3\frac{1}{3}\%$ as a reduced fraction and as a decimal.
(63)
   b. Which form would you use to find $3\frac{1}{3}\%$ of $720.00?

15. Write $\frac{7}{16}$ as a decimal and as a percent.
(63)

16. The regular price was $45, but the item was on sale for $27. What
(58) percent of the regular price was the sale price?

Simplify.

17. $3^1 - 2^2 + 1^3$
(15)

18. $\dfrac{x^4y^3z^2}{x^2y^3z^4}$
(27)

19. $\dfrac{7}{12} - \dfrac{3}{8} \cdot \left(\dfrac{9}{4}\right)^{-1}$
(21, 63)

20. $\dfrac{4.2 + 3}{0.6}$
(24, 25)

Solve.

* 21. $0.8 + 0.3x = 5$
(50)

22. $\dfrac{7}{8}x + \dfrac{1}{2} = \dfrac{7}{8}$
(50)

Model Graph the solutions to Exercises **23–24** on a number line.

* 23. $x + 5 > 1$
(62)

* 24. $2x + 1 \leq -1$
(62)

25. Julie's car has a digital fuel gauge and Julie decided to chart the amount
(41) of fuel she burns when she drives certain numbers of miles. Her data is
shown in the table below. Is the relationship proportional? If so, what is
the constant of proportionality? Why do you think the information in the
table looks the way it does?

Gallons Used	Distance Travelled (miles)
1	18
3	80
9	162
15	450

• Special Right Triangles

Power Up | Building Power

facts | Power Up N

mental math

a. Statistics: Find the median of these numbers: 28.5, 32.9, 17.9, 16.1, 27.2

b. Estimation: $4.9 \times \sqrt{50}$

c. Number Sense: If a number is a perfect square, then its last digit cannot be

    A 0.   **B** 1.   **C** 2.

d. Powers/Roots: $x^2 \cdot x^2$

e. Proportion: $\frac{1}{5} = \frac{x}{15}$

f. Geometry: Find the measure of $\angle x$.

g. Scientific Notation: Write 1.9×10^4 in standard notation.

h. Calculation: 8×2, $\sqrt{}$, $\sqrt{}$, -1, $\sqrt{}$, -1, $\sqrt{}$

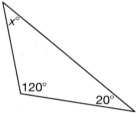

problem solving

The vertices of the larger square are at (0, 0), (2, 0), (2, 2), and (0, 2). The vertices of the smaller square are at (1, 0), (2, 1), (1, 2), and (0, 1). What is the ratio of the area of the smaller square to the area of the larger square?

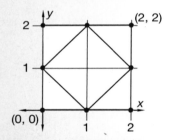

New Concept | Increasing Knowledge

Thinking Skill

Analyze

Can you calculate the measures of the three angles of each triangle? Explain your thinking.

In this lesson we will focus attention on two special right triangles. One triangle is half of a square and the other triangle is half of an equilateral triangle.

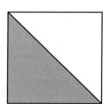 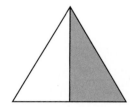

Example 1

Draw the two special right triangles from the beginning of this lesson and on the drawing indicate the measure of each interior angle.

Solution

One of the right triangles has legs of equal length, like two sides of a square. So the triangle is an isosceles triangle.

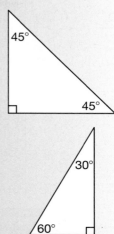

The other right triangle has one leg that measures half the hypotenuse. We begin sketching an equilateral triangle but stop halfway.

We may refer to these special triangles as 45-45-90 triangles and 30-60-90 triangles. Besides knowing the angle measures of these triangles, it is helpful to know their relative side lengths.

Example 2

The length of one leg of each special triangle is given. Find the lengths of the other sides.

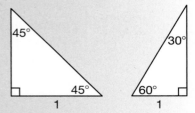

Solution

Since a 45-45-90 triangle is isosceles, we know that the other leg is **1 unit.** Using the Pythagorean Theorem we find that the hypotenuse is $\sqrt{2}$ **units.**

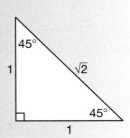

Since a 30-60-90 triangle is half an equilateral triangle, the shortest side is half the hypotenuse. Therefore, the hypotenuse is **2 units.** Using the Pythagorean Theorem we find the remaining side is $\sqrt{3}$ **units.**

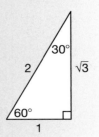

All 45-45-90 triangles have side lengths in the ratio of 1:1:$\sqrt{2}$. All 30-60-90 triangles have side lengths in the ratio of 1:$\sqrt{3}$:2. (Some students remember this ratio as 1, 2, $\sqrt{3}$ noting that the longest side is 2.)

Example 3

The shortest side of a 30-60-90 triangle is 5 inches. What are the lengths of the other sides?

Solution

The side length ratio of all 30-60-90 triangles is 1:$\sqrt{3}$:2. If the shortest side is 5, we simply apply the scale factor 5 to the other two sides. Since 5 times $\sqrt{3}$ is 5$\sqrt{3}$, the other leg is **5$\sqrt{3}$ inches** and the hypotenuse is **10 inches.**

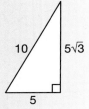

Example 4

A traffic sign in the shape of an equilateral triangle has sides 12 inches long. What is the area of the front of the sign?

Solution

To find the area we multiply the perpendicular base and height. Although we are not given the height, we know that half an equilateral triangle is a 30-60-90 triangle. We can use the side-length ratio 1:$\sqrt{3}$:2 to find that the height is 6$\sqrt{3}$ in.², because 6 times $\sqrt{3}$ is 6$\sqrt{3}$. Now we find the area.

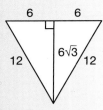

Step:	Justification:
$A = \frac{1}{2}bh$	Area formula for a triangle
$A = \frac{1}{2}(12 \text{ in.})(6\sqrt{3} \text{ in.})$	Substituted
$A = \mathbf{36\sqrt{3} \text{ in.}^2}$	Simplified

To estimate measures involving $\sqrt{2}$ or $\sqrt{3}$, it is helpful to remember these approximations.

$$\sqrt{2} \approx 1.41$$
$$\sqrt{3} \approx 1.73$$

Example 5

A major league baseball diamond is a square with sides 90 feet long. If a base runner tries to "steal" second base, the catcher throws the ball from home plate to second base. How far does the catcher throw the ball?

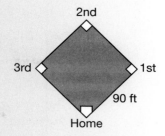

Solution

The throw from home plate to second base is a diagonal of the square, which is the hypotenuse of a 45-45-90 triangle. Since the legs are 90 feet long, the hypotenuse is $90\sqrt{2}$ feet. So the catcher throws the ball **$90\sqrt{2}$ ft** or about **127 ft.**

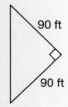

$$90\sqrt{2} \approx 90 \cdot 1.41 \approx 126.9$$

Practice Set

a. Sketch a 45-45-90 triangle and indicate the angle measures and side lengths if one leg measures 1 inch.

b. Sketch a 30-60-90 triangle and indicate the angle measures and side lengths if the shortest side is 1 inch.

c. What are the approximations for $\sqrt{2}$ and $\sqrt{3}$?

d. How many units is it from the origin to (2, 2) on the coordinate plane?

e. If each side of an equilateral triangle is 2 inches long, then what is the area of the triangle?

f. A regular hexagon has sides 2 feet long. What is the exact area of the hexagon? What is the approximate area of the hexagon?

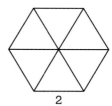

Written Practice   *Strengthening Concepts*

1. **The multipacks of dried fruit contain three bags of banana chips and five bags of raisins. Several multipacks were purchased. If there were 104 bags in all, how many bags of banana chips were there?**
(45)

2. **Sixty percent of the students preferred raisins to banana chips. If there were 95 students in all, how many preferred banana chips?**
(58)

3. **The net weight of a bag of banana chips is 0.9 oz. A bag of raisins has a net weight of 1.1 oz. What is the net weight of a multipack that contains 3 bags of banana chips and 5 bags of raisins?**
(3, 4)

4. **a.** Find the mean, median, mode and range of these measurements:
(53)   2.5 cm, 2.6 cm, 2.5 cm, 2.4 cm, 2.6 cm.

b. Display the data in a line plot.

5. Two of the lines are parallel. Find *x*.
(54)

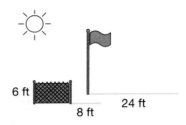

* 6. **Evaluate** A flagpole casts a shadow
(65)   24 feet long while a six-foot fence pole
casts a shadow 8 feet long. Find the height
of the flagpole.

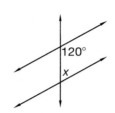

* 7. Write the equation of the line with slope $\frac{1}{2}$ that intercepts the
(56)   *y*-axis at 3.

* 8. **Generalize** Simplify and express using positive exponents.
(51)
$$x^{-2}y^{-1}xy^2z$$

Solve.

9. $\frac{4}{3}x = 28$
(38)

10. $\frac{x^2}{9} = 4$
(14, 36)

11. Find the area of the figure to the right.
(37, 40)   (Use $\pi = 3.14$. Dimensions are mm.)

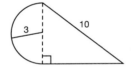

* 12. Write $16\frac{2}{3}\%$ **a** as a decimal and **b** as a reduced fraction.
(63)

* 13. **a.** Write $\frac{2}{3}$ as a decimal and percent.
(63)

b. Select one of these forms to state, "2 out of 3 band members are
also in choir."

* 14. **Evaluate** Evaluate $-b - \sqrt{b^2 - 4ac}$ when $a = 1$, $b = -10$,
(15, 36)   and $c = 9$.

* 15. Blair sees a highway sign that reads "speed limit 100 km/hr." Use a
(64)   unit multiplier to convert 100 km/hr to miles per hour. (1 mi $\approx$ 1.6 km).

* 16. Nora earns $12 per hour. Convert $12 per hour to cents per minute.
(64)

17. Use the Distributive Property to
(21)
a. Factor: $15x^2 - 10x$ **b.** Expand: $7(x - 3)$.

Simplify.

18. $(-3)^2 - 3^2 - 3^0$
(27)

19. $\dfrac{z^3 b^4 r}{x^2 b^2 z^3}$
(27)

20. $\dfrac{9}{16} + \dfrac{1}{16} \cdot \left(\dfrac{2}{3}\right)^{-1}$
(21, 63)

21. $(-6) - (-2) - (-6)(-2)$
(33, 36)

Solve. If there is no solution, write *no solution*.

22. $0.9 + 0.3x = 2.4$
(50)

23. $3(-2x + 1) = 21$
(50)

* 24. **Model** Sketch a square and sketch an equilateral triangle. Draw a
(66) diagonal of the square to divide the square into two congruent right triangles. Then draw a segment in the equilateral triangle that divides it into two congruent right triangles. In all four right triangles, write each angle measure.

* 25. **Formulate** The eighth grade science class noticed that a faucet in the
(41) science lab was leaking. They gathered the following data using a watch with a second hand and a 100-milliliter beaker to measure the amount of water that leaked during a specific time.

Time (Seconds)	10	20	30	40	50	60	70
Water Volume (ml)	8	16	24	32	40	48	56

Is the relationship between elapsed time and water volume proportional? Explain your answer. If the faucet continues to leak at the same rate, how many seconds will it take to fill the 100-milliliter beaker? Write and solve a proportion.

• Percent of Change

facts   Power Up N

mental math

a. Algebra: Evaluate: $\frac{1}{2}(10)(v^2)$ when $v = 3$

b. Estimation: $\sqrt{17} + \sqrt{24}$

c. Fractional Parts: $\frac{2}{3}$ of 99

d. Percent: 25% of $60

e. Rate: How long will it take to read 500 pages at 25 pages per day?

f. Measurement: Find the mass of 7 toy cars.

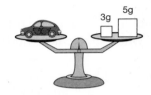

g. Geometry: A cube has how many vertices?

h. Calculation: How much money is half of a tenth of a dollar?

problem solving   Kurt rides his bike 18 miles per hour south. Andrew starts at the same place and rides 17 miles per hour north. If they keep on riding at these rates, about how long will it take them to be 210 miles apart?

We have solved percent problems about part of a whole. Percents are also used to describe a change. A change may be a decrease or an increase.

• The dress was on sale for 30% off the regular price.

• The median price of a house in the county rose 20% last year.

We can use a ratio table to sort the numbers and help us write a proportion. We make three rows, the first for the original number, the second for the change, and the third for the new number. Here is the model.

	%	Actual count
Original	100	
Change (±)		
New		

* 5. ₍₆₇₎ **Analyze** The airline advertised lower fares. The new fare for flying to Austin is $320. If this fare is 20% lower than the original fare, what was the original fare?

6. _(Inv. 2, 35) The two triangles are similar.

    a. Find the lengths x and y.

    b. Are the triangles right triangles? How do you know?

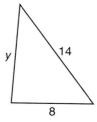

* 7. ₍₅₆₎ Write an equation for a line that has a slope of $\frac{2}{3}$ and passes through the origin.

* 8. ₍₅₆₎ **Represent** Gwen drew a line with a slope of −3 passing through the point (0, −8). Write an equation for the line Gwen drew in slope-intercept form.

9. ₍₄₁₎ Graph the line $y = -2x$. Is the point (−4, 2) on the line?

10. _(Inv. 3) True or false: All parallelograms have two pairs of congruent angles.

11. _(Inv. 2, 39) Find the perimeter of the figure to the right. Round to the nearest meter. (Dimensions are in m. Figure is not to scale.)

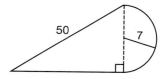

12. ₍₄₈₎ The bank offered a 0.6% interest rate. Find 0.6% of $1300.

* 13. ₍₆₃₎ Write $\frac{5}{9}$ as a **a** decimal and **b** percent.

    c Order these numbers from least to greatest: 0.55, 0.56, $\frac{5}{9}$.

14. _(Inv. 1, 20) The vertices of a triangle are (−6, 1), (6, 1), and (6, 6). Find **a** the perimeter and **b** the area of the triangle.

15. ₍₂₁₎ Use the Distributive Property:

    a. Factor $5x^2 + 5x + 10$

    b. Expand $-3(x + 3)$

16. ₍₃₂₎ Jenny has a deck of 52 alphabet cards (26 uppercase and 26 lower case). Jenny selects one card.

    a. What is the probability that she selects a vowel?

    b. If on the first draw Jenny selects a vowel and does not replace it, how many cards remain? How many are vowels?

    c. Supposing Jenny selected a vowel on her first draw, what is the probability she selects a vowel on her second draw?

* 17. Ryun runs in a race at a rate of 400 meters per minute. Use a unit
(64) multiplier to convert 400 meters per minute to miles per minute.
(1 mi ≈ 1600 m.)

Simplify.

18. a. $(-2)^2 - 4(3) + \left(\sqrt{11}\right)^2$
(15)
   b. $\dfrac{m^2b^4r^3}{r^4b^4m}$
   (27)

19. $\dfrac{5}{12} - \dfrac{3}{4} \cdot \left(\dfrac{9}{5}\right)^{-1}$
(22, 63)

20. Find x, y, and z.
(54)

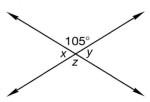

21. A coin is flipped and the spinner is spun.
(32)
   a. What is the sample space of the
      experiment?

   b. What is the probability of tails and a
      consonant (B or C)?

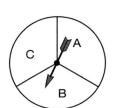

Solve.

22. $5(x - 6) = 40$ **23.** $8x + 3x - 2 = 75$
(50)   (50)

24. $0.007x + 0.28 = 0.7$
(50)

* 25. The purity of gold is described in *karats*.
(41, 48) The graph at right shows the relationship
      between the number of karats a piece of
      gold has and its purity. Is the relationship
      proportional? How many karats is
      100% pure gold? Each karat represents
      what fraction of pure gold? Determine
      the purity of a piece of 14 karat gold
      jewelry.

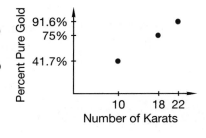

• Probability Multiplication Rule

Power Up

Building Power

facts

Power Up N

mental math

a. **Algebra:** Evaluate: $\sqrt{a^2 + 4^2}$ when $a = 3$

b. **Number Sense:** Compare $\frac{1}{5} + \frac{2}{5} \bigcirc \frac{1}{5} \times \frac{2}{5}$

c. **Powers/Roots:** $x^7 \cdot x$

d. **Probability:** Jackson flipped a coin 20 times and the outcome was tails twelve times. What is the probability that the coin lands tails-up on the next flip?

e. **Ratio:** Every three students shared two calculators. If there are 30 students in the class how many calculators are there?

f. **Geometry:** Find the measure of $\angle m$.

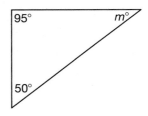

g. **Measurement:** How many inches is 5 ft 2 in.?

h. **Calculation:** How much money is one fifth of one fourth of a dollar?

problem solving

Theo is thinking of two numbers. He gives these two hints: the sum of the numbers is 6, and one number is ten more than the other number. What are the numbers?

New Concept

Increasing Knowledge

When an experiment occurs in stages or has more than one part, we can use a tree diagram to help us find the sample space and count outcomes.

In the experiment of flipping a coin twice (or flipping two coins), there are four outcomes.

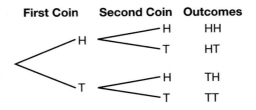

In the first flip, there are two possible outcomes. For each of these outcomes there are two possible outcomes in the second flip. Therefore the total number of outcomes is the product $2 \cdot 2 = 4$.

This experiment illustrates the fundamental counting principle.

Fundamental Counting Principle

If an experiment has two parts, the first part with m possible outcomes and the second part with n possible outcomes, then the total number of possible outcomes for the experiment is the product $m \cdot n$.

Example 1

How many possible outcomes are there for an experiment in which two number cubes are rolled together?

Solution

The first roll has 6 possible outcomes, and the second roll has 6 possible outcomes. Altogether, there are $6 \cdot 6$, or **36 possible outcomes.**

Represent Use a tree diagram to illustrate the number of outcomes when two number cubes are rolled together.

A tree diagram can be used to display and count outcomes of experiments with multiple stages. If an experiment has only two stages, we can use a rectangular table or area model. For example, we can display the sample space of rolling two number cubes this way.

First Cube

	1	2	3	4	5	6
1	(1, 1)	(1, 2)	(1, 3)	(1, 4)	(1, 5)	(1, 6)
2	(2, 1)	(2, 2)	(2, 3)	(2, 4)	(2, 5)	(2, 6)
3	(3, 1)	(3, 2)	(3, 3)	(3, 4)	(3, 5)	(3, 6)
4	(4, 1)	(4, 2)	(4, 3)	(4, 4)	(4, 5)	(4, 6)
5	(5, 1)	(5, 2)	(5, 3)	(5, 4)	(5, 5)	(5, 6)
6	(6, 1)	(6, 2)	(6, 3)	(6, 4)	(6, 5)	(6, 6)

(Second Cube, left axis labels 1–6)

We see that the 6 by 6 rectangle has $6 \cdot 6$ possible outcomes, even though there are only 11 different sums. This sample space display is preferable to listing the 11 different outcomes because the table shows equally-likely outcomes. If we were to simply list the outcomes from 2 to 12, we would not be able to tell which outcomes are more/less likely.

The multiplication counting principle also helps us count ways that an event can occur.

Example 2

Two number cubes are rolled once. How many possible outcomes are there for which the first roll is even and the second roll is even?

Solution

An even outcome (2, 4, or 6) can occur in 3 ways for each number cube, so the total number of possible outcomes for which both outcomes are even is $3 \cdot 3$, which is **9.**

Connect If two number cubes are rolled, what is the probability of rolling an even number on each cube?

To find probabilities of events in experiments with multiple parts, we can count outcomes by multiplying:

$$P(\text{even and even}) = \frac{3 \cdot 3}{6 \cdot 6} = \frac{1}{4}$$

Math Language

In an experiment, two events are **independent** if the occurrence of one does not change the probability of the other.

We can also consider the probability of each part: For the first roll, $\frac{1}{2}$ of the outcomes are even, and for each of these, $\frac{1}{2}$ of the second roll are even. In all, $\frac{1}{4}$ (which is $\frac{1}{2}$ of $\frac{1}{2}$) of the outcomes are two even numbers. That is,

$$P(\text{even and even}) = \frac{1}{2} \cdot \frac{1}{2} = \frac{1}{4}$$

This is an example of a multiplication rule for probability.

Multiplication Rule for Probability

If events A and B are independent, then
$$P(A \text{ and } B) = P(A) \cdot P(B)$$

Example 3

Suppose a bag contains 2 red and 3 blue marbles. Robert selects a marble from the bag, looks at it, then replaces it. Then he selects a second marble from the bag. What is the probability he selects a red marble and then a blue marble?

Solution

$$P(\text{red and blue}) = P(\text{red}) \cdot P(\text{blue})$$
$$= \frac{2}{5} \cdot \frac{3}{5}$$
$$= \frac{6}{25}$$

Verify What is the probability Robert selects a red and blue marble in either order? Calculate the probability then verify the results by drawing a 5 by 5 table.

Example 4

Maribel holds a stack of 52 alphabet cards: 26 uppercase cards (A, B, C, ... Z) and 26 lowercase cards (a, b, c, ... z). She shuffles the stack, and selects one card. What is the probability that it is an uppercase vowel?

Solution

There are five uppercase vowel cards (A, E, I, O, U) in the 52 cards.

$$P(\text{uppercase vowel}) = \frac{5}{52}$$

We can also find this probability by considering the independent events "uppercase" and "vowel". (Even if the letter on the card is uppercase, the probability it is a vowel remains $\frac{5}{26}$.) Since one-half of the cards are uppercase cards, and there are 10 total vowel cards, the probability can be computed:

$$P(\text{uppercase vowel}) = P(\text{uppercase}) \cdot P(\text{vowel})$$

$$P(\text{uppercase vowel}) = \frac{1}{2} \cdot \frac{5}{26}$$

$$P(\text{uppercase vowel}) = \frac{5}{52}$$

Practice Set

In a certain experiment, a coin is tossed and this spinner is spun. Compute the probability of the events in **a** and **b**.

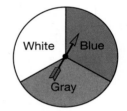

White Blue

Gray

   a. *P*(heads and blue)

   b. *P*(heads and not blue)

   c. Two number cubes are rolled. What is the probability of rolling a number greater than four on each number cube?

   d. A coin will be flipped four times. What is the probability the coin will land heads-up all four times?

Written Practice   *Strengthening Concepts*

   1. In Mateo's aquarium, silver fish outnumber red fish nine to two. If there
   (45)   were twenty-two silver and red fish in all, how many were silver?

   2. Fifteen percent of the teachers wore ties. If 51 teachers did not wear
   (48)   a tie, how many teachers were there in all?

   3. Kurt collected quarters while Robert collected dimes. Together they had
   (3, 4)   30 coins. If ten were quarters, how much money did they have?

* 4. **Analyze** By the middle of the ninth inning, 9% of the fans had left the
(58) stadium. If 18,200 fans remained, how many had left?

5. Imelda held a 6-inch long pencil at arm's length so that it matched the
(65) height of a tree 100 feet away. If the pencil was 24 inches from her eyes,
about how tall was the tree?

6. Mr. Martinez parks at a lot where he is charged a fee as shown in the
(41) table.

Months	1	2	3	4	5
Fee ($)	17	34	51	68	85

If he parks at the lot for 9 months, what will he be charged?

7. If a number cube is rolled once, which is the more likely outcome, a
(32) prime number or a composite number?

8. Graph $y = \frac{2}{5}x - 1$. Is the point (10, 4) on the line?
(41)

* 9. **Model** Two points on the graph of a linear equation are $(-2, 3)$ and
(56) (1, 0). Graph the points and draw a line through the points. Then write
the equation of the line.

10. Find the area of the figure to the right. Round
(Inv. 2, to the nearest square inch. (Units are in
40) inches.)

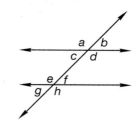

* 11. **Evaluate** Convert $\frac{7}{9}$ to **a** a percent and **b** a decimal, then **c** write these
(63) numbers from least to greatest:

$$\frac{6}{8}, \frac{7}{8}, \frac{7}{9}$$

12. Which two angles are adjacent to $\angle f$?
(54)

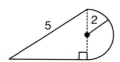

13. Use the Distributive Property to **a** factor: $3x^2 + 3x + 9$, **b** expand
(21) $x(x - 5)$.

* 14. Use a unit multiplier to convert 18 inches per minute to inches per
(64) hour.

* 15. Songhee leisurely rides her bicycle at 10 miles per hour. In about 675
(64) revolutions of the wheels, Songhee travels one mile. Use a unit multiplier
to convert 10 miles per hour to revolutions per hour.

Simplify.

16.  $\dfrac{2m^2ac^3}{ca^2m^2}$
(27)

17. $(-3)(-4) - (-5)(6)$
(36)

Solve.

* **18.** $\dfrac{3}{5}x + \dfrac{7}{10} = \dfrac{19}{10}$
(50)

* **19.** $\dfrac{1}{4} + \dfrac{3}{8}x = 1$
(50)

20. $0.003x - 0.02 = 0.07$
(50)

21. $0.03x - 0.02 = 0.07$
(50)

22. $5x + 2x - 3 = 18$
(50)

Graph the solutions on a number line.

23. $x - 3 < 1$
(62)

24. $3x - 2 \geq 4$
(62)

25. If $\triangle ABC$ is reflected across the
(Inv. 5) line $x = 3$, which of the following will be the coordinates of the image of point A (A')?

A $(1, -1)$ **B** $(2, -1)$

C $(5, -1)$ **D** $(8, -1)$

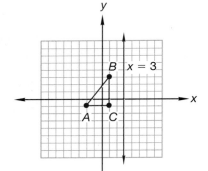

Early Finishers
Real-World Application

The United States Census Bureau provides data about the nation's people and economy. The 2004 U.S. Census reported 1,108 of 4,000 adults aged 25 years and older have a bachelor's degree or higher level of education. What is the chance that the next adult you meet 25-years-old or older has a bachelor's degree or higher level of education?

• Direct Variation

facts   Power Up N

mental math

a. **Statistics:** Find the mode of this data set: 12.5, 13.1, 12.5, 13.0, 12.9

b. **Estimation:** $5\sqrt{17}$

c. **Sequences:** Find the next three terms: $\frac{1}{4}, \frac{1}{2}, \frac{3}{4}, 1, \ldots$

d. **Proportion:** $\frac{3}{7} = \frac{21}{x}$

e. **Measurement:** Find the volume of liquid in this container.

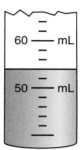

f. **Percent:** 20% of $50

g. **Scientific Notation:** Write 7×10^{-5} in standard notation.

h. **Calculation:** Square 10, $- 1$, $\div 11$, square it, $- 1$, $\div 10$, square it, $- 1$, $\div 9$

problem solving   Some pulsars spin at a rate of 30,000 rotations per minute (rpm). How many rotations per second is that? How long does it take for each rotation?

Math Language
Recall that a **constant** is a number whose value does not change.

Recall that when two variables are proportional the value of one variable can be found by multiplying the other by a constant factor. We call this relationship between the variables **direct variation.**

Direct Variation

$$y = kx$$

In this equation x and y are the variables, and k is the constant multiplier, called the **constant of proportionality.** The constant of proportionality might be a unit rate, a scale factor, a unit multiplier, or an unchanging number that relates the two variables.

Here are some examples of direct variation. Can you find the variables and the constant of proportionality in each example?

• total pay = $12 $\times$ hours worked

• miles traveled = 60 mph $\times$ hours traveled

• perimeter = 4 $\times$ side length

Characteristics of Direct Variation

> **1.** When one variable is zero, the other is zero (the graph of the equation intersects the origin).
>
> **2.** As one variable changes, the other variable changes by a constant factor (the graph is a line).

We can solve the equation $y = kx$ for k by dividing both sides of the equation by x.

$$y = kx \qquad \text{direct variation equation}$$

$$\frac{y}{x} = k \qquad \text{divided by } x$$

This alternate equation shows that the ratio of the two variables in direct variation equals the constant of proportionality. We use the word proportionality because quantities that vary directly are proportional—their ratio is constant. Knowing this we can determine from a table if a relationship is an example of direct variation.

Visit www.SaxonPublishers.com/ActivitiesC3 *for a graphing calculator activity.*

Example 1

Which of the following is an example of direct variation? (The variables are underlined.)

A A taxi company charges three dollars to start the ride plus two <u>dollars</u> per <u>mile</u>.

B The <u>area</u> of a square is the square of the <u>length of its side</u>.

C The <u>perimeter</u> of a square is four times the <u>length of its side</u>.

Solution

We make a table for the three relationships and check for a constant ratio.

$D = 2m + 3$

Miles	Dollars	$\frac{\text{Dollars}}{\text{mile}}$
1	5	$\frac{5}{1}$
2	7	$\frac{7}{2}$
3	9	$\frac{9}{3}$
4	11	$\frac{11}{4}$

$A = s^2$

s	A	$\frac{A}{s}$
1	1	$\frac{1}{1}$
2	4	$\frac{4}{2}$
3	9	$\frac{9}{3}$
4	16	$\frac{16}{4}$

$P = 4s$

s	P	$\frac{P}{s}$
1	4	$\frac{4}{1}$
2	8	$\frac{8}{2}$
3	12	$\frac{12}{3}$
4	16	$\frac{16}{4}$

Of these three, the only example of direct variation is **the relationship between the perimeter of a square and its side length.** In every number pair, the ratio reduces to 4. The relationship is proportional. In the other examples, the ratios vary. The relationships are not proportional.

We can also identify direct variation from a graph. The three relationships from example 1 are graphed below.

A Taxi

$$D = 2m + 3$$

Cost ($) vs Distance (mi)

B Area

$$A = s^2$$

Area vs Side Length

C Perimeter

$$P = 4s$$

Perimeter vs Side Length

Thinking Skill

Analyze

How do $A = s^2$ and $D = 2m + 3$ differ from $y = kx$?

A graph indicates a proportional relationship only if all the points fall along a straight line that intersects the origin. Graph **B** is not linear. Graphs **A** and **C** are linear, but graph **A** does not align with the origin.

Characteristics of Graphs of Direct Variation

Quantities which vary directly have pairs of points which lie on the same line, and the line intersects the origin.

Connect Look at the perimeter graph and follow the line. How many units does the line rise for each unit it moves to the right?

Analyze Are paired quantities whose points are aligned always proportional?

In a direct variation relationship, the value of one variable *depends* on the value of the other. One variable is **independent,** like the number of hours a person works, and the other is **dependent,** like the total pay the person earns. In a table, the independent variable is typically placed in the first column and the dependent variable in the second column. On a graph, the independent variable is typically plotted on the horizontal axis and the dependent variable on the vertical axis. The constant of proportionality is the ratio of the dependent variable to the independent variable.

Example 2

Doubleday doubled every number Eunice said. The table shows four numbers Eunice said and Doubleday's replies. Sketch a graph of all the numbers Eunice and Doubleday could say. Is the relationship of numbers they say an example of direct variation? Write an equation for the relationship. What is the constant of proportionality?

E	D
2	4
1	2
−1	−2
−2	−4

Eunice and Doubleday could go on endlessly saying numbers. The graphed line shows all the numbers they could say that fit their rule. The arrowheads show that the possible pairs of numbers extend without end.

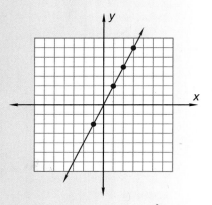

This relationship is an example of direct variation. We see that the possible pairs of numbers form a straight line that passes through the origin. Any two pairs of numbers from the relationship form a proportion.

An equation for the relationship is

$$d = 2e$$

in which e is the number Eunice says and d is the number Doubleday says. The constant of proportionality is **2** because every value of d is 2 times the given value of e. Every d-to-e ratio equals 2.

e	d	$\frac{d}{e}$
2	4	$\frac{4}{2} = 2$
1	2	$\frac{2}{1} = 2$
-1	-2	$\frac{-2}{-1} = 2$
-2	-4	$\frac{-4}{-2} = 2$

Example 3

Jarrod earns \$12 an hour helping a painter. Create a table that shows his pay for 1, 2, 3, and 4 hours of work. Is Jarrod's pay for hours worked an example of direct variation? If so, what is the constant of proportionality?

Solution

We make a table for hours worked and pay. We calculate the ratio of pay to hours worked for each row of our table.

The ratios are equal because they all reduce to $\frac{12}{1}$. The relationship **is an example of direct variation,** and the constant of proportionality is **12.**

Hours	Pay	$\frac{P}{H}$
1	12	$\frac{12}{1}$
2	24	$\frac{24}{2}$
3	36	$\frac{36}{3}$
4	48	$\frac{48}{4}$

Represent Write an equation for the relationship in example 3.

Practice Set

a. Trinny drew some equilateral triangles and measured their side lengths and perimeters. She decided that the perimeter of an equilateral triangle varies directly with its side length. Write an equation for the relationship and identify the constant of proportionality.

s	p
2	6
3	9
4	12
7	21

p = perimeter
s = side length

b. (Analyze) The relationship between Sergio's age and Fernando's age is illustrated in the graph. Are their ages proportional? Is the relationship an example of direct variation? Why or why not?

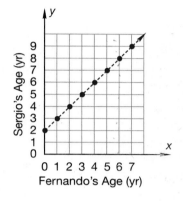

c. (Formulate) The weight of water in a trough is directly proportional to the quantity of water in the trough. If 20 gallons of water weigh 166 pounds, what is the weight of 30 gallons of water? Find the constant of proportionality in pounds per gallon. Then write an equation for the relationship using p for pounds and g for gallons. Start the equation $p =$.

Written Practice   *Strengthening Concepts*

1. The number of home-team fans in a crowd of 980 fans was 630. What
(45) was the ratio of home-team to visiting-team fans?

2. Lani recognized thirty-two percent of the melodies. If she heard
(48) 25 melodies in all, how many did she not recognize?

* 3. The soccer league had 30% more participants this year than last year.
(67) If there are 1170 participants this year, how many participated last year?

* 4. (Analyze) Does the table to the right show
(69) direct variation? If so, state the constant of variation.

x	2	18	30
y	5	45	80

5. Find the lengths of x and y in the similar
(35) triangles to the right.

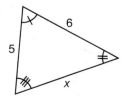

* 6. **Classify** Specifically describe the two triangles in problem **5** as a
(26)   transformation.

7. In the figure, m∠ADC = 60°. Segment CD = 1 in. $m\overline{BC} = m\overline{CD}$
(66)
   a. Find m∠DAC.

   b. Find m∠DBC.

   c. Find DB.

   d. Find DA.

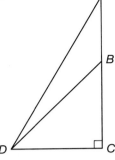

8. Write the equation of the line with slope $\frac{1}{4}$ that passes through the point
(56)   (0, −1).

9. Graph the points (−2, −4) and (6, 0) and draw a line through the points.
(56)   Then write the equation of the line.

10. An atrium was built in the shape shown to
(37)   the right. Find the **a** area and **b** perimeter of
   the atrium. Round to the nearest whole unit.
   (Dimensions are in m.)

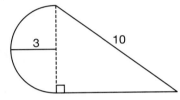

11. Write $\frac{4}{9}$ as **a** a decimal and **b** a percent. **c** Write these three numbers in
(63)   order from least to greatest: 0.4, 0.5, $\frac{4}{9}$.

12. A coin is tossed three times.
(32)
   a. Find the sample space of the experiment.

   b. What is the probability of "at least two heads?"

   c. Name a different event with the same probability as "at least two
   heads."

13. Use the Distributive Property to **a** factor: $5x^2 + 10x + 15$, **b** expand:
(21)   −4(2x + 3).

14. The water leaks from the faucet at a rate of $\frac{1}{2}$ gal/hr.
(69)
   a. Create a table to show how many gallons of water would leak in
   1, 2, 3, and 4 hours.

   b. Is the relationship between the number of hours that elapse and the
   amount of water that leaks an example of direct variation? Explain.

15. Use a unit multiplier to convert the rate described in Exercise **14** to
(64)   gallons per day.

Simplify.

16. $\dfrac{8x^3yz^2}{4xyz^2}$
(27)

17. $\left(-\dfrac{2}{3}\right)^2 \cdot \left(\dfrac{2}{3}\right)^{-1}$
(36, 63)

18. Two of these lines are parallel.
(54)

    a. Compare: $m\angle x \bigcirc m\angle a$.

    b. If it is uncertain whether or not the lines are parallel, can the comparison be made? Explain.

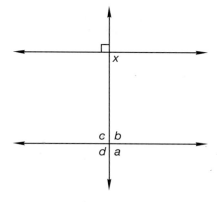

Solve.

19. $\dfrac{3}{8}x + \dfrac{3}{4} = \dfrac{9}{8}$
(50)

20. $\dfrac{1}{3}x - \dfrac{3}{4} = \dfrac{5}{12}$
(50)

21. $0.007x + 0.03 = 0.1$
(50)

22. $\dfrac{x}{5} = \dfrac{1}{2}$
(44)

23. $\dfrac{2}{3} = \dfrac{5}{x}$
(44)

24. Use a unit multiplier to convert 120 miles per hour to miles per minute.
(64)

* 25. *Conclude* The table shows the altitude of a hot-air balloon during the first 5 minutes and 45 seconds of flight. Does the table indicate a proportional relationship? Explain. How can you determine the balloon's altitude 5 minutes into the flight?
(41)

Hot-Air Balloon Altitude

Time (Minutes)	Altitude (Feet)
0	0
1	200
2	400
2.5	500
3	600
4	800
5.75	1150

• Solving Direct Variation Problems

facts Power Up N

mental math

a. Algebra: Evaluate: $\sqrt{6^2 + b^2}$ when $b = 8$

b. Estimation: $2\sqrt{26}$

c. Number Sense: Compare $0.9 \times 0.8 \bigcirc 0.9 + 0.8$

d. Powers/Roots: $x \cdot x \cdot x^2$

e. Rate: Hector sails north at 20 miles per hour. Skip sails south from the same starting location at 25 miles per hour. How far apart are they after 2 hours?

f. Geometry: The triangles are congruent. Use correct symbols to complete this congruence statement: $\triangle WXY \cong \triangle B\_\_$

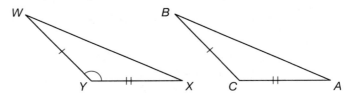

g. Select a Method: To calculate the area of a photo that is $2\frac{1}{2}$ inches long and $1\frac{1}{2}$ inches wide, which would you use?
A mental math
B pencil and paper
C calculator

h. Calculation: Half of half of a dollar is how much more than half of a tenth of a dollar?

problem solving

Find the missing square root. (Hint: What number should be in the tens place of the answer? Since 1521 has a 1 in the ones place, what options do you have for the ones place of the answer?)

$$\sqrt{1600} = 40$$
$$\sqrt{900} = 30$$
$$\sqrt{1521} = \_\_$$

Recall that quantities which vary directly are proportional. The table below shows that the total price of CDs varies directly with the number of CDs a customer purchases.

Number of CDs	Price ($)
1	15
2	30
3	45
4	60

This means their ratio is constant. We often use the letter k for the constant of proportionality.

$$\frac{\text{price}}{\text{number}} = k$$

Every price/number ratio in the table equals the constant, which is 15.

$$\frac{15}{1} = \frac{30}{2} = \frac{45}{3} = \frac{60}{4} = 15$$

One quantity (the price) is determined by multiplying the other variable (the number of CDs) by the constant.

$$\text{price} = 15 \times \text{number}$$

Predict How much will it cost to purchase 6 CDs?

When you know that two quantities vary directly, knowing just one set of paired numbers allows us to solve for missing numbers in other pairs.

Example 1

The amount Ellen charges for banquet food varies directly as the number of people that attend. If Ellen charges $780 for 60 people, how much does she charge for 100 people?

Solution

We know that the ratio of the amount charged to the number of people is constant:

$$k = \frac{\text{charge}}{\text{number}}$$

$$k = \frac{\$780}{60}$$

$$k = \$13$$

The amount Ellen charges is $13 times the number of people who attend:

$$\text{charge} = \$13 \times \text{number}$$

When 100 people attend, Ellen charges **$1300.**

$$\text{charge} = \$13 \times 100 = \$1300$$

An alternate method for solving this problem is to write a proportion. In one case we know the charge and number, $780 and 60. In the second case, the number is known and the charge is unknown. For each case we write a ratio. We know the ratios are equal because we are told the quantities vary directly.

$$\frac{\text{charge}}{\text{number}} \quad \frac{\$780}{60} = \frac{x}{100}$$

Now we can solve using cross products.

$$\$780(100) = 60x$$
$$\$78000 = 60x$$
$$\$1300 = x$$

Example 2

Driving at a constant speed, the distance Maggie travels varies directly with the amount of time she drives. If Maggie drives 75 miles in one and one-half hours, how long does it take her to drive 100 miles?

Solution

Maggie's driving speed is constant. Using k for the constant speed, we write two equivalent equations.

$$k = \frac{\text{distance}}{\text{time}} \qquad \text{distance} = k \times \text{time}$$

The first equation gives us the constant of proportionality, in this situation, Maggie's speed. We express her driving time as 1.5 hours.

$$k = \frac{75}{1.5}$$
$$k = 50 \text{ mph}$$

We have found her speed. We use the second equation to solve for the driving time to drive 100 miles:

$$\text{distance} = 50 \times \text{time}$$
$$d = 50t$$
$$100 = 50t$$
$$2 = t$$

Thinking Skill

Connect

Write a proportion to solve example 2.

It takes Maggie **2 hours** to drive 100 miles.

Example 3

There were 23 people in line in front of Delia waiting to be helped. Three minutes later there were 18 people waiting in front of Delia. Predict how much longer Delia will wait to be helped.

In three minutes, five people (23 − 18 = 5) were helped, so we can find the average length of time it took to help each person.

$$\frac{3 \text{ min}}{5 \text{ people}} = 0.6 \text{ min/person}$$

There is no certainty that the next 18 people will be helped at the same rate, but Delia can use this number to predict how much longer she will wait (*w*) in line.

$$w = 0.6 \text{ min/person} \times 18 \text{ people}$$
$$= 10.8 \text{ min}$$
$$\approx \textbf{11 min}$$

We estimate that Delia will need to wait about 11 more minutes to be helped.

Practice Set

a. The distance a coiled spring stretches varies directly with the amount of weight hanging on the spring. If a spring stretches 2 cm when a 30 gram weight is hung from it, how far will it stretch when a 75 gram weight is hung from the spring?

b. The number of words Tom types varies directly with the amount of time he spends typing. If Tom types 100 words in 2.5 minutes, how many words does he type in 15 minutes?

c. Traffic was dense for the evening commute. Mitchell traveled three miles on the highway in 15 minutes. Predict how much longer it will take Mitchell to reach his exit five miles away.

Written Practice   *Strengthening Concepts*

1. Insects outnumbered humans 1001 to 2 at the campsite. If there were
(45)   10 humans at the campsite, how many insects were there?

2. Sixty-four percent of the coins in the money box were foreign currency.
(48)   If 45 coins were not foreign currency, how many coins were in the money box?

* 3. Robert bought a component for his computer at 15% off. If the sale
(67)   price was $42.50, what was the price before the reduction?

* 4. Does the table to the right show direct
(69)   variation? If so, give the constant of variation.
       If not, explain why.

L	W
6	4
10	8
14	12

5. Find the lengths of x and y in the similar triangles to the right. What is the scale factor from the larger to smaller triangle?
(35)

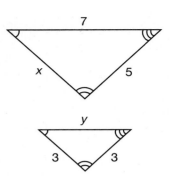

6. Graph $y = \frac{1}{3}x - 1$. Is point (6, 1) a solution?
(56)

* 7. **Analyze** Dale plans to build a balcony similar to the figure. Find the **a** area and **b** perimeter of the balcony. Round to the nearest whole unit. (Dimensions are in feet.)
(37)

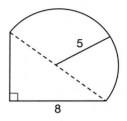

* 8. Write $\frac{1}{6}$ as **a** a decimal and **b** a percent.
(63)

Model Graph each solution on a number line.

* 9. $x - 3 \geq -1$
(62)

* 10. $2x + 3 < 1$
(62)

Simplify.

11. $\dfrac{12w^2x^2y^3}{8x^2y^2}$
(27)

12. $\dfrac{4}{5} - \dfrac{3}{5} \cdot \left(\dfrac{1}{3}\right)^{-1}$
(51)

13. $3x + 3 - x - 3$
(31)

14. $0.2 + 0.3(0.4)$
(24, 25)

Solve.

* 15. $\dfrac{1}{3}x + \dfrac{4}{9} = \dfrac{7}{9}$
(50)

* 16. $\dfrac{4}{5} + \dfrac{2}{15}x = \dfrac{14}{15}$
(50)

17. $0.005x - 0.03 = 0.01$
(50)

18. Use a unit multiplier to convert 30 inches per second to feet per second.
(64)

19. Factor:
(21)
    a. $2x^2 + 6x + 10$ **b.** $x^3 - x^2$

20. There are two black socks and one gray sock in a drawer. Without looking, Phil selects two socks.
(32)
    a. Find the sample space of the experiment

    b. What is the probability that Phil selects two black socks?

21. Two of the lines are parallel. Find x.
(54)

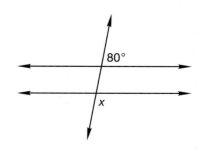

22. A number cube is rolled three times. What is the probability of rolling 6
(68) all three times?

23. Sketch a 30-60-90 triangle. If the longest side of a similar triangle is one
(66) foot, then how many inches long is the shortest side?

24. Solve: $\dfrac{3}{x} = \dfrac{5}{7}$
(44)

25. The cost of renting a riding lawnmower is $17.50, plus $2.50 for every
(69) hour that the mower is out. Create a table for the cost of renting a
lawnmower for the first 5 hours a lawnmower is out. Is the relationship
proportional? Explain why or why not.

Early Finishers
Real-World
Application

The middle-school band is trying to raise money for concert uniforms.
Rodney set up a booth outside the library each day for five days to raise
money. The results of his efforts are recorded in the table below.

   a. On average, how much money did Rodney make per day?

   b. On which day did Rodney make the most dollars per hour?

Day	Donations	Hours
Sunday	$28.55	5
Monday	$26.40	3
Thursday	$22.35	3
Friday	$38.76	4
Saturday	$76.32	8

Focus on
• Probability Simulation

A cereal company claims on its box, "One in three win a free movie pass!" If you decide to purchase this cereal, what is the probability it contains a movie pass?

The cereal box indicates that the probability of selecting a winner is $\frac{1}{3}$. This **theoretical probability** means that, out of all the cereal boxes produced for the contest, one-third contain a free pass. Thus, on average, one in three boxes is a winner.

Does this mean that if you buy three boxes you are guaranteed a winner? We can simulate the purchase of 3 boxes to determine the probability of buying at least one winner.

For the simulation we will need a spinner that looks like this:

Each spin will represent the purchase of one cereal box. The probability of landing on a "winner" is $\frac{1}{3}$, since the winning region fills $\frac{1}{3}$ of the circle.

Activity 1

Probability Simulation

Construct the face of a spinner like the one shown above. Use a compass to draw a circle. Then compute the central angle that intercepts $\frac{1}{3}$ of the circle:

$$\frac{1}{3}(360°) = 120°$$

Use a protractor to help make the central angle. Shade the interior of the 120° central angle, and label the region "winner." You may label the rest of the spinner "not a winner." Place the spinner on a level surface for the simulation.

To keep track of our results, we make and use a table that looks like this:

	Spin 1	Spin 2	Spin 3	At least 1 Winner?
Trial 1				
Trial 2				
Trial 3				
⋮				

Make rows for at least six trials on the table.

Thinking Skill

Explain

If one in three boxes contains a free pass, does buying three boxes guarantee a free pass? Why or why not?

Simulation

1. For each trial, simulate the "purchase" of 3 cereal boxes by spinning the spinner 3 times. For each spin, record W (for winner) or N (for not winner). Then in the "At least one winner?" column write "yes" or "no," depending on whether there were any winners in the three-spin trial.

2. Repeat step 1 until the table is complete. Below are some sample results.

	Spin 1	Spin 2	Spin 3	At Least 1 Winner?
Trial 1	W	N	N	yes
Trial 2	N	N	N	no
Trial 3	N	W	W	yes

3. To compute the **experimental probability** of having at least one winner after purchasing 3 cereal boxes, add the number of successes ("yes" in the last column) and divide by the total number of trials. For our sample results above, this would be:

$$P(\text{at least one winner in 3 boxes}) \approx \frac{2}{3}$$

exercises

Answer the following questions about the simulation.

1. What did one spin of the spinner represent?

2. What did each three-spin trial represent?

3. What did the spinner stopping on "winner" represent?

4. What did the spinner stopping on "not a winner" represent?

5. *Explain* State whether the following tools could have been used instead of a spinner to conduct the simulation. Explain why or why not.

   a. Two red cards and one black card

   b. A coin

   c. Three red marbles

   d. One red, one white, one blue marble

   e. A number cube

6. *Discuss* We use simulation when the theoretical probability is difficult to compute and the experimental probability is impractical to find directly. In this case, why is it desirable to simulate the experiment rather than actually perform the experiment (purchasing the cereal boxes)?

7. Assuming a large number of cereal boxes were made for this contest, one can use probability theory (not simulation) to show that the theoretical probability of having at least one winner after purchasing three cereal boxes is $\frac{19}{27}$ (or about 0.704). What is the difference between the theoretical probability and the experimental probability you found? Why is it important to conduct many trials when computing an experimental probability?

Visit www. SaxonPublishers. com/ActivitiesC3 *for a graphing calculator activity.*

Activity 2

Design and Conduct a Simulation

Materials needed:

- coins
- number cubes
- spinners
- decks of cards
- color tiles or cubes available for students to select

Read each of the following descriptions of events and explain how you would simulate each one. Then choose one of the events and conduct a simulation. Select a simulation tool for the experiment. Draw the chart you would use to organize your results.

1. A basketball player makes an average of 4 in 6 free throws. If he shoots 3 free throws in a game, what is the probability he makes all of them?

2. John has 2 blue socks and 4 black socks in a drawer. On a dark morning, he reaches in and pulls out 2 socks. What is the probability that he has a matching pair?

3. One-half of the chairs at a party have stickers secretly placed under the seats entitling the person sitting in the chair to a door prize. What is the probability that Tina and her two friends will all sit in chairs with stickers under the seats?

• Percent Change of Dimensions

facts | Power Up O

mental math

a. Algebra: Evaluate: $\frac{1}{2}b(10)$ when $b = 2$

b. Fractional Parts: $\frac{5}{6}$ of 48

c. Probability: In a bag of 20 marbles, 4 are blue. What is the probability that one taken at random will not be blue?

d. Proportions: The ratio of DVD players to VCRs was 7 to 2. There were 35 DVD players. How many VCRs were there?

e. Percent: 5% of $40

f. Measurement: Find the temperature indicated on this thermometer.

g. Select a Method: To find the average high temperature for 31 days in August, which would you use?

A mental math
B pencil and paper
C calculator

h. Calculation: Square 10, + 10, + 10, ÷ 2, + 4, $\sqrt{}$, + 1, $\sqrt{}$

problem solving | Danielle has 5 different shirts and 4 different skirts. She wants to wear a different combination each day for as many days as she can before she repeats a combination. How many days can she wear a different combination?

Projectors, copy machines, and graphics software can dilate and reduce the size of images by scale factors that are often expressed as percents.

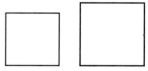

Here we show two squares. The larger square is 20% larger than the smaller square.

Statements about percents of increase or decrease should be read carefully. A small word or single letter can affect the meaning.

> **Dilation:** Percent of original = 100% + percent of increase
>
> **Reduction:** Percent of original = 100% − percent of decrease

Referring to the square above, increasing the size of the smaller square 20% means that the size of the larger square is 120% of the smaller square. Notice that 120% is related to the scale factor. We find that the scale factor from the smaller square to larger square is 1.2 by converting 120% to a decimal.

Example 1

One square is 50% larger than another square.

a. The perimeter of the larger square is what percent of the perimeter of the smaller square?

b. The area of the larger square is what percent of the area of the smaller square?

c. Provide examples of two squares that fit the description.

Solution

Although we are not give the dimensions of the squares, we are given enough information to answer the questions.

a. Since the dimensions of the larger square are 50% greater than the dimensions of the smaller square, each side is 150% of the length of the smaller square. That is, the scale factor from the smaller to larger square is 1.5. Perimeter is a length, so the perimeter of the larger square is 1.5 times, or **150%**, of the perimeter of the smaller square (50% greater).

b. Since area is the product of two dimensions, we square the scale factor to determine the relationship between the areas.

$$(1.5)^2 = 2.25$$

The area of the larger square is 2.25 times the area of the smaller square. In the language of percent, the area of the larger square is **225%** of the area of the smaller square (which means the area is 125% greater).

c. We may select any two side lengths for the squares so that the longer side length is 50% greater than the shorter side length. Here are some sample side lengths: 1 and 1.5, 2 and 3, 10 and 15, 100 and 150. It is

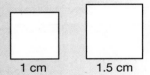

1 cm   1.5 cm

often convenient to choose 1 for the dimension to which the other is compared. The larger square is compared to the smaller, so we choose 1 for the side length of the smaller square. Thus the side length of the larger square is 1.5. We choose to use cm as the unit of measure.

We compare the perimeter of the two squares.

Smaller Square	Larger Square
Perimeter = 4 cm	Perimeter = 6 cm

We see that the perimeter of the larger square is 150% *of* (50% *greater than*) the perimeter of the smaller square.

Now we compare the areas.

Smaller Square	Larger Square
Area = 1 cm²	Area = 2.25 cm²

The area of the larger square is 225% *of* (125% *greater than*) the area of the smaller square.

Example 2

The towel shrunk 10% when it was washed. The area of the towel was reduced by what percent? Provide an example of the "before" and "after" dimensions of the towel.

Solution

Assuming that the towel shrunk uniformly, the length was reduced 10% and the width was reduced 10%. Since the length and width are both 90% of the original size, the scale factor from the original size to the reduced size is 0.9. To find the effect of the contraction on the area we square the scale factor.

$$(0.9)^2 = 0.81$$

This number means that the area of the reduced towel is 81% of the area of the original towel. Thus the area was reduced by **19%.**

To provide a before and after example we select multiples of 10 for the dimensions so that a reduction of 10% is easy to calculate.

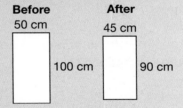

Example 3

The vertices of a square are (4, 4), (4, −4), (−4, −4), and (−4, 4).

a. Sketch the square and its image after a 125% dilation.

b. The perimeter of the dilation is what percent greater than the perimeter of the original square?

c. The area of the dilation is what percent greater than the area of the original square?

Solution

a. The scale factor of the dilation is 1.25 (or $1\frac{1}{4}$). Since 4 × 1.25 is 5, the vertices of the dilation are (5, 5), (5, −5), (−5, −5), and (−5, 5). We sketch the square and its dilation.

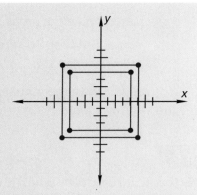

b. The perimeters of the two squares are 20 units and 16 units, so the perimeter of the image is 125% of the perimeter of the original square.

$$\frac{20}{16} = 1.25$$

Therefore, the perimeter of the image is **25%** greater than the perimeter of the original square.

c. The areas of the two squares are 25 square units and 16 square units, so the area of the image is $156\frac{1}{4}$% of the area of the original square.

$$\frac{25}{16} = 1.5625$$

Therefore, the image is $56\frac{1}{4}$% **(or 56.25%)** greater than the area of the original square.

Example 4

If the dimensions of a cube are increased 100%, by what percent is the volume increased?

Solution

We will make the calculations using a specific example. We start with a cube with edges 1 inch long. If the dimensions are increased 100%, then the larger cube has edges 2 inches long.

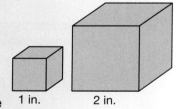

The volume of the smaller cube is 1 in.³. The volume of the larger cube is 8 in.³. The larger cube is 8 times, or 800%, as large as the smaller cube. Therefore, the volume is *increased* **700%.**

1 in.   2 in.

Practice Set

a. There are two concentric circles on the playground. The diameter of the larger circle is 40% greater than the diameter of the smaller circle. The area of the larger circle is what percent greater than the area of the smaller circle?

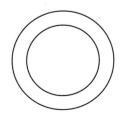

b. Becky reduced the size of the image on the computer screen by 40%. By what percent was the area of the image reduced?

c. If the original square in example 3 were dilated 150%, then by what percent would the perimeter increase, and by what percent would the area increase?

d. Suppose the dimensions of the larger cube in example 4 (the 2-inch cube) were increased by 100%. What would be the edge length of the expanded cube? The volume of the expanded cube would be what percent of the 2-inch cube?

Written Practice   *Strengthening Concepts*

1. Pencils outnumbered pens in the drawer 17 to 3. If there were 80 pens and pencils in all, how many were pens?
(45)

2. Thirty-eight percent of the students watched the ball game. If there are 150 students in all, how many did not watch the ball game?
(48)

* 3. **Analyze** The price of the art piece increased by 90% after the artist became famous. If the new price is $2470, what was the original price?
(67)

* 4. The table shows the perimeter of a polygon given a side length. Does the table show direct variation? If so, state the constant of variation. What type of polygon could this table represent?
(69)

P	s
3	1
6	2
15	5
150	50

* 5. **Analyze** Find the lengths of *x* and *y* in the similar triangles to the right.
(35)

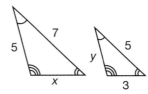

* 6. Graph $y = 2x - 2$. Does the graph indicate direct variation? Explain your answer.
(56, 69)

7. Find the area of the shaded region in the figure to the right. Use $\pi = 3.14$. (Units are in cm.)
(37)

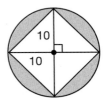

8. When sales tax is 6%, the amount paid for an item is 106% of the price. Write 106% **a** as a decimal and **b** as a fraction. **c** Show how you would use one of these three forms to compute 106% of $21.
(48)

9. Izzy rolls one number cube in a game. He must roll a 3 or more in order to win on his next turn. What is the probability he wins on his next turn?
(32)

10. If the dimensions of a square are increased 10%, by what percent is the area increased?
(71)

11. Graph the solution on a number line. $7 - x \geq 5$
(62)

12. A transversal cuts two parallel lines. Find z.
(54)

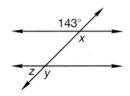

13. Six students estimated the length of the window, in meters: 1.1, 1.3, 1.5, 1.7, 1.9, and 1.5.
(53)

   a. Display the data in a line plot.

   b. Find the mean, median, mode, and range of the estimates.

   c. Which measure indicates the spread of the estimates?

14. Alex swims 100 meters in 50 seconds. Use a unit multiplier to convert 100 meters per 50 seconds to meters per minute.
(64)

* 15. **Model** Sketch $\triangle ABC$ with vertices at $A(-2, -2)$, $B(2, -2)$, and $C(0, 2)$. Then sketch its image $\triangle A'B'C'$ with a scale factor of 3. What are the coordinates of the vertices of $\triangle A'B'C'$?
(Inv. 5)

* 16. In problem **15,** the dimensions of $\triangle ABC$ are what percent of the dimensions of $\triangle A'B'C'$? (Hint: AB is what fraction of $A'B'$?)
(71)

Simplify.

17. $\dfrac{15x^2y^{-1}}{5xy}$
(27)

18. $\dfrac{5}{6} - \dfrac{2}{3} \div \dfrac{4}{5}$
(13, 22)

19. $-3 + (-4)(-5)$
(36)

20. $\dfrac{1.2 + 0.24}{0.3}$
(24, 25)

Solve.

21. $\dfrac{1}{4} - \dfrac{2}{3}x = \dfrac{11}{12}$
(50)

22. $\dfrac{3}{4} = \dfrac{15}{x}$
(44)

23. $0.02x - 0.3 = 1.1$
(50)

24. $6x - 12 = 84$
(50)

25. The frequency table below indicates how many of the 60 Movie Express
(Inv. 6) members rented 1, 2, 3, or 4 movies during a one-week period.

Number of Movies Rented	Tally	Frequency
1	JHT JHT JHT JHT III	23
2	JHT JHT JHT III	18
3	JHT JHT I	11
4	IIII	4

Which is the most reasonable conclusion based on the information in
the chart?

A Most members rent 1 movie per week.

B About $\frac{1}{3}$ of the members rent 3 movies per week.

C Few members rent 2 or 3 movies per week.

D Less than half of the members rent more than 2 movies per week.

Early Finishers
Real-World
Application

A small accounting firm has seven employees. Their annual salaries are as
follows:

$38,500 $41,500 $45,000 $52,000 $43,500 $34,000 $45,000

The employees are voting on two different plans for salary raises for the
coming year. Both plans are one-year proposals and apply to all employees.

Plan 1: Each employee receives a raise of $1,800.

Plan 2: Each employee receives a raise of 4.2%.

a. Make a table of current salaries and proposed salaries under both
plans.

b. Write a brief explanation of how the two plans differ and their
potential financial impact on the highest- and lowest-paid
employees.

• Multiple Unit Multipliers

Power Up | *Building Power*

facts | Power Up O

mental math

a. **Statistics:** Find the range: 27, 26, 26, 24, 28

b. **Estimation:** $3\sqrt{63}$

c. **Number Sense:** Compare $2 \times 0.5 \bigcirc (0.5)^2$

d. **Powers/Roots:** $\sqrt{5 \cdot 5} \cdot \sqrt{3 \cdot 3}$

e. **Proportion:** $\frac{x}{5} = \frac{20}{25}$

f. **Geometry:** Copy and complete this congruence statement: $\triangle ABC \cong$ _____

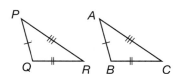

g. **Scientific Notation:** Write 45,000,000 in scientific notation.

h. **Calculation:** $3 \times 3, \times 3, \times 3, \sqrt{\ }, \sqrt{\ }$

problem solving

List the three pairs of one-digit numbers whose product has a 9 in the ones place. Then find the missing digits in the problem below.

$$\begin{array}{r} \_\ \_ \\ \times \_\_ \\ \hline 539 \end{array}$$

New Concept | *Increasing Knowledge*

Math Language

Recall that a **unit multiplier** is a form of 1, because the numerator and denominator are equivalent measures.

We have used single unit multipliers to convert measures and rates. For some conversions we use multiple unit multipliers. Using multiple unit multipliers is similar to multiplying repeatedly by 1.

$$5 \cdot 1 \cdot 1 \cdot 1 = 5$$

We can use two unit multipliers to convert units of area and three unit multipliers to convert units of volume.

Example 1

Thinking Skill

Connect

Does this example involve area or volume? Explain.

The Ortegas want to replace the carpeting in some rooms of their home. They know the square footage of the rooms totals 855 square feet (ft²). Since carpeting is sold by the square yard (yd²), they want to find the approximate number of square yards of carpeting they need. Use unit multipliers to perform the conversion.

To convert square feet (ft^2) to square yards (yd^2), we need to cancel feet twice. We will use two unit multipliers to perform the conversion.

$$855 \text{ ft}^2 \times \underbrace{\frac{1 \text{ yd}}{3 \text{ ft}} \times \frac{1 \text{ yd}}{3 \text{ ft}}}_{\text{unit multipliers}} = \frac{855 \text{ yd}^2}{9} = 95 \text{ yd}^2$$

The Ortegas will need about **95 square yards** of carpet.

To understand visually why it is necessary to cancel feet twice in the example above, consider this diagram of a square yard.

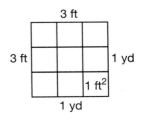

We see that one square yard is equivalent to 9 square feet. Therefore, $855 \text{ ft}^2 = \frac{855}{9} \text{ yd}^2 = 95 \text{ yd}^2$

Analyze What unit multipliers would we use to convert square feet to square miles?

Example 2

Concrete is delivered to a job site in cement trucks and is ordered in cubic yards. Ray is pouring a concrete driveway that is 18 feet wide, 36 feet long, and 4 inches ($\frac{1}{3}$ foot) thick. Ray calculates the volume of concrete he needs is 216 cubic feet. Use unit multipliers to find the number of cubic yards of concrete Ray needs to order.

Solution

To convert cubic feet (ft^3), we need to cancel feet three times. We use three unit multipliers.

$$216 \text{ ft}^3 \times \underbrace{\frac{1 \text{ yd}}{3 \text{ ft}} \times \frac{1 \text{ yd}}{3 \text{ ft}} \times \frac{1 \text{ yd}}{3 \text{ ft}}}_{\text{unit multipliers}} = \frac{216 \text{ yd}^3}{27} = 8 \text{ yd}^3$$

Ray needs to order **8 cubic yards** of concrete.

Using the three unit multipliers in example 2 essentially divides the number of cubic feet by 27 (since $3 \cdot 3 \cdot 3 = 27$). This is reasonable because one cubic yard is equivalent to 27 cubic feet.

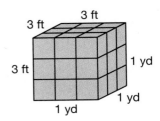

Analyze What unit multipliers would we use to convert cubic inches to cubic feet?

Example 3

Keisha set a school record running 440 yards in one minute (60 seconds). Find Keisha's average speed for the run in miles per hour (1 mi = 1760 yd).

Solution

We are given Keisha's rate in yards per minute. We are asked to find her rate in miles per hour.

We need one unit multiplier to convert yards to miles and another unit multiplier to convert minutes to hours. We select from the following.

$$\frac{1760 \text{ yd}}{1 \text{ mi}} \qquad \frac{1 \text{ mi}}{1760 \text{ yd}} \qquad \frac{1 \text{ hr}}{60 \text{ min}} \qquad \frac{60 \text{ min}}{1 \text{ hr}}$$

We choose the unit multipliers that cancel the units we want to remove and that properly position the units we want to keep.

$$\frac{440 \text{ yd}}{1 \text{ min}} \times \underbrace{\frac{1 \text{ mi}}{1760 \text{ yd}} \times \frac{60 \text{ min}}{1 \text{ hr}}}_{\text{unit multipliers}} = \frac{440 \times 60 \text{ mi}}{1760 \text{ hr}}$$

We perform the arithmetic and find that Keisha's average speed for her run was **15 mph.**

Practice Set

a. *Analyze* Examine problems **b–h.** For which problems will you need to use two unit multipliers? Three unit multipliers? How do you know?

Use multiple unit multipliers to perform each conversion.

b. 9 sq. ft to sq. in.

c. 9 sq. ft to sq. yd

d. 1 m^3 to cm^3

e. $1{,}000{,}000 \text{ mm}^3$ to cm^3

f. 12 dollars per hour to cents per minute

g. 10 yards per second to feet per minute

h. 1 gallon per day to quarts per hour

Written Practice   *Strengthening Concepts*

1. The ratio of fish to ducks at the water's edge was 5 to 2. If there were
(45) 28 in all, how many were fish?

2. Forty-six percent of the pigeons wore bands around one leg. If there
(58) were 81 pigeons without bands, how many pigeons were there in all?

* 3. *Analyze* After the revision, the book had 5% more pages. If the book
(67) now has 336 pages, how many pages did the book have before the revision?

* 4. **Justify** Does the table to the right
(69) show direct variation? If so, state
the constant of variation. If not, state
why not.

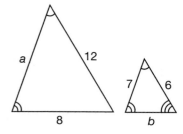

x	y
6	22
15	165
18	176

5. Find the lengths of *a* and *b* in the similar
(35) triangles to the right. What is the ratio of
the perimeter of the smaller triangle to
the perimeter of the larger triangle?

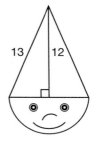

* 6. Gabrielle trains on her bicycle, riding at a rate of 21 miles per hour
(64, 72) during some parts of her training. Use two multipliers to find her rate in
yards per minute (1 mi = 1760 yd).

* 7. Amanda window shops, walking 90 feet per minute. Use unit multipliers
(64, 72) to convert 90 feet per minute to yards per second.

* 8. Graph $y = \frac{1}{2}x$. Does the graph indicate direct variation? Explain.
(56, 69)

9. Stephanie baked a cake according to
(Inv. 2, this plan. In order to know how much
37) batter to use, Stephanie needs to know
the area of the top of the cake. Find
the area of the figure. (Use 3.14 for π.
Dimensions are in inches.)

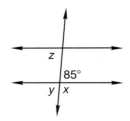

10. Write $33\frac{1}{3}\%$ as a **a** decimal and **b** reduced fraction. **c** Tyrone has read $\frac{2}{3}$
(63) of the book. What percent remains?

11. Two of the lines are parallel. Find *x, y,*
(54) and *z.*

12. A coin will be flipped three times.
(32)
    a. What is the sample space for the experiment?

    b. What is the probability of getting heads two or more times in three
    flips?

    c. **Predict** If a coin were flipped four times, would the probability of
    getting heads two or more times increase or decrease?

13. What is the tax on a $85.97 purchase if the rate is 8%?
(58)

14. The distance from the earth to the moon is approximately
(28) 2.5×10^5 miles. Write that distance in standard form.

15. Factor:
(21)
    a. $7x^2 + 35x - 14$ **b.** $-3x - 15$

Simplify.

16. $\dfrac{8a^2b^{-1}c}{6ab^2c}$ **17.** $-\dfrac{1}{2} - \left(-\dfrac{3}{4}\right)$
(27, 51)   (13, 33)

18. $x + y - x + 3 - 2$ **19.** $0.12 + (0.6)(0.02) + 2$
(31)   (24, 25)

Solve.

20. $2 + \dfrac{1}{3}t = \dfrac{2}{3}$ **21.** $0.9x - 1.3 = 0.5$
(50)   (50)

22. $\dfrac{x}{6} = \dfrac{6}{9}$ **23.** $2(x + 7) - 4 = 24$
(44)   (50)

24. Graph the solution on a number line: $x + 5 \geq 3$
(62)

25. A small appliance company surveyed
(11, 680 consumers to determine which
Inv. 6) of the appliances the consumers had
used within the past month. What
number of consumers had used
neither a blender *nor* a food processor
in the past month?

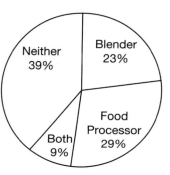

• Formulas for Sequences

facts | Power Up O

mental math

a. **Algebra:** Evaluate: $\frac{1}{2}(8)h$ when $h = 12$

b. **Estimation:** $4\sqrt{99}$

c. **Fractional Parts:** $\frac{2}{7}$ of 42

d. **Percent:** 90% of $30

e. **Rate:** Lance drives west at 65 miles per hour. Courtney starts at the same place and drives west at 50 miles per hour. How far apart are they after 3 hours?

f. **Measurement:** How many quarts is $1\frac{1}{2}$ gallons?

g. **Geometry:** What is the specific name for an equilateral quadrilateral with right angles?

h. **Calculation:** $10 - 9, + 8, - 7, + 6, - 5, + 4, - 3, + 2, - 1$

problem solving

A dentist's drill can spin at 300,000 revolutions per minute. How many revolutions per second is that? A certain carver's drill spins at 450,000 revolutions per minute. How many revolutions per second is that? In ten seconds, how many more times does the carver's drill spin than the dentist's drill?

New Concept *Increasing Knowledge*

Recall that a sequence is an ordered list of numbers that follows a certain rule. In this lesson we will practice using formulas to express the rule.

Also recall that a term in a sequence has a value and a position. We distinguish between the number of the term (its position, such as 1st, 2nd, 3rd) and the value of a term (such as 3, 6, 9).

Number of Term (n)	1	2	3	4	...
Value of Term (a)	3	6	9	12	...

We can write a formula that shows how to find the value of a term if we know its number.

The formula $a_n = 3n$ will generate any term of the sequence in the table.

In place of n we substitute the number of the term we wish to find. Thus, to find the 7th term we substitute 7 for n in the formula.

$$a_7 = 3(7) = 21$$

Example 1

Find the 8th and 9th terms of the sequence generated by the formula $a_n = 3n$.

Solution

The eighth and ninth terms correspond to $n = 8$ and $n = 9$.

$$a_8 = 3(8) \qquad\qquad a_9 = 3(9)$$
$$a_8 = \mathbf{24} \qquad\qquad a_9 = \mathbf{27}$$

When a formula is not given, be creative when looking for the rule or pattern of a sequence. The rule may include addition, subtraction, multiplication, division, powers, arithmetic between several previous terms, or other patterns that are not immediately obvious.

Example 2

Find the next two terms in each sequence.

a. $\{2, 4, 8, 16, 32, \ldots\}$

b. $\{1, 1, 2, 3, 5, 8, \ldots\}$

c. $\{1, -2, 2, -1, 3, 0, 4, \ldots\}$

Solution

a. Each term is twice the previous term.

$\times 2 \ \times 2 \ \times 2 \ \times 2 \ \ \times 2 \ \ \times 2$
2, 4, 8, 16, 32, **64, 128**

b. Each term is the sum of the two previous terms.

1+1 1+2 2+3 3+5 5+8 8+13
1, 1, 2, 3, 5, 8, **13, 21**

c. The pattern alternates between subtracting three and adding four.

−3 +4 −3 +4 −3 +4 −3 +4
1, −2, 2, −1, 3, 0, 4, **1, 5**

Thinking Skill

Analyze

Find another way of describing the pattern of example **2c.**

Example 3

A rule for the sequence in Example 2a is: $a_n = 2^n$

Find the tenth term in the sequence.

Solution

The first term in the sequence corresponds with $n = 1$, so the tenth corresponds with $n = 10$.

$$a_{10} = 2^{10}$$

Visit www. SaxonPublishers. com/ActivitiesC3 *for a graphing calculator activity.*

To use your calculator to compute this power, press

2 y^x 1 0 =

We find that the tenth term is **1024.**

Example 4

Find the first three terms of the sequence given by the formula
$$a_n = 1 + 2n$$

Solution

To find the first three terms we replace n with 1, 2, and 3 and simplify.

$$n = 1 \qquad a_1 = 1 + 2(1) = 3$$
$$n = 2 \qquad a_2 = 1 + 2(2) = 5$$
$$n = 3 \qquad a_3 = 1 + 2(3) = 7$$

The first three terms are **3, 5, and 7.**

Connect One term of the sequence described in example 4 is 37. How could you use an algebraic method to find the number of this term?

Example 5

The numbers in this sequence are sometimes called triangular numbers.
$$\{1, 3, 6, 10, 15, \ldots\}$$
These numbers of objects can be arranged in triangular patterns.

Triangular numbers can be found using this formula.
$$a_n = \frac{n(n + 1)}{2}$$

Find the 10th triangular number.

Solution

Using the formula we replace n with 10 and simplify.

$$a_{10} = \frac{(10)(10 + 1)}{2}$$
$$= \frac{110}{2}$$
$$= \mathbf{55}$$

Practice Set

a. What is the 7th term of this sequence?
$$1, 4, 9, 16, \ldots$$

b. The terms of the following sequence are generated with the formula $a_n = 5n - 2$. Find the tenth term.

$$3, 8, 13, 18, \ldots$$

c. Which formula below generates the terms of the following sequence?

$$0, 3, 8, 15, \ldots$$

A $a_n = n - 1$ **B** $a_n = 2n - 1$ **C** $a_n = n^2 - 1$

d. What are the first four terms in the sequence generated by the formula $a_n = n^2 + 2n + 1$?

e. **Formulate** Can you write another formula that generates the same first four terms in problem **d?** If yes, write the formula.

f. Find the 20th triangular number. (See Example 5.)

Written Practice   *Strengthening Concepts*

1. Renaldo increased the image size on the computer screen by 20%. By
(71) what percent did the area of the image increase?

2. Eighty-eight percent of the students volunteered in their community. If
(48) 110 students volunteered, how many students were there in all?

* 3. **Analyze** The basketball regularly sold for $18. If it is on sale for 20%
(67) off, what is the new price of the basketball?

* 4. **Evaluate** Does the table to the right show
(69) direct variation? If so, state the constant of
variation.

W	D
4	20
7	35
9	45

5. Daniel cut the trapezoid and arranged the
(7) pieces to form a rectangle. What is the length
of the rectangle Daniel formed if it is equal
to the average of the base lengths of the
trapezoid?

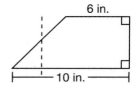

6. Brianne walked 6 yards across the room in 3 seconds. Use a unit
(52) multiplier to write 6 yards in 3 seconds as feet per second.

* 7. Aidan walked 440 feet in one minute. Find his average rate in miles per
(72) hour.

8. Graph $y = -x + 5$.
(56)

9. Find the area of the figure to the right.
(37) (Dimensions are in m.)

10. Write $\frac{1}{12}$ as a **a** decimal and **b** percent. **c** If $\frac{1}{12}$ of the customers pay
(63)   cash, which of these statements is correct:

    A More than one in ten pay cash

    B Less than one in ten pay cash

11. Factor:
(21)
    a. $4x^2 + 12x - 4$ **b.** $-2x - 16$

12. Find the perimeter and area of the
(60)   parallelogram.

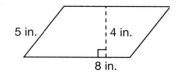

13. Find the perimeter and area of the
(60)   parallelogram.

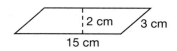

14. The hypotenuse of a right triangle is $\sqrt{39}$ inches. The length of the
(16)   hypotenuse is between:

    A 3 and 4 in.   **B** 4 and 5 in.   **C** 5 and 6 in.   **D** 6 and 7 in.

15. A company charges (*c*) a flat fee of $4 for shipping plus $0.85 per pound
(41)   (*p*). Which equation expresses this relationship?

    A $c = 4p + 0.85$ **B** $c = 0.85p + 4$

    C $c = (4 + 0.85)p$ **D** $4c = 0.85p$

Simplify.

16. $\dfrac{24m^4 b^{-1}}{18m^{-3}b^3}$ **17.** $(-3)^1 - (-2)^2 - 4^0$
(27, 51)   (17, 36)

Solve.

* 18. *Analyze* $\dfrac{2}{3} - \dfrac{1}{6}x = \dfrac{1}{2}$ *** 19.** $\dfrac{4}{5}x + \dfrac{1}{10} = \dfrac{1}{2}$
(50)   (50)

20. $\dfrac{4}{3} = \dfrac{6}{x}$ **21.** $\dfrac{3}{4}x = 12$
(44)   (38)

22. $0.09x + 0.9 = 2.7$ **23.** $3(2x - 1) = 45$
(50)   (50)

* 24. *Generalize* A rule for the following sequence is $a_n = 2n - 3$. Find the
(73)   20th term of the sequence.

$$-1, 1, 3, 5, \ldots$$

*25. Describe a real-life situation illustrated
(41)   by the graph. Explain why the graph
    looks the way it does. Is the relationship
    between the number of rides and the
    cost proportional? If the graph is not
    proportional, how could you change the
    situation to make it proportional?

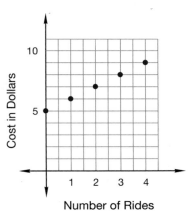

• Simplifying Square Roots

facts   Power Up O

mental math

a. **Algebra:** If $2a = 10$, then a^2 equals what?

b. **Number Sense:** Arrange A, B, C, and D from least to greatest:

   A $\frac{3}{4}$ **B** $\frac{1}{4}$ **C** $\frac{3}{4} + \frac{1}{4}$ **D** $\frac{3}{4} \times \frac{1}{4}$

c. **Powers/Roots:** $x \cdot x^5 \cdot x^3$

d. **Probability:** In a box of blocks, $\frac{3}{4}$ are rectangular. What is the probability that one taken at random will not be rectangular?

e. **Rate:** At 150 mph, how long would it take to fly 750 miles?

f. **Geometry:** Copy and complete this congruence statement:

   $\triangle ABC \cong \triangle D\_\_$

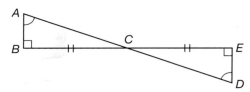

g. **Select a Method:** Howard wants to leave the waiter a 20% tip. Howard will probably

   A estimate an approximate tip.
   B calculate an exact tip.

h. **Calculation:** Square 5, $\times\, 2$, $-\, 1$, $\sqrt{}$, $\times\, 9$, $+\, 1$, $\sqrt{}$, $\times\, 2$, $\sqrt{}$

problem solving   A cube has six faces. Depending on your perspective, you can see different numbers of faces. Draw the following views of a cube, when possible.

1. Only one face visible

2. Exactly two faces visible

3. Exactly three faces visible

4. Exactly four faces visible

We have used the Pythagorean Theorem to find that the diagonal of a 1 by 1 square is $\sqrt{2}$.

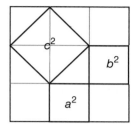

Since the area of c^2 is 2 (see the four half squares), the length of the diagonal is $\sqrt{2}$. The diagonal of a 2 by 2 square is $\sqrt{8}$, but notice that it is also 2 times $\sqrt{2}$. We simplify $\sqrt{8}$ to $2\sqrt{2}$.

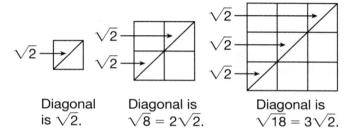

Diagonal is $\sqrt{2}$. Diagonal is $\sqrt{8} = 2\sqrt{2}$. Diagonal is $\sqrt{18} = 3\sqrt{2}$.

The diagonal of a 3 by 3 square is $\sqrt{18}$, but we see it is also 3 times $\sqrt{2}$. We simplify $\sqrt{18}$ to $3\sqrt{2}$. In this lesson we will learn how to simplify some square roots.

We simplify square roots by using the **product property** of square roots, which states that square roots can be multiplied and factored:

$$\sqrt{2}\,\sqrt{6} = \sqrt{12} \qquad \sqrt{12} = \sqrt{4}\,\sqrt{3}$$
$$\text{multiplied} \qquad\qquad \text{factored}$$

We symbolize the product property of square roots this way:

The Product Property of Square Roots
$\sqrt{ab} = \sqrt{a}\,\sqrt{b}$

Above we see that $\sqrt{12}$ has a perfect square factor: $\sqrt{12} = \sqrt{4}\,\sqrt{3}$. Since $\sqrt{4}$ equals 2, we simplify $\sqrt{12}$ to $2\sqrt{3}$.

Step:	Justification:
$\sqrt{12} = \sqrt{4}\,\sqrt{3}$	Factored $\sqrt{12}$
$\sqrt{12} = 2\sqrt{3}$	Simplified $\sqrt{4}$

Similarly, we find perfect square factors in $\sqrt{8}$ and $\sqrt{18}$.

Step:	Justification:	Step:
$\sqrt{8} = \sqrt{4}\,\sqrt{2}$	Factored	$\sqrt{18} = \sqrt{9}\,\sqrt{2}$
$\sqrt{8} = 2\sqrt{2}$	Simplified	$\sqrt{18} = 3\sqrt{2}$

To help us find perfect square factors of a square root, we can write the number under the radical sign as a product of prime factors. Then we look for pairs of identical factors. Each pair is a perfect square. Below we simplify $\sqrt{600}$. First we write the prime factorization of 600.

Step:	Justification:
$\sqrt{600} = \sqrt{2 \cdot 2 \cdot 2 \cdot 3 \cdot 5 \cdot 5}$	Factored 600
$\sqrt{600} = \sqrt{2 \cdot 2}\,\sqrt{5 \cdot 5}\,\sqrt{2 \cdot 3}$	Grouped factors under separate radical signs
$\sqrt{600} = 2 \cdot 5\sqrt{2 \cdot 3}$	Simplified perfect squares
$\sqrt{600} = 10\sqrt{6}$	Multiplied factors

Square roots can be simplified by removing perfect square factors from the radicand. If the perfect square factor is apparent, we can factor in one step.

$$\sqrt{600} = \sqrt{100} \cdot \sqrt{6} = 10\sqrt{6}$$

If the perfect square factors are not apparent, write the prime factorization and look for pairs of common factors.

Example 1

Simplify:

   a. $\sqrt{216}$ **b.** $\sqrt{210}$

Solution

a. We can draw a factor tree to find prime factors. Then we pair identical factors and simplify.

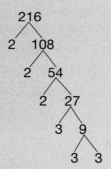

$$\sqrt{216} = \sqrt{2 \cdot 2 \cdot 2 \cdot 3 \cdot 3 \cdot 3} = \sqrt{2 \cdot 2}\,\sqrt{3 \cdot 3}\,\sqrt{2 \cdot 3} = 2 \cdot 3\sqrt{6} = \mathbf{6\sqrt{6}}$$

b. We find the prime factors. There are no pairs of identical factors.

Since 210 has no identical factors, $\sqrt{210}$ cannot be further simplified.

Analyze The radical expression $\sqrt{210}$ is between what two consecutive integers on the number line?

Example 2

Instead of walking 200 yards from point A to point B, Arnold takes a shortcut across a field. How long is the shortcut?

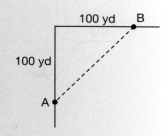

100 yd   B

100 yd

A

Solution

The shortcut is the hypotenuse of a right triangle with legs 100 yd long. We apply the Pythagorean Theorem.

$$a^2 + b^2 = c^2$$
$$100^2 + 100^2 = c^2$$
$$10,000 + 10,000 = c^2$$
$$20,000 = c^2$$
$$\sqrt{20,000} = c$$

Now we simplify the square root.

$$\sqrt{20,000} = \sqrt{100} \cdot \sqrt{100}\,\sqrt{2} = 100\sqrt{2}$$

The length of the shortcut is **$100\sqrt{2}$ yd.** Since $\sqrt{2} \approx 1.41$ the shortcut is about 141 yd.

In example 2 we estimated the length of the shortcut using $\sqrt{2} \approx 1.41$. It is helpful to remember some approximate values of frequently encountered square roots to one or two decimal places.

Approximate Values of Square Roots

Square Root	Approximate Value
$\sqrt{2}$	1.41
$\sqrt{3}$	1.73
$\sqrt{5}$	2.24
$\sqrt{10}$	3.16

Practice Set

Simplify if possible.

a. $\sqrt{20}$ **b.** $\sqrt{24}$ **c.** $\sqrt{27}$

d. $\sqrt{30}$ **e.** $\sqrt{125}$ **f.** $\sqrt{48}$

g. $\sqrt{50}$ **h.** $\sqrt{90}$ **i.** $\sqrt{1000}$

j. Using the numbers in the table in this lesson, calculate to the nearest tenth the values of the square roots in problems **g**, **h**, and **i**.

k. Jenny folded a 10-inch square piece of paper in half diagonally, making a triangle. What is the length of the longest side of the triangle?

l. *Justify* Were there any square roots in problems **a–i** that you could not simplify? Explain why or why not.

Written Practice   *Strengthening Concepts*

1. At basketball practice, every five players shared two basketballs. If there
(45) were 45 players, how many basketballs were there?

* 2. During the year student enrollment increased from 600 to 630. Student
(67) enrollment increased by what percent?

* 3. An item was regularly $90. It was marked 20% off. What was the new
(67) price?

* 4. Does the table show direct
(69) variation? If so, give the constant of variation.

x	0	1	2	3
y	0	3.1	6.2	9.3

For **5** and **6** refer to the two triangles.

5. Find a and b in the similar triangles
(35) to the right.

6. The dimensions of the smaller
(35) triangle are what fraction of the dimensions of the larger triangle? The area of the smaller triangle is what fraction of the area of the larger triangle?

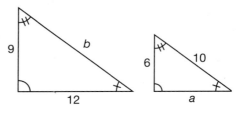

Simplify.

7. $\sqrt{2500}$
(15)

* 8. $\sqrt{18}$
(74)

* 9. $\sqrt{75}$
(74)

10. $\dfrac{36x^3 y^{-1}}{24x^2 y^2}$
(27, 51)

* 11. With a few calculations, Kenan estimated that the tank has a capacity
(72) of about 86,400 in.^{3}. What is its capacity in cubic feet? Use three unit multipliers.

12. Graph $y = -3x + 1$. Is $(2, -5)$ a solution?
(56)

13. The rule of the following sequence is $a_n = n(n + 1)$. Find the 12th term
(73) of the sequence: 2, 6, 12, 20, ...

14. Write $\frac{4}{15}$ **a** as a decimal and **b** as a percent. **c** Select one of these three
(63) forms to state that 16 of the 60 athletes had played in the finals in a
prior season.

15. Factor:
(21)
   a. $6x^2 - 30x - 18$ **b.** $2x^3 + 2x$

16. A 12-inch diameter pizza is sliced into sixths.
(40) Which is the best estimate of the area of the
plate covered by two slices?

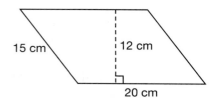

   A 108 in.^{2} **B** 36 in.^{2}

   C 12 in.^{2} **D** 4 in.^{2}

17. If the diameter of a circle is increased 40%, then its perimeter is
(71) increased 40% and its area is increased by what percent?

* 18. Find the perimeter and area of the
(60) parallelogram.

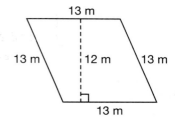

15 cm   12 cm   20 cm

* 19. **a.** Find the area of the parallelogram.
(Inv. 3,
60)   **b.** What kind of parallelogram is this?

13 m   13 m   12 m   13 m   13 m

Solve.

* 20. $\frac{3}{8} - \frac{2}{3}x = \frac{11}{12}$ *** 21.** $\frac{5}{9} = \frac{x}{12}$
(50)   (44)

22. $\frac{4}{3}m = -8$ **23.** $0.07 - 0.003x = 0.1$
(38)   (50)

24. $4x - x - 7 = 5$
(50)

25. Draw the front, top, and right-side views
(Inv. 4) of this figure.

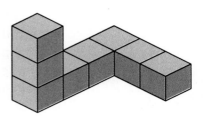

• Area of a Trapezoid

facts | Power Up O

mental math |
a. **Statistics:** Find the mean of these numbers: 13, 15, 15, 15, 12

b. **Estimation:** $5 - \sqrt{3}$

c. **Fractional Parts:** $\frac{4}{5}$ of 60

d. **Percent:** 80% of $20

e. **Proportion:** $\frac{13}{x} = \frac{39}{12}$

f. **Measurement:** Find the length of this object.

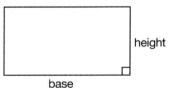

g. **Scientific Notation:** Write 0.00805 in scientific notation.

h. **Calculation:** $5280 \div 10, -8, \div 10, -2, \div 10, -5$

problem solving | Find the missing reduced fraction:

$$\frac{2}{3} \times \frac{7}{4} \times \frac{9}{10} \times \frac{5}{3} \times \frac{?}{?} = 1$$

Thinking Skill

Discuss

How are parallelograms and trapezoids similar? How are they different?

Recall that we multiply the base and height of a parallelogram to find its area. The base and height are perpendicular.

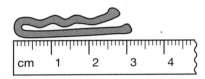

Also recall that the product of multiplying the base and height of a parallelogram is the area of a rectangle. If the parallelogram is not a rectangle, its area is nevertheless equal to the area of a rectangle with the same base and height. This is so because the portion of the parallelogram outside the rectangle matches the "hole" inside the rectangle.

To find the area of a trapezoid we can find the area of a rectangle with the equivalent area. A trapezoid has two bases, which are the parallel sides. The perpendicular distance between the bases is the height. If we multiply either one of the bases by the height the result is a rectangle that is either too small or too large.

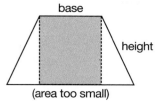

(area too small)

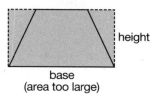

(area too large)

Instead of multiplying one of the bases by the height, we multiply the **average length of the bases** by the height. The average length of the bases is the length halfway between the length of the shorter base and the length of the longer base.

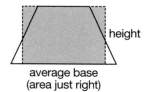

average base
(area just right)

Notice that the parts of the trapezoid outside the rectangle match the "holes" inside the rectangle.

We can express this method for finding the area of a trapezoid with a formula.

Area of trapezoid = average length of the bases · height

$$A = \frac{1}{2}(b_1 + b_2) \cdot h$$

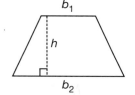

The expression $\frac{1}{2}(b_1 + b_2)$ indicates a way to find the average length of the bases.

Explain Use what you know about averages to explain why the expression $\frac{1}{2}(b_1 + b_2)$ represents the average length of the bases.

Example 1

Erik needs to replace the shingles on the south side of his roof. The section of roof is a trapezoid with the dimensions shown. How many square feet of shingles does he need to cover this section of the roof?

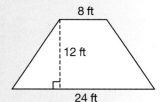

Solution

The parallel sides are the bases. We find the average length of the bases.

$$\text{average length of the bases} = \frac{8 \text{ ft} + 24 \text{ ft}}{2} = 16 \text{ ft}$$

Then we multiply the average of the bases by the height.

Area of trapezoid = average length of the bases · height

$$= 16 \text{ ft} \cdot 12 \text{ ft}$$

$$= 192 \text{ ft}^2$$

The section of roof is covered with **192 ft²** of shingles.

Example 2

The bases of a trapezoid are 12 cm and 18 cm. The height of the trapezoid is 8 cm. Find the area of the trapezoid using the formula $A = \frac{1}{2}(b_1 + b_2) \cdot h$.

Solution

We substitute the values into the equation and then we simplify.

Step:	Justification:
$A = \frac{1}{2}(b_1 + b_2) \cdot h$	Area formula for trapezoid
$A = \frac{1}{2}(12 \text{ cm} + 18 \text{ cm})8 \text{ cm}$	Substituted
$A = \frac{1}{2}(30 \text{ cm})8 \text{ cm}$	Simplified within parentheses
$A = 15 \text{ cm} \cdot 8 \text{ cm}$	Multiplied $\frac{1}{2} \cdot 30$ cm
$A = \mathbf{120 \text{ cm}^2}$	Multiplied 15 cm · 8 cm

Model Make a drawing of this trapezoid with the bases and height labeled.

Practice Set

Find the area of each trapezoid.

a.
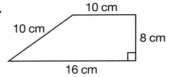

b.
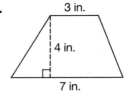

c.
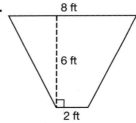

d.
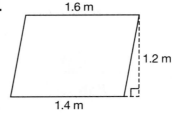

e. The shingles on the south side of Tamika's roof need to be replaced. The shingles she wants come in bundles that cover $33\frac{1}{3}$ square feet. Tamika calculates the area of the section of roof and then determines the number of bundles she needs. She wants to buy two extra bundles to allow for cutting and waste. What is the area of the section of roof and how many bundles should Tamika buy?

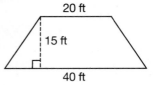

1. Three out of four doctors surveyed recommend the product. If 200 of
(45) the doctors surveyed do not recommend it, how many doctors were
surveyed?

2. Sally bought a collectible at a flea market for $120. She sold it on-line
(67) for $156. The selling price was what percent more than her purchase
price?

3. Sixty-five percent of the tuba players had red cheeks. If there were
(48) fourteen tuba players without red cheeks, how many were there
in all?

* 4. *Analyze* If y varies directly with x, then a and
(69) b in the table equal what numbers?

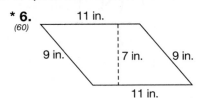

x	2	8	10	a
y	6	24	b	45

Classify For **5** and **6,** name the type of quadrilateral and find its area.

* 5.
(75)

14 m

10 m   10 m

8 m

26 m

* 6.
(60)

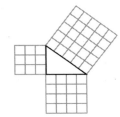

11 in.

9 in.   7 in.   9 in.

11 in.

7. This figure best illustrates
(Inv. 2)
   A the area of a triangle.

   B the perimeter of a triangle.

   C the Pythagorean Theorem.

   D the volume of a prism.

Solve.

8. $0.04 - 0.02x = 0.5$
(50)

9. $\frac{7}{3}x + 1 = \frac{17}{3}$
(50)

10. $\frac{3}{4}r = 33$
(38)

11. $7x + 1 - x = 19$
(50)

12. $\frac{4}{x} = \frac{28}{56}$
(44)

* 13. $\frac{4}{x} = \frac{2.8}{5.6}$
(25, 44)

Generalize Simplify.

* 14. $\sqrt{45}$
(74)

* 15. $\sqrt{50}$
(74)

16. *Formulate* Write $\frac{8}{11}$ as a **a** percent and **b** decimal. **c** Then round the
(63) decimal to the nearest thousandth.

17. If a square is dilated by a scale factor of 2, then the area of the
(26, 71) dilated square is what percent greater than the area of the original
square?

18. During the season a basketball player has made 42 out of 60 free
(59) throws. What is the probability the player will make a free throw?

19. Use unit multipliers to convert 15 miles per hour to feet per second.
(72)

20. Graph $y = \frac{2}{3}x - 3$. Is (9, 3) a solution?
(41)

21. **a.** Find a and b.
(35)

    b. How do we know the triangles are similar?

    c. Classify the triangles by sides.

    d. What is the scale factor from the smaller to the larger triangle?

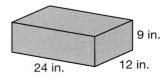

22. A cardboard carton has the given
(42, 43) dimensions. Convert the dimensions to feet
and then find the capacity of the carton in
cubic feet and the surface area of the carton
in square feet.

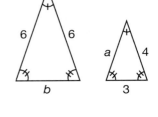

23. A formula for the following sequence is $a_n = 3(n - 1)$. Find the 20th term
(73) of the sequence.

$$0, 3, 6, \ldots$$

24. Sketch a net of this triangular prism.
(36)

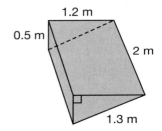

25. Kathy has a prepaid phone card
(69) worth $6.00. Every minute, m, she
uses the phone, $0.02 is deducted
from the phone card balance, b. The
graph indicates this relationship. Is
this relationship an example of direct
variation? If so, what is the constant
of proportionality? If not, why?

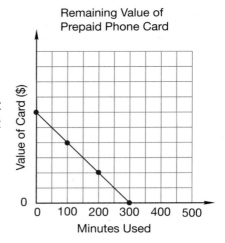

• Volumes of Prisms and Cylinders

Power Up | *Building Power*

facts | Power Up P

mental math

a. Algebra: Simplify: $5 + 2x + 6x$

b. Sequences: A rule of a sequence is $a_n = 5n$. What are the first four terms?

c. Rate: Marvin loads the truck at a rate of 15 boxes per minute. Craig loads 12 boxes per minute. How many boxes can they load together in 3 minutes?

d. Number Sense: Arrange A, B, C and D from least to greatest:

    A 0.04   **B** 0.08   **C** 1.2   **D** 0.016

e. Powers/Roots: $x^4 \cdot x^7 \cdot x^{-1}$

f. Geometry: Rewrite and complete this congruence statement:
$\Box ABCD \cong \Box YX\_\_$

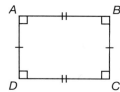 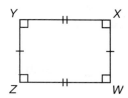

g. Select a Method: Mr. Chen is completing his income tax forms. To perform the calculations, which method is he likely to use?

    A mental math
    B pencil and paper
    C calculator

h. Calculation: How many minutes is a quarter of an hour plus a third of an hour?

problem solving

Theodore counted by threes, saying one number every second. Franklin started 12 seconds later and counted by fives, saying one number every second. Will they ever say the same number at the same time? If so, what will that number be?

> Hint: What number did Theodore say right before Franklin started? How much does Franklin gain on Theodore each second?

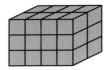

In Lesson 42 we found the volume of a rectangular solid by first calculating the area of the base and then multiplying the area of the base times the height.

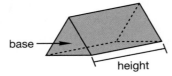

Volume = Area of base · height
$$V = Bh$$

We may use the same process to find the volume of other prisms and cylinders. We find the area of one of the bases, whether it is a polygon or a circle, and multiply that area by the height.

Recall that the bases of a prism or a cylinder are the parallel surfaces at opposite ends of the figure. The height is the perpendicular distance between the bases, whether or not the prism or cylinder is upright.

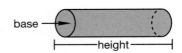

Example 1

Sketch the base of this triangular prism. Find the volume.

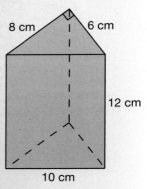

Solution

The base of the triangular prism is a right triangle.

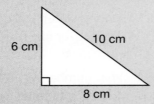

The perpendicular sides are 6 cm and 8 cm, so the area of the triangular base is 24 cm².

Step:	Justification:
$A = \frac{1}{2}bh$	Area formula for triangle
$A = \frac{1}{2}(8 \text{ cm})(6 \text{ cm})$	Substituted
$A = 24 \text{ cm}^2$	Simplified

Thinking Skill

Discuss

How do units used for area differ from units used for volume?

The height of the prism is the distance between the bases, which is 12 cm. We find the volume by multiplying the area of the base times the height.

Step:	Justification:
$V = Bh$	Volume formula
$V = (24 \text{ cm}^2)(12 \text{ cm})$	Substituted
$V = 288 \text{ cm}^3$	Simplified

The volume of the triangular prism is **288 cm³**.

Analyze Notice that in the formulas in example 1, the variable representing *base* appears in both lower-case form (*b*) and upper-case form (*B*). What is the difference in the meanings of *b* and *B*?

Example 2

A double-A battery is about 50 mm long and has a diameter of about 14 mm. Find the approximate volume of a double-A battery. (Use $\frac{22}{7}$ for π.)

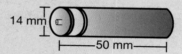

14 mm

50 mm

Solution

The base is a circle. We find the area of the circle substituting $\frac{22}{7}$ for π.

Step:	Justification:
$A = \pi r^2$	Area formula for circle
$A \approx \frac{22}{7}(7 \text{ mm})^2$	Substituted
$A \approx 154 \text{ mm}^2$	Simplified

Reading Math

Recall that the symbol $\approx$ means *is approximately equal to.*

We find the volume of the cylinder by multiplying the area of the base by the height.

Step:	Justification:
$V = Bh$	Volume formula
$V \approx (154 \text{ mm}^2)(50 \text{ mm})$	Substituted
$V \approx 7700 \text{ mm}^3$	Simplified

The volume of a double-A battery is about **7700 mm³**.

Example 3

The figure shows a triangle that is the base of a triangular prism. The height of the prism is 10 cm. Sketch the prism. Then find its volume.

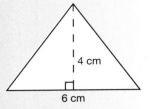

We sketch the prism.

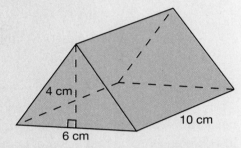

To find the volume we first find the area of the triangular base.

$$A = \frac{1}{2}bh$$

$$A = \frac{1}{2}(6 \text{ cm})(4 \text{ cm})$$

$$A = 12 \text{ cm}^2$$

Then we multiply the area of the base by the height.

$$V = bh$$

$$V = 12 \text{ cm}^2 \cdot 10 \text{ cm}$$

$$V = 120 \text{ cm}^3$$

The volume is **120 cm³**.

Example 4

Buster has two empty cylindrical soup cans. The smaller can is 2 in. in diameter and 3 in. high. The dimensions of the larger can are twice the dimensions of the smaller can.

   a. Find the volume of both cans in terms of π.

   b. What is the scale factor from the smaller can to the larger can?

   c. If the smaller can is filled with water and the contents are poured into the larger can and the process is repeated until the larger can is full, how many small cans of water would be used to fill the larger can?

   a. Smaller can: **3π in.³**

      Larger can: **24π in.³**

b. The dimensions of the larger can are twice the dimensions of the smaller can, so the scale factor is **2.**

c. We cube the scale factor to find the relationship between the volumes of the cans: $2 \cdot 2 \cdot 2 = 8$. **Eight** small cans will fill the larger can.

Practice Set

a. *Analyze* A pup-tent has the dimensions shown. Sketch the base and then find the volume of the tent.

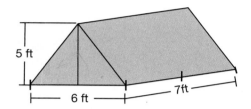

b. A cylindrical backyard pool 3 feet high and 20 feet in diameter holds how many cubic feet of water when full? (Use 3.14 for π and round the answer to the nearest ten cubic feet.)

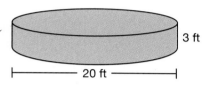

c. The walls of a garage are 8 feet high, and the peak of the gable roof is 12 feet high. The floor of the garage is a 20 foot square. Sketch the base. Then find the volume of the garage.

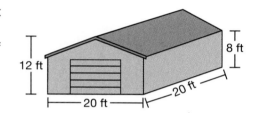

d. A cylindrical candle that is 15 cm high has a diameter of 8 cm. Sketch the shape. Find the volume of the candle (express in terms of π).

e. Brenda is putting cubical boxes that are 6 inches on edge into a larger cubical box with inside dimensions 12 inches on edge. How many of the smaller boxes will fit in the larger box?

Written Practice   *Strengthening Concepts*

* 1. *(67)* *Analyze* A store purchases an item for $80. It marks up the item by 50%. What is the customer's price?

2. *(45)* There were 600 sheep and goats in the hills. If the ratio of sheep to goats is 1 to 2, how many sheep are there?

3. *(48)* Seventy-six percent of the brass instruments in the band leading the parade were trombones. If there were 100 brass instruments, how many were trombones?

* 4. **Explain** In the table are shown the number
(69) of hours Wendy works and the amount of
pay she receives. Does this table show
direct variation? If so, give the constant of
variation. What does the constant mean in
this situation?

Hours	Pay
3	27
4	36
5	45
6	54

* 5. To roughly estimate the volume of the
(76) reservoir pool, Isabel sketches this
diagram. What volume does she estimate?

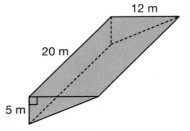

* 6. **Estimate** Suppose a municipal water tank has
(76) these dimensions. Find the volume of the tank.
First express the volume in terms of π. Then
approximate the volume to the nearest whole
unit.

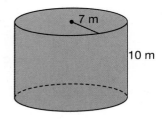

* 7. What is the volume of one cubic yard in cubic feet? Use three unit
(64, 72) multipliers.

8. Find a and b in the similar triangles to
(35) the right.

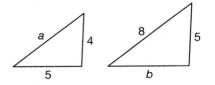

9. **a.** Find the area of the parallelogram.
(Inv. 3, **b.** Name the type of parallelogram.
60)

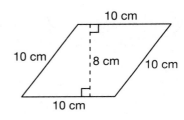

* 10. For **a–c,** refer to the figure.
(Inv. 2,
75) **a.** Classify the quadrilateral.

b. Find its area.

c. Find its perimeter.

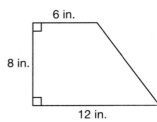

* 11. When fractions with a denominator of 7 are converted to decimals, there
(63)   are six digits that repeat. Write **a** $\frac{1}{7}$ as a decimal and **b** $\frac{2}{7}$ as a decimal.
       c What do you notice about the two decimal numbers?

12. If a cube-shaped box with edges one foot long is stored in a
(42, 48)   cube-shaped box with edges two feet long, then what percent of the
        volume of the larger box is occupied by the smaller box?

* 13. *Generalize* A rule of a sequence is $a_n = 7(2^n)$. Find the first three terms
(73)   of the sequence.

14. Factor:
(21)
     a. $-2x^2 - 2x - 2$ **b.** $5x^2 - 10x$

15. What is the equation of a line passing through the origin with a slope of
(56)   positive one?

Generalize Simplify.

16. $(-3)(-2) - (-1)^2$ *** 17.** $\sqrt{27}$
(36)   (74)

* 18. $\sqrt{32}$ **19.** $\dfrac{12m^3}{18xm^{-2}}$
(74)   (27, 51)

Solve.

* 20. $\dfrac{4.5}{6.3} = \dfrac{x}{7}$ *** 21.** $\dfrac{4}{3}x - \dfrac{2}{7} = -\dfrac{34}{21}$
(25, 44)   (50)

* 22. $\dfrac{x}{4} = \dfrac{0.42}{0.14}$ **23.** $0.03 + 0.011x = 0.36$
(25, 44)   (50)

24. $\dfrac{m}{5} = 2.2$
(25, 38)

25. Describe the transformation
(Inv. 5)   applied to figure
          ABCD to create figure
          A'B'C'D'.

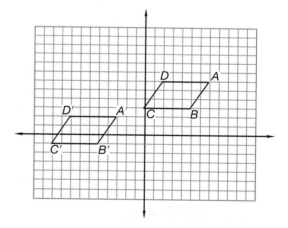

Inequalities with Negative Coefficients

Building Power

facts

Power Up P

mental math

a. **Algebra:** Simplify: $2x + 4 + 7x$

b. **Fractional Parts:** $\frac{3}{8}$ of 56

c. **Probability:** 90% of the students were right handed. What is the probability that one student selected at random will be left handed?

d. **Percent:** 60% of $10

e. **Proportions:** The ratio of runners to cyclists was 9 to 4. If there were 26 all together, how many were cyclists?

f. **Measurement:** What mass is indicated on this scale?

g. **Select a Method:** Before depositing checks at the bank, Imelda fills out a deposit slip. If she is depositing three checks, she will probably use what method?
A mental math
B pencil and paper
C calculator

h. **Calculation:** $100 - 1, \div 9, \times 4, + 1, \div 9, \times 2, - 1, \div 9$

problem solving

Is it possible to remove one block from each side and have the scale still be balanced? If so, which blocks should be removed?

Increasing Knowledge

Solving inequalities is similar to solving equations. However, there is one important difference: when solving inequalities, we reverse the comparison symbol when multiplying or dividing by a negative number.

In the table below we summarize properties of inequalities. Notice the distinction in the table between positive and negative multipliers and divisors.

Properties of Inequalities

If $a > b$, then	
$a + c > b + c$	
$a - c > b - c$	
$ac > bc$	if c is positive
$\dfrac{a}{c} > \dfrac{b}{c}$	if c is positive and $c \neq 0$
$ac < bc$	if c is negative
$\dfrac{a}{c} < \dfrac{b}{c}$	if c is negative and $c \neq 0$

Connect Why can c not equal zero in the inequalities $\frac{a}{c} > \frac{b}{c}$ and $\frac{a}{c} > \frac{b}{c}$?

Example 1

Multiply both sides of this inequality by negative one and explain how to preserve the truth of the inequality.

$$6 > 5$$

Solution

The given inequality is true. Six is greater than five. However, if we multiply both sides of the inequality by -1, then the numbers become -6 and -5. **Since -6 is less than -5, we must reverse the sign of the inequality to preserve the truth of the statement.**

$$-6 < -5$$

Example 2

Solve and graph $-3x - 10 > 8$.

Solution

Step:	Justification:
$-3x - 10 > 8$	Given inequality
$-3x > 18$	Added 10 to both sides
$x < -6$	Divided both sides by -3 and reversed comparison symbol

We graph the result.

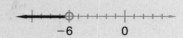

Thinking Skill

Explain

What does the open circle on the graph mean? How would the inequality change if the circle were filled?

Only numbers less than -6 satisfy the original inequality. To check the solution we can substitute a number less than -6 in place of x and simplify. The resulting inequality should be true. We can also substitute a number equal to or greater than -6 in place of x. The resulting inequality should be false.

$$\begin{array}{cc}
\text{Substitute } -7 & \text{Substitute } -5 \\
-3(-7) - 10 > 8 & -3(-5) - 10 > 8 \\
21 - 10 > 8 & 15 - 10 > 8 \\
11 > 8 \text{ (true)} & 5 > 8 \text{ (false)}
\end{array}$$

Justify Is -6 a solution? Explain why or why not.

Example 3

Solve and graph.

$$-5(x - 3) > 5(x - 3)$$

Solution

Step:	Justification:
$-5(x - 3) > 5(x - 3)$	Given inequality
$-5x + 15 > 5x - 15$	Distributive Property
$-10x + 15 > -15$	Subtracted $5x$ from both sides
$-10x > -30$	Subtracted 15 from both sides
$x < 3$	Divided both sides by -10 and reversed the comparison symbol.

Example 4

Marci sees a shirt she likes that costs $50. She will not buy it unless it is $30 or less. To find the percent decrease in the price required, she solves the inequality:

$$50 - 50x \le 30$$

Solve the inequality to find the minimum discount Marci requires.

Solution

First we solve the inequality.

Step:	Justification:
$50 - 50x \le 30$	Given inequality
$-50x \le -20$	Subtracted 50 from both sides
$x \ge \dfrac{20}{50}$	Divided both sides by -50 and reversed comparison symbol

Now we express the result as a percent.

$$x \ge 40\%$$

Marci will not buy the shirt unless it is discounted **40%** or more.

Practice Set

Model Solve. Then graph each inequality to show your work.

a. $-5x - 6 < -1$ **b.** $7x + 2 \le 8x + 4$

c. $-5x + 25 > 5(x - 5)$ **d.** $-2x + 3 \ge -(x - 7)$

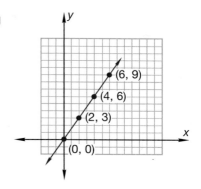

Written Practice *Strengthening Concepts*

1. The ratio of broken to working piano keys was 2 to 9. If there were
(45) 88 keys on the old piano, how many were broken?

2. Last month the snake had a mass of 6.6 kilograms. Its mass is now 60%
(67) more than it was last month. What is the snake's mass now?

* 3. Dena ordered a $8.40 meal and Jules ordered a $9.20 meal. If the sales
(58) tax is 8% and they leave a 20% tip on the price of the meal including
tax, then what is the total?

4. Rosa watched the distance markers along
(69) the highway. She noticed that she had
traveled 3 km in 2 min, 6 km in 4 min,
and 9 km in 6 min. Does this graph show
direct variation? If so, give the constant
of proportionality. How far will she have
traveled in 10 min?

* 5. A calling card charges 50¢ for each call placed plus 3¢ per minute
(41, 69) on the call.

a. Make a table that lists possible durations of phone calls and the
corresponding charges.

b. Write an equation for the function.

c. Is the function an example of direct variation? Why or why not?

* 6. *Connect* O'Grady averages 5 baskets for every 8 free throw attempts.
(Inv. 6) If you were to construct a spinner to simulate O'Grady's free throw
attempts, how many degrees would represent free throws made? Free
throws missed?

Solve these inequalities and graph their solutions on a number line.

* 7. $-x + 15 > 0$ *** 8.** $-2x \ge 2$
(77) (77)

Analyze Solve. If there is no solution, write, "no solution."

* 9. $\dfrac{4}{x} = \dfrac{0.22}{0.55}$ **10.** $\dfrac{3}{4} - \dfrac{2}{3}x = \dfrac{11}{12}$
(25, 44) (50)

11. $12 = 16 - 4x$ **12.** $\dfrac{7}{8}x = 49$
(14) (38)

13. If the dimensions of a 3-4-5 right triangle are increased 50%, by what
(71) percent is the area increased?

14. The rule for a sequence is $a_n = n^2 - n$. The first three terms are 0, 2,
(73) and 6. What is the tenth term?

Classify For problems **15** and **16,** name the type of quadrilateral and find its
area.

* 15.
(75)

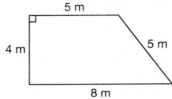

5 m

4 m

5 m

8 m

* 16.
(60)

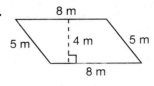

8 m

5 m   4 m   5 m

8 m

Simplify.

* 17. $\sqrt{54}$
(74)

18. $\dfrac{7x^{-3}y^3z^3}{14x^4y^3z^2}$
(27, 51)

19. Estimate each number between a range of two consecutive integers.
(16)
   a. $\sqrt{90}$ **b.** $\sqrt{80}$ **c.** $\sqrt{70}$

20. Geneka watched the bug scuttle 10 inches in 2 seconds. Use unit
(72) multipliers to convert 10 inches in 2 seconds to feet per minute.

21. Graph $y = -\frac{1}{2}x + 2$. Is (2, 1) a solution?
(56)

Find the volume of each solid. Express your answer in terms of π.

* 22.
(76)

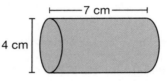

7 cm

4 cm

* 23.
(76)

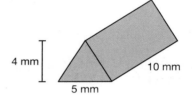

4 mm

5 mm

10 mm

24. Write $\frac{4}{11}$ as a **a** decimal and **b** percent.
(63)

25. The table shows the favorite fruit of the eighth-grade class at Canyon
(11, Vista Middle School.
Inv. 6)

Favorite Fruit	Percent of Class
strawberries	13.3
oranges	23.3
bananas	36.7
blueberries	6.7
grapes	20.0

   a. What percent of the class preferred strawberries or bananas?

   b. If there are 30 eighth-graders in Mrs. Li's class, what is a good
   estimate for the number of students who prefer blueberries?
      A 21   **B** 18   **C** 7   **D** 2

   c. What fraction of eighth-graders do not prefer grapes?

• Products of Square Roots

facts | Power Up P

mental math

a. Statistics: Find the median of these numbers: 13, 15, 15, 15, 12

b. Estimation: $6 - \sqrt{15}$

c. Number Sense: Arrange A, B, C, and D from least to greatest:

  A 0.2   **B** 0.8   **C** 0.2 + 0.8   **D** 0.2 × 0.8

d. Powers/Roots: $x^4 \cdot x^{-3}$

e. Proportion: $\frac{x}{12} = \frac{2}{3}$

f. Geometry: Draw an equilateral triangle. Does it have a right angle?

g. Scientific Notation: Write 4.01×10^8 in standard notation.

h. Calculation: $16 \div 2, \div 2, \div 2, \div 2, \div 2, \div 2$

problem solving

Karina and Kristina are facing each other. Karina turns around once every 3 seconds. Kristina turns around once every 4 seconds. How long will it take for Karina and Kristina to face each other again? If they turn at this rate for a minute, how many times will they face each other?

New Concept | *Increasing Knowledge*

Math Language

The number under the radical symbol is called the **radicand.** In the expression $\sqrt{12}$, 12 is the radicand.

Recall the product property of square roots:

$$\sqrt{ab} = \sqrt{a}\sqrt{b}$$

This property means that square roots can be factored.

$$\sqrt{3} = \sqrt{} \cdot \sqrt{} = 4\sqrt{}$$

This property also means square roots can be multiplied. Here is one example.

$$\sqrt{4} \cdot \sqrt{} = \sqrt{3}$$

$$2 \cdot 3 = 6$$

In this lesson we will practice multiplying square roots.

Example 1

Find the length and width of this rectangle. Then find its area.

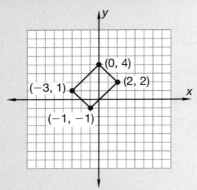

Solution

Using the Pythagorean Theorem we find that the length is $\sqrt{18}$, which simplifies to $\mathbf{3\sqrt{2}}$, and the width is $\sqrt{8}$, which simplifies to $\mathbf{2\sqrt{2}}$. To find the area we multiply the length and width.

$$\sqrt{18}\,\sqrt{8} = 3\sqrt{2} \cdot 2\sqrt{2}$$
$$\sqrt{144} = 6\sqrt{4}$$
$$\mathbf{12 = 12}$$

Using either the unsimplified or the simplified square roots, we find that the area is **12 square units.**

> *Verify*   How can we determine by the illustration that the area is 12 square units?

Example 2

The area of □*PQRS* is what percent of the area of □*ABCD*?

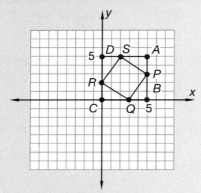

Solution

Square *ABCD* is 25 square units. To find the area of square *PQRS* we square the length of its side. We use the Pythagorean Theorem to find the length.

$$s^2 = 2^2 + 3^2$$
$$s^2 = 4 + 9$$

$$s^2 = 13$$
$$s = \sqrt{13}$$

Each side of $\square PQRS$ is $\sqrt{13}$, so the area of $\square PQRS$ is

$$(\sqrt{13})^2 = \sqrt{13} \cdot \sqrt{13} = \sqrt{169} = 13$$

Square $PQRS$ is $\frac{13}{25}$ of the area of $\square ABCD$. We convert the fraction to a percent.

$$\frac{13}{25} \cdot \frac{4}{4} = \frac{52}{100} = 52\%$$

We find the area of $\square PQRS$ is **52%** of the area of square $ABCD$.

Model Was it necessary to find the length of one side of $\square PQRS$ in order to find its area? Sketch $\triangle PBQ$ and draw squares on each side to represent the squares of the legs and hypotenuse. Does square $PQRS$ appear in your sketch? Now look at the solution steps for finding the length of a side. Does the area of the square appear in the solution steps? Explain.

Example 3

Simplify:

   a. $\sqrt{3} \cdot \sqrt{12}$ b. $\sqrt{42} \cdot \sqrt{24}$ c. $3\sqrt{6} \cdot \sqrt{15}$

Solution

a.

Step:	Justification:
$\sqrt{3} \cdot \sqrt{12}$	Given
$\sqrt{36}$	Multiplied $\sqrt{3}$ and $\sqrt{12}$
6	Simplified

b.

Step:	Justification:
$\sqrt{42} \cdot \sqrt{24}$	Given
$\sqrt{42 \cdot 24}$	Multiplied $\sqrt{42}$ and $\sqrt{24}$
$\sqrt{2 \cdot 3 \cdot 7 \cdot 2 \cdot 2 \cdot 2 \cdot 3}$	Factored 42 and 24
$\sqrt{2 \cdot 2}\sqrt{2 \cdot 2}\sqrt{3 \cdot 3}\sqrt{7}$	Grouped pairs
$2 \cdot 2 \cdot 3\sqrt{7}$	Simplified radicals
$12\sqrt{7}$	Multiplied $2 \cdot 2 \cdot 3$

c.

Step:	Justification:
$3\sqrt{6}\sqrt{15}$	Given
$3\sqrt{6 \cdot 15}$	Multiplied $\sqrt{6}$ and $\sqrt{15}$
$3\sqrt{2 \cdot 3 \cdot 3 \cdot 5}$	Factored 6 and 15
$3\sqrt{3 \cdot 3}\sqrt{2 \cdot 5}$	Grouped pair of 3s
$3 \cdot 3 \cdot \sqrt{10}$	Simplified radical
$9\sqrt{10}$	Multiplied $3 \cdot 3$

Practice Set | Simplify.

     a. $\sqrt{2}\sqrt{14}$ **b.** $\sqrt{15}\sqrt{3}$

     c. $3\sqrt{5}\sqrt{5}$ **d.** $5\sqrt{3}\cdot 2\sqrt{3}$

     e. Find the length of a side of the square with vertices at (5, 4), (1, 5), (0, 1), and (4, 0). Then find the area of the square.

     f. Find the area of a rectangle with vertices at (0, 5), (3, −1), (−1, −3), and (−4, 3).

     g. *Explain* Describe two ways to find the area of the shaded square, one way using radicals and one way without using radicals.

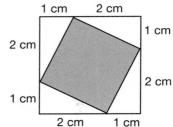

Written Practice   *Strengthening Concepts*

* 1. *(71)* **Evaluate** A perpendicular cut through a water pipe reveals a circular cross section. The diameter of a $\frac{3}{4}$-inch pipe is 50% greater than the diameter of a $\frac{1}{2}$-inch pipe. The cross-sectional area of the larger pipe is what percent greater than the cross-sectional area of the smaller pipe?

* 2. *(67)* **Analyze** This year there was a thirty-five percent reduction in spending from last year's budget. If last year's budget was $1,200,000, what was this year's spending?

3. *(48)* About twelve percent of the 4000 roses had already opened. About how many of the roses had opened?

4. *(70)* The number of apples on display varies directly as the number of boxes the produce manager unloads. If the manager unloads 2 boxes there are 60 apples. How many apples are there when he unloads 3 boxes?

Simplify.

* 5. *(74, 78)* $\sqrt{8}\sqrt{6}$ *** 6.** *(74, 78)* $\sqrt{21}\sqrt{3}$

* 7. *(74, 78)* $\sqrt{6}\sqrt{12}$ **8.** *(27, 51)* $\dfrac{8m^4 b^3}{4m^2 b^{-1}}$

Solve.

* 9. *(25, 44)* $\dfrac{0.32}{0.56} = \dfrac{8}{m}$ **10.** *(50)* $0.5 - 0.02x = 0.1$

11. *(50)* $\dfrac{6}{7} - \dfrac{1}{2}x = -\dfrac{1}{7}$ **12.** *(38)* $\dfrac{2}{3}m = 6$

For problems **13** and **14,** name the type of quadrilateral and find its area. Dimensions are in cm.

* 13.
(75)

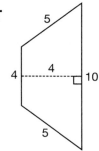

* 14.
(60)

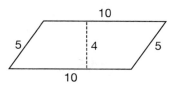

15. Estimate each number between a range of two consecutive integers.
(16)
   a. $\sqrt{60}$ **b.** $\sqrt{50}$ **c.** $\sqrt{40}$

16. Use unit multipliers to convert $15 per hour to cents per minute.
(72)

17. Graph $y = -\frac{1}{2}x - 5$. Is $(-10, 0)$ a solution?
(56)

18. A five-foot tall person casts a six-foot shadow when a nearby tree casts
(65) a 30 foot shadow. Sketch a diagram for the description and find the approximate height of the tree.

Analyze For problems **19** and **20** find the volume of each solid. For problem **20** leave in terms of π.

* 19.
(76)

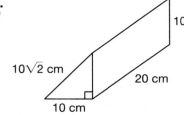

* 20.
(76)

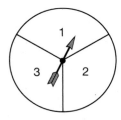

21. A machine shrink-wraps 100 packages every 4 minutes. Which
(34) equation shows how many packages the machine can shrink-wrap in 10 minutes?

  A $\dfrac{100}{4} = \dfrac{x}{10}$ **B** $\dfrac{100}{x} = \dfrac{10}{4}$ **C** $\dfrac{100}{10} = \dfrac{x}{4}$ **D** $100 \cdot 4 = 10 \cdot x$

22. A number cube is rolled and the spinner
(63, 68) is spun. What is the probability that both outcomes will be even numbers? Express the probability as a ratio, as a percent, and as a decimal number rounded to the nearest hundredth.

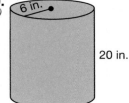

Solve these inequalities and graph their solutions on a number line.

23. $4(x + 1) > 0$ **24.** $2(3 - x) \le 2$
(77)   (77)

25. (Inv. 4) **Analyze** Use this figure to answer the following questions.

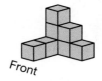

Front

a. Which of the following is the top view of the figure?

A

B

C

D

b. Which of the following is the front view of this figure?

A

B

C

D

• Transforming Formulas

Power Up Building Power

facts Power Up P

mental math

a. Algebra: Simplify: $12x - x - 9x$

b. Sequences: A rule of a sequence is $a_n = 6n$. What are the first four terms?

c. Fractional Parts: $\frac{3}{7}$ of 63

d. Percent: 0% of $400

e. Proportions: 10 push-ups in 20 seconds is a rate of how many per minute?

f. Measurement: Find the volume indicated by this figure.

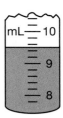

g. Geometry: What is the specific name for an equilateral quadrilateral that is not a square?

h. Calculation: Square 5, $\times$ 2, $-$ 2, $\div$ 2, $+$ 1, $\sqrt{}$

problem solving Patty, Laura and Barbara traveled by plane, train, and automobile. Patty did not take the train, and Barbara didn't travel on the ground. Which form of transportation did each woman use?

New Concept *Increasing Knowledge*

We have used formulas to solve problems about rates and about geometric figures. Standard formulas are expressed with one variable isolated. For example, a rate equation states that distance traveled equals rate of travel times the amount of time traveled.

$$\text{distance} = \text{rate} \cdot \text{time}$$
$$d = rt$$

If we know the rate and time we substitute those values for r and t and then simplify to find d.

$$d = (60 \text{ mph})(2 \text{ hours})$$

However, if we know distance and rate, then we need to solve the equation to find the time.

$$240 \text{ mi} = (60 \text{ mph})t$$

Instead of solving the equation *after* substituting values, we can rearrange the formula *before* we substitute. Since we know we want to find time, we transform the formula to isolate t.

Step:	Justification:
$d = rt$	Distance formula
$\dfrac{d}{r} = \dfrac{rt}{r}$	Divided both sides by r
$\dfrac{d}{r} = t$	Simplified
$t = \dfrac{d}{r}$	Symmetric property of equality

Example 1

The illustration shows that w equals the sum of x and b.

$$w = x + b$$

Solve this equation for x.

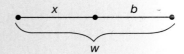

Solution

Step:	Justification:
$w = x + b$	Given equation
$w - b = x$	Subtracted b from both sides
$x = w - b$	Symmetric property of equality

The equation, solved for x, is $x = w - b$.

Referring to the diagram, we find x by subtracting length b from length w.

Example 2

If we measure the circumference of a basketball, then we can calculate its diameter using the formula $C = \pi d$. Transform the formula to solve for d. Then describe the meaning of the transformed formula.

Solution

We isolate d by dividing both sides by π.

Step:	Justification:
$C = \pi d$	Given equation
$\dfrac{C}{\pi} = \dfrac{\overset{1}{\cancel{\pi}}d}{\underset{1}{\cancel{\pi}}}$	Divided both sides by π
$\dfrac{C}{\pi} = d$	Simplified
or $d = \dfrac{C}{\pi}$	Symmetric property of equality

The formula means that **to find the diameter of a circle (or of the basketball), we divide the circumference by π.**

Example 3

The formula for the area of a rectangle is $A = lw$. Transform the formula to solve for l. Then use the transformed formula to find the length of a rectangle with area 42 and width 3.

l

w

Solution

We isolate l by dividing both sides by w.

Step:	Justification:
$A = lw$	Given equation
$\dfrac{A}{w} = \dfrac{l\cancel{w}}{\cancel{w}}$	Divided both sides by w
$l = \dfrac{A}{w}$	Symmetric property of equality

If $A = 42$ and $w = 3$, then $l = \dfrac{A}{w} = \dfrac{42}{3} = $ **14 units.**

Example 4

Solve for r: $A = \pi r^2$. Then find the radius of a circle with an area of 154 cm². Use $\dfrac{22}{7}$ for π.

Solution

Thinking Skill

Explain

Explain the formula $r = \sqrt{\dfrac{A}{\pi}}$ in words.

Isolate r by dividing both sides by π, then taking the square root.

Step:	Justification:
$A = \pi r^2$	Given equation
$\dfrac{A}{\pi} = \dfrac{\overset{1}{\cancel{\pi}} r^2}{\underset{1}{\cancel{\pi}}}$	Divided both sides by π
$\dfrac{A}{\pi} = r^2$	Simplified
$\sqrt{\dfrac{A}{\pi}} = r$	Square root of both sides
$r = \sqrt{\dfrac{A}{\pi}}$	Symmetric property of equality

We can use this transformed formula to find the radius. We substitute 154 for area and $\dfrac{22}{7}$ for π.

$$r \approx \sqrt{\dfrac{154}{\dfrac{22}{7}}}$$

$$r \approx \sqrt{154 \cdot \dfrac{7}{22}}$$

$$r \approx \sqrt{49} = 7$$

The radius is about **7 cm.**

Practice Set

a. The formula for the area of a rectangle is $A = lw$. Solve the formula for w. Then describe the meaning of the transformed formula.

b. Solve the equation $P = a + b + c$ for a.

c. The Pythagorean Theorem is

$$a^2 + b^2 = c^2$$

Solve for c.

d. The formula for the perimeter of a rectangle is $P = 2l + 2w$. Solve the formula for l. Then use your formula to find the length of a rectangular field with a perimeter of 620 feet and a width of 140 feet.

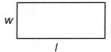

Written Practice   *Strengthening Concepts*

1. Last year's meeting had 18,000 attendees. This year 20,700 attended.
(67) This was an increase of what percent?

2. If 46% of the 150 people responded to the survey, how many
(48) did not?

* 3. **Analyze** One cube has edges 2 ft long. Another cube has edges 50%
(43, 71) longer. The surface area of the large cube is what percent greater than the surface area of the smaller cube?

* 4. Solve for r. $C = 2\pi r$
(79)

* 5. Two red marbles and one blue marble are in a bag. One marble is
(32) selected then placed back in the bag, then a marble is selected again.
   a. Find the sample space.
   b. What is the probability of selecting the blue marble twice?

6. Write $\frac{5}{9}$ as a **a** percent and **b** decimal. **c** If 5 out of 9 people surveyed
(63) favor an initiative, is the fraction of those who favor the initiative closer to one half or two thirds?

* 7. A portion of a wall that runs along a highway
(75) has this shape and will be painted on one side. Name the shape to be painted and find its area.

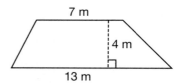

Solve.

8. $0.03x + 0.1 = 0.7$
(50)

9. $\frac{2}{3}x - \frac{1}{2} = \frac{1}{6}$
(50)

* 10. $3(x - 2) = x + 6$
(38)

* 11. $2x + 3 = x - 5$
(79)

Generalize Simplify.

* 12. $\sqrt{30}\ \sqrt{3}$
(74, 78)

* 13. $\sqrt{8}\ \sqrt{5}$
(74, 78)

* 14. $(-2)(-3) + (-2) - (-3)$
(31, 36)

15. $\dfrac{32r^2m^3}{16r^2m^{-1}}$
(27, 51)

* 16. Name the shape and find its area.
(60)

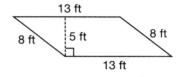

17. The collect call will cost Trini 50¢ per minute. Use unit multipliers to convert 50¢ per minute to dollars per hour.
(72)

18. Graph $y = \frac{1}{3}x$. Is (15, 3) a solution?
(41)

* 19. Carmen designed the company's logo using similar triangles like these. Find x and y. (Hint: Draw the three triangles separately.)
(65)

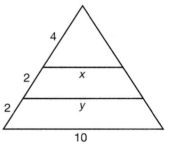

20. A slab of concrete with these dimensions will support a gazebo at the park. What volume of concrete is needed? (Use $\pi \approx 3.14$)
(76)

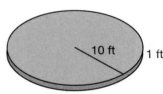

* 21. Concrete is ordered by the cubic yard. How many whole cubic yards should be ordered for the slab in problem **20?**
(72)

* 22. Find the volume of the triangular prism.
(76)

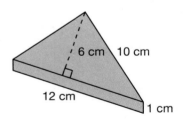

* 23. The weight of an object on the moon varies directly with the object's weight on earth. If an astronaut who weighs 150 pounds on earth weighs 25 pounds on the moon, what is the weight on the moon of an astronaut who weighs 180 pounds on earth?
(51)

* 24. Solve the inequality and graph the solution.
(77, 79)

$$2x + 2 \geq x + 6$$

25. **a.** Figure *WXYZ* is rotated 90° clockwise
(Inv. 5) about point *Z*. Which are the
coordinates of the vertices of its
image?

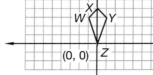

  A *W'* (−3, −1), *X'* (−4, 0), *Y'* (−3, 1),
   Z' (0, 0)

  B *W'* (3, 1), *X'* (4, 0), *Y'* (3, −1), *Z'* (0, 0)

  C *W'* (1, 3), *X'* (3, 3), *Y'* (−1, 3), *Z'* (0, 0)

  D *W'* (1, −3), *X'* (0, −4), *Y'* (1, −3), *Z'* (0, 0)

  b. The coordinates for points *Z* and *Z'* are (0, 0). Explain why.

Early Finishers
Real-World
Application

The local fuel plant has cylindrical holding tanks that are each 85 feet tall with
a diameter of 32 feet. One million cubic feet of oil is being shipped to the
plant for storage. How many holding tanks will the plant need to store all the
oil?

• Adding and Subtracting Mixed Measures
• Polynomials

facts | Power Up P

mental math

a. Algebra: Simplify: $x^2 + 3x - 2x + 1$

b. Number Sense: Arrange from least to greatest:

$$\frac{1}{5} \qquad \frac{2}{5} \qquad \frac{1}{5} + \frac{2}{5} \qquad \frac{1}{5} \times \frac{2}{5}$$

c. Powers/Roots: $x^{10} \cdot x^{10}$

d. Probability: A spinner has eight congruent sections labeled 1–8. Find the probability that the spinner stops on the number 3 or 4.

e. Proportions: The ratio of ducks to geese was 12 to 5. If there were 48 ducks, how many geese were there?

f. Geometry: What is the measure of each angle of an equilateral triangle?

g. Select a Method: The parade officials want to know how many spectators came to watch the parade. The officials will probably

A estimate an appropriate number.
B count the exact number.

h. Calculation: 12×3, $\sqrt{}$, $\times 5$, -10, $\times 5$, $\sqrt{}$, -1, $\sqrt{}$

problem solving

The first three terms of the arithmetic sequence $a_n = 10n$ are 10, 20, and 30. The first three terms of the sequence of squares, $a_n = n^2$, are 1, 4, and 9.

The first term of the first sequence is 10 times the first term of the second sequence (10 to 1). The second term of the first sequence is 5 times the second term of the second sequence (20 to 4).

a. Which numbered term of the two sequences is equal?

b. For which numbered term is the number in the second sequence twice the number in the first sequence?

c. For which numbered term is the number in the second sequence three times the number in the first sequence?

adding and subtracting mixed measures

Mixed measures are measurements with different units combined by addition. We combine the units separately and then make adjustments to the totals as we see in the following examples.

Example 1

Competing in a triathlon Regina swam for 27 minutes 15 seconds, she biked for 48 minutes 28 seconds, and she ran for 42 minutes 53 seconds. What was her total time for the triathlon?

Solution

Each length of time is expressed with two units, minutes and seconds. We add the units separately.

$$\begin{array}{r} 27 \text{ min } 15 \text{ sec} \\ 48 \text{ min } 28 \text{ sec} \\ + 42 \text{ min } 53 \text{ sec} \\ \hline 117 \text{ min } 96 \text{ sec} \end{array}$$

Now we adjust the units. Since 96 seconds is 1 min 36 sec, we add 1 minute to the total leaving 36 in the seconds column.

$$118 \text{ min } 36 \text{ sec}$$

Since 118 minutes is 1 hr 58 min, we express Regina's total time as

1 hr 58 min 36 sec

When we subtract we often need to regroup units.

Example 2

Junior weighed 7 pounds 12 ounces when he was born. Now he weighs 12 pounds 5 ounces. How much weight has Junior gained since he was born?

Solution

In order to subtract ounces we rename 1 pound as 16 ounces. Combining the 16 ounces and 5 ounces, we get 21 ounces. Then we subtract.

$$\begin{array}{r} \overset{\displaystyle\frown \; 1 \; pound \; \frown}{} \\ \overset{11}{} \qquad \overset{21}{} \\ \cancel{12} \text{ pounds } \cancel{5} \text{ ounces} \\ - \; 7 \text{ pounds } 12 \text{ ounces} \\ \hline 4 \text{ pounds } \;\; 9 \text{ ounces} \end{array}$$

Junior has gained **4 pounds 9 ounces.**

Example 3

Last year during physical fitness testing, Marissa completed a mile in 12 minutes 15 seconds. This year she finished in 9 minutes 48 seconds. Marissa improved her time by how many minutes and seconds?

Solution

We compare the times by subtracting. We need to rename 1 minute as 60 seconds. Combining 60 seconds and 15 seconds totals 75 seconds. Then we subtract.

$$
\begin{array}{r}
\overset{\overset{\displaystyle 1\ min}{\frown}}{}\\[-2pt]
\overset{11}{\cancel{12}}\ min\ \overset{75}{\cancel{15}}\ sec\\
-\ \ \ 9\ min\ 48\ sec\\
\hline
2\ min\ 27\ sec
\end{array}
$$

Marissa improved her time by **2 minutes and 27 seconds.**

---

polynomials

A **monomial** is a constant or the product of constants and/or variables with exponents that are whole numbers. Examples of monomials are:

$$2x^3 \qquad -5 \qquad x^2y \qquad 7xyz \qquad z$$

Math Language

The prefixes in the words *monomial, binomial,* and *trinomial* indicate the number of terms in algebraic expressions. *Mono-* means one, *bi-* means two, and *tri-* means three.

A monomial consists of one term. A **binomial** has two terms. Examples of binomials are:

$$x + 5 \qquad xy - 7y \qquad x^2 - 4$$

A **trinomial** has three terms. Examples of trinomials are:

$$x^2 + 7x + 10 \qquad x + 2xy + y$$

Monomials, binomials, and trinomials are all types of **polynomials** and can have any number of terms. The **degree of a polynomial** is the highest degree of any of its terms. If a term has one variable, the **degree of the term** is the exponent of the variable.

y exponent is 1, first degree term (linear)

b^2 exponent is 2, second degree term (quadratic)

p^3 exponent is 3, third degree term (cubic)

If a term has more than one variable, the degree of the term is the sum of the exponents.

$4ac$ exponents are $1 + 1 = 2$, the degree of the term is 2

$6xyz$ exponents are $1 + 1 + 1 = 3$, the degree of the term is 3

(Classify) What is the degree of these terms: $a^2 - 4$, $m + 3$, and $5abc + 2$?

Thus we find the degree of each of the following polynomials by looking at the highest degree of the terms in the polynomial.

Example 4

Name the degree of each polynomial.

a. $4x^3 + 2x^2 + x - 3$

b. $x^2 - 2x + 3$

c. $x + y + 3$

Solution

For each polynomial we find the term with the highest degree by finding the sum of the exponents of variables.

a. The term of highest degree is $4x^3$, which is **3rd degree.**

b. The term of the highest degree is x^2, so the polynomial is **2nd degree.**

c. Both x and y are first degree terms and the polynomial is **1st degree.**

Notice the order of the three terms in this polynomial.

$$x^2 - 2x + 3$$

We say that the terms are arranged in **descending order** because the degrees of the terms are 2 then 1 then 0.

Example 5

Arrange the terms of this polynomial in descending order.

$$x + 2x^3 + 2 + x^2$$

Solution

We focus our attention on the variables and their exponents.

$$2x^3 + x^2 + x + 2$$

Example 6

Add the binomials $x + 6$ and $2x - 3$.

Solution

We collect like terms. One way to do this is to align like terms and add.

$$\begin{array}{r} x + 6 \\ 2x - 3 \\ \hline 3x + 3 \end{array}$$

Practice Set

Add.

a.　3 hr 32 min 45 sec
　　+ 1 hr 43 min 27 sec

b.　5 lb 8 oz
　　+ 8 lb 9 oz

c. *Model* Neil needs to buy molding to mount around the room, so he makes a sketch of the room recording its length and width. What is the perimeter of the room?

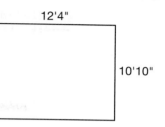

12'4"

10'10"

Perform each indicated subtraction.

d.   3 hr 15 min 12 sec
     − 1 hr 42 min 30 sec

e.   6 ft 2 in.
     − 4 ft 7 in.

f. Tony cut 2 ft $7\frac{1}{2}$ in. length from an 8 ft long board. What is the remaining length of the board?

Classify Identify each polynomial below as a monomial, binomial, or trinomial and name its degree.

g. $x^2 - 25$ **h.** $x^2 - 6x + 7$ **i.** $2x + 3y - 4$

j. Arrange the terms in descending order.
$$5 - 4x^2 + x^3 - 6x$$

k. Add the trinomials $x + y - 2$ and $x - y + 4$.

Written Practice   *Strengthening Concepts*

1.
(67)
The item was discounted 20% from the original price of $90.

a. What was the sale price?

b. If the sale price was reduced 30% for closeout, what was the closeout price?

2.
(48)
The farmer planted crops on 84% of his land. If 12 acres were left unplanted, how many acres of land did the farmer have?

3.
(29, 32)
The bag contained only red and green marbles in the ratio of 2 to 7. If one marble is drawn from the bag, what is the probability the marble will be red?

4.
(Inv. 2)
The sides of a triangle are 5 in., 7 in., and 9 in. Is the triangle a right triangle? Explain your answer.

* 5.
(79)
Analyze Solve for r: $A = \pi r^2$

* 6. Find the volume of this cylindrical container
(76) to the nearest cubic foot.

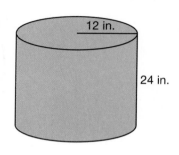
12 in.

24 in.

Analyze   Solve.

* 7. $7x - 3 = 2x - 33$
(79)

8. $0.003x - 0.2 = 0.01$
(50)

9. $\frac{4}{5}x - \frac{2}{3} = \frac{1}{3}$
(50)

* 10. $3(x - 2) = 2(2x - 1)$
(79)

11. **a.**   1 yd 2 ft 7 in.
(80) $\underline{+\ 2\ yd\ 1\ ft\ 10\ in.}$

b.   2 hr 15 min
$\underline{-\ 1\ hr\ 37\ min}$

12. **a.** What type of polynomial is $3x + 2$?
(80)

b. What is the sum of $3x + 2$ and $x - 1$?

Simplify.

* 13. $\sqrt{1000}$
(74)

* 14. $\sqrt{40}\ \sqrt{10}$
(74, 78)

* 15. $\sqrt{3}\ \sqrt{24}$
(74, 78)

16. $\dfrac{12x^3y^4}{3x^4y^{-3}}$
(27, 51)

* 17. **Analyze**   The figure shows the city limits of
(75) a town. The town covers how many square
miles?

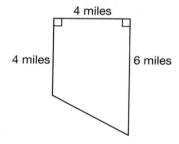
4 miles

4 miles

6 miles

18. Find the area of the parallelogram.
(60)

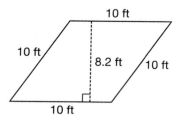
10 ft

10 ft

8.2 ft

10 ft

10 ft

19. Solve this inequality. Graph its solutions.
(77)

$$2x + 1 \le x + 2$$

20. Aziel ordered carpeting installed for $27 per square yard. Use unit
(64, 72) multipliers to convert $27 per square yard to dollars per square foot.

21. Graph $y = -2x + 6$. Is (3, 0) a solution?
(56)

22. Tedman plans a larger version of his model
(65) airplane. A diagram of the tail of the airplane
and the larger version are shown here. Find
a, b, and c.

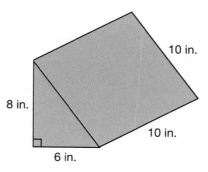

* 23. [Model] Find the volume of the triangular prism, then sketch the net of
(76) the prism.

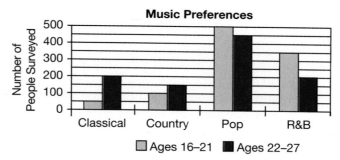

24. Write $\frac{2}{9}$ **a** as a decimal and **b** as a percent.
(63)

25. The bar graph shows the music preferences of two different age groups.
(Inv. 6)

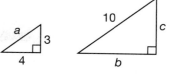

a. How many people in the survey are in the 16–21 age group?

b. How many more people in the 22–27 age group prefer classical
music than in the 16–21 age group?

c. How many people in the survey preferred R&B?

Focus on

• Scatterplots

Mr. Lopez's physical education class did laps around the track, and each student recorded the total time exercising and the number of laps he or she completed. The students' data are shown below.

Number of Laps	Time (Minutes)
3	9
4	8
2	3
6	12
5	15
4	7
2	4
6	15
3	7
5	11
6	7

We make a **scatterplot** of these paired data by plotting these data points on the coordinate plane. The number of laps is the independent variable and is marked on the horizontal axis. Elapsed time is the dependent variable and is marked on the vertical axis. Thus, the first set of numbers, "3 laps in 9 minutes" is plotted as the point (3, 9) on the graph.

Number of of Laps	Time (min)
3	9
4	8
2	3
6	12
5	15
4	7
2	4
6	15
3	7
5	11
6	7

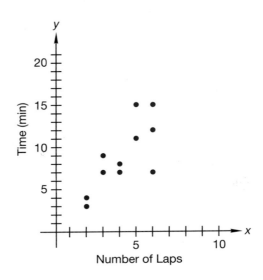

Although the points do not lie on a straight line, we see that the points have a somewhat linear relationship. That is, a line can be drawn on the graph that passes near all of the points.

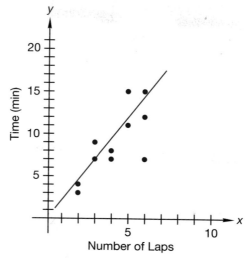

This means that there is a **correlation** between the number of laps completed and the total time completing laps. That is, there is a linear relationship between the quantities. As we follow the line above we see that as the number of laps increase, the total time increases.

Many lines can be drawn passing near these points. We choose a line that minimizes the distance between data points and the line; we call it a **best-fit line** or line of best fit. For now, we choose a best-fit line by visual inspection.

There is one data point that lies apart from the rest. The point (6, 7) does not seem to follow the trend of the other data points. It represents 6 laps in 7 minutes, a very fast run compared to the rest of the class. Points like these that lie away from the others are called **outliers** and are not taken into consideration when plotting a best-fit line. In this situation, the slope of the best-fit line indicates the average running pace of the students. That is, it indicates the average rate in minutes per lap.

Example 1

Draw a best-fit line that fits the data above. Find the equation of the line.

Plot the points on graph paper and use a ruler to find a best-fit line.

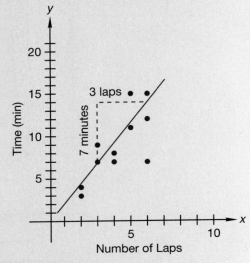

The slope of this line is about:

$$m = \frac{7 \text{ min}}{3 \text{ laps}} = 2\frac{1}{3} \text{ min/lap}$$

The *y*-intercept is zero, so with *x* representing the number of laps and *y* representing the time, the equation of the line is:

$$y = 2\frac{1}{3}x$$

This means the ratio of time to laps is about $2\frac{1}{3}$ min/lap (2:20/lap).

Predict Use the graph of the equation to predict how long it would take the students to run 7 laps.

It is probably not a surprise to us that the number of laps one completes and the elapsed time are correlated. However, some relationships are not as obvious as distance run and time. Researchers can create scatterplots to pair two sets of data in order to see if they are related or to investigate how they are related.

Quantities can be positively correlated (meaning the slope of the best-fit line is positive), negatively correlated (slope is negative), or not correlated (no linear relationship).

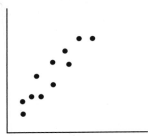

Positive correlation
as one variable increases,
so does the other

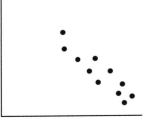

Negative correlation
as one variable increases,
the other decreases

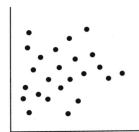

Zero correlation
no linear relationship

Example 2

Determine whether the wages of college graduates and the wages of high-school-only graduates in the listed jobs are correlated. If so, find the equation of a best-fit line.

Median Weekly Earnings, 2000

	High School Diploma ($)	Bachelor Degree or Higher ($)
Cashier	303	384
Cook	327	396
Computer Programmer	864	1039
Data-Entry Keyers	475	514
Designers	633	794
Electricians	714	976
Police	726	886
Legal Assistants	563	725
Real Estate Sales	614	918

Solution

There is a **positive correlation** between the wages of a high school graduate and a college graduate in the given jobs.

The slope of a best fit line is $m = \dfrac{\$500}{\$400} = 1.25$

The equation of the line is $y = 1.25x$

This means that in these jobs, the wages of a college graduate are about 1.25 times the wages of a high-school-only graduate. In other words, college graduates in these jobs earn about 25% more than high-school-only graduates.

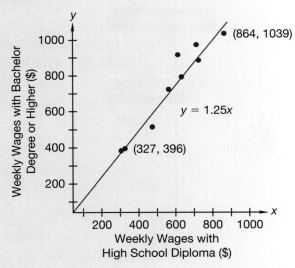

Scatterplots and best-fit lines can be easily graphed using a graphing calculator or a spreadsheet computer program. Calculators are programmed to use "least-squares fitting" to determine the line that best represents the data. Rather than multiple lines that are possible with an inspection method, *exactly one best-fit line* is computed for a given data set.

To use a graphing calculator (TI-83+) to compute the best-fit line, follow the steps below:

Visit www. SaxonPublishers. com/ActivitiesC3 *for a graphing calculator activity.*

1. **Enter the data**

   (STAT), choose "edit", then type the data points into the lists L_1 and L_2.

2. **Make a scatterplot**

   (2nd) StatPlot, choose "Plot1:on", scatterplot (the graph with points only), "X list:L_1", "Y list:L_2", "Mark:0".

   (GRAPH) If the window does not display all the points, select (WINDOW), then type 0 for *X*Min and *Y*Min and 1100 for *X*Max and *Y*Max. Type 100 for *X*Scl and *Y*Scl. (GRAPH)

3. **Compute best-fit line**

   (STAT), choose "CALC," "LinReg ($ax + b$)," enter L_1, L_2, Y by pressing (2nd) L_1 (,) (2nd) L_2 (,) (VARS), "Y Vars," "1. Function," Y_1.

   (ENTER) Write down the equation and numbers you see.

4. **Graph the scatterplot and best-fit line together.**

   (GRAPH)

Example 3

Use a graphing calculator to find the best-fit line (using least squares fitting) for the data in example 2.

Solution

> Lin Reg
> $y = ax + b$
> $a = 1.291547346$
> $b = -12.06506654$

The equation of the best-fit line is $y = ax + b$, with $a = 1.3$ and $b = -12.1$ (rounding to one decimal place).

The equation of the line is $y = 1.3x - 12.1$

Make a Scatterplot and Graph a Best-fit Line

Materials needed:

- multiplication worksheet
- stopwatch

Are the time you spend working and the number of math problems you complete positively correlated?

To study this correlation, conduct the following experiment:

5. Complete a multiplication worksheet while a timer keeps track of how long you work. Work in order—you may skip problems but not go back to complete previously skipped problems. The timer will say "time" at 30-second intervals and "stop" after 3 minutes. When the timer says "time," circle the most recently completed problem. When the timer says "stop," put down your pencil.

6. Correct your work and record data. After checking your answers, fill in the chart to indicate how many problems (in total) you completed correctly after $\frac{1}{2}$ min, 1 min, $1\frac{1}{2}$ min, etc.

7. **Model** Make a scatterplot and if the quantities are correlated, determine the equation of a best-fit line. The slope is indicative of how many problems per minute you completed.

8. Freddy ran for eight minutes wearing a heart rate monitor. He checked his pulse at every minute of his run. Freddy's heart rate at each minute is charted below:

Minutes Run	Heart Rate
1	95
2	110
3	133
4	142
5	150
6	158
7	170
8	175

Make a scatterplot of the data.

Analyze Which axis should we use for minutes run?

If there is a correlation, then find a best-fit line by visual inspection, and state whether the correlation is positive or negative. What does the slope of the line represent? What does the intercept represent?

Explain Do you think Freddy's heart rate would continue to increase at the same rate if he ran for 30 minutes? Explain your answer.

9. Shayna conducted a survey to find out how much money her customers are willing to pay for a new item. She charted the number of people willing to buy the item for each proposed price.

Proposed Price ($)	Number of Purchasers
1.00	10
1.50	9
2.00	6
2.50	5

Represent Make a scatterplot of the data. Are the quantities positively or negatively correlated? Sketch a best-fit line.

• Central Angles and Arcs

facts | Power Up Q

mental math |
a. **Statistics:** Find the mode: 13, 15, 15, 15, 12, 11

b. **Estimation:** $4 - \sqrt{10}$

c. **Fractional Parts:** $\frac{3}{5}$ of 20

d. **Measurement:** Find the temperature indicated on this thermometer.

e. **Percent:** 10% more than $70

f. **Proportion:** $\frac{9}{12} = \frac{x}{8}$

g. **Scientific Notation:** Write 41,600,000 in scientific notation.

h. **Calculation:** Three fourths of an hour is how many minutes more than two thirds of an hour?

problem solving | Find the next three numbers in this sequence: 1, 1, 2, 4, 7, 13, 24, 44 (Hint: Look at the previous terms.)

An angle whose vertex is the center of a circle is called a **central angle.** Below we see four central angles of a circle, each measuring 90°. Notice each angle intercepts $\frac{1}{4}$ of the circle, and the sum of the four angle measures is 360°.

For any circle, the sum of its central angle measures is 360°.

Conclude What is the greatest number of obtuse central angles into which a circle can be divided? Explain your answer.

Example 1

Draw a circle with central angles that divide the circle evenly into thirds.

Solution

The three angle measures are equal and total 360°.

$$\text{angle measure} = 360° \div 3$$

$$\text{angle measure} = 120°$$

Use a compass to draw a circle. Then use a protractor as a guide to sketch a 120° angle with its vertex at the center of the circle. Rotate the paper or protractor to draw the next angle.

Example 2

The court of a playground game is painted on the blacktop. The court is formed by a circle with two intersecting diameters, forming four central angles. One of the central angles measures 50°. Find the measures of the other angles.

Solution

The court looks like this:

Thinking Skill

Explain

Describe another way to find the measures of the other angles in the circle without using a protractor.

In the figure, one acute and one obtuse angle together form a straight angle (180°). Thus, the other angle measures are **130°, 50°, and 130°**.

An **arc** is a portion of a circle intercepted by a central angle. One way to describe an arc is by the portion of the circle it represents. An arc that forms one half of a circle is a semicircle. An arc less than a semicircle is a minor arc. An arc greater than a semicircle is a major arc.

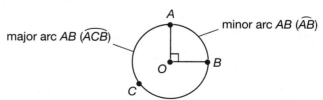

major arc AB ($\overarc{ACB}$)   minor arc AB ($\overarc{AB}$)

In the figure on the previous page, the quarter circle from A to B is arc AB (abbreviated $\overset{\frown}{AB}$) and is a minor arc. The arc from A to B through C is named major arc AB ($\overset{\frown}{AB}$) or arc ACB ($\overset{\frown}{ACB}$). One way to describe the size of an arc is by the measure of its central angle. Arc AB ($\overset{\frown}{AB}$) measures 90°.

Conclude Major arc AB ($\overset{\frown}{AB}$) measures how many degrees?

We may also refer to the length of an arc. The length of an arc is proportional to the fraction of the circumference of the circle the arc represents. For example, if the circumference of the circle above is 100 cm, then minor $\overset{\frown}{AB}$ is 25 cm long and major $\overset{\frown}{AB}$ is 75 cm long.

Example 3

What is the measure of $\overset{\frown}{AD}$ in degrees? What fraction of the circumference is the arc intercepted by $\angle ABD$? If the radius of the circle is 6 cm, what is the length of the arc in terms of π?

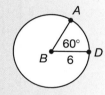

Solution

The measure of the arc equals the measure of its central angle, so $\overset{\frown}{AD}$ measures 60°. The arc is $\frac{60°}{360°} = \frac{1}{6}$ of the circle. The arc length is $\frac{1}{6}$ of the circumference.

$$C = \pi d$$
$$C = 12\pi \text{ cm}$$
$$\text{Arc} = \frac{1}{6}(12\pi \text{ cm})$$
$$= 2\pi \text{ cm}$$

Explain What is the measure of a central angle that intercepts $\frac{1}{9}$ of a circle? How do you know?

Practice Set

For **a–d** refer to the figure.

a. A circle is divided into eighths. What is the measure of each acute central angle?

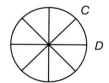

b. What is the measure of arc CD?

c. What is the measure of major arc CD?

d. **Justify** If the diameter of the circle is 20 cm, what is the length of arc CD? Use 3.14 for π. Describe the steps you follow to solve the problem.

Written Practice   Strengthening Concepts

1. Between the starting line and finish line there was an 8% decrease in
(67)   the number of runners. If there were 75 runners at the starting line, how many made it to the finish line?

2. The weight of the load varies directly with the number of crates that
(70) make up the load. If a load of 15 crates weighs 2 tons, how many crates
would weigh 10 tons?

3. Draw quadrilateral *ABCD* with points *A* (2, 2), *B* (2, −2), *C* (−2, −2), and
(Inv. 5) *D* (−2, 2). Then draw its image after a dilation of scale factor 1.5.

* 4. **a.** Find the perimeter of the two squares you drew in problem **3.**
(8,
Inv. 5) **b.** Does the scale factor apply to the perimeters? Justify your answer.

c. Find the area of the two squares.

d. How does the scale factor apply to the areas?

* 5. **Evaluate** A circle with a diameter of 12 inches is divided into
(40, 81) 6 congruent sectors. Write answers to **b** and **c** in terms of π.

a. Find the measure of one of the central angles.

b. Find the length of the arc of one sector.

c. Find the area of one sector.

* 6. Expand: $3d(d + 4)$
(21, 36)

Generalize Simplify.

* 7. $\sqrt{50}$ *** 8.** $5\sqrt{20}$
(74)   (74)

9. $\dfrac{5x^4m^{-1}}{3mx}$ **10.** $\dfrac{(-5)(-10)}{(-5) - (-10)}$
(25, 51)   (36)

* 11. **Model** Solve and graph the solution on a number line: $x - 2x - 1 < 3$
(77)

* 12. Transform this equation to solve for *y*: $y + 5 = 2x$
(79)

13. Can a right triangle have sides of length 1, $\sqrt{3}$, and 2? Explain.
(Inv. 2)

14. Suppose a plumber charges $45 per hour. Use unit multipliers to
(64, 72) convert this rate to cents per minute.

15. The following (*x, y*) points represent the number of baked goods Tina
(69) orders (*x*) and the corresponding price in dollars (*y*).

(2, 5), (4, 10), (6, 15), (0, 0)

Does this set of points show direct variation? If so, what is the constant
of variation?

16. Write $\frac{4}{15}$ as **a** a decimal and **b** a percent. **c** Is $\frac{4}{15}$ closer to $\frac{1}{3}$ or $\frac{1}{4}$?
(63)

17. Find the area of the trapezoid.
(75)

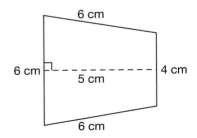

* 18. Preston's pup tent has the dimensions shown. Find the volume of the
(76) tent.

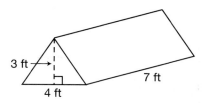

3 ft

7 ft

4 ft

19. Find *a, b,* and *c* in the similar triangles below. (Hint: First find side *a*
(35) using the Pythagorean Theorem.)

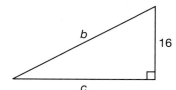

b

16

a

17

15

c

Solve.

20. $\frac{1}{3}m - \frac{2}{3} = 0$
(23, 38)

21. $3x - 5 = x$
(79)

22. **a.** Factor: $10x^2 - 15x$ **b.** Expand: $10(2x + 1.5)$
(21, 36)

23. Mrs. Ortiz is 5 ft 4 in. tall. Her four-year-old son is 3 ft $8\frac{1}{2}$ in. tall.
(80) Mrs. Ortiz is how much taller than her son?

24. Which rule below best describes the sequence 2, 5, 10, 17, ...?
(73)

   A $a_n = 2n$ **B** $a_n = n^2 + 1$ **C** $a_n = n + (n + 2)$

   What is the 10th term of the sequence?

25. The circle graph shows the preferred
(Inv. 6) juice drinks among the students in
Meerca's class.

   a. List the juices in order of
   preference, starting with the juice
   that is preferred by the most
   students.

   b. About what percent of the class
   prefers orange juice?

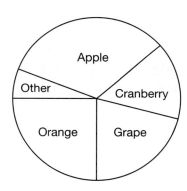

• Graphing Equations Using Intercepts

facts

Power Up Q

mental math

a. **Algebra:** Simplify: $6x - 2y - x + 2y$

b. **Estimation:** $10\sqrt{120}$

c. **Number Sense:** $5^2, 15^2, 25^2, 35^2, \ldots$ What are the last two digits of these squares?

d. **Rate:** 3000 revolutions per minute is how many revolutions per second?

e. **Powers/Roots:** $\frac{x^7}{x^2}$

f. **Geometry:** If two angles are supplementary, the sum of their measures is 180°. $\angle A$ and $\angle B$ are supplementary. $m\angle A = 100°$. Find $m\angle B$.

g. **Select a Method:** To calculate the approximate distance he can drive on a tank of gas, Gilbert will probably use which method?

A mental math **B** pencil and paper **C** calculator

h. **Calculation:** $8 + 8, \sqrt{}, \times 12, + 1, \sqrt{}, \times 9, + 1, \sqrt{}, + 1, \sqrt{}$

problem solving

Pente gazed at the clock and wrote this sequence:

5, 10, 3, 8, 1, 6, 11, 4, 9, ...

What are the next three numbers in the sequence? What is the rule of the sequence?

Previously we have graphed equations by making a table of (x, y) pairs that satisfy the equation and graphing the corresponding points. We have also graphed linear equations using the slope and y-intercept. In this lesson we will graph equations that are expressed in a form called the **standard form** of a linear equation.

Standard Form

$$Ax + By = C$$

We could transform equations in this form to slope-intercept form, but there is an easier way to graph these equations. A quick way to graph linear equations in standard form is to choose zero for each variable and then solve for the other variable. We illustrate this with the following equation:

$$2x + 3y = 12$$

x	y
0	
	0

If we choose 0 for x, then $3y = 12$, so $y = 4$. If we choose 0 for y, then $2x = 12$, so $x = 6$.

$2(0) + 3y = 12$	$2x + 3(0) = 12$
$3y = 12$	$2x = 12$
$y = 4$	$x = 6$
$(0, 4)$	$(6, 0)$

x	y
0	4
6	0

Math Language
Recall that the **x-intercept** is the point where a graph intersects the x-axis, and the **y-intercept** is the point where a graph intersects the y-axis.

We find that two solutions to the equation are $(0, 4)$ and $(6, 0)$. Notice that **by choosing 0 for x and 0 for y we have found the x- and y-intercepts of the equation's graph.** The equation is linear, so we graph all the solutions to the equation by drawing a line through these two points.

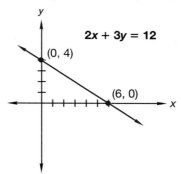

Visit www.
SaxonPublishers.
com/ActivitiesC3
*for a graphing
calculator activity*

Example 1

Find the x- and y-intercepts of 3x + 4y = 24. Then graph the equation.

Solution

The y-intercept is where $x = 0$.

$$3(0) + 4y = 24$$
$$4y = 24$$
$$y = 6$$

The y-intercept is **(0, 6)**.

x	y
0	6
8	0

The x-intercept is where $y = 0$.

$$3x + 4(0) = 24$$
$$3x = 24$$
$$x = 8$$

The x-intercept is **(8, 0)**.

Once we know the intercepts of a line, we can graph the line by connecting the two points.

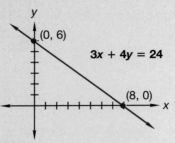

Example 2

Martin is thinking of two numbers. He gives these hints. The sum of the numbers is 3. If you double the first number and add the second number, the sum is 8. Graph this system of equations to find the two numbers.

$$\begin{cases} x + y = 3 \\ 2x + y = 8 \end{cases}$$

Solution

We can graph each equation by finding the *x*- and *y*-intercepts. We can mentally calculate the value of each variable when the other is zero.

x + y = 3			2x + y = 8	
x	**y**		**x**	**y**
0	3		0	8
3	0		4	0

Now we graph both equations by drawing lines through the *x*- and *y*-intercepts for each equation.

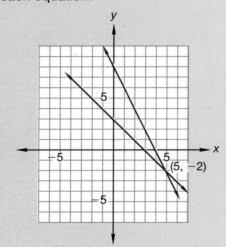

Thinking Skill

Verify

Look back at the original problem. Do 5 and −2 satisfy the conditions described in the hints?

Every point on a line represents a pair of numbers that satisfy the equation. The point where the lines intersect represents the pair of numbers that satisfies both equations. We see that the pair of numbers that satisfies both equations is (5, −2). This means that one of Martin's numbers is **5** and the other is **−2**.

Practice Set

Formulate Find the *x*- and *y*-intercepts of the following equations, then graph them.

a. $2x + y = 8$

b. $3x - 2y = 12$

c. $-x + 3y = 6$

d. $x + 2y = -4$

e. Tanisha is thinking of two numbers. The sum of the numbers is 5. If the lesser number is subtracted from the greater number the result is 7. Graph this system of equations to find the two numbers.

$$\begin{cases} x + y = 5 \\ x - y = 7 \end{cases}$$

Written Practice   *Strengthening Concepts*

1. A book was lengthened by 8 pages for the second edition. If the book
(67)   was originally 200 pages long, what was the percent increase?

2. If measured at the same moment, the length of a person's shadow varies
(71)   directly with the height of the person. If a 3 foot child casts a 5 foot shadow, what is the length of the shadow of a 6 foot adult?

* 3. **Evaluate** On a coordinate plane graph two points representing the
(47)   height (*x*) and shadow lengths (*y*) of the two people described in problem **2**. Then draw a line through the two points. What are the coordinates of the point on the line that corresponds to a shadow length of $2\frac{1}{2}$ ft? What do those numbers mean for this function?

4. The beans were packed in a can that had a height of 10 cm and a radius
(76)   3 cm. Sketch the cylinder and find its volume in terms of π.

For problems **5** and **6** refer to the similar triangles. Units are cm.

5. Draw the two similar triangles separately. Then find *x*, *y*, and *z*.
(Inv. 2,
35)

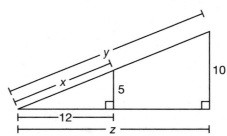

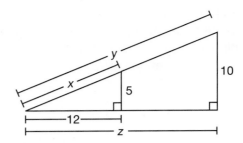

6. **a.** What is the scale factor from the smaller to larger triangle in
(20, 35)   problem **5**?

b. Find the area of each triangle.

c. The area of the larger triangle is how many times the area of the smaller triangle?

* 7. *Analyze* A circle with a diameter of 15 inches is divided into fifths.
(81)   **a** What is the degree measure of each central angle? **b** What is the length of one arc? Use 3.14 for π.

* 8. *Model* Graph the equation $2x + 3y = 6$ using the method taught in this
(82)   lesson.

9. During middle school Ting read seventy percent of the 110 books on
(48)   the recommended list. How many of the recommended books did she read?

Generalize Simplify.

* 10. $3\sqrt{27}$
(78)

* 11. $\sqrt{2}\sqrt{22}$
(78)

12. $\dfrac{6xm}{12x^2}$
(27, 51)

* 13. $\left(\dfrac{2}{3}\right)^2\left(\dfrac{1}{3}\right)^2\left(\dfrac{4}{9}\right)^{-2}$
(23, 63)

* 14. Find the area of the trapezoid. Is the area greater than or less
(75)   than 2 cm^2?

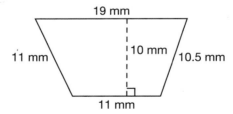

* 15. Can a right triangle have sides of length 2, 2, and $2\sqrt{2}$? Explain.
(Inv. 2, 78)

* 16. Sam's car averages 30 miles per gallon on the highway. Traveling
(64, 72)   at 60 miles per hour, Sam's car uses how many gallons per hour?

17. Write $1\frac{1}{6}$ as a **a** decimal and **b** percent. **c** Round the decimal form to the
(63)   nearest thousandth.

* 18. Transform this formula to solve for s. $A = \pi rs$
(79)

* 19. *Evaluate* The Abbot, Benitez, and Cheung mailboxes are side by side.
(32) A postal worker with a bundle of mail for each of the three families
places a bundle in each box without checking the recipient's name.

    a. List the possible arrangements of the bundles.

    b. What is the probability the postal worker placed the bundles in the
    correct boxes?

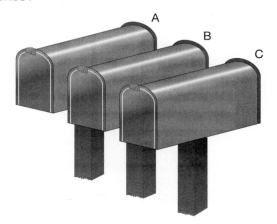

20. Lilly plans a circular garden that she will
(40) border with pavers as in the shaded region of
this figure. What is the area that the pavers
will cover? (Round to the nearest whole
unit.)

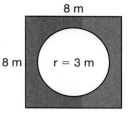

21. Use the distributive property to factor -3 from $-3x^2 - 9x - 42$.
(21)

Solve.

22. $\frac{1}{2}w - \frac{2}{3} = \frac{2}{3}$
(23, 38)

23. $0.3m = 0.4m - 0.5$
(38, 79)

24. Arrange each trinomial in descending order and then find their sum.
(80)
$$5 + x^2 + 2x \qquad x + x^2 - 1$$

25. Sandy's Boat Rental rents paddleboats for $8 per hour. Which graph
(69) represents the relationship of number of hours rented to the cost? Is
this a proportional relationship?

A

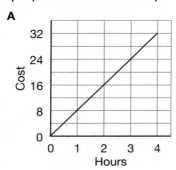

B

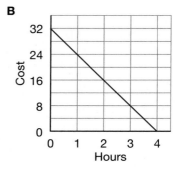

C

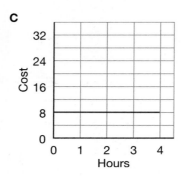

D

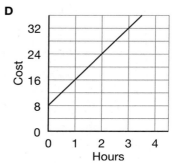

• Probability of Dependent Events

facts | Power Up Q

mental math

a. Algebra: Simplify: $5x + 3x + 1$

b. Statistics: What is the mean of 24 and 42?

c. Probability: A spinner has five equal sectors labeled 1–5. Find the probability that the spinner stops on an even number.

d. Percent: 25% more than $80.

e. Rate: Micah swam laps at a rate of 2 laps per minute. How long does it take Micah to swim 5 laps?

f. Measurement: The odometer read 2150.5 miles at the beginning of the trip and 3876.5 miles at the end. How long was the trip?

g. Select a Method: To find the 7.75% sales tax on a $46.85 purchase, a retailer would probably use
  A mental math.
  B pencil and paper.
  C a calculator.

h. Calculation: $6 \times 11, -2, \sqrt{}, +1, \sqrt{}, \times 12, \sqrt{}, -7$

problem solving

The vertices of each square are the midpoints of the sides of the next larger square. If the area of the inner square is 1 cm², what are the perimeter and area of each of the four squares? Make a table listing the perimeters and areas in order. Then use the table to predict the perimeter and area of a square drawn inside the inner square.

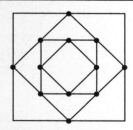

New Concept | Increasing Knowledge

We know that when events are independent, the probability that both will occur is the product of their probabilities.

$$P(A \text{ and } B) = P(A) \cdot P(B)$$

In this lesson, we will compute probabilities of events that are **not independent.** Events that are not independent are sometimes called **dependent** events because the probability of one event *depends* on the other event.

Example 1

Two eighth graders and one seventh grader want to go on the trip, but only two students will be selected. If their names are drawn at random, what is the probability that the two eighth graders will get to go?

Solution

We make a tree diagram to find the probability.

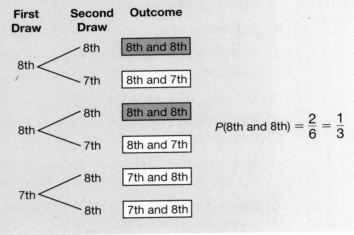

$$P(\text{8th and 8th}) = \frac{2}{6} = \frac{1}{3}$$

Rather than counting outcomes in example 1, we can consider probabilities at each stage of the experiment. In the first stage (the selection of the first student), $\frac{2}{3}$ of the students are eighth graders.

After an eighth grader is selected, there is a 50% chance (or probability of $\frac{1}{2}$) that the second eighth grader will be selected.

The probability that two eighth graders are selected is $\frac{2}{3} \cdot \frac{1}{2}$, which is $\frac{1}{3}$.

Stage 1:

2 out of 3 are eighth graders

Stage 2:

1 out of 2 are eighth graders

This illustrates a multiplication rule for events A and B that are **not independent**.

$$\begin{array}{ll} P(A) & \text{under initial conditions} \\ \times \ P(B) & \text{under new conditions} \\ \hline P(A \text{ and } B) \end{array}$$

The problem above is an example of **conditional probability.**

The multiplication rule would look like this:

$$P(\text{8th and 8th}) = P(\text{eighth first draw}) \cdot P(\text{eighth second draw})$$

$$= \frac{2}{3} \cdot \frac{1}{2}$$

$$= \frac{1}{3}$$

The selection of the first eighth grader and the second are not independent events. When one name is selected, the probability of selecting the next name changes.

Thinking Skill

Predict

In an experiment, two marbles are drawn from a bag with two white marbles and three blue marbles. Do you think you will be more likely to select two blue marbles if the marble is replaced after the first draw, or if the marble is not replaced after the first draw? Explain your reasoning.

This differs from independent events in which one event does not change the probability of future events. For example, consider an experiment in which a marble is selected from a bag, its color is recorded, and it is replaced. Then another marble is drawn, recorded, and replaced. The events, "blue" (for the first marble) and "blue" (for the second marble), are independent because every time a marble is selected the probability of selecting a blue marble remains $\frac{3}{5}$.

However, if a blue marble is drawn and not replaced, the conditions have changed and the probability of selecting another blue marble changes to $\frac{1}{2}$.

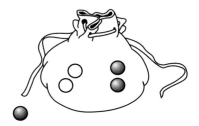

Example 2

A bag contains two white marbles and three blue marbles. If two marbles are selected, what is the probability of selecting two blue marbles if the marbles are selected a. with replacement? b. without replacement?

Solution

a. The events are independent:

$$P(\text{blue and blue}) = P(\text{blue}) \cdot P(\text{blue})$$

$$= \frac{3}{5} \cdot \frac{3}{5}$$

$$= \frac{9}{25}$$

The probability is $\frac{9}{25}$, which is 0.36.

b. The events are not independent.

$$P(\text{blue and blue}) = P(\text{blue}) \cdot P(\text{blue after one blue was selected})$$

$$= \frac{3}{5} \cdot \frac{\overset{1}{\cancel{2}}}{\underset{2}{\cancel{4}}}$$

$$= \frac{3}{10}$$

The probability is $\frac{3}{10}$, which is 0.30.

Practice Set

a. If a spinner is spun twice, are the events independent or not independent? Justify your answer.

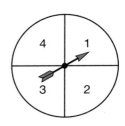

b. Four different cards lettered *A, B, C,* and *D* are face down on a table with the order scrambled. As a card is turned over it is left letter side up. Find these probabilities.
- Turning over the *A* card first.
- Turning over *A,* then *B.*
- Turning over *A,* then *B,* then *C.*
- Turning over *A,* then *B,* then *C,* then *D.*

c. The four letter cards in **b** are face down on the table. Michelle turns over one card and finds that it is *A.* If she turns over the remaining cards one at a time, what is the probability that she turns them over in alphabetical order?

d. Recall that the probabilities of an event and its complement total 1. In a bag are six marbles: 3 red, 2 white, and 1 blue. Three marbles are drawn one at a time without replacement. What is the probability of drawing red, then white, then blue? What is the probability of not drawing red then white, then blue?

e. **Analyze** Gerry's drawer has two black socks and four blue socks. If he selects two socks without looking, what is the probability they are both black?

Written Practice   *Strengthening Concepts*

1. The number of calories of a food item varies directly with the size of the
(71) portion. If a 2-inch slice of a certain delicacy contains 170 calories, how many calories are in a 3-inch slice?

* 2. **Explain** Greg rode his bike 30 miles to the summit averaging 10 miles
(49) per hour. Then, he rode back down averaging 30 miles per hour. How far did he ride? How long did it take him? What was his average speed? Explain why Greg's average speed is not 20 miles per hour.

3. The sweater was normally $50. It was on sale for $12 off. What was the
(67) percent of discount?

* 4. **Analyze** a. Find the measure of the central
(81) angle *x.*

b. Find the length of the arc *AB.*

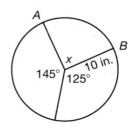

* 5. A square washer 4 cm on a side has a 2 cm diameter hole. When the washer is bolted in place, what is the area of the washer against the object to which it is bolted? Round your answer to the nearest square centimeter.
(8, 40)

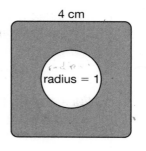
4 cm
radius = 1

←—2 cm—→

* 6. **a.** How many cubic feet of concrete is needed to build a walkway 36 feet long, 3 feet wide, and 3 inches thick?
(42, 72)

  b. The number of cubic feet needed equals how many cubic yards?

7. Tim bikes at a park that has two triangular paths, as shown here. Some of the path lengths are marked. Assuming the triangles are similar, find x and y. Then find the ratio of the perimeter of the shorter path to the longer path.
(65)

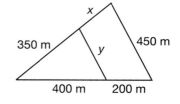
x
350 m
y
450 m
400 m
200 m

* 8. **Model** Solve and graph the solution set on a number line: $5x - 7x \leq -4 - (-4)$
(77)

* 9. **Justify** Can a right triangle have sides of 1, 2, and $\sqrt{5}$? Justify your answer.
(Inv. 2, 78)

* 10. Sue swims 400 m in 6 minutes. Which equation could be used to find how far Sue swims in 9 minutes?
(49)

  A $\dfrac{400}{x} = \dfrac{6}{9}$ **B** $\dfrac{400}{9} = \dfrac{x}{6}$ **C** $\dfrac{400}{6} = \dfrac{9}{x}$ **D** $400x = 9 \cdot 6$

11. Uriel buys a one-acre lot (4,840 yd^2). Use unit multipliers to convert this area to square feet.
(72)

* 12. **Model** Sketch $\triangle ABC$ with vertices $A(0, 0)$, $B(0, 5)$, and $C(2, 3)$. Then sketch the image of $\triangle ABC$ after a translation of $(-2, -3)$.
(Inv. 5)

13. Identify the type of quadrilateral and find the area.
(Inv. 3, 75)

5 cm
3 cm
2.5 cm
7 cm

Simplify.

* 14. $6\sqrt{12}$
(78)

* 15. $5\sqrt{5}\sqrt{5}$
(78)

16. $\dfrac{9x^2y^4 2m^{-1}}{2x^2y^3 9m^{-1}}$
(27, 51)

17. $\dfrac{(-15)(10)}{-15 - 10}$
(36)

* 18. Find the capacity (volume) of a cylindrical cup with an inside diameter of 6 cm and a height of 9 cm. Use 3.14 for π.
(76)

* 19. In the metric system, small quantities of liquid are measured in milliliters
$_{(6)}$ (mL) or cubic centimeters (cm^3 or cc). A liter is 1000 mL or 1000 cc. The
capacity of the cup in problem **18** is about what fraction of a liter?

A $\frac{1}{2}$ **B** $\frac{1}{4}$ **C** $\frac{1}{5}$ **D** $\frac{2}{5}$

* 20. *Explain* Graph $y = -x - 1$ using slope-intercept. Then on another
$_{(56, 82)}$ coordinate plane graph $x + y = -1$ using the standard form
method. What do you notice about the two graphs? Explain your
observation.

* 21. Arc *AB* measures 100°. What is the measure
$_{(81)}$ of major arc *AB*?

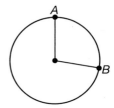

Solve.

22. $\frac{0.2}{0.5} = \frac{8}{x}$
$_{(25, 44)}$

23. $\frac{2}{3}x = \left(\frac{3}{2}\right)^{-1}$
$_{(38, 68)}$

* 24. Find the sum of the trinomials $2x + 3y - 4$ and $x - 3y + 7$.
$_{(80)}$

25. The graph shows the growth in the populations of California and New
$_{(Inv. 6)}$ York from 1920 to 1990.

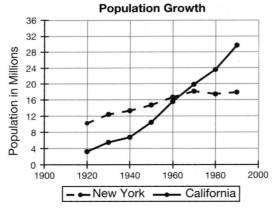

a. About how much did the population of California increase between
1980 and 1990?

b. In which decade were the populations of California and New York
about the same?

    A 1920–1930

    B 1940–1950

    C 1960–1970

    D 1970–1980

• Selecting an Appropriate Rational Number

facts | Power Up Q

mental math

a. Statistics: Find the range of this data: 13, 15, 15, 15, 12

b. Estimation: $2 \times \$1.47 + \$12.97$

c. Number Sense: The numbers 2^2, 12^2, 22^2, 32^2, etc., all have the same digit in the ones place. What is that digit?

d. Powers/Roots: $\frac{x^4}{x^3}$

e. Proportion: $\frac{7}{8} = \frac{56}{x}$

f. Geometry: Angle P and $\angle R$ are supplementary. The measure of $\angle P$ is $50°$. Find $m\angle R$.

g. Scientific Notation: Write 4.2×10^{-1} in standard notation.

h. Calculation: Two dozen $\div\ 3$, $\times\ 4$, $+\ 1$, $\div\ 3$, $\times\ 5$, $+\ 1$, $\div\ 7$

problem solving

A rectangle has vertices at $(5, 7)$, $(5, -1)$, $(1, 7)$, and $(1, -1)$. How many units and in which directions should the rectangle be translated (moved) so that the center of the rectangle is at the origin?

Consider how you would calculate the answers to these two questions.

• The tax rate is 7%. How much money is 7% of $12.60?

• The discount is $33\frac{1}{3}\%$. How much money is $33\frac{1}{3}\%$ of $12.60?

To find a percent of a number, we usually convert the percent to a decimal or fraction before we multiply. Sometimes we choose to use the decimal form and sometimes we choose to use the fraction form. The better choice depends upon the situation. To answer the two questions above, many people would convert 7% to a decimal because it is relatively easy to multiply by 0.07. Since $33\frac{1}{3}\%$ is a repeating decimal, many people would convert it to the fraction $\frac{1}{3}$ to perform the calculation.

Example 1

a. How much is 40% of $\frac{1}{2}$ of $12.60?

b. How much is $66\frac{2}{3}\%$ of $\frac{5}{6}$ of $12.60?

a. Both 40% and $\frac{1}{2}$ easily convert to decimal form. We rewrite the numbers as decimals and perform the calculations.

Step:	Justification:
40% of $\frac{1}{2}$ of $12.60	Given
$0.4 \times 0.5 \times \$12.60$	Converted 40% and $\frac{1}{2}$ to decimals
$0.2 \times \$12.60$	Multiplied 0.4 and 0.5
$2.52	Multiplied

b. Both $66\frac{2}{3}\%$ and $\frac{5}{6}$ convert to repeating decimals. We choose to convert the percent to a fraction before performing the calculations.

Step:	Justification:
$66\frac{2}{3}\%$ of $\frac{5}{6}$ of $12.60	Given
$\overset{1}{\cancel{\frac{2}{3}}} \times \frac{5}{\underset{3}{\cancel{6}}} \times \$12.60$	Convert to fraction
$\frac{5}{\underset{1}{\cancel{9}}} \times \$\overset{1.40}{\cancel{12.60}}$	Multiplied $\frac{2}{3}$ and $\frac{5}{6}$
$7.00	Multiplied

Example 2

In a public opinion survey, 180 of the 280 women surveyed agree or strongly agree with the mayor, while 150 of the 275 men surveyed agree or strongly agree with the mayor. Which choice listed below is the most appropriate way to describe the results of the survey?

A $\frac{9}{14}$ of the women and $\frac{6}{11}$ of the men agree with the mayor.

B 0.643 of the women and 0.545 of the men agree with the mayor.

C 64% of the women and 55% of the men agree with the mayor.

Solution

Thinking Skill

Discuss

The question says that 180 of the 280 women and 150 of the 275 men surveyed "agree or strongly agree" with the mayor. Each answer choice only says "agree." Why might we combine these two categories?

When stating comparisons it is appropriate to use forms of numbers that are easy to compare and are readily understood.

- Choice A uses two fractions with different denominators, which makes it difficult to compare.

- Choice B states the comparison with decimals, which is a less common way to express a comparison.

- **Choice C** expresses the comparison to the nearest percent. The numbers are easy to compare and understand.

Practice Set

a. Find 20% of $\frac{1}{4}$ of $12.00.

b. Find $33\frac{1}{3}$% of $\frac{3}{8}$ of $12.00.

c. **Explain** To find 80% of $21.50, would you convert 80% to a fraction or a decimal? Why?

d. **Explain** To find 25% of $24, would you convert 25% to a fraction or decimal? Why?

e. Which expression is the most appropriate rational number to announce a discount of $14 from a regular price of $40?

A save $\frac{7}{20}$ **B** save 0.35   **C** save 35%

Written Practice   *Strengthening Concepts*

1. At a certain store, there are about 11 CD's sold for every 4 cassettes.
(45) If the store sells 270 CD's and cassettes in all, about how many are CD's?

2. Irina made 85% of her free-throw attempts. If she missed only six
(48) free throws, how many did she attempt?

3. Robert saved 15% on his backpack. If he paid $17, what was the
(67) original price?

* 4. **Analyze**   **a.** How many degrees is arc *RS*?
(81)
     b. Find the length of the arc *RS*.

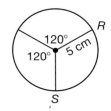

* 5. If the diameter of a circle is reduced 20%, then its area is reduced by
(71) what percent?

6. Refer to the triangles to the right to
(65) answer parts **a–c**.

     a. How do we know the triangles are similar?

     b. Find *x* and *y*.

     c. What is the scale factor from $\triangle DEF$ to $\triangle ABC$?

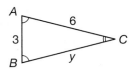

* 7. This number line shows the solution set
(77) to which of these inequalities?

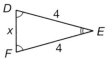

     A $x > 3$ **B** $-x < 3$ **C** $x + 1 \geq 2$ **D** $1 - x \leq -2$

* 8. **Classify** Classify the following polynomial by the number of terms and
(80) by degree. Then arrange the terms in descending order.

$$5 + x^2 - 2x$$

* 9. **Analyze** Find the surface area of a 1-yard cube in square yards and in
(43, 52)   square feet.

Generalize Simplify.

10. $\sqrt{6}\sqrt{8}$
(78)

11. $3x + 2x^2 + x^2 - x$
(31)

12. $\dfrac{4x^2y^{-1}2x^3}{6x^4y^2}$
(27, 51)

* 13. 50% of $\dfrac{2}{3}$ of $1.20
(84)

A triangle has vertices at (0, 0), (3, 3), and (−4, 4). Use this information in
problems **14** and **15.**

* 14. **a.** Classify the triangle by angles.
(Inv. 2,
20)
   b. What is the area of the triangle?

* 15. **a.** What is the length of the longest side of the triangle?
(Inv. 2,
74)
   b. What is the perimeter of the triangle?

16. Write $1\frac{1}{3}$ as **a** a decimal and **b** a percent.
(63)

17. When U.S. coins were silver the weight of the coin was proportional to
(29)   its value.

   a. What is the ratio of a quarter to a dime?

   b. What percent of a quarter is a dime?

18. A piece has been cut from a sheet
(75)   of construction paper. The remaining
   portion covers how many square
   inches?

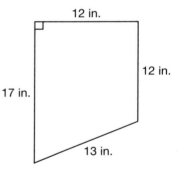

19. Graph $y = x$ and graph $x + y = 0$.
(56, 82)

20. The rule of a sequence is $a_n = 2n^2 - 1$. What is the 10th term of the
(73)   sequence?

Solve.

21. $\dfrac{x}{5} = \dfrac{0.42}{0.3}$
(25, 44)

22. $3x + 2 = x + 14$
(79)

23. $\dfrac{3}{4}x = \left(\dfrac{4}{3}\right)^{-2}$
(38, 63)

24. $-\dfrac{x}{5} = 10$
(23, 38)

25. The paint store uses the amounts shown on the table to make a gallon of a certain shade of paint. Use the data in the table to make a graph. Is the relationship between the amount of yellow paint and the amount of red paint per gallon proportional? If it is a proportional relationship, write an equation that describes the relationship and state the constant of proportionality. (see below.)

	Parts Red Paint (x)	Parts Yellow Paint (y)
1 gal	3	5
2 gal	6	10
3 gal	9	15
4 gal	12	20

Early Finishers
Real-World Application

The Marquez family has four children born in different years.

a. List all possible permutations of boy and girl children that the Marquez family could have.

b. Find the probability that the Marquez family has three boys and one girl in any order.

• Surface Area of Cylinders and Prisms

facts | Power Up Q

mental math

   a. Algebra: Simplify: $5x - 6x + 3$

   b. Geometry: If a right triangle is isosceles, then each acute angle measures how many degrees?

   c. Fractional Parts: $\frac{4}{7}$ of 49

   d. Rate: Kurt and Dash clear debris from fields. If Kurt cleans 1.5 acres per week and Dash cleans 1 acre per week, how much can they clean together in 4 weeks?

   e. Measurement: Find the length of this object.

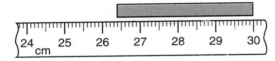

   f. Percent: 50% more than $90.

   g. Select a method: While grocery shopping, Reagan keeps track of the total price of the items in the basket. Reagan probably uses
   A mental calculation.
   B mental estimation.
   C pencil and paper.
   D a calculator.

   h. Calculation: $100 \div 5, + 5, \div 5, + 5, \div 5, + 5$

problem solving | There are 3 numbers whose sum is 77. The second number is twice the first, and the third number is twice the second. What is the second number?

New Concept   *Increasing Knowledge*

To make a soup can, the manufacturer assembles two circles and one rectangle of metal in the shape of a cylinder.

The area of the material used on one soup can represents the **surface area** of the cylinder. We compute the surface area of an object by adding the areas of all its faces or surfaces.

Example 1

Compute the surface area of this cylinder in terms of π.

We find the area of each face or surface of the cylinder. The curved surface of the cylinder is equivalent to a rectangle with a length equal to the circumference of the circular bases, which is $2\pi r$, or 4π units. The rectangle's height is equal to the height of the cylinder (5 units)

Thinking Skill

Analyze

What parts of the cylinder does 4π represent in the expression? What does 20π represent?

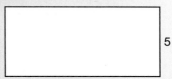

Surface Area $= A_1 + A_2 + A_3$

Surface Area $= 4\pi + 4\pi + 20\pi$

Surface Area $= \mathbf{28\pi}$ **square units**

Example 2

A tent in the shape of a triangular prism is shown. What is the surface area of the tent in square feet?

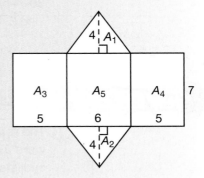

The triangular prism has five faces.

$A = A_1 + A_2 + A_3 + A_4 + A_5$

$A = 12 + 12 + 35 + 35 + 42$

$A = 136$

The tent is made with **136 ft²** of material.

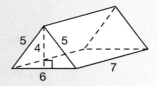

The **lateral surface area** of a prism is the area of the faces between its bases. The lateral surface area of a cylinder is the area of the surface between its bases. For example, the lateral surface area of a soup can is represented by the label but not by the circular top and bottom of the can. A quick way to find the lateral surface area of a cylinder or prism is to multiply the *perimeter* of the base by the *height* of the cylinder or prism.

Example 3

Hector wants to paint the exterior walls of a
shed. Find the lateral surface area of the shed.

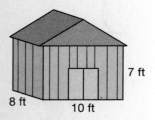

7 ft

8 ft   10 ft

Solution

Instead of finding and adding the areas of the four walls, we multiply the
perimeter of the base by the height.

$$\text{lateral surface area} = \text{perimeter of base} \cdot \text{height}$$
$$= (8 + 10 + 8 + 10) \text{ ft} \cdot 7 \text{ ft}$$
$$= 36 \text{ ft} \cdot 7 \text{ ft}$$
$$= \mathbf{252 \text{ ft}^2}$$

Practice Set

Find the total surface area of these figures.

a.

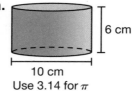

6 cm

10 cm

Use 3.14 for π

b. 10 in.

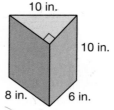

10 in.

8 in.   6 in.

c. What is the lateral surface area of the cylinder in problem **a?**

d. What is the lateral surface area of the triangular prism in **b?**

e. **Analyze** Find the lateral surface area of a
30-ft-high water tower with a 40 ft diameter.
Use 3.14 for π.

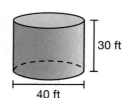

30 ft

40 ft

Written Practice   *Strengthening Concepts*

1. The ratio of cyclists to pedestrians on a certain trail was 2 to 7. If there
(45) were 27 people on the trail, how many more pedestrians were there than
cyclists?

* 2. **Connect** A jogger ran three miles down the trail then turned around and
(72) ran back, completing the roundtrip in 45 minutes.

   a. The jogger averaged how many minutes per mile?

   b. The jogger's average rate was how many miles per hour?

3. The traffic to the website increased $16\frac{2}{3}$% during the month. If there
$(67, 84)$ were 3,000,000 visitors per day at the beginning of the month, how many were there by the end of the month? What rational number did you use to help you find the answer? Why did you use that number?

* 4. (Analyze) A circle is divided into ten equal sectors.
(81)
    a. Find the measure of each central angle.

    b. If the diameter is 50 in., find the length of an arc of one of the sectors.

5. The thickness of a dime is about one millimeter, which is 0.001 meters.
(51) Write 0.001 in scientific notation.

6. Refer to the similar triangles to the right to
(65) answer parts **a–c.**

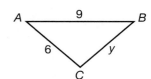

    a. Find x and y in the similar triangles.

    b. Which angle has the same measure as $\angle C$?

    c. What is the scale factor from the smaller to larger triangle?

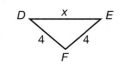

* 7. (Justify) Will a 9 inch-by-13 inch picture
$(Inv.\ 2)$ frame fit completely inside a box of this size? Justify your answer.

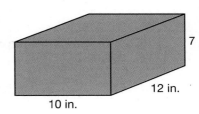

* 8. A rectangle's length is $(x + 4)$ and width is $(x + 3)$. By adding binomials
(80) give its perimeter.

9. Solve the inequality and graph the solution on a number line:
(77)
$$-2x + 1 > 1$$

* 10. The label of a can represents the lateral surface area of a cylinder. What
(85) is the lateral surface area of a can of beans with a diameter of 7 cm and a height of 11 cm? (Use $\frac{22}{7}$ for π.)

11. Expand $-7x(5 - x)$ and arrange the terms of the product in descending
$(36, 80)$ order.

12. There are 2 red marbles and 3 green marbles in a bag. If two marbles
(83) are drawn at the same time from the bag, what is the probability both will be green?

(Generalize) Simplify.

* 13. $\sqrt{500}$
(74)

* 14. $2\sqrt{5}\sqrt{20}$
(78)

* 15. Ming is playing a board game with a red number cube and a green
(59)   number cube. She needs to roll a sum of 10 or more with the two
        number cubes to win on this turn.

   a. Of the 36 possible combinations, how many result in a sum of 10 or
        more? List them.

   b. What is the probability she will win on this turn?

16. *Evaluate* A 10-foot ladder leans against a building. The foot of the
(Inv. 2) ladder is 3 feet from the building. Which choice below best indicates
        how far up the building the ladder reaches?

   A Between 7 and 8 feet   **B** Between 8 and 9 feet

   C Between 9 and 10 feet   **D** Between 10 and 11 feet

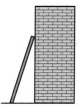

17. Laura walked the 440 feet from history class to math class in
(64, 72) 2 minutes. Use unit multipliers to convert 440 feet in 2 minutes to
        miles per hour.

* 18. Graph $y = \frac{1}{2}x + 2$. Then graph $x + y = 5$. What are the coordinates of
(56, 82) the point at which the graphed lines intersect?

19. *Explain* One wall of an attic room is shown. The window is 4 feet wide
(75)   and 3 feet high. Fernando wants to paint the wall. Find the area of the
        wall to be painted. Explain how you found your answer.

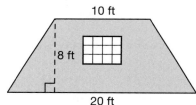

10 ft

8 ft

20 ft

20. Use your ruler to find the scale of the drawing in problem **19.** If the
(35)   length of the room is 25 feet, which would be the length of the room in a
        drawing at this scale?

Solve.

21. $\dfrac{x}{3} = \dfrac{1.5}{0.5}$
(25, 44)

22. $\dfrac{2}{3}x + 2 = \left(\dfrac{1}{3}\right)^{-1}$
(38, 63)

23. $0.2x - 0.02x = 0.018$
(38, 50)

24. $2x^2 - 2 = 48$
(35, 50)

25. Describe the transformation(s) needed to align figure 1 with figure 2.
(Inv. 5)

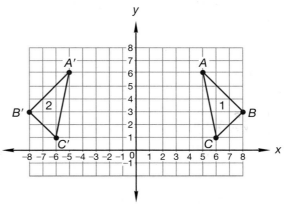

Sherman just purchased a dance hall. One of the hall's badly worn dance surfaces is shaped like an isosceles trapezoid. The trapezoid's bases measure 40 feet and 60 feet, and its height measures 30 feet. If hardwood flooring costs $2.25 per square foot, what will it cost Sherman to install new floors on four such dance surfaces?

• Volume of Pyramids and Cones

facts | Power Up R

mental math

a. Algebra: If x is 10, then what is the product of $x + 1$ and $x - 1$?

b. Number Sense: The perfect squares 9^2, 19^2, 29^2, 39^2, etc. all have what digit in the one's place?

c. Ratio: The ratio of statues to live creatures was 6 to 7. If there were 39 in all, how many were statues?

d. Number Sense: The sum of two numbers is 8. Their product is 12. What are the two numbers?

e. Probability: The probability that his name will be drawn is $\frac{1}{1000}$. Find the probability that his name will not be drawn.

f. Geometry: Angle X and $\angle Y$ are supplementary. If m$\angle Y$ is 90°, find m$\angle X$.

g. Scientific notation: Light travels at about 300,000 km/sec. Express this number in scientific notation.

h. Calculation: 7×20, $+ 4$, $\sqrt{}$, $+ 4$, $\sqrt{}$, $+ 4$

problem solving

A shoe manufacturer loads boxes of shoes onto pallets and ships them to a retailer. Each pallet can hold a stack of shoeboxes 3 ft high, 3 ft wide, and 4 ft long. If each shoebox is 12 inches long, 6 inches wide, and 6 inches high, how many shoeboxes fit on each pallet?

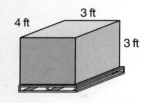

We know how to compute the volume of cylinders and prisms.

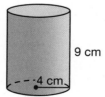

The volume
of this cylinder is:
$V = B \cdot h$
$V = \pi(4 \text{ cm})^2 \cdot 9 \text{ cm}$
$= 144\pi \text{ cm}^3$

The volume
of this prism is:
$V = B \cdot h$
$V = (6 \text{ in.} \cdot 8 \text{ in.}) \cdot 10 \text{ in.}$
$= 480 \text{ in.}^3$

In this lesson we will learn the formulas to calculate the volumes of cones and pyramids, beginning with the formula for the volume of a cone.

Suppose we have a cone-shaped container and a cylindrical can with the same diameters and heights.

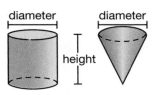

To fill the can with water using the cone, we would have to fill and empty the cone three times.

volume of cylinder   =   volume of 3 cones

The volume of a cone is $\frac{1}{3}$ the volume of a cylinder with the same base diameter and height.

volume of cone $= \frac{1}{3} B \cdot h$

$V = \frac{1}{3}\pi r^2 h$

Thinking Skill

Explain

Why are $V = \frac{1}{3}Bh$ and $V = \frac{1}{3}\pi r^2 h$ equivalent formulas for the volume of a cone?

Example 1

Aidan built two cones out of sand on the beach. The diameter of the smaller cone is 6 in. and its height is 5 in. The diameter of the larger cone is 12 in. and its height is 10 in. How many cubic inches of sand did Aidan use for each cone? How many times greater is the volume of the larger cone than the volume of the smaller cone? (Express π as π.)

Solution

We draw diagrams of the two cones, one with a radius of 3 in. and a height of 5 in., the other with a radius of 6 in. and a height of 10 in.

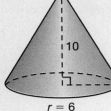

$r = 3$ $r = 6$

Smaller Cone		**Larger Cone**
Step:	Justification:	Step:
$V = \frac{1}{3}\pi r^2 h$	Formula	$V = \frac{1}{3}\pi r^2 h$
$V = \frac{1}{3}\pi(3)^2(5)$	Substituted for r and h	$V = \frac{1}{3}\pi(6)^2(10)$

$$V = \frac{1}{3}\pi \cdot 9 \cdot 5 \qquad \text{Evaluated } 3^2 \text{ and } 6^2 \qquad V = \frac{1}{3}\pi \cdot 36 \cdot 10$$

$$V = 15\pi \text{ in.}^3 \qquad \textbf{Simplified} \qquad\qquad V = 120\pi \text{ in.}^3$$

Dividing 120π in.^{3} by 15π in.^{3}, we find that the larger cone has a volume **8 times as large** as the smaller cone.

$$\frac{120 \, \cancel{\pi} \text{ in.}^{\cancel{3}}}{15 \, \cancel{\pi} \text{ in.}^{\cancel{3}}} = 8$$

The volumes of prisms and pyramids are similarly related. From a block of clay in the shape of a prism, we can form 3 pyramids with the same base and height as the original prism. This is shown with a square base below, but it is true for bases of any shape.

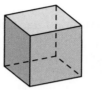 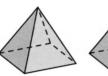

volume of 1 prism =   volume of 3 pyramids

The volume of a pyramid is $\frac{1}{3}$ the volume of a prism with the same base and height.

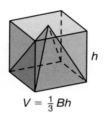

$$V = \frac{1}{3}Bh$$

Example 2

Find the volume of this square pyramid.

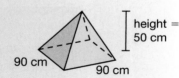

90 cm   90 cm   height = 50 cm

Solution

The volume is $\frac{1}{3}$ the product of the area of the square base and the height.

Step:	Justification:
$V = \dfrac{1}{3}Bh$	Given
$V = \dfrac{1}{3}(90 \text{ cm})^2(50 \text{ cm})$	Substituted for B and h
$V = \dfrac{1}{3}(8100 \text{ cm}^2)(50 \text{ cm})$	Squared 90 cm
$\boldsymbol{V = 135{,}000 \text{ cm}^3}$	Simplified

Example 3

A triangular pyramid is shown. The area of its base is $36\sqrt{3}$ units. Its height is 10 units. What is its volume?

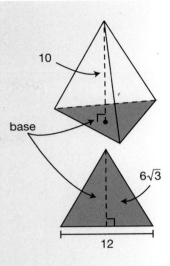

Solution

We are given the height and the area of the base. We substitute these values for B and h in the formula and perform the calculations.

Step:	Justification:
$V = \frac{1}{3}Bh$	Formula
$V = \frac{1}{3}36\sqrt{3} \cdot 10$	Substituted
$V = 120\sqrt{3}$	Simplified

Practice Set

Find the volume.

a. Refer to the pyramid in example 2. Find the volume of a pyramid with dimensions $\frac{1}{10}$ of the pyramid in example 2. The volume of the smaller pyramid is what fraction of the volume of the pyramid in example 2?

> **Connect** The volume of the smaller pyramid is what percent of the volume of the pyramid in example 2?

b. The area of the hexagonal base of a pyramid is $18\sqrt{3}$. The height of the pyramid is 12. What is the volume?

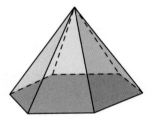

c. The diameter of the base of the cone is 20 inches. The height is 21 inches. What is the volume of the cone? Express the answer in terms of π and again rounded to the nearest cubic inch.

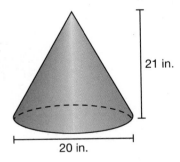

d. Using a 12 inch cube of clay, Lucian makes a 12-inch high pyramid with a 12-inch square base. What is the volume of the pyramid? Does Lucian have enough clay from the original cube to make another pyramid the same size as the first?

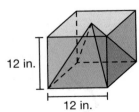

e. *Connect* A party hat arrived in a cylindrical container in which it fit perfectly. If the volume of the container is 99 in.^{3}, what is the volume of the cone-shaped hat?

Written Practice   *Strengthening Concepts*

1. In two months a puppy's weight increased from 5 lb 15 oz to 8 lb 5 oz.
(80)   Find the amount of increase.

2. The infant's weight increased from 15 to 21 pounds in three months.
(71)   What was the percent increase?

* 3. *Evaluate* A 12-inch diameter wall clock is divided into twelfths by
(81)   equally spaced numbers.

  a. Find the measure of the central angle between consecutive numbers.

  b. Find the length of the arc between two consecutive numbers on the circle.

4. A pyramid with a square base is 4 in. high.
(86)   What is the volume of the pyramid?

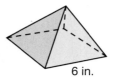

6 in.

5. Find the lateral surface area of the cylinder to
(85)   the nearest square *centimeter*.

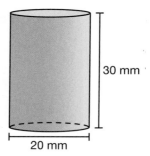

30 mm

20 mm

* 6. *Analyze* The sun casts shadows
(65)   as shown. How tall is the flag pole?
    (Use similar triangles.)

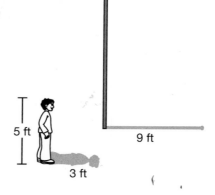

5 ft

3 ft

9 ft

7. Find the sum of the polynomials $3x - 2y - 4$ and $x + 2y - 7$.
(80)

* 8. Errin bought stock in XYZ Corporation for $100 per share. One month
(67) later the stock was up 10%. After another month the stock fell 10%.

    a. What was the price of the stock after the first month?

    b. What was the price of the stock after the second month?

9. There were 11 girls in a class of 25. What percent of the students were
(58) boys?

* 10. If a car can travel 192 miles when its gas tank is 75% full, then how far
(48) can it travel at the same mileage rate on a full tank?

11. Write the first four terms of the sequence that follows the rule
(73) $a_n = 5n - 2$.

Generalize Simplify.

* 12. $\sqrt{5}\sqrt{15}$
(74, 80)

13. $3(-1) - 2\left(-\dfrac{1}{2}\right)$
(36)

* 14. Find $33\frac{1}{3}\%$ of $\frac{4}{5}$ of $4.50.
(84)

* 15. Solve for y: $y - 7 = 2x - 3$
(79)

16.   Evaluate Demonstrate that a right triangle
(Inv. 2) can have the dimensions shown.

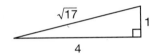

17. There are three winning cards out of twenty-one placed in a hat.
(83)

    a. What is the probability that the first person to draw a card will select
a winning card?

    b. If the first person selects a winning card and does not replace it, what
is the probability that the second person will select a winning card?

18. Use unit multipliers to convert 60 miles per hour to feet per minute.
(64, 72)

19. **a.** Whatever number Xavier says, Yanos says the opposite.
(56, 82)   Graph $y = -x$.

    b. The numbers Xena and Yolanda say always total 4. Graph $x + y = 4$.

    c. Describe the relationship of the two graphed lines.

Solve.

20. $\dfrac{2.7}{5.4} = \dfrac{3}{x}$
(25, 44)

21. $1.2 - 3.4x = -5.6$
(25, 36)

22. $\dfrac{4}{7} - \dfrac{3}{7}x = 1$
(22, 38)

23. $17 - 2x = x - 7$
(78)

* 24. Cubic boxes are stacked as shown. Sketch
(Inv. 4) the front, top, and right side views.

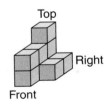

25. If each box in problem **24** has edges two feet long, then what is the
(42) volume of the stack of boxes?

• Scale Drawing Word Problems

Power Up *Building Power*

facts | Power Up R

mental math

a. **Statistics:** Find the mean of this data: 11, 11, 17, 16, 15

b. **Intercepts:** The sum of two numbers is 7. Their product is 12. What are the two numbers?

c. **Fractional Parts:** $\frac{2}{5}$ of 65

d. **Proportion:** $\frac{42}{x} = \frac{6}{5}$

e. **Measurement:** Find the mass of the bag.

f. **Percent:** 100% more than $60

g. **Scientific Notation:** Write 0.00009 in scientific notation.

h. **Calculation:** What is the average of the first five positive odd numbers?

problem solving Is it possible to remove a block from each side and still have a balanced scale? If so, which blocks can be removed?

New Concept *Increasing Knowledge*

A map is a small representation of a larger physical region. If the map is **drawn to scale,** then small units of measure on the map correspond with larger units of measure in the actual region. Therefore, the lengths on the map are proportional to those in the actual region.

For example, if the key on a map shows the scale relationship 1 inch = 10 miles, then a one-inch length on the map corresponds with a 10-mile distance in the region represented by the map. Thus, a road 2.25 inches long on the map is really 22.5 miles long.

Most architectural drawings are also created to scale. People who look at architectural drawings or blueprints can easily calculate the actual sizes of the structures.

We use a scale factor to convert measures on the scale drawing to the actual measures they represent and vice versa. The scale factor is the ratio of a length on the drawing to the length it represents on the actual object. Since the scale factor is a ratio, we can solve scale drawing problems using proportions.

Example 1

Mary sketched a scale drawing of her dream garden using 1 inch to represent 2 feet.

a. If the drawing of the garden is 8 inches long, how long will Mary's garden be?

b. If Mary's garden is to be 12 feet wide, how wide should the drawing be?

Solution

We may find the answer to both questions by solving proportions. The scale of the drawing determines the ratio.

Thinking Skill

Explain

Explain two different methods we can use to solve a proportion.

a.

	Scale	Measure
Drawing, in inches	1	8
Garden, in feet	2	L

$\rightarrow \dfrac{1}{2} = \dfrac{8}{L}$

We solve the proportion and find that the length of the garden is **16 feet**.

b.

	Scale	Measure
Drawing, in inches	1	W
Garden, in feet	2	12

$\rightarrow \dfrac{1}{2} = \dfrac{W}{12}$

We solve the proportion and find that the drawing of the garden should be **6 inches wide**.

Represent Draw Mary's garden with a scale of 1 inch = 4 feet.

Example 2

Edmund wants to convert the attic to living space. He has made a scale drawing (1 inch = 16 feet) of his plans. What are the dimensions of Bedroom 2 excluding the closet?

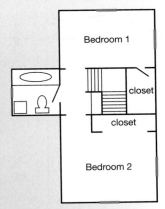

Solution

Using a ruler to measure, we find that Bedroom 2 is 1 inch long and $\frac{3}{4}$ inch wide. Since 1 inch on the drawing represents 16 feet, the room is **16 feet long**. An alternative to writing and solving a proportion is multiplying by the scale factor. To compute the width of Bedroom 2, we multiply the width on the drawing by the scale factor.

$$\underbrace{\frac{3}{4} \text{ in.}}_{\text{drawing}} \times \underbrace{\frac{16 \text{ ft}}{1 \text{ in.}}}_{\substack{\text{scale} \\ \text{factor}}} = \underbrace{12 \text{ ft}}_{\text{measure}}$$

We find the room is **12 ft wide.**

Predict Would a drawing of Bedroom 2 with a scale of 1 inch = 1 foot fit on this page? Explain.

Practice Set

a. A map is drawn with a scale of 1 inch = 8 miles. Two towns $2\frac{3}{4}$ inches apart on the map are how many miles apart?

b. **Justify** Mariah is making a scale drawing of her apartment. Her apartment measures 36 feet long and 30 feet wide. She wants the drawing to fit on an 8.5 in.-by-11 in. piece of paper. Which of the following would be a good scale for Mariah to use? Why?

A 1 in. = 2 ft   **B** 1 in. = 3 ft   **C** 1 in. = 4 ft   **D** 1 in. = 6 ft

c. Refer to the floor plan in Example 2 to find the actual length and width of Bedroom 1, excluding the closet.

d. **Model** Create a scale drawing of a building, classroom, playground, or landscape area. Specify the scale factor you used for your drawing.

Written Practice   *Strengthening Concepts*

* 1. **Analyze** Arthur is drawing the floor plan of his home. The scale is 1 inch
(87)   equals 2 feet.

    a. If he draws his hallway 5 inches long, how long is the actual hallway?

    b. If the kitchen is 12 feet wide, how wide should the drawing of the kitchen be?

* 2. **Analyze** Kay is sketching her dream house. The scale is 1 inch equals
(87)   5 feet.

    a. If her dining room is to be 25 feet long, how long should the drawing be?

    b. If she draws the bookcase 2 inches long, how long is the actual bookcase?

3. Mariya charged three purchases on her credit card. The charges were
(17) $19.53, $12.99, and $26.90. She estimated to total by adding $20, $13, and $27. Without performing the addition, explain how you know whether the actual total is greater than or less than her estimate.

* 4. **Connect** If this net is folded to make a prism, what is its surface
(55, 85)   area?

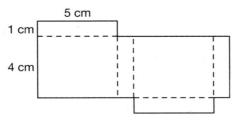

   A 20 cm²   **B** 25 cm²

   C 50 cm²   **D** 58 cm²

* 5. In terms of π, what is the volume of
(75, 86)   the cylinder and the volume of the
cone?

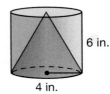

6. In terms of π, what is the total surface area of the cylinder in problem **5?**
(85)

* 7. **Model** Yoshi drew plans for a rectangular table that would measure 4
(87)   feet by 6 feet. On his drawing, $\frac{1}{2}$ in. represented 1 foot. Sketch the plan
of the table top that Yoshi drew and label the side lengths.

* 8. Three vertices of a square are (1, 1), (0, −1), and (−2, 0).
(Inv. 1,
Inv. 2)   **a.** What are the coordinates of the fourth vertex?

  b. What is the length of each side?

  c. What is the area of the square?

9. In an alphabet tile game there are six tiles face down on the table, and
(83)   two of the tiles are A. If Gabbie selects two tiles randomly, what is the
probability that both tiles will be A?

* 10. A rectangle has sides with length $(x + 3)$ and $(x + 4)$. Find the perimeter
(80)   by adding binomials.

* 11. **Generalize** Expand and add like terms: $x(x + 3) - 4(x + 3)$
(31, 36)

Simplify.

* 12. $3x^{-1}y^2x^3y^{-1}$
(51)

* 13. $\sqrt{8}\sqrt{10}$
(78)

14. $\dfrac{(-3) - (2)(-4)}{(-2) - (-1)}$
(36)

* 15. 50% of $33\frac{1}{3}$% of 1.2
(84)

16. Nigel takes a shortcut on his walk to school.
(Inv. 2)   Instead of going to the corner at *B,* he cuts
across the field from *A* to *C.* If the distance
from *A* to *B* is 60 yards, and from *B* to *C* is
80 yards, how many yards does he cut by
taking the short cut?

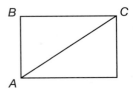

* 17. A circle is divided into nine equal sectors.
(81)

    a. Find the measure of the central angle for one of the sectors.

    b. If the circumference of the circle is 27 in., find the length of the arc of one of the sectors.

* 18. **a.** A meter is about 1.1 yards. A cubic meter is about how many cubic yards?
(72)

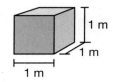

    b. A cubic yard is 27 cubic feet, so a cubic meter is about how many cubic feet? Round to the nearest cubic foot.

19. Graph $y = x$. Then graph $x + 2y = 6$. What are the coordinates of the point where the lines intersect?
(56, 82)

Solve.

20. $\dfrac{2.1}{1.4} = \dfrac{x}{6}$
(25, 44)

21. $0.3x - 0.9 = 0.3$
(25, 50)

22. $\dfrac{4}{5} - \dfrac{3}{5}x = \dfrac{3}{5}$
(23, 50)

23. $\left(\dfrac{7}{2}\right)^{-1} = -\dfrac{1}{7}x$
(38, 63)

24. $4(x - 3) + 5 = x + 2$
(21, 78)

25. Klaudia can swim five laps in the city pool every two and a half minutes. If she swims for half an hour every day, how many laps does she swim every four days?
(49)

Early Finishers
Real-World Application

The Pantheon was built over 2000 years ago under the Roman Empire. It is the best-preserved of all Roman buildings due to its thickness and the quality of its materials. The main room of the Pantheon resembles a giant cylinder. The diameter of the cylindrical room is 43.3 meters and the height of the cylinder is 14 meters. What is the lateral surface area of the Pantheon's cylindrical room? (Use 3.14 for π.)

• Review of Proportional and Non-Proportional Relationships

Building Power

facts

Power Up R

mental math

a. Algebra: If x is 10, then what is the product of $x + 2$ and $x - 2$?

b. Intercepts: The sum of two numbers is 6. Their product is 8. What are the two numbers?

c. Number Sense: What digit is in the ones place in the product 34×2593?

d. Powers/Roots: $\frac{x^7}{x^5}$

e. Function: The mystery man made a marvelous machine. When Ricky typed 6, the machine output 35. When Albert typed 7, the machine output 48. When Jackie typed 8, the machine output 63. Wally's brother Timmy typed a number. The machine output 143. What did Timmy type?

f. Geometry: Angle B and $\angle C$ are supplementary. If $m\angle B$ is 20°, find $m\angle C$.

g. Sequences: What is the next term in this sequence?

penny, dime, dollar bill, ...

h. Calculations: $102 - 3, \div 3, \div 3, - 3, \div 2, \sqrt{\ }$

problem solving

Brad and William start from the same point and fly in the same direction. Brad flies at 500 miles per hour, while William flies at 200 miles per hour. If William has a 150 mile head start, how long will it take Brad to catch up?

New Concept *Increasing Knowledge*

Let us review three important ideas about proportional relationships.

1. Proportional relationships involve **two variables**—two quantities that can change values—such as the distance traveled and the time spent traveling or the quantity of items and the total cost. As one variable increases, the other changes by a constant factor k: $y = kx$.

2. In a table of values for the two variables, all pairs form **equal ratios.** That is, in a proportional relationship, the ratio is constant. This idea is expressed by the equation $\frac{y}{x} = k$, in which k (think k for "konstant") is the **constant of proportionality.**

3. A graph of a proportional relationship is **linear and aligns with the origin** (if one variable is zero, the other variable is zero). The graph may be a continuous line or may simply be aligned points.

Example 1

State whether each graph below depicts a proportional relationship. Justify your decision.

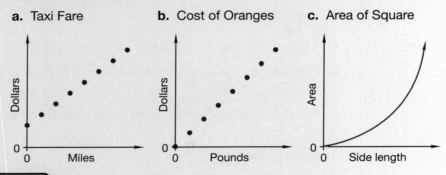

a. Taxi Fare

b. Cost of Oranges

c. Area of Square

Solution

a. The data points are aligned but not with the origin, thus the relationship is **not proportional.**

b. The data points are aligned and include the origin, thus the relationship is **proportional.**

c. The origin is included but the graph is not linear, thus the relationship is **not proportional.**

Example 2

The following tables record the effect of dilating on the perimeter and area of a square.

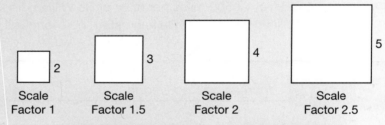

| Scale Factor 1 | Scale Factor 1.5 | Scale Factor 2 | Scale Factor 2.5 |

Side Length	Perimeter		Side Length	Area
2	8		2	4
3	12		3	9
4	16		4	16
5	20		5	25

State whether each table displays a proportional relationship and justify your decision.

Why is the word "proportional" used to describe a relationship between two variables whose ratios are constant?

To determine if paired numbers represent a proportional relationship, we check the ratios formed by the pairs. If the ratio is constant, the relationship is proportional.

Side Length	Perimeter	P/s
2	8	$\frac{8}{2} = 4$
3	12	$\frac{12}{3} = 4$
4	16	$\frac{16}{4} = 4$
5	20	$\frac{20}{5} = 4$

Side Length	Area	A/s
2	4	$\frac{4}{2} = 2$
3	9	$\frac{9}{3} = 3$
4	16	$\frac{16}{4} = 4$
5	25	$\frac{25}{5} = 5$

We see that as a square is dilated, the ratio of side length to perimeter does not change. However, the ratio of side length to area does change. **Thus, side length and perimeter are proportional, but side length and area are not proportional.**

Example 3

The heights of similar triangles are proportional to their bases. As a triangle is dilated, its base and height increase proportionally.

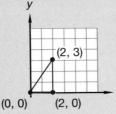

Base	Height
2	3
3	4.5
4	6
5	7.5

(0, 0)   (2, 0)

Predict the height of the triangle shown when its base is 15.

Solution

We will show two methods for finding the height. Since the relationship is proportional, one way to find the unknown height is to write and solve a proportion. We select any row of numbers from the table to write one ratio for the proportion. We use 15 as the base for the second ratio and h for the unknown height.

$$\frac{\text{base}}{\text{height}} \quad \frac{2}{3} = \frac{15}{h}$$

We solve the proportion using cross products.

$$\frac{2}{3} = \frac{15}{h}$$

$$2h = 3 \cdot 15$$

$$h = \frac{45}{2}$$

$$h = 22.5$$

Another way to find the height is to multiply the base by the constant of proportionality. Notice that the ratio of height to base is constant.

Therefore, we can multiply the base (15) by the constant of proportionality (1.5) to find the height.

height = 1.5 × base

$h = 1.5 \times 15$

$h = 22.5$

Base	Height	$\dfrac{\text{height}}{\text{base}}$
2	3	$\frac{3}{2} = 1.5$
3	4.5	$\frac{4.5}{3} = 1.5$
4	6	$\frac{6}{4} = 1.5$
5	7.5	$\frac{7.5}{5} = 1.5$

Notice the equation we wrote at the end of example 3.

height = 1.5 × base

$h = 1.5b$

This form of equation is typical of proportional relationships. Here we show the general form of the equation. Look at it carefully.

$$y = kx$$

We see two variables (x and y) on either side of the equals sign with a constant (k) multiplying one variable. We do not see addition, subtraction, or exponents.

Example 4

State whether each equation below indicates a proportional relationship. Justify your decisions.

a. $c = \pi d$ **b.** $A = \pi r^2$ **c.** $y = 2x + 1$

Solution

a. We see two variables on either side of the equals sign. The constant, π, multiplies one variable. The relationship is **proportional.**

b. We see two variables on either side of the equals sign. The constant, π, multiplies one variable, but the variable r has an exponent of 2. The relationship is **not proportional.**

c. We see two variables on either side of the equals sign. The constant, 2, multiplies one variable. The number 1 is added to one side of the equation. That means if one variable is zero, the other is not zero. The relationship is **not proportional.**

Example 5

Which of these problems can be solved with proportional thinking?

A It takes 2 workers 5 hours to complete the job. How long would it take 4 workers to complete the job?

B The shipping charge is $2 per item plus a $3 flat fee, so it costs $13 to ship 5 items. What is the shipping charge on 10 items?

C Each student read the same number of books. Thirty students read a combined total of 150 books. How many books were read by 60 students?

Only in **C** are the quantities related by a constant factor. We can find the number of books read by 60 students by solving the proportion:

$$\frac{30}{150} = \frac{60}{x}$$

$$x = 300$$

Practice Set

a. *Explain* In a proportional relationship, there are how many variables? The graph of the relationship is what shape? When one variable is zero, what number is the other variable?

b. An equation in the form $y = kx$ indicates a direct proportion. As x changes by 1, y changes by a constant factor, k. Which of these equations indicates a direct proportion?

A $y = 5 + x$ **B** $P = 4s$ **C** $A = bh$

c. The equation $y = kx$ transforms to $\frac{y}{x} = k$, which means that the ratio of y to x is constant. If $\frac{y}{x} = 3$, then which pair of numbers does not fit the relationship?

A (4, 12) **B** (6, 18) **C** (9, 3)

d. *Conclude* Which of the following is not an example of a proportional relationship?

A Distance and time when driving at a constant speed

B Quantity and price when the unit price remains the same

C Hours awake and hours asleep when the number of hours in a day remains the same

e. *Classify* Which graph below shows a proportional relationship?

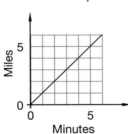 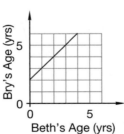 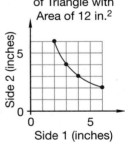

A Distance Traveled at 60 mph

B Ages of Bry and Beth

C Length and Width of Triangle with Area of 12 in.²

f. Which equation is an example of a proportional relationship?

A $d = 30t$ **B** $A = s^2$ **C** $P = 2l + 2w$

* 1. **Analyze** Michael drew a scale drawing of a building. The scale is 0.5
(87)   inches equals 1 foot.

    a. If he drew the hallway 2.5 inches wide, how wide is the actual
     hallway?

    b. The building itself is 60 feet wide. How wide is it in Michael's
     drawing?

* 2. **Estimate** To estimate the area of a pond
(75)   Gerard added the areas of two trapezoids.
   What is the approximate area of the pond?
   (Dimensions are in yards.)

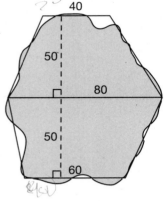

3. When the teacher returned the tests, the number of smiles increased by
(67)   forty percent. If there had been 15 smiles in the room before the tests
   were returned, how many were there after the tests were returned?

* 4. **Analyze** A rectangle has sides with lengths $(x + 5)$ and $(x + 6)$. Find
(80)   the perimeter.

* 5. Expand and write the product in descending order. $-2x(3 + x - 3x^2)$
(36, 80)

6. Find the lateral surface area of this
(85)   triangular prism with an equilateral
   base.

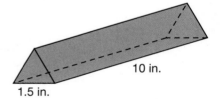

10 in.

1.5 in.

7. The illustration shows a pyramid with
(86)   the same base and height as the cube.

    a. What is the volume of the cube?

    b. What is the volume of the pyramid?

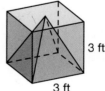

3 ft

3 ft

For **8–11** refer to the circle. Segment *AE* and *BD*
are diameters. Units are in cm.

* 8. **Classify** **a.** Name two obtuse central
(81)   angles

    b. Name two acute central angles.

9. **a.** What is the area of each triangle?
(20)

    b. Classify $\triangle ABC$ by sides.

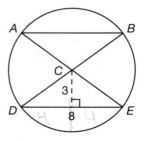

10. The dashes showing the height divide $\triangle CDE$ into two congruent right triangles.
(Inv. 2, 39)

    a. Use the Pythagorean Theorem to find CD.

    b. What is the diameter of the circle?

11. In terms of π, the two triangles occupy what fraction of the area of the circle? Is the fraction more or less than $\frac{1}{3}$? How do you know?
(20, 40)

12. a. Find x and y in the similar triangles to the right.
(40, 45)

    b. Find the ratio of the perimeter of $\triangle ABC$ to the perimeter of $\triangle DEF$.

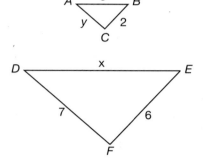

Simplify.

* 13. $\sqrt{98}$
(74)

* 14. $4\sqrt{10}\sqrt{20}$
(78)

15. $\left(\frac{2}{3}\right)^{-2} + \left(\frac{3}{4}\right)^{-1}$
(22, 63)

16. 150% of $\frac{3}{10}$ of 0.2
(84)

* 17. **Evaluate** Joshua completed tasks and recorded the time it took to complete them. Does this table show a proportional relationship? If so, give the constant of proportionality. Use words to describe the relationship between the number of tasks and the time necessary to complete them.
(88)

Tasks	Time
3	18 sec.
4	24 sec.
5	30 sec.
6	36 sec.

18. Siti registered to run in a 10k (a 10 km race). Use a unit multiplier to write the distance in miles. (One mile is just over 1.6 km.) Round down to the tenth of a mile.
(64, 72)

19. Write $\frac{5}{9}$ as **a** a percent, **b** a decimal, and **c** a decimal rounded to the nearest hundredth.
(63)

20. Graph the equations $y = x$ and $y = x + 2$. Describe the relationship of the two lines.
(41, 56)

Solve.

21. $\frac{2}{3}x + \frac{3}{8} = \frac{3}{4}$
(23, 50)

22. $\frac{2.7}{1.8} = \frac{x}{2}$
(25, 44)

23. $0.4 - 0.02x = 1.2$
(25, 50)

24. $\frac{2}{5}n = -18$
(24, 38)

25. The table below show the amount of Mighty Clean Detergent needed to wash loads of laundry.
(69)

Amount of Detergent	$\frac{3}{4}$ C	$1\frac{1}{2}$ C	$2\frac{1}{4}$ C	3 C	$3\frac{3}{4}$ C
Number of Laundry Loads	1	2	3	4	5

Graph the data. Is the number of cups of detergent proportional to the number of loads of laundry? If so, write an equation that describes the graph and state the constant of proportionality. If not, explain why.

Early Finishers
Real-World Application

The Great Pyramid of Giza was built as a tomb for the Egyptian king Khufu nearly 5000 years ago. The pyramid is made from blocks of limestone and granite, with each block weighing between 2 and 2.5 tons. If the side length of the base of this square pyramid is about 745 feet and the height of the pyramid is about 449 feet, what is the approximate volume of the Great Pyramid of Giza? (Round your answer to the nearest thousand cubic feet.)

• Solving Problems with Two Unknowns by Graphing

facts | Power Up R

mental math

a. **Algebra:** If $3x = 9$, then what is the product of x^3 equal?

b. **Fractional Parts:** $\frac{9}{8}$ of 480

c. **Probability:** Ten of the 11 girls completed the assignment. Twelve of 13 boys completed the assignment. If the teacher calls on a student at random, what is the probability that the student has the assignment completed?

d. **Percent:** What number is 75% more than $40?

e. **Ratio:** The ratio of lions to tigers to bears was 2 to 3 to 5. If there were 45 bears, how many tigers were there?

f. **Measurement:** Find the volume of liquid in this graduated cylinder:

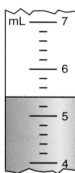

g. **Select a Method:** With a tape measure you find the dimensions of a room. To find its perimeter you would probably use

   A   mental math.

   B   pencil and paper.

   C   a calculator.

h. **Calculation:** $6 \times 5, + 2, \div 2, \sqrt{\ }, \times 7, - 1, \div 3, \sqrt{\ }$

problem solving | When $\frac{85}{99}$ is written as a decimal, what is the 237th digit to the right of the decimal?

Theo is thinking of two numbers. He says that the sum of the two numbers is 6. He also says that one number is 10 more than the other number. What are the two numbers?

We could solve this problem by guessing and checking, but there are other strategies we can use.

Notice we are given two pieces of information about two numbers. One piece of information is that their sum is 6. We can write that information as an equation. We will use x and y to represent the two numbers.

$$x + y = 6$$

The other piece of information is that one number is ten more than the other number. We can write an equation for this information also. We let y represent the larger number.

$$y = x + 10$$

We have written two equations using the same two variables. Together the equations form a **system of equations,** which is two or more equations with common variables. We often use a brace to indicate a system of equations:

$$\begin{cases} x + y = 6 \\ y = x + 10 \end{cases}$$

Systems of equations are sometimes called **simultaneous equations.** We solve a system of equations by finding the values that satisfy both equations simultaneously. There are several ways to solve a system of equations. The method we will practice in this lesson is solving by graphing.

Visit www.
SaxonPublishers.
com/ActivitiesC3
*for a graphing
calculator activity.*

Example

Theo is thinking of two numbers. He says that the sum of the numbers is 6. He also says that one number is 10 more than the other number. What are the two numbers? Solve this system of equations by graphing.

$$\begin{cases} x + y = 6 \\ y = x + 10 \end{cases}$$

Solution

By graphing the two equations, we will "see" the answer to the problem. We will graph a line for each equation. There are many pairs of numbers that satisfy $x + y = 6$ such as $6 + 0$, $3 + 3$, and $0 + 6$. We can make a table of ordered pairs, plot the points, and draw a line to represent all possible pairs of numbers whose sum is 6.

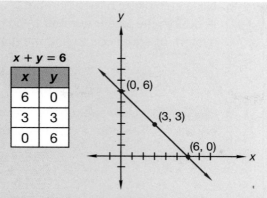

Now we consider the second equation, $y = x + 10$. Some (x, y) pairs that satisfy this equation are $(-10, 0)$, $(-5, 5)$, and $(0, -10)$. We graph this equation on the same coordinate plane as the first equation.

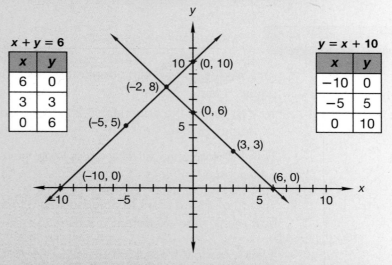

Thinking Skill

Infer

Each equation in this example has an infinite number of solutions. How do we know that the system of equations has only one solution?

This line represents all pairs of numbers in which one number is 10 more than the other number. We see that there is one point, or one pair of numbers, that is a solution to both equations. The lines intersect at $(-2, 8)$. This means that $x = -2$ and $y = 8$ is the one pair of values that satisfies both equations and both conditions of the problem. Therefore, **Theo's two numbers are –2 and 8.**

Verify Look back. Do these two numbers satisfy the condition of the original problem?

Practice Set

Solve each problem by graphing the system of equations and checking the solution.

a. Together Xena and Yolanda have $12. Yolanda has 6 dollars more than Xena. How much money does each person have? Graph this system of equations to find the answer.

$$\begin{cases} x + y = 12 \\ y = x + 6 \end{cases}$$

b. Nikki is thinking of two numbers. Their sum is 12. The greater number is double the lesser number. What are the two numbers? Graph this system of equations to find the numbers.

$$\begin{cases} x + y = 12 \\ y = 2x \end{cases}$$

c. (Formulate) Use inspection to predict the solutions to this system of equations. Then graph the equations to verify your prediction.

$$\begin{cases} x + y = 6 \\ y = x \end{cases}$$

d. Describe two numbers represented by this system of equations. Then graph to find the two numbers.

$$\begin{cases} y = x - 6 \\ x + y = 0 \end{cases}$$

Written Practice   *Strengthening Concepts*

* 1. Andy is designing a large doghouse. His drawing has the scale 1 inch
(87) equals 4 feet.

  a. If he draws the doghouse 3.25 inches long, how long is the actual doghouse?

  b. If the doghouse is to be 6 feet wide, how wide must Andy draw the doghouse?

2. The ratio of exuberant to deflated was 6 to 1. If there were 567 in all,
(45) how many were exuberant?

3. If value of a $100 stock drops 10% one year, then increases 10% the
(67) next year, what is the value of the stock after the increase?

4. An hour is 3.6×10^3 seconds. A common year is 8.76×10^3 hours.
(46) Find the number of seconds in a common year. Express the answer in scientific notation.

* 5. Using whole numbers of inches only, give the widths and lengths of six
(8, 9) different rectangles that have an area of 90 square inches.

6. A one cubic meter shipping container can hold how many 1-cm cubes?
(72) Use unit multipliers.

7. Add the trinomials $x + 2y - 4$ and $-x - y + 2$.
(80)

Analyze The illustration shows a building with a gabled roof. Refer to the illustration for problems **8–10.**

* 8. What is the volume of the
(76) building?

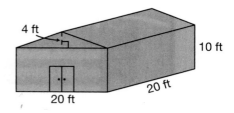
4 ft
10 ft
20 ft
20 ft

* 9. Enough paint must be purchased
(43) to cover the four sides of the building. What is the total surface area of the four sides?

10. The roof overhangs the building by about one foot on all four sides.
(Inv. 2, 85) Estimate the total area of the two sloping sections of roof. (Hint: first use the Pythagorean Theorem to estimate the distance from the peak to the edge of the roof.)

11. **Evaluate** On a sunny day Jason sees that the shadow of a flag pole is
(40, 45) about four times as long as his own shadow. If Jason is about $5\frac{1}{2}$ feet tall, about how tall is the flagpole?

Simplify.

* 12. $2\sqrt{6}\sqrt{15}$
(78)

13. $\sqrt{720}$
(54)

14. $(-12) + (-3)(-4) - (-5)$
(36)

15. $\frac{2}{3} \times 0.15$
(84)

* 16. Can a right triangle have sides of length 1, 7, and $5\sqrt{2}$? Justify your
(Inv. 2, 78) answer.

* 17. Alex's phone company charges these rates
(88) for collect calls. Does this table show a proportional relationship? If so, state the constant of proportionality.

Minutes	Charge
1	$1.10
2	$1.20
3	$1.30
4	$1.40

18. Graph the lines $y = -x$ and $y = x$. Describe the relationship between
(41) the two lines. Where do the lines intersect?

19. **Model** Melissa is thinking of two numbers. She gives two hints: their
(89) sum is 1, and their difference is 5. Solve this system of equations by graphing to find her two numbers.

$$\begin{cases} x + y = 1 \\ x - y = 5 \end{cases}$$

20. Write $\frac{8}{9}$ as **a** a percent, **b** a decimal, and **c** a decimal rounded to two
(63) places.

Solve.

21. $\dfrac{0.5}{2.5} = \dfrac{x}{10}$
(25, 44)

22. $0.007x - 0.07 = 0.7$
(25, 50)

23. $x - 4 = 3x + 2$
(14, 36)

24. $\dfrac{5}{9}x - \dfrac{1}{3} = \dfrac{2}{3}$
(23, 50)

25. Sketch the top, front, and right side views of this object.
(Inv. 4)

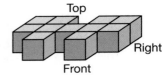

Early Finishers
Real-World Application

Cartography is the art of making maps. Cartographers create large-scale maps and small-scale maps. A large-scale map shows a small area with a large amount of detail. A small-scale map shows a large area with a small amount of detail. To understand the relationship between the distances on a map and the corresponding distances on the Earth's surface, you must understand the map's scale factor. On an certain globe, one inch represents five hundred miles on the Earth's surface. If a region is 660 miles wide, how many inches wide is the region on the globe?

• **Sets**

facts   Power Up R

mental math

a. **Statistics:** Find the median of this data: 11, 11, 17, 16, 15

b. **Estimation:** $19.89 + $21.89

c. **Number Sense:** What digit is in the ones place in the product 1528 × 432?

d. **Powers/Roots:** $\frac{m^5}{m}$

e. **Proportion:** $\frac{3}{2} = \frac{x}{4}$

f. **Geometry:** Angle P and $\angle Q$ are supplementary. If $m\angle P = 95°$, find $m\angle Q$.

g. **Scientific Notation:** Write 6.02×10^{-2} in standard notation.

h. **Calculation:** $8 \times 7, - 20, \sqrt{\ }, \times 8, + 1, \sqrt{\ }, \div 2$

problem solving   How many different sizes of triangles are there in this figure? How many triangles of each size are there? How many triangles all together?

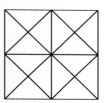

New Concept   *Increasing Knowledge*

A set is a collection of elements. Almost anything can be an element, for example, numbers, ordered pairs, variables, or geometric figures. The elements of a set are often listed in braces, { }. The set of whole numbers can be expressed this way:

$$\{0, 1, 2, 3, ...\}$$

Thinking Skill

Explain

Why does the set of whole numbers not begin with an ellipsis?

The ellipsis (...) indicates that the elements of the set continue in the same pattern without end. The set of integers can be expressed this way:

$$\{..., -3, -2, -1, 0, 1, 2, 3, ...\}$$

Example 1

Use set notation to denote the set of even integers.

Solution

$$\{..., -4, -2, 0, 2, 4, ...\}$$

To indicate an element of a set we use a symbol like a curved E.

$3 \in \{\text{integers}\}$

Read: 3 is an element of the set of integers

$\frac{1}{2} \notin \{\text{integers}\}$

Read: $\frac{1}{2}$ is not an element of the set of integers

Example 2

Use set notation to indicate whether each number is an element of the set of rational numbers.

a. 2

b. $\frac{1}{2}$

c. $\sqrt{2}$

Solution

a. $2 \in \{\text{rational numbers}\}$

b. $\frac{1}{2} \in \{\text{rational numbers}\}$

c. $\sqrt{2} \notin \{\text{rational numbers}\}$

Connect What other math symbol uses a slash mark to indicate the word *not*?

Example 3

Graph the elements of each set on a number line.

a. the set of integers

b. $\{\sqrt{1}, \sqrt{2}, \sqrt{3}, \sqrt{4}\}$

Solution

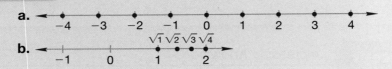

a.

b.

Sets can be related to each other in various ways. If all of set A is contained in set B, then set A is a **subset** of set B. The subset symbol is similar to a ∪ rotated 90° clockwise.

$$A \subset B$$

"A is a subset of B"

For example, the set of rectangles is a subset of the set of quadrilaterals, because every rectangle is a quadrilateral.

$$\text{rectangles} \subset \text{quadrilaterals}$$

Example 4

Use the subset symbol to show the relationship between the set of natural numbers ℕ and the set of integers ℤ.

Recall from Lesson 1 that natural numbers are the numbers we say when we count by ones starting from 1: {1, 2, 3, …}. Integers include all the natural numbers, their opposites, and zero. Thus, the set of natural numbers is a subset of the set of integers.

$$\mathbb{N} \subset \mathbb{Z}$$

Symbols like the N and Z in example 4 are used in some math books to indicate certain sets of numbers. Here are the symbols that are used.

$\mathbb{N}$ Natural (counting) numbers

$\mathbb{Z}$ Integers

$\mathbb{Q}$ Rational numbers (quotients)

$\mathbb{R}$ Real numbers

We can use **Venn diagrams** to illustrate the relationships among sets. This diagram shows that set S (squares) is a subset of set R (rectangles).

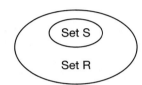

Sets A and B overlap if some elements of set A are also elements of set B. The shaded region of the Venn diagram illustrates the common elements. This region is called the **intersection** of sets A and B and is shown with the symbol ∩. The intersection of A and B is the set of elements that are in both sets.

$$A \cap B$$

"The intersection of sets A and B."

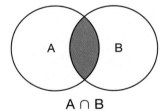

A ∩ B

For example, the intersection of the set of rectangles and the set of rhombuses is the set of squares.

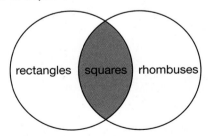

We use the word **union** and the symbol ∪ to indicate the combining of sets. The shaded Venn diagram below illustrates the union of sets A and B. The union is the set of all the elements of sets A and B combined.

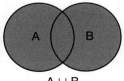

A ∪ B

For example, the union of the set of boys in the classroom and the set of girls in the classroom is the set of students in the classroom.

Example 5

Sets A and B are shown below. Find A ∩ B and A ∪ B.

$$A = \{1, 2, 3, 4\} \qquad B = \{2, 4, 6, 8\}$$

Solution

$$A \cap B = \{2, 4\}$$
$$A \cup B = \{1, 2, 3, 4, 6, 8\}$$

If there are no elements in the intersection of two sets, we say the intersection is an **empty set**: ∅.

Formulate Name two sets in which the intersection is the empty set.

In the table below we list some symbols commonly used to refer to relationships among sets.

Set Notation Symbols

{ }	The set of
∈	Is an element of
⊂	Is a subset of
∩	The intersection of
∪	The union of
∅	Empty set

Example 6

Use a Venn Diagram to illustrate the sets. Find A ∩ B.

a. A = {1, 2, 3}   b. A = {2, 4, 6}
   B = ℕ   B = {1, 3, 5}

Solution

a.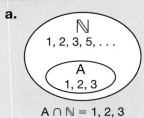

A ∩ ℕ = 1, 2, 3

b.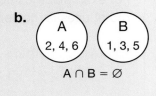

A ∩ B = ∅

Practice Set

Indicate if the number is an element of the set of integers.

a. −1   b. 2   c. √3

Use the subset symbol to show the relationship between A and B.

d. A = {all triangles}; B = {all polygons}

e. $A = \mathbb{R}; B = \mathbb{Q}$

f. A = {vowels}; B = {letters in the alphabet}

Use a Venn Diagram to illustrate the relationship between the sets.
Find A ∩ B.

g. A = {3, 6, 9, 12}
B = {5, 12, 15, 20}

h. A = {all parallelograms}
B = {all trapezoids}

Use a Venn Diagram to illustrate the relationship between the sets. Then
indicate the union of the sets.

i. {10, 20, 30} ∪ {20, 30, 40}

j. {2, 4, 6, 8, ...} ∪ {4, 8, 12, ...}

Written Practice   *Strengthening Concepts*

* 1. James drew a village to the scale 1 inch equals 10 feet. **a** If he drew
(87)   one building 2.1 inches long, how long was the actual building?
b If a building was actually 18 feet wide, how wide was the building on
James' drawing?

* 2. **Formulate** One atom of mercury has a mass of 3.33×10^{-22}g. What
(46)   is the mass of 6.02×10^{23} atoms of mercury? Express your answer in
standard form.

3. The $45 price was increased by 8% for sales tax. What was the total
(67)   price including tax?

* 4. What is the measure of the central angle formed by the hour and
(81)   minute hands on the face of a clock at the following times?
a. 6:00   **b.** 3:00   **c.** 4:30   **d.** 1:30

5. If a coin is flipped five times, what is the probability of getting heads all
(68)   five times?

6. Describe the rule of this sequence and express the rule with an
(73)   equation. Then use your rule to find the 20th term.

$$5, 10, 15, 20, \ldots$$

7. On a coordinate plane draw a rectangle with these vertices: (3, 1),
(Inv. 2)   (2, −2), (−1, −1), and (0, 2). Then find the length of each side.

8. Find the area of the rectangle you drew in exercise **9**.
(Inv. 2,
78)
* 9. Some pulsars spin at 60,000 rotations per minute (rpm). How many
(64)   rotations per second is that? How long does it take for each rotation?

Generalize Simplify.

10. $(-3)^1 - 3^2 - (-3)^3$
(33, 51)

11. 75% of $\frac{2}{3}$ of $1.44
(84)

12. $\sqrt{20,000}$
(74)

13. $2x^{-2}y^{-3}xy$
(51)

14. At the airport, Andre surveyed passengers and asked how many trips
(53) in the last year they had taken by air. The data he collected is shown
below:

$$3, 1, 2, 1, 5, 6, 2, 2, 3, 4$$

a. Find the mean, median, mode, and range of the data.

b. Which of these measures would you report to someone interested
in knowing the most common number of trips people at this airport
take in a year?

15. Rene recorded the time and average
(41) speed she traveled on different trips
from her home to work.

a. Graph the pairs of points on
coordinate axes.

b. Use the graph to predict how long
the trip might take Rene if she travels
55 mph.

Rate (mph)	Time (min)
30	90
45	60
50	54
60	45

* 16. **Model** Bingham and Gridley each thought of a number. Bingham's
(89) number was 5 less than Gridley's number. The sum of their numbers
was 3. Find their two numbers by graphing this system of equations.

$$\begin{cases} y = x - 5 \\ x + y = 3 \end{cases}$$

* 17. **Explain** Adam counted by twos to twelve. Beverlee counted by threes
(90) to twelve.

$$A = \{2, 4, 6, 8, 10, 12\}$$
$$B = \{3, 6, 9, 12\}$$

a. Find $A \cap B$ and describe what the answer means.

b. Find $A \cup B$ and describe what the answer means.

18. The driving speed limit of some highways is 60 mph. Use unit multipliers
(64, 72) to convert 60 miles per hour to feet per second. (1 mi = 5280 ft)

19. Write $\frac{2}{11}$ **a** as a percent and **b** as a decimal, and **c** as a decimal rounded
(63) to three decimal places.

20. a. Factor: $5m^2 - 40m + 5$ **b.** Expand: $7x(x - 3)$
(21)

Solve.

21. $\frac{1.5}{2.5} = \frac{3}{x}$
(25, 44)

22. $0.05 + 0.02x = 1.95$
(25, 50)

23. $\frac{4}{5} - \frac{2}{3}x = \frac{14}{15}$
(23, 50)

24. $-\frac{m}{3} = 3$
(23, 38)

25. The graph below shows the change in the total average daily
attendance in all public schools in one state over a period of time.
Use the graph to answer the following questions.

(Inv. 6)

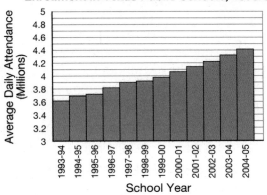

Enrollment in Texas Public Schools, 1993-2005

a. What kind of graph is this?

b. The data for which state is displayed in the graph?

c. What is the interval used for data categories on the *x*-axis?

d. In which school year was enrollment about 15% greater than the
enrollment in 1993-94?

e. Describe the trend of the data, if one exists.

Early Finishers

*Real-World
Application*

Mr. Tham needs to rent a car for an upcoming vacation. Cars R Us charges
an initial fee of $10.00 plus $15.00 a day. Roger's Rental charges an initial fee
of $15.00 plus $10.00 a day.

a. If *d* represents the number of days one car is rented and *y* represents
the total cost of renting the car, write one equation for each rental car
company to represent the cost of renting a car.

b. Graph both equations for values of *d* from 1–5. Which rental car
company offers the best price for a five-day rental?

c. On which day do both shops charge the same amount?

Focus on
• Sampling Methods

The U.S. Census Bureau reported that for a one-way trip to work, the average nationwide commute time was 24.3 minutes in 2003. The American Community Survey randomly selected households from across the nation to participate in a similar commute study. Individual state and city averages vary as follows.

Nationwide	New York	Nebraska	Chicago	Tulsa
24.3 min	30.4 min	16.5 min	33.2 min	17.1 min

Every ten years, the U.S. Census Bureau conducts a study, the intent of which is to survey every resident in the nation. This study, in which every member of the population is included, is called a *census study*. With a census study, it is possible to accurately describe characteristics of the population.

Math Language

In a **random sample,** each member of the population has an equal probability of being selected.

A census of an entire population requires a large amount of time and money, so it is generally not practical to conduct a census study. Instead we study a **sample,** or small group, of the population. A sample should be **representative** of, or very similar to, the larger population. If a large enough sample is selected **randomly** from the entire population, we can be reasonably sure that the sample is representative.

Analyze How might the results of the survey above differ if the sample was selected from only one city or state? How representative would the results be for the entire nation?

When we conduct a study of a population, we randomly select a sample of sufficiently large size. To do this, we can assign a number to each member of the population to be studied and randomly select numbers to choose sample subjects.

Random Number Generators

Thinking Skill

Discuss

What population does the health food store want to survey? Why might the store want to survey this particular population?

A health food store wants to test its consumers' response to a new product. The researcher assigned a number to each person in the customer database and used a random number generator to select a sample of customers.

1. Use a random number table to imitate the selection of sample subjects. A random number table is a random listing of integers 0 to 9. (A random number table is provided for this exercise in *Saxon Math Course 3 Instructional Masters* Investigation Activity 8, or you may create your own.) Start at any number on the table, then record all the numbers that follow in the same row or column. Write the list as pairs of numbers, but discard pairs that do not fall in the range 00–19. List the corresponding customer names with each remaining number pair. Repeat these steps, if necessary, until five subjects are chosen.

Customers

00	Strickler	10	Smith
01	Sanchez	11	Zamora
02	Tsing	12	McNelly
03	LaBrado	13	Jensen
04	Poth	14	Gomez
05	Barrell	15	Rathod
06	Jo	16	O'Grady
07	Joslen	17	Barreto
08	Cheung	18	Davis
09	Clark	19	Lee

Our results may look like this:

0781351205941609

Write the list as pairs of numbers, and eliminate numbers that do not apply.

07   8̶1̶   3̶5̶   12   05   9̶4̶   16   09

Then find the corresponding name:

07: Joslen

12: McNelly

05: Barrell

16: O'Grady

09: Clark

2. Now we will repeat the selection process using a random number generator instead of a table. A graphing calculator generates one random number with these commands:

MATH , ◄ , ENTER

The word "rand" is displayed on the screen above a decimal number between 0 and 1. Take the first two digits displayed, and discard numbers not in the range 00–19. Record the corresponding customer names.

If the random number is .9518983326, we discard it.

If the random number is .1081104624, we write:

10: Smith

Press ENTER to produce more random numbers until the sample size reaches 5.

In the previous activity, we used *random sampling* methods. An alternative is *systematic sampling,* in which subjects are chosen from a list with equal spacing between selected subjects. For example, to choose five subjects from the customer list, we could have chosen every fourth customer: 03, 07, 11, 15, 19. With either method, a certain amount of *sampling error* is introduced. Error is minimized when we use objective methods to ensure an equal probability for any population member to be selected. In an actual study, we would select a much larger sample of a population.

Example 1

Identify a problem with each sampling method described below. Describe a better sampling method for each situation.

a. Matthew wants to know the percentage of students at his middle school who favor school uniforms. He decides to ask everyone in his English class whether or not they favor school uniforms.

b. In response to requests for a new softball diamond, the city council would like to know how important building a softball diamond is to the residents. They ask researchers to conduct a phone survey by randomly selecting 20 names from the city's phone book.

Solution

a. Matthew's survey is not a random sampling of the entire school. This type of sampling is sometimes called convenience sampling, because subjects are chosen for their convenience rather than randomly. Instead, he could arrange the names of all of the students in his school alphabetically, number them, then use a random number generator to choose subjects for his sample. He could also use another sampling method called *stratified sampling*. With this method, the researcher identifies groups within the population and randomly chooses a sample within each group, called a **stratum.** Matthew could take a random sample from each of the grades (6th, 7th, and 8th) of his middle school.

b. Surveying only 20 people in a large city would not give a highly reliable result. The greater the sample size, the greater confidence we have that the results from the sample would match the results from the desired population. For example, it would be better for the city to sample 200 people.

Once a representative sample is selected, care is taken to assure that information is gathered without bias. For a study involving a **survey,** we choose questions carefully to avoid bias. The survey questions should not sway the persons answering them one way or another.

Example 2

Consider the following questions proposed for a survey about a new softball diamond. Select the most appropriate (least biased) question and explain your choice.

1. **Wouldn't you like a new softball diamond at Filly Park?**

2. **Which of these describes your desire for a softball diamond at Filly Park: I do not want one / I feel indifferent to the idea / I would like one / I want one built soon.**

3. **The city pays for many services under a tight budget, like street lighting, crossing guards in front of schools, and park maintenance. How important is it to you that the city build a softball diamond at Filly Park?**

Solution

Only **question 2** would be an appropriate survey question. With question 1, many people would agree to a new softball diamond. Question 3 sets a tone that may dissuade respondents from stating that they want a new softball diamond.

Formulate Write a question that Matthew could ask in example 1a that would be an appropriate survey question. Then write a question that he could ask that may sway the responses.

As a class, decide on an opinion survey you want to conduct. Answer these questions to help you plan:

- You will measure the percentage of a certain opinion on what issue?
- What population will you study?
- What will be the sample size?
- How will you select your sample?
- What question will you ask?
- How will you collect responses?
- How will you report and display the results?

After you have planned your survey, answer these questions.

- Can you think of other populations in which the outcome would be different?
- Can you think of a different way of asking the question that would sway the responses?

• Effect of Scaling on Perimeter, Area, and Volume

facts | Power Up S

mental math

a. **Algebra:** Solve: $5x = 15$

b. **Estimation:** $\$1.48 \times 2 + 3 \times \$7.98$

c. **Fractional Parts:** $\frac{5}{7}$ of 56

d. **Percent:** 20% more than $50

e. **Rate:** Three rings per hour is how many rings per day?

f. **Geometry:** A loop of string in the shape of a square encloses 36 in.^{2} If the same loop of string is arranged to make an equilateral triangle, then each side of the triangle will be how long?

g. **Measurement:** Find the temperature indicated on this thermometer.

h. **Calculation:** 9×4, $\sqrt{}$, $\times 7$, $+ 7$, $\sqrt{}$, $- 8$

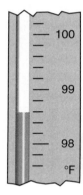

problem solving | A room is 12 ft wide and 18 ft long. How many square yards of flooring is required to cover the floor?

In many lessons and problems, we have noted the effect of changing dimensions on the perimeter, area, surface area, and volume of various figures. These effects are summarized in the table below. If the dimensions (length, width, height) of a figure or object are changed proportionally (by a scale factor), then these ratios between the two figures apply:

Scale Factor

ratio of perimeters = scale factor

ratio of areas = (scale factor)^2

ratio of volumes = (scale factor)^3

In this lesson we will review these relationships with a variety of examples.

Example 1

Vince was asked to paint a mural near the city park based on a painting he made in art class. The mural was to be ten times the length and width of his original canvas painting. Vince bought ten times as much paint as he used for his original painting, yet he ran out of paint after only a fraction of the mural was painted. Why did Vince run out of paint so quickly? How much paint did he need? Explain.

Solution

The dimensions of his painting are increased by a factor of 10, so the area of the mural is 100 times (10^2) the area of the canvas. He needs to buy 100 times as much paint as he used for his original painting.

We can say, "Vince is increasing *two* dimensions (both length and width) by a factor of 10. This results in an increase of the area by a factor of 10 × 10, which is 100. Therefore, painting the mural is like painting 100 canvases."

When the dimensions of a figure are proportionally increased or decreased by a scale factor, the area is increased or decreased by the square of the scale factor.

Scale Factor

ratio of lengths = scale factor

ratio of areas = (scale factor)^2

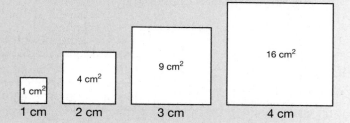

1 cm²   4 cm²   9 cm²   16 cm²

1 cm   2 cm   3 cm   4 cm

Example 2

Irene works with stained glass. She wants to make a replica of a stained glass window. Her replica will be related to the original window by a scale factor of $\frac{1}{3}$. The original window is made of 3456 square inches of glass and has a perimeter of 240 inches. How much glass will Irene use in the replica? What will be the perimeter of her replica?

Thinking Skill

Explain

Do you need to know the shape of a figure to find its perimeter and area when its dimensions are scaled by a factor?

Irene's replica will have $\frac{1}{3}$ the linear dimensions of the original window, and $(\frac{1}{3})^2$, or $\frac{1}{9}$ the area. We multiply the amount of glass in the original window by the square of the scale factor to find the amount of glass Irene will use:

$$\frac{1}{9}(3456 \text{ in.}^2) = \textbf{384 in.}^2$$

Now we consider the perimeter of the replica. The side lengths of the window are scaled down by a factor of $\frac{1}{3}$. Regardless of the shape of the window, each corresponding side of the replica is $\frac{1}{3}$ the length of the side of the window. Therefore, the sum of the lengths of the sides will be $\frac{1}{3}$ the perimeter of the window.

$$\frac{1}{3}(240 \text{ in.}) = \textbf{80 in.}$$

Example 3

Nelson built this figure with four one-inch cubes. How many one-inch cubes are needed to build a similar figure with the dimensions doubled?

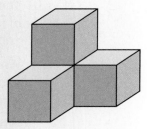

Solution

Doubling every dimension means every distance is twice as large. We are asked how changing the dimensions by a scale factor of 2 affects the volume.

Scale Factor

ratio of lengths = scale factor
ratio of volumes = (scale factor)^3

Since the scale factor is 2, the ratio of the volumes is 8.

$$(2)^3 = 8$$

That means 8 times as many cubes are needed to build a similar figure with double the dimensions of the original figure.

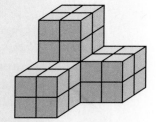

$$8 \cdot 4 = 32$$

We find that **32 cubes** are needed.

Practice Set

a. If you know the scale factor between two polygons, describe how you can find the ratios of their perimeters and areas.

b. The dimensions of a wallet-size photo are about half the dimensions of a 5 inch-by-7 inch photo. The area of a wallet-size photo is about what fraction of the area of a 5-by-7 photo?

c. A triangle with vertices at (0, 0), (1, 0), and (0, 2) is dilated by a scale factor of 4. The area of the dilated image is how many times the area of the original triangle?

d. Find the scale factor of the smaller rectangle to the larger rectangle. Find the ratios of the perimeters and areas of the larger rectangle to the smaller rectangle.

4 in. 6 in. 2 in.

e. Find the surface area and volume of a 1-inch cube and a 10-inch cube. What is the ratio of the surface areas? What is the ratio of the volumes?

f. One square has sides 10 cm long. A second square has sides 50% longer. How long are the sides of the larger square? What is the scale factor from the smaller to larger square written as a decimal? What is the ratio of the areas of the smaller to larger square written as a decimal?

g. *Justify* To celebrate her pizza shop's 10th anniversary, Blanca is offering "Double Your Dollar Days." During this time, customers can get twice as much pizza for the price of a 10-inch pizza. What is the approximate diameter of a pizza twice the area of a 10-inch pizza? (Assume the thicknesses of the all pizzas are the same.) Justify your answer.

A 12 in.   **B** 14 in.   **C** 16 in.   **D** 20 in.

h. If a 6-inch diameter balloon holds a pint of water, how much water does a 6-inch radius balloon hold?

1. The weight of a quantity of water varies directly with its volume. If a
(70)   pint (2 cups) of water weighs one pound, how much does one gallon
(16 cups) of water weigh?

2. The shoes were marked up 90% after being autographed by the sports
(67)   star. If they were originally $70, what was the new price?

* 3. **Estimate** The regular price was $23.95. This discount was $33\frac{1}{3}$%.
(84)   Estimate the sale price.

* 4. **Analyze** A farmer drew a picture of his fields to the scale of 1 inch
(87)   equals 250 feet.

    a. If the corn field is 1000 feet long, how long is the drawing of the
field?

    b. If the drawing of the pumpkin patch is 0.5 inches wide, how wide is
the actual patch?

5. Baxter filled a 15 cm tall cylindrical container
(79)   with 750 cm^3 of water.

    a. How can Baxter find the area of the base
of the container using this information?

    b. Transform the formula $V = Bh$ to solve for B.

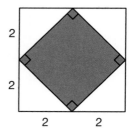
750 cm³   15 cm

6. If Baxter uses the water in the container to fill cones with the same base
(86)   and height as the cylinder, how many cones can he fill?

7. Expand and simplify this expression and arrange the terms of the
(36, 80) product in descending order. $3(x - 4) + x(x - 4)$

* 8. **Evaluate** For a decorative table top, Misti
(Inv. 2) will make a mosaic in the shaded portion of
the square as shown in the diagram. Find the
area of Misti's mosaic. (Dimensions are in
feet.)

    2

    2

         2   2

9. Given circle A with a radius of x inches and circle B with a radius of
(91) $2x$ inches, find the following ratios.

    a. circumference of circle A to circumference of circle B

    b. area of circle A to area of circle B

10. The letter cards A, E, I, O, U are face down on a table. If cards are
(83) selected one at a time and held, what is the probability of selecting all
five cards in alphabetical order?

11. **a.** What is the area of the triangle?
(20, 40)

b. What is the area of the circle in terms
of π?

c. What fraction of the area of the circle is
the area of the triangle? Express in terms
of π.

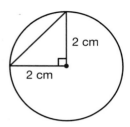

Generalize Simplify.

12. $3xyx - 2xy^2 + yx^2$
(31)

* 13. Simplify: $2\sqrt{8}\sqrt{18}$
(74, 78)

14. The number of ride tickets sold at a carnival in a week is shown below.
(Inv. 6)

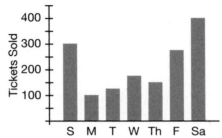

a. On which day were the most tickets sold? the least?

b. How many tickets were sold on Friday and Saturday combined?

* 15. *Evaluate* A sign maker paints two signs with proportional dimensions.
(91)

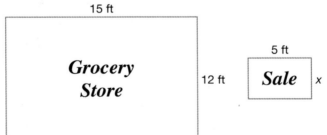

a. Find x.

b. Find the ratio of the perimeter of the larger sign to the perimeter of
the smaller sign.

c. Find the ratio of the area of the larger sign to the smaller sign.

d. If it takes a pint of paint to paint the small sign, how much paint is
needed to paint the large sign?

A triangle has vertices at (0, 0), (2, 1) and (−2, 4). Sketch the triangle on a
coordinate plane to answer to answer problems **16** and **17**.

* 16. *Model* **a.** Classify the triangle by angles.
(Inv. 2,
78) **b.** What is the area of the triangle?

* 17. **a.** What is the length of the longest side of the triangle? (Hint: Tracing
(Inv. 1, Inv. 2) grid lines make a right triangle with the same longest sides.)

b. What is the perimeter of the triangle?

* 18. The days that Zachary attends soccer practice (*S*) and music lessons
(90) (*M*) are expressed in set notation below.

$$S = (M, W, F)$$

$$M = (T, W)$$

a. Find $S \cap M$ (The days Zachary attends both soccer practice and
music lessons.)

b. Find $S \cup M$ (The days Zachary attends either soccer practice or
music lessons or both.)

19. Under standard conditions one kilogram of water occupies 1000 cm^3.
(64, 72) Use unit multipliers to convert 1000 cm^3 to cubic inches. (Either use a
calculator to find the volume to the nearest cubic inch or leave the result
undivided.)

20. **a.** Factor: $4x^2 + 4x$ **b.** Expand: $-(x - 3)$
(21)

21. Write $\frac{5}{11}$ **a** as a percent, **b** as a decimal, and **c** as a decimal rounded to
(63) the nearest thousandth.

Solve.

22. $\dfrac{1.2}{1.6} = \dfrac{9}{x}$ **23.** $\dfrac{5}{6}x = \dfrac{1}{2}x + \dfrac{1}{3}$
(25, 44)   (23, 78)

24. $0.1 + 0.001x = 1.01$
(41)

25. Study this tile pattern and describe how it
(Inv. 2) illustrates the Pythagorean Theorem. Support
your description with an illustration.

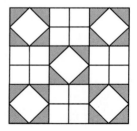

• Areas of Rectangles with Variable Dimensions
• Products of Binomials

facts   Power Up S

mental math

a. **Algebra:** Solve: $\frac{1}{2}x = 20$

b. **Number Sense:** What digit is in the one's place in the product 97×97

c. **Powers/Roots:** $\frac{m^4}{m^2}$

d. **Probability:** Three dozen blocks were in the bag, of which four were red. What is the probability that a block removed at random will not be red?

e. **Ratio:** In an old town, 50 of the 70 residents were miners. What was the ratio of miners to non-miners?

f. **Select a Method:** To find the approximate height of a tall tree, what method would you probably use?
   A Climb the tree and measure directly.
   B Use similar figures and a proportion.
   C Measure the circumference and divide by π.

g. **Geometry:** If two angles are complementary, the sum of their measures is 90°. If $\angle C$ and $\angle D$ are complementary, and $m\angle C = 60°$, find $m\angle D$.

h. **Calculation:** $-1 - 2, + 4, \times 5000, \div 10, \div 10$

problem solving

A polygon with 3 sides has interior angles with a sum of 180°. A polygon with 4 sides has interior angles with a sum of 360°. A polygon with 5 sides has interior angles with a sum of 540°. If the sum of the interior angles of a polygon is 1800°, how many sides does it have?

New Concepts *Increasing Knowledge*

areas of rectangles with variable dimensions

Suppose the floor of a square closet with sides measuring 3 ft 6 in. will be covered with one-foot square tiles. We can sketch a plan to show a way the tiles can be arranged and use the sketch to help us find the area of the floor.

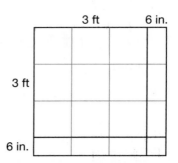

We see one part of the area is covered with full tiles, two parts are covered with half tiles, and one part is covered with a quarter tile. We count the tiles in each part.

9 full tiles + 3 half tiles + 3 half tiles + 1 quarter tile

We can combine the two groups of half tiles.

9 full tiles + 6 half tiles + 1 quarter tile

Six half tiles equal three full tiles. We add to find that our sketch shows 12 full tiles and one quarter tile, meaning the area of the floor is $12\frac{1}{4}$ ft.

Sketching can also help us visualize the areas of rectangles with dimensions expressed as variables.

Example 1

Suppose a rectangular rug measures $2\frac{1}{2}$ ft by $4\frac{1}{4}$ ft. We can represent its area by writing:

$$\left(2 + \frac{1}{2}\right)\left(4 + \frac{1}{4}\right)$$

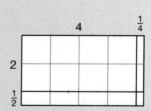

a. **Sketch the rectangle on a grid to depict the product.**

b. **What is the total area of the rug?**

Solution

a.

The rectangle is $(4 + \frac{1}{4})$ ft long and $(2 + \frac{1}{2})$ ft wide. The area of each rectangular portion is represented by a different pair of factors from the expression $(2 + \frac{1}{2})(4 + \frac{1}{4})$.

b. We multiply each term in one set of parentheses by each term in the other set of parentheses to find the total area.

	Step:	**Justification:**

$$\text{Area} = (2 \cdot 4) + \left(2 \cdot \frac{1}{4}\right) + \left(\frac{1}{2} \cdot 4\right) + \left(\frac{1}{2} \cdot \frac{1}{4}\right)$$ Area of four regions

$$= 8 + \frac{1}{2} + 2 + \frac{1}{8}$$ Multiplied

$$= 10\frac{5}{8}$$ Added

As we see in the diagram, the area of the rug is $10\frac{5}{8}$ **ft²**.

Example 2

A rectangle is $x + 3$ units long and $x + 2$ units wide. What is the area of the rectangle?

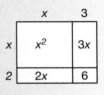

Solution

This rectangle has side lengths expressed with binomials. We sketch the rectangle and divide the length and width into two parts, representing the two terms of each binomial. This divides the area of the rectangle into four smaller rectangular parts. We find the area of each part.

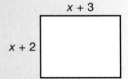

	Step:	**Justification:**
	$\text{Area} = x^2 + 3x + 2x + 6$	Added areas of parts
	$\text{Area} = x^2 + 5x + 6$	Combined $3x$ and $2x$

products of binomials

We can find the area of the rectangle in example 2 by multiplying the binomials. In this part of the lesson we will practice techniques for multiplying binomials.

The product of two binomials has four terms, some of which may be like terms. Each term of one binomial is multiplied by each term of the other binomial.

We will use an arithmetic model to show the multiplication of two binomials. Breaking out the partial products of the problem on the left, we show that multiplying 43×42 is equivalent to multiplying $(40 + 3) \times (40 + 2)$.

$$
\begin{array}{r}
43 \\
\times\ 42 \\
\hline
86 \\
1720 \\
\hline
1806
\end{array}
\quad
\begin{array}{l}
= \\
= (40 + 3)2 \\
= (40 + 3)40
\end{array}
\quad
\begin{array}{l}
= \\
=
\end{array}
\quad
\begin{array}{r}
(40 + 3) \\
\times\ (40 + 2) \\
\hline
80 + 6 \\
1600 + 120 \\
\hline
1600 + 200 + 6 = 1806
\end{array}
$$

To find the first partial product on the right, we multiply $(40 + 3)$ by 2 (the "ones place" of the bottom factor) and record the result. To find the second partial product on the left, we multiply $(40 + 3)$ by 40 (the "tens place" of the bottom factor) and record the result. We add all the products to find the total product. We will use this method in example 3.

Example 3

Expand: $(x + 3)(x + 2)$

Solution

First we will use an arithmetic model to find the product.

$$\begin{array}{r} (x + 3) \\ (x + 2) \\ \hline 2x + 6 \\ x^2 + 3x \quad\ \\ \hline x^2 + 5x + 6 \end{array}$$

Connect   Compare the results of this multiplication to the solution to example 2.

We must pay attention to the signs when multiplying binomials. Multiplying binomials involves multiplying terms. Every term has a sign. Like signs result in a positive product. Unlike signs result in a negative product. Refer to Lesson 36 to review the method for properly multiplying terms.

The acronym **FOIL** can help us remember the four multiplications that occur when multiplying two binomials.

product of two binomials = products of **F**irst + **O**utside + **I**nside + **L**ast

Example 4

Expand: $(x - 3)(x - 4)$

Solution

We can multiply using the algebraic method (and its acronym, FOIL). First we multiply both terms in the second binomial by x. Then we multiply both terms by -3.

Thinking Skill

Connect

Does reversing the order of the binomials change the product?

Step:	Justification:
$(x - 3)(x - 4)$	Given expression
$(x - 3)(x - 4)$	Applied **FOIL**
$x^2 - 4x - 3x + 12$	Multiplied
$x^2 - 7x + 12$	Simplified

Practice Set

Refer to this rectangle to answer problems **a** and **b.**

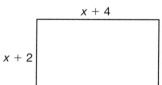

a. *Analyze*   What is the area of the rectangle? (Make a sketch and check by multiplying.)

b. *Justify*   What is the area of a square with a side length $x + 4$? List and explain each step you used to find the answer.

c. Find the product: $(x + 2)(x + 5)$. Depict the product with a rectangle.

Expand.

d. $(x + 1)(x + 3)$ **e.** $(x + 4)(x + 7)$

f. $(x + 3)^2$ **g.** $(x + 10)^2$

h. $(x + 8)(x - 3)$ **i.** $(x - 5)(x + 2)$

j. $(x + 2)(x - 2)$ **k.** $(20 + 2)(20 - 2)$

Written Practice   *Strengthening Concepts*

1. The old fee of $12 was increased by 175%. What is the new fee?
(67)

* 2. **Analyze** Joseph's workshop is 12 feet by 7 feet. If the drawing of the
(87) workshop has a scale of 1 inch = 2 feet, what are the dimensions of the
drawing?

* 3. Three siblings buy a soccer goal for $29.99 plus 6% tax. Which
(67) expression shows the total paid if they split the cost evenly?

     A ($29.99 × 0.06) ÷ 3   **B** ($29.99 + 0.06) ÷ 3

     C ($29.99 × 1.06) ÷ 3   **D** ($29.99 + 1.06) ÷ 3

* 4. **Model** On a number line graph all the real numbers greater than negative
(1, 16) 2 that are also less than or equal to 5.

* 5. What central angle is formed by the hands of a clock at 4:00? The
(11, 81) sector of the arc between the hands at 4:00 is what percent of a full
circle?

* 6. **Analyze** The set of the first five counting numbers is {1, 2, 3, 4, 5}. The
(90) set of the first five even counting numbers is {2, 4, 6, 8, 10}.

     a. What is the intersection of the two sets?

     b. What is the union of the two sets?

For problems **7** and **8**, write an expression representing the area of each
rectangle.

7.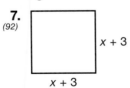
(92)

     $x + 3$

     $x + 3$

8. $x - 1$
(62)

     $4x + 2$

9. The vertices of a rectangle are at $(1, -3)$, $(4, 0)$, $(0, 4)$, and $(-3, 1)$.
(Inv. 1, Inv. 2) Sketch the rectangle and find the length and width of the rectangle. Use
the information for problems **10** and **11**.

10. What is the area of the rectangle in problem **9**?
(78)

11. What is the length of a diagonal of the rectangle?
(78, Inv. 2)

12. Write a variable expression for the perimeter of the square with side
$\quad$ $2x + 1$.
(80)

* 13. **Evaluate** A triangle with a base and height
(20, 40) of 12 in. is enclosed in a circle which has a
diameter of 20 in.

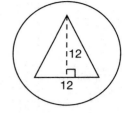

$\quad$ **a.** Visually estimate the percent of the area
$\quad\quad$ of the circle occupied by the triangle.

$\quad$ **b.** Find the area of the triangle and the area
$\quad\quad$ of the circle. (Use 3.14 for π.)

$\quad$ **c.** Calculate to the nearest percent the portion of the circle occupied by
$\quad\quad$ the triangle.

14. Which of the following combinations of lengths *cannot* construct a
(20) triangle?

$\quad$ **A** $2, 3, 4$ $\quad\quad$ **B** $2, 3, 5$ $\quad\quad$ **C** $3, 4, 5$ $\quad\quad$ **D** $3, 4, 6$

15. Can a right triangle have the
(Inv. 2) side lengths shown? Justify your
answer.

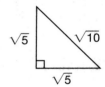

16. If the foundation of a new house requires 240 cubic feet of concrete,
(72) then how many cubic yards of concrete should be ordered?

17. Find the perimeter and area of the
(75) trapezoid.

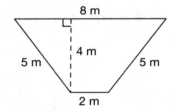

18. One of the 25 students was absent. Write $\frac{1}{25}$ as **a** a percent and
(12) **b** a decimal.

Solve.

19. $\dfrac{2.4}{1.8} = \dfrac{x}{6}$ $\quad\quad\quad\quad\quad\quad$ **20.** $0.05 + 0.01x = 0.4$
(25, 44) $\quad\quad\quad\quad\quad\quad\quad\quad\quad\quad\quad\quad\quad$ *(25, 36)*

21. $\dfrac{5}{3} + \dfrac{3}{4}x = \dfrac{5}{4}$ $\quad\quad\quad\quad\quad$ **22.** $3(x + 2) = 2x + 15$
(22, 38) $\quad\quad\quad\quad\quad\quad\quad\quad\quad\quad\quad$ *(50)*

23. Find the volume of each prism. The two rectangular prisms are similar
(42) figures. Units are in inches. Refer to these figures for problems **24–25.**

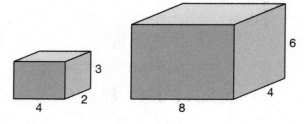

* 24. a. *Analyze*   What is the scale factor from the smaller to larger
(91)   prism?

    b. What is the ratio of the area of one face of the larger prism to the
    area of the corresponding face of the smaller prism?

    c. What is the ratio of the volume of the larger prism to the volume of
    the smaller prism?

25. Sketch and label a net of the smaller prism.
(Inv. 4,
55)

Rita and Saul are building their first house on a city lot. To provide a proper
foundation or their yard, they want to deposit fertilizer-rich topsoil 2-feet
deep over their entire yard. Their front yard is rectangular and measures
200 feet by 125 feet.

    a. What is the area of Rita and Saul's front yard? How much topsoil do
    they need (volume)?

    b. If topsoil sells for $2.95 per cubic yard, about how much should Rita
    and Saul budget to spend on topsoil?

• Equations with Exponents

Power Up   *Building Power*

facts | Power Up S

mental math

   a. Estimation: $52.80 − $33.05

   b. Fractional Parts: $\frac{2}{9}$ of 81

   c. Percent: 0% more than $30

   d. Proportion: $\frac{x}{15} = \frac{9}{5}$

   e. Scientific Notation: Compare 7.3×10^{-4} ◯ 0.00073

   f. Statistics: Find the mode of this data: 11, 11, 17, 16, 15

   g. Measurement: The odometer read 58,254.7 miles at the beginning of the trip. At the end of the trip it read 60,355.9 miles. How long was the trip?

   h. Calculation: $50 + 13$, $\div 3$, $\div 3$, $\times 7$, $\sqrt{\ }$, $\times 9$

problem solving | The original price was $100. For a sale, an employee marked the item down 20%. After the sale, the employee marked the item up 20% from its sale price. What is its current price?

New Concept   *Increasing Knowledge*

An equation containing a variable with an exponent of 2 (but no greater exponents) is called a **quadratic equation.** The following is a quadratic equation:

$$x^2 = 25$$

Both $x = 5$ and $x = -5$ are solutions of the equation above, because 5^2 is 25 and $(-5)^2$ is 25. Every quadratic equation has up to two real solutions. When the solutions are opposites, as in the equation above, we can write them as ± 5, which is read "plus or minus five".

Formulate   Write a quadratic equation with a solution of $x = \pm 4$.

Example 1

Solve: $2x^2 + 7 = 25$

Math Language

Recall that **inverse operations** are operations that "undo" each other. For example, multiplication is the inverse operation of division.

We isolate the x by applying inverse operations. The inverse operation of squaring is finding the square root.

Step:	Justification:
$2x^2 + 7 = 25$	Given equation
$2x^2 = 18$	Subtracted 7 from both sides
$x^2 = 9$	Divided both sides by 2
$x = \pm 3$	Found square roots of 9

Check by testing both solutions in the original equation.

Check $x = 3$	Justification	Check $x = -3$
$2(3)^2 + 7 = 25$	Substituted	$2(-3)^2 + 7 = 25$
$2(9) + 7 = 25$	Simplified	$2(9) + 7 = 25$
$25 = 25$ (true)	Simplified	$25 = 25$ (true)

There are two solutions of this equation because every positive number has two square roots.

$$\text{If } x^2 = 9, \text{ then } x = \pm 3.$$

Example 2

Solve: $2x^2 = 16$

Solution

Step:	Justification:
$2x^2 = 16$	Given
$x^2 = 8$	Divided both sides by 2
$x = \pm \sqrt{8}$	Found square roots of 8
$x = \pm 2\sqrt{2}$	Simplified $\sqrt{8}$

Note that the radical sign indicates only the positive square root, also called the **principal square root.**

$$\sqrt{9} = 3$$

In Examples 1 and 2 we used $\pm$ to indicate both the positive and negative solutions.

Example 3

The two rectangles are similar. Find x by solving this proportion.

Solve: $\dfrac{9}{x} = \dfrac{x}{16}$

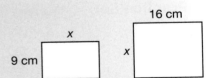

9 cm x 16 cm x

Solve for x.

Step:	**Justification:**
$\dfrac{9}{x} = \dfrac{x}{16}$	Given equation
$x^2 = 9 \cdot 16$	Cross multiplied
$x^2 = 144$	Multiplied 9 and 16
$x = \pm 12$	Found the square roots of 144

Both 12 and -12 are solutions of the proportion.

Check $x = 12$	Justification	Check $x = -12$
$\dfrac{9}{12} = \dfrac{12}{16}$	Substituted	$\dfrac{9}{-12} = \dfrac{-12}{16}$
$\dfrac{3}{4} = \dfrac{3}{4}$	Simplified	$-\dfrac{3}{4} = -\dfrac{3}{4}$

For applications of quadratic equations, often only one of the solutions will make sense. We must consider physical reality when reporting solutions. In this situation, only $+12$ applies because lengths are positive. The dimension is **12 cm.**

Example 4

Jaime is making a mosaic using 100 one-inch square tiles. If the shape of his mosaic is to be a square, about how long should each side be?

Solution

Since the area of Jaime's mosaic will be 100 in.^{2}, we can use the area formula to solve for the length of each side.

Step:	**Justification:**
$A = s^2$	Formula for area of a square
$100 \text{ in.}^2 = s^2$	Substituted 100 in.^{2} for A
$\pm 10 \text{ in.} = s$	Found square roots of 100 in.^{2}

Although we find two solutions of the equation, only the principal square root makes sense in this situation—a side length must be positive. Jamie should make his mosaic **10 in. on each side.**

Discuss Why is it a good idea to reread the problem after solving for x?

Practice Set

Evaluate Solve and check both solutions to each quadratic equation.

a. $2x^2 - 3 = 29$

b. $5x^2 + 6 = 16$

c. $-7x^2 + 8 = -13$

d. $8 = 4w^2 - 24$

e. $\dfrac{4}{x} = \dfrac{x}{9}$

f. $\dfrac{x}{3} = \dfrac{4}{x}$

g. Yoli arranges 121 square tiles into one larger square. How many tiles are along each edge of the larger square?

1. Seven of every 10 doctors recommend brand *X*. If 30 doctors were
(45) surveyed, how many did not recommend brand *X*?

2. An item sold for $10, plus 8.25% tax. What was the total?
(67)

* 3. *Analyze* The scale on a drawing is 1:24. If a building is drawn 10 inches
(87) wide, how many inches wide is it? How many feet wide is it?

* 4. *Model* The vertices of a quadrilateral are at $(-2, -1)$, $(2, -1)$, $(1, 1)$ and
(Inv.5, $(-1, 1)$. Sketch the quadrilateral and its image after a dilation of scale
91) factor 2. What is the ratio of the area of the dilated image to the area of
the original quadrilateral?

* 5. On a number line, graph all the real numbers that are greater than or
(1,16) equal to -1 that are also less than 5.

* 6. *Evaluate* From an original deck of 26 alphabet cards, the *X*-card is
(32, 83) missing.

    a. What is the probability of drawing *A, E, I, O,* or *U* from the deck?
Express the probability as a decimal.

    b. If one of the five vowels is drawn but not replaced, then what is
the probability of drawing one of the four remaining vowels on the
second draw? Express the probability as a fraction.

7. Which point is in Quadrant IV?
(Inv. 1)
    A $(3, 2)$ **B** $(3, -2)$ **C** $(-3, 2)$ **D** $(-3, -2)$

* 8. The set of even counting numbers is {2, 4, 6, 8,...}. The set of odd
(90) counting numbers is {1, 3, 5, 7, ...}.

    a. What is the intersection of these two sets?

    b. What is the union of these two sets?

9. Find two solutions of the equation $x^2 + 1 = 50$.
(93)

Expand.

10. $(x - 7)^2$
(92)

11. $(2x + 4)(x - 1)$
(92)

12. Find the area of the rectangle.
(78)

$\sqrt{10}$

$2\sqrt{10}$

13. Write expressions using a variable for the area and perimeter of a
(8, 36) square with side 2*x*.

14. In terms of π, what fraction of area of the
(7, 40) square is the area of the circle?

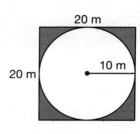

20 m

20 m

10 m

* 15. [Generalize] Simplify: $2\sqrt{12} + 4\sqrt{75}$
(74)

* 16. Solve and graph the solutions on a number line: $-2x - 4 \geq x + 5$
(77)

* 17. [Model] Graph these equations to find the x and y values that make both
(89) equations true: $\begin{cases} x + 2y = 4 \\ x - y = 1 \end{cases}$

18. Walter replaced a 10-gallon-per-minute shower head with a new
(64, 72) 2.5 gallon-per-minute shower head. If the shower is used an average of
20 minutes per day, how many gallons per week are saved by using the
new shower head?

19. Write 0.875 as **a** a percent and **b** as a reduced fraction.
(12)

* 20. [Connect] A formula for the surface area of a cube with edges of
(79) length s is $A = 6s^2$. Solve this formula for s.

* 21. A cube that contains a liter of water has a surface area of 600 cm^2.
(43)
    a. What is the area of each face?

    b. What is the length of each edge?

    c. What is the volume of the cube?

Solve.

22. $1.2x - 0.5 = x$
(79)

23. $\frac{2}{3}x + \frac{3}{4} = \frac{3}{2}$
(50)

24. Solve this proportion by cross multiplying and then solving the resulting
(44, 79) equation. We show the first step.

$$\frac{x}{3} = \frac{x + 1}{2}$$

$$2x = 3(x + 1)$$

* 25. A Major League baseball diamond
(Inv. 2) is actually a square with sides
90 feet long. If a catcher throws a
ball from home to second base,
about how far is the throw?

    A 90 ft   **B** 127 ft

    C 157 ft   **D** 180 ft

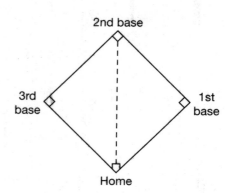

2nd base

3rd base

1st base

Home

• Graphing Pairs of Inequalities on a Number Line

facts   Power Up S

mental math

a. Algebra: Solve: $\frac{x}{3} = 6$

b. Estimation: $230 − $58.99

c. Number Sense: What digit is in the ones place in the product 183 × 183?

d. Powers/Roots: $\frac{m^5}{m^5}$

e. Rate: Six classes per day is how many classes per 5-day week?

f. Statistics: What is the mode, median, and range of the number of days in the 12 months of a common year?

g. Geometry: Angle X and $\angle Z$ are complementary. If $m\angle X = 80°$, find $m\angle Z$.

h. Calculation: 63 ÷ 3, ÷ 7, × 33, + 3, × 2, + 1

problem solving

Can one item be removed from each side of the scale and have it remain balanced? If so, which items can be removed? If not, why?

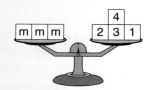

New Concept   *Increasing Knowledge*

We have graphed inequalities on a number line. In some situations, we consider pairs of inequalities. For example, suppose we know that the heights of professional male basketball players range from 5′5″ (65 in.) to 7′6″ (90 in.). This can be expressed on a number line and with an inequality expression.

$$60 \quad 65 \quad 70 \quad 75 \quad 80 \quad 85 \quad 90 \quad 95$$
$$65 \leq h \leq 90$$

The inequality expression can be read "h is greater than or equal to 65 and less than or equal to 90." The expression can be written:

$$h \geq 65 \text{ and } h \leq 90$$

Notice the graph of the inequality contains only the points that are *both* greater than or equal to 65 *and* less than or equal to 90. The graph is the **intersection** (overlap) of the graphs of $h \geq 65$ and $h \leq 90$.

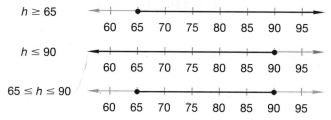

$h \geq 65$

$h \leq 90$

$65 \leq h \leq 90$

Explain What do the closed circles at 65 and 90 indicate?

Example 1

Use a number-line graph and an inequality expression to describe the statement:

"The length of one side (*s*) of a triangle is greater than 2 and less than 8."

Solution

The side length cannot equal 2 or 8, so we use open circles.

Since $s > 2$ and $s < 8$, we write:

$$2 < s < 8$$

Example 2

Graph the solution set of the inequality $-1 \leq x < 5$.

Solution

We graph the numbers greater than or equal to -1 and less than 5.

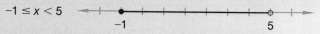

$-1 \leq x < 5$

Predict How would the graph look if the statement were $-1 < x < 5$?

Suppose school officials cancel outdoor physical education classes when the outside temperature is below 40°F or above 100°F. If *t* represents the outside temperature, this rule can be expressed:

$$t < 40 \text{ or } t > 100$$

Example 3

Graph this pair of inequalities:

$$t < 40 \text{ or } t > 100$$

We use the word "or" to indicate that t must meet one *or* the other condition, not both. Therefore, the graph is the **union** (or combination) of the graphs of each inequality.

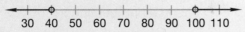

To summarize, when we graph a pair of solutions with an "and" statement we graph the intersection.

Thinking Skill

Analyze

Can we graph the intersection of the solutions of $x < 1$ and $x > 2$? Why or why not?

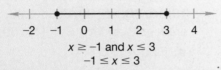

$x \geq -1$ and $x \leq 3$
$-1 \leq x \leq 3$

When we graph a pair of solution with an "or" statement, we graph the union.

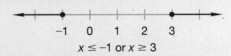

$x \leq -1$ or $x \geq 3$

Discuss How is an intersection different from a union?

Practice Set

Graph these inequalities on a number line.

a. $-3 < x < 3$

b. $x \leq -3$ or $x \geq 3$

c. $5 < x \leq 10$

d. $x < 0$ or $x \geq 5$

e. *Represent* A college will cancel a lab class if less than 12 students enroll. No more than 30 students can enroll in one lab class. Write an inequality that shows the numbers of students (s) that could be in a lab class and illustrate this inequality on a number line.

Written Practice   *Strengthening Concepts*

1. About 30,000 prospectors made it to Dawson City during the Klondike
(48)   Gold Rush. If this total was about 30% of those participating in the Gold Rush, about how many of the participants did not make it to Dawson City?

* 2. *Estimate* An item that regularly sold for $199 was marked down to
(67) $139. Estimate the percent discount.

3. The scale of a model car is 1 in. = 1.5 feet. If the model is 8 inches long,
(87)   how long is the actual car?

* 4. *Analyze* If you roll two number cubes at the same time, what is the
(32)   probability that both number cubes will show odd numbers?

5. Square *ABCD* has vertices *A* (0, 2), *B* (2, 0), *C* (0, −2), *D* (−2, 0).
(9,
Inv. 4)   **a.** Sketch *ABCD*.

   b. A rotation of how many degrees shows that *ABCD* has rotational symmetry?

6. Refer to the square in problem **5.**
(Inv. 2, 78)

    a. What is the length of each side?

    b. What is its area?

* 7. ⟨Model⟩ Use words to describe this inequality. Then graph the solution
(94) set on a number line.

$$x < -1 \text{ or } x \geq 0$$

* 8. ⟨Analyze⟩ Set A contains multiples of fives that are less than fifty. $A = \{5,$
(90) $10, 15, 20, 25, 30, 35, 40, 45\}$; Set B contains multiples of fifteen that
are less than fifty. $B = \{15, 30, 45\}$

    a. Find $A \cup B$

    b. Find $A \cap B$

Expand.

* 9. $(x + 10)^2$
(92)

* 10. $(x - 10)^2$
(92)

* 11. $(x + 10)(x - 10)$
(92)

* 12. Find the area of this rectangle.
(78)

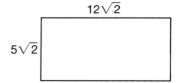

13. Write an expression using a variable for the area and perimeter of a
(92) square with side $2x - 2$.

14. The figure shows a triangle and a circle with
(20, 40) dimensions in inches. Refer to the figure
for **a–c**.

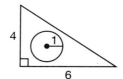

    a. What is the area of the triangle?

    b. What is the area of the circle in
       terms π?

    c. Use your answers to **a** and **b** to write the fraction of the triangle
       occupied by the circle. Is the fraction more or less than $\frac{1}{4}$? How do
       you know?

15. Simplify: $2\sqrt{27} + \sqrt{12}$
(74)

* 16. Solve: $x^2 + 1 = 50$
(93)

* 17. Cheryl draws a triangle with sides 2 in., 3 in., and 4 in. long. Which
(20, Inv. 2) choice best describes the triangle? Justify your choice.

    A acute   **B** right   **C** obtuse

18. During the storm, Larry filled 8 quarts of water in an hour from leaks in
(64, 72) the barn. Use unit multipliers to convert 8 quarts per hour to cups per
minute. (1 quart = 4 cups). About how many minutes would it take to fill
one cup at that rate?

19. Use the distributive property to:
(21)

   a. Factor: $5x^2 - 25x + 60$

   b. Expand: $-3x(3x - 1)$

20. A net of a square pyramid is shown.
(55,
Inv. 4) **a.** Find the area of the net, which is the
surface area of the pyramid.

   b. Sketch the pyramid.

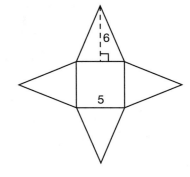

For problems **21–23,** refer to the figure of the similar trapezoids

* 21. **a.** Find the scale factor from the smaller to
(91)   larger figure.

   b. If you are told the perimeter of the
smaller trapezoid, how could you find the
perimeter of the larger trapezoid?

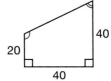

* 22. Find the ratio of the areas of the larger to
(91) smaller trapezoid. Express the ratio as a
decimal. Explain how you can find the ratio
without finding the areas of the trapezoids.

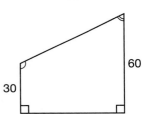

23. The area of the larger trapezoid is what
(71) percent greater than the area of the smaller
trapezoid?

24. Write $2\frac{2}{3}$ **a** as a percent, **b** as a decimal, and **c** as a decimal rounded to
(63) the nearest hundredth.

25. Solve: $2.3 - 0.02x = 0.5$
(50)

• Slant Heights of Pyramids and Cones

facts   Power Up S

mental math

a. **Algebra:** Solve: $10 = \frac{x}{5}$

b. **Fractional Parts:** $\frac{2}{11}$ of 110

c. **Percent:** 30% more than $20

d. **Probability:** Half of the target was painted red. Half of the remaining portion was painted blue. An object hits the target. What is the probability that it hits the unpainted part?

e. **Ratio:** Sixteen of the 30 students were boys. What was the ratio of the number of boys to the number of girls?

f. **Statistics:** What is the mean number of days in the 12 months of a leap year?

g. **Measurement:** Millie measures the length of a rectangle using the middle of the ruler. Find the length of the rectangle.

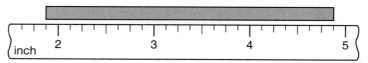

h. **Calculation:** 205×2, $\div\ 10$, $-\ 1$, $\div\ 10$, $-\ 4$, $\times\ 9$

problem solving

D and *E* are digits. Find what *D* and *E* represent:

$$
\begin{array}{r}
DD \\
\times\ DD \\
\hline
EE \\
EE \\
\hline
108E
\end{array}
$$

New Concept — *Increasing Knowledge*

The term *height* in geometry generally means the perpendicular distance from the highest point of a figure to its base. The heights of the pyramid and cone shown below are equal. Both are the distance from the apex of the solid to the point on the base directly below the apex. We use the height to calculate the volume.

For pyramids and cones we are also interested in another measure called the **slant height,** which is the diagonal distance along the surface from the apex to the base. We use the slant height to calculate surface area. We will learn more about the surface areas of pyramids and cones in Lesson 100.

If we look at a vertical **cross-section** of a pyramid or cone at the apex, we see that the cross-section is an isosceles triangle. The height divides the cross-section into two right triangles. The slant height is to the hypotenuse of one of the right triangles.

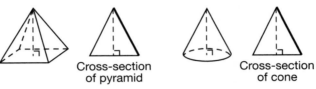

Cross-section of pyramid

Cross-section of cone

If we know the dimensions of the base and the height of the solid, then we know the measures of two legs of the right triangles. We can use the Pythagorean Theorem to find the hypotenuse.

Example 1

Calculate the slant height of the cone.

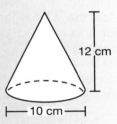

12 cm

10 cm

Solution

The apex of the cone is centered above the circular base. So the radius of the base is one leg of a right triangle and the height is the other leg. We apply the Pythagorean Theorem to find the slant height.

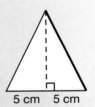

5 cm   5 cm

$$a^2 + b^2 = c^2$$
$$5^2 + 12^2 = c^2$$
$$25 + 144 = c^2$$
$$169 = c^2$$
$$13 = c$$

We find that the slant height is **13 cm.**

Example 2

Arman is building a model of the Great Pyramid to approximate proportions. The base of the pyramid will be a square 30 cm on each side. He wants the height of the pyramid to be 20 cm. What should be the slant height of the triangular faces? Sketch a net of the Great Pyramid model Arman is building.

Solution

The slant height of a pyramid is not an edge, but rather the shortest distance from the apex to a side of the base.

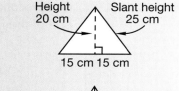

The apex is centered above the square base, so the height divides the cross-section of the pyramid into two right triangles and is one leg of the right triangles. The other leg of each triangle is 15 cm. We recognize the dimensions as multiples of a 3-4-5 triangle. Thus the slant height is 25 cm.

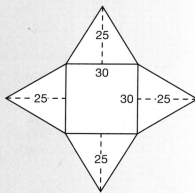

Arman should make the triangular faces of the pyramid using a slant height of **25 cm.** He can use these dimensions to sketch and label a net.

Thinking Skill

Analyze

Use the labeled net in this example to calculate the surface area of the pyramid by finding the area of each face separately.

Connect If an actual size net is folded to form a pyramid, what would be the height of the pyramid?

Practice Set

a. *Explain* Describe the difference between the height of a cone and its slant height.

b. Compared to the height of a pyramid, the slant height of the pyramid is

  A longer   **B** shorter   **C** the same

Evaluate For problems **c** and **d,** find the slant height. Sketch a right triangle to show the dimensions you used for your calculations.

c.

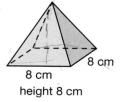

8 cm
8 cm
height 8 cm

d. ├──4 in.──┤

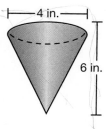

6 in.

1. *Analyze*   Kerry used a 20% off coupon on $45 purchase. How much did
(58)   she save?

2. The scale on the map was 1 inch = 2 miles. If the cities are $13\frac{1}{2}$ inches
(87)   apart on the map, how many miles apart are they?

3. In the United States we measure fuel economy in miles per gallon. At
(4)   the beginning of a trip a car's odometer read 23,450 miles, and the fuel
tank was full. At 23,742 miles the fuel tank was refilled with 9.87 gallons
of gas. Estimate the number of miles per gallon the car averaged on that
portion of the trip.

* 4. *Analyze*   In Europe, fuel economy is measured in liters per 100
(4)   kilometers. On a 2500 km road trip a car used 150 liters of fuel. The car
used an average of how many liters per 100 km?

* 5. You roll two number cubes at the same time. What is the probability that
(32)   both will show the number "1"?

6. Use words to state this inequality. Then graph the solution set.
(94)
$$-4 < x \le -1$$

* 7. *Analyze*   Consider these two sets of numbers:
(90)
$$E = \{\dots, -4, -2, 0, 2, 4, 6, 8,\dots\}$$
$$F = \{\dots, -3, -1, 1, 3, 5, 7,\dots\}$$

a. How many elements are in E ∩ F?

b. E ∪ F is what set of numbers?

8. Rita looks at the floor plan of a room that was drawn with scale factor
(91)   1 inch equals 4 feet. She multiplied the perimeter of the drawing by four
and ordered that many feet of molding. She multiplied the area of the
drawing by four and ordered that many square feet of carpeting. When
the orders arrived, Rita had enough molding, but the carpet covered
only $\frac{1}{4}$ of the room. Explain the error.

* 9. The heating and cooling system forces
(40)   air through cylindrical ducts. A 10-inch
diameter line branches into an 8-inch
and 6-inch line. Find the area of each
circular opening in terms of π and
describe the relationship of the areas.

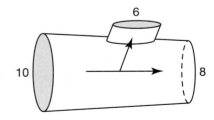

10. Two of the lines are parallel. Find x, y
(20, 54)   and z.

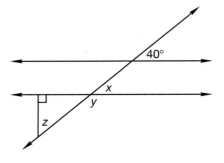

11. The taxi charges $2 plus $1.60 per mile.
(41)
  a. Make a chart of three possible mile and charge combinations.

  b. Write an equation for the function.

Expand.

12. $(2x + 5)^2$
(92)

13. $(2x - 5)^2$
(92)

14. $(2x + 5)(2x - 5)$
(92)

15. Find the area of the rectangle to the
(78)   right.

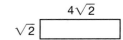

16. Write expressions for the area and perimeter of a square with sides
(92) $x + 10$ using a variable.

17. Simplify: $2\sqrt{20}\sqrt{15}$
(78)

18. Write $\frac{1}{2}$% as a fraction and as a decimal.
(63)

For problems **19** and **20** refer to this drawing of a
circular washer. Units are mm.

19. What fraction of the circular piece of metal
(40)   was removed when the hole was made?

20. When the washer is bolted into place, what
(40, 72)   is the area of the washer that is pressed
  against the object to which it is bolted?
  Use 3.14 for π. Express your answer in square millimeters and again in
  square centimeters.

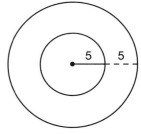

* 21. The cone in problem **23** has a slant height of
(Inv. 2,   20 cm and a circular base with a radius of
95)   5 cm. Use the Pythagorean Theorem to
  find the height of the cone. If a calculator is
  available, find the height to the nearest tenth
  of a centimeter.

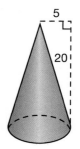

22. Fabian walked across the length and width of a room and estimated
(72) that it covered 40 yd². Use unit multipliers to write the area in square
feet.

* 23. *Represent* The net of a cone is shown here.
(40, 81)

a. Sketch the cone.

b. Find the surface area of the cone in terms of π.

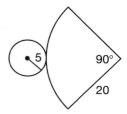

Solve.

* 24. $2x^2 = 50$
(93)

25. A music download website charges its non-members a $2.50 fee to
(69) download plus $1.25 for every song. The total download fee (f) for
non-members is a function of the number of downloads (n):

$$f = \$2.50 + \$1.25n$$

a. Create a table of prices for the first 6 downloads.

b. Graph the function on axes like the ones shown below.

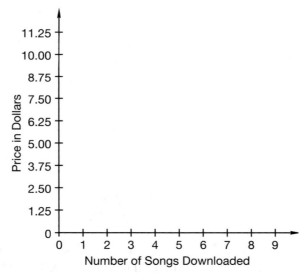

c. Is the relationship a proportion? Justify your answer.

• Geometric Measures with Radicals

facts Power Up T

mental math

a. **Estimation:** Estimate the product in hours: 27 minutes × 12

b. **Number Sense:** What digit is in the ones place of 1^2, 9^2, 11^2, 19^2, 21^2, 29^2, etc.?

c. **Powers/Roots:** $\frac{m^2}{m^3}$

d. **Proportion:** $\frac{12}{x} = \frac{4}{7}$

e. **Scientific Notation:** Compare $4.5 \times 10^5 \bigcirc 45{,}000$

f. **Statistics:** Find the range of this data: 11, 11, 17, 16, 15

g. **Geometry:** Angle R and $\angle Q$ are complementary. If $m\angle R = 45°$, find $m\angle Q$.

h. **Calculation:** $0 + 90$, $+ 19$, $+ 2$, $- 10$, $- 1$, $\sqrt{}$

problem solving

Use the relationships shown by the equations to fill in the comparison symbols (assume all variables represent positive numbers):
A $\bigcirc$ B $\bigcirc$ C $\bigcirc$ D.

$$\frac{1}{A} + \frac{1}{B} + \frac{1}{C} = \frac{1}{D}$$

$$\frac{1}{A} + \frac{1}{B} = \frac{1}{C}$$

$$\frac{1}{A} = \frac{1}{B}$$

We must add and multiply radicals to find the perimeters and areas of some geometric figures. We have practiced multiplying radicals. Below we show how to add radicals.

Adding radicals is similar to collecting like terms. We group common radicals with a coefficient. The perimeter of the square at right is therefore:

$$\sqrt{2} + \sqrt{2} + \sqrt{2} + \sqrt{2} = 4\sqrt{2}$$

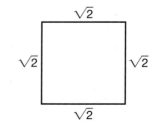

Example 1

Find a. the perimeter and b. the area of this rectangle.

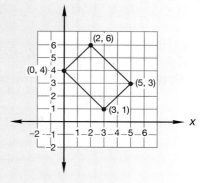

Solution

We use the Pythagorean Theorem to find that the length of the rectangle is $3\sqrt{2}$ and the width is $2\sqrt{2}$.

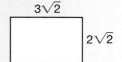

Explain Explain how you calculated the length and width of the rectangle.

a. We add the lengths of the four sides to find the perimeter.

$$3\sqrt{2} + 2\sqrt{2} + 3\sqrt{2} + 2\sqrt{2} = \mathbf{10\sqrt{2} \text{ units}}$$

b. We multiply the length and width of the rectangle to find the area.

$$(3\sqrt{2})(2\sqrt{2}) = 6\sqrt{4} = \mathbf{12 \text{ units}^2}$$

Verify Explain how we can use the grid on the coordinate plane to find the perimeter and the area of the rectangle in example 1.

Example 2

Refer to the illustrated right triangle to find:

a. the length of the hypotenuse.

b. the perimeter of the triangle.

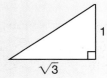

Solution

a. We use the Pythagorean Theorem to find the hypotenuse.

Step:	Justification:
$a^2 + b^2 = c^2$	Pythagorean Theorem
$1^2 + (\sqrt{3})^2 = c^2$	Substituted for a and b
$1 + 3 = c^2$	Simplified
$4 = c^2$	Simplified
$2 = c$	Found square root of 4

The length of the hypotenuse is **2 units**.

b. To find the perimeter we add the lengths of the three sides.

Thinking Skill

Generalize

What is the square of the square root of any number? Why?

Step:	Justification:
Perimeter $= 1 + 2 + \sqrt{3}$	Added three sides
$= \mathbf{3 + \sqrt{3}}$	Simplified

Practice Set

Find the perimeter and area of each figure in units.

a.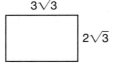
$3\sqrt{3}$

$2\sqrt{3}$

b.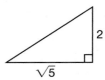
2

$\sqrt{5}$

c. *Evaluate* What is the perimeter and area of a square with vertices at $(2, 0)$, $(0, 2)$, $(-2, 0)$, and $(0, -2)$?

d. What is the perimeter and area of this square?

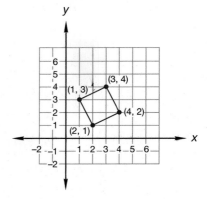

Written Practice · *Strengthening Concepts*

1. Only thirty percent of the ten contestants had the correct answer. How many did not have the correct answer?
(58)

2. The local bookstore increased their inventory by 22%. If they had 5000 titles before the increase, how many did they after the increase?
(67)

3. On the scale drawing of the sailboat, the scale is 2 inches represents 3 feet. If the mast is 12 feet tall, how tall is it in the drawing?
(87)

* 4. *Analyze* Find the lateral surface area of the prism.
(85)

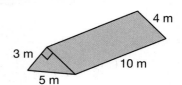
4 m

3 m
10 m
5 m

* 5. *Analyze* Find the total surface area of the cylinder. Use 3.14 for π.
(85)

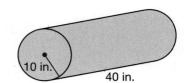
10 in.
40 in.

• Relations and Functions

facts | Power Up T

mental math

a. Algebra: Solve: $y + 10 = 7$

b. Number Sense: What digit is in the ones place of 2^2, 8^2, 12^2, 18^2, 22^2, 28^2, etc.?

c. Powers/Roots: $\frac{m^5}{m^7}$

d. Probability: The children dyed 4 eggs out of the dozen, then returned them to the carton. If one egg is selected at random, what is the probability of selecting an egg that has not been dyed?

e. Ratio: Eight of the 20 silkworms had turned into moths. Find the ratio of moths to worms.

f. Select a Method: The owner of a small business is depositing 23 checks received for sales. To find the total deposit, the owner will probably use

  A mental math.

  B pencil and paper.

  C a calculator.

g. Geometry: Angles P and $\angle Y$ are complementary. If $m\angle Y$ is 20°, find $m\angle P$.

h. Calculation: $36 \div 4$, $\sqrt{}$, $+ 8$, $\times 7$, $+ 4$, $\sqrt{}$

problem solving | A standard number cube is designed so that the sum of the numbers on the opposite faces will always be 7. If you fold this net into a cube, will the sum of the numbers on each pair of opposite faces be 7? If not, use this net as a guide to draw a possible net for a standard number cube.

			2	
1	3	6	4
			5	

A **relation** is a pairing of two sets of information. In a phone book names are paired with phone numbers. The name-number pairing is a relation. For a basketball game, a player's jersey number might be paired with the number of points the player scored. This number-number pairing is also a relation. Relations may be described in set notation, with a table of pairs, or with an equation or graph.

Below are different ways of describing the same relation, the correspondence between the number of tickets purchased and the purchase price.

t	p
0	0
1	7
2	14
3	21

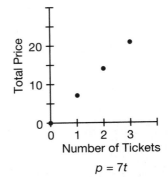

$p = 7t$

In this relation, the set of input values, or domain, consists of the number of tickets purchased. The set of output values, or range, are the prices.

This relation is an example of a function. As we discussed in Lesson 41, a function takes one input value and assigns it exactly one output value.

Not all relations are functions. Consider the relation of the number of minutes that Preston arrives before school starts and the number of students at school at that time.

m	n
4	50
10	20
12	12
10	25
6	40
4	60

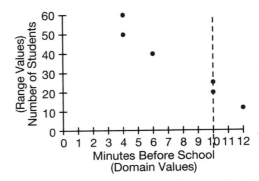

In this case, each input value is not paired with exactly one range value. On the days Preston arrived 10 minutes before school, he saw 20 students the first time and 25 students the second time. We see this on the table as the same input value paired with different output values: (10, 20) and (10, 25). Notice these points lie on a vertical line on the graph since they share the same input value.

To check if the graph of a relation represents a function, we can use the vertical line test:

Vertical Line Test

If a vertical line can be drawn that passes through more than one point on the graph of a relation, then the relation is not a function.

Thinking Skill

Analyze

Find another pair of points that demonstrate this relation is not a function.

Example 1

Different relations are graphed below. Which of these are also functions?

A 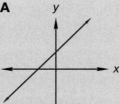   B 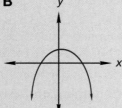   C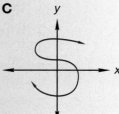

Solution

Both A and B are functions—no vertical line passes through more than one point. Relation C is not a function.

Example 2

Sketch a circle on the coordinate plane with its center on the origin and a radius of 5. The circle's graph is a relation. List four pairs of the points in the relation. Is the relation a function?

Solution

The circle and a table of four points are shown below.

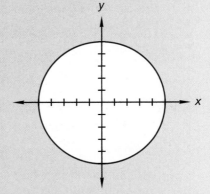

x	y
−5	0
0	5
5	0
0	−5

With the vertical line test, we can see that this relation is **not a function.**

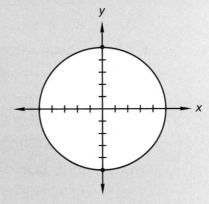

x	y
−5	0
0	5
5	0
0	−5

For the same input value there are two different output values.

In Investigation 8, we saw that a scatterplot is useful to convey paired data. Similarly, we can use scatterplots to present the data in relations. In fact, recall that sometimes plotting data in a scatterplot helps quickly reveal the **correlation,** or relationship, between the sets of data. To determine whether there is a correlation among the data, we draw a best-fit line. The best-fit line can help identify a trend in the data stemming from a relation that we can then use to make predictions.

Example 3

Near the beginning of the lesson, we saw that Preston gathered data on the number of students at school and the time he arrived. Look again at the graph where we plotted the data. Because the entire set of data can be conveyed as distinct points on the graph, this is the same as a scatterplot. To sketch a graph of the function that models the data, draw a best-fit line on the scatterplot. Now you can use the best-fit line, sometimes called a *trend line*, to make predictions about the data set.

   a. According to your graph, how many students might Preston see 10 minutes before school starts?

   b. How many students might Preston expect to see 8 minutes before school starts?

Solution

See student work.

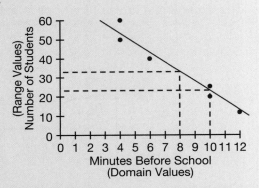

   a. Ten minutes before school starts, Preston might expect to see 23 students.

   b. Eight minutes before school starts, Preston might expect to see 33 students.

Practice Set For **a–d,** state whether each graph or table is a function.

   a.

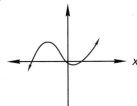

   b.

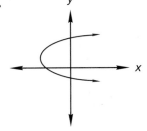

c.

x	y
3	5
2	6
1	7
2	8
3	9

d.

x	y
2	5
3	4
4	5

Written Practice   *Strengthening Concepts*

1. Martha washed two out of every three dishes. Her sister washed the
(34) rest. If Martha washed 12 dishes, how many did they wash all together?

2. The amount received was 20% less than the amount requested. If the
(67) amount received was $800, how much was requested?

* 3. A bag contains two colors of marbles; 8 are blue and 2 are green.
(68, 83)
    a. What is the probability of drawing a blue marble, replacing it, and
    drawing a blue marble again?

    b. What is the probability of drawing two blue marbles in two draws if
    the marble drawn first is not replaced?

* 4. **Classify** Is this relation a function? Is it linear
(98) or non-linear?

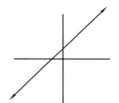

* 5. Is this relation a function? Is it linear or
(98) non-linear?

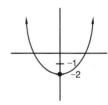

6. Find $\sqrt[3]{1000}$. Then sketch a cube that has a volume of 1000 cm³. Label
(15, 42) the side lengths.

7. For a field trip fund raiser each $30
(41) raised pays for one student to go on
the trip. The table shows the number of
students (s) whose way is paid based
on the number of dollars (d) raised.
Write an equation for the function.
Begin the equation $s =$.

d	s
30	1
60	2
90	3
120	4

For **8** and **9** refer to this triangular prism.

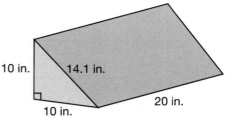

8. What is the volume of the prism?
(76)

* 9. What is the lateral surface area of the prism?
(85)

* 10. **Evaluate** Use words to describe this inequality: $x < 4$ or $x \geq 6$. Then graph the solution set.
(94)

* 11. **Model** By graphing, find the one pair of x and y values that make both equations true.
(89)

$$\begin{cases} 4x + 2y = 8 \\ y = x + 1 \end{cases}$$

12. Quan placed an ad in the paper and for 6 days the number of interested callers were: 4, 7, 14, 11, 9, 9.
(53)

    a. Find the mean, median, and mode of the data.

    b. Which of these measures would change if on the seventh day there were 3 interested callers?

* 13. The ratio of corresponding sides of similar triangles is 5 to 3. What is the ratio of their areas?
(91)

Solve.

14. $\dfrac{3.4}{4.4} = \dfrac{x}{22}$
(25, 44)

15. $0.5x = 1.5 + 0.8x$
(79)

16. $\dfrac{2}{3}x - \dfrac{1}{2} = -\dfrac{1}{6}$
(50)

17. $3x^2 + 2 = 50$
(93)

18. A recursive rule for a sequence is $\begin{cases} a_1 = 5 \\ a_n = 2(a_{n-1}) \end{cases}$.
(97)

    If the fourth term of the sequence is 40, then what is the fifth term?

19. Write an expression using a variable for the area and perimeter of a rectangle with length $x + 6$ and width $x - 1$.
(92)

Simplify.

20. $2\sqrt{3}\,\sqrt{3}$
(78)

21. $\dfrac{(-25)(-20)}{(-25) - (-20)}$
(33, 36)

22. Simplify: $\dfrac{wx^3y^4}{x^2y^2z^{-2}}$
(27)

23. The radius of the circle is 10. The circle occupies about what fraction of the triangle?
(20, 40)

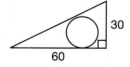

    A $\dfrac{1}{2}$ **B** $\dfrac{1}{3}$

    C $\dfrac{1}{4}$ **D** $\dfrac{1}{9}$

24. Can any triangle have sides of length 3, 6, and 9? Justify your answer.
(20)

25. Terry designs a logo by performing transformations on this triangle. She contracts the triangle by a factor of $\frac{1}{2}$ then contracts the image by a factor of $\frac{1}{2}$. Sketch all three triangles.
(Inv. 5)

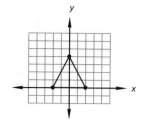

Early Finishers
Real-World Application

A square-pyramid-shaped roof needs to be shingled. If the side length of the base of the roof is 18 feet and the height of the roof is 12 feet, determine the slant height of the roof. Then calculate the roof's lateral surface area to determine the area that needs to be shingled.

• Inverse Variation

facts Power Up T

mental math
a. **Estimation:** $2090 ÷ 3

b. **Fractional Parts:** $\frac{4}{9}$ of 63

c. **Percent:** 90% more than $100

d. **Proportion:** $\frac{3}{5} = \frac{x}{20}$

e. **Scientific Notation:** Compare 6.3×10^{-5} ◯ 630,000

f. **Statistics:** Find the mean of this data: 9, 4, 9, 5, 3

g. **Measurement:** Find the volume of liquid in this graduated cylinder.

h. **Calculation:** 9×7, $+ 1$, $\sqrt{}$, $- 8$, $\times 12$, $÷ 4$

problem solving Some institutions use a four-digit number to denote time of day. Using this system, 9:30 a.m. is "0930 hours," and 1:00 p.m. is "1300 hours." If someone tells you the time is "2222 hours," what time is it?

New Concept *Increasing Knowledge*

In Lesson 69 we studied direct variation in which the *quotient* of two variables is a constant. In most real-world applications, as one variable increases, the other increases.

$$\boxed{\text{direct variation: } \frac{y}{x} = k}$$

In this lesson we will study **inverse variation** in which the *product* of two variables is a constant. In most real-world applications, as one variable increases, the other decreases.

$$\boxed{\text{inverse variation: } xy = k}$$

Consider the following scenario:

Every week Elizabeth bakes a loaf of cornbread for the family meal. The cornbread is a favorite— each person receives an equal portion, and it is always eaten completely. The size of each slice is small when many people are present but large when few are there. The table at right shows the fraction of the cornbread each person receives, depending on the number of people present.

Number of People	Size of Slice
2	$\frac{1}{2}$
4	$\frac{1}{4}$
10	$\frac{1}{10}$

The relationship between serving size and the number of servings is an example of **inverse variation, or inverse proportion.** The product of the variables is constant. In this case, the constant is one loaf and the product of the variables (serving size and number of servings) is 1 for every number pair. As one variable changes, the other changes in such a way that their product is constant.

An equation that describes the relation above is:

$$n \cdot s = 1$$

In this equation, n is the number of people present, s is the size of each slice, and 1 represents the whole loaf.

Example 1

Trinh travels 120 miles to visit his friend. If he averages 60 miles per hour, then the trip takes 2 hours.

a. **What is the quantity that remains constant in this problem?**

b. **Fill in this chart with other possible rate and time combinations.**

c. **Show that the rates and times for Trinh's trips are inversely proportional.**

Rate (mph)	Time (hr)
60	2
40	
30	
20	

Solution

a. The distance is constant, **120 miles.**

b. Traveling at 40 mph, it would take Trinh 3 hours to travel 120 miles, since

$$40\frac{\text{miles}}{\text{hour}} \times 3 \text{ hours} = 120 \text{ miles}$$

The other rate and time pairs are similarly computed.

Rate (mph)	Time (hr)
60	2
40	3
30	4
20	6

c. To show that the variables are inversely proportional, we show that their product is constant.

Rate (mph)	Time (hr)	Rate × Time
60	2	120
40	3	120
30	4	120
20	6	120

The product is 120 for every pair of numbers, so the variables are inversely proportional.

An equation that describes the rate and time relationship from example 1 is:

$$r \cdot t = 120$$

More rate and time values are graphed below.

Thinking Skill

Explain

How do graphs of direct and inverse variation differ? How do the relationships between their number pairs differ?

Rate (mph)	Time (hr)
2	60
4	30
6	20
10	12
12	10
20	6
30	4
60	2

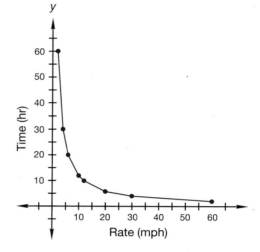

This graph is characteristic of inverse variation. We see that the function is non-linear. Any variables x and y that are inversely proportional have a similar equation and graph.

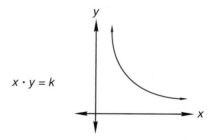

$$x \cdot y = k$$

Visit www. SaxonPublishers. com/ActivitiesC3 for a graphing calculator activity.

Example 2

Robots are programmed to assemble widgets on an assembly line. The number of robots working (*n*) and the amount of time (*t*) it takes to assemble a fixed number of widgets are inversely proportional. If it takes 10 robots 8 hours to assemble a truckload of widgets, how long would it take 20 robots to assemble the same number of widgets?

Solution

We know that the product of the number of robots (*n*) and the time it takes to complete the task (*t*) is constant. We know one pair of numbers (*n* = 10 and *t* = 8). We multiply to find the constant product.

$$10 \cdot 8 = 80$$

The other pair has the same product. We substitute 20 for *n* and solve for *t*.

$$20 \cdot t = 80, \; t = 4$$

We see that it would take 20 robots **4 hours** to assemble a truckload of widgets.

Analyze In the example, we see that doubling the number of robots reduces the working time to one-half the original time. If the number of robots were tripled, what fraction of the original time would it take the robots to complete the job?

Practice Set

Describe each equation or graph in **a–d** as inverse variation or direct variation.

a. $\dfrac{y}{x} = k$

b. $x \cdot y = k$

c.

d.

e. *Justify* Determine if the variables in tables 1 and 2 are inversely proportional and explain how you made that determination.

1.

x	y
1	6
2	3
3	2

2.

m	n
1	3
2	2
3	1

f. *Analyze* If it takes 3 people, working at an equal rate, 5 hours to complete a job, how long would it take 5 people to complete the same job?

g. The volume of air in a balloon and the air pressure in the balloon are inversely proportional. If the balloon is squeezed to $\frac{1}{2}$ its original volume, by what factor does the air pressure increase?

Written Practice *Strengthening Concepts*

1. The amount of vitamin C in the breakfast cereal varies directly with the
(69) serving size. If $1\frac{1}{4}$ cups of cereal provides 10% of the recommended daily value of vitamin C, what percent of the recommended daily value of vitamin C is in 2 cups of cereal?

* 2. The time it takes Javier to walk to school and his walking rate are
(99) inversely proportional. If it takes Javier 12 minutes walking at 3 miles per hour, how long does it take walking at 4 miles per hour?

* 3. *Analyze* If you roll two number cubes at the same time, what is the
(68) probability of rolling a 3 or an even number on both cubes?

For problems **4** and **5** refer to the illustration of the cone.

* 4. The machine dredging the channel deposited sand on a barge. The
(*Inv. 2,*
95) accumulating sand assumed the shape of a cone. If the diameter of the
base is 30 ft and the height is 8 ft, then the slant height is about:

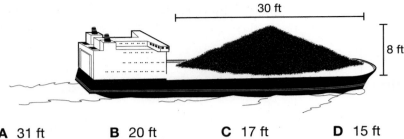

30 ft

8 ft

A 31 ft   **B** 20 ft   **C** 17 ft   **D** 15 ft

* 5. Find the volume of the cone in cubic feet. If the sand weighs about
(*86*) 100 lb per ft^3, then what is the approximate weight of the sand on the
barge?

* 6. *Analyze* If it takes 5 workers 6 hours to complete a task, how long will
(*86*) it take 3 workers to do the task?

7. The number of students visiting the nurse each day in a week were:
(*53*)
$$5, 7, 12, 7, 4$$

a. Find the mean, median, mode, and range of the data.

b. Which measure gives an indication of how the number of students
visiting the nurse varies?

8. Use the distributive property.
(*21*)
a. Factor: $48x^2 - 24x - 36$

b. Expand: $24\left(\frac{3}{8}x + \frac{5}{6}\right)$

* 9. *Model* State this inequality with words. Then graph the solution set.
(*94*)
$$7 \leq x < 10$$

* 10. Find the lateral surface area: Use 3.14 for π
(*96*) and round the answer to the nearest square
centimeter.

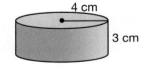

4 cm

3 cm

11. Sketch a triangle with lengths of sides 4, 5, and 6 units. Is the triangle
(*Inv. 2*) acute, right, or obtuse? How do you know?

* 12. *Justify* The number of laps in the walkathon that the participants
(*Inv. 6*) completed is listed below:
$$4, 3\frac{1}{2}, 5, 2\frac{3}{4}, 3, 4\frac{1}{2}, 3\frac{3}{4}$$

Which type of graph would best represent the data?

A Bar Graph   **B** Histogram   **C** Circle Graph

* 13. *(Analyze)* Bill plans a stained glass window in the shape of a regular
(91)   pentagon. The length of the side in his drawing is 6 inches, but the
window will measure 12 inches on a side.

  a. What will be the perimeter of the window?

  b. The area of the window will be how many times the area of the
  drawing?

Solve.

14. $\dfrac{2.8}{4.8} = \dfrac{7}{x}$
(25, 44)

15. $0.0004x - 0.05 = -0.04$
(50)

16. $\dfrac{5}{8} - \dfrac{1}{2}x = \dfrac{1}{2}$
(50)

* 17. $\dfrac{x^2}{4} = 16$
(93)

Expand.

* 18. $(2x + 1)(x - 1)$
(92)

* 19. $(x - 7)^2$
(92)

* 20. Write an expression using a variable for the area and perimeter of a
(92)   rectangle with length $2x + 1$ and width $2x - 1$.

Simplify.

21. $\sqrt{8}\sqrt{6}$
(78)

22. $2x + y + x - y$
(31)

23. $\dfrac{m^{-3}xz^5}{m^3xm}$
(27, 51)

24. A recursive rule for a sequence is
(97)
$$\begin{cases} a_1 = 1 \\ a_n = 2a_{n-1} + 1 \end{cases}$$

If the sixth term is 63 what is the seventh term?

25. The product of the velocity of an object going around the sun and its
(99)   distance from the sun is constant.

  a. What is the constant product for a comet which travels 6.0×10^3 mi/hr
  when it is 9.0×10^8 mi from the sun?

  b. How fast is the comet moving when it is 3.0×10^7 mi from the sun?
  (Hint: Remember, the product is a constant.)

• Surface Areas of Right Pyramids and Cones

Building Power

facts

Power Up T

mental math

a. Algebra: Solve: $z - 8 = -1$

b. Estimation: $1603 \div 4$

c. Number Sense: Write the digit in the ones place of each perfect square below. For example, since 8^2 is 64, write 4.

$$1^2, 2^2, 3^2, 4^2, 5^2, 6^2, 7^2, 8^2, 9^2, 10^2$$

d. Powers/Roots: $\frac{m^9}{m^{12}}$

e. Rate: 2 miles per day is how many miles per common year?

f. To be sure you have enough money to pay for the items you put in your basket, you can estimate the total after rounding each price

  A down to the dollar.

  B up to the dollar.

  C to the nearest dollar.

g. Geometry: Angle W and angle B are complementary. If m$\angle W$ is 40°, find m$\angle B$.

h. Calculation: $0 \div 10, + 8, \times 7, - 1, \div 5, - 8, - 4$

problem solving

If an object spins 720° per second, how many revolutions per minute is that?

Increasing Knowledge

Below are a pyramid and its net. This pyramid is a right square pyramid because it has a square base and its apex is directly above the center of its base.

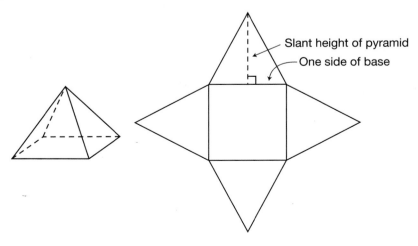

Slant height of pyramid

One side of base

The lateral surface area of a pyramid is composed of triangles. To find the area of one triangular face we multiply the length of a side of the base and the slant height and divide the product by two.

$$\text{Area of triangular face} = \frac{\text{length of side of base} \cdot \text{slant height}}{2}$$

To find the lateral surface area we have a choice. For this pyramid, we could find the area of one face and then multiply it by four. Alternately, we could multiply the perimeter of the base by the slant height (s) and divide the product by two.

$$\text{Lateral surface area} = \frac{4 \cdot \text{length of side of base} \cdot s}{2} = \frac{\text{perimeter} \cdot s}{2}$$

We may use this general formula for the lateral surface area of a pyramid.

**Lateral Surface Area
of a Right Pyramid**

$$A_s = \frac{ps}{2} \text{ or } \frac{1}{2}ps$$

A_s is lateral surface area
p is perimeter of base
s is slant height

Example 1

Find the lateral surface area of a right pyramid with a slant height of 10 cm and a regular hexagonal base with sides 5 cm long.

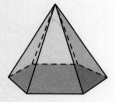

Solution

The base is a regular hexagon with sides 5 cm long, so the perimeter of the base is 30 cm (6 × 5 cm). We show two ways to find the lateral surface area:

Area of each face:

$$A = \frac{bh}{2}$$

$$A = \frac{(5 \text{ cm})(10 \text{ cm})}{2}$$

$$A = 25 \text{ cm}^2$$

Total area of 6 faces:

$$6 \times 25 \text{ cm}^2$$

$$\mathbf{150 \text{ cm}^2}$$

Lateral surface area:

$$A_s = \frac{ps}{2}$$

$$A_s = \frac{(30 \text{ cm})(10 \text{ cm})}{2}$$

$$A_s = \mathbf{150 \text{ cm}^2}$$

To find the lateral surface area of a right cone, we may use the same concept we used to find the lateral surface area of a pyramid. For a cone with a circular base, the perimeter is the circumference.

$$\text{Lateral Surface area} = \frac{\text{circumference} \cdot \text{slant height}}{2}$$

We can substitute for circumference in this formula. Recall that there are two formulas for circumference, $C = \pi d$ and $C = 2\pi r$. We will use the second formula.

$$A_s = \frac{2\pi r \cdot s}{2}$$

Reducing gives us the following formula:

**Lateral Surface Area
of a Right Cone**

$$A_s = \pi r s$$

A_s is lateral surface area
r is the radius of the base
s is slant height

Example 2

Find the lateral surface area and the total surface area of this cone.

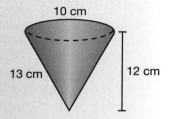

Find the lateral surface area and the total surface area of this cone.

Solution

We will use the formula to find the lateral surface area. The figure indicates the diameter of the base is 10 cm, so the radius is 5 cm. The slant height is 13 cm.

$$A_s = \pi r s$$
$$= \pi(5\text{ cm})(13\text{ cm})$$
$$= \mathbf{65\pi\ cm^2}$$

The total surface area is the lateral surface area plus the area of the base. We find the area of the circular base.

$$A = \pi r^2$$
$$= \pi(5\text{ cm})^2$$
$$= 25\pi\text{ cm}^2$$

Now we add this area to the lateral surface area.

$$\text{total surface area} = 65\pi\text{ cm}^2 + 25\pi\text{ cm}^2$$
$$= \mathbf{90\pi\ cm^2}$$

We have found the surface area in terms of π for each calculation. We may select an approximation for π to express the area as a rational number.

$$90\pi\text{ cm}^2 \approx 90(3.14)\text{ cm}^2$$
$$\approx \mathbf{282.6\ cm^2}$$

Practice Set

a. Describe how to find the lateral surface area of a right pyramid.

b. Describe how to find the lateral surface area of a right circular cone.

c. Find the total surface area of a square pyramid with base sides 5 inches long and a slant height of 4 inches.

d. Find the lateral surface area of a cone with a diameter of 8 cm and a slant height of 12 cm. Express it in terms of π.

Written Practice   *Strengthening Concepts*

1. Find the volume and surface area of a cone
(86, 100) with a diameter of 12 in., a height of 8 in., and a slant height of 10 in. Express both measures in terms of π.

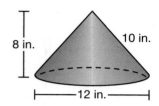

2. Forty-four percent of the drive was bumpy. If there were 11 bumpy
(48) miles, how long was the drive?

3. Attendance was down 8 percent from last year. If attendance was 460
(67) this year, what was the attendance last year?

* 4. A bag contains 2 red marbles, 3 white marbles, and 5 blue marbles.
(68, 83)
   a. If Doug draws a marble, replaces it, and draws again, what is the probability of drawing a red marble both times?

   b. What is the probability of drawing red both times if the first marble is not replaced?

* 5. A toy truck is a 1:18 scale replica of the original. What is the ratio of the
(87) surface areas of the trucks?

* 6. The corresponding sides of a pair of similar hexagons are 3 inches and
(91) 15 inches. What is the ratio of their perimeters? What is the ratio of their areas?

* 7. $\sqrt{12} + \sqrt{27}$
(74)

* 8.   **Justify**   Is this relation a function? Justify
(98) your answer.

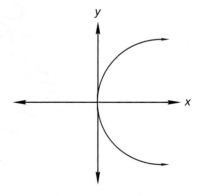

* 9. Find the total surface area rounded to the nearest square inch.
(85)

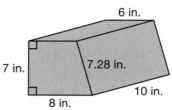

6 in.

7 in.   7.28 in.

8 in.   10 in.

10. Use words to describe this inequality. Then graph the solution set.
(94) $x < -12$ or $x \geq -6$

* 11. $\boxed{\text{Evaluate}}$ **a.** Find the slant height of the
(Inv. 2,   pyramid.
100)

b. Use the answer to **a** and the information
in the diagram to find the area of
one of the triangular faces of the
pyramid.

c. What is the total surface area of the pyramid?

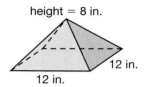

height = 8 in.

12 in.

12 in.

12. A spinner is constructed so that the winning sector covers $\frac{1}{3}$ of the
(68)   circle. The spinner is spun twice. What is the probability of spinning a
win on both spins?

13. **a.** Set S is the set of perfect squares less than 100. List the members of
(90)   Set S.

b. Set C is the set of positive cubes less than 100. List the members of
set C.

c. Find $S \cap T$.

d. Find $S \cup T$

Solve.

14. $\dfrac{7.2}{4.8} = \dfrac{3}{x}$
(25, 44)

15. $0.07 - 0.007x = 0.7$
(50)

Expand.

16. $(3x - 4)(3x + 4)$
(92)

17. $(x - 11)^2$
(92)

18. Write expressions using variables for the area and perimeter of the
(92)   rectangle with length $3x + 2$ and width x.

19. Find the mean, median, and mode(s) of 1, 0.1, 0.01, 0.01, 0.1,
(7)   and 1.

Simplify.

20. $\dfrac{72m^4x^3b^{-1}}{36m^3x^3b^2}$
(27, 51)

21. $\dfrac{(-21) + (-24)}{(-21) - (-24)}$
(36)

22. $2x + 3y + 1 + x - 3y + 4$
(31)

23. 50% of $\frac{2}{3}$ of 1.5
(84)

* 24. The volume of air in a balloon is inversely proportioned to the pressure.
(99) If the volume increases as the balloon rises, what happens to the pressure?

25. The chart below shows the number of bytes to bit in a binary computing
(69) system. What is the constant of proportionality? Use it to write a
function that describes the relationship between bytes, y, and bits, b.
Then graph the function on axes similar to the ones shown here.

Conversion of Bytes to Bits

b	y
1	8
2	16
3	24
4	32

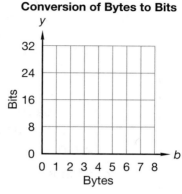

Early Finishers
*Real-World
Application*

The intensity I of light follows an "inverse variation squared" law. This means
that if P is the brightness of a light source (measured in units called lumens)
and D is the distance from the light source to an object (measured in feet),
then the intensity of the light reaching the object is given by the equation

$$I = \frac{P}{D^2}$$

If a light source in a photography studio has a brightness of 100 lumens,
what is the intensity of light that hits an object 5 feet away from this
source? (Note: The units of your answer should be lumens per square foot.)

Focus on
• Compound Interest

If you deposit money in a bank, the bank does not simply hold your money in a safe. It loans your money to others in order to make more money. In some cases, the bank pays you a percentage of the money deposited. The amount of money you deposit is called the **principal.** The amount of money the bank pays you is called **interest.**

There is a difference between **simple interest** and **compound interest.** Simple interest is paid on the principal only, not on any accumulated interest. For instance, if you deposit $100 in an account that pays 5% simple interest, you would be paid $5 (5% of $100) each year your $100 was in the bank. If you take your money out after three years, you would have a total of $115.

Most interest-earning accounts, however, pay compound interest. In a compound interest account, interest is paid on accumulated interest as well as on the principal. If you deposit $100 in an account with 5% annual percentage rate, the amount of interest you would be paid each year increases. If the earned interest is left in the account, after three years you would have a total of $115.76. We show the calculations here:

Compound Interest

$100.00	principal
+ $5.00	first-year interest (5% of $100.00)
$105.00	total after one-year
+ $5.25	second-year interest (5% of $105.00)
$110.25	total after two years
+ $5.51	third-year interest (5% of $110.25)
$115.76	total after three years

Notice that in three years, $100.00 grows to $115.00 at 5% simple interest, while it grows to $115.76 at 5% compound interest. The difference is only 76¢ in three years, but as this table shows, the difference can become much larger over time. What is the difference of the final amounts after 50 years?

Total Value of $100 at 5% Interest

Number of Years	Simple Interest	Compound Interest
3	$115.00	$115.76
10	$150.00	$162.89
20	$200.00	$265.33
30	$250.00	$432.19
40	$300.00	$704.00
50	$350.00	$1146.74

A demonstration of the effect of compound interest can be seen in a plan created by Benjamin Franklin. In his will Franklin gave 1000 pounds, about $4500, to both Boston and Philadelphia. He stipulated that the funds were to be loaned at interest to assist young people starting in business. After 100 years, $\frac{3}{4}$ of the accumulated funds were to be used for public works and $\frac{1}{4}$ was to continue funding new businesses. After another 100 years, the remaining portion of Ben Franklin's gifts had grown to $2\frac{1}{4}$ million in Philadelphia and nearly $5 million in Boston.

Example 1

Over the last century, the stock market's average annual rate of return has been about 10%. Make a table that shows the value of a $1000 investment growing 10% each year after 1, 2, 3, 4, and 5 years.

Solution

At a 10% growth rate, the end-of-year value of an investment is 110% of the value at the end of the prior year. Therefore, to find the value of an investment year by year, we multiply the previous year's balance by 110%, or 1.1.

Total Value of $1000 at 10% Interest

Number of Years	Compound Interest
1	$1100.00
2	$1210.00
3	$1331.00
4	$1464.10
5	$1610.51

We can express periodic compound interest using recursive rules for sequences. In example 1, the first term in the sequence is the original deposit of $1000. Each succeeding term is the product of 1.1 and the previous term.

$$\begin{cases} a_1 = 1000 \\ a_n = 1.1a_{n-1} \end{cases}$$

Calculating Interest and Growth

1. Most calculators can calculate compound interest. Perform the calculation in example 1 by using the multiplication sequence shown below.

$$1000 \times 1.1 \times 1.1 \times 1.1 \times 1.1 \times 1.1 =$$

Do the numbers displayed after each multiplication match the numbers in the table?

Notice that 1.1 occurs repeatedly as a factor. We can group these factors using an exponent to find the value of an investment after a specific number of years. To find the value of the investment in example 1 after five years, we could use this expression:

$$1000(1.1)^5$$

2. A calculator with an exponent key y^x enables us to calculate compound interest for any number of years with just a few keystrokes. We can use this formula.

$$V = P(1 + i)^n$$

V is the current value.

P is the principal.

i is the interest (expressed as a decimal).

n is the number of interest compounding periods.

Find the value of $1000 investment growing 10% each year after 5 years using the exponent key.

or (1) (.) (1)

What number is displayed?

3. Use a calculator with an exponent key for the following problem:

Although the average rate of return for the stock market has been about 10%, the market rises more some years and falls in others. Thus, stocks are usually considered a long term investment. If a 25-year-old invested $1000 in a mutual fund that averaged 10% return per year, what would be the value of the investment when the person was 65? Round the answer to the nearest thousand dollars.

Use the compound interest formula. Substitute 1000 for the principal (*P*), 0.1 for the interest (*i*), and 40 for the number of compounding periods (*n*).

$$V = P(1 + i)^n \qquad \text{Compound interest formula}$$
$$V = 1000(1.1)^{40} \qquad \text{Substituted}$$

The keystroke sequence is:

[1] [0] [0] [0] [×] [1] [.] [1] [yˣ] [4] [0] [=]

- What number is displayed?
- What is the value rounded to the nearest thousand dollars?

Use a calculator with an exponent key for exercises **4** and **5.**

4. Ben Franklin thought the funds he left to Boston and Philadelphia would earn 5% interest annually. Find the value of a $4500 investment earning 5% interest compounded annually after 100 years. Round the answer to the nearest thousand dollars.

5. Find the difference in value, after 30 years, of a $5000 investment that earns 6% interest compounded annually to an investment that earns 5% interest compounded annually.

The **Rule of 72** is a rule of thumb we can use to find how quickly an investment will double at a given interest rate. Though the rule is not exact, we can use it to estimate the doubling time of various growth rates. We divide 72 by the interest rate to find the approximate doubling time.

Rule of 72

$$\frac{72}{\text{Growth rate}} = \text{Doubling time}$$

Example 2

Apply the rule of 72 to estimate how quickly an investment will double at 6% interest.

Solution

When we divide 72 by 6 the quotient is 12, which means the investment doubles about every 12 years.

Approximate Value of $1000 Investment at 6% Compounded Annually

Growing at 6% for (Years)	0	12	24	36
Value of Account	1000	2000	4000	8000

6. Considering example 2, predict the value of a $1000 investment growing at 12% for 36 years.

7. Apply the Rule of 72 to find the approximate value of a $1000 investment growing at 12% for 36 years.

8. Compare the growth rates between 6% in example 2 and exercise 7.

Compound interest is an example of **exponential growth.** The graph of exponential growth is a curve. The increase may seem small at first but becomes greater as it builds upon itself. Direct variation ($y = kx$) is an example of **linear growth.** Over a period of time, exponential growth surpasses linear growth.

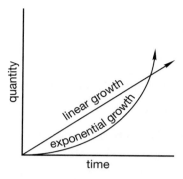

9. A property owner was going away on business for three weeks and wanted to hire a person to care for the property. Two people applied for the job. The first person wanted to be paid $100 per day. The second person wanted a penny the first day, two cents the second day, four cents the third day, and so on, doubling every day for three weeks. Which person should the property owner hire? Explain your choice.

Biology offers other examples of compound growth. Human population grew exponentially for a long period of time.

10. At a steady growth rate of 1.2%, about how long would it take for the world's population to double?

The growth of population and world economies place demands on energy supplies and food production.

11. If a country's energy needs increase at 3% per year, then its energy needs will double in about how many years?

• Geometric Probability

facts | Power Up U

mental math

a. Algebra: Solve: $z - 1 = -8$

b. Fractional Parts: $\frac{3}{11}$ of 99

c. Percent: 10% less than $70

d. Probability: Fourteen of the 20 students walked to school. If a student is selected at random, what is the probability that the student did not walk to school?

e. Ratio: Three thousand people attended the concert, and 500 stood. Find the ratio of standers to sitters.

f. Geometry: The large room is 20-feet-by-20-feet. The small room is 10-feet-by-10-feet. The area of the large room is how many times the area of the small room?

g. Measurement: Find the temperature indicated on this thermometer:

h. Calculation: $-1 + 40, \div 3, + 3, \sqrt{}, + 7, \times 9$

problem solving

A security camera rotates a certain number of degrees in one minute. The next minute, it rotates twice as many degrees. In the third minute, it rotates half as many degrees as it did in the second minute. Altogether, the camera rotated one full revolution in three minutes. How many degrees did it rotate the first minute? Hint: Write an equation and solve.

When playing a target game, we usually aim for the center, which increases the probability of a hit near the center than very close to the edges. Suppose instead there is an equal probability of a hit anywhere on the target. In this case, the probability of hitting the bull's-eye depends on the size of the bull's-eye relative to the size of the target. The smaller the bull's-eye, the smaller the probability of hitting it. If the target is hit, we measure the probability of hitting the bull's-eye in this way:

$$\text{probability of hitting bull's-eye} = \frac{\text{area of bull's-eye}}{\text{area of target}}$$

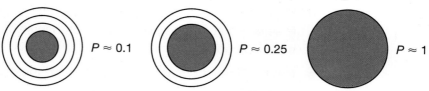

Notice that the probability is zero if the area of the bull's-eye is zero. The probability increases as the size of the bull's-eye increases. When the bull's-eye is the size of the target, the probability of hitting the bull's-eye (if the target is hit) is 1.

Example 1

If a 12-inch diameter target is hit, what is the probability that the 4-inch diameter bull's-eye is hit?

Solution

We find the area of each circle in terms of π. The radius of the target is six inches and the radius of the bull's-eye is two inches.

Area of Target	Area of Bull's-Eye
$A = \pi r^2$	$A = \pi r^2$
$A = \pi(6 \text{ in.})^2$	$A = \pi(2 \text{ in.})^2$
$A = 36\pi \text{ in.}^2$	$A = 4\pi \text{ in.}^2$

$$\text{Probability} = \frac{\text{area of bull's-eye}}{\text{area of target}}$$

$$= \frac{4\pi \text{ in.}^2}{36\pi \text{ in.}^2}$$

$$= \frac{4}{36}$$

$$= \frac{1}{9} \approx 0.11$$

Thinking Skill

Analyze

What is the probability that the bull's-eye is not hit?

Probability based on the area of the regions is called **geometric probability**. Assuming an object lands with equal probability at any point in a known region, the geometric probability that it lands in a specific region is:

$$\text{Probability} = \frac{\text{area of specific region}}{\text{area of known region}}$$

Example 2

What is the probability that an object that lands in the circle also lands in the square?

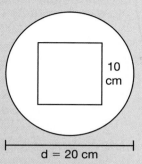

10 cm

d = 20 cm

Solution

Using $\pi = 3.14$		Leaving π as π
Probability $= \dfrac{s^2}{\pi r^2}$	Formulas	Probability $= \dfrac{s^2}{\pi r^2}$
$\approx \dfrac{(10 \text{ cm})^2}{3.14(10 \text{ cm})^2}$	Substituted	$= \dfrac{(10 \text{ cm})^2}{\pi (10 \text{ cm})^2}$
$\approx \dfrac{100}{314} \approx 0.32$	Simplified	$= \dfrac{100}{100\pi} = \dfrac{1}{\pi}$

Model The probability that an object will land in a square inside another square is $\frac{1}{4}$. Draw the two squares and label your drawing with possible lengths of sides.

Practice Set

Find the probability that an object that lands in the larger figure will also land in the smaller figure.

a.

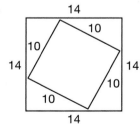

4
3
6
8

b.

10
6 10
8

c.

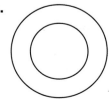

14
10
10
14
14
10
10
14

d.

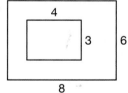

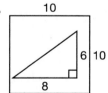

radius of larger circle = 10
radius of smaller circle = 6
(leave π as π)

e. **Analyze** Which is greater, the probability that an object lands in the square or the probability that an object lands in the circle outside the square?

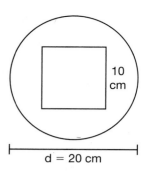

10 cm

d = 20 cm

1. Your textbook has a cover of about
(Inv. 4, 42) these dimensions. Sketch a prism with this base and with height 2 in. Then find the volume.

11 in.

8 in.

2. A sandbag target consists of a
(101) 12-inch diameter circle within a 24-inch diameter circle. If the center of a sandbag lands within the large circle, what is the probability that it also lands within the small circle?

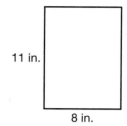

3. Due to inclement weather, the deliveries decreased by 22%. If there had
(67) been 150 deliveries before the decrease, how many were made after the decrease?

* 4. **Analyze** A container holds 4 green marbles and 3 blue marbles. If two
(83) marbles are drawn without replacement,

    a. What is the probability that the first one is green, and the second one is blue?

    b. What is the probability that the first one is blue and the second one is green?

    c. What is the probability that the two marbles will be blue and green in either order?

* 5. The diameter of one spherical ball is 12 in. The diameter of another
(91) spherical ball is 3 inches. From larger to smaller, **a** what is the ratio of their diameters? **b** What is the ratio of their surface areas?

* 6. The students have five dozen tiles that
(99) they arrange in different ways to form rectangles. The dimensions of the rectangles are shown in the table.

x	y
2	30
3	20
4	15
5	12

a. Does this table show inverse variation? How can you tell?

b. A rectangle of 6 rows would have how many tiles in each row?

7. Find the mean, median, and mode of 40, 45, 50, 55, and 60. How can
(53) we find the mean of these numbers by inspection?

* 8. *Analyze* Which of the following sampling methods is *least* likely to
(Inv. 9) produce a sample that is representative of a population?

    A random sampling   **B** convenience sampling

    C systematic sampling

In this figure, two squares are inside a circle. The
shaded triangles are isosceles and congruent.
Units are cm. Refer to this figure for problems
9 and **10**.

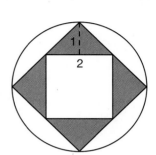

9. **a.** What is the diameter of the circle?
(40)

    b. What is the radius of the circle?

    c. What is the area of the circle in terms
    of π?

10 **a.** What is the area of each shaded triangle?
(40)

    b. In terms of π, the shaded region of the circle is what fraction of the
    area of the whole circle?

    c. Is more or less than $\frac{1}{3}$ of the circle shaded? How do you know?

* 11. *Connect* If 6.022×10^{23} atoms of nitrogen have a mass of 14.01 g,
(46) what is the mass of one nitrogen atom?

* 12. Write expressions using a variable for the area and perimeter of a
(92) rectangle with length $w + 7$ and width w.

* 13. Two number cubes are rolled. Find the probability of rolling a number
(83) greater than zero and less than 7 on each cube.

* 14. Two of the lines are parallel.
(20, 54) Find x.

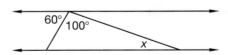

Solve.

15. $\dfrac{4.1}{5.1} = \dfrac{82}{x}$
(25, 44)

16. $0.04 - 0.003x = 0.4$
(25, 50)

17. $-\dfrac{1}{8} + \dfrac{1}{2}x = 0$
(24, 50)

Expand.

18. $(5x - 5)(5x + 5)$
(92)

19. $(x + 9)^2$
(92)

Simplify.

20. $(-2)^2(4)^{-1}$
(36, 51)

21. $\dfrac{3x^3}{2x^{-1}y}$
(27, 51)

22. $\dfrac{(-2)(-3)(-4)}{-2 + (-3) - (-4)}$
(31, 36)

* 23. ⬭Evaluate⬭ In terms of π, what is the probability that an object that lands in the circle to the right will also land in the triangle. The radius is 10, and the base of the triangle is a diameter.
(101)

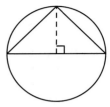

Mary walks every day for exercise. The number of miles she walks and the amount of time it takes her on different days are shown in the scatterplot, along with a line of best fit. Refer to the scatterplot for problems **24** and **25.**

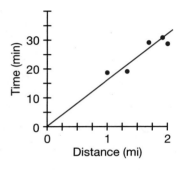

24. **a.** Predict the amount of time it would take Mary to walk 1 mi.
(Inv. 8)

    b. Write the equation of this best-fit line.

    c. Interpret the equation of the best-fit line with words.

* 25. **a.** ⬭Classify⬭ What type of variation is displayed by the scatterplot?
(70)

    b. Which of the following is the constant displayed?

        A Walking rate   **B** Walking time

        C Walking distance   **D** Walking fit

• Growth and Decay

Power Up Building Power

facts Power Up U

mental math

a. **Estimation:** $27,900 ÷ 4

b. **Number Sense:** A certain perfect square has a 4 in the ones place. The square root of that number has either an 8 or a __ in the ones place.

c. **Powers/Roots:** $(x^2)^3$

d. **Proportion:** $\frac{4}{3} = \frac{24}{x}$

e. **Scientific Notation:** Compare $8.04 \times 10^2 \bigcirc 80,400$

f. **Statistics:** Find the median of these values: 9, 4, 9, 5, 3

g. **Geometry:** The sum of the measures of two complementary angles is ___.

h. **Calculation:** $99 - 100, \times 8, + 11, - 13, \times 10$

problem solving

A planner designs a city block in the shape of a hexagon, as shown. If a car drives around the block, how many degrees has the car turned in total? How does that compare to a rectangular city block?

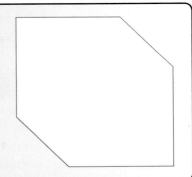

New Concept Increasing Knowledge

In geometric sequences the ratio of consecutive terms is constant.

$$1, 2, 4, 8, 16, 32, \ldots$$

In this sequence the constant ratio is 2.

$$\frac{2}{1} = 2, \frac{4}{2} = 2, \frac{8}{4} = 2, \text{ and so on.}$$

The constant ratio is the multiplier from term to term.

$$\overset{\times 2 \quad \times 2 \quad \times 2 \quad \times 2}{1, \ 2, \ 4, \ 8, \ 16, \ldots}$$

If the terms of a geometric sequence become smaller, then the constant ratio is a number between 0 and 1.

$$\overset{\times\frac{1}{2} \quad \times\frac{1}{2} \quad \times\frac{1}{2} \quad \times\frac{1}{2}}{128, \ 64, \ 32, \ 16, \ 8, \ldots}$$

We can apply geometric sequences in the real world to study **growth** and **decay.**

We often see examples of growth and decay in the financial realm. The effect of compound interest on the value of a bank account is an example of growth. The effect of inflation on the value of the dollar in the world market is an example of decay.

In biological sciences we find growth and decay in increasing or decreasing populations of living organisms.

Example 1

The population of a county has grown 2% per year for the last five years. If the county's population is now 1,200,000, what will the population be in three years at the current growth rate? (Round to the nearest thousand.) Using the Rule of 72, about how many years will it take for the county population to double at this rate?

Solution

Thinking Skill

Explain

Describe the Rule of 72 in your own words.

To find the population year-by-year we can add 2% to each year's population.

After 1 Year	After 2 Years	After 3 Years
1,200,000	1,224,000	1,248,480
× 1.02	× 1.02	× 1.02
1,224,000	1,248,480	1,273,449.6

After three years, the county's population would be about **1,273,000**. Dividing 72 by 2 (the growth rate percentage), we find that the population would double in about **36 years** at the current rate.

Example 2

Carbon-14 is a radioactive isotope occurring in nature. It is used to date materials containing carbon, such as once-living plants and animals. Carbon-14 has a half-life of about 5730 years. If a bone is found containing about $\frac{1}{4}$ the amount of Carbon-14 as modern-day bones, about how old is the bone?

A 11,000 years   B 5500 years   C 2700 years

Solution

The half-life of radioactive material is the length of time it takes for half the isotopes to lose their radioactivity. When a plant or animal dies, it stops taking in carbon-14. After about 5730 years half of the carbon-14 atoms become nitrogen. In another 5730 years half of the remaining carbon-14 has become nitrogen. Thus, in two half-lives, one fourth (half of one half) of the number of carbon-14 atoms remain. Therefore the best choice is **A 11,000 years,** which is about two half-lives.

Practice Set

a. Find the next term in this growth sequence. What is the constant ratio?

16, 24, 36, 54, …

b. Find the next term in this decay sequence. What is the constant ratio?

256, 192, 144, 108, …

c. (Analyze) The schools in an older community experienced a decline in the student population of 3% per year. If the decline continues at the same rate, and there are 5,000 students in the community this year, about how many students will be in the community in three years? (Round to the nearest hundred.)

d. The number of bacteria in a culture increased from about 2000 to about 4000 in one hour. Assuming the bacteria population grows exponentially, how much longer will it take the number to increase to 16,000?

Written Practice   *Strengthening Concepts*

1. Holley is waiting in a line in which about 3 people are helped every
(4)   5 minutes. If there are 20 people in front of Holley, how long will she have to wait in line before she is helped?

2. As of last week, 56% of the players had never even seen the game
(48) played. If there were 25 players in all, how many had not seen it played?

3. The number of players increased by 135%. If there had been 20 players
(67) before the increase, how many are there now?

* 4. **a.** Find the first 3 terms of the sequence generated by $a_n = 3n - 1$.
(73, 97)

   b. Find the first three terms of the sequence.
$$\begin{cases} a_1 = 2 \\ a_n = a_{n-1} + 3 \end{cases}$$

* 5. A cup contains 5 red marbles and 4 yellow marbles. Two marbles are
(83) removed one at a time without replacement.

   a. **(Evaluate)** Find the probability of selecting a red marble and then a yellow marble.

   b. Find the probability of selecting a yellow marble, then a red marble.

   c. Find the probability of selecting a red marble and yellow marble in either order.

   d. What is the probability of not drawing one red and one yellow.

* 6. **(Connect)** Two wooden cubes sit in a carpenter's shop. One is twice as
(85) tall as the other. If the small cube will take one ounce of paint to cover it completely, how much paint is needed to cover the larger cube? Hint: What is the ratio of their surface areas?

7. Find the perimeter and area of a square with the length of a side of
(78, 96) $\sqrt{3}$ units.

8. Is this relation a function? Is it linear or
(98)
non-linear?

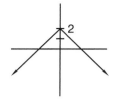

9. Two cards are drawn one at a time with replacement from a deck of
(68)
12 cards. Find the probability of drawing a green card and then a gray
card if there are 4 green, 2 yellow, and 6 gray cards.

* 10. A cone-shaped cup is 12 cm deep. Its open end is 10 cm in diameter.
(95, 100)
    a. Find the slant height of the cone.

    b. Find the lateral surface area of the cone. (Use 3.14 for π.)

11. Write expressions using variables for the area and perimeter of a
(65)
rectangle with length $2w + 9$ and width w.

12. What part of $1000 is one penny? Express your answer using
(51)
exponents.

13. Consider the two solids formed by four
(42, 43)
cubes.

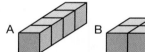

    a. Find and compare their volumes.

    b. Find and compare their surface area.

14. Sheila sketches the floor plan of a house which measures 40 ft by 60 ft
(87)
at its greatest dimension. She makes her sketch on grid paper which is
marked 30 units wide and 40 units long. What scale factor should she
choose that would allow her to make the largest possible drawing?

* 15. **Analyze** The number of students who knew the secret grew
(102)
exponentially. If four students knew the secret at 8 a.m. and if eight
students knew at 9 a.m., then how many students knew the secret by
noon?

* 16. **Formulate** A rule for a sequence is $a_n = 2n$. Write the first four terms
(73, 97)
of the sequence. Then write a recursive formula for the sequence.

Solve.

17. $-\dfrac{4}{3} + \dfrac{1}{12}x = -\dfrac{3}{4}$
(23, 50)

18. $4x^2 = 64$
(93)

19. **a.** Factor: $32x^2 - 16x + 8$
(21, 36)

    b. Expand: $\dfrac{1}{2}(3x - 6)$

Expand.

20. $(2x + 1)^2$
(92)

21. $(2x + 1)(2x - 1)$
(92)

* 22. $\dfrac{6x^4y^9}{3x^{-2}y^3}$
(27, 51)

* 23. $33\dfrac{1}{3}\%$ of $\dfrac{3}{4}$ of 0.12
(78, 96)

Simplify.

24. *(99)* (Analyze) Six people can craft a large quilt in 8 hours. How long will it take four people to craft a large quilt?

25. *(69)* Which of the following graphs represents a direct variation between *x* and *y*?

A
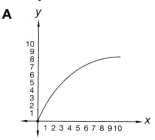

B
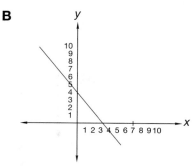

C
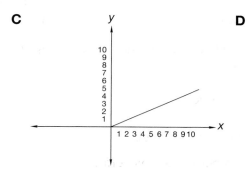

D
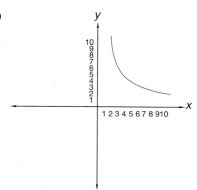

- **Line Plots**
- **Box-and-Whisker Plots**

Power Up | *Building Power*

facts | Power Up U

mental math

a. **Algebra:** Solve: $\frac{1}{p} = \frac{1}{2}$

b. **Estimation:** $47,899 ÷ 12$

c. **Fractional Parts:** $\frac{8}{7}$ of 77

d. **Percent:** 20% less than $60

e. **Rate:** A plant grows 2.4 inches per day. How many inches per hour is that?

f. **Select a Method:** A building contractor is preparing a bid that will state how much the contractor will charge to build a house. To find the total charge, the contractor will probably use
   A mental math.
   B pencil and paper.
   C a calculator.

g. **Measurement:** The odometer read 631.4 miles at the beginning of the trip. When there was exactly one mile left on the trip, the odometer read 971.8 miles. How long was the trip?

h. **Calculation:** $-100 + 50, ÷ 5, + 20, × 5, + 5$

problem solving | In calculus, we sometimes find the derivative of an expression. The derivative of x^4 is $4x^3$. The derivative of x^9 is $9x^8$. Find the derivative of x^3.

New Concepts | *Increasing Knowledge*

line plots | A **line plot** is a visual display of data based on a number line. Each value of a data set is indicated by an X above that value on the number line.

Example 1

During physical fitness testing, students count the number of curl-ups they can do in one minute. The following totals were collected for the class. Graph the data on a line plot.

32, 45, 36, 40, 29, 50, 39, 45, 52

26, 38, 48, 55, 40, 38, 51, 35, 40

53, 42, 39, 46, 43, 40, 38, 45, 41

We draw a portion of a number line that includes the span of the values in the data. For each value in the data set we write an X above the corresponding value on the number line.

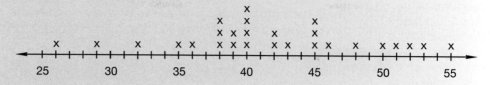

A line plot displays the spread of a data set. Another way to illustrate the spread of data is with a **box-and-whisker plot,** which identifies the median, quartiles, and the extremes.

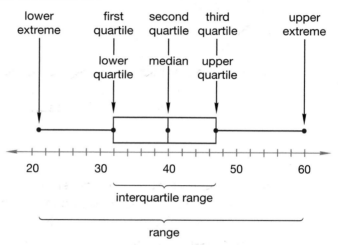

The **quartiles** are the quarter groupings of the data. To divide a data set into quartiles (quarters), we first find the median. Then we find the middle values between the median and the extremes. The median, quartiles, and extremes of the data from example 1 are indicated below.

LE | Q1 | Q2 M | Q3 | UE

26 , 29, 32, 35, 36, 38, 38 , 38, 39, 39, 40, 40, 40, 40 , 42, 42, 43, 45, 45, 45, 46 , 48, 50, 51, 52, 53, 55

Example 2

Visit www.
SaxonPublishers.
com/ActivitiesC3
*for a graphing
calculator activity.*

Create a box-and-whisker plot for the data in example 1.

Solution

Above the number line we place five dots, one at each extreme, at the median, and at the first and third quartiles. We box in the middle half of the values and draw "whiskers" for the values below the lower quartile and above the upper quartile.

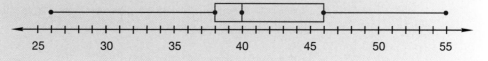

Example 3

Thinking Skill

Connect

Although all measures of central tendency are considered averages, which one is commonly referred to as "the average?" Describe how it is found.

In this lesson, we constructed a line plot and a box-and-whisker plot of the same data.

a. Which plot would you use to convey quickly the mode of the data?

b. Which plot would you use to convey quickly the median of the data?

Solution

a. The mode is easily seen on a **line plot.** It is the number with the greatest number of X's above it. (In this case, the mode is 40.)

b. The median is seen in a **box-and-whisker plot.** It is represented by a dot within the box, which in this case is 40.

Practice Set

Connect In fitness testing, students did push-ups and recorded the number they were able to complete. The students' totals are recorded below.

12, 16, 21, 17, 23, 13, 23, 24, 35, 36,

15, 25, 25, 36, 38, 14, 26, 30, 40

a. Make a line plot of the data.

b. Create a box-and-whisker plot of these data.

c. Which of these graphs quickly conveys the median? The mode?

Written Practice

Strengthening Concepts

1. A wallet-sized photo is usually $2\frac{1}{2}$ in. by $3\frac{1}{2}$ in. What scale factor is used
(91) to enlarge it to a 5 in. by 7 in. photo? The area of a 5-by-7 photo is how many times the area of a wallet-sized photo?

2. Seventy-six percent of the drivers passed the driving test. If 6 did not
(48) pass the test, how many drivers took the driving test?

3. What is the total price if 8.25% tax is added to a $100 item?
(67)

For problems **4** and **5,** use the following information. A gym charges $35 per month for membership plus $2 for each class a member takes. A formula the gym uses to find the monthly charge is

$$y = 2x + 35$$

where x represents the number of classes taken in a month and y is the monthly charge.

* 4. **a.** Create a table of total monthly charges for 1, 2, 3, and 4 classes.
(98)

b. **Evaluate** Are the number of classes taken and the monthly charge directly proportional?

* 5. **a.** Is the gym's formula a function?
(98)

b. Is the relationship between fee and classes linear or non-linear?

* 6. Consider the sequence {1, 3, 5, 7, ...}
(73)

   a. Describe this sequence in words.

   b. Find the next three numbers in the sequence.

   c. A formula for determining the nth number in this sequence is
$a_n = 2n - 1$. Find the 100th term in the sequence.

7. Lydia drinks water from a small cylindrical cup with a diameter of 5 cm
(76) and height of 8 cm. What is the volume of the cup?

* 8. *Analyze* A hat contains 7 slips of paper numbered 1 to 7. If two slips
(83) are removed without replacement, find the probability that both are even
numbers.

* 9. Jaime painted a model of an Egyptian pyramid and prepared to paint
(91, 100) another similar but larger pyramid. The corresponding edges of the
similar pyramids measure 60 cm and 30 cm. If the smaller pyramid
required 6 ounces of paint, how many ounces of paint are needed for
the larger pyramid?

10. **a.** Sketch a square pyramid.
(Inv. 4, 55)

   b. Sketch the net of the square pyramid from part **a.**

* 11. *Evaluate* Refer to the cone to answer **a** and **b.**
(Inv. 2, 95) Units are inches.

   a. What is the slant height of the cone from
the apex to the base?

   b. If the top 12 inches of the cone are
removed, what is the slant height of the
portion of the cone that remains?

   c. What is the diameter of the circle at
the top of the portion of the cone that
remains?

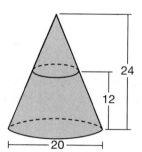

12. It is recommended that we each drink eight 8-oz glassfuls of water a
(72) day. About how many 150 cm³ cups of water are equivalent to eight
8-ounce glassfuls of water? (1 cm³ equals about 1 mL, and 1 ounce is
about 30 mL.)

13. A coin is tossed and a number cube is rolled. What is the probability of
(9, 68) heads and a composite number?

14. Find the mean, median, and mode of 7, 2.5, 0.5, 7, and 3.
(7)

Solve.

15. $\dfrac{3.6}{2.6} = \dfrac{9}{x}$
(25, 44)

16. $0.04x - 0.44 = 0.02x$
(25, 50)

17. $-\frac{1}{3}x - \frac{3}{5} = \frac{1}{3}$
(23, 50)

18. $x^2 - 16 = 9$
(93)

Expand.

19. $(3x + 1)(x + 5)$
(92)

20. $(2x - 2)^2$
(92)

Simplify.

21. $(6.0 \times 10^8)(5.0 \times 10^{-3})$
(46)

22. Simplify: $\dfrac{3x^{-2}y^5}{9x^3y^5}$
(27, 41)

23. 75% of $\frac{5}{6}$ of 4.8
(84)

24. Write expressions using variables for the area and perimeter of the
(92) rectangle with length $2x + 3$ and width $2x - 3$.

25. The line plot shows the number of points scored by 11 players in a
(103) basketball game. Convert the line plot to a box-and-whiskers plot.

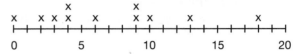

Early Finishers
Real-World Application

The Louvre Museum in Paris, France is one of the largest and most famous museums in the world. The museum holds priceless art from all over the world, including one of the world's most famous paintings, the Mona Lisa. A new entrance to the Louvre was created in 1989 by architect I. M. Pei. The entrance is a large glass square pyramid based on the Pyramid of Giza. The side length of the base measures 35.42 meters and the slant height measures 27.95 meters. What is the total surface area of the Louvre's entrance?

• Volume, Capacity, and Mass in the Metric System

Building Power

facts

Power Up U

mental math

a. **Algebra:** Solve: $2p + 5 = 17$

b. **Number Sense:** A perfect square has a 6 in the ones place. Its square root has a 4 or a ___ in the ones place.

c. **Powers/Roots:** $(m^4)^2$

d. **Probability:** Is the probability of rolling a 5 with a number cube greater than, less than, or equal to the probability of rolling a 4?

e. **Ratio:** If 50 states voted and 20 states opposed the amendment, what was the ratio of supporters to opponents?

f. **Select a Method:** To find the number of gallons of paint needed to paint the interior of her house, Trinh will probably use
   A mental math.
   B pencil and paper.
   C a calculator.

g. **Geometry:** The sum of the measures of two supplementary angles is _____.

h. **Calculation:** $55 - 44, \times 8, - 7, \sqrt{}, \times 6$

problem solving

If each egg is the same weight and we ignore the weight of the egg carton, how many eggs must be removed to balance the scale if one "x" block is removed?

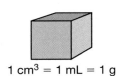

2 Dozen Eggs

Increasing Knowledge

Units of volume, capacity, and mass are closely related in the metric system. The relationships between these units are based on the physical characteristics of water under certain standard conditions. We state two commonly used relationships.

For Water Under Standard Conditions

Volume		Capacity		Mass
1 cm^3	=	1 mL	=	1 g

One cubic centimeter contains 1 milliliter of water, which has a mass of 1 gram.

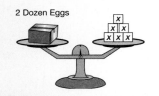

$1 \text{ cm}^3 = 1 \text{ mL} = 1 \text{ g}$

For Water Under Standard Conditions

Volume	Capacity	Mass
1000 cm^3 =	1 L =	1 kg

One thousand cubic centimeters can contain 1 liter of water, which has a mass of 1 kilogram.

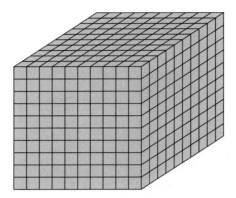

Example 1

Ray has a fish aquarium that is 50 cm long and 20 cm wide. The aquarium is filled with water to a depth of 40 cm.

a. **How many liters of water are in the aquarium?**

b. **What is the mass of the water in the aquarium?**

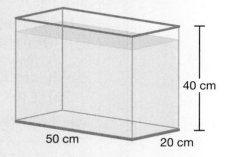

40 cm

50 cm

20 cm

Solution

First we find the volume of the water in the aquarium.

$$(50 \text{ cm})(20 \text{ cm})(40 \text{ cm}) = 40{,}000 \text{ cm}^3$$

a. Each cubic centimeter of water is 1 milliliter. Forty thousand milliliters is **40 liters.**

b. Each liter of water has a mass of 1 kilogram, so the mass of the water in the aquarium is **40 kilograms.** (Since a 1-kilogram mass is about 2.2 pounds on Earth, the water in the aquarium weighs about 88 pounds.)

Example 2

Malia wants to find the volume of a vase. She fills a 1-liter beaker with water and then uses all but 280 milliliters to fill the vase.

a. **What is the volume of the vase?**

b. **If the mass of the vase is 640 grams, what is the mass of the vase filled with water?**

1 L

beaker

vase

a. The 1-liter beaker contains 1000 mL of water. Since Malia used 720 mL (1000 mL − 280 mL), the volume of the inside of the vase is **720 cm³**.

b. The mass of the water (720 g) plus the mass of the vase (640 g) is **1360 g**.

Example 3

On a fishing boat, live bait is kept in a cubic tank measuring 1 meter on each edge.

  a. **Find the volume of the tank.**
  b. **Find the capacity of the tank in liters.**
  c. **Find the approximate mass of a tank full of water in kilograms and the approximate equivalent weight in pounds. (Use the approximation 1 kg ≈ 2.2 lb.)**

Solution

a. Since the shape of the tank is a cube, we cube the length of the edge to find the volume. (1 m = 100 cm)

$$(100 \text{ cm})^3 = \textbf{1,000,000 cm}^3$$

b. The capacity of a cubic centimeter is one milliliter. To find the capacity of the tank in liters, we multiply 1,000,000 mL by a unit multiplier.

$$1{,}000{,}000 \text{ mL} \times \frac{1 \text{ L}}{1000 \text{ mL}} = \textbf{1000 L}$$

Notice that a 1-cubic-meter tank has a capacity of one thousand liters.

Thinking Skill

Analyze

Which is heavier, a ton or a metric ton? By about how many pounds?

c. Under standard conditions the mass of a liter of water is one kilogram. The mass of a tank full of water is about **1000 kg,** which is about **2200 pounds.**

Example 3 illustrates another volume-capacity-mass equivalent for the metric system.

For Water Under Standard Conditions

Volume		Capacity		Mass
1 m³	=	10⁶ mL	=	10³ kg

Example 4

To increase the community water supply, the water company will replace the existing water tank. The new tank has dimensions 30% greater. If the current cylindrical tank is 10 meters high and 20 meters in diameter, then the new tank will hold about how many liters of water?

  A 3.4×10^5 L   B 4.1×10^8 L   C 6.9×10^6 L

> **Solution**
>
> The volume of the current tank is relatively easy to calculate.
>
> $$V = \pi r^2 h$$
> $$= 3.14(10 \text{ m})^2(10 \text{ m})$$
> $$= 3140 \text{ m}^3$$
>
> The capacity of a cubic meter is 1000 liters, so the maximum capacity of the current tank is about 3,140,000 liters (or about 3.14×10^6 liters).
>
> We could calculate the capacity of the larger tank, but we realize that increasing the size of the tank by 30% applies a scale factor of 1.3, which a little more than doubles the volume.
>
> $$(1.3)^3 = 2.197$$
>
> Since the volume is more than doubled, the capacity is more than doubled. The best choice is **C. 6.9×10^6 L.**

Practice Set

a. What is the mass of 3 liters of water?

b. What is the volume of 2 liters of water?

c. When the bottle was filled with water, the mass increased by 1 kilogram. How many milliliters of water were added?

d. An aquarium that is 25 cm long, 10 cm wide, and 12 cm deep can hold how many liters of water?

e. If the aquarium in **d** weighs about 6 pounds when empty, then about how much would it weigh when it is $\frac{3}{4}$ full of water? (Use 1 kg ≈ 2.2 lb.)

Written Practice   *Strengthening Concepts*

1. Transform the formula $c = \pi d$ to solve for d. Use the transformed
(79, 39) formula to find the approximate diameter of a basketball that has a circumference of 77 cm. (Use $\frac{22}{7}$ for π.)

2. What percent of the circle
(40, 48) is not shaded?

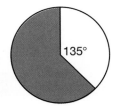
135°

3. The $80 jacket is on sale for 25% off. What is the sale price?
(67)

* **4.** **Analyze** Find the volume of the cone in
(86) terms of π.

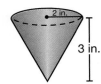
2 in.
3 in.

* 5. **Model** A carnival charges \$1 admission, plus \$2 per ride. This is
(56)   described by the equation $y = 2x + 1$. Graph the function. What point
    on the graph corresponds to the cost for five rides?

* 6. The following sequence is expressed by the equation $a_n = 50 - 2n$.
(73)   Find the tenth and twentieth terms of the sequence.

$$48, 46, 44, 42, \ldots$$

* 7. **Connect** An architect's scale model of a building is $\frac{1}{100}$ of the height of
(87)   the actual building. What fraction of the surface area of the building is
    the surface area of the model?

* 8. Destiny can take two balloons from a paper sack containing 4 green
(83)   balloons, 3 red balloons, and 5 white balloons. If she does not replace
    the balloons, what is the probability of drawing

   a. a red balloon and then a green balloon?

   b. a green balloon and then a red balloon?

   c. a red balloon and a green balloon in either order?

   d. What is the probability she will not draw exactly one red and one
      green balloon?

* 9. On a coordinate plane draw a square with vertices at (2, 2), (−2, 2),
(Inv. 1)   (−2, −2) and (2, −2). On the same plane draw a square with vertices
    at (2, 1), (−1, 2), (−2, −1), and (1, −2). Then shade the four regions
    outside the smaller square but inside the larger square.

* 10. **Evaluate** Refer to the answer to problem 9 for **a–d.**
(Inv. 2,
101)   **a.** What is the area of the larger square?

   b. What is the area of the smaller square?

   c. What is the probability that an object that lands in the larger square
      also lands in the smaller square?

   d. What percent of the larger square is shaded?

11. A recursive formula for the growth of a \$1000 investment at 5%
(97,
Inv. 7)   interest is

$$\begin{cases} a_1 = 1000 \\ a_n = 1.05a_{n-1} \end{cases}$$

In this formula a_1 means the value at the beginning of the first year.
Find the value of the investment at the end of the first and second
years.

12. Cokie drew a parallelogram with an area of 5 square inches. If she draws
(60)   another parallelogram that is similar to the first but with sides four times
    as long, what will be the area of the new parallelogram?

Solve.

13. $\frac{4.2}{3.5} = \frac{x}{10}$
(25, 44)

14. $0.03x - 0.5 = 0.01$
(25, 50)

15. $\frac{2}{5}x + \frac{1}{2} = \frac{7}{10}$
(23, 50)

16. $\frac{1}{2}x^2 = 50$
(79)

Expand.

17. $(2x + 1)(x + 2)$
(92)

18. $(x - 10)^2$
(92)

Simplify.

19. $\dfrac{2x^3 m^{-1}}{6m^4 x^2}$
(27, 51)

20. $\dfrac{(-2)(-3) - (-4)}{(-2) + (-3)(-4)}$
(33, 36)

21. 150% of $\frac{3}{4}$ of 2.4
(84)

22. $2\sqrt{6}\sqrt{12}$
(78)

23. Write expressions using variables for the area and perimeter of the
(92) rectangle with length $2x - 1$ and width $x + 5$.

* 24. **Evaluate** Assuming standard
(104) conditions, find the volume of this
aquarium in cubic centimeters, its
capacity in liters, and the mass of
the water in the aquarium when it is
full.

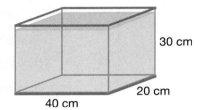

30 cm

20 cm

40 cm

25. **a.** **Classify** Which graph illustrates direct variation?
(69, 99)

b. Which graph illustrates inverse variation?

A

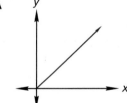

B

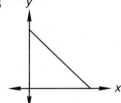

C

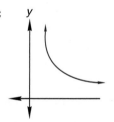

D

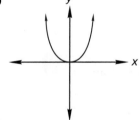

• Compound Average and Rate Problems

Power Up — *Building Power*

facts | Power Up U

mental math
  a. **Estimation:** $23.75 ÷ 6
  b. **Fractional Parts:** $\frac{5}{9}$ of 54
  c. **Percent:** 25% less than $40
  d. **Proportion:** $\frac{x}{110} = \frac{9}{10}$
  e. **Scientific Notation:** Compare 5.1×10^{-2} ◯ 0.00051
  f. **Statistics:** Find the mode of this data: 9, 4, 9, 5, 3
  g. **Measurement:** Find the length of the rectangle.

  h. **Calculation:** $54 + 8, + 2, \sqrt{}, - 9, + 54$

problem solving | The earth rotates one revolution per day. How many degrees does it rotate per hour?

New Concept — *Increasing Knowledge*

In this lesson we will solve problems that include more than one average or rate.

Example 1

Blanca bought 200 shares of XYZ stock at an average price of $54 per share. Later she bought 100 more shares at $60 per share. What was the average price per share Blanca paid for all 300 shares of XYZ?

Solution

Blanca purchased different quantities of stock at different prices, so we cannot simply find the average of $54 and $60. Instead we find the amount Blanca paid for all the shares and divide by the total number of shares.

$$\begin{array}{r}
200 \text{ shares at } \$54 = \$10{,}800 \\
+\ 100 \text{ shares at } \$60 = \$\ \ 6{,}000 \\
\hline
300 \text{ shares at } x = \$16{,}800
\end{array}$$

We find that 300 shares cost $16,800. Now we can find the average cost (x) per share.

$$x = \frac{\$16,800}{300} = \$56$$

The average cost per share was **$56.**

Example 2

After three games of bowling, Gabe's average score was 88. What score does Gabe need to average in his next two games to raise his average to 90?

Solution

Although we do not know the specific scores in Gabe's first three games, we know that his total number of pins must be 264 for his average to be 88.

$$3 \times 88 = 264$$

To have a five-game average of 90, Gabe's pin total needs to reach 450.

$$5 \times 90 = 450$$

We subtract to find that Gabe needs 186 pins in his next two games to reach 450.

$$450 - 264 = 186$$

Therefore, he needs to average **93** in his next two games to raise his average score to 90.

$$186 \div 2 = 93$$

Many rate problems can be solved like average problems as we see in this example.

Example 3

Because of the heavy traffic, the 120-mile trip to San Diego took 3 hours. The traffic was lighter on the return trip, which took 2 hours. Find the average speed for each trip and for the round trip.

Solution

Thinking Skill

Analyze

Suppose the traffic on the drive to San Diego was jammed and took 6 hours. What would be the average speed of the drive to San Diego? The average speed round trip?

The average speed on the trip to San Diego was **40 miles per hour.**

$$\frac{120 \text{ mi}}{3 \text{ hr}} = 40 \text{ mph}$$

The average speed for the return trip was **60 miles per hour.**

$$\frac{120 \text{ mi}}{2 \text{ hr}} = 60 \text{ mph}$$

To find the average speed for the round trip we divide the total distance by the total time.

$$\frac{120 \text{ mi} + 120 \text{ mi}}{3 \text{ hr} + 2 \text{ hr}} = \frac{240 \text{ mi}}{5 \text{ hr}} = 48 \text{ mph}$$

The average speed for the round trip was **48 miles per hour.**

Practice Set

a. Jarrod earns $12 per hour helping a painter and $15 per hour helping a carpenter. What is Jarrod's average hourly rate for the week if he helps the painter for 10 hours and the carpenter for 20 hours?

b. Kaitlyn has a four-game average of 87. What does she need to average in her next two games to have a six-game average of 90?

c. Sonia covered 30 km for the fund raiser. She jogged the first 15 km in two hours and walked the rest of the distance in three hours. Find her average jogging speed, average walking speed, and average speed for the full 30 km.

Written Practice   *Strengthening Concepts*

1. Bryon rides his bike to the bus stop. It takes Bryon 40 seconds to ride
(3, 4)   his bike one block. His bus arrives in 3 minutes, and he is 4 blocks away. Will Bryon make it in time?

2. For every person who walked, there were forty-four who drove. If 270
(45)   made the trip, how many walked?

3. A store purchases a chair for $20 from the manufacturer. What is the
(67)   selling price if it sells for 80% more?

4. Bridgette is thinking of two numbers. She gives these hints. The sum of
(89)   the numbers is 4. If you double the first number and then subtract the second number from the sum the difference is 2. Make a table of (x, y) values for each equation below using integers near zero. Then plot the points and graph the two equations to find the numbers.

$$\begin{cases} x + y = 4 \\ 2x - y = 2 \end{cases}$$

Analyze A class built a model of the Great Pyramid with these dimensions. Refer to this model for problems **5** and **6**.

* 5. Find the volume of the model.
(86)

* 6. The ratio of the dimensions of the model to
(87, 100)   the dimensions of the Great Pyramid is about 1 to 38. The surface area of the real pyramid is about how many times the surface area of the model?

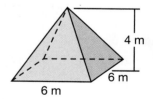

4 m

6 m

6 m

* 7. **Model** The cost of the pizza is $10, plus $1 for each topping. This is
(83)   modeled by $y = 1x + 10$. Graph this function. What are the coordinates of the point that shows the cost of a four-topping pizza?

8. Julius did a standing long jump six times. The distances he jumped in feet were: 3.5, 4.1, 4.9, 5.6, 6.4, and 4.9.
(53)

  a. Find the mean, median, mode and range of the data.

  b. Which measure describes the variation in the distance of his jumps?

* 9. Eight slips of paper, numbered 1 to 8, are placed in a hat.
(68, 83)

  a. If two numbers are drawn without replacement, what is the probability that both numbers drawn are less than 3?

  b. If the numbers are drawn one at a time and the first draw is replaced in the hat, then what is the probability that both numbers drawn are less than 3?

  c. How do the two probabilities compare?

* 10. (*Analyze*) The circle is divided into five congruent sectors.
(81)

  a. Arc *AB* measures how many degrees?

  b. The diameter of the circle is 10 inches. Find the length of arc *AB* to the nearest inch.

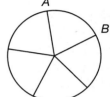

* 11. (*Explain*) A carnival game is played in a way that a small pebble is tossed into a bucket with an inside diameter of 10 inches. A triangle is painted at the bottom of the bucket as shown here. Before tossing the pebble, the player selects "inside the triangle" or "outside the triangle" as the winning landing location. Which would you select? Explain.
(41, 47)

* 12. Jenna pedaled the 24 miles to the lake against the wind and averaged 8 miles per hour. On the return trip she averaged 12 miles per hour. Find Jenna's average speed for the round trip.
(105)

Solve.

13. $\dfrac{5.1}{1.7} = \dfrac{3}{x}$
(25, 44)

14. $0.02x + 0.1 = 1$
(25, 50)

15. $5x^2 - 2 = 3$
(93)

16. $\dfrac{3}{7} + \dfrac{1}{3}x = \dfrac{10}{21}$
(23, 50)

Simplify.

17. 10% of $\dfrac{1}{8}$ of 4.24
(84)

18. $\dfrac{16z^4 m^3}{56m^{10}z^{-1}}$
(27, 51)

19. $\sqrt{1,000,000}$
(15)

20. $\sqrt{8} + \sqrt{8}$
(54, 64)

* 21. (*Evaluate*) Jonya would like to bring her average golf score down to 99. Her three round average is 103. What does she need to average in her next four rounds of golf to have a seven-round average of 99?
(105)

* 22. **Evaluate** Consider the following sequence: 5, 12, 19, 26, ...
(73)

    a. Describe the rule for this sequence.

    b. Find the next 3 terms.

    c. Which equation below generates this sequence?

    A $a_n = n + 7$ **B** $a_n = 7n + 2$ **C** $a_n = 7n - 2$ **D** $a_n = 5 + 7n$

Expand.

23. $(3x + 1)^2$
(92)

24. Write expressions using variables for the area and perimeter of a
(92) rectangle with length $2x + 3$ and width $2x - 3$.

25. From the graph, make a table of values for x and y. Then find the
(69) constant of proportionality of the two variables.

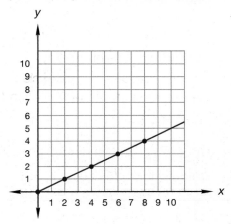

• Reviewing the Effects of Scaling on Volume

facts | Power Up V

mental math

a. **Algebra:** Solve: $90p + 5 = 50$

b. **Estimation:** 697 miles + 206 miles + 151 miles

c. **Number Sense:** A perfect square has a 5 in the ones place. Its square root has a __ in the ones place.

d. **Powers/Roots:** $(y^3)^3$

e. **Rate:** Two minutes per page is how many pages per minute?

f. **Statistics:** The number that appears most often in a set of data is the

A mean.   B median.

C mode.   D range.

g. **Geometry:** The sum of the measures of the angles in a triangle is ___.

h. **Calculation:** $53 - 9, \div 4, \times 7, + 2, - 19$

problem solving | If $0 < r < 1$, arrange these numbers from least to greatest: $r, r^2, 2r$.

New Concept | Increasing Knowledge

We have considered the effects of scaling on perimeter, area, surface area, and volume. The shape of the figure does not affect the relationship.

Mabel collects marbles. Her marble jar is full, so to make room for more, she finds another jar with twice the height and diameter of the first. She figures that when she transfers the marbles, the new jar will be half full, yet she is surprised to find that it is only $\frac{1}{8}$ full.

Recall that the volume of a scaled object does not increase proportionally to changes in edge lengths.

Consider these cubes. The large cube has 3 times the length, width, and height of the small cube. We must stack 27 small cubes to create a cube the size and shape of the large one, which means the volume of the large cube is 27 times the volume of the small cube.

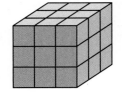

Since the three dimensions—length, width, and height—were multiplied by 3, the volume was multiplied by $3 \times 3 \times 3$, or 27.

We can make the following statement about any scaled object:

When the dimensions of an object are multiplied by a factor, the volume of the object is multiplied (or divided) by the cube of that factor.

Scale Factor

Ratio of lengths = scale factor
Ratio of areas = (scale factor)^2
Ratio of volumes = (scale factor)^3

Example

The large brick has 1.5 times the length, width, and height of the small brick. What is the ratio of the volume of the larger brick to the smaller brick? If the large and small bricks are made of the same mix of materials, what is the weight of the large brick?

Thinking Skill

Formulate

Find the weight of the larger brick by writing and solving a proportion using the ratios of volumes and weights. (Hint: The ratio of volumes is expressed as a decimal.)

3 lb

Solution

Although we do not know the actual dimensions of the two bricks, we can calculate the ratio of the volumes by cubing the scale factor.

$$\text{Ratio of volumes} = (\text{scale factor})^3$$
$$= \mathbf{(1.5)^3} \text{ or } \mathbf{3.375}$$

effect on volume $\times (1.5)^3$

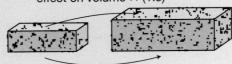

scale factor $\times$ 1.5

The weight of the large brick is $(1.5)^3(3 \text{ lb}) = \mathbf{10.125 \ lbs.}$

Practice Set

a. This 3-by-3 cube is constructed of 27 blocks. How many blocks are needed to construct a cube with twice the dimensions of the original cube?

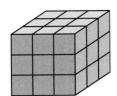

b. An architect builds a $\frac{1}{50}$ scale model of a proposed building for a client. The volume of the model is what fraction of the volume of the proposed building?

c. The larger cylindrical container has dimensions 25% greater than the smaller container. The larger container will hold about what percent more flour than the smaller container?

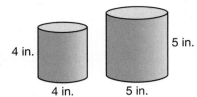

4 in.   5 in.

4 in.   5 in.

   A 25%   **B** 50%

   C 100%   **D** 200%

d. A crate with the given dimensions will hold 24 cubes with 12-inch sides. How many 6-inch cubes will the crate hold?

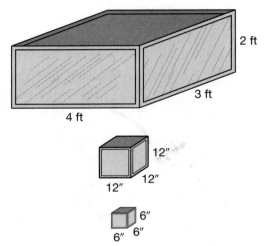

2 ft

3 ft

4 ft

12″

12″   12″

6″

6″ 6″

Written Practice   *Strengthening Concepts*

1. A soup can has a paper label around the curved surface. The label overlaps itself as shown. What is the length and width of the label when it is laid flat? (use 3.14 for π)
(85)

4 cm

0.38 cm

9.8 cm

* 2. **Evaluate** Winifred earns 2 dollars for each carnation bouquet she delivers and 3 dollars for each rose bouquet. This equation shows that she earns 18 dollars: $2c + 3r = 18$. The following equation shows she delivers 8 bouquets: $c + r = 8$. Use guess and check or graphing to find the number of each type of bouquet she delivers.
(89)

3. Twenty-five percent of the students had forgotten the year of the Declaration of Independence. If 57 students had not forgotten, how many students were there in all?
(48)

4. The number of celebrants at the Independence Day party had increased
(67) by 25% over last year. If there were 2220 people gathered this year, how many were there last year?

5. Martin is thinking of two numbers. He gives these hints. The sum
(89) of the numbers is 3. If you double the first number and add the second number, the sum is 8. Graph this system of equations to find the two numbers.

$$\begin{cases} x + y = 3 \\ 2x + y = 8 \end{cases}$$

Expand.

6. $(5y - 4)(5y + 4)$
(92)

7. $(5y + 4)^2$
(92)

8. Write expressions using a variable for the area and perimeter of a
(92) rectangle with length $5x$ and width $2x + 7$.

* 9. *Analyze* Hector buys potted plants for the nursery. He purchased
(105) 40 potted plants from one supplier for \$1.20 each, and he bought 60 potted plants from another supplier for \$1.10 each. What was the average price Hector paid for the plants?

* 10. *Analyze* A cup contains 1 kidney bean, 2 green beans, 2 pinto beans,
(93) and 3 lima beans. Two beans are removed from the cup at random and without replacement. What is the probability that a lima bean and pinto bean are removed?

A $\dfrac{3}{28}$ **B** $\dfrac{5}{8}$ **C** $\dfrac{3}{32}$ **D** $\dfrac{5}{32}$

11. Find the slant height of this right
(95) circular cone by using the Pythagorean Theorem.

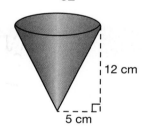

12 cm

5 cm

* 12. Use your answer to problem **11** to
(80, 100) find the lateral surface area of the cone.

Solve.

13. $3x^2 + 7 = 34$
(93)

14. $0.05x + 0.2 = 2$
(25, 50)

15. $\dfrac{1}{3} + \dfrac{3}{4}x = \dfrac{1}{8}$
(23, 50)

16. $\dfrac{3.2}{2.4} = \dfrac{6}{x}$
(25, 44)

Simplify.

17. $\sqrt{50} + \sqrt{50}$
(74, 96)

18. $33\dfrac{1}{3}\%$ of $\dfrac{5}{8}$ of 2.4
(84)

19. $1\dfrac{2}{3} \cdot \dfrac{3}{8} + 2\dfrac{1}{2}$
(13, 23)

20. $\dfrac{42a^2bc}{49a^{-2}b^{20}}$
(27, 51)

21. If you roll two number cubes at the same time, what is the probability of
(68) rolling an odd number greater than 4 on both cubes?

* 22. *Generalize* This block is built with eight
(91)   cubes. A block with the dimensions
      doubled would require how many
      cubes?

Evaluate For **23–25**, refer to the two cones shown here. Units are inches.

* 23. **a.** What is the scale factor from the smaller
(91)   cone to the larger cone?

b. What is the ratio of the area of the larger
cone to the smaller cone?

c. What is the ratio of the volume of the
larger cone to the smaller cone?

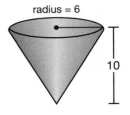

radius = 6

10

radius = 3

5

* 24. **a.** Find the volume of the larger cone in terms of π.
(86)
b. Find the volume of the smaller cone in terms of π.

* 25. If the lower 5 inches of the larger cone
(86)   were removed, this bowl shape would
      remain.

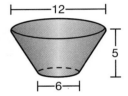

—12—

5

—6—

a. What is the volume of the remaining solid
in terms of π?

b. Calculate the volume to the nearest
ten cubic inches using 3.14 for π.

• Volume and Surface Area of Compound Solids

facts | Power Up V

mental math

    a. Algebra: Solve: $40j + 8 = 128$

    b. Fractional Parts: $\frac{9}{10}$ of 70

    c. Percent: 50% less than $30

    d. Probability: One fourth of the athletes wore jerseys. The rest wore sweats. Find the probability that an athlete selected at random is wearing sweats.

    e. Ratio: Last night 6 of the 8 frogs had enough to eat. What was the ratio of full to hungry frogs?

    f. Statistics: Benton has read 30 pages of a 310-page book. To finish the book in one week, Benton needs to read an average of how many pages per day?

    g. Measurement: Find the mass of the mystery box.

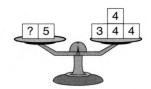

    h. Calculation: $60 - 47, + 1, \div 2, \times 6, - 3$

problem solving

In calculus, we sometimes find the derivative of an expression. The derivative of $2x^3$ is $6x^2$. The derivative of $3x^5$ is $15x^4$. The derivative of $4x^6$ is $24x^5$. Find the derivative of $3x^4$.

New Concept | Increasing Knowledge

Structures and other objects in the world around us are often complex shapes. However, many complex shapes can be broken down into combinations of simpler shapes. A small building might be composed of a rectangular prism and a triangular prism. A large building might be a combination of rectangular prisms and a pyramid. A round table might combine a wide, short cylinder and four narrow, long cylinders.

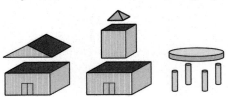

Breaking complex figures into combinations of simpler forms helps us approximate or calculate their volumes and surface areas.

Example

The diagram shows a desk with dimensions in inches. Convert the dimensions to feet. Then calculate the volume and exposed surface area (excluding the part of the desk that rests on the floor).

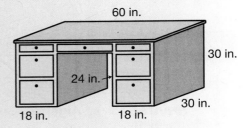

Solution

Thinking Skill

Verify

Explain how the volume of the desk can be found by calculating the volume of a larger rectangular solid, then subtracting the knee space. Use this subtraction method to find the volume of the desk.

We divide each given dimension by 12 to convert it to feet, which we express in decimal form. We divide the figure into three rectangular prisms and indicate the dimensions.

The volume of the top portion is:

$$V = lwh$$
$$= 5(2.5)(0.5)$$
$$= 6.25 \text{ ft}^3$$

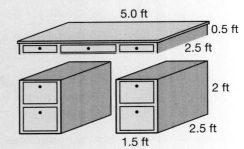

The two pillars have identical dimensions. The volume of each one is:

$$V = lwh$$
$$= 2.5(1.5)(2)$$
$$= 7.5 \text{ ft}^3$$

We combine the three parts to find the total volume.

Top portion	6.25 ft³
Left pillar	7.5 ft³
Right pillar	+ 7.5 ft³
	21.25 ft³

We can also use the divided figure to help us calculate the surface area. The exposed surface area of the desk includes the total surface area of the top portion, minus the area that rests on the pillars, plus the lateral surface area of the pillars.

Total surface area of top portion	32.5 ft²
− Area resting on pillars	7.5 ft²
Exposed area of top portion	25 ft²
+ Lateral surface area of pillars	32 ft²
	57 ft²

Calculate the Volume of an Object Built with Solids

Illustrated is an object built with several solids. Assemble this object or another object using solids or other manipulatives. Sketch the object you built. Then calculate (or estimate) the volume of the object using the appropriate formula for each component of the object.

Practice Set

a. This figure is composed of 1-inch cubes. Find its volume and exposed surface area (excluding the base).

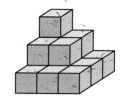

b. Find the volume of this building. Then find the total area of the building's rectangular walls. Dimensions are in feet.

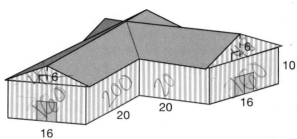

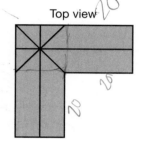

c. **Estimate** Find a building at school or in the neighborhood that is a combination of simple shapes. Sketch the building and its estimated dimensions. Then estimate the volume of the building.

Written Practice   *Strengthening Concepts*

1. Jo Ann sold 30 cups of lemonade in 45 minutes. Which of these is an equivalent rate?
(49)

   A 15 cups in $\frac{1}{2}$ hr   **B** 35 cups in 1 hr

   C 60 cups in $1\frac{1}{2}$ hr   **D** 50 cups in 1 hr

2. Seven out of eleven students surveyed preferred the longer lunch break. If there were 132 students surveyed, how many did not prefer the longer lunch break?
(45)

3. A used car is priced at 85% more than the value of the trade-in. If the car is priced at $7400, what is the value of the trade-in?
(69)

* 4. **Model** The following equation is written in standard form. Find the intercepts of the line, then graph the equation. $6x - 5y = 30$
(82)

* 5. **Model** Scott is thinking of two numbers. He says their sum is 4 and
(82, 89) their difference is 6. Graph these equations to find the two numbers.
$$\begin{cases} x + y = 4 \\ x - y = 6 \end{cases}$$

6. It takes Charles 15 minutes to ride his bike to school at 8 miles per hour.
(99) How fast does he need to ride to get to school in 10 minutes?

7. Solve this inequality and graph the solution set on a number line.
(62)
$$2(x + 3) < x + 7$$

* 8. **Evaluate** Geraldine is thinking of two whole numbers. If you multiply
(82, 89) the first number by two and the second number by three, the sum of the
products is 6. If you add the second number to the opposite of the first,
the sum is −3. Graph this system of equations to find the two numbers.
$$\begin{cases} 2x + 3y = 6 \\ -x + y = -3 \end{cases}$$

9. Find the volume of this square pyramid:
(86) The height is 2 m.

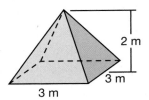

10. a. Find the number of cubic inches of steel in this barbell in terms
(107) of π.

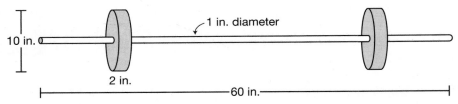

1 in. diameter

10 in.

2 in.

60 in.

b. Estimate the weight of the barbell using 0.3 pounds per cubic inch
for the weight of the steel.

The base of a right circular cone has a diameter of 14 feet and a height of
24 feet. Refer to this cone for problems **11–14**, and use $\frac{22}{7}$ for π.

11. Use the Pythagorean Theorem to find the slant
(Inv. 2, 95) height of the cone.

* 12. Find the lateral surface area of the cone. Use $\frac{22}{7}$
(95, 106) for π.

7

24

* 13. a. Find the volume of the cone. Use $\frac{22}{7}$
(86, 106) for π.

b. If a cut is made parallel to the base, halfway up the cone (at 12 ft),
then the upper cone would have dimensions $\frac{1}{2}$ the dimensions of the
original cone. What would be the volume of the upper cone?

* 14. **a.** Draw the shape that remains if the upper 12 ft of the cone is
(Inv. 4, 86) removed.

b. Find the volume of the remaining figure by subtracting the volume of the removed cone from the volume of the original cone.

Solve.

15. $5m^2 + 3 = 83$
(50, 93)

16. $0.1r + 2.5 = 3.1$
(25, 50)

17. $\dfrac{1}{3} + \dfrac{1}{4}x = \dfrac{5}{6}$
(23, 50)

18. $\dfrac{2.4}{1.8} = \dfrac{6}{x}$
(25, 44)

Simplify.

19. $\dfrac{35w^5 m^{-4}}{14wm^{-4}}$
(27, 51)

20. $66\dfrac{2}{3}\%$ of $\dfrac{3}{4}$ of \$0.72
(84)

Expand.

21. $(x + 3)(x - 3)$
(92)

22. $(x - 3)^2$
(92)

23. Write expressions using variables for the area and perimeter of a
(92) rectangle with a length $3x + 1$ and width $x - 2$.

Evaluate For problems **24** and **25,** refer to these similar triangular prisms.
Units are centimeters.

* 24. **a.** What is the lateral surface area of
(85, 100) the smaller prism?

b. What is the scale factor from the smaller to larger prism?

c. What is the ratio of the surface area of the larger prism to the surface area of the smaller prism?

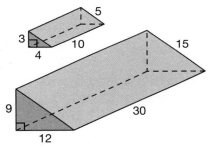

* 25. **a.** What is the volume of the larger prism?
(76, 106)

b. What is the volume of the smaller prism?

c. What is the ratio of the volumes of the larger to smaller prism?

• Similar Solids

facts | Power Up V

mental math

a. **Estimation:** 5220 + 6980 + 2100 + 5900

b. **Number Sense:** A perfect square has a 0 in the ones place. Its square root has a ___ in the ones place.

c. **Powers/Roots:** $(x^3)^4$

d. **Proportion:** $\frac{5}{x} = \frac{10}{12}$

e. **Scientific Notation:** Compare 0.4044 ◯ 4.044 × 10⁴

f. **Statistics:** Find the range of this data: 9, 4, 9, 5, 3

g. **Geometry:** An equilateral triangle has three angles with the same measure. Find the measure of each angle.

h. **Calculation:** 39 + 40, − 3, − 14, ÷ 2

problem solving | Peter and his sisters were chased to the plateau. They left 1 hour before their pursuers and jogged at 5 miles per hour. Their pursuers rode at 7 miles per hour. The plateau was 15 miles away. Were Peter and his sisters caught before they reached the plateau?

New Concept | Increasing Knowledge

Recall that **similar figures** have the same shape, which means that their corresponding dimensions are proportional.

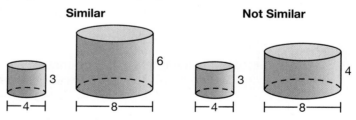

The two cylinders on the left are similar because their corresponding dimensions are proportional.

$$\begin{array}{c} \text{diameter} \\ \text{height} \end{array} \quad \frac{4}{3} = \frac{8}{6}$$

The two cylinders on the right are not similar.

$$\begin{array}{c} \text{diameter} \\ \text{height} \end{array} \quad \frac{4}{3} \neq \frac{8}{4}$$

Recall that the dimensions of similar figures are related by a **scale factor.** The scale factor applies to all measures, including circumferences, perimeters, surface areas and volumes of figures.

Relationship of Scale Factor to Area and Volume

> ratio of dimensions = scale factor
> ratio of surface areas = (scale factor)^2
> ratio of volumes = (scale factor)^3

Example 1

An architect builds a scale model of a building she is designing. The scale of the model is 1 inch = 5 feet.

   a. What is the scale factor from the model to the building?

   b. The surface area of the building is how many times the surface area of the model?

   c. The volume of the building is how many times the volume of the model?

Solution

Thinking Skill

Estimate

Estimate the building's height in inches if the scale model is 13.2 in. tall.

Scale models are designed to be similar to the objects they model.

   a. To find the scale factor we express the measures in the same units. Since 5 feet is 60 inches, the scale is 1 inch on the model equals 60 inches on the building. Therefore, the scale factor is **60.**

   b. The surface areas of similar figures are related by the square of the scale factor. The surface area of the building is 60^2, or **3600** times the surface area of the model.

   c. The volumes of similar figures are related by the cube of the scale factor. The volume of the building is 60^3, or **216,000** times the volume of the model.

Example 2

To create a cast bronze sculpture, the ancient artist carved the subject in beeswax. Then the wax was covered in liquid clay and heated. As the cast was heated, the clay hardened and the wax melted away. Melted bronze was then poured into the hardened clay mold. When the metal cooled the clay was removed, leaving the durable bronze figure. If the artist was commissioned to create a similar solid bronze figure ten times larger than the original sculpture, how many times as much bronze would be used?

Since both objects are solid and not hollow, we compare their volumes. We know the scale factor is 10. That means the length, width, and height of the new figure are 10 times greater. These three dimensions are multiplied to calculate a volume, so the volumes are related by the cube of the scale factor (10^3). Therefore, the larger sculpture would use **1000** times as much bronze.

Practice Set

For problems **a–c,** use the scale factor to help you answer the question.

a. The two cylinders are similar. If the lateral surface area of the smaller cylinder is 20π in.^{2}, then what is the lateral surface area of the larger cylinder?

b. The volume of a model car built to a 1:20 scale is what fraction of the volume of the car?

c. A toy doll modeled after an infant is a 1:3 scale model. The amount of material required to clothe the doll is about what fraction of the amount of material needed to clothe an infant?

Written Practice   *Strengthening Concepts*

1. Fifteen percent of the school days were blustery. If there were 180 school days in all, how many were not blustery?
(48)

2. Reginald is charged $7\frac{1}{2}\%$ for sales tax on a \$400 item. What is the amount of the sales tax only?
(67)

* 3. **Analyze** If $\triangle ABC$ is reflected across the line $y = 1$, what will be the coordinates of the image of point C?
(Inv. 5)

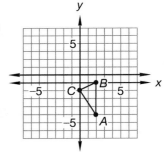

Model Give x and y intercepts, then graph problems **4** and **5**.

* 4. $3x + 5y = 15$
(82)

* 5. $2x + 4y = -12$
(82)

* 6. **Evaluate** You draw two marbles from a bag containing 21 red marbles, 5 blue and 10 green. Find the following probabilities:
(83)

   a. drawing red then blue without replacement

   b. drawing blue then red without replacement

   c. drawing red and blue in either order without replacement

   d. not drawing exactly one red and one blue without replacement

* 7. Evaluate Marla is Jill's older sister. The sum of their ages is 12. The
(82, 89) difference of their ages is 4. Graph this system of equations to find their
ages. How old is Marla?

$$\begin{cases} x + y = 12 \\ y - x = 4 \end{cases}$$

Refer to the two cones for problems **8** and **9**. Units are inches.

* 8. **a.** What is the scale factor from the
(91, 100)   larger to smaller cone?

b. What is the ratio of the surface
areas of the larger to smaller
cones?

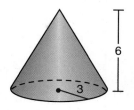

c. What is the ratio of the volumes
of the larger to smaller cones?

* 9. **a.** Find the volume of the larger cone in terms of π.
(86)

b. Find the volume of the smaller cone.

10. Find the first 6 terms in the sequence generated by $a_n = 12 - 2n$.
(73)

11. This box-and-whisker plot indicates the distribution of raw scores on
(103) a standardized test from a sample of 500 students. Refer to the plot to
answer the questions that follow.

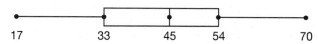

17   33   45   54   70

a. What was the median score?

b. What was the range of scores?

c. The scores of the middle 250 students range between what two
scores?

d. Name the two measures of central tendency that we cannot identify
from the plot.

* 12. Brenda averaged 8 points per game in her first 3 basketball games.
(105) How many points per game does she need to average in her next two
games to have a five game average of 10 points per game?

* 13. In 2 hours Barton drives 120 miles from his home near Lubbock to visit
(105) a friend near Amarillo. Due to bad weather the return trip takes 3 hours.

a. Find Barton's average speed from Lubbock to Amarillo.

b. Find his average speed on the return trip.

c. Find his average speed for the round trip.

14. A 15-foot ladder is placed against a building
(Inv. 2)   as shown. About how high up the building
does the ladder reach?

A 10 ft   **B** 12.1 ft

C 14.1 ft   **D** 15.8 ft

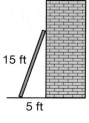

Solve.

15. $12x^2 + 2 = 50$
(93)

16. $0.008m + 0.12 = 1$
(25, 50)

17. $\dfrac{1}{8} + \dfrac{1}{3}x = \dfrac{11}{12}$
(23, 50)

18. $\dfrac{5}{x} = \dfrac{2.7}{8.1}$
(25, 44)

19. Write expressions using variables for the area and perimeter of a
(92)   rectangle with length $5x + 9$ and width 2.

Simplify.

20. $\dfrac{20x^4m^{-2}}{4x^3m^{-2}}$
(27, 51)

21. 250% of $\dfrac{5}{6}$ of $0.48
(84)

A juice carton has these approximate dimensions.
Refer to this diagram for problems **22–24.**

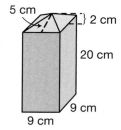

* 22. Find the volume of the carton.
(107)

* 23. Find the total surface area of the carton.
(107)

24. The capacity of the carton is about how
(104)   many liters?

* 25. The graph below shows the number of visitors to the library each day
(Inv. 6)   for a week. Represent the data in a table.

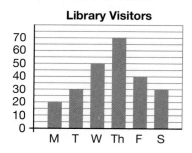

Library Visitors

• Consumer Interest

facts

Power Up V

mental math

a. Algebra: Solve: $2h + h = 36$

b. Estimation: $678 + 297 - 399$

c. Fractional Parts: $\frac{4}{11}$ of 88

d. Percent: 100% less than $25

e. Rate: How many hours per mile is 4 miles per hour?

f. Select a Method: To find the average height of the five starters on the basketball team, the coach will probably use
A mental math.
B pencil and paper.
C a calculator.

g. Measurement: What is the volume of the liquid in this graduated cylinder?

h. Calculation: $31 + 49, \div 2, \div 2, \div 2, \div 2, \div 2$

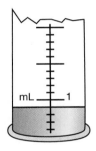

problem solving

The lady thought of a number. The knight asked if the number was a multiple of three. The lady said, "Yes." The knight asked if the number was prime. The lady answered. The knight then knew what the number was. What was the lady's answer? What was the number?

It costs money to borrow money. Borrowing money to buy a home, a car, or even a jacket includes additional costs, and sometimes those additional costs are high. We will consider three situations.

Thinking Skill

Explain

Why are the payments on a 15-year loan not twice the payments on a 30-year loan?

Mortgage: The money borrowed to buy a home is called a mortgage. If a person borrows $300,000 at about 7% fixed interest for 30 years, then the borrower makes equal monthly payments of about $2000. The monthly payment pays current interest and a small amount of the $300,000 principal. If the borrower continues to make the mortgage payment for 30 years (360 monthly payments) until the loan is repaid, then the sum of the payments will be about $720,000. This total includes the original $300,000 loan plus about $420,000 in finance charges (interest, etc.). A 15-year loan repays the original loan in half the time and may have a lower interest rate, so the total finance charge is less. However, the monthly payments on a 15-year loan are higher (not twice as high but perhaps a third higher).

Auto loans: Auto loans are usually at a higher interest rate than home loans and are for a shorter duration, such as 36 months or 60 months. A person who borrows $20,000 for 36 months at 7.5% annual percentage rate (APR) pays about $622 per month. The interest paid totals about $2,400. Borrowing $20,000 at 7.5% for 60 months reduces the monthly payment to about $400 but increases the total interest paid to over $4000.

Credit Card Interest: Credit card companies offer a line of credit to consumers and require monthly payments for purchases made with the credit card. If the charges are paid in full each month, no interest is charged. If the charges are not paid in full then an interest charge is added to the bill at the end of the monthly billing cycle. The monthly charge is typically $1\frac{1}{2}$ to 2% of the unpaid balance.

Visit www.SaxonPublishers.com/ActivitiesC3 *for a graphing calculator activity.*

Example

Jason has a credit card balance of $1200. The card company charges 2% interest on the unpaid balance each month.

   a. What is the interest charge on Jason's current balance?

   b. Jason has a choice of paying the balance plus interest in full, or making a minimum payment of $50. If Jason makes only the minimum payment each month, what balance will he carry to the next month?

   c. If Jason makes only the minimum payment and makes an additional $800 in purchases during the next month, what will be Jason's approximate balance after the next month's interest is applied?

Solution

   a. To find the interest charge we multiply the $1200 by 2%.

$$0.02 \times \$1200 = \textbf{\$24}$$

   b. The $24 charge is added to $1200 making the total $1224. If Jason makes a $50 payment the balance carried to next month is

$$\$1224 - \$50 = \textbf{\$1174}$$

   c. If Jason makes the minimum payment, carries over the $1174 balance, and adds $800 in charges, then his balance is $1174 + $800 + interest. The interest charge is 2% of $1174 plus 2% from the date of purchase for new purchases.

$$2\% \text{ of } \$1174 = \$23.48$$
$$\text{up to } 2\% \text{ of } \$800 = \$16.00$$

Since we do not know when the purchases were made, we know only that the total interest charges will range from about $24 to about $40.

$$\$1174 + \$800 + \$24 = \$1998$$
$$\$1174 + \$800 + \$40 = \$2014$$

Jason's balance will be **about $2000.**

Practice Set

a. Two families repaid their 30-year $300,000 mortgages. One loan was repaid at about $1800 per month at 6%, and the other at about $2000 per month at 7%. What is the difference in interest charges over the life of the loans?

b. Monthly mortgage payments vary with the length of the loan. The table below shows the approximate monthly payment on a $300,000 mortgage for 15 years and for 30 years at 7%.

Payback Period	Approximate Payment
15 yr	2700
30 yr	2000

What is the difference in interest charges over the life of the loans?

c. *Evaluate* Two friends each had an auto loan of $20,000. One was repaid at 8% in 3 years at about $625 per month, and the other at 8% in 5 years at about $405 per month. The total of the payments for the 5-year loan is how much more than the total for the 3-year loan?

d. The table shows the approximate monthly payments on a 5-year auto loan of $20,000 at two different interest rates.

Rate	Payment
8%	405
9%	415

Over the life of the loan, how much more is paid at 9% than at 8%?

e. Find the monthly interest charge on a $12,000 credit card balance at 2% per month. If the $12,000 balance is carried for a year, what is the total of the interest charges?

Written Practice *Strengthening Concepts*

* 1. *Predict* A random survey of 400 out of 20,000 likely voters found that the ballot measure was favored by 220 people. Based on these results, predict the number of people who will vote in favor of the ballot measure.
(34, Inv. 9)

* 2. **Classify** If two people can set up all the chairs in 12 minutes, how long
(99) will it take

    a. three people?

    b. four people?

    c. Are the variables in this situation directly or inversely related?

* 3. When Fred started the novel he read an average of 35 pages a day for
(105) four days. To finish the book he averaged 42 pages a day for three days.
What was the average number of pages he read per day for the seven
days?

* 4. Find the perimeter of a room that is 12 ft 9 in. long and 11 ft 4 in.
(80) wide.

* 5. **a.** Factor: $6x^2 - 9x$ **b.** Expand: $-2x(x - 3)$
(21)

6. What is the probability that an object that
(60, 101) lands in the parallelogram also lands in
triangle *ABC*?

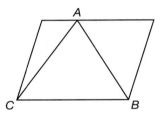

7. Graph these two equations and find the *x, y* pair that satisfies both
(82) equations.

$$\begin{cases} -2x + 3y = 18 \\ x + y = -4 \end{cases}$$

* 8. The difference between length and width of a rectangle is 3 inches. The
(82) perimeter is 10 inches. Find the length and width by graphing.

$$\begin{cases} x - y = 3 \\ 2x + 2y = 10 \end{cases}$$

For problems **9–11,** refer to the similar pyramids. Units are inches.

* 9. **a.** **Model** Find the volume of the larger
(55, 86)   pyramid.

    b. Draw a net of the larger pyramid.

* 10. **a.** Use the Pythagorean Theorem to find the
(95, 100)   slant height of the larger pyramid.

    b. Use the slant height to help you find
    the lateral surface area of the larger
    pyramid.

* 11. **a.** Find the scale factor from the smaller to larger pyramid.
(91, 106)

    b. Find the ratio of the surface areas of the larger to smaller
    pyramids.

    c. Find the ratio of the volumes of the larger to smaller pyramids.

12. **a.** What is the measure of the central angle formed by the hands of a clock at 5 o'clock?
(81)

    b. If the diameter of the clock is 12 inches, what is the length of the arc intercepted by the angle in **a?** Use 3.14 for π and round to the nearest inch.

* 13. Sonya has a $4,500 balance on her credit card. The finance charge is 2% per month. If she makes only the minimum payment of $100 this month, how much of her payment will pay interest and how much will be used to reduce the balance she owes?
(109)

14. A bag contains three red marbles and six blue marbles. Find the probability of drawing a red marble and then a red marble in two draws
(68, 83)

    a. with replacement.   **b.** without replacement.

15. Find the mean of 2.5, 3.6, 4.5, and 5.5. Round to the appropriate number of decimal spaces.
(7)

Simplify.

16. $(-1)^2 - (-1)^3 - (-1)^4$
(36)

17. 20% of 108% of $20.00
(84)

18. $\dfrac{2.4 \times 10^6}{4.0 \times 10^{-2}}$
(27, 51)

19. $\dfrac{36m^4b^2}{48m^{-1}b^5}$
(51)

Expand.

20. $(3x + 1)(x - 3)$
(92)

21. $(x - x)^2$
(92)

22. Write expressions using variables for the area and perimeter of a rectangle with length $5x - 4$ and width $4x - 5$.
(92)

Solve.

23. $7x^2 - 50 = 650$
(93)

24. $0.04r - 0.2 = 1$
(25, 50)

* 25. Draw the solid figure that has these front, top, and side views.
(Inv. 4)

      Front   Top   Right Side

• Converting Repeating Decimals to Fractions

facts Power Up V

mental math

a. Algebra: Solve: $\frac{1}{2}(50n) = 125$

b. Number Sense: A perfect square has a 9 in the ones place. Its square root has a ___ or a ___ in the ones place.

c. Powers/Roots: $(r^2)^4$

d. Probability: A random number generator gives a random number from 1 to 20. Rob designs a game where he wants the probability of a special event to be 25%. He wants special events to happen when the random number generator gives low numbers. What numbers should be assigned special events?

e. Ratio: In the game of capture the flag, there were 32 participants. If 12 players were your opponents, what was the ratio of your opponents to your teammates?

f. Statistics: The photographer lined up 21 students in order of height. The height of the 11th student in line represents the
   A mean.
   B median.
   C mode.
   D range.

g. Geometry: A right triangle has one angle that measures 90°. What is the sum of the other two angles?

h. Calculation: $2\frac{1}{2} \times 2, \times 2, \times 2, \times 2, \times 2, \times 2$

problem solving

Use the given information to answer **a** and **b**.

a. Two brothers are 5 years apart, and the sum of their ages is 41. What are the ages of the brothers?

b. A girl is 3 times the age of her younger sister, and the sum of their ages is 20. What are the ages of the sisters?

Recall from Lesson 30 that some fractions convert to repeating decimals.

$$\frac{3}{11} \approx 11\overline{)3.00000}^{.27272}$$

One way to express a repeating decimal is by writing a bar over the repetend.

$$\frac{3}{11} = 0.\overline{27}$$

To convert a repeating decimal to a fraction we write three equations using algebra to remove the repeating part of the decimal.

1. To write the first equation we choose a variable, like f (for fraction), and make it equal to the decimal.

$$f = 0.\overline{27}$$

2. To make the second equation we first count the numbers of digits that repeat. Then we multiply both sides of the first equation by a power of 10 using the number of repeating digits as the exponent. For example, the decimal $0.\overline{27}$ has two digits that repeat ($0.272727...$), so we multiply by 10^2, or 100.

$$100f = 100(0.\overline{27}) \qquad \text{Multiplied both sides of first equation by 100}$$

$$100f = 27.\overline{27} \qquad \text{Simplified}$$

Notice that the decimal in this second equation has the same repetend as the decimal in the first equation.

3. To make the third equation, we subtract the first equation from the second equation.

$$\begin{array}{lr} \text{Second equation} & 100f = 27.\overline{27} \\ -\ \text{First equation} & f = 0.\overline{27} \\ \hline \text{Third equation} & 99f = 27.0 \end{array}$$

Since the repetends are identical, subtracting $0.\overline{27}$ from $27.\overline{27}$ results in a third equation without a repetend. We solve the third equation and reduce to find the equivalent fraction.

$$99f = 27$$

$$f = \frac{27}{99} = \frac{3}{11}$$

Example 1

Convert $0.1\overline{6}$ to a fraction.

Solution

We write three equations. This time we show an alternate way to indicate a repetend, using three dots (called an ellipsis) to show in this case that the pattern of repetition continues.

$$\text{First equation} \qquad f = 0.16666...$$

One digit repeats ($0.1666...$) so we multiply both sides by 10, which shifts the decimal point one place.

$$\text{Second equation} \quad 10f = 1.6666...$$

Now we make a third equation by subtracting the first from the second.

$$
\begin{array}{llr}
\text{Second equation} & 10f = 1.6666\ldots \\
-\ \text{First equation} & f = 0.1666\ldots \\
\hline
\text{Third equation} & 9f = 1.5
\end{array}
$$

The third equation is $9f = 1.5$. We solve the equation.

$$9f = 1.5$$

$$f = \frac{1.5}{9}$$

Since the fraction has a decimal, we multiply the fraction by $\frac{10}{10}$ to shift the decimal point one place and reduce the fraction.

$$\frac{1.5}{9} \cdot \frac{10}{10} = \frac{15}{90}$$

$$\frac{15}{90} = \frac{1}{6}$$

Example 2

The theoretical probability of rolling 12 with two number cubes is $0.02\overline{7}$. About how many times would you expect to roll 12 in 180 rolls?

Solution

We have a choice of rounding $0.02\overline{7}$ or converting to an equivalent fraction. Since rounding introduces small errors we choose to convert to a fraction. One digit repeats so we multiply by 10.

$$f = 0.02\overline{7}$$

$$10f = 0.27\overline{7}$$

Now we subtract the first equation from the second equation and reduce.

$$
\begin{array}{lr}
& 10f = 0.27\overline{7} \\
& -f = 0.02\overline{7} \\
\hline
& 9f = 0.25
\end{array}
$$

$$f = \frac{0.25}{9}$$

To clear the decimal we multiply by $\frac{100}{100}$, forming an equivalent fraction without decimals. We reduce this fraction to find the answer.

$$\frac{0.25}{9} \cdot \frac{100}{100} = \frac{25}{900} = \frac{1}{36}$$

The decimal probability $0.02\overline{7}$ is equivalent to the fraction $\frac{1}{36}$. This means we would expect 12 as the outcome $\frac{1}{36}$ of the times we roll a pair of number cubes. We find $\frac{1}{36}$ of 180.

$$\frac{1}{36} \cdot 180 = \frac{180}{36} = 5$$

In 180 rolls we would expect to roll 12 **five times.**

Convert each repeating decimal to a reduced fraction.

a. $0.\overline{8}$ **b.** $0.\overline{72}$ **c.** $0.8\overline{3}$

d. Write $33.\overline{3}\%$ as a repeating decimal, then as a reduced fraction.

e. Write $12.\overline{12}\%$ as a repeating decimal, then as a reduced fraction.

f. (Predict) The theoretical probability of rolling 8 with a pair of number cubes is $0.13\overline{8}$. About how many times would you expect to roll 8 in 180 rolls?

Written Practice   *Strengthening Concepts*

* 1. (Classify) Describe each of the following relationships as either direct
(69, 99) variation or inverse variation.

  a. The relationship between the length and width of various rectangles whose areas are 24 in.^{2}

  b. The relationship between the length of a side and perimeter of a square.

* 2. A 1:16 scale model car stands $3\frac{1}{2}$ inches high. How high is the actual
(87) car?

3. You draw two marbles from a bag containing 4 green marbles and
(81) 6 blue marbles without replacement. What is the probability that they are both green?

4. If a \$20,000 automobile depreciates (declines in value) 20% per year,
(102) then what is its value after three years?

Figure *ADEF* is a rectangle. $AB = BD$, $BC = CD$,
$AF = AC$, and $FE = 12$ cm. Refer to this figure for
problems **5–7.**

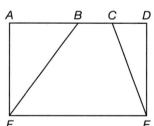

* 5. Find:
(20, 35)
  a. *AB*   **b.** *BC*

  c. *AF*   **d.** Area of $\triangle ABF$

6. Find:
(75)
  a. The area of trapezoid *BCEF*

  b. The fraction of the rectangle occupied by trapezoid *BCEF*.

7. Find:
(Inv. 2)
  a. *CE*   **b.** The perimeter of $\triangle CDE$

8. Which point is in Quadrant II?
(Inv. 1)
  A $(2, 5)$ **B** $(-2, 5)$

  C $(2, -5)$ **D** $(-2, -5)$

* 9. *(Evaluate)* Find the probability in terms of π
(101) that an object that lands in the triangle to the
right will also land in the circle.

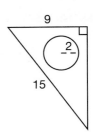

* 10. The probability found in problem **9** is closest
(16) to

    A $\frac{1}{12}$ **B** $\frac{1}{4}$ **C** $\frac{1}{2}$ **D** $\frac{3}{4}$

* 11. *(Formulate)* Convert $0.\overline{36}$ to a fraction.
(110)

12. Which of the following sets of numbers could be the lengths of sides of
(Inv. 2) a right triangle?

    A 2, 3, 4   **B** 6, 8, 10   **C** 5, 10, 15   **D** 4, 5, 6

(Generalize) Simplify.

* 13. 125% of $\frac{5}{6}$ of 0.48   **14.** $\sqrt{8}\sqrt{12}$
(84)   *(73)*

15. $\dfrac{(-12) - (-5)(6)}{(-2)(3)}$ *** 16.** 11 lbs 7 oz − 7 lbs 11 oz
(33, 36)   *(80)*

17. The graph of $2x - 5y = 20$ passes through which of these points?
(82)

    A (0, 4)   **B** (4, 0)   **C** (0, −4)   **D** (−4, 0)

Expand.

18. $(2x + 5)(2x - 5)$ **19.** $(2x - 5)^2$
(92)   *(92)*

20. Write expressions using variables for the perimeter and area of a
(92) rectangle with length $6x + 1$ and width $2x + 5$.

Solve.

21. $3x^2 - 8 = 100$ **22.** $0.08 - 0.02x = 1$
(93)   *(25, 50)*

23. $3(x + 6) - 4x = 20$ **24.** $\dfrac{1.6}{3.6} = \dfrac{x}{27}$
(21, 79)   *(25, 44)*

25. Draw the front, top, and right side views of
(Inv. 4) this figure.

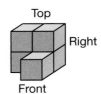

Focus on
• Non-Linear Functions

Suppose we took photographs at half-second intervals of a ball that is pushed off a high ledge. The result might look like this.

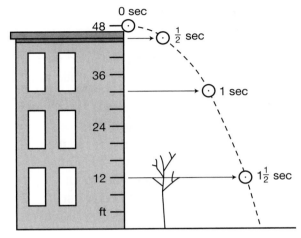

The ball does not follow a straight-line path to the ground. The path is curved. We call this kind of curve a **parabola.** We can write a function that models the height of the ball at any given time. The graph of the function is a parabola.

Activity 1

Modeling Freefall

1. Use the diagram above to chart the height (*h*) of the ball at each instant of time (*t*) shown.

t	h

2. Use the function below to complete the chart. Then graph the points.

$$y = -16x^2 + 48$$

x	y
0	
$\frac{1}{2}$	
1	
$1\frac{1}{2}$	

We see that the graph of the height of the falling ball versus time and the graph of the function are the same.

t	h
0	48
$\frac{1}{2}$	44
1	32
$1\frac{1}{2}$	12

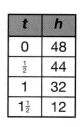

x	y
0	48
$\frac{1}{2}$	44
1	32
$1\frac{1}{2}$	12

Math Language

The term **quadratic** comes from the Greek word *quad,* meaning *square.*

Notice that the function has an x^2 term. We call this type of function a **quadratic function.** The quadratic function models the height (y) of the ball at any time (x).

Discuss Describe this graph of $y = -16x^2 + 48$ for x-values less than zero. In the situation of the falling ball, why are we only concerned about the portion of the graph in the first quadrant?

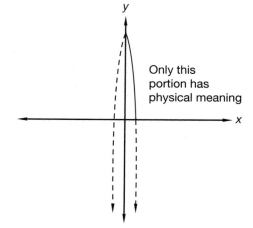

Only this portion has physical meaning

Visit www. SaxonPublishers. com/ActivitiesC3 *for a graphing calculator activity.*

The quadratic function $y = -16x^2 + 48$ comes from the study of physics and the force of gravity. When we only consider the force of gravity and ignore other forces such as wind and friction, the height of any falling object on the earth can be described by a quadratic equation.

The graphed functions on the previous page are **non-linear functions**—their points do not lie on a straight line. The height of the ball is *not* subject to a constant rate of change. The ball moves faster as time goes by.

In the first half-second, the ball falls 4 ft, but in the second half-second it falls 8 ft.

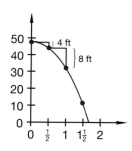

Using the Graph of a Quadratic Function

Leslie held a softball 5 ft above the ground and threw it straight up at 32 ft/sec. She has studied physics and predicted that, considering only gravity, the height (h) in feet of the ball at (t) seconds could be described by this function:

$$h = -16t^2 + 32t + 5$$

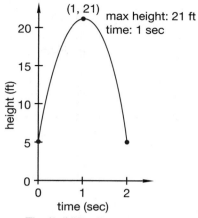

max height: 21 ft
time: 1 sec

The ball hits the ground shortly after 2 seconds

3. Complete the table to find the height of the ball at 0, 1, and 2 seconds.

t	h
0	
1	
2	

4. Plot the points (t, h). What is the maximum height of the ball? When does it reach this height? Approximately when does the ball hit the ground?

Discuss For Activities 1 and 2, describe the difference between the *paths* that the balls follow and the *graphs* of height versus time.

The graph of every quadratic function is a parabola with a **vertex** that indicates the maximum or minimum value of the function.

Maximization

Uriel has 20 yards of fencing, and he would like to use it to enclose a rectangular plot of land for a garden. He wonders what the rectangle's dimensions would have to be to contain the maximum area with his length of fence.

To solve the puzzle, Uriel drew a rectangle and named one side length x. Since half the perimeter of the rectangle is 10 yards, he reasoned that the other side length, must be $10 - x$.

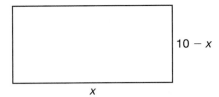

Using x and $10 - x$ as the dimensions of the rectangle, we can write an equation for its area:

Step:	Justification:
$A = lw$	Area formula
$A = x(10 - x)$	Substituted
$A = 10x - x^2$	Distributive property

We see that the equation is a quadratic function.

Analyze What x-values make sense in this situation? Write an inequality to show the values x can take.

5. Use the area function to complete the table and find the areas of rectangles with different lengths. Sketch each rectangle. What dimensions give maximum area?

x	A
2	16
3	21
4	
5	
6	
7	
8	

16 8

2

21 7

3

6. Graph the points (x, A) from your chart. Indicate the vertex on your graph.

facts | Power Up W

mental math

a. **Estimation:** $118 \div 3$

b. **Fractional Parts:** $\frac{7}{9}$ of 45

c. **Percent:** Ten percent less than $90 is $90 − $9 = __. Ninety percent of $90 is also __. So 10% less than a number is 90% of the number.

d. **Proportion:** $\frac{1}{7} = \frac{x}{49}$

e. **Scientific Notation:** Compare $0.000008 \bigcirc 8 \times 10^{-6}$

f. **Statistics:** Which measure of central tendency could be used to find a home price that is less than or equal to the prices of half of the homes sold?

g. **Measurement:** Find the temperature indicated on this thermometer.

h. **Calculation:** $160 \div 10, -6, \div 10, -1, -8$

problem solving

In calculus we sometimes find the derivative of an expression. The derivative of $9x^2$ is $18x$. The derivative of $9x$ is 9. The derivative of 9 is 0. What is the derivative of $23x$?

There is a special relationship between the volume of a cylinder, the volume of a cone, and the volume of a sphere. Picture two identical cylinders whose heights are equal to their diameters. Inside one cylinder is a cone with the same diameter and height as those of the cylinder. Inside the other cylinder is a sphere with the same diameter as that of the cylinder.

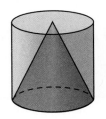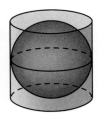

We learned in Lesson 86 that the volume of the cone is $\frac{1}{3}$ the volume of the cylinder. The volume of the sphere is twice the volume of the cone, or $\frac{2}{3}$ the volume of the cylinder. Here we use a balance scale to represent this relationship in another way.

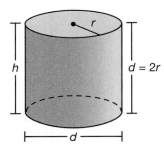

Imagine the objects on the balance scale are solid and composed of the same material. The diameters of all three solids are equal. The heights of the cylinder and cone equal their diameters. The balance scale shows that the combined masses of the cone and sphere equal the mass of the cylinder. The cone's mass is $\frac{1}{3}$ the cylinder's mass, and the sphere's mass is $\frac{2}{3}$ the cylinder's mass.

We develop the formula for the volume of a sphere by multiplying the formula for the volume of a cylinder by $\frac{2}{3}$. In this diagram, we have labeled the diameter (d), the radius (r), and the height (h).

To calculate the volume of a cylinder, multiply the area of its circular base times its height.

$$\text{volume of cylinder} = \text{area of circular base} \cdot \text{height}$$
$$V = \pi r^2 \cdot h$$

For a cylinder whose height is equal to its diameter, we can replace h in the formula with d or with $2r$, since two radii equal the diameter.

Step:	Justification:
$V = \pi r^2 \cdot h$	Formula for volume of a cylinder
$V = \pi r^2 \cdot 2r$	Replaced h with $2r$, which equals the height of the cylinder
$V = 2\pi r^3$	Simplified

This formula, $V = 2\pi r^3$, gives the volume of a cylinder whose height is equal to its diameter. The volume of a sphere is $\frac{2}{3}$ the volume of a cylinder with the same diameter. We multiply this formula for the volume of a cylinder by $\frac{2}{3}$ to find the formula for the volume of a sphere.

Step:	Justification:
$V = \frac{2}{3} \cdot 2\pi r^3$	Formula for volume of a sphere
$= \frac{4}{3}\pi r^3$	Multiplied $\frac{2}{3}$ by 2

We have found the formula for the volume of a sphere.

Volume of a sphere $= \dfrac{4}{3}\,\pi r^3$

Example 1

A ball with a diameter of 20 cm has a volume of how many cubic centimeters? (Round your answer to the nearest hundred cubic centimeters.)

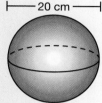

$\longmapsto$ 20 cm $\longmapsto$

Use 3.14 for π

Solution

The diameter of the sphere is 20 cm, so its radius is 10 cm. We use the formula for the volume of a sphere, substituting 3.14 for π and 10 cm for r.

Step:	Justification:
$V = \frac{4}{3}\pi r^3$	Formula
$\approx \frac{4}{3}(3.14)(10\text{ cm})^3$	Substituted 3.14 for π and 10 cm for r
$\approx \frac{4}{3}(3.14)(1000\text{ cm}^3)$	Cubed 10 cm
$\approx \frac{4}{3}(3140\text{ cm}^3)$	Multiplied 3.14 by 1000 cm^3
$\approx 4186\frac{2}{3}\text{ cm}^3$	Multiplied $\frac{4}{3}$ by 3140 cm^3
$\approx \mathbf{4200\text{ cm}^3}$	Rounded to nearest hundred cubic centimeters.

If we cut a grapefruit in half, the **cross section** is a circle. We can calculate the area of the cross-section with the familiar formula for the area of a circle.

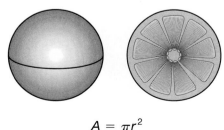

$A = \pi r^2$

Thinking Skill

Analyze

Is the radius of a great circle greater than, equal to, or less than the radius of its sphere?

The largest cross-section of a sphere is called its **great circle**. A great circle of a sphere includes the sphere's center.

The surface area of a sphere is four times the area of a great circle of the sphere.

$$\text{Surface Area of a Sphere} = 4\pi r^2$$

Example 2

The diameter of a softball is nearly 10 cm. Find the surface area of a softball. Express the answer twice, once in terms of π, and once using 3.14 for π.

Solution

The diameter is about 10 cm, so the radius is about 5 cm. We substitute 5 cm for r in the formula.

Step:	Justification:
$A = 4\pi r^2$	Surface area formula
$A = 4\pi(5 \text{ cm})^2$	Substituted 5 cm
$A = 4\pi \cdot 25 \text{ cm}^2$	Simplified
$A = \mathbf{100\pi \text{ cm}^2}$	Simplified
$A = 100(3.14) \text{ cm}^2$	Substituted 3.14 for π
$A = \mathbf{314 \text{ cm}^2}$	Simplified

In terms of π, the surface area of a softball is about $100\pi \text{ cm}^2$, or about 314 cm^2.

Practice Set

Pictured at right is a cylinder whose height is equal to its diameter. In the cylinder is the largest sphere it can contain. Packing material is used to fill all the voids in the cylinder not occupied by the sphere. Use this information to answer problems **a** and **b**.

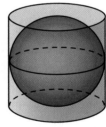

a. What fraction of the volume of the cylinder containing the sphere is occupied by the volume of the sphere?

b. What fraction of the volume of the cylinder containing the sphere is occupied by the volume of the packing material?

Find the volume and surface area of each sphere.

c. |———— 6 in. ————|

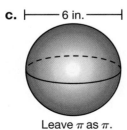

Leave π as π.

d.   d = 30 cm

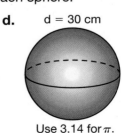

Use 3.14 for π.

A spherical "snow globe" is shipped in a cylindrical container in which it just fits. Refer to the figure for problems **1** and **2**.

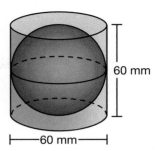

60 mm

├──60 mm──┤

1. **a.** What is the volume of the cylindrical container in terms of π?
(76, 111)

    b. What is the volume of the sphere in terms of π?

    c. What fraction of the volume of the cylinder is occupied by the volume of the sphere?

2. **a.** What is the area of the base of the cylinder in terms of π?
(40, 111)

    b. What is the surface area of the sphere in terms of π?

    c. The surface area of the sphere is how many times the area of the base of the cylinder?

3. The elephant's weight increased by 8% after it walked through the peanut field. If it gained 80 pounds, how much did it weigh before it walked though the field?
(67)

Add.

* 4. $(3a + 2b - 5) + (4a + 5b - 2)$
(80)

5. Convert $0.8\overline{3}$ to a reduced fraction.
(110)

* 6. Graph $-4x + 3y = 24$. When x is 12, what is y?
(82)

7. Find the product: $(x + 3)(x - 1)$
(92)

* 8. **Analyze** Factor: $36x^2y - 24xy^2$
(21, 36)

* 9. Find the probability that an object that lands in the triangle to the right will also land in the circle. The radius is 26. (Express in terms of π.)
(101)

35   21   26

10. Can a right triangle have sides of length 6, 1, and $2\sqrt{10}$?
(Inv. 2)

* 11. **Model** Graph $y = -\dfrac{1}{16} x^2$.
(Inv. 11)

* 12. **Evaluate** Draw a line through the points $(2, 5)$ and $(-3, -10)$. Then find the equation of the line in slope-intercept form.
(56)

Solve.

13. $5x^2 - 3 = 42$
(93)

14. $0.14 - 0.02x = 0.88$
(25, 50)

15. $\dfrac{3}{4}x + \dfrac{1}{3} = \dfrac{5}{6}$
(23, 50)

16. $\dfrac{2.42}{0.11} = \dfrac{44}{x}$
(25, 44)

17. Solve: $2x - 5 < 3x - 7$
(77, 79)

18. Solve: $4(2x - 3) = 4x - 4$
(79)

Simplify.

19. $\dfrac{45x^5 m}{9\,m^{-1}x}$
(27, 51)

20. $\dfrac{(-2)(-3) - (-2)}{(-2) - (-3)}$
(33, 36)

21. Find the mean, median, and mode of 2.7, 3.6, 4.5, 4.6, and 2.7. Round
(7) the mean to one decimal place.

22. Find the perimeter and area of a rectangle with vertices at (3, 0), (1, −2),
(96) (−2, 1), and (0, 3)?

23. What are the coordinates of the image of the rectangle in exercise **22**
(Inv. 5) after a translation of 2 units to the right and 3 units up?

24. Write expressions using variables for the area and perimeter of a
(92) rectangle with length $3x + 1$ and width $3x - 1$.

25. Use the table to complete **a** and **b**.
(69)

    a. Write an equation for the relationship between A and B.

    b. Complete the table and find the constant of proportionality.

A	B	B/A
1	3	
2	6	
4	12	
5	15	

Early Finishers
Real-World
Application

Isaam, a swimming pool repairman, has been hired to fill a rectangular wading pool that is 10 yards long and 20 feet wide with specially treated water. If the pool holds 2,100 cubic feet of water, what is its depth? If Isaam charges $0.20 per cubic foot for the specially treated water, how much will Isaam charge to fill the wading pool?

•Ratios of Side Lengths of Right Triangles

facts | Power Up W

mental math

a. Algebra: Solve: $125h = \frac{125}{2}$

b. Estimation: $69 + $73 + $71

c. Number Sense: A perfect square has a 1 in the ones place. Its square root has a __ or a __ in the ones place.

d. Powers/Roots: $\frac{(x^2)^3}{x}$

e. Rate: Sixty miles per hour is how many minutes per mile?

f. Statistics: The photographer lined up 21 students in order of height. The difference between the first and last student in line represents the
A mean.
B median.
C mode.
D range.

g. Geometry: A triangle has sides with lengths 12 and 8. The length of the third side must be between what two numbers?

h. Calculation: $-8 + 20, + 5, + 8, \sqrt{}, + 9$

problem solving

Reggie read the first half of the book on Monday. On Tuesday, he read $\frac{2}{3}$ of the second half. Reggie read 175 pages altogether on Monday and Tuesday. How many pages are in the book? (Hint: You could write a proportion to solve the problem.) What fraction of the book has he read?

Trigonometry is a branch of mathematics that explores the relationships between the angles and sides of triangles. Trigonometry is used in surveying, navigation, and engineering. Global Positioning Systems use trigonometric calculations to determine a receiver's location.

In this lesson we will learn some math language used in trigonometry. We know that a right triangle has two acute angles and two legs. For each one of the acute angles, one of the legs is the **opposite side** and the other leg is the **adjacent side.** The sides of an acute angle of a right triangle are its adjacent side and the hypotenuse. Its opposite side is across the triangle from the angle.

- From the perspective of ∠A, $\overline{BC}$ is the opposite side and $\overline{AC}$ is the adjacent side.
- From the perspective of ∠B, $\overline{AC}$ is the opposite side and $\overline{BC}$ is the adjacent side.

Trigonometry deals with the ratios of the sides of triangles. In this lesson's examples and exercises, we will find the ratios of lengths of sides for selected angles.

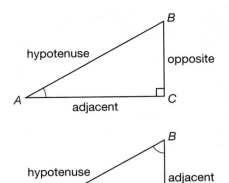

Example 1

a. **Find the ratio of the length of the side opposite ∠A to the length of the hypotenuse. Express the ratio as a fraction and as a decimal.**

b. **Find the ratio of the length of the side opposite ∠X to the length of the hypotenuse. Express the ratio as a fraction and as a decimal.**

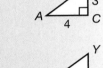

Solution

Thinking Skill

Generalize

For △ABC, the side opposite ∠A is side $\overline{BC}$. For △XYZ, the side opposite ∠X is side $\overline{YZ}$. If a triangle is named △DEF, name the sides opposite ∠D, ∠E, and ∠F, respectively.

a. The side opposite ∠A is $\overline{BC}$, and the hypotenuse is $\overline{AB}$. The ratio of the length of the opposite side to the length of the hypotenuse is

$$\frac{\text{opposite}}{\text{hypotenuse}} = \frac{3}{5} = 0.6$$

b. The side opposite ∠X is $\overline{YZ}$. The hypotenuse is $\overline{XY}$. The ratio of the side opposite to hypotenuse is:

$$\frac{\text{opposite}}{\text{hypotenuse}} = \frac{6}{10} = 0.6$$

Notice that the ratio we found for △XYZ reduces to the same ratio we found for △ABC because the triangles are similar.

Example 2

Find the ratio of the length of the side opposite ∠R to the length of the adjacent side. Then use a calculator to find the ratio rounded to 3 decimal places.

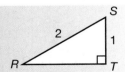

The side opposite $\angle R$ is $\overline{ST}$ and its length is 1.
The side adjacent $\angle R$ is $\overline{RT}$, and its length is not
given. We use the Pythagorean Theorem to find
$\overline{RT}$ is $\sqrt{3}$. The ratio of the length of the opposite
side to the length of the adjacent side is:

$$\frac{\text{opposite}}{\text{adjacent}} = \frac{1}{\sqrt{3}} = \frac{\sqrt{3}}{3}$$

We can use a calculator to find the ratio expressed as a decimal.

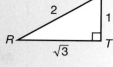

Rounded to three decimal places, the ratio is **0.577**.

Practice Set

Find each ratio for the illustrated right triangle.
Express each ratio as a fraction and as a decimal
rounded to three decimal places.

With respect to $\angle A$, find the ratio of the lengths:

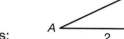

   a. Opposite side to adjacent side

   b. Opposite side to hypotenuse

   c. Adjacent side to hypotenuse

With respect to $\angle B$, find the ratio of the lengths:

   d. Opposite side to adjacent side

   e. Opposite side to hypotenuse

   f. Adjacent side to hypotenuse

Written Practice   *Strengthening Concepts*

1.
(3, 4)
The party was expensive. It cost $1200 to rent the facility and $20 for
each guest. If the total cost of the party was $3880, how many people
attended?

* 2.
(39, 111)
Connect A classroom globe of the earth has a diameter of 12 inches.

   a. On the globe how many inches long is the equator to the nearest
inch?

   b. What is the surface area of the globe to the nearest square
inch?

* 3.
(111)
Which is the best estimate of the volume of the globe described in
problem **2**?

   A 450 in.^{2} **B** 900 in.^{3} **C** 1200 in.^{3} **D** 7000 in.^{3}

* 4. The top, front, and side views of a roof are shown. From the edge of the
(20, 75) roof to the peak is 11 ft. Find the surface area of the roof.

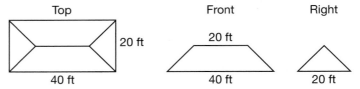

| Top | Front | Right |

* 5. Convert $0.\overline{81}$ to a reduced fraction.
(110)

6. Simplify: $(3r + 5p - 4) + (2r - 4p + 4)$
(89)

* 7. For **a** and **b**, refer to triangle ABC.
(112)

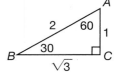

   a. What is the ratio of the length of the
side opposite $\angle B$ to the length of the
hypotenuse?

   b. What is the ratio of the side adjacent to
$\angle B$ and the length of the hypotenuse?

   c. What is the ratio of the length of the side opposite $\angle A$ to the length
of the side adjacent to $\angle A$?

* 8. **a.** Find the volume of the smaller cone
(86, 91)   (in terms of π). Dimensions are in
inches.

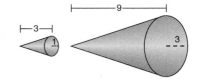

   b. Find the ratio of the surface areas.

   c. Find the ratio of the volumes.

* 9. Graph these equations and find the common solution.
(82)
$$\begin{cases} x + y = 1 \\ -5x + 10y = -20 \end{cases}$$

10. A company has $4 flat fee for shipping, plus $2 for each item ordered.
(41, 56) This can be represented by $y = 2x + 4$. Graph this equation. Find the
coordinates of the point representing an order of 4 items.

11. Find the probability that an object that
(101) lands inside the triangle to the right will
also land in the circle. (Express in terms
of π.)

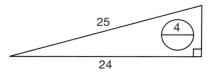

12. Simplify: $\dfrac{24x^5y^{-9}}{32xy^{10}}$
(27, 51)

13. You roll two number cubes at the same time. Find the probability of
(68) rolling a number less than 5 on both cubes.

14. Draw a line through points (2, 8) and (8, −4) and find the equation of the
(56) line in slope-intercept form.

Solve.

15. $7x^2 + 5 = 68$
(93)

16. $0.4 - 0.03x = 1$
(25, 50)

17. $\frac{5}{8}x - \frac{1}{3} = \frac{1}{2}$
(23, 50)

18. $\frac{9.3}{6.3} = \frac{x}{21}$
(25, 44)

19. Factor: $44x^2 - 22x + 33$
(21)

20. Solve: $3(x - 7) \leq 5x + 1$
(77, 79)

21. Solve: $5x - (-2) = 10x - 3(x + 1)$
(79)

22. Does the figure illustrate a possible right triangle? Explain.
(Inv. 2)

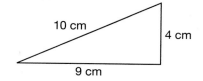

10 cm   4 cm   9 cm

* 23. Graph $y = -0.8x^2$
(Inv. 11)

24. Write a variable expression for the area and perimeter of the rectangle with length $5x + 4$ and width $4x - 5$.
(92)

25. Although it is said that 1 dog year is the equivalent of 7 human years, some people use the following rule: A dog ages 10.5 human years for both of its first two years and 4 human years for each year afterwards. Using this rule, calculate in human years the ages of dogs that are the following ages.
(105)

a. 3 dog years

b. 10 dog years

Early Finishers
Real-World Application

The average monthly temperatures in Tampa, Florida, are below. If Tampa's average yearly temperature is 72°F, what is July's average monthly temperature?

January	59°F	July	?°F
February	61°F	August	82°F
March	66°F	September	80°F
April	71°F	October	74°F
May	77°F	November	67°F
June	81°F	December	62°F

• Using Scatterplots to Make Predictions

facts | Power Up W

mental math

a. **Algebra:** Solve: $5a + 5 = 555$

b. **Fractional Parts:** $\frac{7}{10}$ of 80

c. **Percent:** Twenty percent less than $60 is the same as what percent of $60?

d. **Probability:** The probability of a matching object being selected at random is $\frac{2}{9}$. What is the probability that the item selected is not a matching object?

e. **Ratio:** Of the 12 bushes Laura planted, 8 bloomed. What was the ratio of blooming bushes to bushes that did not bloom?

f. **Select a Method:** A structural engineer wants to find the size of concrete and steel beams needed to support the floor in a parking structure. The engineer will probably use
   A mental math.
   B pencil and paper.
   C a calculator.

g. **Measurement:** The odometer read 2250.3 miles at the beginning of the trip. Half-way through the trip, the odometer read 3251.5 miles. How long was the trip?

h. **Calculation:** $14 + 12$, $\div 2$, $\times 3$, $+ 1$, $\div 2$, $\div 2$

problem solving

Daniel has 19 feet of fence. He wants to enclose a garden in the corner of his yard. If he wants a garden with an area of 78 ft², what should the length and width of the garden be?

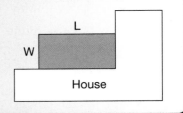

New Concept | *Increasing Knowledge*

Recall that scatterplots are graphs of points representing a relationship between two variables. Below we review common relationships displayed in three sample scatterplots.

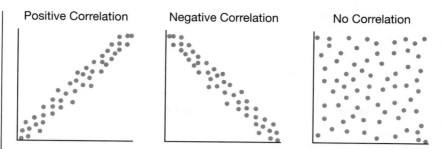

Positive Correlation   Negative Correlation   No Correlation

The first two scatterplots indicate a strong relationship between the variables plotted. A **positive correlation** means that as one variable increases the other variable also increases. For example, the number of hours a student spends studying and the student's grade point average might have a positive correlation.

A **negative correlation** means that as one variable increases the other decreases. For example, the number of hours a student spends watching TV and the student's grade point average might have a negative correlation.

If the graph shows no meaningful connection between variables, we say that they have **no correlation.** For example, the last digit of a student's address and the student's grade point average have no correlation.

Researchers use these terms to describe relationships between variables. They also describe these relationships algebraically to determine the predictability of the relationship.

Example

Each student in a science lab receives a specimen of a certain mineral and measures its volume and mass. These measures for each specimen are listed below, and each data point is graphed on the scatterplot.

Volume (cm³)	Mass (g)
2.0	5.3
4.5	12.0
5.7	16.1
10.0	27.0
7.5	20.9
6.9	16.2
3.1	9.0
9.4	26.0

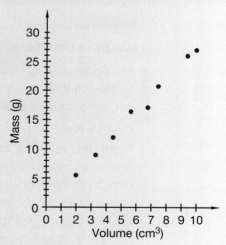

a. Draw a best-fit line and find its equation.

b. Use the best-fit line to predict the mass of a specimen of this mineral with 100 cm³ volume.

Thinking Skill

Analyze

How do we know that the best-fit line intersects the origin?

a. A best–fit line is graphed below.

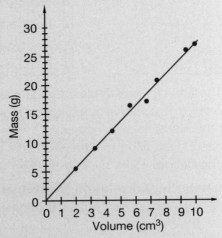

We see that it passes through the origin and the data point (10.0, 27.0), so its slope is 2.7. Therefore, the equation is:

$$y = 2.7x$$

In this equation, x represents volume and y represents mass.

b. The mass of a 100 cm³ specimen can be predicted by substituting 100 for x in the equation:

$$y = 2.7x$$
$$y = 2.7 \cdot 100$$
$$y = 270 \text{ g}$$

Activity

Using a Scatterplot to Make Predictions

In this activity you and your classmates will research the number of vowels as related to the total number of letters appearing in various books. Each of you should choose a different book before beginning the activity.

Open a book of your choice to any page of text and select a paragraph. From the first letter of the paragraph, count every letter in order (both vowels and consonants) through one to four sentences. Do not count spaces or punctuation. Record the total and note where your count ended. (One to four sentences generally contain between 50 and 200 letters.) Then go back to the beginning of the paragraph and count the total number of vowels (a, e, i, o, u) up to but not past the letter where the first count ended. Express your two counts as an ordered pair (all letters, vowels).

As a class, draw a first-quadrant graph on grid paper. Mark on the horizontal axis the total number of letters, and on the vertical axis the number of vowels. Let each grid line represent five letters. Plot each student's results on the graph. When all points are plotted, the graph might look similar to the one on the following page. (We broke the scales so the graph will fit on the page. You do not need to break your scales.)

Frequency of Vowels in English

- Describe the relationship indicated on your scatterplot.

- On your scatterplot, sketch a best-fit line. Should the line pass through the origin? Why?

- Find the slope of your best-fit line. Then convert the slope to a percent. What does the slope mean in this situation?

- Is there predictive value to your results? Select another paragraph from a book and count the number of letters in a few sentences. Based on your scatterplot, predict the number of vowels there are in the same sentences. Then count the vowels. Was your prediction close?

Practice Set Because this lesson's activity is time-consuming, there is no practice set.

Written Practice   *Strengthening Concepts*

1. The number of pretzels in the jar had been decreased by 15%. If there
(67) were 12 fewer pretzels than before, how many were left?

* 2. **Analyze** Nelson ran for 30 minutes at 8 miles per hour, and then he
(105) jogged for 30 minutes at 6 miles per tour. What was his average speed for the hour of running and jogging?

* 3. **Evaluate** To calculate the volume of a rock, Brenda submerged the
(104) rock in a graduated cylinder of water. Before
the rock was submerged, the water level
was 150 mL. After the rock was submerged
the water level rose to 213 mL. What is the
approximate volume of the rock in cubic
centimeters?

213 mL (After submerged)
150 mL (Before submerged)

Simplify problems **4–7.**

4. $(3x + 4y - 5) + (4x + 3y + 2)$
(80)

5. $(4m - 5y - 6) + (2m - 4y - 6)$
(80)

6. $\left(-\dfrac{2}{3}\right)^2$
(63)

7. $\left(-\dfrac{2}{5}\right)^0$
(27, 63)

8. Find the product: $(x - 1)(x - 10)$
(92)

9. A shade for a lamp appears to be part of a cone. Refer to the cone to answer the following questions in terms of π.
(100)

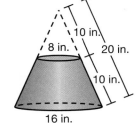

   a. What is the lateral surface area of the cone with the 16-in. diameter base?

   b. What is the lateral surface area of the cone with the 8-in. diameter base?

   c. What is the lateral surface area for the part of the cone between the 16-inch diameter base and the 8-inch diameter base?

10. **a.** What is the scale factor from the smaller to larger cone in problem **9**?
(91, 100)

   b. What is the ratio of the surface areas of the larger to smaller cone?

* 11. *Evaluate* Find the area of the shaded region in terms of π. The radius of the circle is 1. Units are cm.
(37, 40)

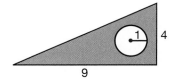

12. **a.** Find the ratio of the length of the side opposite $\angle A$ to the length of the hypotenuse.
(112)

   b. Find the ratio of the length of the side opposite $\angle A$ to the length of the side adjacent to $\angle A$.

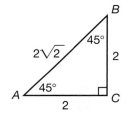

* 13. *Model* Graph $y = -0.5x^2$.
(Inv. 11)

14. Betty drew a point at $(0, 8)$. From that point she counted down two units and right one unit and drew another point, then she drew a line through both points. What is the equation of the line Betty drew in slope-intercept form?
(56)

Analyze Solve problems **15–18.**

15. $-2x^2 + 11 = 3$
(93)

16. $0.3 - 0.02x = 0.4$
(25)

17. $\dfrac{3}{4}x + \dfrac{1}{3} = \dfrac{7}{12}$
(22, 50)

18. $\dfrac{2.8}{3.5} = \dfrac{x}{5}$
(25, 44)

19. Factor: $27x^2 + 18x - 81$.
(21)

Solve.

20. $7(3 - x) < 3 - x$
(77, 79)

21. $2x - (-3x) = 7x - 2$
(79)

22. Simplify: $\dfrac{32m^4 b^{-4}}{20b^4 m}$
(27, 51)

23. Convert $0.41\overline{6}$ to a reduced fraction.
(110)

24. Find the mean and median of 10.4, 10.6, 10.8, 10.0 and 9.8. Round
(7) each to one decimal place.

25. Suppose you have enough money to buy either one 12-inch pizza or
(91) two 8-inch pizzas. Which choice gives you more pizza? Explain your
answer.

Early Finishers
Real-World
Application

The city council funded a new spherical water tank that is 200 feet wide
to top the town's water tower. Before putting it to use, the city will paint its
surface and fill the tank to capacity. Determine the approximate area to be
painted and volume of the spherical tank. (Use 3.14 for π.)

• Relative Sizes of Sides and Angles of a Triangle

facts Power Up W

mental math

a. **Algebra:** Solve: $2k + 2 = 170$

b. **Estimation:** 205×390

c. **Fractional Parts:** $\frac{5}{11}$ of 77

d. **Percent:** 25% less than $120 is the same as what percent of $120?

e. **Rate:** $100 per week is about how many dollars per year?

f. **Geometry:** The Pythagorean Theorem applies to all
   A acute triangles.
   B right triangles.
   C obtuse triangles.

g. **Measurement:** How long would 10 bolts be?

h. **Calculation:** $81 \div 3, \div 3, \div 3, \div 3, \div 3$

problem solvings The number 1 can be written as the difference of two prime numbers (3 − 2). The numbers 2 and 3 can also be written as the difference of two prime numbers (5 − 3 and 5 − 2, respectively). If possible, write each of the numbers 4, 5, 6, and 7 as the differences of two prime numbers.

Refer to △*ABC* below. Listed in order of size, the angles of △*ABC* are ∠*A*, then ∠*B*, and ∠*C*. Listed in order of length, the sides of △*ABC* are side *BC*, side *CA*, and side *AB*.

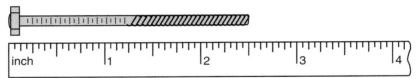

$$m\angle A < m\angle B < m\angle C$$
$$BC < CA < AB$$

Notice that the side opposite the smallest angle is the shortest side (side *BC* and ∠*A*). Likewise, the side opposite the longest angle is the longest side (side *AB* and ∠*C*).

The order of lengths of sides of a triangle when written from shortest to longest corresponds to the order of the sides' opposite angles when written from smallest to largest.

Example 1

List the sides of this triangle in order from shortest to longest.

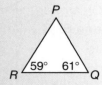

Solution

The sum of the interior angles of a triangle is 180°, so m∠P = 60°. Thus, the angles in order from smallest to largest are ∠R, ∠P, and ∠Q. The lengths of the sides opposite these angles follow the same order. The side opposite ∠R is side $\overline{PQ}$, which is the shortest side. We list the three sides in order of length.

$$\text{side } \overline{PQ} \text{ (or } \overline{QP}\text{), side } \overline{QR} \text{ (or } \overline{RQ}\text{), side } \overline{RP} \text{ (or } \overline{PR}\text{)}$$

Example 2

The lengths of sides of △ABC are given. List the angles in order of size from smallest to largest.

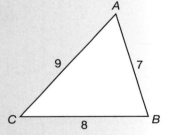

Solution

The shortest side is opposite the smallest angle. Since side $\overline{AB}$ is the shortest side, the smallest angle is ∠C. We list the angles in order from smallest to largest.

$$\angle C, \angle A, \angle B$$

Example 3

If we know the length of side *AC*, how could we find the length of side *WY*? Find *WY* if *AC* = 6 cm.

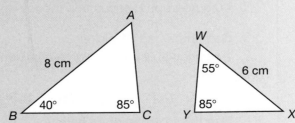

The sum of the angle measures of a triangle is 180°, so we know that $m\angle A = 55°$ and $m\angle X = 40°$. We see that the three angles of $\triangle WXY$ are congruent to the corresponding angles of $\triangle ABC$. Therefore, the two triangles are similar and their lengths of sides are proportional.

The scale factor from the larger triangle to the smaller triangle is $\frac{6}{8}$ (or $\frac{3}{4}$). **If we know $\overline{AC}$, we can find $\overline{WY}$ by multiplying $\overline{AC}$ by the scale factor or by solving a proportion.** We show both methods for finding $\overline{WY}$ when $\overline{AC}$ is 6 cm.

Thinking Skill

Explain

How do we determine the scale factor for example 3?

Multiplying by Scale Factor:

$$\overline{WY} = \text{scale factor} \times \overline{AC}$$

$$= \frac{3}{4} \cdot 6$$

$$= 4.5 \text{ cm}$$

Solve Proportion:

$$\frac{8}{6} = \frac{6}{x}$$

$$8x = 36$$

$$x = 4.5 \text{ cm}$$

Practice Set

For **a** and **b** refer to figure *ABCD*.

  a. Find the measure of $\angle CBD$, $\angle DBA$, and $\angle DAB$.

  b. Arrange these segments in order from shortest to longest.

    $\overline{AB}$, $\overline{AD}$, $\overline{BD}$, $\overline{BC}$, $\overline{DC}$,

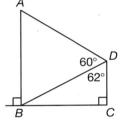

  c. *Classify* Which angle in $\triangle WXY$ is the smallest angle?

  d. *Analyze* Refer to the triangles in example **3.** Find *WY* if *AC* is 6.4 cm.

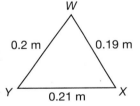

  1. Connie was confused and filled forty-four percent of the flower pots with flour. If she did not fill 28 of the pots, how many pots were there in all?
  (48)

  2. The local art museum purchased one of Connie's flour-filled pots for $700. It is a hit, so Connie makes $5 per autograph. If she signs 40 autographs, how much money does she make in all?
  (3, 4)

  * 3. Samantha had 21 nickels and dimes. They totaled $1.60. Two equations that show this information are shown below.
  (89)

$$\begin{cases} u + d = 21 \\ 5u + 10d = 160 \end{cases}$$

Use this information to help you find and check the numbers of nickels and dimes Samantha has.

* 4. *Analyze* Find the area of $\triangle FBD$ by
$_{(20)}$ subtracting the areas of the three smaller triangles from the area of rectangle *ACDE*.

* 5. *List* Name the sides of FBD in order from
$_{(115)}$ least to greatest. Then list the angles of FBC in order from least in measure to greatest.

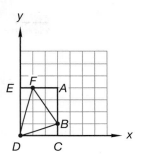

Expand.

6. $4m(3m^2 - 2)$
$_{(95)}$

7. $(2x - 4)^2$
$_{(62)}$

Simplify.

8. $(4x^2)^3$
$_{(94)}$

9. $\left(-\dfrac{2}{5}\right)^3$
$_{(93)}$

10. $\left(\dfrac{-2}{1.1}\right)^0$
$_{(93)}$

11. $\dfrac{25m^4b^2}{35b^{-1}m}$
$_{(27,\,51)}$

12. Where does the graph of the equation $-2x + 5y = -20$ intercept the x
$_{(82)}$ and y-axis?

* 13. Find the perimeter and area in units of a rectangle with vertices at $(0, 0)$,
$_{(96)}$ $(-2, -2)$, $(-3, -1)$, and $(-1, 1)$.

* 14. *Evaluate* The rectangle in problem **13** is rotated 90° clockwise about
$_{(Inv.\,5)}$ the origin. What are the coordinates of the vertices of the image?

15. Write expressions using variables for the area and perimeter of a
$_{(92)}$ rectangle with length $5w - 6$ and width w.

16. Demonstrate that a right triangle can have sides with lengths
$_{(Inv.\,2)}$ 7, 24, 25.

17. A line has a y-intercept of 4 and a slope of -2. What is its equation in
$_{(56)}$ slope-intercept form?

* 18. Graph $y = -\dfrac{1}{9}x^2$.
$_{(Inv.\,11)}$

* 19. Write $1.1\overline{6}$ as a mixed number with a reduced fraction.
$_{(110)}$

* 20. Solve: $-x^2 - 9 = -25$
$_{(93)}$

21. Solve: $\dfrac{1}{3}x - \dfrac{1}{2} = \dfrac{1}{12}$
$_{(48)}$

22. Solve: $\dfrac{1.2}{3.2} = \dfrac{3}{x}$
$_{(44)}$

23. Solve: $3x + 5 > 4x$
$_{(77)}$

24. Which type of borrowing usually has the highest interest rate?
$_{(109)}$
    A home loan   **B** auto loan   **C** credit card loan

25. Assume that the Earth's circumference is exactly 25,000 miles. How
$_{(39,\,52)}$ many hours and minutes will it take a jet traveling 600 mph to fly around the world?

• Division by Zero

facts | Power Up X

mental math

a. Algebra: Solve: $3e + 2 = 22 - 5$

b. Number Sense: Can a perfect square have a 3 in the ones place?

c. Powers/Roots: $\frac{(x^5)^2}{x}$

d. Probaillity: The probability of an event happening is 0.09. What is the probability that the event won't happen?

e. Ratio: Bob has 6 fruits and 3 vegetables. Larry has 2 fruits and 9 vegetables. Altogether, what is the ratio of fruits to vegetables?

f. Number Sense: Arrange these numbers from least to greatest:
$$-\frac{2}{3}, -\frac{3}{2}, -1$$

g. Geometry: $\triangle ABC \cong \triangle DEF$. Therefore, $\angle C \cong \angle$ ___.

h. Calculation: $\frac{1}{3} + \frac{2}{3}$, $\times 256$, $- 6$, $\div 10$, $- 5$

problem solving

In calculus we sometimes find the derivative of an expression. The derivative of $x^3 + 5x + 3$ is $3x^2 + 5$. The derivative of $2x^3 + 4x^2 + 3x$ is $6x^2 + 8x + 3$. The derivative of $x^4 + x^2 + 1$ is $4x^3 + 2x$. Find the derivative of $x^2 + 6x + 9$.

New Concept   *Increasing Knowledge*

When performing algebraic operations, we guard against dividing by zero. For example, the following expression reduces to 3 only if x is not zero:

$$\frac{3x}{x} = 3 \text{ if } x \neq 0$$

We substitute 0 for x and find that when x is 0, the expression equals $\frac{0}{0}$. What is the value of $\frac{0}{0}$? Try the division with a calculator. Notice that the calculator displays an error message when division by zero is entered. The display is frozen and other calculations cannot be performed until the erroneous entry is cleared.

We say that division by zero is **undefined.** In this lesson we will consider why division by zero is not possible.

As we know, zero lies on the number line with negative numbers to its left and positive numbers to its right.

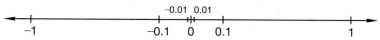

In the following example, we will consider the quotient $\frac{10}{x}$ as the values of x approach zero. We will divide 10 by numbers closer and closer to zero and examine the quotients for any pattern.

Example 1

Find each set of quotients. As the divisors become closer to zero, do the quotients become closer to zero or farther from zero?

a. $\dfrac{10}{1}, \dfrac{10}{0.1}, \dfrac{10}{0.01}$

b. $\dfrac{10}{-1}, \dfrac{10}{-0.1}, \dfrac{10}{-0.01}$

Solution

a. 10, 100, 1000

b. −10, −100, −1000

As the divisors get closer to zero, the quotients get **farther from zero.**

Reading Math

Recall that ∞ (which looks like 8 laid on its side) is the symbol for infinity. "Infinity" refers to a very great (limitless) amount. It would take forever to count an infinite number of objects.

As the divisors approach zero from the positive side, the quotients get greater and greater toward positive infinity ($+\infty$). However, as the divisors approach zero from the negative side, the quotients become lesser and lesser toward negative infinity ($-\infty$). In other words, as the divisors of a number approach zero from opposite sides, the quotients do not become closer, but grow infinitely farther apart.

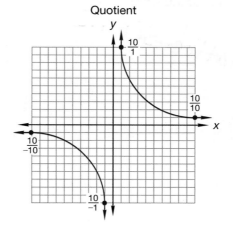

Quotient

Considering this growing difference in quotients as divisors approach zero from opposite sides can help us understand why division by zero is not possible.

Considering the relationship between multiplication and division can also help us see why division by zero is not possible. Recall that multiplication and division are inverse operations. The numbers that form a multiplication fact may be arranged to form two division facts. For the multiplication fact $4 \times 5 = 20$, we may arrange the numbers to form two division facts.

$$\frac{20}{4} = 5 \qquad \frac{20}{5} = 4$$

We see that if we divide the product of two factors by either factor, the result is the other factor.

$$\frac{\text{product}}{\text{factor}_1} = \text{factor}_2 \qquad \frac{\text{product}}{\text{factor}_2} = \text{factor}_1$$

This relationship between multiplication and division breaks down when zero is one of the factors.

Example 2

The numbers in the multiplication fact $2 \times 4 = 8$ can be arranged to form two division facts.

$$\frac{8}{4} = 2 \qquad \frac{8}{2} = 4$$

If we attempt to form two division facts for the multiplication fact $2 \times 0 = 0$, one of the arrangements is not a fact. Which arrangement is not a fact?

Solution

The product is 0 and the factors are 2 and 0. So the possible arrangements are:

$$\text{Fact:} \qquad \text{Not a Fact:}$$
$$\frac{0}{2} = 0 \qquad \frac{0}{0} = 2$$

Division by 0 is not defined. The arrangement $0 \div 0 = 2$ **is not a fact.**

The multiplication fact $2 \times 0 = 0$ does not imply $0 \div 0 = 2$. Similarly, $3 \times 0 = 0$ does not imply $0 \div 0 = 3$. This breakdown in the inverse relationship between multiplication and division when zero is one of the factors is another indication that division by zero is not defined.

Example 3

If we were asked to graph the following equation, what number could we not use in place of x when generating a table of ordered pairs?

$$y = \frac{12}{x + 2}$$

Solution

This equation involves division. Since division by zero is not defined, we need to guard against the divisor, $x + 2$, being zero. When x is 0, the expression $x + 2$ equals 2, so we may use 0 in place of x. However, when x is -2, the expression $x + 2$ equals zero.

Step:	Justification:
$y = \dfrac{12}{x + 2}$	Equation
$y = \dfrac{12}{(-2) + 2}$	Replaced x with -2
$y = \dfrac{12}{0}$	Simplified (Not permitted)

We may not use -2 in place of x in this equation. Using symbols, we can write our restriction this way:

$$x \neq -2$$

Practice Set

a. Use a calculator to divide several different numbers of your choosing by zero. Remember to clear the calculator between calculations. What answers are displayed?

b. If we attempt to form two division facts for the multiplication fact $5 \times 0 = 0$, one of the arrangements is not a fact. Which arrangement is not a fact and why?

c. *Explain* Under what circumstances does the following expression not equal 1? Why?

$$\frac{x-1}{x-1}$$

For the following expressions, find the number or numbers that may not be used in place of the variable.

d. $\dfrac{5}{x}$

e. $\dfrac{2}{x-3}$

f. $\dfrac{6}{2w}$

g. $\dfrac{y+3}{2y-8}$

h. $\dfrac{16}{x^2-4}$

i. $\dfrac{3ab}{c}$

Written Practice　　*Strengthening Concepts*

1. There was a 30% increase in productivity when the teacher walked
(67) up and down the rows of students. If the productivity rate had been 50 before the increase, what was it after the increase.

* 2. *Analyze* Beau cut a paper cone along the slant height and unrolled the
(40, 100) cone. Answer **a** and **b** in terms of π.

a. What was the lateral surface area of the cone?

b. What was the circumference of the base of the cone?

12 cm
120°

3. Quentin had $1.10 in nickels and quarters. He had 10 nickels and quarters
(89) in all. Write equations to show this information. Use this information to help you find the number of quarters and nickels Quentin has.

Generalize Simplify problems **4–9.**

4. **a.** $\left(\dfrac{3}{2}\right)^{-2}$
(63)

b. $\left(\dfrac{3}{5}\right)^{-1}$

c. $(5)^{-2}$

d. $\left(\dfrac{1}{2}\right)^{-3}$

5. $\dfrac{24x^5b^{-1}}{18xb^4}$
(27, 51)

6. $\left(-\dfrac{3}{5}\right)^{0}$
(27)

7. $(5r^3)^2$
(27, 36)

8. $4x(2x^3+3)$
(36)

* 9. $(7+3m-5x)+(2x+3-3m)$
(80)

10. In the following expression, the value of x cannot be which numbers?
(116)

$$\frac{2y}{x(2 - x)}$$

11. Expand: $(x + 2)(x - 6)$
(92)

12. Where does the graph of $-5x + 2y = 30$ intercept the x and y-axes?
(82)

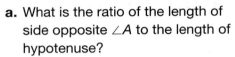 For problems **13** and **14** refer to the figure ABC on the coordinate plane.

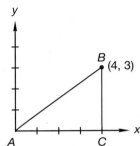

* 13. Express each answer as a decimal form.
(112)

    a. What is the ratio of the length of side opposite $\angle A$ to the length of hypotenuse?

    b. What is the ratio of the length of side adjacent $\angle A$ to the length of hypotenuse?

    c. What is the ratio of the length of side opposite $\angle A$ to the side adjacent?

* 14. Rotate $\triangle ABC$ $180°$ about the origin. What are the coordinates of the vertices of $\triangle A'B'C'$?
(Inv. 5)

* 15. **Formulate** Write expressions using a variable for the area and perimeter of a rectangle with length $2x + 7$ and width $2x - 7$.
(92)

* 16. **Model** Graph $y = x^2 - 4$.
(Inv. 11)

17. A line intercepts the y-axis at the point $(0, 8)$. From there the line moves up four units for every one unit it moves to the right. What is the equation of the line in slope-intercept form?
(56)

18. Solve: $-2x^2 + 5 = -3$
(93)

19. Solve: $\frac{1}{5} + \frac{1}{2}x = \frac{3}{20}$
(21)

20. Factor: $32x^2 - 8x + 24$
(21)

Solve.

21. $0.2 + 0.02x = 3$
(25, 50)

22. $4x > 6x + 6$
(77)

23. $3x - 7 = 5x + 3(x - 4)$
(50, 79)

24. The base of the triangle is a diameter of the circle. Find the area of the shaded region in terms of π. Units are cm.
(37, 40)

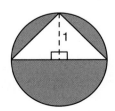

25. Eric's bill at a restaurant was $38.50, and he left a 20% tip. How much did he spend in all?
(67)

• Significant Digits

facts | Power Up X

mental math

a. Estimation: 18 × $307

b. Fractional Parts: $\frac{10}{9}$ of 36

c. Percent: Thirty percent less than a number is what percent of the number?

d. Proportion: $\frac{x}{18} = \frac{20}{12}$

e. Scientific Notation: Compare 0.007 ◯ 7×10^{-2}

f. Statistics: Which measure of central tendency would be used to find the equal share of a rental fee for a car used by three people?

g. Measurement: Find the mass of 7 mystery bags.

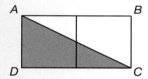

h. Calculation: 20 + 40, ÷ 2, + 9, + 30, + 1

problem solving | There are two squares in this diagram. The side length of each is 2 units. What is the area of the shaded triangle? What is the area of □ABCD? (Hint: First find the fraction of □ABCD that is shaded.)

The length of this segment is 8.9 centimeters.

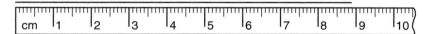

We use two digits to write the length. We use one decimal place because the ruler is divided into tenths. Using this ruler, we may be able to estimate the digit in the hundredths place of the length, but it would only be a rough guess. We cannot estimate the digit in the thousandths place.

We distinguish between counts and measures. We can count the exact number of students in the classroom or the exact number of books in the library. Measures, however, are not exact. A measure can be expressed only to a degree of precision.

Visit www.
SaxonPublishers.
com/ActivitiesC3
*for a graphing
calculator activity.*

To measure the segment in the beginning of the lesson, we compare its length to a standard scale and use a number of units to refer to its length. We are confident that the segment is about 9 cm long. We are also confident that the segment is about 8.9 cm long. However, we cannot say with confidence that the segment is 8.916 cm long, because our measuring tool does not measure with that type of precision. The number of digits we use to express a measure with confidence is called the number of **significant digits.** For example, the length of the segment above, 8.9 cm, has two significant digits.

Below are rules for counting significant digits.

1. Count all non-zero digits. (The number 2.54 has 3 significant digits.)

2. Count zeros between non-zero digits. (The number 10.7 has 3 significant digits.)

3. Count trailing zeros in a decimal number. (The decimal 1.20 has 3 significant digits.)

4. Do not count zeros if there is not a non-zero digit to the left. (The number 0.023 has 2 significant digits.)

5. Trailing zeros of whole numbers are ambiguous. The number of significant digits in 1200 might be 2, 3, or 4. Expressing a number in scientific notation removes the ambiguity. (The number 1.20×10^3 has 3 significant digits.)

6. Exact numbers have an unlimited number of significant figures, so we do not count any digits when dealing with exact numbers. Exact numbers include counts and exact conversion rates like 3 ft/yd and 2.54 cm/in.

Example 1

State the number of significant digits in a–d.

   a. The diameter of a quarter is about 2.5 cm.

   b. A ruler is about 30.5 cm long.

   c. Earth's average distance from the sun is about 9.3×10^7 miles.

   d. A mile is 5280 feet.

Solution

   a. two

   b. three

   c. two

   d. unlimited (exact)

When performing calculations with measures, scientists are careful not to include more digits in the results than are justified. We follow this rule to determine the correct number of significant digits when we multiply and divide.

> **A product or quotient of two measurements may only have as many significant digits as the measurement with the least number of significant digits.**

Example 2

If the segment at the beginning of the lesson is divided into 4 equal segments, what would be the length of each segment?

Solution

We divide 8.9 into four equal parts.

$$8.9 \div 4 = 2.225$$

The number 4 is an exact number and does not affect the count of significant digits. The measure 8.9 cm has two significant digits, so the quotient should have no more than two significant digits. We round the quotient to two significant digits. Each small segment is **2.2 cm long.**

Example 3

What is the volume of a cube with edges 1.2 mm long?

Solution

Thinking Skill

Explain

Why is 1.728 mm^3 reduced to 1.7 mm^3?

We calculate the volume by finding the cube of the length of the edge.

$$V = (1.2 \text{ mm})^3$$
$$= 1.728 \text{ mm}^3$$

We report the volumes as **1.7 mm^3**

We follow this rule to determine the correct number of significant digits when we add and subtract.

> **A sum or difference may show no more decimal places than the measurement with the fewest number of decimal places.**

Example 4

The test tube contained 6.24 mL of liquid before it was heated and 4.6 mL after it was heated. How many mL of liquid evaporated when the test tube was heated?

We subtract 4.6 from 6.24

$$\begin{array}{r} 6.2 \,|\, 4 \\ -\ 4.6 \,| \\ \hline 1.6 \,|\, 4 \end{array}$$

Since 4.6 mL has only one decimal place, the difference should have only one decimal place. Thus **1.6 mL** of liquid evaporated.

Practice Set

Write the number of significant digits in the following numbers.

a. 10.68 sec

b. 0.063 L

c. 4.20 kg

d. 1.609×10^3 m

Perform each calculation and express each answer with the correct number of significant digits.

e. 3.6 cm · 4.2 cm

f. $\dfrac{16.4 \text{ m}}{1.2 \text{ sec}}$

g. 1.24 g + 6.4 g + 5.1 g

h. 0.249 L − 0.12 L

Written Practice   *Strengthening Concepts*

1. Sales were up by 30% over last year. If last year's sales totaled $2500, what were this year's sales?
(67)

2. At the rented cabin, the Bunyan family decided to watch movies. The local video store charges $15 for a DVD Player rental, plus $4 per DVD. If the family spends $27 how many DVDs did they rent?
(3, 4)

* 3. **Classify** State whether the following relationships are examples of direct variation, inverse variation, or neither.
(69, 99)

a. The number of workers painting a building and the hours required to paint the building

b. The surface area to be painted and the quantity of paint needed to paint the building

c. The hourly pay rate of the painters and the cost per gallon of paint

* 4. **Analyze** The contractor bought 10 gallons of paint from one supplier at $18.00 per gallon and 6 gallons of paint from another supplier at $20.00 per gallon. What was the average price per gallon for all 16 gallons?
(105)

5. Factor: **a.** $9a^2 - 12a$ **b.** $8x^3 + 12x^2$
(21, 36)

6. Simplify: **a.** $(4)^{-2}$ **b.** $\left(\dfrac{1}{3}\right)^{-3}$ **c.** $\left(\dfrac{2}{5}\right)^{-2}$
(51, 63)

7. $\left(-\dfrac{1}{9}\right)^0$
(27)

8. $(3x^5)^3$
(27)

9. Expand: $7m^2(3m + 5)$
(21)

10. Simplify: $(6x + 5m + 3) + (3m + 1 + 4x)$
(80)

* **11.** An inflated playground ball is about 8 inches in diameter. What is the
(111)   surface area of the ball rounded to the nearest ten square inches?

* **12.** What is the volume of air in the playground ball described in problem **11**
(111)   to the nearest ten cubic inches?

13. *Formulate* Write expressions using variables for the area and perimeter
(92)   of a rectangle with length $2x + 9$ and width $x + 7$.

* **14.** *Evaluate* Refer to the
(112)   diagram of the roof for
   problems **a** and **b**. Express
   each answer as a decimal.
   Round repeating decimals to
   the thousandth place.

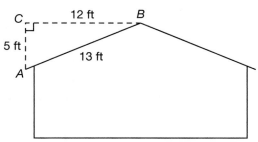

   a. What is the ratio of the
      side opposite $\angle A$ to the
      side adjacent $\angle A$?

   b. What is the ratio of the side opposite $\angle A$ to the hypotenuse?

* **15.** *Conclude* In the following expression, what number cannot be
(116)   substituted for x?
$$\frac{x + 4}{2x - 6}$$

16. Graph $y = -x^2$
(96)

17. Write the equation of a line with a slope of $\frac{2}{3}$ and a y-intercept of $\frac{19}{3}$.
(56)

18. Find the mean, median, and mode of these numbers:
(7)   98.2, 98.6, 99.0, 98.6, 98.6.

Solve.

19. $-3x^2 + 50 = -25$
(73)

20. $\frac{4}{5} + \frac{1}{3}x = \frac{23}{30}$
(48)

21. $\frac{2.4}{4.4} = \frac{x}{33}$
(25, 44)

22. $3x < 5x - 8$
(77)

Expand.

23. $(x + 3)(x - 1)$
(92)

24. $(2x - 5)^2$
(92)

25. Suppose you could buy ice cream in either a cylindrical cup 3 inches
(86)   high or a cone 6 inches high. If the cup and the cone have the same
   radius, then which could hold more ice cream? Explain.

• Sine, Cosine, Tangent

facts | Power Up X

mental math |
a. **Algebra:** Solve: $-2s + 5 = 17$

b. **Estimation:** $2019 \div 51$

c. **Number Sense:** Can a perfect square have a 7 in the ones place?

d. **Powers/Roots:** $\frac{(m^4)^2}{m^2}$

e. **Rate:** Twenty dollars per month is how many dollars per year?

f. **Statistics:** On the table were three stacks of math books of varying heights. Shannon wrote down the number in each stack and rearranged the books so that the stacks were the same height. The number in each stack now represents the
A mean.
B median.
C mode.
D range.

g. **Geometry:** An equilateral quadrilateral with equal angles is also known as a ___.

h. **Calculation:** $70 - 1, \div 3, + 2, \sqrt{}, \times 10$

problem solving | In the land of Infinitum, there is a road that goes on and on forever. How long would it take you to walk half the length of the road? How long would it take you to walk 1% of the length of the road? Does it matter how fast you walk?

In Lesson 112, we practiced finding the ratios of lengths of sides of right triangles. These ratios have special names.

Trigonometry Ratios

Name	Ratio
sine	$\dfrac{\text{opposite leg}}{\text{hypotenuse}}$
cosine	$\dfrac{\text{adjacent leg}}{\text{hypotenuse}}$
tangent	$\dfrac{\text{opposite leg}}{\text{adjacent leg}}$

Example 1

Find the sine, cosine, and tangent of ∠A.

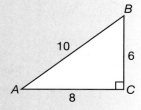

Solution

Math Language

"Trig" is short for trigonometry, which is a branch of mathematics dealing with triangles. It comes from the greek words *tri* (meaning "three") and *meter* (meaning "measure").

From the perspective of ∠A, the opposite leg is $\overline{BC}$ and the adjacent leg is $\overline{AC}$.

$$\text{sine } \angle A = \frac{\text{opposite leg}}{\text{hypotenuse}} = \frac{6}{10} = \textbf{0.6}$$

$$\text{cosine } \angle A = \frac{\text{adjacent leg}}{\text{hypotenuse}} = \frac{8}{10} = \textbf{0.8}$$

$$\text{tangent } \angle A = \frac{\text{opposite leg}}{\text{adjacent leg}} = \frac{6}{8} = \textbf{0.75}$$

Notice that we express each ratio as a decimal. The "trig" ratios for this angle happen to be rational numbers. Since many right triangles involve irrational numbers, the trig ratios for many angles are irrational. We express the irrational ratios rounded to a selected number of decimal places.

Example 2

Find the sine, cosine, and tangent of ∠R.

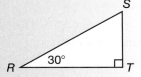

Solution

We recognize the triangle as a 30-60-90 triangle. We recall that all 30-60-90 triangles have lengths of sides in the ratio $1:\sqrt{3}:2$.

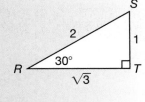

$$\text{sine } 30° = \frac{1}{2} = \textbf{0.5}$$

$$\text{cosine } 30° = \frac{\sqrt{3}}{2} \approx \textbf{0.866}$$

$$\text{tangent } 30° = \frac{1}{\sqrt{3}} \approx \textbf{0.577}$$

We can find the decimal values of trig ratios on a calculator. For the ratios in example 2, a calculator with a square root key is sufficient. For other angles we can use the trig functions on a scientific or graphing calculator, which are often abbreviated with three letters.

Name	Abbreviation
sine	sin
cosine	cos
tangent	tan

If you have a calculator with these three keys, practice using them for the triangle in Example 2.

To find the sine of a 30° angle use this key sequence:

The display should read 0.5. Use a similar key sequence to find the cosine and tangent of a 30° angle. Compare the results to the solutions in Example 2.

Visit www. SaxonPublishers. com/ActivitiesC3 *for a graphing calculator activity.*

Example 3

Use a calculator to find the sine and cosine of ∠A.

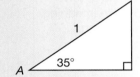

Solution

We enter 35 and press ⓢⒾⓃ. The display fills with digits. Rounded to the nearest thousandth, **the sine of ∠A is 0.574.** Following a similar process, we find **the cosine of ∠A is about 0.819.**

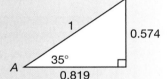

$$\text{sine } 35° = \frac{\text{opposite leg}}{\text{hypotenuse}} = \frac{0.574}{1}$$

$$\text{cosine } 35° = \frac{\text{adjacent leg}}{\text{hypotenuse}} = \frac{0.819}{1}$$

Thinking Skill

Evaluate

What would be the lengths of the legs of a 35°-55°-90° triangle whose hypotenuse measures 5 cm?

Trig ratios can help us calculate measures we cannot perform directly. Suppose we want to find the height of a tree. If the angle of elevation from 100 ft away is 22°, we can find the height of the tree using the tangent function.

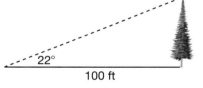

$$\text{tangent } 22° = \frac{\text{opposite leg}}{\text{adjacent leg}} = \frac{\text{tree height}}{100 \text{ ft}} = \frac{0.404}{1}$$

We let t stand for the height of the tree and solve the proportion.

$$\frac{t}{100} = \frac{0.404}{1}$$

$$1t = 40.4$$

The height of the tree is about 40.4 ft.

Practice Set

Classify Name the trig function for each ratio.

a. $\dfrac{\text{opposite leg}}{\text{adjacent leg}}$ **b.** $\dfrac{\text{opposite leg}}{\text{hypotenuse}}$ **c.** $\dfrac{\text{adjacent leg}}{\text{hypotenuse}}$

For **d–f** refer to this triangle. Write each ratio in fraction form.

d. What is the sine of ∠B?

e. What is the cosine of ∠B?

f. What is the tangent of ∠B?

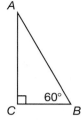

Use a calculator with trig functions to find the following. Round to the nearest thousandth.

g. What is the sine of ∠A?

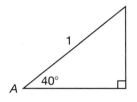

h. What is the tangent of ∠B?

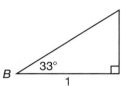

i. The height of the kite is the side opposite the 42° angle. The known side (100 ft) is the side adjacent. Find the height of the kite.

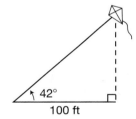

Written Practice   *Strengthening Concepts*

1. The number of parrots in the neighborhood increased by 90% over the
(67) last year's number. If there are 3800 parrots this year, how many parrots were there last year?

2. Gerry and Kay were 438 miles apart and decided to meet part way.
(105) They started driving toward each other at the same time. Gerry drove 60 miles per hour. After 4 hours, they met. What was Kay's average speed?

3. How many hours, minutes, and seconds is it from 9:42:30 a.m. to
(80) 1:17:15 p.m.?

4. The balance on the loan is $1000. The monthly interest rate is 1.5%.
(109) The minimum payment for the month is $20.

   a. What is the interest paid this month?

   b. What is the principal paid this month?

   c. What is the new balance?

Simplify.

5. $(3x^4y^5)(-2x^2y^4)$
(27, 36)

6. $(5x^3m)(-5x^3m)$
(27, 36)

7. $5x^2(3x + 2)$
(21, 36)

8. $(4x^5)^2$
(27)

9. $(2x + 3y - 7) + (x - 3y + 2)$
(80)

10. **a.** $(3)^{-3}$
(51, 63)

b. $\left(\dfrac{1}{5}\right)^{-2}$

11. **a.** $\left(-\dfrac{2}{5}\right)^{0}$
(27, 63)

b. $\left(\dfrac{3}{7}\right)^{-2}$

Analyze For **12–14** refer to this illustration. The illustration shows a platform. The front and back are parallel congruent trapezoids. The top and bottom are parallel rectangles.

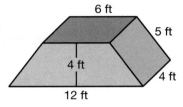

12. Draw the front top and right side views.
(Inv.4)

* **13.** Find the volume of the platform.
(107)

* **14.** **a.** The six faces of the platform are covered with plywood. How many square feet of plywood are used to cover the platform?
(107)

b. Plywood is available in 4 ft by 8 ft sheets. What is the minimum number of sheets needed to cover the platform?

15. Write expressions using variables for the area and perimeter of a rectangle with length $2x - 1$ and width $x + 2$.
(92)

* **16.** Represent Bryan drew a line with a slope of positive one passing through the point $(0, -6)$. What is the equation of Bryan's line in slope-intercept form?
(56)

17. Factor using the greatest common factor: $8x^3 + 4x^2 + 12x$
(21)

18. For questions **a–c** use 1.732 for $\sqrt{3}$ if it is involved in the calculations.
(118)

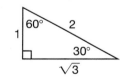

a. What is the sine of 60°?

b. What is the tangent of 60°?

c. What is the cosine of 60°?

Solve.

19. $5x^2 + 3 = 48$
(93)

20. $\dfrac{1}{2} + \dfrac{1}{6}x = \dfrac{1}{3}$
(48)

21. $0.2 - 0.03x = 2$
(41)

22. $2x + 3 = 5x + 6$
(50, 57)

23. Give the x and y intercepts, then graph $-5x + y = 10$.
(82)

24. Graph $y = 3x^2$.
(Inv.11)

25. Which of the following is equivalent to 7^3?
(15, 27)

A $7 + 7 + 7$ **B** $(4 + 3)^3$ **C** $7^5 - 7^2$ **D** $7^2 + 7$

• Complex Fractions

facts | Power Up X

mental math

a. Algebra: Solve: $-6r + 5 = 41$

b. Fractional Parts: $\frac{3}{10}$ of 900

c. Percent: Forty percent less than a number is what percent of the number?

d. Probability: The multiple-choice question has options A, B, C, D, and E. Find the probability of selecting the correct answer at random.

e. Ratio: Twenty-four students attended an aerobics class. Three students lazed about while the rest learned a new workout. What was the ratio of the number of students who worked out to the number of students who lazed about?

f. Select a Method: To find the approximate number of math problems you have solved in this course, you would probably use
A mental math.
B mental estimation.
C pencil and paper.
D a calculator.

g. Measurement: Find the volume of liquid in this graduated cylinder.

h. Calculation: $50 \times 10, - 10, \div 7, \div 7, \times 6$

problem solving

Which two of these statements are possible? Give examples of both.
A The product of two primes can be a prime number.
B The sum of two prime numbers can be a prime number.
C The quotient of two prime numbers can be a prime number.
D The difference of two prime numbers can be a prime number.

A **complex fraction** is a fraction containing one or more fractions in the numerator or denominator. Each of the following is a complex fraction:

$$\frac{\frac{1}{5}}{\frac{2}{3}} \qquad \frac{16\frac{2}{3}}{100} \qquad \frac{15}{6\frac{2}{3}}$$

It is customary to express the numerator and denominator of a fraction with integers when possible. We will describe two methods for simplifying a complex fraction. One method is to use the identity property of multiplication (multiply the numerator and denominator of the fraction by a common denominator).

Example 1

Simplify: $\dfrac{\frac{1}{5}}{\frac{2}{3}}$

Solution

Thinking Skill

Analyze

What is the result of multiplying the numerator and denominator by the common denominator?

The complex fraction is composed of the fractions $\frac{1}{5}$ and $\frac{2}{3}$ which have a common denominator of 15. We multiply the complex fraction by $\frac{15}{15}$. We perform the multiplication in the numerator and the multiplication in the denominator separately.

$$\frac{\overset{3}{\cancel{15}}}{\underset{5}{\cancel{15}}} \cdot \frac{\overset{1}{\cancel{\frac{1}{5}}_{1}}}{\underset{\cancel{\frac{2}{3}}_{1}}{}} = \frac{3}{10}$$

Notice that we multiplied the complex fraction by a fraction equivalent to 1, so the resulting fraction is equal to the original complex fraction.

An alternative method for simplifying some complex fractions is to treat the complex fraction as a fraction division problem. We can express the division problem with a division sign instead of a fraction bar.

$$\begin{array}{l}\textbf{dividend} \\ \textbf{divisor}\end{array} \quad \frac{\frac{1}{5}}{\frac{2}{3}} \rightarrow \frac{1}{5} \div \frac{2}{3}$$

We simplify the division using the method described in Lesson 22.

$$\frac{1}{5} \div \frac{2}{3} = \frac{1}{5} \cdot \frac{3}{2} = \frac{3}{10}$$

Example 2

Simplify: $\dfrac{\frac{1}{2} + \frac{1}{3}}{\frac{1}{3} + \frac{1}{4}}$

Solution

We cannot cancel because the fractions in the numerator and denominator are added, not multiplied. First we simplify the numerator and denominator by adding the fractions.

$$\frac{\frac{1}{2} + \frac{1}{3}}{\frac{1}{3} + \frac{1}{4}} = \frac{\frac{5}{6}}{\frac{7}{12}}$$

Now we can express the complex fraction as a division of fractions.

$$\frac{5}{6} \div \frac{7}{12}$$

$$\frac{5}{6} \cdot \frac{12}{7} = \mathbf{\frac{10}{7}}$$

We could have used the method of multiplying by the common denominator (12) even before adding.

$$\frac{12}{12} \cdot \frac{\left(\frac{1}{2} + \frac{1}{3}\right)}{\left(\frac{1}{3} + \frac{1}{4}\right)} \qquad \text{Multiplied by } \frac{12}{12}$$

$$\frac{{}^{6}\cancel{12}\left(\frac{1}{2}\right) + {}^{4}\cancel{12}\left(\frac{1}{3}\right)}{{}^{4}\cancel{12}\left(\frac{1}{3}\right) + {}^{3}\cancel{12}\left(\frac{1}{4}\right)} \qquad \text{Distributive property}$$

$$\frac{6 + 4}{4 + 3} \qquad \text{Simplified}$$

$$\mathbf{\frac{10}{7}} \qquad \text{Simplified}$$

Example 3

Change $28\frac{4}{7}\%$ to a fraction and simplify.

Solution

A percent is a fraction that has a denominator of 100. Thus $28\frac{4}{7}\%$ is

$$\frac{28\frac{4}{7}}{100}$$

Next, we write both the numerator and denominator as fractions.

$$\frac{\frac{200}{7}}{\frac{100}{1}}$$

Now we multiply the complex fraction by $\frac{7}{7}$.

$$\frac{{}^{1}7\left(\frac{200}{7}\right)}{7\left(\frac{100}{1}\right)} = \frac{200}{700} = \mathbf{\frac{2}{7}}$$

Practice Set

Simplify.

a. $\dfrac{\frac{2}{5}}{\frac{2}{3}}$

b. $\dfrac{30}{\frac{5}{6}}$

c. $\dfrac{\frac{1}{2} - \frac{1}{3}}{\frac{1}{2} + \frac{1}{4}}$

Change each percent to a fraction and simplify:

d. $14\frac{2}{7}\%$

e. $8\frac{1}{3}\%$

f. $4\frac{1}{6}\%$

1. There were 120 employees. Now there are 168. What was the percent
(67) increase?

2. The monthly payment on a $300,000 30 year mortgage at 6% annual
(109) interest is $1798.65.

    a. What is the monthly interest rate?

    b. What is the interest payment for the first month?

    c. How much of the principal do you pay in the first month?

    d. Estimate how much of the principal is left after the first payment.

3. Freddy and Elizabeth went for a run on the trail. They started at the
(95) same time and ran in the same direction. Elizabeth ran 9 miles per hour.
After $1\frac{1}{2}$ hours, they were 3 miles apart. How fast was Freddy running?

4. **Connect** A square pyramid is carved
(86) from a cube of wood with each edge
6 inches long.

    a. What is the volume of the cube?

    b. What is the volume of the pyramid?

    c. What is the fraction of the cube carved away to make the pyramid?

* 5. Calculate the slant height of the pyramid in problem **4.** Then find the
(95, 100) lateral surface area of the pyramid.

6. Factor using the greatest common factor:
(21)
$$6x^3 + 3x^2 + 9x$$

Generalize Simplify.

7. $(4x^5y)(-2x)$
(27, 36)

8. $(-2m^2b^5)(-m^3)$
(27, 36)

9. $-2x^3(x + 1)$
(21, 36)

10. $(4x^5)^3$
(27)

* 11. **a.** $\dfrac{1\frac{1}{2}}{1\frac{3}{4}}$
(119)

b. $\dfrac{16\frac{2}{3}}{100}$

12. **a.** $\left(\dfrac{1}{9}\right)^0$
(27, 63)

b. $\dfrac{6x^5m^2}{9x^5m}$

13. $(1 - 2x + 3m) + (4m + 5x - 6)$
(80)

14. Expand: $(3m + 1)^2$
(92)

15. *Analyze* For **a–c** use 1.414 for $\sqrt{2}$ if it is
(118) involved in the calculation. (You may use a
calculator).

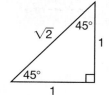

a. What is the tangent of 45°?

b. What is the sine of 45°?

c. What is the cosine of 45°?

16. Write expressions using variables for the area and perimeter of the
(92) rectangle with length $7w + 9$ and width w.

Solve.

17. $-5x^2 + 9 = -36$
(93)

18. $3(x - 1) = 4x - 9$
(50, 79)

19. $\dfrac{3}{7}x - \dfrac{1}{14} = \dfrac{1}{2}$
(48)

20. $\dfrac{3.5}{1.4} = \dfrac{x}{4}$
(25, 44)

21. Write equation of the line which includes points $(7, -14)$ and $(0, 7)$.
(56)

22. Find the pair of x, y values that makes both equations true.
(82)

$$\begin{cases} 3x - 4y = 12 \\ x + y = 4 \end{cases}$$

23. Find the perimeter and area of square $ABCD$ with vertices at $A\ (3, 0)$,
(96) $B\ (4, -3)$, $C\ (1, -4)$ and $D\ (0, -1)$.

24. If square $ABCD$ in problem **23** is translated so that the coordinates of
(Inv. 5) A' are $(0, 3)$, then what are the coordinates of B', C', and D'?

25. A pendulum swings 30 degrees to the right and to the left of vertical.
(81) If the pendulum is 9 feet long, how long is the arc it makes? Explain.

• Rationalizing a Denominator

facts | Power Up X

mental math

a. **Estimation:** $586 - 119 + 408$

b. **Number Sense:** Which of these numbers cannot be in the ones place of a perfect square?

    A 4   **B** 6   **C** 8   **D** 0

c. **Powers/Roots:** $\frac{(m^5)^3}{m^{10}}$

d. **Proportion:** $\frac{15}{x} = \frac{10}{2}$

e. **Scientific Notation:** Compare $4360 \bigcirc 4.36 \times 10^3$

f. **Statistics:** What would you use to find how much of a difference there is between the hottest and coldest temperatures in a certain city: range, median, mode, or mean?

g. **Geometry:** $\triangle ABC \cong \triangle DEF$. Therefore, $\overline{AB} \cong$ __.

h. **Calculation:** $60 - 5, \times 2, + 10, + 13, - 13$

problem solving

If a pulsar spinning 100 times per second is 10 miles in diameter, how many miles does point A travel in one second? (Hint: What is the circumference of the sphere? How many times does point A circle the sphere in one second?) Use $\pi = 3.14$.

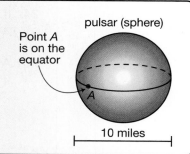

pulsar (sphere)

Point A is on the equator

10 miles

There are customs we follow in mathematics, such as reducing fractions to lowest terms and writing common fractions with integers. Another custom we follow is expressing fractions with rational denominators. Specifically, if a fraction's denominator contains a radical, we take steps to form an equivalent fraction that does not. (Note: Radicals in the numerator are acceptable.)

Recall that we rename a fraction by multiplying it by a fraction equivalent to 1. In the following example we multiply the given fraction by the fraction $\frac{\sqrt{2}}{\sqrt{2}}$.

Example 1

An isosceles right triangle with a hypotenuse of 1 has legs of the length shown. Express the length with a rational denominator.

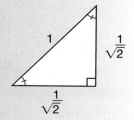

Solution

Thinking Skill

Formulate

Use a calculator to find the approximate number for $\frac{1}{\sqrt{2}}$ and for $\frac{\sqrt{2}}{2}$. Round each answer to three decimal places.

We use the identity property of multiplication to multiply the fraction by a fraction equivalent to 1. Since $\sqrt{2} \cdot \sqrt{2} = 2$, we choose to multiply by the fraction $\frac{\sqrt{2}}{\sqrt{2}}$.

$$\frac{1}{\sqrt{2}} \cdot \frac{\sqrt{2}}{\sqrt{2}} \qquad \text{Identity property of multiplication}$$

$$\frac{\sqrt{2}}{2} \qquad \text{Multiplied}$$

Example 2

Rationalize the denominator of each fraction.

   a. $\dfrac{3}{\sqrt{5}}$ **b.** $\dfrac{6}{\sqrt{3}}$ **c.** $\dfrac{\sqrt{3}}{\sqrt{2}}$

Solution

a. We multiply the fraction by $\frac{\sqrt{5}}{\sqrt{5}}$.

$$\frac{3}{\sqrt{5}} \cdot \frac{\sqrt{5}}{\sqrt{5}} \qquad \text{Identity property of multiplication}$$

$$\frac{3\sqrt{5}}{5} \qquad \text{Rationalized denominator}$$

b. We rationalize the denominator and then reduce the fraction.

$$\frac{6}{\sqrt{3}} \cdot \frac{\sqrt{3}}{\sqrt{3}} \qquad \text{Identity property of multiplication}$$

$$\frac{6\sqrt{3}}{3} \qquad \text{Rationalized denominator}$$

$$2\sqrt{3} \qquad \text{Reduced}$$

c. We multiply by $\frac{\sqrt{2}}{\sqrt{2}}$.

$$\frac{\sqrt{3}}{\sqrt{2}} \cdot \frac{\sqrt{2}}{\sqrt{2}} = \frac{\sqrt{6}}{2} \qquad \text{Rationalized denominator}$$

Practice Set

Rationalize the denominator of each fraction in **a–h.**

   a. $\dfrac{3}{\sqrt{2}}$ **b.** $\dfrac{4}{\sqrt{2}}$

   c. $\dfrac{1}{\sqrt{3}}$ **d.** $\dfrac{2}{\sqrt{3}}$

e. $\dfrac{8}{\sqrt{8}}$

f. $\dfrac{5}{\sqrt{10}}$

g. $\dfrac{\sqrt{2}}{\sqrt{3}}$

h. $\dfrac{2\sqrt{3}}{\sqrt{2}}$

Written Practice   Strengthening Concepts

1. Accidents were down 44% from last year. This year there were
(67) 140 accidents. This is how many fewer than last year?

2. A car sold for $20,000 is financed with a 3 year loan at a 6% annual
(109) interest rate. The payment for the first month is $608.44.

  a. What is the monthly interest rate?

  b. What is the interest payment for the first month?

  c. How much of the principal is paid the first month?

  d. Estimate how much of the principal remains to be paid after the first payment.

* 3. (Analyze) Mariya and Irina folded laundry. Mariya folded 10 items per
(105) minute. After 8 minutes, they had folded 164 items. How many items per minute did Irina fold?

(Generalize) Simplify.

4. $\dfrac{x^{-5}m^3}{m^2z^{-3}}$
(51)

5. $\dfrac{4x^2r^{-2}}{2x^{-1}r^3}$
(51)

6. $3x^2y(-2y^3)$
(27, 36)

7. $-2y^3(3y + 1)$
(21, 36)

8. **a.** $(4)^{-1}$
(27, 51)

  b. $(5r^4)^3$

9. **a.** $\left(\dfrac{1}{5}\right)^0$
(27, 63)

  b. $\left(\dfrac{2}{5}\right)^{-2}$

10. $5 - 4x - (2x - 1)$
(31, 36)

* 11. **a.** $\dfrac{\frac{3}{4}}{6}$
(119)

  b. $\dfrac{3\frac{1}{3}}{100}$

* 12. **a.** $\dfrac{1}{\sqrt{2}}$
(120)

  b. $\dfrac{1}{\sqrt{3}}$

* 13. Approximate $\dfrac{1}{\sqrt{2}}$ as a decimal. Use 1.414 for $\sqrt{2}$.
(120)

14. Factor using the greatest common factor.
(21)
$$10x^3 - 8x^2 - 6x$$

15. Expand: $(2x + 9)^2$
(63)

16. Solve: $4x^2 - 9 = 91$
(93)

17. $0.2x + 0.8 = 5$
(25, 50)

18. Solve: $\dfrac{1}{8}x + \dfrac{3}{4} = 2$
(23, 50)

19. $-2(x + 3) = -x$
(50, 79)

20. Write a variable expression for the area and perimeter of a rectangle
(92) with length x and width $2x - 13$.

21. What is the slope-intercept equation of a line with a slope of negative
(56) one-half and a *y*-intercept of positive two?

* 22. **Formulate** Express the sine of a 30° angle as
(118) a fraction and as a decimal.

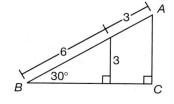

* 23. **Analyze** Find the length of segment $\overline{AC}$.
(35, 118) (Units are inches.)

24. Find the area of the shaded region of the
(40) figure to the right in terms of π. (Hint: the
unshaded part is one fourth of a circle.) The
radius of this circle is 2 cm.

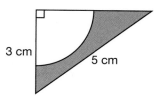

25. If a driver's average reaction time to perceive danger and to react to
(64, 72) it is about 0.75 seconds and if the driver's car is traveling at 60 miles
per hour, then how many feet does the car travel before the driver
begins to react when he perceives danger?

Early Finishers
Real-World
Application

Lisa wants to build a scale model of Manhattan, a borough of New York
City, for a class geography project. Manhattan is an island approximately
13 miles long and 2.3 miles wide. If Lisa plans on using a scale factor of
1 inch to .25 mile, what will be the approximate area (in square feet) of
Lisa's scale model?

Focus on
• Proof of the Pythagorean Theorem

When mathematicians want to demonstrate that a particular idea is true, they construct a **proof.** Following logical steps, a proof describes how certain given information leads to a certain conclusion. Indeed, virtually all of the structure of mathematics is built upon conclusions reached by a proof. One extremely important and useful conclusion about our world is that in a right triangle the lengths of the legs (a and b) and the length of the hypotenuse (c) are related in the following way:

$$a^2 + b^2 = c^2$$

Recall that this conclusion is called the Pythagorean Theorem. Mathematicians have constructed literally hundreds of proofs of the Pythagorean Theorem. In fact, one of America's presidents, James A. Garfield, is credited with providing an original proof of this theorem. In this investigation we will develop one of the many proofs of the Pythagorean Theorem. Work through the exercises in this investigation as a whole class with the guidance of your teacher.

The following proof is based upon the characteristics of similar triangles. Recall that the corresponding angles of similar triangles are congruent and that the lengths of their corresponding sides are proportional.

1. Begin by sketching a right triangle. Name the vertices *A, B,* and *C,* with $\angle C$ being the right angle. This triangle is a generic right triangle, so the measures of the acute angles do not affect the outcome of the proof.

2. It is customary to refer to the lengths of the sides of a right triangle by the lowercase form of the letter of the opposite vertex. So along segment *AB* write a lower case *c,* along segment *BC* write a lowercase *a,* and along segment *CA* write a lowercase *b.* Remember these lowercase letters refer to the lengths of the sides. Since this is a generic right triangle, we do not know the lengths of the sides. However, it is not necessary to know the lengths of the sides in order to establish the relationship among the sides.

3. Next draw a segment from vertex *C* across the triangle to $\overline{AB}$ so that the segment is perpendicular to $\overline{AB}$. Name the point where the segment intersects $\overline{AB}$ point *D.*

4. Point *D* divides $\overline{AB}$ into two shorter segments. We have already labeled the distance from *A* to *B* as *c.* Now label the distance from *B* to *D* as *x.* The distance from *D* to *A* is the rest of *c,* which can be found by subtracting *x* from *c.* So label the length of $\overline{AD}$ as $c - x$.

Check your drawing with the following description: If you have performed steps 1–4 correctly, you should have drawn a right triangle ABC and divided the right triangle into two smaller right triangles named $\triangle BCD$ and $\triangle ACD$. The hypotenuse and one leg of $\triangle BCD$ should be labeled a and x. The hypotenuse and one leg of $\triangle ACD$ should be labeled b and $c - x$. The hypotenuse of $\triangle ABC$ should be labeled c and is equal to length x plus length $c - x$. (Note that you may need to position c on your drawing to indicate that it represents the entire length of $\overline{AB}$.)

We use this figure and what we know about similar triangles to prove the Pythagorean Theorem. Before proceeding with the algebraic proof, we need to be convinced that the three triangles in the figure are similar. If their corresponding angles are congruent, then the triangles are similar.

5. What is the sum of the measures of the two acute angles of a right triangle? Why?

6. If $\angle B$ of $\triangle ABC$ measures m degrees, then how many degrees is the measure of $\angle A$?

7. If the number of degrees in the measure of $\angle B$ is m and the number of degrees in the measure of $\angle A$ is $90° - m$, then how many degrees are in the measure of

    a. $\angle BCD$?   b. $\angle ACD$?

8. Can we conclude that all three triangles in the figure are similar? Why?

Since similar triangles have sides whose measures are proportional, we can write proportions that relate the lengths of the sides of the triangles. Because we are referring to the lengths of the sides, we will use the lowercase letters on the diagram: a, b, c, x, and $c - x$.

Recall that we began with the largest triangle, $\triangle ABC$, and that we divided this triangle into two smaller triangles. We will write two proportions that each relate the largest triangle to one of the smaller triangles.

9. Write a proportion that relates the hypotenuse of $\triangle ABC$ and $\triangle BCD$ to corresponding legs of $\triangle ABC$ and $\triangle BCD$.

$$\begin{array}{cc} & hyp \quad leg \\ \triangle ABC & \dfrac{c}{a} = \dfrac{\square}{\square} \\ \triangle BCD & \end{array}$$

Thinking Skill

Explain

Explain how to find the cross products of a proportion.

10. Write a proportion that relates the hypotenuses of $\triangle ABC$ and $\triangle ACD$ to corresponding legs of $\triangle ABC$ and $\triangle ACD$.

$$\begin{array}{cc} & hyp \quad leg \\ \triangle ABC & \dfrac{c}{b} = \dfrac{\square}{\square} \\ \triangle ACD & \end{array}$$

11. Cross multiply the proportions you wrote in problems 9 and 10.

12. If you have finished cross multiplying the proportion from problems 9 and 10, you should see the term cx in both cross products. (Check with your teacher if you do not see cx in both cross products.) In the first cross product we see that cx equals a^2. This means that we can replace cx with a^2 in the second cross product. Replace cx with a^2 in the second cross product and write the result.

13. Transform the equation you gave for your answer to problem 12 by solving for c^2. What do we call this equation?

This completes an algebraic proof of the Pythagorean Theorem for all right triangles.

Some proofs of the Pythagorean Theorem involve the dividing up and rearranging areas of squares built on the sides of a right triangle. These proofs are based on the concept that a^2, b^2, and c^2 represent areas of squares, as shown below.

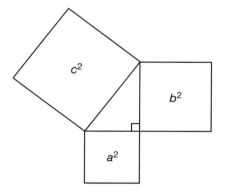

This algebra appendix provides the opportunity to augment Course 3 with additional algebra topics. It contains supplements to selected lessons and written practice sets, as outlined in the table of contents. **Each supplementary lesson is taught in addition to the core lesson containing the same lesson number.**

Algebra appendix lessons A61–A99 provide thorough preparation in foundational algebra skills. Use the supplemental written practice problems numbered between 26 and 30 to reinforce these lessons.

For students who wish to continue their study of algebra, lessons A101–A120 cover more advanced algebra topics. Use the written practice problems numbered between 31 and 35 to reinforce these lessons.

LESSON A61

New Concept

Simplifying Equations with Decimals

We may use the Distributive Property and a power of 10 to convert decimal numbers to whole numbers in an equation. Recall that powers of 10 are 10, 100, 1000, and so on.

Example 1

Solve: $-0.2 + 0.03x = 0.004$

Solution

First we find the number in the equation with the most places after the decimal point. In this equation it is 0.004, which has three decimal places. We know that multiplying a number by 1000 shifts the decimal point three places to the right, so we multiply both sides of the equation by 1000. To multiply more than one term, we apply the Distributive Property.

Step:	Justification:
$-0.2 + 0.03x = 0.004$	Given
$1000(-0.2 + 0.03x) = 1000(0.004)$	Multiplied both sides by 1000
$-200 + 30x = 4$	Simplified
$30x = 204$	Added 200 to both sides
$x = \dfrac{204}{30}$	Divided both sides by 30
$x = 6.8$	Simplified

Multiplying both sides of the equation by 1000 created an equivalent equation that we could solve using whole numbers. To check the solution we substitute 6.8 for x in the original equation and simplify.

Step:	Justification:
$-0.2 + 0.03(6.8) = 0.004$	Substituted
$-0.2 + 0.204 = 0.004$	Multiplied
$\checkmark 0.004 = 0.004$	Added

A ratio in a proportion may be changed to an equivalent ratio without changing the truth of the proportion.

Connect How do we change a ratio with decimals to an equivalent ratio with whole numbers?

Example 2

Solve: $\dfrac{5.1}{0.3} = \dfrac{34}{x}$

Solution

Since both 5.1 and 0.3 have one decimal place, multiplying by $\frac{10}{10}$ results in an equivalent ratio with whole numbers.

$$\frac{5.1}{0.3} \cdot \frac{10}{10} = \frac{51}{3}$$

We use this equivalent ratio to solve the proportion.

Step:	Justification:
$\dfrac{51}{3} = \dfrac{34}{x}$	Equivalent proportion
$51x = 3 \cdot 34$	Multiplied cross products
$x = \dfrac{3 \cdot 34}{51}$	Divided both sides by 51
$x = 2$	Simplified

We check the solution using the original proportion.

$\dfrac{5.1}{0.3} = \dfrac{34}{2}$	Substituted 2 for x
$\checkmark\, 17 = 17$	Divided

The ratios are equal. The solution is correct.

Practice Set

Solve the following equations.

a. $0.25 - 0.5x = 0.05$

b. $0.03 + 0.3x = 3$

c. $4 = 0.1x - 0.02$

d. $0.009x - 0.4 = -0.31$

e. $\dfrac{x}{11} = \dfrac{1.8}{6.6}$

f. $\dfrac{3.2}{2.4} = \dfrac{x}{18}$

g. $\dfrac{15}{x} = \dfrac{0.3}{0.12}$

h. $\dfrac{0.17}{0.51} = \dfrac{3}{x}$

i. *Justify* Show how to check your solution to problem **g.**

LESSON A61

Written Practice

26. $0.4x + 1.3 = 1.5$
(A61)

27. $0.002x + 0.03 = 0.92$
(A61)

LESSON A62

Written Practice

26. $1.3x - 0.32 = 0.98$
(A61)

27. $0.001x + 0.09 = 1.1$
(A61)

LESSON A63

New Concept

Simplifying Equations with Fractions

Recall that a first step we may take when solving equations with decimals is to convert all decimal numbers to whole numbers. We may use a similar method to remove fractions from an equation. We multiply both sides of the equation by a common denominator of the fractions to convert them to whole numbers.

Example 1

Solve: $\dfrac{5}{3}x - \dfrac{1}{2} = \dfrac{3}{4}$

Solution

Math Language

We commonly refer to the **coefficient,** a number that multiplies the variable(s) in an algebraic term. In the term $\dfrac{5}{3}x$, the coefficient is $\dfrac{5}{3}$.

Thinking Skill

Represent

Use the variables a, b, and c to state the Distributive Property in equation form.

We can form an equivalent equation without fractions by multiplying both sides of the equation by a number that is a common denominator of the fractions. A common denominator of the fractions is 12, so we begin by multiplying both sides by 12. To multiply the left side by 12, we group the two terms with parentheses and use the Distributive Property.

Step:	Justification:
$\dfrac{5}{3}x - \dfrac{1}{2} = \dfrac{3}{4}$	Given equation
$12\left(\dfrac{5}{3}x - \dfrac{1}{2}\right) = \left(\dfrac{3}{4}\right)12$	Multiplied both sides by 12
$20x - 6 = 9$	Simplified
$20x = 15$	Added 1 to both sides
$x = \dfrac{15}{20}$	Divided both sides by 20
$x = \dfrac{3}{4}$	Reduced

Discuss What are other common denominators of the fractions? Using another common denominator, solve the equation. Is the value of x the same?

Example 2

Solve: $\dfrac{5}{3} = 2x - \dfrac{1}{2}$

Solution

We can choose to solve this equation by adding $\dfrac{1}{2}$ to both sides and then dividing both sides by 2. We can also choose to make and solve an equivalent equation without fractions. We choose to multiply both sides of the equation by 6 to form an equivalent equation without fractions.

Step:	Justification:
$6\left(\dfrac{5}{3}\right) = \left(2x - \dfrac{1}{2}\right)6$	Multiplied both sides by 6
$10 = 12x - 3$	Simplified
$13 = 12x$	Add 3 to both sides
$\dfrac{13}{12} = x$	Divided both sides by 12

Check by substituting $\dfrac{13}{12}$ for x in the original equation.

Practice Set

Solve each equation below. For the first step, eliminate fractions by multiplying both sides of the equation by a common denominator.

a. $\dfrac{1}{5}x + \dfrac{1}{3} = \dfrac{2}{15}$

b. $\dfrac{1}{3} + \dfrac{5}{6}x = \dfrac{3}{4}$

c. $1 = \dfrac{1}{6} + \dfrac{3}{4}x$ **d.** $\dfrac{7}{8} = 2 - \dfrac{1}{2}x$

e. *Justify* For each problem **a–d,** state the common denominator you used.

LESSON A63

Written Practice

26. $0.02x + 0.1 = 1$
(A61)

27. $1.2 + 0.2x = 0.8$
(A61)

28. $\dfrac{3}{5}x + \dfrac{7}{10} = \dfrac{19}{10}$
(A63)

29. $\dfrac{1}{4} + \dfrac{3}{8}x = 1$
(A63)

LESSON A64

Written Practice

26. $2.1 + 0.7x = 9.8$
(A61)

27. $0.8 + 0.3x = 5$
(A61)

28. $\dfrac{3}{8}x + \dfrac{3}{4} = \dfrac{9}{8}$
(A63)

29. $\dfrac{1}{3}x - \dfrac{3}{4} = \dfrac{5}{12}$
(A63)

LESSON A65

Written Practice

26. $0.9 + 0.3x = 2.4$
(A61)

27. $0.007x + 0.28 = 0.7$
(A61)

28. $\dfrac{1}{3}x + \dfrac{4}{9} = \dfrac{7}{9}$
(A63)

29. $\dfrac{4}{5} + \dfrac{2}{15}x = \dfrac{14}{15}$
(A63)

LESSON A66   Writing the Equation of a Line Given the Slope and a Point on the Line

New Concept

Math Language

Recall that **slope** is the ratio of a line's change in the vertical direction (rise) to its change in the horizontal direction (run).

Remember the slope-intercept form of a linear equation is

$$y = mx + b$$

Recall that the letter m stands for the slope and b stands for the y-intercept. In this lesson we write an equation given the slope and a point on the line. We replace m with the slope number, and then we find the y-intercept. We will show two ways to find the y-intercept.

Example 1

Write the equation of a line that passes through (1, −2) with a slope of −3.

Solution

The slope is −3, so we can write −3 in place of m.

Step:	Justification:
$y = mx + b$	Slope-intercept form
$y = -3x + b$	Substituted -3 for m

To complete the equation we need to find the y-intercept.

Method 1: Graph the line to see the y-intercept. One way to find the intercept is to graph the given point, and then draw a line with a slope of -3 through the point.

Thinking Skill

Analyze

How can you tell from just looking at the graph that the slope of the line is negative?

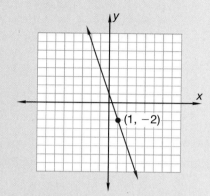

We see that the line passes through the y-axis at $+1$. So the equation of the line is:

$$y = -3x + 1$$

Method 2: Substitute for x and y and solve for b. The graphing method is not always accurate enough. A more accurate method uses substitution. The given point has values for x and y that satisfy the equation since the coordinates of any point on the line make the equation true. We substitute these x- and y-values in the equation and solve for b. For this equation, the slope is -3 and the x and y values are 1 and -2.

Step:	Justification:
$y = mx + b$	Slope-intercept equation
$y = -3x + b$	Replaced m with -3
$-2 = -3(1) + b$	Replaced x, y with 1, -2, respectively

Now b is the unknown. We solve for b.

$-2 = -3 + b$	Multiplied -3 and 1
$1 = b$	Added 3 to both sides

Since b is 1, the equation for the line is

$y = -3x + 1$	Substituted -3 for m and 1 for b

One method can serve as a check for the other method.

Example 2

The slope of a line is $-\frac{1}{2}$. A point on the line is (3, -1). Find the y-intercept and write the equation of the line in slope-intercept form.

We will use substitution to find the *y*-intercept and graphing to check our work. We begin with the slope-intercept equation. We replace *m* with $-\frac{1}{2}$. We replace *x* and *y* with 3 and -1 respectively.

$$y = mx + b$$
$$(-1) = \left(-\frac{1}{2}\right)(3) + b$$

Next we solve for *b*.

Step:	**Justification:**
$(-1) = \left(-\frac{1}{2}\right)(3) + b$	Substituted
$-1 = -\frac{3}{2} + b$	Simplified
$\frac{1}{2} = b$	Added $\frac{3}{2}$ to both sides

We have found *b*. The *y*-intercept is $\frac{1}{2}$. Now we write the slope-intercept equation using $-\frac{1}{2}$ for *m* and using $\frac{1}{2}$ for *b*.

$$y = mx + b \qquad \text{Slope-intercept equation}$$
$$y = -\frac{1}{2}x + \frac{1}{2} \qquad \text{Substituted } -\frac{1}{2} \text{ for } m \text{ and } \frac{1}{2} \text{ for } b$$

We can check the reasonableness of the answer by graphing a line with the original conditions, with a slope of $-\frac{1}{2}$ passing through $(3, -1)$.

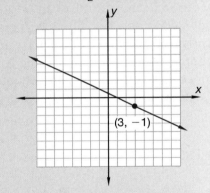

(3, −1)

We see that the line intercepts the *y*-axis between 0 and 1, so our answer is reasonable.

Practice Set

Represent Write an equation in slope-intercept form for the lines described in problems **a–e**. Graph or use substitution to solve.

a. slope = 1; point (5, −2)

b. slope = 3; point (−3, −2)

c. slope = $\frac{1}{5}$; point (−5, 3)

d. slope = $-\frac{3}{4}$; point (8, −1)

e. slope = −5; point (−1, −8)

LESSON A66

26. $0.003x - 0.02 = 0.07$
(A61)

27. $\frac{1}{4} - \frac{2}{3}x = \frac{11}{12}$
(A63)

28. $2 + \frac{1}{3}t = \frac{2}{3}$
(A63)

29. Write an equation for the line with slope $\frac{1}{2}$ through the point $(-4, 1)$.
(A66)

30. Write an equation for the line with slope -2 through the point $(-3, 5)$.
(A66)

LESSON A67

26. $0.03x - 0.02 = 0.07$
(A61)

27. $\frac{2}{3} - \frac{1}{6}x = \frac{1}{2}$
(A63)

28. $\frac{4}{5}x + \frac{1}{10} = \frac{1}{2}$
(A63)

29. Write an equation for the line with slope $\frac{2}{3}$ through the point $(-3, -2)$.
(A66)

30. Write an equation for the line with slope -3 through the point $(-3, 1)$.
(A66)

LESSON A68

New Concept

Finding a Slope from Two Given Points

Recall that the slope of a line is the ratio of its rise to run. If we are given the coordinates of two points on a line, we can find the slope of the line. Here we show two ways to find the slope of a line through $(-2, 1)$ and $(6, -3)$.

Method 1: Graph the points and count units to find the rise and run. The number of horizontal units between the points is the run, and the number of vertical units is the rise. Beginning at the point on the left, we count horizontally to the right until we are directly above or below the second point. Then we count units up (positive) or down (negative) to the second point. We write the ratio and reduce if possible.

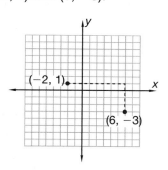

$$\text{slope} = \frac{\text{rise}}{\text{run}} = \frac{\text{change in } y}{\text{change in } x} = \frac{-4}{8} = -\frac{1}{2}$$

Thinking Skill

Conclude

How can we determine from the sign of the slope whether the line rises or falls from left to right?

Method 2: Use subtraction to find the rise and run. Notice that the rise is the change in the y numbers and the run is the change in the x numbers. To calculate the change we subtract the coordinates of one point from the coordinates of the other point.

$$\boxed{\text{slope} = \frac{\text{rise}}{\text{run}} = \frac{\text{change in } y}{\text{change in } x}}$$

The change in y is $(-3) - (1)$, or -4.

The change in x is $(6) - (-2)$, or 8.

The slope is the change in y over the change in x.

$$\text{slope} = \frac{-4}{8} = -\frac{1}{2}$$

Example 1

Use the second method to find the slope of a line through points (−2, 1) and (6, −3) by subtracting the coordinates (6, −3) from (−2, 1).

Solution

Changing the order of subtraction does not change the slope. However, the subtractions to find the rise and run must start from the coordinates of the same point.

$$
\begin{array}{cc}
x & y \\
(-2, & 1) \\
-(6, & -3) \\
\hline
-8 & 4
\end{array}
$$

$$\text{slope} = \frac{\text{rise}}{\text{run}} = \frac{\text{change in } y}{\text{change in } x} = \frac{4}{-8} = -\frac{1}{2}$$

Explore Reverse the order of subtraction. How does the answer change? Does the slope change?

Infer If the y-coordinates of two points on a line are equal, what is the slope of the line? Explain.

Example 2

Suppose a six-mile taxi ride costs $15, and a 12-mile taxi ride costs $27. Use the data points (6, 15) and (12, 27) to find the slope, which is the cost per mile.

Solution

We may graph the points, or we may subtract the coordinates to find the slope. We choose to subtract.

$$
\begin{array}{ccc}
& x & y \\
\textbf{Point B} & (12, & 27) \\
\textbf{Point A} & (6, & 15) \\
\hline
& 6 & 12
\end{array}
$$

$$\text{slope} = \frac{\text{change in } y}{\text{change in } x} = \frac{12}{6} = 2$$

For this situation, y-values are numbers of dollars and x-values are numbers of miles. Dividing y by x results in dollars per mile. Thus, the slope **2** means the charge per mile is $2. Taxi service usually has a start charge (in this case $3) as well as a distance charge. Because of the start charge, we cannot find the charge per mile simply by dividing $15 by 6 miles or $27 by 12 miles.

Practice Set Use both methods described in this lesson to find the slope of a line through the given points.

a. (−1, 2) and (2, −4)

b. (−1, −2) and (5, 1)

c. (−5, −7) and (3, −1)

d. (−2, 3) and (3, 3)

e. (3, −2) and (3, 4)

f. (Connect) Water freezes at 0° on the Celsius scale and at 32° on the Fahrenheit scale. Water boils at 100° on the Celsius scale and 212° on the Fahrenheit scale. If a line is drawn to represent the relationship between Fahrenheit and Celsius through the data points (0, 32) and (100, 212), then what would be the slope of the line? What is the meaning of this slope?

LESSON A68

Written Practice

26. $0.007x + 0.03 = 0.1$
(A61)

27. $\dfrac{3}{8} - \dfrac{2}{3}x = \dfrac{11}{12}$
(A63)

28. Write an equation for the line with slope $-\dfrac{4}{3}$ that passes through point $(-3, 5)$.
(A66)

29. Find the slope of the line that passes through $(-2, 4)$ and $(3, -1)$.
(A68)

30. Find the slope of the line that passes through $(0, 0)$ and $(2, 3)$.
(A68)

LESSON A69

Written Practice

26. $0.005x - 0.03 = 0.01$
(A61)

27. $\dfrac{4}{3}x - \dfrac{2}{7} = -\dfrac{34}{21}$
(A63)

28. Write an equation for the line with slope 1 that passes through point $(-5, -7)$.
(A66)

29. Find the slope of the line that passes through $(4, 3)$ and $(-3, 4)$.
(A68)

30. Find the slope of the line that passes through $(-2, 5)$ and $(3, 5)$.
(A68)

LESSON A70

Written Practice

26. $0.9x - 1.3 = 0.5$
(A61)

27. $\dfrac{7}{3}x + 1 = \dfrac{17}{3}$
(A63)

28. Write an equation for the line with slope $\dfrac{1}{4}$ that passes through point $(5, 4)$.
(A66)

29. Find the slope of the line that passes through $(-2, -4)$ and $(3, 5)$.
(A68)

30. Find the slope of the line that passes through $(1, -5)$ and $(2, -3)$.
(A68)

LESSON A71

New Concept

Finding the Equation of a Line Given Two Points

Two points determine a line. In this lesson we describe two methods to find the equation of a line given two points.

Method 1: Graph the points, draw the line, and visually determine the slope and *y*-intercept. Then substitute the slope and *y*-intercept into the slope-intercept equation.

$$y = mx + b$$

Method 2: A more accurate method is to use the two points to calculate the slope of the line. Then substitute the coordinates of one of the points for x and y in the equation to calculate the y-intercept. In this lesson we will practice this method.

Example

Write the equation of the line through (1, 2) and (−3, −4).

Solution

First we calculate the slope (m) by subtracting the points' coordinates.

$$\text{slope} = \frac{\text{change in } y}{\text{change in } x} = \frac{(-4) - (2)}{(-3) - (1)} = \frac{-6}{-4} = \frac{3}{2}$$

Next we substitute $\frac{3}{2}$ for m and substitute the coordinates of one of the given points for x and y in the equation. Then we solve the equation to find b. We will use the point (1, 2), but either point may be used.

Step:	Justification:
$y = mx + b$	Slope-intercept form
$2 = \frac{3}{2}(1) + b$	Substituted for m, x, and y
$2 = \frac{3}{2} + b$	Simplified
$\frac{1}{2} = b$	Subtracted $\frac{3}{2}$ from both sides

We found that the y-intercept is $\frac{1}{2}$. Now that we know both the slope (m) and the y-intercept (b), we are ready to write the equation of the line.

$$y = \frac{3}{2}x + \frac{1}{2} \qquad \text{Substituted for } m \text{ and } b$$

One way to check our work is by substituting the coordinates of the second point into the equation to see if the x- and y-values satisfy the equation.

Step:	Justification:
$y = \frac{3}{2}x + \frac{1}{2}$	Given equation
$(-4) = \frac{3}{2}(-3) + \frac{1}{2}$	Substituted (−3, −4)
$-4 = \frac{-9}{2} + \frac{1}{2}$	Simplified
$-4 = -4$	Simplified

We see that both points satisfy the equation of the line.

Another way to check our work is by graphing the points and sketching a line through the points (Method 1). The graph allows us to visually check the slope and *y*-intercept.

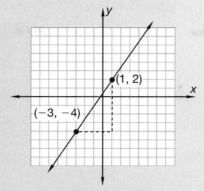

Justify How can we use this graph to show that our equation is reasonable?

Explain When you are given two points on a line, how can you find the equation of the line?

Practice Set | **Formulate** Find the equation of the line which includes the given pairs of points in **a–d**. Check your work.

a. $(-4, 5)$ and $(-2, 2)$ **b.** $(3, -4)$ and $(-3, 4)$

c. $(8, 5)$ and $(2, 5)$ **d.** $(-9, 5)$ and $(6, 0)$

LESSON A71

Written Practice

26. $0.09x + 0.9 = 2.7$
(A61)

27. $\dfrac{3}{4} - \dfrac{2}{3}x = \dfrac{11}{12}$
(A63)

28. Write an equation for the line with slope 5 that passes through point $(1, -2)$.
(A66)

29. Write an equation for the line that passes through $(3, 4)$ and $(-3, 0)$.
(A71)

30. Write an equation for the line that passes through $(1, 3)$ and $(-1, -1)$.
(A71)

LESSON A72

Written Practice

26. $0.07 - 0.003x = 0.1$
(A61)

27. $\dfrac{6}{7} - \dfrac{1}{2}x = -\dfrac{1}{7}$
(A63)

28. Write an equation for the line that passes through $(-4, 0)$ with a slope of $\dfrac{1}{2}$.
(A66)

29. Write an equation for the line that passes through $(-2, 0)$ and $(2, -2)$.
(A71)

30. Write an equation for the line that passes through $(-4, 5)$ and $(4, -5)$.
(A71)

Graphing Sequences

Consider the sequence:

$$5, 10, 15, 20, \ldots$$

The table of the number of each term and its value is shown below.

Number of Term (n)	Value of Term (a)
1	5
2	10
3	15
4	20

We can graph the sequence by graphing the pairs of numbers (n, a) from the table.

Notice that the sequence values are graphed along the y-axis. Reading the graph from left to right helps us visualize the change in value of the sequence at each step. From this graph we see that each term is 5 greater than the previous term.

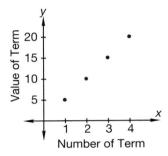

Example 1

Graph the first five terms of the sequence:

$$1, \frac{1}{2}, \frac{1}{3}, \frac{1}{4}, \frac{1}{5}, \ldots$$

Solution

We make a table of value pairs.

Number of Term (n)	Value of Term (a)
1	1
2	$\frac{1}{2}$
3	$\frac{1}{3}$
4	$\frac{1}{4}$
5	$\frac{1}{5}$

Then we graph the pairs of numbers from the table.

We see that the values decrease more and more gradually. If we continued graphing values, we would see that the points approach the x-axis but never touch it.

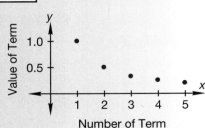

Example 2

Graph the first three terms of the sequence $a_n = 3^n$.

Solution

We make a table of the sequence, then graph the value pairs.

number of term (n)	value of term (a)
1	3
2	9
3	27

These values increase rapidly. The next term, 81, is not on the range of this graph.

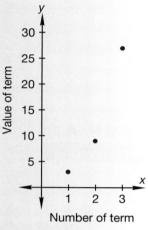

Calculator Activity Sequences are easily graphed with a graphing calculator. The following instructions are for the TI-83+:

See Graphing Calculator Activities *for a graphing calculator activity.*

1. Press MODE, ↓, ←, and ENTER to select "Seq" (for sequence mode).

2. Press Y= to enter the equation of the sequence. For the nMin, enter 1. For u(n), enter the equation of the sequence, 3", using ∧ and XTON.

3. Press WIN to change the numbers of the terms you want graphed (nMin and nMax) and to change the graphing window. For example, to graph the first 3 terms, set nMin to 1, nMax to 3, PlotStart to 1, and PlotStep to 1. Then set Xmin to −1, Xmax to 4, Ymin to −5 and Ymax to 30.

4. Press GRAPH to view the graph.

5. Press TRACE and ← or → to view how the pairs from the graph match to the value pairs in table of the sequence.

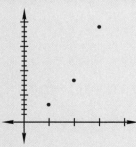

Graph each sequence in **a–c.**

    a. 3, 6, 9, 12, . . .

    b. $a_n = 2^n$

    c. $2, \dfrac{3}{2}, \dfrac{4}{3}, \dfrac{5}{4}, \ldots$

LESSON A73

Written Practice

26. $0.03 + 0.011x = 0.36$
(A61)

27. $\dfrac{2}{3}x - \dfrac{1}{2} = \dfrac{1}{6}$
(A63)

28. Write an equation for the line with slope $-\dfrac{4}{5}$ that passes through point
(A66) $(5, -5)$.

29. Write an equation for the line that passes through $(-4, 1)$ and $(-8, 6)$.
(A71)

30. Graph the following sequence: $a_n = 2n + 1$
(A73)

LESSON A74

Written Practice

26. $0.04 - 0.02x = 0.5$
(A61)

27. $\dfrac{4}{5}x - \dfrac{2}{3} = \dfrac{1}{3}$
(A63)

28. Write an equation for the line with slope $-\dfrac{3}{2}$ that passes through $(-2, 5)$.
(A66)

29. Write an equation for the line that passes through $(-2, 8)$ and $(2, 4)$.
(A71)

30. Graph the following sequence: 16, 8, 4, 2, . . .
(A73)

LESSON A75

Written Practice

26. $0.007 + 0.001x = 0.4$
(A61)

27. $\dfrac{1}{3}m - \dfrac{2}{3} = 0$
(A63)

28. Write an equation for the line that passes through point $(3, -4)$ with
(A66) slope $-\dfrac{4}{3}$.

29. What is the equation of the line that intercepts the y-axis at 2 and the
(A71) x-axis at -4?

30. Graph the following sequence: $\dfrac{1}{2}, \dfrac{2}{3}, \dfrac{3}{4}, \dfrac{4}{5}, \ldots$
(A73)

LESSON A76

Written Practice

26. $0.5 - 0.02x = 0.1$
(A61)

27. $0.03x + 0.1 = 0.7$
(A61)

28. $\dfrac{1}{2}w - \dfrac{2}{3} = \dfrac{2}{3}$
(A63)

29. Write an equation for the line that passes through $(2, 6)$ with slope 2.
(A66)

30. Write an equation for the line that passes through $(-2, -5)$ and $(6, 3)$.
(A71)

LESSON A77

Written Practice

26. $0.003x - 0.2 = 0.01$
(A61)

27. $\dfrac{5}{3}x = \dfrac{2}{3}x + \dfrac{1}{3}$
(A63)

28. Write an equation for the line that passes through $(2, 5)$ with slope $-\dfrac{1}{2}$.
(A66)

29. Write an equation for the line that passes through $(-10, 0)$ and $(0, 5)$.
(A71)

30. Graph the following sequence: $a_n = n^2$
(A73)

LESSON A78

Written Practice

26. $0.03 - 0.01m = 0.5$
(A61)

27. $0.3m = 0.4m - 0.5$
(A61)

28. $\dfrac{2}{7} - \dfrac{4}{7}x = 6$
(A63)

29. Write an equation for the line that passes through $(2, 10)$ with slope $-\dfrac{1}{2}$.
(A66)

30. Write an equation for the line that passes through $(-3, 1)$ and $(6, -2)$.
(A71)

LESSON A79

Written Practice

26. $0.48 - 0.2x = 0.18$
(A61)

27. $\dfrac{5}{6} - \dfrac{1}{3}x = \dfrac{1}{2}$
(A63)

28. Write an equation for the line that passes through $(3, 9)$ with slope $-\dfrac{5}{3}$.
(A66)

29. Write an equation for the line that passes through $(-2, 5)$ and $(5, -9)$.
(A71)

30. Graph the following sequence: $a_n = 2n - 1$
(A73)

LESSON A80

New Concept

Adding and Subtracting Polynomials

Recall that to add polynomials, we add like terms.

$$(x^2 - 3x + 5) + (2x^2 + x - 4)$$
$$= 3x^2 - 2x + 1$$

One way to do this is to write one polynomial below the other, aligning like terms:

$$\begin{array}{r} x^2 - 3x + 5 \\ + 2x^2 + x - 4 \\ \hline 3x^2 - 2x + 1 \end{array}$$

Example 1

Add $(3x^2 - x + 1) + (x^2 - 5)$

$$
\begin{array}{r}
3x^2 - x + 1 \\
x^2 \quad\;\; - 5 \\
\hline
4x^2 - x - 4
\end{array}
$$

We can subtract polynomials by rewriting the subtraction as addition. Consider the example below:

$$(x + 5) - (2x - 3)$$

First, we use the Distributive Property to rewrite the second polynomial, multiplying each term by -1. This simply changes the sign of each term. The problem is now an addition problem.

$$(x + 5) + (-1)(2x - 3)$$
$$(x + 5) + (-2x + 3)$$

Then we add the polynomials:

$$-x + 8$$

Example 2

Subtract the polynomials

$$(x^2 - 5x + 1) - (x^2 + 2x - 5)$$

First, we write the problem as addition by changing the signs of the second polynomial.

$$(x^2 - 5x + 1) + (-x^2 - 2x + 5)$$

Then we add like terms.

$$
\begin{array}{r}
x^2 - 5x + 1 \\
-x^2 - 2x + 5 \\
\hline
-7x + 6
\end{array}
$$

Practice Set

a. $(x + y + 1) + (2x - 2y - 2)$ **b.** $(3x^2 - x + 3) + (-5x^2 + 3x - 1)$

c. $(x^2 + x - 7) + (x^2 - 3)$ **d.** $(x - 5) - (3x + 5)$

e. $(4x^2 - 3x + 2) - (x^2 - 3x + 4)$ **f.** $(x^2 - x + 6) - (7x + 4)$

LESSON A80

Written Practice

26. $0.2 - 0.02x = 0.08$
(A61)

27. Write an equation for the line that passes through (4, 6) with slope 12.
(A66)

28. Write an equation for the line that passes through (2, 0) and (−4, 0).
(A71)

29. $(x + 7) - (2x - 3)$
(A80)

30. $(6x^2 - 3x) - (5x^2 + 1)$
(A80)

LESSON A81

Written Practice

26. $\dfrac{4}{7} - \dfrac{3}{7}x = 1$
(A63)

27. Write an equation for the line that passes through (8, 6) with slope 3.
(A66)

28. Write an equation for the line that passes through (5, 2) and (−5, 4).
(A71)

29. $(4x^2 - 3x + 7) + (-x^2 - 5x + 2)$
(A80)

30. $(9x - 6y + 2) + (x + y + 3)$
(A80)

LESSON A82

Written Practice

26. $1.2 - 3.4x = -5.6$
(A61)

27. Write an equation for the line that passes through (3, 1) with slope $-\frac{1}{4}$.
(A66)

28. Write an equation for the line that passes through (3, −4) and (−1, 6).
(A71)

29. Graph the following sequence: $a^n = 4n - 3$
(A73)

30. $(x^2 + 5x + 6) + (x^2 - 5x - 6)$
(A80)

LESSON A83

New Concept

Proportions with Unknowns in Two Terms

Cross products are particularly useful when proportions have variables in more than one term.

$$\frac{x + 3}{x} = \frac{3}{2}$$

The equation above can be cross-multiplied, then solved using familiar steps.

Step:	Justification:
$2(x + 3) = 3 \cdot x$	Cross-multiplied
$2x + 6 = 3x$	Distributive Property
$6 = x$	Subtracted $2x$ from both sides

We check the solution by substituting 6 for x in the original proportion.

Step:	Justification:
$\dfrac{6 + 3}{6} = \dfrac{3}{2}$	Substituted 6 for x
$\dfrac{9}{6} = \dfrac{3}{2}$	Added $6 + 3$
$\dfrac{3}{2} = \dfrac{3}{2}$ ✓	Reduced $\dfrac{9}{6}$

The ratios are equal, so the solution checks.

Example

Solve for x. $\dfrac{x + 1}{7} = \dfrac{x}{6}$

Solution

Cross multiply first.

Step:	Justification:
$6(x + 1) = 7 \cdot x$	Cross-multiplied
$6x + 6 = 7x$	Distributive Property
$6 = x$	Subtracted $6x$ from both sides

Check:

$\dfrac{6 + 1}{7} = \dfrac{6}{6}$	Substituted 6 for x
$\dfrac{7}{7} = \dfrac{6}{6}$	Added $6 + 1$
$1 = 1$ ✓	Simplified

Practice Set

Solve and check.

a. $\dfrac{3x}{4x + 1} = \dfrac{2}{3}$

b. $\dfrac{x}{12} = \dfrac{x - 3}{8}$

c. $\dfrac{1}{x - 1} = \dfrac{2}{x + 5}$

d. $\dfrac{1}{3m} = \dfrac{2}{3m + 1}$

LESSON A83

Written Practice

26. $\dfrac{2}{3}x + \dfrac{3}{8} = \dfrac{3}{4}$
(A63)

27. Write an equation for the line that passes through (2, 8) with slope $-\dfrac{1}{2}$.
(A66)

28. Write an equation for the line that passes through (−5, 10) and (0, 2).
(A71)

29. $(2x^2 - x + 3) - (5x^2 - 4x - 3)$
(A80)

30. Solve: $\dfrac{2x + 3}{x + 4} = \dfrac{1}{3}$
(A83)

LESSON A84 Adding Radical Expressions

We simplify expressions containing radicals much like we do expressions with variables. For example, to find the perimeters of the equilateral triangles shown below we add like terms.

$$x + x + x = 3x \qquad \text{Collected like terms}$$
$$\sqrt{2} + \sqrt{2} + \sqrt{2} = 3\sqrt{2} \qquad \text{Collected like terms}$$

Unlike terms cannot be combined. We can only indicate the addition.

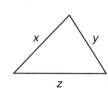

$$x + y + z \qquad \text{Cannot be simplified}$$
$$1 + \sqrt{2} + \sqrt{3} \qquad \text{Cannot be simplified}$$

Math Language

Recall that a **radicand** is a number, term, or expression under a radical sign. The radicand in $\sqrt{2}$ is 2.

Radical expressions with the same radicands are like terms and can be combined.

Step:	Justification:
$3\sqrt{2} + \sqrt{2}$	Given expression
$4\sqrt{2}$	Added $3\sqrt{2} + \sqrt{2}$

Radical expressions with different radicands are unlike terms and cannot be combined. Radical terms also cannot be combined with constants.

$$5 + 5\sqrt{3} + 5\sqrt{2} \qquad \text{Cannot be simplified}$$

Example 1

Simplify:

$$\sqrt{125} + \sqrt{5} + 5$$

Solution

Thinking Skill

Explain

Write the steps to show how to simplify $\sqrt{125}$.

Step:	Justification:
$\sqrt{125} + \sqrt{5} + 5$	Given expression
$5\sqrt{5} + \sqrt{5} + 5$	Simplified: $\sqrt{125} = 5\sqrt{5}$
$(5\sqrt{5} + \sqrt{5}) + 5$	Grouped like terms (Associative Property)
$6\sqrt{5} + 5$	Added $5\sqrt{5} + \sqrt{5}$

Example 2

$$\sqrt{12} + \sqrt{18} + 2\sqrt{2}$$

Solution

Step:	Justification:
$\sqrt{12} + \sqrt{18} + 2\sqrt{2}$	Given expression
$2\sqrt{3} + 3\sqrt{2} + 2\sqrt{2}$	Simplified $\sqrt{12}$ and $\sqrt{18}$
$2\sqrt{3} + (3\sqrt{2} + 2\sqrt{2})$	Grouped like terms
$\mathbf{2\sqrt{3} + 5\sqrt{2}}$	Added $3\sqrt{2} + 2\sqrt{2}$

Practice Set   Simplify and combine like terms.

a. $\sqrt{8} + \sqrt{18}$ **b.** $\sqrt{50} + \sqrt{75}$

c. $\sqrt{12} + \sqrt{48}$ **d.** $\sqrt{98} - \sqrt{32}$

e. $3\sqrt{2} + \sqrt{8}$ **f.** $\sqrt{20} + \sqrt{40}$

g. Create   Write a radical expression with two or more terms that can be combined. Then simplify the expression.

LESSON A84

Written Practice

26. $0.3x - 0.9 = 0.3$
(A61)

27. Write an equation for the line that passes through $(-3, -5)$ and
(A71) $(-1, -1)$.

28. $(3x + 2y + 6) + (-2x - y + 3)$
(A80)

29. Solve: $\dfrac{4}{x} = \dfrac{x}{9}$
(A83)

30. Simplify: $6 + 3\sqrt{18}$
(A84)

LESSON A85

Written Practice

26. $\dfrac{5}{9}x - \dfrac{1}{3} = \dfrac{2}{3}$
(A63)

27. Write an equation for the line that passes through $(-2, 5)$ and $(1, -1)$.
(A71)

28. $(2x + y - 5) - (x + y + 2)$
(A80)

29. Solve: $\dfrac{x + 1}{x} = \dfrac{10}{8}$
(A83)

30. Simplify: $\sqrt{75} + \sqrt{12}$
(A84)

LESSON A86

Written Practice

26. $0.4 - 0.02x = 1.2$
(A61)

27. Graph the following sequence: $a_n = 2^n$
(A73)

28. $(3x - 4y + 1) + (-3x + 2y - 3)$
(A80)

29. Solve: $\dfrac{x-5}{2} = \dfrac{x-2}{3}$
(A83)

30. Simplify: $\sqrt{18} + 2\sqrt{8}$
(A84)

Solving Equations with Two Variables Using Substitution

New Concept

The equation shown below has two variables. Can we find x if we also know that $y = x + 2$?

$$x + y = 24$$

If we know the value of one variable, we can substitute its value into the equation to find the other variable.

Step:	Justification:
$x + y = 24$	Given
$x + (x + 2) = 24$	Substituted $x + 2$ for y
$x + x + 2 = 24$	Cleared parentheses
$2x + 2 = 24$	Collected like terms
$2x = 22$	Subtracted 2 from both sides
$x = 11$	Divided both sides by 2

Since $y = x + 2$, we also know that $y = 11 + 2 = 13$.

Example 1

If $x = y - 3$, find y in $2x + 3y = 39$. Then find x.

Solution

We write the given equation and substitute $y - 3$ for x.

Step:	Justification:
$2x + 3y = 39$	Given equation
$2(y - 3) + 3y = 39$	Substituted $y - 3$ for x
$2y - 6 + 3y = 39$	Distributive Property
$5y - 6 = 39$	Commutative and Associative Properties
$5y = 45$	Added 6 to both sides
$y = 9$	Divided both sides by 5

We have found that $y = 9$. Since we are given that $x = y - 3$, we can find x by subtracting 3 from 9.

Step:	Justification:
$x = y - 3$	Given
$x = 9 - 3$	Substituted 9 for y
$x = 6$	Simplified

By substitution we find that **$y = 9$** and **$x = 6$.**

Practice Set

Use substitution to solve the following equations.

a. If $y = x - 2$, find x in $4x + 3y = 78$. Then find y.

b. If $y = x - 5$, find x in $3x - 2y = 12$. Then find y.

c. If $x = 3y$, find y in $y = 2x + 10$. Then find x.

d. If $x = y + 2$, find y in $y = 3x - 4$. Then find x.

LESSON A87

Written Practice

26. Write an equation for the line that passes through (4, 5) and (−5, −4).
(A71)

27. $(x^2 - 3x + 5) + (2x^2 - x - 7)$
(A80)

28. Solve: $\dfrac{25}{x} = \dfrac{x}{4}$
(A83)

29. Simplify: $\sqrt{50} + \sqrt{8}$
(A84)

30. If $y = x - 2$, find x in $2x + 3y = 19$. Then find y.
(A87)

LESSON A88

Written Practice

26. $0.007x - 0.07 = 0.7$
(A61)

27. $\dfrac{4}{5} - \dfrac{2}{3}x = \dfrac{14}{15}$
(A63)

28. Write an equation for the line that passes through (3, 8) with slope $-\dfrac{1}{3}$.
(A66)

29. Simplify: $\sqrt{75} + 2\sqrt{27}$
(A84)

30. If $y = 2x + 1$, find x in $2x + y = 13$. Then find y.
(A87)

LESSON A89

Written Practice

26. Write an equation for the line that passes through (−7, 0) and (−3, 12).
(A71)

27. $(x + y + 1) - (x - y - 1)$
(A80)

28. Solve: $\dfrac{x + 1}{x} = \dfrac{75}{50}$
(A83)

29. Simplify: $\sqrt{18} + \sqrt{72}$
(A84)

30. If $x = 5 - y$, find y in $2x + 3y = 11$. Then find x.
(A87)

LESSON A90

Written Practice

26. Write an equation for the line that passes through (4, 8) with slope $\dfrac{3}{4}$.
(A66)

27. Write an equation for the line that passes through (−1, 0) and (5, 0).
(A71)

28. Solve: $\dfrac{x - 2}{3} = \dfrac{x - 4}{2}$
(A83)

29. Simplify: $\sqrt{20} + 2\sqrt{45}$
(A84)

30. Find x in $2x - y = 9$ when $y = x - 1$. Then find y.
(A87)

LESSON A91

Written Practice

26. Write an equation for the line that passes through $(-2, 5)$ and $(4, 2)$.
(A71)

27. $(3x - y + 7) + (4x + y + 3)$
(A80)

28. Solve: $\dfrac{3}{x} = \dfrac{x}{27}$
(A83)

29. Simplify: $\sqrt{98} + \sqrt{8}$
(A84)

30. Find x in $3x + 2y = -12$ when $y = -\frac{1}{2}x - 2$. Then find y.
(A87)

LESSON A92

New Concept

Solving Systems of Equations by Substitution, Part 1

Recall that a system of equations is a set of two or more equations that describe different aspects of the same situation. We can use a system of equations to solve problems with more than one variable.

> *In the cash register there were $3.50 in nickels and dimes. If there were 10 more nickels than dimes, how many dimes were there?*

We use two equations. One equation is about the number of coins. The other equation is about the value of the coins.

We will let n stand for the number of nickels and d stand for the number of dimes. We are told that there are ten more nickels than dimes. We write this relationship as an equation about the number of coins.

$$n = d + 10$$

The value of n nickels is $5n$ cents, and the value of d dimes is $10d$ cents. We are given that their total value is 350 cents, so we write this equation about the value of the coins.

$$5n + 10d = 350$$

Together, these two equations form a **system of equations** about the nickel and dime problem. We can show that these equations are linked using a brace:

$$\begin{cases} n = d + 10 \\ 5n + 10d = 350 \end{cases}$$

Linking the equations means that n has the same meaning and value throughout the system of equations, and d has the same meaning and value.

To solve the nickel and dime problem, we can guess combinations of nickels and dimes to find a pair of numbers that satisfy the condition of the problem. However, we do not need to guess if we use algebra. Instead we can solve the system of equations to find the number of nickels and dimes.

Example 1

Solve: $\begin{cases} n = d + 10 \\ 5n + 10d = 350 \end{cases}$

There are two equations. In the first equation we see that n equals $d + 10$. Therefore, we can substitute $d + 10$ in place of n in the second equation.

Step:	Justification:
$n = d + 10$	Given first equation
$5n + 10d = 350$	Given second equation
$5(d + 10) + 10d = 350$	Substituted $d + 10$ for n
$5d + 50 + 10d = 350$	Distributive Property
$15d + 50 = 350$	Added $5d$ and $10d$
$15d = 300$	Subtracted 50 from both sides
$d = 20$	Divided both sides by 15

We find that d is 20. Now we can find n by substituting 20 in place of d. We choose the first equation to do this.

Step:	Justification:
$n = d + 10$	First equation
$n = (20) + 10$	Substituted 20 for d
$n = 30$	Simplified

We find that n is 30. We check our solution by substituting both values in the second equation.

Step:	Justification:
$5n + 10d = 350$	Second equation
$5(30) + 10(20) = 350$	Substituted 30 for n and 20 for d
$150 + 200 = 350$	Simplified
$350 = 350 \checkmark$	Simplified

Using substitution to solve this system of equations, we find that $d = 20$ and $n = 30$. Applying this system of equations to the nickel and dime problem, we find that there were 30 nickels and 20 dimes in the cash register.

We use the following steps to solve a system of two equations with two unknowns using substitution.

1. Isolate one of the variables in one of the equations.

2. Substitute for that variable in the other equation and solve.

3. Substitute the found value of the variable in one of the equations and solve for the other variable.

4. Check the solution pair in both of the equations.

Practice Set Solve each system of equations by substitution. Check solutions in both equations.

a. $\begin{cases} 4x - 3y = 9 \\ x = 3 \end{cases}$

b. $\begin{cases} 2b + 2h = 28 \\ b = h + 2 \end{cases}$

$$\text{c.} \quad \begin{cases} 3x - 2y = -7 \\ y = 2 \end{cases} \qquad\qquad \text{d.} \quad \begin{cases} p = q - 4 \\ 3p - 2q = 8 \end{cases}$$

e. *Evaluate* Denise participated in a marathon—a race of just over 26 miles. She ran part of the way averaging 7 miles per hour and walked part of the way averaging 3 miles per hour. It took Denise 6 hours to complete the marathon. For how many hours did Denise run and for how many hours did Denise walk? Solve this system of equations to find the answer.

$$\begin{cases} w = 6 - r \\ 3w + 7r = 26 \end{cases}$$

LESSON A92

Written Practice

26. Write an equation for the line that passes through $(7, -2)$ and $(-1, 6)$.
(A71)

27. $(x^2 - 5) + (x + 4)$
(A80)

28. Solve: $\dfrac{x - 3}{5} = \dfrac{x - 7}{10}$
(A83)

29. Simplify: $4\sqrt{5} + \sqrt{180}$
(A84)

30. Find x by substitution: $\begin{cases} 5x - 4y = 20 \\ y = 5 \end{cases}$
(A92)

LESSON A93

Written Practice

26. Write an equation for the line that passes through $(8, 2)$ with slope $\frac{1}{2}$.
(A66)

27. Graph the following sequence: $a_n = n^2 - 1$
(A73)

28. Solve: $\dfrac{12}{x} = \dfrac{x}{27}$
(A83)

29. Simplify: $2\sqrt{72} + \sqrt{50}$
(A84)

30. Find x by substitution: $\begin{cases} 2x + 3y = -1 \\ y = -5 \end{cases}$
(A92)

LESSON A94

Written Practice

26. Write an equation for the line that passes through $(1, 4)$ with slope $-\frac{3}{2}$.
(A68)

27. Write an equation for the line that passes through $(-2, -6)$ and $(-1, 0)$.
(A71)

28. $(3x^2 - 5x + 7) + (2x^2 - 4)$
(A80)

29. Simplify: $\sqrt{28} + 2\sqrt{63}$
(A84)

30. Find y by substitution: $\begin{cases} 4x + 3y = 1 \\ x = 4 \end{cases}$
(A92)

LESSON A95

New Concept

Products Equal to Zero

If the product of two numbers is zero, then at least one of the numbers must be zero.

Zero Product Property

If $ab = 0$, then $a = 0$ or $b = 0$.

We apply this fact to solve the following equation.

$$(x + 2)(x + 3) = 0$$

This equation has two factors whose product is 0. The equation is true only if a factor is zero. Therefore, either $x + 2 = 0$ or $x + 3 = 0$. Thus, to solve the original equation, we set each factor equal to zero and solve. We will show the steps used to solve each equation on either side of each step's justification.

Step:	Justification:	Step:
$x + 2 = 0$	Factor = 0	$x + 3 = 0$
$x = -2$	Solved	$x = -3$

We find the solutions $x = -2$ and $x = -3$. We test the solutions in the original equation.

Check $x = -2$:	Justification:	Check $x = -3$:
$(x + 2)(x + 3) = 0$	Equation	$(x + 2)(x + 3) = 0$
$(-2 + 2)(-2 + 3) = 0$	Solved	$(-3 + 2)(-3 + 3) = 0$
$(0)(1) = 0$	Simplified	$(-1)(0) = 0$
$0 = 0$ ✓	Simplified	$0 = 0$ ✓

Our check confirms that if x is -2 or -3, the product of the binomials is zero.

Example 1

Solve: $(x - 2)(x + 4) = 0$

Solution

The product of two factors is 0. Therefore, one factor or the other equals zero.

Step:	Justification:
$(x - 2)(x + 4) = 0$	Given equation
$x - 2 = 0$ or $x + 4 = 0$	Zero Product Property
$x = 2$ or $x = -4$	Solved

Example 2

Solve: $2w(w - 3) = 0$

Solution

The three factors are 2, w, and $w - 3$. For their product to be 0, one of the factors must be zero.

Step:	Justification:
$2w(w - 3)$	Given equation
$2w = 0$ or $w - 3 = 0$	Zero Product Property
$w = 0$ or $w = 3$	Solved

Example 3

Solve: $(2x + 3)(4x - 5) = 0$

At least one binomial above equals zero. Thus, for the equation above to be true, either $2x + 3 = 0$ or $4x - 5 = 0$. We solve each equation.

Step:	**Justification:**	**Step:**
$2x + 3 = 0$	Zero Product Property	$4x - 5 = 0$
$2x = -3$	Isolated x-term	$4x = 5$
$x = -\dfrac{3}{2}$	Isolated x	$x = \dfrac{5}{4}$

We can check the solutions $x = -\frac{3}{2}$ and $x = \frac{5}{4}$ in the original equation.

Check $x = -\dfrac{3}{2}$:	**Justification:**	**Check $x = \dfrac{5}{4}$:**
$(2x + 3)(4x - 5) = 0$	Equation	$(2x + 3)(4x - 5) = 0$
$\left[2\left(-\dfrac{3}{2}\right) + 3\right]\left[4\left(-\dfrac{3}{2}\right) - 5\right] = 0$	Substituted	$\left[2\left(\dfrac{5}{4}\right) + 3\right]\left[4\left(\dfrac{5}{4}\right) - 5\right] = 0$
$(-3 + 3)(-6 - 5) = 0$	Simplified	$\left(\dfrac{5}{2} + 3\right)(5 - 5) = 0$
$(0)(-11) = 0$	Simplified	$\left(\dfrac{11}{2}\right)(0) = 0$
$0 = 0 \checkmark$	Simplified	$0 = 0 \checkmark$

We have confirmed that $-\frac{8}{2}$ and $\frac{5}{4}$ are solutions to the equation.

Practice Set

Find two solutions for each equation.

a. $(x - 3)(x - 4) = 0$ **b.** $(x + 2)(x - 5) = 0$

c. $3x(x - 1) = 0$ **d.** $(x + 3)(x + 4) = 0$

e. $(2x - 5)(4x + 1) = 0$

f. $(9x - 1)(2x + 3) = 0$

g. $(4x + 5)(2x - 1) = 0$

LESSON 95A

Written Practice

26. Find an equation for the line that passes through $(8, 9)$ with slope $\frac{1}{5}$.
(A68)

27. Simplify: $3\sqrt{8} + 2\sqrt{18}$
(A84)

28. Simplify: $4\sqrt{18} + \sqrt{50}$
(A84)

29. Find x by substitution: $\begin{cases} x - 2y = 5 \\ y = -3 \end{cases}$
(A92)

30. Find both solutions: $(x + 3)(x - 1) = 0$
(A95)

26. Write an equation for the line that passes through $(-1, 5)$ with slope $-\frac{7}{3}$.
(A68)

27. $(9x - 8y + 7) + (6y - 5 + 4x)$
(A80)

28. Solve: $\dfrac{x - 2}{5} = \dfrac{2x + 3}{15}$
(A83)

29. Find y by substitution: $\begin{cases} 3x - 4y = -2 \\ x = 2 \end{cases}$
(A92)

30. Find both solutions: $(x + 4)(x - 5) = 0$
(A95)

LESSON A97

New Concept

Writing Equations with Two Variables

In this lesson we will practice solving problems by writing a system of equations. Suppose we know that John sent 19 pieces of mail, some letters and some postcards. Then we can write an equation about the number of letters and postcards he sent.

$$x + y = 19$$

In this equation, x represents the number of letters and y represents the number of postcards John sent. This single equation has two variables.

Discuss Can we solve the equation to find how many letters John sent and how many postcards he sent?

To find how many letters or postcards John sent we need more information.

Suppose we also know that John spent 39¢ postage for each letter, 24¢ for each postcard, and $6.06 in all. We can use this information to write the following equation about the cost to mail the letters and postcards.

$$39x + 24y = 606$$

This equation is about money. The term $39x$ means 39¢ times x. The term $24y$ means 24¢ times y, and 606 means 606¢.

Thinking Skill

Explain

What is one way to solve a system of equations?

Now we have two different equations about the same situation. We have two equations in two unknowns, or a **system of equations.** If we have as many equations as we have unknowns, we can solve the system of equations.

Analyze Test possible solution pairs to determine the number of postcards and the number of letters John sent.

Example

On weekends Jose mows lawns and washes cars to earn spending money.

a. One month the number of lawns Jose mowed equaled the number of cars he washed. Write an equation about the number of lawns he mowed and the number of cars he washed.

b. Jose earns eight dollars for mowing a lawn and seven dollars for washing a car. Write an equation that shows he earned ninety dollars that month mowing lawns and washing cars.

c. Find the number of cars washed and lawns mowed.

a. We let m stand for the number of lawns mowed and w for the number of cars washed during the month. We are told that the numbers were equal, so we write this equation.

$$m = w$$

b. Jose earns 8 dollars for each lawn mowed, so $8m$ is how much he earns for m lawns. He earns 7 dollars for each car washed, so he earns $7w$ for w cars. He earned 90 dollars in the month, so we can write this equation about money. $8m + 7w = 90$

c. One way to find the number of jobs Jose did is to try some numbers and adjust the choices, that is, guess and check. We know the number of lawns and cars are equal so we could try 10 and 10.

$$8(10) + 7(10) \neq 90$$

We see 10 is way too many. We try 5.

$$8(5) + 7(5) \neq 90$$

Five is too few, but is close. We try 6.

$$8(6) + 7(6) = 90$$

We find **Jose washed 6 cars and mowed 6 lawns.**

Connect Show how to solve this system of equations by substitution.

Practice Set

Write a system of equations for each problem.

a. Sally sells shells for 70¢ each and tumbled rocks for 20¢ each. Write an equation showing she sells a total of 60 shells and rocks. Write another equation that shows her sales total $33.

b. **Formulate** During the basketball game Kwan made 14 of his 24 shots. His long shots were 3 points each. His short shots were 2 points each. Write an equation that shows how many shots Kwan made. Write another equation that shows he made 32 points. Then guess and check some pairs of numbers to find the number of each kind of shot Kwan made.

c. On the shelf were large and small soup cans. The large soup cans weigh 26 ounces each and the small soup cans weigh 12 ounces each. Write an equation that shows there were 12 soup cans. Write another equation that shows the total weight was 172 ounces. Then find the number of large and small cans by guessing and checking.

d. Xavier says a number Yoli triples it. Write an equation relating the numbers Xavier and Yoli say. Write another equation that shows that one time Xavier and Yoli's numbers added to 8. Find the two numbers by graphing the two equations.

26. Write an equation for the line that passes through (50, 50) and
(A71) $(-10, -10)$.

27. Simplify: $2\sqrt{12} + 4\sqrt{75}$
(A84)

28. Solve by substitution: $\begin{cases} 5x - 4y = 20 \\ y = x - 5 \end{cases}$
(A92)

29. Find both solutions: $(x - 1)(x - 10) = 0$
(A95)

30. Sven earns three points for each correct answer, and loses one point for
(A97)   each incorrect answer. Write an equation that shows he earns 78 points.
Write another equation that shows he answered all 30 questions.

LESSON A98

New Concept

Function Notation

Consider the following scenario:

Fernando manufactures crayons. His machine produces twelve crayons for
every block of crayon mixture he puts into the machine. To show how many
crayons his machine produces with one, two, and three blocks of mixture,
Fernando writes:

1 block into Fernando's machine	$f(1) = 12$	produces 12 crayons
2 blocks into Fernando's machine	$f(2) = 24$	produces 24 crayons
3 blocks into Fernando's machine	$f(3) = 36$	produces 36 crayons

Fernando writes this general rule to represent that when x blocks are put into
the machine $12x$ crayons are produced.

$$\underset{\uparrow}{\text{input}} \quad \overset{\text{output}}{\overbrace{}}$$
$$f(x) = 12x$$

Fernando's expression is an example of a function, and it is written in
customary function notation. The letter f is the name of Fernando's function.
To read the function we say, "f of x equals $12x$." Using function notation we
replace y with the name of the function.

Example 1

Complete the table using Fernando's function: $f(x) = 12x$

x	f(x)
1	12
2	24
3	36
4	
5	

Solution

To find the numbers for the $f(x)$ column we substitute numbers from the x
column for x in $12x$. We find $f(x)$ for $x = 4$ and for $x = 5$.

$$f(4) = 12(4) \qquad\qquad f(5) = 12(5)$$
$$= 48 \qquad\qquad\qquad\quad = 60$$

Thinking Skill

Explain

How do we read each line of this function table?

x	f(x)
1	12
2	24
3	36
4	48
5	60

To graph Fernando's function we replace $f(x)$ with y. That is, we graph the points $(x, f(x))$. At right is a graph of Fernando's function. Since only whole numbers of blocks can be put into Fernando's machine, the domain of the function that makes physical sense is the set of the whole numbers. Therefore the graph is aligned points rather than a line.

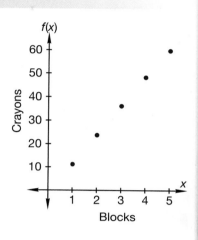

Example 2

Olga drives east on Interstate 10 from El Monte, which is 13 miles east of Los Angeles. She drives 60 mph, and her freeway distance from Los Angeles after *x* hours of driving is given by the function:

$$d(x) = 60x + 13$$

Find her distance from Los Angeles after $\frac{1}{4}$ hr and $\frac{1}{2}$ hr of driving.

Solution

This function is read, "*d* of *x* equals 60*x* plus 13."

We find $d(x)$ for $x = \frac{1}{4}$, which is Olga's distance from Los Angeles after a quarter hour of driving.

Step:	Justification:
$d\left(\frac{1}{4}\right) = 60\left(\frac{1}{4}\right) + 13$	Replaced x with $\frac{1}{4}$
$d\left(\frac{1}{4}\right) = 15 + 13$	Simplified
$d\left(\frac{1}{4}\right) = 28$	Simplified

We have found the distance after $\frac{1}{4}$ hour. Now we find $d(x)$ for $x = \frac{1}{2}$.

Step:	Justification:
$d\left(\frac{1}{2}\right) = 60\left(\frac{1}{2}\right) + 13$	Replaced x with $\frac{1}{2}$
$d\left(\frac{1}{2}\right) = 30 + 13$	Simplified
$d\left(\frac{1}{2}\right) = 43$	Simplified

After $\frac{1}{4}$ hr, Olga is **28 miles** from Los Angeles. After $\frac{1}{2}$ hr, she is **43 miles** from Los Angeles.

Generalize Find $d(x)$ for $x = 0$. What does this distance mean in the context of the problem?

a. Use words to show how this expression is read.

$$f(x) = 2x - 1$$

Evaluate each function for the indicated value.

b. For $f(x) = 4x + 1$, find $f(5)$.   **c.** For $g(x) = \frac{1}{2}x + 2$, find $g(10)$.

d. The parking lot is open from 7 a.m. to 6 p.m. The lot charges $1.50 per hour or fraction thereof. That is, the time in the lot is rounded up to the next hour.

$$f(x) = 1.5x$$

Find $f(3)$.

LESSON A98

Written Practice

26. Graph the following sequence: $a_n = 2n - 2$
(A73)

27. Solve by substitution: $\begin{cases} 3x + 2y = 16 \\ x = 2y \end{cases}$
(A92)

28. Find both solutions: $(x - 7)(x) = 0$
(A95)

29. Lucy earns 5¢ for selling one cup of lemonade and 10¢ for giving advice. Write an equation that shows she earns $2.10. Write another equation that shows she sells lemonade the same number of times as she gives advice.
(A97)

30. Use the function $f(x) = 3x - 1$ to find $f(10)$.
(A98)

LESSON A99

New Concept

Solving Systems of Equations by Elimination, Part 1

We have simplified equations by adding equal quantities to both sides of an equation.

Step:	Justification:
$x - 7 = 12$	Given equation
$x - 7 + 7 = 12 + 7$	Added 7 to both sides
$x = 19$	Simplified

In the second step above, adding 7 to both sides of the equation maintained the equality. Recall that this step employs the Addition Property of Equality, which is expressed symbolically this way:

$$\text{If } a = b,$$

$$\text{then } a + c = b + c$$

Informally we could say it this way: If equals are added to equals, the sums are equal.

We may use this property as an alternative method to solve systems of equations. We add two equations to form a third equation that we can solve.

Step:	Justification:

$$\begin{cases} 2x + 3y = 16 \\ 2x - 3y = 4 \end{cases}$$

First equation

Second equation

$$4x = 20$$

Third equation (sum of first and second)

Notice that the y-term does not appear in the third equation. It was **eliminated** by adding the first two equations because $3y$ and $-3y$ are opposites. In the third equation there is one unknown which is easily found.

Below we use pictures of balance scales to illustrate the concept of adding equations. We see that $2x + 3y$ and 16 are balanced (equal). We also see that $2x - 3y$ and 4 are equal.

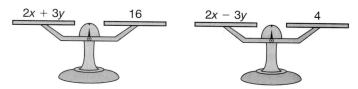

Now we combine these quantities on one balance scale. The scale remains balanced because we are adding equals to equals.

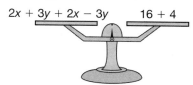

We simplify the quantities on both sides of the scale.

The y-term has been eliminated and we can find x.

$$4x = 20$$

$$x = 5$$

Now that we know the value of one of the variables, we can substitute its value into either the first or second equation and solve for y.

Step:	Justification:
$2x + 3y = 16$	Given equation
$2(5) + 3y = 16$	Substituted 5 for x
$3y = 6$	Subtracted 10 from each side.
$y = 2$	Divided both sides by 3

Thinking Skill

Verify

How can you check that (5, 2) is the solution of the system of equations?

We find that y is 2. Thus (5, 2) is the x, y pair that satisfies both of the original equations.

The strategy of solving a system of equations by adding equations to eliminate a variable term is called **solving by elimination.**

We follow four steps to solve a system of two equations in two unknowns by elimination.

1. Add two equations to form a third equation that has one variable.

2. Solve the third equation.

3. Substitute the solution of the third equation into one of the original equations and solve for the other variable.

4. Check the solution in the other equation.

Example

Solve by elimination: $\begin{cases} 2y + 3 = 7 \\ -2y - x = -9 \end{cases}$

Solution

Step 1: We add the two equations to make a third equation. Since the coefficients of the y-terms are opposites, the sum of the y-terms is zero, and the y-term is eliminated.

Step:	Justification:
$\begin{cases} 2y + 3x = 7 \\ -2y - x = -9 \end{cases}$	First equation Second equation
$\qquad\quad 2x = -2$	Third equation

Step 2: We solve the third equation.

$2x = -2$	Third equation
$x = -1$	Divided by 2

Step 3: Now we substitute -1 in place of x in either the first or second equation. Here we show substitution in the first equation.

Step:	Justification:
$2y + 3x = 7$	First equation
$2y + 3(-1) = 7$	Substituted -1 for x
$2y - 3 = 7$	Simplified
$2y = 10$	Added 3 to both sides
$y = 5$	Divided both sides by 2

We find that x is -1 and y is 5. We can write the solution this way: $(-1, 5)$.

Step 4: We check the system by substituting -1 for x and 5 for y in the other original equation.

Step:	Justification:
$-2y - x = -9$	Second equation
$-2(5) - (-1) = -9$	Substituted $(-1, 5)$ for x and y
$-10 + 1 = -9$	Simplified
$-9 = -9$	Simplified

Discuss In a previous lesson, we solved systems of equations by using substitution. This method involves solving one equation for a variable and substituting this solution into the second equation. When we are given a system of equations to solve, how might we decide whether to use elimination or substitution?

Practice Set

Solve by elimination:

a. $\begin{cases} 2x + 5y = 9 \\ -2x + 7y = 3 \end{cases}$

b. $\begin{cases} 6x + y = 15 \\ 4x - y = 5 \end{cases}$

c. $\begin{cases} 4x - 2y = -4 \\ -x + 2y = -2 \end{cases}$

d. $\begin{cases} -x + y = -10 \\ x - 8y = 3 \end{cases}$

LESSON A99

Written Practice

26. Solve: $\dfrac{2}{3x} = \dfrac{8}{2x + 10}$
(A83)

27. Find both solutions: $(x + 5)(x + 1) = 0$
(A95)

28. Basic items (x) cost \$4. Deluxe items (y) cost \$8. Daisy spends \$24 on the items. Show this in an equation. She buys 5 items. Show this in another equation.
(A97)

29. Find $f(-4)$ for the function $f(x) = x^2 + 3$.
(A98)

30. Solve by elimination. $\begin{cases} 3x - 4y = 5 \\ x + 4y = 3 \end{cases}$
(A99)

LESSON A100

Written Practice

26. Write an equation for the line that passes through $(-2, -2)$ and $(4, 1)$.
(A71)

27. Solve by substitution: $\begin{cases} -5x + y = 8 \\ y = 3x - 2 \end{cases}$
(A92)

28. Joseph pays \$2 for each widget and \$3 for each gadget. Write an equation that shows he spends \$30. Write another equation that shows he buys 12 items.
(A97)

29. Find $f(50)$ for the function $f(x) = \frac{1}{5}x - 4$.
(A98)

30. Solve by elimination. $\begin{cases} -x + 3y = 7 \\ x + 2y = -2 \end{cases}$
(A99)

LESSON A101

New Concept

Factoring Quadratics

Consider the following problems.

- Max is thinking of two numbers. Their product is six and their sum is five. What are the two numbers?

- Briana is thinking of two numbers. Their product is six and their sum is negative seven. What are the two numbers?

- The product of two numbers is negative six and their sum is negative one. What are the two numbers?

The mental process we use to solve these problems can also be used to factor quadratic expressions.

Since Lesson 92 we have multiplied binomials. Often the products have been quadratic equations. For example,

$$(x + 1)(x + 2) = x^2 + 3x + 2$$

In this lesson we will reverse the process. We will find the two binomials that produce the quadratic expression.

$$(\,?\,)(\,?\,) = x^2 + 3x + 2$$

To factor quadratics we need to understand how the terms of the quadratic are formed. If the coefficient of the x^2 term is 1, then the following applies.

Constant is product of last terms
Coefficient of x is sum of last terms
Product of first terms

$$x^2 + 3x + 2$$

Example 1

Factor: $x^2 + 5x + 6$

Solution

The task is to find the two binomials whose product is $x^2 + 5x + 6$. The x^2 term is the product of x and x, so we begin by writing

$$(x \quad)(x \quad)$$

For the last terms of the binomials, we find two numbers whose product is positive 6 and whose sum is positive 5.

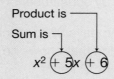

Product is
Sum is

$$x^2 + 5x + 6$$

Justify Which factor pair for 6 should we choose, 6 and 1 or 3 and 2?

We complete the two binomials.

$$(x + 3)(x + 2)$$

We can check the factors by multiplying the binomials to see if the product is the original quadratic expression.

Step:	Justification:
$(x + 3)(x + 2)$	Binomial factors
$x^2 + 2x + 3x + 6$	Multiplied
$x^2 + 5x + 6$	Simplified

Example 2

Factor: $x^2 - 7x + 6$

Solution

The first terms of the binomial factors are x and x.

$$(x\ )(x\ )$$

The product of the last terms is positive 6 and the sum is negative 7.

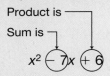

The product of two negative numbers is a positive number. For the sum to be -7, the factors must be -6 and -1.

$$(x - 6)(x - 1)$$

Example 3

Factor: $x^2 - x - 6$

Solution

We begin with the factors of x^2.

$$(x\ )(x\ )$$

The product of the last terms is -6. The coefficient of the x term is -1, so the sum of the factors is -1.

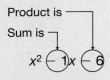

For the product to be -6, one factor is negative and the other is positive. Our choices are

| -6 and $+1$ | $+6$ and -1 | -3 and $+2$ | $+3$ and -2 |

The pair whose sum is -1 is -3 and $+2$.

$$(x - 3)(x + 2)$$

Generalize Write a rule that describes how the sign of the last term in a quadratic polynomial determines whether the last terms in the binomial factors are both positive, both negative, or one negative and one positive.

Example 4

Factor: $x^2 - 16$

Solution

Notice that there is no x term. This means that the coefficient of x is zero.

$$x^2 + 0x - 16$$

We look for factors of -16 whose sum is zero. The factors are $+4$ and -4.

$$(x + 4)(x - 4)$$

Practice Set

Evaluate Factor each quadratic polynomial. Check your work by multiplying the binomials.

a. $x^2 + 7x + 10$ **b.** $x^2 + 12x + 35$

c. $x^2 - 3x + 2$ **d.** $x^2 - 8x + 12$

e. $x^2 - 4x - 5$ **f.** $x^2 + 2x - 8$

g. $x^2 - 9$ **h.** $x^2 - 100$

LESSON A101

Written Practice

26. Solve: $\dfrac{5}{2x} = \dfrac{3}{x + 1}$
(A83)

27. Solve by substitution: $\begin{cases} -2x + 3y = 9 \\ x = y + 4 \end{cases}$
(A92)

28. Gertrude pays \$3 for carnations and \$5 for roses. Write an equation
(A97) that shows she spends \$45. Write another equation that shows she purchases 10 flowers.

29. Find $f(3)$ in $f(x) = x^2 + x - 5$
(A98)

30. Solve by elimination. $\begin{cases} 2x - 5y = 11 \\ -2x - 6y = 0 \end{cases}$
(A99)

31. Factor: $x^2 + 8x + 12$
(A101)

32. Factor: $x^2 - 5x + 6$
(A101)

33. Factor: $x^2 + 8x + 15$
(A101)

34. Factor: $x^2 - 81$
(A101)

35. Factor: $x^2 - 7x + 12$
(A101)

LESSON A102

New Concept

Solving Systems of Equations by Substitution, Part 2

We have solved systems of equations with the substitution method. For some systems of equations we need to manipulate an equation before we can substitute. We begin by solving one of the equations for one of the variables. Then we substitute for this variable in the other equation.

Example 1

Solve: $\begin{cases} 2x + 8y = 12 \\ x - 3y = -8 \end{cases}$

We decide to solve the second equation for x because its coefficient is 1.

Step:	**Justification:**
$x - 3y = -8$	Second equation
$x = -8 + 3y$	Added $3y$ to both sides
$x = 3y - 8$	Commutative Property

Next, since x equals $3y - 8$, we substitute $3y - 8$ in place of x in the first equation. This gives us one equation with one unknown.

Step:	**Justification:**
$2x + 8y = 12$	First equation
$2(3y - 8) + 8y = 12$	Substituted $3y - 8$ for x
$6y - 16 + 8y = 12$	Distributive Property
$14y - 16 = 12$	Combined $6y$ and $8y$
$14y = 28$	Added 16 to both sides
$y = 2$	Divided both sides by 14

We have found that y equals 2. Now to find x we substitute 2 for y in either equation.

Step:	**Justification:**
$x - 3y = -8$	Second equation
$x - 3(2) = -8$	Substituted 2 for y
$x - 6 = -8$	Simplified
$x = -2$	Added 6 to both sides

We have solved for both x and y. We can write the solution **(−2, 2).**

To check that $(-2, 2)$ is a solution to both equations, we can substitute -2 for x and 2 for y in both equations.

Example 2

Solve: $\begin{cases} -x + 2y = -1 \\ -x - y = 2 \end{cases}$

Thinking Skill

Conclude

Could we chose to solve the equation $-x + y = 2$ for x instead of for y? Would this have affected the solution of the system?

We choose to solve the second equation for y because its coefficient is positive 1.

Step:	Justification:
$-x + y = 2$	Second equation
$y = x + 2$	Added x to both sides

Next we substitute $x + 2$ for y in the other equation.

Step:	Justification:
$-x + 2y = -1$	First equation
$-x + 2(x + 2) = -1$	Substituted $x + 2$ for y
$-x + 2x + 4 = -1$	Distributive Property
$x + 4 = -1$	Combined $-x$ and $2x$
$x = -5$	Subtracted 4 from both sides

Now we substitute -5 for x in either equation.

Step:	Justification:
$-x + y = 2$	Second equation
$-(-5) + y = 2$	Substituted -5 for x
$5 + y = 2$	Simplified $-(-5)$
$y = -3$	Subtracted 5 from both sides

The solution to the system of equations is **(−5, −3)**.

Practice Set

Solve the following systems.

a. $\begin{cases} 2x + y = 4 \\ -x + 3y = -2 \end{cases}$

b. $\begin{cases} x + 4y = 3 \\ 2x - y = 15 \end{cases}$

c. $\begin{cases} 2x + 6y = -8 \\ x + 4y = -4 \end{cases}$

d. $\begin{cases} x - 7y = 15 \\ -2x + 5y = -3 \end{cases}$

e. **Justify** How do you know that your solution of the system in problem **d** is correct?

LESSON A102

Written Practice

26. Write an equation for the line that passes through $(-2, 3)$ with slope $-\frac{5}{7}$.
(A68)

27. Simplify: $2\sqrt{27} + \sqrt{12}$
(A84)

28 Frank earns 5 points for each correct answer in a game and loses a point for each incorrect answer. He earned a total of 138 points. He answered all 30 questions. Write two equations showing this.
(A97)

29. Solve by elimination. $\begin{cases} 3x + 4y = 5 \\ 2x - 4y = 15 \end{cases}$
(A99)

30. Simplify: $(3x + 2y - 5) - (2y + 4x - 3)$
(A80)

31. Factor: $x^2 + 9x + 20$
(A101)

32. Factor: $x^2 - 3x - 4$
(A101)

33. Factor: $x^2 + x - 30$
(A101)

34. Solve by substitution. $\begin{cases} 5x - 2y = 3 \\ -2x + y = 5 \end{cases}$
(A92, A102)

35. Solve by substitution. $\begin{cases} -x + 2y = -1 \\ -x + y = 2 \end{cases}$
(A92, A102)

LESSON A103

Written Practice

26. Write an equation for the line that passes through $(-2, 5)$ and $(3, -5)$.
(A71)

27. Solve: $\dfrac{4}{3 + x} = \dfrac{6}{5 + x}$
(A83)

28. Find both solutions: $(x + 2)(x - 6) = 0$
(A95)

29. Angie carries nickels and dimes. Write an equation that shows she has 12 coins. Write another equation that shows she has \$1.05.
(A97)

30. Solve by elimination. $\begin{cases} 4x + 3y = 9 \\ 7x - 3y = -9 \end{cases}$
(A99)

31. Factor: $x^2 - 10x + 25$
(A101)

32. Factor: $x^2 + 6x + 9$
(A101)

33. Factor: $x^2 + 11x + 24$
(A101)

34. Solve by substitution. $\begin{cases} x - 4y = 1 \\ 2x - 3y = 7 \end{cases}$
(A92, A102)

35. Solve by substitution. $\begin{cases} 2x + 5y = 90 \\ 4x + y = 0 \end{cases}$
(A92, A102)

LESSON A104

New Concept

Solving Systems of Equations by Elimination, Part 2

The following system of equations can be solved using graphing, substitution, or elimination. In this lesson we will learn more about solving systems of equations by elimination. Consider the following equations.

$$\begin{cases} x + y = 5 \\ 2x + y = 7 \end{cases}$$

Notice that adding these equations in their current form does not eliminate a variable. However, if we multiply both sides of one of the equations by -1, the y terms will be opposites. Then we can add the equations and eliminate the y term. We choose to multiply the first equation by -1. The result is $-x - y = -5$.

Step:	Justification:
$x + y = 5$	First equation
$-x - y = -5$	Multiplied both sides by -1

Now we add this modified equation and the second equation.

$$\begin{cases} -x - y = -5 \\ 2x + y = 7 \end{cases}$$ First equation modified
Second equation
$$\overline{ x = 2}$$ Added equations; y is eliminated

We find that x is 2. Substituting 2 for x in either original equation gives a y-value of 3. Instead of eliminating y, we could have eliminated x by multiplying the first equation by -2 and adding.

Step:	Justification:
$x + y = 5$	First equation
$\begin{cases} -2x - 2y = -10 \\ 2x + y = 7 \end{cases}$	Multiplied both sides by -2 Second equation
$-y = -3$	Added modified first and second equation
$y = 3$	Divided both sides by -1

By substituting 3 for y in either original equation, we find $x = 2$. Using either method, we find the solution to the system of equations is (2, 3).

There are many ways to eliminate one variable in a system of equations. The strategy is to make the coefficients of one variable opposites by multiplying both sides of one or both equations by selected factors.

Example

Solve by elimination.

$$\begin{cases} 2x + 5y = 7 \\ -x - 3y = -1 \end{cases}$$

Solution

We study the two equations to decide which variable to eliminate. If we multiply the second equation by 2, the x term becomes $-2x$, which is the opposite of $2x$ in the first equation.

Step:	Justification:
$-x - 3y = -1$	Second equation
$\begin{cases} -2x - 6y = -2 \\ 2x + 5y = 7 \end{cases}$	Multiplied both sides by 2 First equation
$-y = 5$	Added equations
$y = -5$	Divided both sides by -1

We found y. Now, we substitute -5 for y in either equation to find x.

Step:	Justification:
$-x - 3y = -1$	Second equation
$-x - 3(-5) = -1$	Substituted -5 for y
$-x + 15 = -1$	Simplified
$-x = -16$	Subtracted 15 from both sides
$x = 16$	Divided both sides by -1

Eliminate a variable to solve the following systems of equations.

a. $\begin{cases} 3x + y = 21 \\ -x + y = -7 \end{cases}$ **b.** $\begin{cases} x - 3y = 14 \\ -2x - 2y = 4 \end{cases}$

c. $\begin{cases} 3x + 5y = 4 \\ x - y = 4 \end{cases}$ **d.** $\begin{cases} -2x - y = 3 \\ 7x + 8y = 12 \end{cases}$

LESSON A104

Written Practice

26. Graph the following sequence: $a_n = n(n - 1)$
(A73)

27. Solve: $\dfrac{3x}{2 + x} = \dfrac{9}{4}$
(A83)

28. Find both solutions: $(x + 3)(x - 1) = 0$
(A95)

29. George found 48 nickels and dimes on the ground. They totaled \$4.40 in value. Write two equations that express this.
(A97)

30. Find $f(0)$ in $f(x) = 12x^3 + 4x^2 - 3x + 5$
(A98)

31. Factor: $x^2 + 4x - 5$
(A101)

32. Factor: $x^2 - 6x + 8$
(A101)

33. Solve by substitution. $\begin{cases} -3x + y = 1 \\ 2x + 2y = 10 \end{cases}$
(A92, A102)

34. Solve by elimination. $\begin{cases} y = -x + 3 \\ y = 2x - 3 \end{cases}$
(A99, A104)

35. Solve by elimination. $\begin{cases} 2x + 2y = 4 \\ 3x - 2y = 1 \end{cases}$
(A104)

LESSON 105A

New Concept

Solving Quadratic Equations by Factoring, Part 1

We have factored quadratic equations since Appendix Lesson 101. We can use factoring to solve some quadratic equations.

Example 1

Solve by factoring: $x^2 + 5x + 6 = 0$

Solution

One side of the equation is zero. We factor the other side.

Step:	Justification:
$x^2 + 5x + 6 = 0$	Given equation
$(x + 3)(x + 2) = 0$	Factored

The product of the two binomials is 0. Therefore, one binomial or the other must equal zero.

either $x + 3 = 0$	Zero Product Property
or $x + 2 = 0$	
$x = -3$ or $x = -2$	Solved

We check the two solutions by substituting each solution in the original equation. We will show the steps using both solutions on either side of each step's justification.

Check $x = -3$:	Justification:	Check $x = -2$:
$x^2 + 5x + 6 = 0$	Equation	$x^2 + 5x + 6 = 0$
$(-3)^2 + 5(-3) + 6 = 0$	Substituted	$(-2)^2 + 5(-2) + 6 = 0$
$9 + (-15) + 6 = 0$	Simplified	$4 - 10 + 6 = 0$
$0 = 0 \checkmark$	Simplified	$0 = 0 \checkmark$

Example 2

Solve $x^2 - 6x + 9 = 0$ by factoring. Check your work.

Solution

Step:	Justification:
$x^2 - 6x + 9 = 0$	Given equation
$(x - 3)(x - 3) = 0$	Factored

Since the binomial factors are identical, we set one to zero and solve for x.

$x - 3 = 0$	Zero Product Property
$x = 3$	Added 3 to both sides

To check the solution we replace x with 3.

Check $x = 3$:	Justification:
$x^2 - 6x + 9 = 0$	Equation
$(3)^2 - 6(3) + 9 = 0$	Substituted 3 for x
$9 - 18 + 9 = 0$	Simplified
$0 = 0$	Simplified

Example 3

Solve $x^2 + 3x - 4 = 0$ by factoring.

Solution

Step:	Justification:
$x^2 + 3x - 4 = 0$	Given equation
$(x + 4)(x - 1) = 0$	Factored
$x + 4 = 0$ or $x - 1 = 0$	Zero Product Property
$x = -4$ or $x = 1$	Solved

Verify How can we be sure that both -4 and 1 are solutions?

Practice Set   Solve each equation by factoring.

a. $x^2 + 7x + 10 = 0$

b. $x^2 - 3x + 2 = 0$

c. $x^2 + 12x + 35 = 0$

d. $x^2 - 11x + 24 = 0$

e. $x^2 - x - 6 = 0$ **f.** $x^2 + 5x - 6 = 0$

LESSON A105

Written Practice

26. Simplify: $2\sqrt{20} + \sqrt{45}$
(A84)

27. Solve by substitution: $\begin{cases} 3x - y = 7 \\ x = y - 1 \end{cases}$
(A92)

28. Samantha had 21 nickels and dimes. They totaled $1.60. Write two
(A97) equations that show this.

29. Find $f(-1)$ for $f(x) = 2x^2 + 3x + 4$
(A98)

30. Simplify: $(x^2 - 5x) - (x^2 - 10)$
(A80)

31. Solve by substitution. $\begin{cases} 3x + 2y = 8 \\ x - 3y = -1 \end{cases}$
(A92, A102)

32. Solve by elimination. $\begin{cases} 4x + y = 7 \\ -2x + 3y = 7 \end{cases}$
(A99, A104)

33. Solve by elimination. $\begin{cases} 2x + 5y = 7 \\ 3x + 5y = 8 \end{cases}$
(A99, A104)

34. Solve by factoring: $x^2 + 6x + 5 = 0$
(A105)

35. Solve by factoring: $x^2 + 4x + 4 = 0$
(A105)

LESSON 106A

New Concept

Graphing Linear Inequalities on the Coordinate Plane, Part 1

Consider the following situation.

When Xavier says a number, Yoli says a greater number. Graph all the pairs of numbers Xavier and Yoli could say.

Some pairs of numbers Xavier and Yoli could say are shown below.

Xavier	Yoli
3	7
5	200
$\frac{1}{2}$	27
-3	-1

This list could continue endlessly because any pair of numbers in which Yoli's number is greater than Xavier's number meets the conditions described. If we let y equal Yoli's number and x equal Xavier's number, we can write the relationship this way:

$$y > x$$

We can graph all the pairs of numbers that satisfy this inequality on a coordinate plane. We begin by graphing the equation $y = x$.

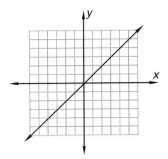

The line shows all the number pairs in which y is equal to x. Since we want to graph all pairs of numbers (x, y) in which y is greater than x, we shade the region above the line $y = x$.

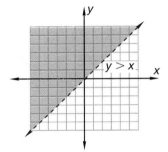

Every point in the shaded region above the line has a y value greater than its x-value. Thus, the coordinates of any point in the shaded region above the line are a pair of numbers that Xavier and Yoli could say.

Since we only want to show y values greater than x, we need to make a change to the line $y = x$. We want this line to serve as a border to the $y > x$ region, but we do not want to include the points where $y = x$, so we make the line broken instead of solid. A broken line indicates that points on the line are not included in the graph of the solution. The graph below represents the **solution set** of the inequality $y > x$.

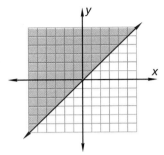

Verify Select a point above the line in the shaded region and another below the line in the unshaded region. For each point, substitute the x- and y-values in the inequality $y > x$. For which point chosen, is the inequality true? For which is it false?

Example 1

Graph $y \leq 3x - 6$.

Solution

First we graph $y = 3x - 6$. The line is solid and not broken because points on the line are included.

Identify Describe how you know that the line is solid and not broken.

Then we determine the region where $y = 3x - 6$. We find y-values **less than** $3x - 6$ **below** the line. We shade the region below the line and then check our solution.

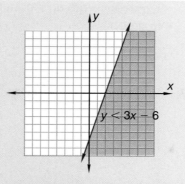

We can check our results by selecting points inside or outside the shaded region to determine whether or not the (x, y) values satisfy the inequality.

The point $(0, 0)$ is not shaded, so we should get a false statement when we substitute 0 for x and 0 for y in the inequality. The point $(4, 2)$ is shaded, so we should get a true statement substituting 4 for x and 2 for y.

Check (0, 0):	**Justification:**	**Check (4, 2):**
$y \leq 3x - 6$	Given inequality	$y \leq 3x - 6$
$0 \leq 3(0) - 6$	Substitute	$2 \leq 3(4) - 6$
$0 \leq -6$ False	Simplified	$2 \leq 6$ True

The results support the solution.

Example 2

Graph on a coordinate plane:

    a. $x > -3$ **b.** $x \leq 0$

Solution

    a. $x > -3$ **b.** $x \leq 0$

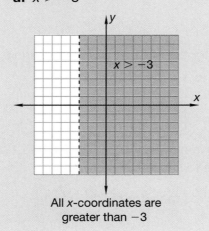 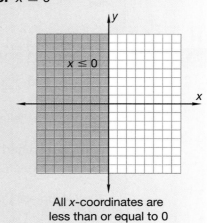

All x-coordinates are   All x-coordinates are
greater than -3 less than or equal to 0

Practice Set

Graph the inequalities in **a–d.**

    a. $y < 2x + 4$ **b.** $y \geq \frac{1}{3}x + 1$

    c. $x \geq 2$ **d.** $y < 4$

e. *Formulate* Write an inequality that describes this graph.

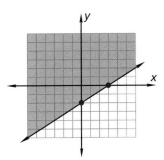

LESSON A106

Written Practice

26. Write an equation for the line that passes through $(10, -3)$ and $(-2, 3)$.
(A71)

27. Simplify: $\sqrt{196} - \sqrt{169}$
(A84)

28. Find both solutions: $(3x + 2)(4x - 5) = 0$
(A95)

29. Mr. Allen earns 3 dollars for every car he parks correctly, and loses $10 for every car that he parks crookedly. He parks 30 cars. He makes $64. Write two equations that show this information.
(A97)

30. Find $g(8)$ if $g(x) = 5 + \frac{1}{2}x$
(A98)

31. Solve by substitution. $\begin{cases} x + 4y = -1 \\ 2x - y = 7 \end{cases}$
(A92, A102)

32. Solve by elimination. $\begin{cases} -2x - y = 26 \\ -2x + 3y = 18 \end{cases}$
(A99, A104)

33. Solve by factoring: $x^2 + 9x + 8 = 0$
(A105)

34. Graph this inequality on a coordinate plane: $y \geq x - 1$
(A106)

35. Graph $y \geq 2x - 1$
(A106)

LESSON A107

Written Practice

26. Write an equation for the line that passes through $(-7, 4)$ and $(7, -2)$.
(A71)

27. Solve: $\dfrac{5 + x}{x - 5} = 11$
(A83)

28. Solve by elimination: $\begin{cases} x - 3y = 14 \\ 6x + 3y = 0 \end{cases}$
(A99)

29. Quentin had $6.40 in nickels and dimes. He had 78 nickels and dimes in all. Write equations to show this information.
(A97)

30. Complete the function table for $h(x) = \frac{1}{2}x^2 + 1$
(A98)

x	h(x)
0	1
1	
2	

31. Solve by substitution. $\begin{cases} 2x - 5y = 28 \\ -x + y = 10 \end{cases}$
(A92, A102)

32. Solve by elimination. $\begin{cases} 5x - 2y = 9 \\ 4x - 2y = 6 \end{cases}$
(A99, A104)

33. Solve by factoring: $x^2 - x - 20 = 0$
(A105)

34. Graph $y \leq \frac{2}{3}x + 3$
(A106)

35. Graph $y > 5x - 5$
(A106)

LESSON A108

New Concept

Graphing Linear Inequalities on the Coordinate Plane, Part 2

To graph an inequality with variables x and y, we may first need to transform the inequality to slope-intercept form.

Step:	**Justification:**
$2x + 3y > 6$	Given inequality
$3y > -2x + 6$	Subtracted $2x$ from both sides
$y > \frac{1}{3}(-2x + 6)$	Multiplied both sides by $\frac{1}{3}$
$y > -\frac{2}{3}x + 2$	Simplified to slope-intercept form

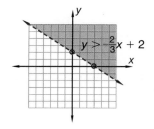

As a check to our work, we substitute the coordinates of any point in the *original inequality*. The point (0, 0) is not shaded, so we should get a false statement when we check it in the inequality.

Check (0, 0):	**Justification:**
$2(0) + 3(0) > 6$	Substituted (0, 0)
$0 > 6$ False	Simplified

Example 1

Graph the solution set of the inequality $5x - 2y \leq 4$.

Solution

We isolate y, being careful to reverse the inequality when we multiply or divide by a negative number.

Step:	**Justification:**
$5x - 2y \leq 4$	Given inequality
$-2y \leq -5x + 4$	Subtracted $5x$ from both sides
$y \geq -\frac{1}{2}(-5x + 4)$	Multiplied both sides by $-\frac{1}{2}$
$y \geq \frac{5}{2}x - 2$	Simplified

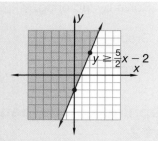

$$y \geq \frac{5}{2}x - 2$$

Check: The point (0, 0) is shaded and so should yield a true statement.

Check (0, 0):	Justification:
$5(0) - 2(0) \leq 4$	Substituted (0, 0)
$0 \leq 4$ True	Simplified

Example 2

Carlos wants to make individually sized pizzas for his friends—some cheese and some vegetarian. If he can fit only a total of 12 pizzas in his oven, we can use c for the number of cheese pizzas and v for the vegetarian pizzas he makes, and write:

c	+	v	≤	12
number of cheese pizzas		number of vegetarian pizzas		is less than or equal to 12

Graph this inequality. $c + v \leq 12$

Solution

These equations do not contain x and y. We are looking for (c, v) pairs. We make v the vertical axis and solve for v.

To graph $c + v \leq 12$, isolate the v:

$$v \leq -c + 12$$

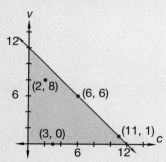

Any point in the shaded region satisfies the inequality, but only shaded points with whole-number coordinates make sense. Carlos cannot make a negative pizza, and he probably would not make a $\frac{2}{3}$ pizza. Some possible (c, v) pairs are shown on the graph.

Practice Set

Model Graph the inequalities.

a. $x + y \geq 5$

b. $x - y \leq 0$

c. $6x + 4y < 8$

d. $-3x - 7y \leq 14$

State whether the point $(0, 0)$ is in the solution set of the inequality.

e. $3x + 2y \geq 10$ **f.** $9x - 4y \leq 3$

LESSON A108
Written Practice

26. Write an equation for the line that passes through $(-2, 3)$ with slope -1.
(A68)

27. Simplify: $\sqrt{8} + \sqrt{8}$
(A84)

28. Franci has 19 nickels and dimes totaling $1.65. Write two equations to illustrate this.
(A97)

29. Find $g(3)$ for $g(x) = x^2 - 2x + 1$
(A98)

30. $(x^2 - 3x + 4) - (1 + 2x - 5x^2)$
(A80)

31. Solve by substitution. $\begin{cases} x + y = 13 \\ 4x + 2y = 40 \end{cases}$
(A92, A102)

32. Solve by elimination. $\begin{cases} 9x - y = 0 \\ x + y = 10 \end{cases}$
(A99, A104)

33. Solve by factoring: $x^2 + 14x + 33 = 0$
(A105)

34. Graph: $5y \geq x - 5$
(A108)

35. Graph: $3x + y \geq 7$
(A108)

LESSON A109
Written Practice

26. Write an equation for the line that passes through $(2, 8)$ and $(-4, 5)$.
(A71)

27. Solve: $= \dfrac{32}{3x} = \dfrac{8}{x - 1}$
(A83)

28. Simplify: $\sqrt{50} + \sqrt{50}$
(A84)

29. Pedro earns 4 points for every correct answer in a game, and loses one point for every incorrect answer. He answers 40 questions and earns 110 points. Write two equations that illustrate this.
(A97)

30. Find $g(6)$ for $g(x) = 2x^2 - 3x + 1$.
(A98)

31. Solve by substitution. $\begin{cases} 2x + y = 0 \\ 3x + 2y = 5 \end{cases}$
(A92, A102)

32. Solve by elimination. $\begin{cases} 2x - y = -1 \\ -x + 2y = 11 \end{cases}$
(A99, A104)

33. Solve by factoring: $x^2 + x - 6 = 0$
(A105)

34. Graph: $x + 2y < 4$
(A108)

35. Graph: $4x - 5y \leq 0$
(A108)

LESSON A110
New Concept

Solving Quadratic Equations by Factoring, Part 2

We have used factoring to solve quadratic equations in which the coefficient of the x^2 term was 1.

$$x^2 + 5x - 6 = 0$$

coefficient is 1

$$(x + 6)(x - 1) = 0$$

$$x = -6 \text{ or } x = 1$$

In this lesson we will factor some quadratic equations in which the x^2 term has a coefficient greater than 1.

Example 1

Solve by factoring. $2x^2 + 3x - 2 = 0$

Solution

The first terms of the binomial factors multiply to equal $2x^2$, so we make one x and the other $2x$.

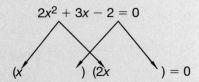

We need to be thoughtful when finding the second terms of the binomials. Since the last term of the quadratic is -2, one factor is positive and the other is negative. The following pairs of numbers produce -2 for the third term of the quadratic.

$$+1 \text{ and } -2 \qquad -1 \text{ and } +2 \qquad +2 \text{ and } -1 \qquad -2 \text{ and } +1$$

We need to identify the pair of numbers that results in $+3x$ for the middle term of the quadratic. Recall that the middle term is the sum of the inner and outer products when the binomials are multiplied. Below we try the four options.

$(x + 1)(2x - 2) = 0$	$(x - 1)(2x + 2) = 0$
Middle term is $-2x + 2x = 0$	Middle term is $+2x - 2x = 0$
$(x + 2)(2x - 1) = 0$	$(x - 2)(2x + 1)$
Middle term is $-x + 4x = +3x$	Middle term is $+x - 4x = -3x$

We find that the factors are $(x + 2)$ and $(2x - 1)$.

$2x^2 + 3x - 2 = 0$	Given Equation
$(x + 2)(2x - 1) = 0$	Factored
$x + 2 = 0 \text{ or } 2x - 1 = 0$	Zero Product Property
$x = -2 \text{ or } x = \dfrac{1}{2}$	Solved

We check both solutions in the original equation.

Check $x = -2$:	**Justification:**	**Check $x = \dfrac{1}{2}$:**
$2x^2 + 3x - 2 = 0$	Equation	$2x^2 + 3x - 2 = 0$
$2(-2)^2 + 3(-2) - 2 = 0$	Substituted	$2\left(\dfrac{1}{2}\right)^2 + 3\left(\dfrac{1}{2}\right) - 2 = 0$
$8 - 6 - 2 = 0$	Simplified	$\dfrac{1}{2} + \dfrac{3}{2} - 2 = 0$
$0 = 0 \checkmark$	Simplified	$0 = 0 \checkmark$

Example 2

Solve by factoring: $4x^2 - 25 = 0$

Solution

There are many possible factors that produce a first term of $4x^2$ and third term of -25. However, only one pair of factors gives a second term of 0. Notice that $4x^2$ and 25 are both squares whose difference is zero. Such quadratics are sometimes referred to as a **difference of squares.** We can factor difference-of-squares equations by writing the square root of the first term of the quadratic as the first terms of the binomials plus and minus the square root of the second term of the quadratic.

$$4x^2 - 25 = 0 \qquad \text{Given equation}$$

$$(2x + 5)(2x - 5) = 0 \qquad \text{Factored}$$

Notice that $2x$ is a square root of $4x^2$ and that 5 is a square root of 25. For the equation to be true, one of the binomials must equal zero.

$$2x + 5 = 0 \text{ or } 2x - 5 = 0 \qquad \text{Zero Product Property}$$

$$x = -\frac{5}{2} \text{ or } x = \frac{5}{2} \qquad \text{Solved}$$

Remember that not all quadratic equations can be solved by factoring. Equations that cannot be solved by factoring can be solved by other methods such as by using the quadratic formula.

Practice Set

Solve by factoring.

   a. $2x^2 + 7x + 3 = 0$

   b. $3x^2 - x - 2 = 0$

   c. $9x^2 - 16 = 0$

   d. $16x^2 - 9 = 0$

LESSON A110

Written Practice

26. Graph the following sequence: $a_n = (n - 1)(n - 1)$
(A73)

27. Simplify: $\sqrt{12} + \sqrt{147}$
(A84)

28. Forty nickel and penny coins total 64¢. Write two equations that illustrate this.
(A97)

29. Solve by elimination: $\begin{cases} 5n + p = 64 \\ -n - p = -40 \end{cases}$
(A99)

30. Find $f(3)$ if $f(x) = x^2 + 4x - 4$
(A98)

31. Solve by elimination. $\begin{cases} 4x - 3y = 5 \\ 2x - 3y = -5 \end{cases}$
(A99, A104)

32. Graph: $-x + 2y > 6$
(A108)

33. Graph: $2x + 3y \geq 6$
(A108)

34. Solve by factoring: $2x^2 + 9x + 10 = 0$
(A110)

35. Solve by factoring: $2x^2 + 7x + 3 = 0$
(A110)

LESSON A111

26. Write an equation for the line that passes through $(2, -3)$ and $(-4, 12)$.
(A71)

27. Solve: $\dfrac{5}{2x + 1} = \dfrac{7}{3x + 1}$
(A83)

28. Simplify: $\sqrt{225} - \sqrt{625}$
(A84)

29. Penelope's pennies and Nancy's nickels are 70 in number and 4.30 in value. Write two equations to express this.
(A97)

30. $(12x - 5y) - (10x + 5y)$
(A80)

31. Solve by substitution. $\begin{cases} 5x + y = 8 \\ 2x + 3y = -2 \end{cases}$
(A92, A102)

32. Solve by elimination. $\begin{cases} 4x + 3y = 2 \\ 3x + 3y = 3 \end{cases}$
(A99, A104)

33. Graph: $-5x + y < 3$
(A108)

34. Solve by factoring: $3x^2 + 5x + 2 = 0$
(A110)

35. Solve by factoring: $5x^2 + 7x + 2 = 0$
(A110)

LESSON A112

Thinking Skill

Recall

What two symbols tell us to use a dashed line when we graph solutions to inequalities? What two symbols tell us to use a solid line?

Graphing Systems of Inequalities

Recall that we graph linear inequalities by shading the part of the plane in which we find pairs of numbers that satisfy the inequality. We illustrate the set of solutions for a system of inequalities by finding where the graphs of the inequalities intersect (overlap). Consider the following situation:

A principal assigns students to primary classrooms so that the number of boys and girls in each classroom is equal to or less than 20.

$$b + g \le 20$$

The principal also tries to have at least 16 children in each classroom.

$$b + g \ge 16$$

These two inequalities are both satisfied by pairs of numbers in the region where the graphs of the two inequalities intersect (overlap).

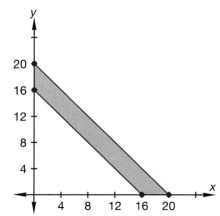

Graph the solution of $\begin{cases} y \le -2x + 5 \\ y > x - 2 \end{cases}$

Solution

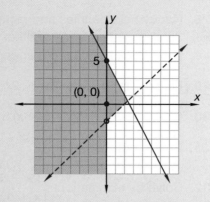

The solution is the overlap of the two shaded regions. To check the solution we select a point from the graphed region and test the x, y values in the original inequalities.

Step:	Justification:	Step:
$y \le -2x + 5$	Original equations	$y > x - 2$
$0 \le -2(0) + 5$	Substituted (0, 0)	$0 > (0) - 2$
$0 \le 5$ (true)	Simplified	$0 > -2$ (true)

Practice Set

Evaluate Graph each system of inequalities.

a. $\begin{cases} y < 2x - 1 \\ y \ge -x - 3 \end{cases}$ **b.** $\begin{cases} y \ge x \\ y \ge -\frac{1}{2}x + 1 \end{cases}$

LESSON A112

Written Practice

26. Write an equation for the line that includes points (0, −1) and (5, 5).
(A71)

27. Simplify: $\sqrt{288} - \sqrt{50}$
(A84)

28. The 33 quarters and nickels totaled $4.25. Write two equations to represent this.
(A97)

29. Solve by substitution: $\begin{cases} n = 33 - q \\ 25q + 5n = 425 \end{cases}$
(A92)

30. Find $h(-3)$ for $h(x) = x^2 - 3x - 9$
(A98)

31. Solve by substitution. $\begin{cases} x + 2y = 4 \\ 2x - y = -7 \end{cases}$
(A92, A102)

32. Solve by elimination. $\begin{cases} 2x + 3y = -9 \\ -x - y = 4 \end{cases}$
(A99, A104)

33. Solve by factoring: $x^2 - 3x - 18 = 0$
(A105)

34. Graph this system. $\begin{cases} y \le x - 2 \\ y \ge -x + 2 \end{cases}$
(A112)

35. Graph this system. $\begin{cases} y \ge x - 2 \\ y \le -x \end{cases}$
(A112)

26. Solve: $\dfrac{x}{12} = \dfrac{2x-1}{20}$
(A83)

27. Find both solutions: $(2x + 7)(3x - 5) = 0$
(A95)

28. Nancy held pennies and nickels. Write an equation that shows she has
(A97, A102) 30 all together. Write another that shows she has 70 cents. Then guess
and check to find the number of pennies and nickels Nancy held.

29. Find $f(-1)$ for $f(x) = 3x^2 - 2x - 1$
(A98)

30. $(x - 7y + 1) - (3y + 5 + x)$
(A80)

31. Solve by substitution. $\begin{cases} 3x + y = 7 \\ 2x + y = 6 \end{cases}$
(A92, A102)

32. Solve by elimination. $\begin{cases} 2x + 5y = -3 \\ x + 5y = -1 \end{cases}$
(A99, A104)

33. Solve by factoring: $x^2 - x - 90 = 0$
(A105)

34. Solve by factoring: $3x^2 + 5x - 2 = 0$
(A110)

35. Graph this system. $\begin{cases} y \geq -\frac{2}{3}x + 2 \\ y \leq x - 2 \end{cases}$
(A112)

LESSON A114

New Concept

Solving Systems of Inequalities from Word Problems

Suppose Brandon has at most one dollar in nickels and dimes in his pocket.
An inequality describing the value of the coins would be written:

$$5x + 10y \leq 100$$

In this inequality, x is the number of nickels Brandon has, and y is the number
of dimes.

Also suppose Brandon has no more than five dimes in his pocket.

$$y \leq 5$$

Below is a graph of these inequalities.

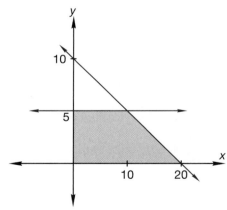

Notice that only points in the first quadrant are shaded. This is because
Brandon cannot have a negative number of dimes or nickels. That is, $x \geq 0$
and $y \geq 0$ are two additional inequalities that affect the graph. Although we
shaded a region, only the points in the region with whole-number coordinates
meet the given conditions.

Example 1

By looking at the plot of possible nickel-dime combinations, tell whether or not the following combinations, of coins meet the conditions given.

    A two nickels, six dimes

    B ten nickels, five dimes

    C fourteen nickels, two dimes

    D sixteen nickels, four dimes

Solution

The combinations of coins correspond to points on the graph. For example, "two nickels, six dimes" is the point (2, 6) and "ten nickels, five dimes" is (10, 5). The listed combinations are plotted below.

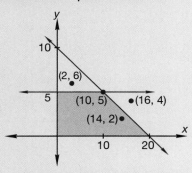

Only (10, 5) and (14, 2) lie within the shaded region, so Brandon may have **B ten nickels, five dimes** or **C fourteen nickels, two dimes.** He does not have **A** two nickels and six dimes or **D** sixteen nickels and four dimes.

Analyze At which point on the graph do we find the maximum amount of money possible with the maximum number of dimes?

Example 2

Joey sells bags of peanuts and popcorn for two dollars each at the game. Peanut bags take up one unit of space and popcorn bags fill two units of space on his tray.

    a. Write an inequality to show that Joey can carry up to thirty dollars in products.

    b. Write an inequality to show that Joey can carry a maximum of twenty units on his tray.

    c. Plot the inequalities and tell which of the following combinations Joey can carry: ten peanut bags and four popcorn bags; fifteen peanut bags and no popcorn; five peanut bags, nine popcorn bags.

    d. Which combination fills his tray with a maximum number of peanuts?

Solution

    a. If we let x be the number of peanut bags and y be the number of popcorn bags that Joey carries on his tray, then the value in dollars of peanuts and popcorn Joey can carry is $2x + 2y \leq 30$.

b. Joey can carry 20 units, but popcorn bags take two units. Therefore, $x + 2y \le 20$.

c. To graph these inequalities, we will first write them in slope-intercept form.

First Inequality:	Justification:	Second Inequality:
$2x + 2y \le 30$	Given equation	$x + 2y \le 20$
$2y \le -2x + 30$	Isolated y-term	$2y \le -x + 20$
$y \le -x + 15$	Divided both sides by 2	$y \le -\frac{1}{2}x + 10$

Now we graph these inequalities. Only the first quadrant is shaded since the number of bags is not negative, that is $x \ge 0$ and $y \ge 0$.

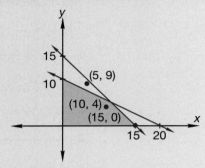

The points (10, 4) and (15, 0) lie within the shaded region, so Joey can carry **ten peanut bags and four popcorn bags** or **fifteen peanut bags and no popcorn.** He cannot carry 5 peanuts and 9 popcorns.

d. Joey's tray is filled when $x + 2y = 20$. The combinations are represented by the graph of this line. The greatest number of peanuts he can carry with his tray filled is the furthest right point on that line, (10, 5). He can carry **ten bags of peanuts and five bags of popcorn.**

Practice Set

a. Hank loaded $4 sandwiches and $1 containers of juice on his lunch truck. Letting x equal the number of sandwiches and y the number of containers of juice he loads, write an inequality that shows he loads at most $100 worth of sandwiches and containers of juice. Write another inequality that shows he loads at most 36 containers of juice.

b. At a collectibles show Martina pays $2 for each photograph and $3 for each poster. Let x equal the number of photographs and y the number of posters she buys. Write an inequality that shows she spends at most $30. Write another inequality that shows she buys at most 12 items. Since she cannot buy a negative number of items, show this in two or more inequalities. Graph the system of inequalities.

LESSON A114

Written Practice

26. Write an equation for the line that passes through $(-12, -5)$ with slope $-\frac{5}{7}$.
(A68)

27. Write an equation for the line that includes points (5, −2) and (1, 6).
(A71)

28. Simplify: $\sqrt{150} + \sqrt{24}$
(A84)

29. Find both solutions: $(3x + 9)(3x - 2) = 0$
(A95)

30. Eric earns 4 points for each correct answer, and loses one point for
(A97, A99) each incorrect answer. Write an equation that shows he earns 70 points. Write another equation that shows he answers 20 questions. Then find the number of questions Eric answered correctly.

31. Solve by substitution. $\begin{cases} 2x - 3y = -1 \\ x - 2y = -2 \end{cases}$
(A92, A102)

32. Solve by elimination. $\begin{cases} x + 5y = 3 \\ 2x - y = 17 \end{cases}$
(A99, A104)

33. Solve by factoring: $x^2 + 6x + 9 = 0$
(A105)

34. Graph this system. $\begin{cases} y \geq 2x - 1 \\ -2x + y \leq 1 \end{cases}$
(A108, A112)

35. Bill earns $5 commission for each piccolo he sells, and $10 for each
(A114) tuba. Write an inequality that shows he can earn no more than $60 total. Write an inequality that shows he has only three tubas to sell. Write inequalities that show he cannot sell a negative number of either instrument. Graph the system. What is the fewest number of total instruments Bill can sell to earn the $60?

LESSON A115

Written Practice

26. Write an equation for the line that includes points $(-3, 1)$ and $(6, -2)$.
(A71)

27. Solve: $\dfrac{7}{2x + 1} = \dfrac{2}{3x - 7}$
(A83)

28. Find both solutions: $(x + 1)(3x - 3) = 0$
(A95)

29. Martin earns 5 cents for every plastic bottle and 7 cents for every glass
(A97, A92) bottle he returns. Write an equation that show he returns six more glass bottles than plastic bottles. Write an equation that shows he earns 90¢. Find the number of bottles of each type Martin returned.

30. Find $g(5)$ for $g(x) = 2x^3 - 3x^2 - x + 12$
(A98)

31. Solve by substitution. $\begin{cases} 2x + y = -4 \\ 3x + 3y = -6 \end{cases}$
(A92, A102)

32. Solve by elimination. $\begin{cases} 2x + 7y = 60 \\ x + 6y = 50 \end{cases}$
(A99, A104)

33. Solve by factoring: $10x^2 - x - 11 = 0$
(A110)

34. Graph this system. $\begin{cases} 2x + 3y > 12 \\ -4x + y < 4 \end{cases}$
(A108, A112)

35. Clear glass panes cost $2. Stained glass panes cost $9. Write an
(A114) inequality that shows Gerry can spend no more than $72. Write an inequality that shows that he can buy a maximum of 9 clear panes. Write inequalities that show he cannot buy a negative number of either type of pane. Graph the system. What is the maximum number of total panes he can buy?

The Quadratic Formula, Part 1

Recall that a quadratic equation contains a second degree polynomial (exponent is 2). Quadratic equations can be written in this form with the right side of the equal sign 0.

$$ax^2 + bx + c = 0$$

The letters a, b, and c represent numbers. Here are two examples.

$$x^2 + 3x - 4 = 0 \qquad 2x^2 - 50 = 0$$

In the equation on the left, a, b, and c are 1, 3, and -4 respectively. In the equation on the right a is 2, b is zero because there is no x term, and c is -50. Notice that the signs of a, b, and c are the signs of the terms.

Example 1

If the equation is quadratic, identify a, b, and c.

a. $x^2 - 5x + 6 = 0$ **b.** $3x^2 - 27 = 0$

c. $x + 3 = 0$ **d.** $x^3 + 2x^2 + 4 = 0$

Solution

a. Quadratic: $a = 1$, $b = -5$, $c = 6$

b. Quadratic: $a = 3$, $b = 0$, $c = -27$

c. Not quadratic, since it contains no squared variables.

d. Not quadratic, since it contains an x^3 term.

To solve a quadratic equation we find the value or values of the variable that satisfy the equation (make it true). We have solved quadratic equations like **b** in example 1 by isolating the x^2 term, and we have solved equations like **a** by factoring. Another way to solve quadratic equations is with the **quadratic formula.** Later, you will understand how it is derived. For now you should memorize the formula.

Thinking Skill

Analyze

Why is there a $\pm$ symbol in the quadratic formula?

The solutions to an equation $ax^2 + bx + c = 0$ are

$$x = \frac{-b \pm \sqrt{b^2 - 4ac}}{2a}$$

Recall that most quadratic equations have two solutions. The $\pm$ symbol in the formula means that for one solution we add the radical and for the other solution we subtract.

Example 2

Use the quadratic formula to find the solutions to $x^2 - 5x + 6 = 0$.

Solution

We see that $a = 1$, $b = -5$, and $c = 6$. We substitute these numbers into the quadratic formula.

Steps:	Justification:

$$x = \frac{-(-5) \pm \sqrt{(-5)^2 - 4(1)(6)}}{2 \cdot 1}$$

Substituted for a, b, and c

$$= \frac{5 \pm \sqrt{25 - 24}}{2}$$

Simplified under the radical

$$= \frac{5 \pm 1}{2}$$

Simplified radical

Now we apply the $\pm$ sign and find the two solutions.

$$x = \frac{5 + 1}{2} \text{ or } x = \frac{5 - 1}{2}$$

$x = 3$ or 2

We check the answers by replacing x with each number and simplifying.

Check $x = 3$:	Justification:	Check $x = 2$:
$(3)^2 - 5(3) + 6 = 0$	Substituted	$(2)^2 - 5(2) + 6 = 0$
$9 - 15 + 6 = 0$	Simplified	$4 - 10 + 6 = 0$
$0 = 0$ ✓	Simplified	$0 = 0$ ✓

Each answer results in a true equation.

Practice Set

Classify For problems **a** and **b** state why the equation is not quadratic.

a. $y = 3x - 2$

b. $x^2 - 3x + 2 = x^3$

Use the quadratic formula to solve the problems **c–f**.

c. $2x^2 - 12x + 10 = 0$

d. $x^2 + 7x + 12 = 0$

e. $3x^2 - 3x - 18 = 0$

f. $6x^2 - x - 1 = 0$

LESSON A116

Written Practice

26. Write an equation for the line that includes points $(5, -5)$ and $(2, 1)$.
(A71)

27. Graph the following sequence: $a_n = \left(\frac{1}{n}\right)^{-1}$
(A73)

28. Find the perimeter of the rectangle.
(A84)

$$2\sqrt{7}$$
$$\sqrt{7}$$

29. Find both solutions: $(3x + 12)\left(\frac{1}{2}x - 2\right) = 0$
(A95)

30. Solve by substitution. $\begin{cases} x = y + 1 \\ 9x - 10y = 0 \end{cases}$
(A92)

31. William receives one point for each correct answer in a quiz game
(A97, A104) and loses two points for each incorrect answer. Write an equation that shows he receives 10 points. Write another equation that shows he answers 16 questions. Find the number of questions he answers correctly. What strategy did you use to find the number correct?

32. Solve by factoring: $x^2 + 7x + 6 = 0$
(A105)

33. Graph this system. $\begin{cases} 2x + 3y \leq 6 \\ 2x + 3y \geq -6 \end{cases}$
(A108, A112)

34. Use the quadratic formula to solve: $x^2 + 5x + 4 = 0$.
(A116)

35. Use the quadratic formula to solve: $x^2 - 7x + 12 = 0$.
(A116)

---

26. Write an equation for the line that includes points $(-5, 2)$ and $(10, -1)$.
(A71)

27. Solve: $\dfrac{2x - 1}{30} = \dfrac{x + 1}{18}$
(A83)

28. Simplify: $(x^2 - 5x + 4) - (-5x + x^2 + 4)$
(A80)

29. Find $h(-1)$ for $h(x) = 3x^3 - x^2 - 3x + 9$
(A98)

30. Solve by substitution. $\begin{cases} 5x - 3y = -11 \\ x = y - 5 \end{cases}$
(A92)

31. Albert's football team scores 3 points for every field goal and 7 points
(A97, A102) for each touchdown, because they always make the extra point. Write
an equation showing they score 33 points. Write another equation
showing they score 7 times. Then find the number of field goals and
touchdowns the team scored.

32. Samuel carries 12 bottles per trip and Joseph carries 8 bottles per trip.
(A114) Write an inequality that shows the two can move as many as all 96
bottles. Write an inequality that shows Joseph can take only 6 trips.
Write inequalities that show that neither of the two can take a negative
number of trips. Graph the system. What is the maximum number of
trips Samuel can take?

33. Use the quadratic formula to solve: $2x^2 - 18 = 0$
(A116)

34. Use the quadratic formula to solve: $5x^2 - 6x + 1 = 0$
(A116)

35. Use a calculator to find the length of side BC to
(A117) the nearest inch.

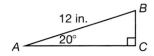

The Quadratic Formula, Part 2

Recall that an equation that can be written in the form

$$ax^2 + bx + c = 0$$

Thinking Skill

Analyze

Why can we use the
quadratic formula
when b and c equal
zero, but not when a
equals zero?

is a **quadratic equation.** The letters a, b, and c in the formula represent
numbers, and a does not equal zero. Any quadratic equation can be solved
with the quadratic formula.

$$x = \frac{-b \pm \sqrt{b^2 - 4ac}}{2a}$$

Sometimes, quadratic equations must be manipulated to determine the
coefficients a, b, and c.

Example 1

Determine a, b, and c for the equation $x^2 = -15 - 8x$. Then use the quadratic formula to solve the equation.

Solution

We begin by expressing the equation in the standard form for a quadratic equation. Be sure that one side equals zero.

Step:	Justification:
$x^2 = -15 - 8x$	Given equation
$x^2 + 8x + 15 = 0$	Added $8x$ and 15 to both sides

Now the equation is in standard form. We see that $a = 1$, $b = 8$, and $c = 15$. We substitute these values into the quadratic formula and simplify to find the solution.

Step:	Justification:
$x = \dfrac{-8 \pm \sqrt{8^2 - 4 \cdot 1 \cdot 15}}{2 \cdot 1}$	Substituted
$= \dfrac{-8 \pm \sqrt{4}}{2}$	Simplified under radical
$= \dfrac{-8 \pm 2}{2}$	Simplified radical
$= -3 \text{ or } -5$	Simplified

Justify How can we check our work?

Example 2

A rectangular field is 20 yards longer than it is wide. Its area is 2400 square yards. What are the dimensions of the field?

Solution

We will begin by drawing a sketch. We label the width w and the length $w + 20$.

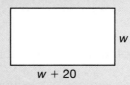

w

$w + 20$

The area of the rectangle is 2400 yd^2 which is the product of the length and width.

Step:	Justification:
$lw = A$	Formula for area
$(w + 20)(w) = 2400$	Substituted for l, w, and A.
$w^2 + 20w = 2400$	Distributive Property
$w^2 + 20w - 2400 = 0$	Subtracted 2400 from both sides

We have expressed the quadratic equation in standard form. We see that a is 1, b is 20, and $c = -2400$. Now we substitute these numbers into the quadratic formula.

$$w = \frac{-b \pm \sqrt{b^2 - 4ac}}{2a} \qquad \text{Quadratic formula}$$

$$w = \frac{-20 \pm \sqrt{(20)^2 - 4(1)(-2400)}}{2(1)} \qquad \text{Substituted for } a, b, \text{ and } c$$

$$w = \frac{-(20) \pm \sqrt{400 + 9600}}{2} \qquad \text{Simplified}$$

$$w = \frac{-(20) \pm 100}{2} \qquad \text{Simplified}$$

$$w = \frac{80}{2} \text{ and } w = \frac{-120}{2} \qquad \text{Simplified}$$

$$w = 40 \text{ and } w = -60$$

We see there are two solutions to the equation, 40 and -60. Remember we are finding the width of a rectangular field, so only the positive solution applies to this problem. Thus the width is **40 yards.** Since the length is 20 yards greater than the width, the length is **60 yards.**

Practice Set

Express each of the following equations in $ax^2 + bx + c = 0$ form. Then use the quadratic formula to solve.

a. $3x^2 - 2x = 1$

b. $4x + 2 = -2x^2$

c. $-2 = 5x - 3x^2$

d. $4x^2 = 3 + x$

e. *Formulate* Write the following quadratic equation in $ax^2 + bx + c = 0$ form. Identify and list the numbers you will use for a, b, and c in the quadratic formula. Then solve.

$$x^2 = x + 6$$

LESSON A118

Written Practice

26. Write an equation for the line that includes points $(-1, 2)$ and $(1, -1)$.
(A71)

27. Graph the following sequence: $a_n = \dfrac{4n + 2}{\sqrt{4}}$
(A73)

28. Three vertices of a square are $(1, 1)$, $(0, -1)$, and $(-2, 0)$. What is the perimeter of the square?
(A84)

29. Find both solutions: $(5x + 12)(2x - 6) = 0$
(A95)

30. Solve by elimination. $\begin{cases} -2x + -3y = -6 \\ 2x + 7y = 6 \end{cases}$
(A99)

31. Jaime found a dime under every chair and a quarter under every cushion. Write an equation that shows he found 95 cents. Let $d =$ the number of dimes and $q =$ the number of quarters. Write another equation that shows he found 5 coins. How many of each coin did Jaime find?
(A97, A102)

32. Solve by factoring: $3x^2 + 3x - 6 = 0$
(A110)

33. Graph this system. $\begin{cases} 3x - 7y \geq 21 \\ 2x + 3y \leq 0 \end{cases}$
(A108, A112)

34. Find the base, *b*, of the triangle to the nearest foot.
(117)

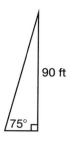

90 ft

75°

35. Use the quadratic formula to solve: $4x^2 = 5x - 1$
(A118)

LESSON A119

New Concept

Finding the Equation of a Line Parallel to a Given Line through a Given Point

Find the slope of lines *s* and *t* graphed below. What do you notice about the relationship of the lines and their slopes?

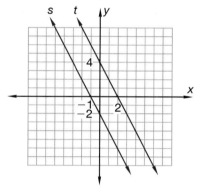

Activity   On a coordinate plane, draw two parallel lines. Find the slopes of the lines. Try this for pairs of parallel lines that run in several different directions. What can you conclude about the slopes of parallel lines?

Example 1

Find the slope of a line parallel to

   a. the *x*-axis   **b. the line $y = 2x - 5$**

Solution

   a. The *x*-axis has zero slope, so any line parallel to it has a slope of **zero.**

   b. The line has a slope 2, so any line parallel to it has slope **2.**

There are many lines parallel to the lines mentioned above. Illustrated on the following graph are some of those lines.

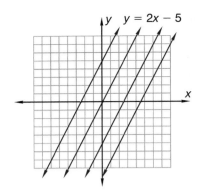

If we know that a line is parallel to another, and we know one point on the line, we can find the equation of the line.

Example 2

Write the equation of the line parallel to $y = 2x - 5$ that passes through (2, 1).

Solution

The line has a slope of 2, so its equation takes the form

$$y = 2x + b$$

To complete the equation we only need to find is b, which is the y-intercept. We will show two ways. One way is to graph a line through (2, 1) that has a slope of 2. Then we inspect the graph to find the y-intercept of the line.

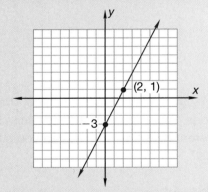

Since the y-intercept is -3, the equation of the line is $y = 2x - 3$.

Another way to find b is to substitute (2, 1) for x and y in the equation and solve for b. Since (2, 1) satisfies the equation, we substitute 2 for x and 1 for y and solve for b.

Step:	Justification:
$y = 2x - 5$	Given equation
$y = 2x + b$	Equation of parallel line
$1 = 2(2) + b$	Substituted (2, 1) for x and y
$1 = 4 + b$	Simplified
$-3 = b$	Solved for b

We find that b is -3, so the equation of the line is

$$y = 2x - 3$$

Write the equation of a line parallel to $y = \frac{3}{4}x + 1$ that passes through the point $(1, -2)$.

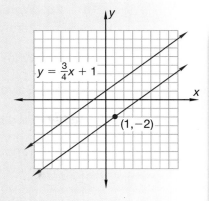

$y = \frac{3}{4}x + 1$

$(1, -2)$

A line parallel to $y = \frac{3}{4}x + 1$ has a slope of $\frac{3}{4}$. We want to find the y-intercept. A line through $(1, -2)$ parallel to the given line intercepts the y-axis between -2 and -3. To find the y-intercept we substitute the x- and y-values from $(1, -2)$ into the equation of the parallel line.

Step:	Justification:
$y = \frac{3}{4}x + 1$	Given equation
$y = \frac{3}{4}x + b$	Equation of parallel line
$-2 = \frac{3}{4}(1) + b$	Substituted $(1, -2)$ for x and y
$-\frac{8}{4} = \frac{3}{4} + b$	Rewrote -2 as an improper fraction
$-\frac{11}{4} = b$	Subtracted $\frac{3}{4}$ from both sides

This y-intercept value is reasonable because $-\frac{11}{4}$ is between -2 and -3 as we saw on the graph. Now we write the equation.

$$y = \frac{3}{4}x - \frac{11}{4} \quad \text{Substituted } -\frac{11}{4} \text{ for } b$$

Practice Set

Write an equation for a line through the given point that is parallel to the given equation.

a. $(3, 7)$; $y = 6x + 8$

b. $(10, 0)$; $y = -\frac{2}{5}x + 1$

c. $(8, 24)$; $y = -\frac{7}{2}x + 16$

d. $(2, 3)$; $y = \frac{1}{3}x + 1$

26. Write an equation for the line that passes through (0, 4) with a slope of
(A68) zero.

27. Write an equation for the line that includes points $(-1, 2)$ and $(0, -1)$.
(A71)

28. Solve: $\dfrac{3x - 1}{2x} = \dfrac{4}{3}$
(A83)

29. Find both solutions: $(4x + 9)(2x - 8) = 0$
(A95)

30. Timothy earns five dollars for mowing small lawns and seven dollars for
(A97) mowing large lawns. Write an equation that shows he earns
34 dollars. Let s = the number of small yards and l = the number of
large yards. Write another equation that shows he mows 6 lawns. Then
find the number of each size lawn Timothy mowed.

31. Solve by factoring: $x^2 - x - 2 = 0$
(A105)

32. Graph this system. $\begin{cases} 4x - 6y \le 12 \\ 3x + 2y \ge -4 \end{cases}$
(A108, A112)

33. The parking meter charges 5¢ for every 6 minutes of parking. Sage says
(A114) he has only nickels and dimes in his cup holder. Write an inequality that
shows the maximum amount of money he can insert into the meter is
$1.00. Write an inequality that shows Sage has only 4 dimes to spend.
Write inequalities that show Sage cannot spend a negative number of
nickels or dimes. Graph the system. What is the fewest number of coins
Sage can use to buy a full two hours?

34. Use the quadratic formula to solve: $3x + 2x^2 = 2$
(A118)

35. Write an equation for the line through the point $(6, -2)$ and parallel to
(A119) the equation $y = \frac{4}{3}x - 5$

LESSON A120

Finding the Equation of a Line Perpendicular to a Given Line through a Given Point

Recall that two lines which intersect to form a right angle are perpendicular
lines.

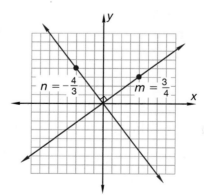

The slopes of perpendicular lines are related as we see in the following
examples.

Example 1

The slope of the line shown is $\frac{3}{2}$. Draw this line on a coordinate plane. Then draw the line through $(-1, 2)$ that is perpendicular to the line shown. Estimate its slope.

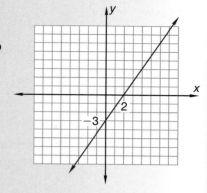

Solution

Use the edge of a ruler to draw the perpendicular line through $(-1, 2)$.

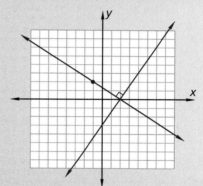

The slope appears to be $-\frac{2}{3}$.

The slopes of perpendicular lines are negative reciprocals. In example 1, the slope of the first line is $\frac{3}{2}$, and the slope of the perpendicular is the negative reciprocal of $\frac{3}{2}$, which is $-\frac{2}{3}$.

Example 2

a. **What is the slope of a line perpendicular to $y = 2x - 5$?**

b. **Write the equation of the line through $(0, 1)$ which is perpendicular to the line $y = 2x - 5$.**

Solution

a. The slope of the first line is 2, so the slope of any line perpendicular to it is $-\frac{1}{2}$.

b. We will calculate the y-intercept and then confirm the answer by graphing. The equation of a line perpendicular to the given line is

$$y = -\frac{1}{2}x + b$$

Since the graph of the equation passes through $(0, 1)$, we know that $x = 0$ and $y = 1$ satisfy the equation.

Step:	Justification:
$y = -\frac{1}{2}x + b$	Equation
$1 = -\frac{1}{2}(0) + b$	Substituted (0, 1) for x and y
$b = 1$	Simplified

Thus, the y-intercept is 1 and the equation is $y = -\frac{1}{2}x + 1.$

The graph of this equation is perpendicular to $y = 2x - 5$ and passes through (0, 1).

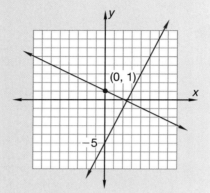

Example 3

Write the equation of the line perpendicular to $y = -\frac{1}{5}x + 7$ that passes through (5, 8).

Solution

The slope of the line perpendicular to the given line is the negative reciprocal of $-\frac{1}{5}$; that is, 5.

Step:	Justification:
$y = 5x + b$	Equation of perpendicular

We substitute 5 for x and 8 for y.

$8 = 5(5) + b$	Substituted (5, 8) for x and y
$8 = 25 + b$	Simplified
$-17 = b$	Subtracted 25 from both sides

The y-intercept is -17, so the equation of the perpendicular through (5, 8) is

$y = 5x - 17$	Substituted -17 for b

Practice Set

Write an equation for the line that includes the given point and is perpendicular to the given line.

a. (0, 0); $y = x$

b. (2, 2); $y = 2x$

c. $(2, 1); y = -\frac{1}{2}x + 3$

d. $(4, 2); y = -\frac{4}{3}x + 1$

LESSON A120

Written Practice

26. Write an equation for the line that includes points $(-3, 2)$ and $(15, -1)$.
(A71)

27. $(x^2 + 2x + 3) - (x^2 + 3x + 2)$
(A80)

28. The vertices of a rectangle are at $(1, -3)$, $(4, 0)$, $(0, 4)$, and $(-3, 1)$. What
(A84) is the perimeter of the rectangle?

29. Winifred earns 2 dollars for each carnation bouquet she delivers, and
(A97) 3 dollars for each rose bouquet. Write an equation that shows she earns 17 dollars. Write another equation that shows she delivers 8 bouquets. Find the number of each type of bouquet she delivers.

30. Find $f(-2)$ for $f(x) = x^2 - 13x - 9$
(A98)

31. Solve by factoring: $2x^2 - 4x - 30$
(A110)

32. Use the quadratic formula to solve: $3x^2 - 5x = 2$
(A118)

33. Use the quadratic formula to solve: $x = 2 - 6x^2$
(A118)

34. Write the equation of the line through $(3, -2)$ that is parallel to the
(A119) equation $y = \frac{1}{3}x + 1$.

35. Write an equation for the line that includes the point $(-3, 2)$ and is
(A120) perpendicular to the line $y = \frac{5}{3}x - 4$.

A

absolute value
valor absoluto
(1)

The distance from the graph of a number to the origin on a number line. The symbol for absolute value is a vertical bar on each side of a numeral or variable, e.g., $|-x|$.

$$|+3| = |-3| = 3$$

*Since the graphs of −3 and +3 are both 3 units from the number 0, the **absolute value** of both numbers is 3.*

---

acute angle
ángulo agudo
(18)

An angle whose measure is between 0° and 90°.

*An **acute angle** is smaller than both a right angle and an obtuse angle.*

---

acute triangle
triángulo acutángulo
(20)

A triangle whose largest angle measures between 0° and 90°.

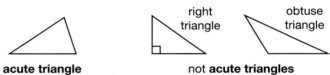

---

addition rule
regla de suma
(32)

A counting principle for "or", the number of ways that one of two or more events can occur is the sum of the ways each event can occur.

Darren has five casual shirts and two dress shirts. Darren has 5 + 2 choices of a casual shirt or a dress shirt.

In probability, the likelihood that one or the other of two mutually exclusive events will occur is the sum of the probabilities of each. $P(A \text{ or } B) = P(A) + P(B)$

The probability of rolling an even number or a 5 with one roll of a number cube is: $\frac{1}{2} + \frac{1}{6} = \frac{2}{3}$

---

additive identity
identidad aditiva
(1)

The number 0. *See also **identity property of addition.***

$$7 + 0 = 7$$

additive identity

*We call zero the **additive identity** because adding zero to any number does not change the number.*

---

additive inverse
inverso aditivo
(31)

A number with the same absolute value as another number but with the opposite sign.

*−3 is the **additive inverse** of 3*

The sum of a number and its **additive inverse** is zero.

GLOSSARY

adjacent angles *ángulos adyacentes* *(54)*	Two angles that have a common side and a common vertex but do not overlap. The angles lie on opposite sides of their common side.

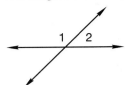

*∠1 and ∠2 are **adjacent angles**. They share a common side and a common vertex.*

adjacent sides *lados adyacentes* *(Inv. 3)*	In a polygon, two sides that intersect to form a vertex.

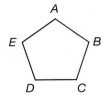

*$\overline{AB}$ and $\overline{BC}$ are **adjacent sides**. They meet at vertex B.*

alternate exterior angles *ángulos alternos externos* *(54)*	A special pair of angles formed when a transversal intersects two lines. Alternate exterior angles lie on opposite sides of the transversal and are outside the lines.

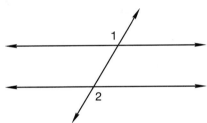

*∠1 and ∠2 are **alternate exterior angles**. When a transversal intersects parallel lines, as in this figure, **alternate exterior angles** have the same measure.*

alternate interior angles *ángulos alternos internos* *(54)*	A special pair of angles formed when a transversal intersects two lines. Alternate interior angles lie on opposite sides of the transversal and are between the lines.

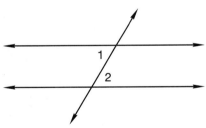

*∠1 and ∠2 are **alternate interior angles**. When a transversal intersects parallel lines, as in this figure, **alternate interior angles** have the same measure.*

altitude *altura* *(20)*	The perpendicular distance from the base of a triangle to the opposite vertex. *See also* **height**.

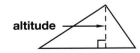

altitude

angle *ángulo* (18)	Two rays with the same endpoint. The endpoint of the rays is the vertex of the angle. *These rays form an **angle.***
apex *ápice* *(Inv. 4)*	The vertex at the top of a cone or pyramid is called the **apex.** 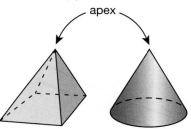
arc *arco* (40)	A portion of a circle intercepted by a central angle. 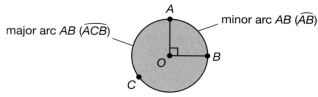 *An **arc** whose measure is less than 180° is a **minor arc.** An **arc** whose measure is more than 180° is a **major arc.***
area *área* (8)	The size of the surface of a flat shape. The area of a surface is measured in square units. 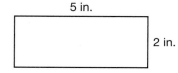 *The **area** of this rectangle is 10 square inches.*
arithmetic sequence *secuencia aritmética* (61)	A sequence that has a constant difference between terms. *3, 6, 9, 12, ...* *This sequence is an **arithmetic sequence** with a constant difference of 3. Each term is 3 more than the previous term.*
associative property of addition *propiedad asociativa de la suma* (1)	The grouping of addends does not affect their sum. In symbolic form, $a + (b + c) = (a + b) + c$. Unlike addition, subtraction is not associative. $(8 + 4) + 2 = 8 + (4 + 2)$ $(8 - 4) - 2 \neq 8 - (4 - 2)$ *Addition is **associative.*** *Subtraction is not **associative.***
associative property of multiplication *propiedad asociativa de la multiplicación* (1)	The grouping of factors does not affect their product. In symbolic form, $a \times (b \times c) = (a \times b) \times c$. Unlike multiplication, division is not associative. $(8 \times 4) \times 2 = 8 \times (4 \times 2)$ $(8 \div 4) \div 2 \neq 8 \div (4 \div 2)$ *Multiplication is **associative.*** *Division is not **associative.***

average *promedio* (7)	The number found by dividing the sum of the elements of a set by the number of elements; one of several measures of central tendency, also called *mean*. *To find the **average** of the numbers 5, 6, and 10, add.* $$5 + 6 + 10 = 21$$ *There were three addends, so divide the sum by 3.* $$21 \div 3 = 7$$ *The **average** of 5, 6, and 10 is 7.*

B

base *base* (20)	**1.** A designated side (or face) of a geometric figure. 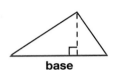 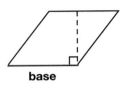 base base base **2.** The lower number in an expression x^a. base $\longrightarrow 5^3 \longleftarrow$ exponent 5^3 *means* $5 \times 5 \times 5$, *and its value is 125.*
best-fit line *línea de mejor ajuste* (Inv. 8)	*See* **line of best fit.**
bias *sesgo* (Inv. 6)	A slant toward a particular point of view caused by a sampling procedure. *To avoid any **bias** in a survey about the candidates for mayor, the researchers chose a representative sample and used neutral wording for the survey questions.*
binomial *binomio* (80)	A polynomial with two terms. *The expression 2x + 1 is a **binomial.***
bisector *bisectriz* (maintenance)	The midpoint of a line segment or ray that divides an angle into two equal angles. 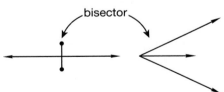

box-and-whisker plot *gráfica de frecuencias acumuladas* *(103)*	A graph that displays the median, quartiles, and extremes of a set of data. The quantities are the three numbers that divide the set of data into quarters.

Box-and whisker-plot

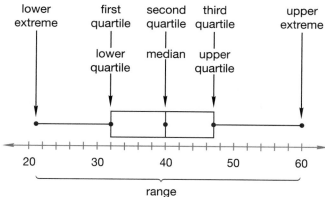

C

center *centro* *(maintenance)*	The point inside a circle or sphere from which all points on the circle or sphere are equally distant.

 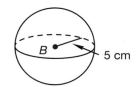

*The **center** of circle A is 2 inches from every point on the circle.*
*The **center** of sphere B is 5 centimeters from every point on the sphere.*

central angle *ángulo central* *(40)*	An angle whose vertex is the center of a circle.

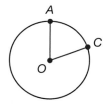

∠AOC is a **central angle.**

chance *posibilidad* *(32)*	A way of expressing the likelihood of an event; the probability of an event expressed as a percent.

*The **chance** of snow is 10%. It is not likely to snow.*
*There is an 80% **chance** of rain. It is likely to rain.*

chord *cuerda* *(maintenance)*	A segment whose end points lie on a circle.

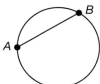

$\overline{AB}$ is a **chord** of the circle.

circle	A closed, curved shape in which all points on the shape are the same
círculo	distance from its center.
(maintenance)	

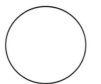

circle

---

circle graph	A method of displaying data, often used to show information about relative
gráfica circular	sizes of parts of a whole. A circle graph is made of a circle divided into
(Inv. 6)	sectors.

Class Test Grades

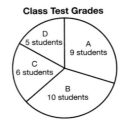

*This **circle graph** shows data for a class's test grades.*

---

circumference	The distance around of a circle.
circunferencia	
(39)	

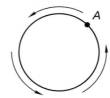

*If the distance from point A around to point A is 3 inches, then the **circumference** of the circle is 3 inches.*

---

coefficient	**1.** In common use, the number that multiplies the variable(s) in an algebraic
coeficiente	term. If no number is specified, the coefficient is 1.
(28, 31)	

In the term $-3x$*, the **coefficient** is* -3*.*
In the term y^2*, the **coefficient** is 1.*

2. In scientific notation, the number that is multiplied by a power of 10.

3.0×10^8 *The number 3.0 is the **coefficient**.*

---

commutative property of addition	Changing the order of addends does not change their sum. In symbolic form, $a + b = b + a$. Unlike addition, subtraction is not commutative.
propiedad conmutativa de la suma	
(1)	

$8 + 2 = 2 + 8$	$8 - 2 \neq 2 - 8$
*Addition is **commutative**.*	*Subtraction is not **commutative**.*

---

commutative property of multiplication	Changing the order of factors does not change their product. In symbolic form, $a \times b = b \times a$. Unlike multiplication, division is not commutative.
propiedad conmutativa de la multiplicación	
(1)	

$8 \times 2 = 2 \times 8$	$8 \div 2 \neq 2 \div 8$
*Multiplication is **commutative**.*	*Division is not **commutative**.*

compass *compás* *(maintenance)*	A tool used to draw circles and arcs. 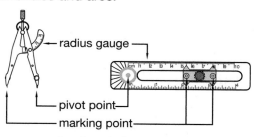 two types of **compasses**
complement of an event *complemento de un evento* *(32)*	The set of outcomes in the sample space that are not included in the event. *If a bag contains three marbles, red, blue, and yellow, the* ***complement*** *of choosing a blue marble is choosing* ***not*** *blue, that is, a red or a yellow marble.*
complement of a set *complemento de un conjunto* *(90)*	The elements not contained in a given set or subset. *In the universe of students in a school, the* ***complement of the set*** *of boys is the set of girls.*
complementary angles *ángulos complementarios* *(54)*	Two angles whose measures total 90°. 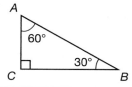 $\angle A$ *and* $\angle B$ *are* ***complementary angles.***
complex fraction *fracción compleja* *(119)*	A fraction that contains one or more fractions in its numerator or denominator. $\dfrac{\frac{3}{5}}{\frac{2}{3}}$ $\dfrac{25\frac{2}{3}}{100}$ $\dfrac{15}{7\frac{1}{3}}$ $\dfrac{\frac{a}{b}}{\frac{b}{c}}$ $\dfrac{1}{2}$ $\dfrac{12}{101}$ $\dfrac{xy}{z}$ **complex fractions** not **complex fractions**
composite number *número compuesto* *(9)*	A counting number greater than 1 that can be expressed as a product of prime numbers. Every composite number has three or more factors. *9 is divisible by 1, 3, and 9. It is* ***composite.*** *11 is divisible by 1 and 11. It is not* ***composite.***
compound interest *interés compuesto* *(Inv. 10)*	Interest that pays on accumulated interest as well as on the principal. The formula for computing the current value V of an investment with compound interest is $V = P(1 + i)^n$ where P is the principal, i is the interest rate, and n is the number of compounding periods.

Compound Interest

	$100.00	principal
+	$6.00	first-year interest (6% of $100.00)
	$106.00	total after one year
+	$6.36	second-year interest (6% of $106.00)
	$112.36	total after two years

Simple Interest

	$100.00	principal
	$6.00	first-year interest
+	$6.00	second-year interest
	$112.00	total after two years

congruent *congruente* *(19)*	Having the same size and shape. *These polygons are* ***congruent.*** *They have the same size and shape.*

constant	A number whose value does not change.
constante (14)	In the expression $2\pi r$, the numbers 2 and π are **constants,** while r *is a variable.*

constant of proportionality	The number that relates two variables in a direct variation. *See also* **direct variation.**
constante de proporcionalidad (69)	In the function $y = kx$, k *is the* **constant of proportionality.**

continuous function	A type of function whose graph is not interrupted with gaps.
función continua (47)	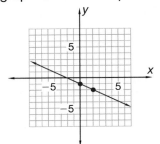

This graph shows that the function $y = -\frac{1}{2}x - 1$ is a **continuous function.**

coordinate(s)	**1.** A number used to locate a point on a number line.
coordenada(s) (Inv. 1)	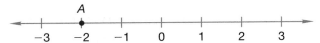

The **coordinate** of point A is -2.

2. An ordered pair of numbers used to locate a point in a coordinate plane.

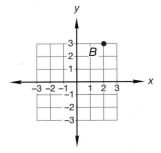

The **coordinates** of point B are (2, 3). The x-coordinate is listed first, the y-coordinate second.

coordinate plane	A grid on which any point can be identified by an ordered pair of numbers.
plano coordenado (Inv. 1)	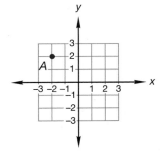

Point A is located at $(-2, 2)$ on this **coordinate plane.**

correlation *correlación* (Inv. 8)	A measure of the degree of relationship between two variables.

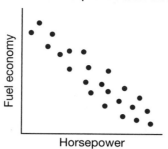

*The scatterplot indicates a negative **correlation** between the horsepower and fuel economy of a car.*

corresponding angles *ángulos correspondientes* (54)	A special pair of angles formed when a transversal intersects two lines. Corresponding angles lie on the same side of the transversal and are in the same side of each of the lines.

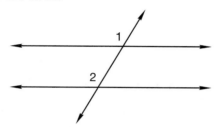

*∠1 and ∠2 are **corresponding angles.** When a transversal intersects parallel lines, as in this figure, **corresponding angles** have the same measure.*

corresponding parts *partes correspondientes* (35)	Sides or angles of similar polygons that occupy the same relative positions.

$\overline{BC}$ **corresponds** to $\overline{YZ}$.
∠A **corresponds** to ∠X.

cosine *coseno* (118)	In a right triangle, the ratio of the adjacent leg to the hypotenuse.

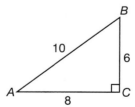

$$\cos \angle A = \tfrac{8}{10} = 0.8$$

counting numbers *números de conteo* (1)	The numbers used to count; the members of the set {1, 2, 3, 4, 5, ...}. Also called *natural numbers.*

*1, 24, and 108 are **counting numbers.***
*−2, 3.14, 0, and $2\tfrac{7}{9}$ are not **counting numbers.***

cross products *productos cruzados* *(44)*	The product of the numerator of one fraction and the denominator of another. 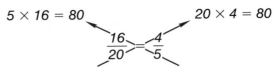 $5 \times 16 = 80$ $20 \times 4 = 80$ $$\frac{16}{20} = \frac{4}{5}$$ *The* **cross products** *of equal ratios are equal.*
cube root *raíz cúbica* *(15)*	One of three equal factors of a number. The symbol for the cube root of a number is $\sqrt[3]{}$. *The cube root of 27 is 3 because* $3 \times 3 \times 3 = 27$.

D

decay *decaimiento* *(102)*	A pattern of change that results in a decrease in value over time. *The effect of inflation on the value of the dollar is an example of* **decay.**
degree (°) *grado* *(18)*	**1.** A unit for measuring angles.  *There are 90* **degrees** *There are 360* **degrees** *(90°) in a right angle.* *(360°) in a circle.* **2.** A unit for measuring temperature. 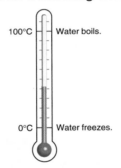 *There are 100* **degrees** *between the freezing and boiling points of water on the Celsius scale.*
denominator *denominador* *(5)*	The bottom term of a fraction. $\dfrac{5}{9}$ ← numerator ← **denominator**
dependent event *evento dependiente* *(83)*	An event is *dependent* if the outcome of another event affects the probability that the other event will occur. *Whitney draws a card from a regular deck and does not replace it. Then Chad draws a card from the same deck. The probability of Chad's drawing the queen of hearts is* **dependent** *on what card Whitney draws. If Whitney draws the queen of hearts, the probability of Chad's drawing it is 0. If Whitney draws a different card, the probability of Chad's drawing the queen of hearts is* $\frac{1}{51}$.

dependent variable *variable dependiente* *(69)*	The variable in an equation whose value is determined by the value chosen for one or more other variables. *See also* **independent variable.** *In* y = 2x + 1, *the* **dependent variable** *is* y.
diagonal *diagonal* *(maintenance)*	A line segment, other than a side, that connects two vertices of a polygon. 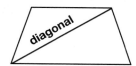
diameter *diámetro* *(39)*	The distance across a circle measured through its center. *The* **diameter** *of this circle is 3 inches.*
dilation *dilatación* *(19)*	An enlargement of a figure. 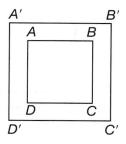 *Square* A′B′C′D′ *is a* **dilation** *of square* ABCD.
direct proportion *proporción directa* *(47)*	A type of function whose graph is linear, includes the origin, and has a positive slope. *See also* **direct variation.** 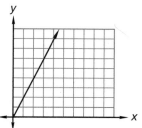 *This graph shows that the function* y = 2x *is a* **direct proportion.**
direct variation *variación directa* *(69)*	A proportional relationship between two variables in which one variable is a constant multiple of the other, that is, the quotient of the two variables is a constant. *See also* **direct proportion** and **constant of proportionality.** y = kx *represents a* **direct variation** *between* x *and* y *with a* **constant of proportionality** k. *When* x *is zero,* y *is zero. As* x *increases,* y *increases by the factor* k.

distributive property
 propiedad distributiva
 (21)
	Expressed symbolically: a(b + c) = ab + ac
	
	        a(b + c) expands to ab + ac
	        ab + ac factors to a(b + c)
	
	 8 × (2 + 3) = (8 × 2) + (8 × 3)
	 Multiplication is **distributive** *over addition.*

divisible divisible (9)	Able to be divided by a counting number without a remainder. $$\frac{5}{4)\overline{20}}$$ The number 20 is **divisible** by 4, since $20 \div 4$ has no remainder. $$\frac{6\ R\ 2}{3)\overline{20}}$$ The number 20 is not **divisible** by 3, since $20 \div 3$ has a remainder.
domain dominio (47)	The set of inputs of a function. *In the function* y = 3x, *the values of* x *are the* **domain.**

E

edge arista (Inv. 4)	A line segment formed where two faces of a polyhedron intersect. 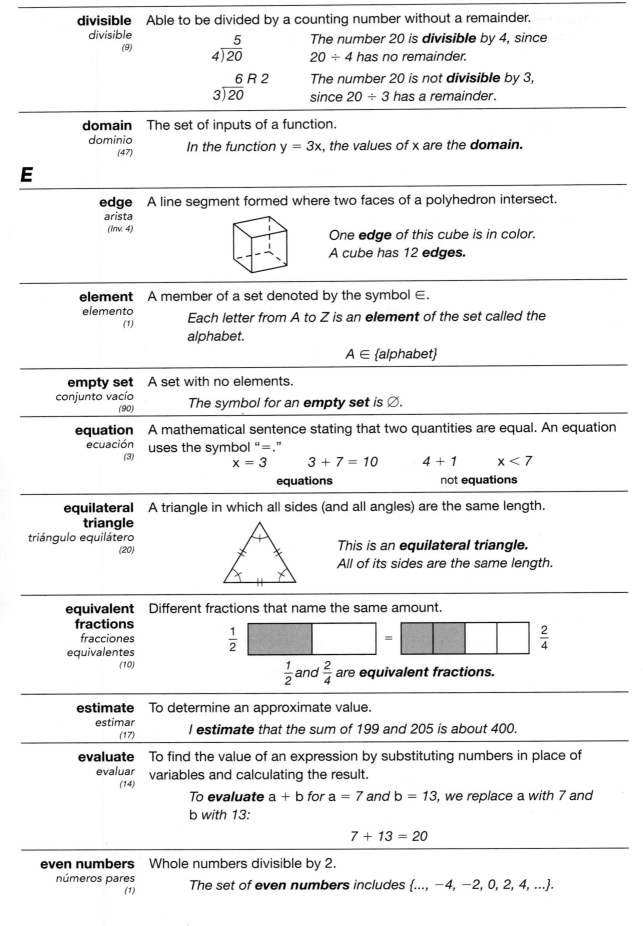 One **edge** of this cube is in color. A cube has 12 **edges.**
element elemento (1)	A member of a set denoted by the symbol ∈. *Each letter from A to Z is an* **element** *of the set called the alphabet.* $$A \in \{alphabet\}$$
empty set conjunto vacío (90)	A set with no elements. *The symbol for an* **empty set** *is* ∅.
equation ecuación (3)	A mathematical sentence stating that two quantities are equal. An equation uses the symbol "=." $x = 3$ $3 + 7 = 10$ $4 + 1$ $x < 7$ **equations** not **equations**
equilateral triangle triángulo equilátero (20)	A triangle in which all sides (and all angles) are the same length. *This is an* **equilateral triangle.** *All of its sides are the same length.*
equivalent fractions fracciones equivalentes (10)	Different fractions that name the same amount. $\frac{1}{2}$ = $\frac{2}{4}$ $\frac{1}{2}$ *and* $\frac{2}{4}$ *are* **equivalent fractions.**
estimate estimar (17)	To determine an approximate value. *I* **estimate** *that the sum of 199 and 205 is about 400.*
evaluate evaluar (14)	To find the value of an expression by substituting numbers in place of variables and calculating the result. *To* **evaluate** a + b *for* a = 7 *and* b = 13, *we replace* a *with 7 and* b *with 13:* $$7 + 13 = 20$$
even numbers números pares (1)	Whole numbers divisible by 2. *The set of* **even numbers** *includes* {..., −4, −2, 0, 2, 4, ...}.

experimental probability probabilidad experimental (32)	Probability of an outcome that is determined statistically. Experimental probability is the ratio of the number of times an event occurs to the number of trials. *If you toss heads 7 times out of 10 tosses, the **experimental probability** of tossing heads again is $\frac{7}{10}$.*

exponent exponente (15)	The upper number in an expression x^a. If the exponent is a counting number, it indicates how many times the base is to be used as a factor. base $\longrightarrow 5^3 \longleftarrow$ **exponent** 5^3 *means $5 \times 5 \times 5$, and its value is 125.*

exterior angle ángulo externo (maintenance)	In a polygon, the supplementary angle of an interior angle. 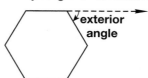

F

face cara (Inv. 4)	A flat surface of a geometric solid. *One **face** of the cube is shaded.* *A cube has six **faces**.*

factor factor (21)	**1.** Noun: One of two or more numbers that are multiplied. $\quad 5 \times 6 = 30 \quad$ *The **factors** in this mathematical statement are 5 and 6.* **2.** Noun: A whole number that divides another whole number without a remainder. *The numbers 5 and 6 are **factors** of 30.* **3.** Verb: To write as a product of factors. *We can **factor** the number 6 by writing it as 2×3.*

factor tree árbol de factores (9)	A method of finding all the prime factors of a number. The numbers on each branch of the tree are factors of the number. Each number at the end of a branch is a prime factor. 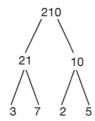 *This **factor tree** shows that the prime factors of 210 are 2, 3, 5, and 7.*

formula fórmula (3)	An equation used to calculate a desired result. *The **formula** for the area of a circle is $A = \pi r^2$, where A is the area and r is the length of the radius.*

fraction *fracción* (5)	A number that names part of a whole.

$\frac{1}{4}$ of the circle is shaded.

$\frac{1}{4}$ is a **fraction.**

function *función* (41)	A mathematical rule that identifies a relationship between two sets of numbers. Each number from the first set (an input) is used to calculate another number from the second set (an output). Each input produces only one output.

$$y = 3x$$

x	y
3	9
5	15
7	21
10	30

There is exactly one resulting number for every number we multiply by 3. Thus, y = 3x is a **function.**

G

geometric probability *probabilidad geométrica* (101)	Probability based on the area of regions.

 $P \approx 0.1$ $P \approx 0.25$ $P \approx 1$

The probability of hitting the shaded region of each circle =

$$\frac{area\ of\ shaded\ region}{area\ of\ whole\ circle}.$$

geometric sequence *secuencia geométrica* (61)	A sequence that has a constant ratio between terms.

3, 9, 27, 81, ...

This sequence is a **geometric sequence** *with a constant ratio of 3. Each term is 3 times the previous term.*

geometric solid *sólido geométrico* (Inv. 4)	A three-dimensional geometric figure.

geometric solids   not **geometric solids**

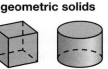   

cube   cylinder   circle   rectangle   hexagon

geometry *geometría* (maintenance)	A major branch of mathematics that deals with shapes, sizes, and other properties of figures.

Some figures we study in **geometry** *are angles, circles, and polygons.*

greatest common factor (GCF) *máximo común divisor (MCD)* (10)	The largest whole number that is a factor of two or more indicated numbers.

The factors of 12 are 1, 2, 3, 4, 6, and 12.
The factors of 18 are 1, 2, 3, 6, 9, and 18.
The **greatest common factor** *of 12 and 18 is 6.*

growth *crecimiento* (102)	A pattern of change that results in an increase in value over time.

The effect of compound interest on an investment is an example of **growth.**

H

height
altura
(20)

The perpendicular distance from the base to the opposite side of a parallelogram or trapezoid; from the base to the opposite face of a prism or cylinder; or from the base to the opposite vertex of a triangle, pyramid, or cone. *See also* **altitude.**

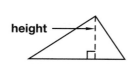 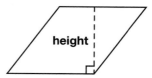 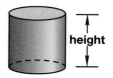

histogram
histograma
(53)

A method of displaying a range of data. A histogram is a special type of bar graph that displays quantitative data in intervals of equal size with no space between bars.

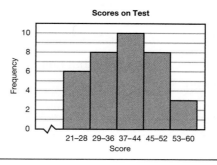

histogram

hypotenuse
hipotenusa
(Inv. 2)

The longest side of a right triangle.

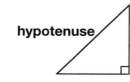

*The **hypotenuse** of a right triangle is always the side opposite the right angle.*

I

identity property of addition
propiedad de identidad de la suma
(1)

The sum of any number and 0 is equal to the initial number. In symbolic form, $a + 0 = a$. The number 0 is referred to as the *additive identity*.

*The **identity property of addition** is shown by this statement:*
$$13 + 0 = 13$$

identity property of multiplication
propiedad de identidad de la multiplicación
(1)

The product of any number and 1 is equal to the initial number. In symbolic form, $a \times 1 = a$. The number 1 is referred to as the *multiplicative identity*.

*The **identity property of multiplication** is shown by this statement:*
$$94 \times 1 = 94$$

independent variable
variable independiente
(69)

A variable in an equation whose value can be freely chosen.
See also **dependent variable.**

*In $y = 2x + 1$, the **independent variable** is x.*

improper fraction
fracción impropia
(10)

A fraction with a numerator equal to or greater than the denominator.

$\frac{12}{12}$, $\frac{57}{3}$, and $2\frac{13}{2}$ *are **improper fractions.***
*All **improper fractions** have an absolute value greater than or equal to 1.*

independent events *eventos independientes* *(68)*	Two events are *independent* if the outcome of one event does not affect the probability that the other event will occur. *If a number cube is rolled twice, the outcome of the first roll does not affect the probability of any outcome on the second roll. The first and second rolls are **independent events.***
index *índice* *(15)*	A number associated with a radical sign that indicates a root of a number. *The small 5 in $\sqrt[5]{32}$ is the **index** of the root. $\sqrt[5]{32}$ means the fifth root of 32, which equals 2.*
indirect measure *medida indirecta* *(65)*	A measurement technique that uses a proportional relationship between given measures to calculate an unknown measure. ***Indirect measurement** is used when an object or distance cannot be measured with a measurement tool such as a ruler. For example, we can use an **indirect measure** to find the height of a flag pole or the distance across a lake.*
inequalities *desigualdades* *(62)*	A mathematical statement that has $<$, $>$, $\leq$, or $\geq$ as a symbol of comparison. $x \leq 4 \qquad 2 < 7 \qquad\qquad x = 2 \qquad 9 + 10$ **inequalities** $\qquad\qquad\qquad$ not **inequalities**
inscribed *inscrito* *(maintenance)*	In geometry, one figure drawn within another figure; an angle with its vertex on a circle and with sides that are chords. **inscribed** $\qquad$ **inscribed** **hexagon** $\qquad$ **angle**
integers *números positivos, negativos y el cero* *(1)*	The set of counting numbers, their opposites, and zero; the members of the set $\{\dots, -2, -1, 0, 1, 2, \dots\}$. *$-57$ and 4 are **integers**. $\frac{15}{8}$ and -0.98 are not **integers**.*
intercept *intersección* *(56)*	The point at which the graph of an equation intersects an axis, especially the *y*-intercept.
interest *interés* *(Inv. 10)*	An additional amount added to a loan, account, or fund, usually based on a percentage of the principal; the difference between the principal and the total amount owed (with loans) or earned (with accounts and investment funds). *If we borrow $500.00 from the bank and repay the bank $575.00 for the loan, the **interest** on the loan is $575.00 − $500.00 = $75.00.*
interest rate *tasa de interés* *(Inv. 10)*	A percent that determines the amount of interest paid on a loan over a period of time. *If we borrow $1000.00 and pay back $1100.00 after one year, our **interest rate** is* $$\frac{\$1100.00 - \$1000.00}{\$1000.00} \times 100\% = 10\% \text{ per year}$$

| **interior angle**
ángulo interno
(maintenance) | An angle that opens to the inside of a polygon.

<div align="center">**interior angle**</div> |

| **International System**
Sistema internacional
(6) | *See* **metric system.** |

| **intersection**
intersección
(Inv. 1, 90) | **1.** A common point or points shared by two geometric figures.

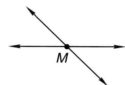
M

*These two lines **intersect**.*
They share the common point M.

2. The common elements of two or more sets denoted by the symbol ∩.

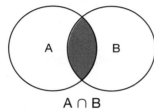
A   B
 A ∩ B
*This diagram shows the **intersection** of sets A and B.* |

| **inverse operations**
operaciones inversas
(38) | Operations that "undo" one another.

$a + b - b = a$ *Addition and subtraction are*
$a - b + b = a$ ***inverse operations.***

$a \times b \div b = a$ $(b \neq 0)$ *Multiplication and division are*
$a \div b \times b = a$ $(b \neq 0)$ ***inverse operations.***

$\sqrt{a^2} = a$ $(a \geq 0)$ *Raising to powers and finding*
$(\sqrt{a})^2 = a$ $(a \geq 0)$ *roots are **inverse operations.*** |

| **inverse variation**
variación inversa
(99) | A relationship between two variables in which the product of the two variables is a constant.

$xy = k$ *represents an **inverse variation** between x and y. As one variable increases, the other variable decreases so that the product k is always the same.*

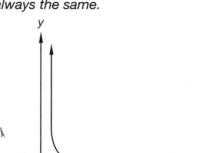 |

irrational numbers *números irracionales* (16)	Numbers that cannot be expressed as a ratio of two integers. Their decimal expansions are nonending and nonrepeating. π *and* $\sqrt{3}$ *are* ***irrational numbers.***
isosceles triangle *triángulo isósceles* (20)	A triangle with at least two sides of equal length. At least two angles of an isosceles triangle are also equal. *Two of the sides of this* ***isosceles triangle*** *have equal lengths.*

L

lateral surface area *área superficial lateral* (43)	The combined area of the surfaces on the sides of a geometric solid excluding the areas of the bases. 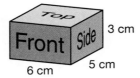 *Area of front* $= 3\ cm \times 6\ cm = 18\ cm^2$ *Area of back* $= 3\ cm \times 6\ cm = 18\ cm^2$ *Area of side* $= 3\ cm \times 5\ cm = 15\ cm^2$ $+$ *Area of side* $= 3\ cm \times 5\ cm = 15\ cm^2$ **Lateral surface area** $= 66\ cm^2$
laws of exponents *leyes de exponentes* (27)	Rules that describe relationships between exponents for certain operations. **Laws of Exponents for Multiplication and Division** $x^a \cdot x^b = x^{a+b}$ $\dfrac{x^a}{x^b} = x^{a-b}$ for $x \neq 0$ $(x^a)^b = x^{a \cdot b}$ $x^0 = 1$ for $x \neq 0$ $x^{-a} = \dfrac{1}{x^a}$ for $x \neq 0$
least common denominator (LCD) *mínimo común denominador (mcd)* (13)	The least common multiple of the denominators of two or more fractions. *The* ***least common denominator*** *of* $\frac{5}{6}$ *and* $\frac{3}{8}$ *is the least common multiple of 6 and 8, which is 24.*
least common multiple (LCM) *mínimo común múltiplo (mcm)* (maintenance)	The smallest whole number that is a multiple of two or more given numbers. *Multiples of 6 are 6, 12, 18, 24, 30, 36, ...* *Multiples of 8 are 8, 16, 24, 32, 40, 48, ...* *The* ***least common multiple*** *of 6 and 8 is 24.*

legs *catetos* *(Inv. 2)*	The two shorter sides of a right triangle that form a 90° angle at their intersection. 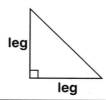 *Each **leg** of this right triangle is shorter than the hypotenuse.*
like terms *términos semejantes* *(31)*	Algebraic terms with identical variable factors. $-3x^2y \quad 2yx^2 \qquad\qquad -2x^2y \quad 3xy^2$ **like terms** \qquad\qquad not **like terms**
line *línea* *(18)*	A straight path that extends without end in both directions. $A \qquad\qquad\qquad\qquad B$ **line** *AB* or **line** *BA*
linear equation *ecuación lineal* *(41)*	An equation whose graph is a line. 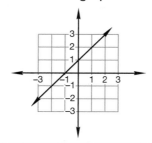 $y = x + 1$ is a ***linear equation*** *because its graph is a line.*
linear inequality *desigualdad lineal* *(A106)*	A mathematical sentence using $<$, $>$, $\le$, or $\ge$ whose graph is bounded by a straight line.
linear pair *par lineal* *(18)*	Two adjacent angles that together form a straight line. 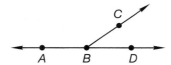 $\angle ABC$ and $\angle CBD$ are a ***linear pair.*** *The sum of their measures is 180°.*
line of best fit *línea de mejor ajuste* *(Inv. 8)*	A straight line on a scatter plot that best approximates the relationship between two sets of data. 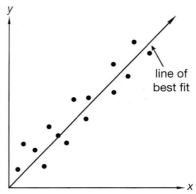
line of symmetry *eje de simetría* *(Inv. 3)*	*See* **Reflective symmetry.**

line plot *diagrama de puntos* (53)	A graph that displays the spread of a data set on a number line.

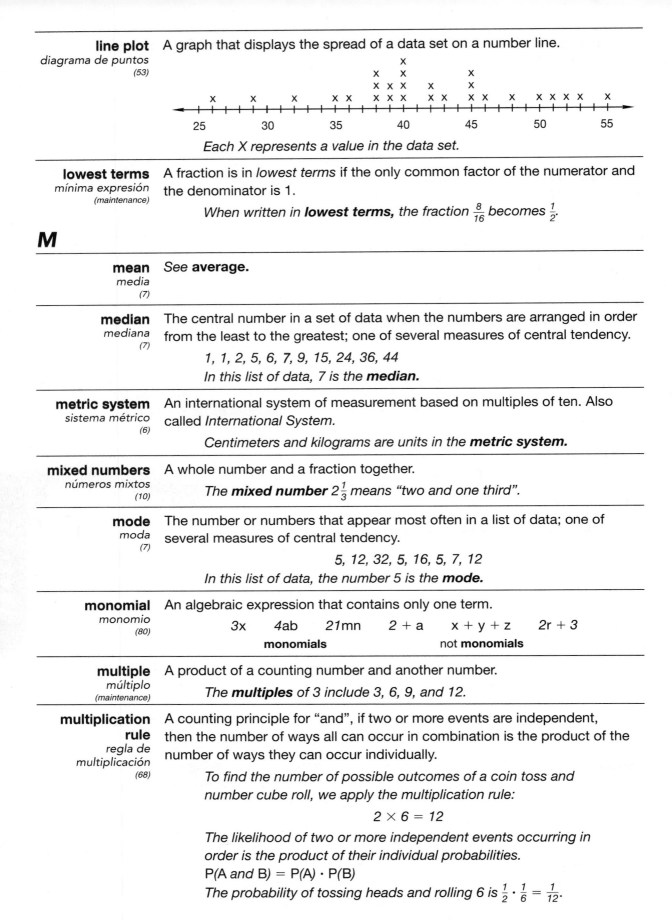

Each X represents a value in the data set.

lowest terms *mínima expresión* (maintenance)	A fraction is in *lowest terms* if the only common factor of the numerator and the denominator is 1. *When written in **lowest terms,** the fraction $\frac{8}{16}$ becomes $\frac{1}{2}$.*

M

mean *media* (7)	*See* **average.**
median *mediana* (7)	The central number in a set of data when the numbers are arranged in order from the least to the greatest; one of several measures of central tendency. *1, 1, 2, 5, 6, **7**, 9, 15, 24, 36, 44* *In this list of data, 7 is the **median.***
metric system *sistema métrico* (6)	An international system of measurement based on multiples of ten. Also called *International System.* *Centimeters and kilograms are units in the **metric system.***
mixed numbers *números mixtos* (10)	A whole number and a fraction together. *The **mixed number** $2\frac{1}{3}$ means "two and one third".*
mode *moda* (7)	The number or numbers that appear most often in a list of data; one of several measures of central tendency. *5, 12, 32, **5**, 16, **5**, 7, 12* *In this list of data, the number 5 is the **mode.***
monomial *monomio* (80)	An algebraic expression that contains only one term. $3x \qquad 4ab \qquad 21mn \qquad\qquad 2+a \qquad x+y+z \qquad 2r+3$ **monomials** $\qquad\qquad\qquad$ not **monomials**
multiple *múltiplo* (maintenance)	A product of a counting number and another number. *The **multiples** of 3 include 3, 6, 9, and 12.*
multiplication rule *regla de multiplicación* (68)	A counting principle for "and", if two or more events are independent, then the number of ways all can occur in combination is the product of the number of ways they can occur individually. *To find the number of possible outcomes of a coin toss and number cube roll, we apply the multiplication rule:* $$2 \times 6 = 12$$ *The likelihood of two or more independent events occurring in order is the product of their individual probabilities.* *P(A and B) = P(A) · P(B)* *The probability of tossing heads and rolling 6 is $\frac{1}{2} \cdot \frac{1}{6} = \frac{1}{12}$.*

multiplicative identity *identidad multiplicativa* *(1)*	The number 1. *See also* **identity property of multiplication.** $$-2 \times 1 = -2$$ ↑ **multiplicative identity** *The number 1 is called the **multiplicative identity** because multiplying any number by 1 does not change the number.*
multiplicative inverse *inverso multiplicativo* *(22)*	*See* **reciprocal.**

N

natural numbers *números naturales* *(1)*	*See* **counting numbers.**
net *malla* *(55)*	A two-dimensional image of the surfaces of a solid. 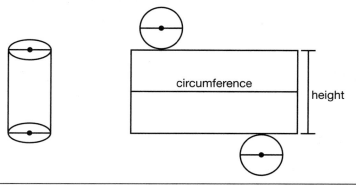
numerator *numerador* *(5)*	The top term of a fraction. $\dfrac{9}{10}$ ← **numerator** ← denominator

O

obtuse angle *ángulo obtuso* *(18)*	An angle whose measure is between 90° and 180°. 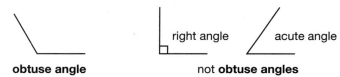 *An **obtuse angle** is larger than both a right angle and an acute angle.*
obtuse triangle *triángulo obtusángulo* *(20)*	A triangle whose largest angle measures between 90° and 180°. 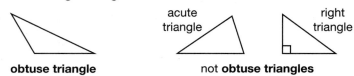
odd numbers *números impares* *(1)*	Whole numbers not divisible by 2. *The set of **odd numbers** includes {..., −3, −1, 1, 3, ...}.*

odds *posibilidad* *(32)*	A way of describing the likelihood of an event; the ratio of favorable outcomes to unfavorable outcomes. *If you roll a number cube, the **odds** of getting a 3 are 1 to 5, or 1:5.*
opinion survey *operaciones* *aritméticas* *(Inv. 6)*	A survey in a free-response format without predetermined choices. *An **opinion survey** about food preferences might ask, "What is your favorite food?"*
opposite angles *ángulos opuestos* *(54)*	*See **vertical angles.***
opposites *opuestos* *(1)*	Two numbers whose sum is zero; a positive number and a negative number whose absolute values are equal. $$(-3) + (+3) = 0$$ *The numbers +3 and −3 are **opposites.***

origin *origen* *(1)*	**1.** The location of the number 0 on a number line.  **origin** on a number line **2.** The point (0, 0) on a coordinate plane. 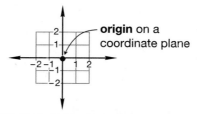

outlier *valor lejano* *(Inv. 8)*	A number in a list of data that is distant from the other numbers in the list. *1, 5, 4, 3, 6, 28, 7, 2* *In this list, the number 28 is an **outlier** because it is distant from the other numbers in the list.*

P

parabola *parábola* *(Inv. 11)*	A U-shaped curve characteristic of the graphs of quadratic equations. 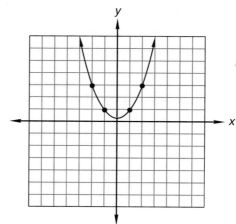 *The graph of $y = x^2$ is a **parabola.***

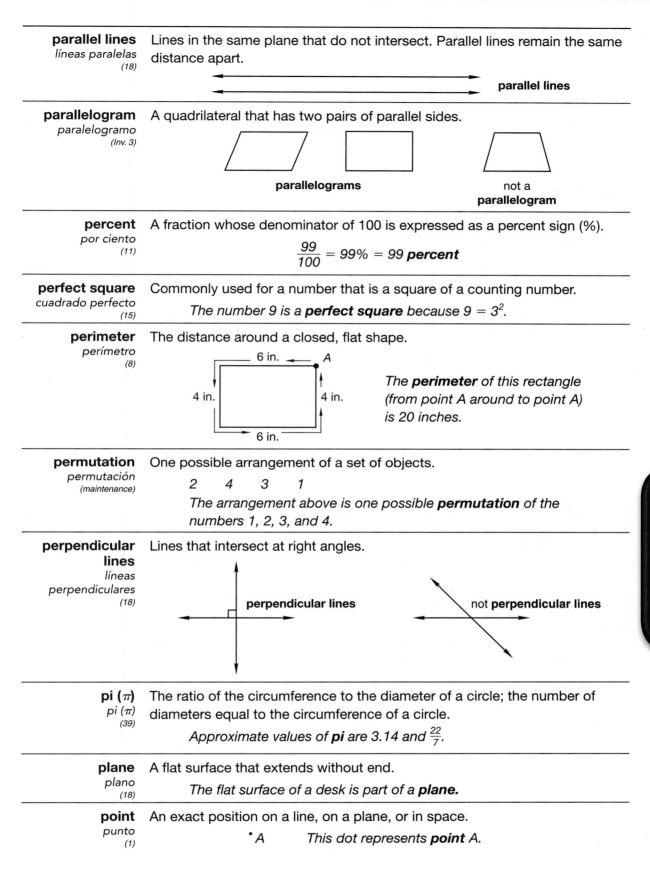

parallel lines *líneas paralelas* (18)	Lines in the same plane that do not intersect. Parallel lines remain the same distance apart. parallel lines
parallelogram *paralelogramo* (Inv. 3)	A quadrilateral that has two pairs of parallel sides. **parallelograms** not a **parallelogram**
percent *por ciento* (11)	A fraction whose denominator of 100 is expressed as a percent sign (%). $$\frac{99}{100} = 99\% = 99\ \textit{percent}$$
perfect square *cuadrado perfecto* (15)	Commonly used for a number that is a square of a counting number. *The number 9 is a **perfect square** because $9 = 3^2$.*
perimeter *perímetro* (8)	The distance around a closed, flat shape. 6 in. *A* 4 in. 4 in. 6 in. *The **perimeter** of this rectangle (from point A around to point A) is 20 inches.*
permutation *permutación* (maintenance)	One possible arrangement of a set of objects. 2 4 3 1 *The arrangement above is one possible **permutation** of the numbers 1, 2, 3, and 4.*
perpendicular lines *líneas perpendiculares* (18)	Lines that intersect at right angles. **perpendicular lines** not **perpendicular lines**
pi (π) *pi (π)* (39)	The ratio of the circumference to the diameter of a circle; the number of diameters equal to the circumference of a circle. *Approximate values of **pi** are 3.14 and $\frac{22}{7}$.*
plane *plano* (18)	A flat surface that extends without end. *The flat surface of a desk is part of a **plane**.*
point *punto* (1)	An exact position on a line, on a plane, or in space. •*A* *This dot represents **point** A.*

GLOSSARY

Glossary   **881**

polygon *polígono* (19)	A closed, plane figure with straight sides. polygons not **polygons**
polyhedron *poliedro* (Inv. 4)	A geometric solid whose faces are polygons. **polyhedrons** not **polyhedrons** cube triangular prism pyramid sphere cylinder cone
polynomial *polinomio* (80)	An algebraic expression that has one or more terms. *The expression $3x^2 + 13x + 12y$ is a **polynomial.***
population *población* (Inv. 6)	A group of people that researchers want to study to make predictions or draw conclusions. *To analyze the food preferences of teenagers, researchers would study the **population** of people between the ages of 13 and 19.*
power *potencia* (15)	**1.** The value of an exponential expression. *16 is the fourth **power** of 2 because $2^4 = 16$.* **2.** An exponent. *The expression 2^4 is read "two to the fourth **power**."*
prime factorization *factorización prima* (9)	The expression of a composite number as a product of its prime factors. *The **prime factorization** of 60 is $2 \times 2 \times 3 \times 5$.*
prime factors *factores primos* (74)	The factors of a number that are prime numbers. *The factors of 45 are 1, 3, 5, 9, 15, and 45. Its **prime factors** are 3 and 5.*
prime number *número primo* (9)	A counting number greater than 1 whose only two factors are the number 1 and itself. *7 is a **prime number**. Its only factors are 1 and 7.* *10 is not a **prime number**. Its factors are 1, 2, 5, and 10.*
principal *capital* (Inv. 10)	The amount of money borrowed in a loan, deposited in an account that earns interest, or invested in a fund. *If we borrow $750.00, our **principal** is $750.00.*
prism *prisma* (Inv. 4)	A polyhedron with two congruent parallel bases. rectangular **prism** triangular **prism**

probability *probabilidad* (32)	The likelihood that a particular event will occur. *See also* **theoretical probability** and **experimental probability.** The **probability** of rolling a 3 with a standard number cube is $\frac{1}{6}$.
proof *prueba* (Inv. 12)	A method that uses logical steps to describe how certain given information can lead to a certain conclusion. *For an example of a **proof,** please see the **proof** of the Pythagorean theorem in Investigation 12.*
proportion *proporción* (34)	A statement that two ratios are equal. $$\frac{6}{10} = \frac{9}{15}$$ *These two ratios are equal, so this is a **proportion.***
protractor *transportador* (18)	A tool that is used to measure and draw angles. 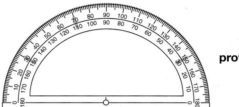 protractor
Pythagorean theorem *teorema de Pitágoras* (Inv. 2)	The area of a square constructed on the hypotenuse of a right triangle is equal to the sum of the areas of squares constructed on the legs of the right triangle expressed as a formula, $c^2 = a^2 + b^2$. 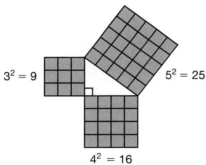 $3^2 = 9$ $5^2 = 25$ $4^2 = 16$ $5^2 = 4^2 + 3^2$ $25 = 16 + 9$ $25 = 25$

Q

quadrant *cuadrante* (Inv. 1)	A region of a coordinate plane formed when two perpendicular number lines intersect at their origins. 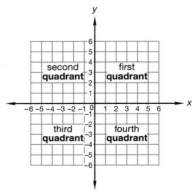

quadratic equation *ecuación cuadrática* (93)	An equation containing a variable with an exponent of 2 (but no greater exponents). $2x^2 + 7x = 45$ and $x^2 = 25$ are **quadratic equations.**
qualitative data *datos cualitativos* (Inv. 6)	Data that falls into categories. *People's preferences for types of restaurants, books, or movies are examples of **qualitative data**.*
quantitative data *datos cuantitavos* (Inv. 6)	Data that is numerical. *The number of people who see a movie each night of the week is an example of **quantitative data**.*
quartile *cuartil* (103)	*See* **box-and-whisker plot.**

R

radicand *radical* (15)	The number under a radical sign. *See also* **radical expression.**
radical expression *expresión radical* (15)	An expression that indicates the root of a number. A radical expression contains a radical sign, $\sqrt{\ }$. 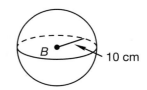 $$\sqrt{15^2} \qquad\qquad \sqrt{9} \qquad\qquad 2 + 4 \qquad\qquad 16$$ $$\sqrt{x} \qquad\qquad 2 + \sqrt{13} \qquad\qquad xy \qquad\qquad 4133$$ **radical expressions** not **radical expressions**
radius *radio* (39)	(Plural: *radii*) The distance from the center of a circle or sphere to a point on the circle or sphere. *The **radius** of circle A is 2 inches.* *The **radius** of sphere B is 10 centimeters.*
random *al azar* (Inv. 9)	Descriptive of a selection process for a sample in which each member of the population has an equal likelihood of being chosen. *Participants in the survey were chosen at **random**.*
range *intervalo* (7, 47)	**1.** The difference between the greatest number and least number in a set of data. *5, 17, 12, 34, 29, 13* *To calculate the **range** of this list, we subtract the least number from the greatest number. The **range** of this list is 29 because $34 - 5 = 29$.* **2.** The set of outputs of a function. *In the function $y = 3x$, the values of y are the **range**.*
rate *tasa* (7)	A division relationship between two measures. *If a car travels 240 miles in 4 hours, its average **rate** is 240 miles ÷ 4 hours, which equals 60 miles per hour (mph).*

rate of change *tasa de cambio* *(44)*	How quickly one variable changes as a related variable changes; the slope. *The inflation rate is the **rate of change** of the cost of living per year.*
ratio *razón* *(29)*	A comparison of two numbers by division. △ △ △ ☆ ☆ ☆ ☆ ☆ ☆ *There are 3 triangles and 6 stars. The **ratio** of triangles to stars is 3:6 (or 1:2), which is read as "3 to 6" (or "1 to 2").*
rational numbers *números racionales* *(10)*	All numbers that can be written as a ratio of two integers. $\frac{15}{16}$ *and 37 are **rational numbers.*** $\sqrt{2}$ *and π are not **rational numbers.***
ray *rayo* *(18)*	A part of a line that begins at a point and extends without end in one direction. A ———————————— B **ray** *AB*
real numbers *números reales* *(16)*	All the numbers that can be represented by points on a number line. *The set of **real numbers** is composed of all rational and irrational numbers.*
reciprocal *recíproco* *(22)*	The result of inverting a number; also called **multiplicative inverse.** *The **reciprocal** of $\frac{3}{4}$ is $\frac{4}{3}$.* *The product of a number and its **reciprocal** is 1.* $\frac{3}{4} \times \frac{4}{3} = \frac{12}{12} = 1$
rectangle *rectángulo* *(Inv. 3)*	A quadrilateral that has four right angles. **rectangles** not **rectangles**
recursive formula *fórmula recursiva* *(97)*	A formula for a sequence that expresses the value of any term by reference to the preceding term. $A_n = a_{n-1} + 3$ *is a **recursive formula** because each term equals the preceding term plus 3.*
reduce *reducir* *(10)*	To rewrite a fraction in lowest terms. *If we **reduce** the fraction $\frac{9}{12}$, we get $\frac{3}{4}$.*
reflection *reflexión* *(26)*	A transformation by flipping a figure to produce a mirror image. A reflection does not change the size of a figure. **reflection**

GLOSSARY

reflective symmetry *simetría de reflexión* *(Inv. 3)*	A characteristic of a figure that can be divided into mirror images by a line or plane; also known as line symmetry. Has **reflective symmetry**　　　Does not have **reflective symmetry**
regular polygon *polígono regular* *(19)*	A polygon in which all sides have equal lengths and all angles have equal measures. **regular polygons**　　　　　not **regular polygons**
relation *relación* *(98)*	The pairing of two sets of data. Set notation, a table of pairs of values, an equation, or a graph can express a relation. *A table of values that pairs the number of tickets purchased with a purchase price shows a **relation.***
repetend *período decimal* *(30)*	The repeating digits of a decimal number. The symbol for a repetend is an overbar. $$0.83333333... = 0.8\overline{3}$$ *In the number above, 3 is the **repetend.***
representative *representativõ* *(Inv. 9)*	Having the same characteristics as a larger population. *A **representative** sample of a community would include a range of participants proportional to the range in the population.*
rhombus *rombo* *(Inv. 3)*	A parallelogram with all four sides of equal length. 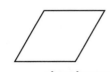 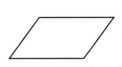 **rhombuses** (or rhombi)　　　　　not **rhombuses**
right angle *ángulo recto* *(18)*	An angle that forms a square corner and measures 90°. It is often marked with a small square. **right angle**　　　　　not **right angles** *A **right angle** is larger than an acute angle and smaller than an obtuse angle.*
right solid *sólido recto* *(maintenance)*	A geometric solid whose sides are perpendicular to its base. 　**right solids**

right triangle *triángulo rectángulo* *(20)*	A triangle whose largest angle measures 90°.

right triangle   not **right triangles**

root *raíz* *(15)*	A value of a radical expression. $$\sqrt{16} = 4$$ *4 is a **root** of this radical expression.*
round *redondear* *(17)*	To find a nearby number that ends with one or more zeros. The mathematical procedure for rounding involves place value. *When we **round** 188 to 190, we are **rounding** to the nearest ten.*
rotation *rotación* *(26)*	A transformation by turning a figure about a specified point called the *center of rotation*. A rotation does not change the size of a figure.

rotation

rotational symmetry *simetría rotacional* *(Inv. 3)*	A characteristic of a figure that its image reappears more than once as the figure is rotated a full turn.

*Every parallelogram has **rotational symmetry**.*

S

sample *ejemplo* *(Inv. 6)*	A smaller part of a larger population that is representative of the larger population. *Identifying every 10th person who enters a theater gives a **sample** of the entire audience.*
sample space *espacio muestral* *(32)*	The collection of all the possible outcomes of an event. *The **sample space** of outcomes when flipping a coin consists of heads and tails.*
scale drawing *dibujo a escala* *(87)*	A drawing on which small units of measure correspond to larger units of measure on an actual object or physical area. *If the **scale drawing** of a city shows a distance that is 2 inches long, and the scale is 1 inch = 1 mile, then the actual distance is 2 miles long.*

scale factor	The constant that relates the ratios formed by corresponding sides of similar
factor de escala	polygons.
(26)	

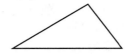

25 mm

10 mm

4 mm

10 mm

*The **scale factor** from the smaller rectangle to the larger rectangle is 2.5.*

scalene triangle	A triangle with three sides of different lengths. All three angles also have
triángulo escaleno	different measures.
(20)	

*All three sides of this **scalene triangle** have different lengths.*

scatterplot	A graph showing data points on a coordinate plane that represent pairs of
diagrama de	related values from two sets of data.
dispersión	
(Inv. 8)	

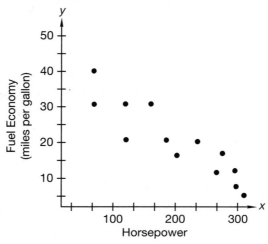

*This **scatterplot** shows data collected on the number of mid-size cars relating to the horsepower of the engine and the gas milage of the car in city driving.*

scientific notation	A method of writing a number as a product of a decimal number between
notación científica	1 and 10 and a power of 10.
(28)	

*In **scientific notation**, 34,000 is written 3.4×10^4.*

sector	A portion of the interior of a circle enclosed by an arc and two radii of
sector	a circle.
(40)	

*This circle is divided into 3 **sectors**.*

segment	A part of a line with two distinct endpoints.
segmento	
(18)	

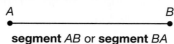

A   B

segment *AB* or **segment** *BA*

semicircle
semicírculo
(40)

An arc that forms one half of a circle.

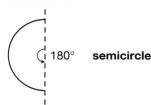

180° **semicircle**

*A **semicircle** is an arc whose measure is 180°.*

---

sequence
secuencia
(1)

An ordered list of numbers, called terms, that follow a certain pattern or rule.

*The numbers 2, 4, 6, 8, ... form a **sequence.** The rule is "count up by twos."*

---

set
conjunto
(1)

A collection of elements, such as numbers, ordered pairs, variables, or polygons denoted with bracess { }.

*The **set** of whole numbers is expressed as {0, 1, 2, 3, 4, ...}.*

---

significant digits
cifras significativas
(117)

The number of digits that can be used with confidence when expressing or calculating with measurements; indicating the precision of a measurement.

---

similar
semejante
(19)

Having the same shape but not necessarily the same size. Corresponding parts of similar figures are proportional.

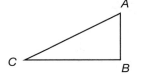

*△ABC and △DEF are **similar.** They have the same shape, but not the same size.*

---

simple interest
interés simple
(Inv. 10)

Interest paid on the principal only.

*The formula for calculating **simple interest** is I = P × r × t, where P is the principal, r is the rate of interest, and t is the time in years.*

---

sine
seno
(118)

In a right triangle, the ratio of the opposite leg to the hypotenuse.

$$\textbf{sin}\ \angle A = \frac{6}{10} = 0.6$$

---

skew lines
lineas al sesgo
(18)

Lines that are in different planes that do not intersect.

---

slant height
altura inclinada
(100)

In a pyramid or cone, the diagonal distance along the surface from the apex to the base.

slant heights

slope *pendiente* *(44)*	The number that represents the slant of the graphed line. The slope is the ratio of the **rise** to the **run** between any two points on the graph and indicates the rate of change.

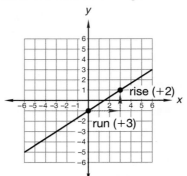

*Every vertical increase (rise) of 2 units leads to a horizontal increase (run) of 3 units, so the **slope** of this line is $\frac{2}{3}$.*

slope-intercept form *forma pendiente-intersección* *(56)*	The form $y = mx + b$ for a linear equation. In this form m is the slope of the graph of the equation, and b is the y-intercept. *In the equation y = 2x + 3, the **slope** is 2 and the **y-intercept** is 3.*
solid *sólido* *(1)*	See **geometric solid**.
solution *solución* *(14)*	The number or numbers that satisfy an equation (make the equation true). *In the equation 4x = 20, the **solution** is 5 because 4 × 5 = 20.*
solution set *conjunto solución* *(1)*	All possible numbers that satisfy an inequality. *The **solution set** for 2x > 4 includes all numbers greater than 2. So 3 is part of the **solution set** and 2 is not.*
sphere *esfera* *(111)*	A round geometric surface whose points are all an equal distance from its center.

sphere

square *cuadrado* *(maintenance)*	**1.** A rectangle with all four sides of equal length.

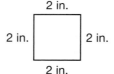

*All four sides of this **square** are 2 inches long.*

2. The product of a number and itself.

*The **square** of 4 is 16.*

square root *raíz cuadrada* *(15)*	One of two equal factors of a number. The symbol for the principal, or positive, square root of a number is $\sqrt{}$. *A **square root** of 49 is 7 because 7 × 7 = 49.*
standard form *forma estándar* *(82)*	The form $Ax + By = C$ for a linear equation. *The equation −2x + y = 5 is in **standard form.***

statistics *estadística* *(Inv. 6)*	The science of collecting data and interpreting the data in order to draw conclusions or make predictions.
stem-and-leaf plot *diagrama de tallo y hojas* *(maintenance)*	A method of graphing a collection of numbers by placing the "stem digits" (or initial digits) in one column and the "leaf" digits (or remaining digits) out to the right.

Stem	Leaf
2	1 3 5 6 6 8
3	0 0 **2** 2 4 5 6 6 8 9
4	0 0 1 1 1 2 3 3 5 7 7 8
5	0 1 1 2 3 5 8

*In this **stem-and-leaf plot**, 3|2 represents 32.*

step function *función escalón* *(47)*	A type of function whose graph has a stair step pattern.

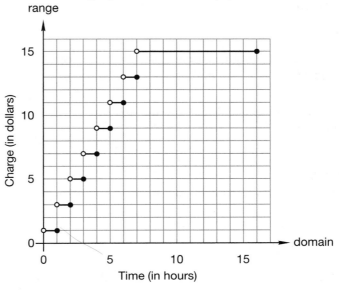

*This graph shows charges for certain time intervals as a **step function.***

straight angle *ángulo llano* *(18)*	An angle that measures 180° and thus forms a straight line.

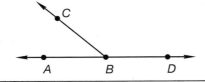

Angle ABD is a **straight angle.**
Angles ABC and CBD are
not **straight angles.**

subset *subconjunto* *(90)*	A part of a set denoted with the symbol ⊂.

*The set of rectangles is a **subset** of the set of quadrilaterals.*
{rectangles} ⊂ {quadrilaterals}

supplementary angles *ángulos suplementarios* *(54)*	Two angles whose measures total 180°.

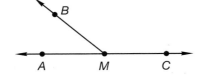

∠AMB and ∠CMB are
supplementary.

surface area *área superficial* *(43)*	The combined area of the surfaces of a geometric solid.

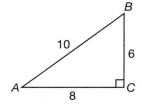

Area of top = 5 cm × 6 cm = 30 cm²
Area of bottom = 5 cm × 6 cm = 30 cm²
Area of front = 3 cm × 6 cm = 18 cm²
Area of back = 3 cm × 6 cm = 18 cm²
Area of side = 3 cm × 5 cm = 15 cm²
+ Area of side = 3 cm × 5 cm = 15 cm²
Total **surface area** = 126 cm²

---

symmetry *simetría* *(Inv. 3)*	*See* **reflective symmetry** and **rotational symmetry.**

---

system of equations *sistema de ecuaciones* *(89)*	Two or more equations with common variables. $\begin{cases} x + y = 2 \\ 3x + 2y = 5 \end{cases}$ **system of equations**

T

---

tangent *tangente* *(118)*	In a right triangle, the ratio of the opposite leg to the adjacent leg. $\tan \angle A = \frac{6}{8} = 0.75$

---

term *término* *(31, 61)*	**1.** A number that serves as a numerator or denominator of a fraction. $\dfrac{5}{6}$ > **terms** **2.** One of the numbers in a sequence. *1, 3, 5, 7, 9, 11, ...* *Each number in this sequence is a* **term.** **3.** A constant or variable expression composed of one or more factors in an algebraic expression. *The expression* 2x + 3xyz *has two* **terms,** 2x *and* 3xyz.

---

theoretical probability *probabilidad teórica* *(32)*	Probability of an outcome that is found by analyzing a situation; the ratio of favorable outcomes to all possible outcomes. *The* **theoretical probability** *of tossing heads with a single coin toss is 1 out of 2 because there are 2 outcomes, heads and tails, and both are equally likely, but only 1 is a favorable outcome.*

---

transformation *transformación* *(26)*	The changing of a figure's position or form. *See also* **rotation, reflection,** and **translation.**

Transformations

Movement	Name
flip	reflection
slide	translation
turn	rotation

translation *traslación* *(26)*	A transformation by sliding a figure from one position to another without turning or flipping the figure. A translation does not change the size of a figure.

transversal *transversal* *(54)*	A line that intersects one or more other lines in a plane.

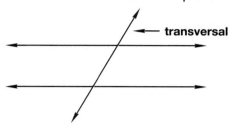

trapezoid *trapecio* *(Inv. 3)*	A quadrilateral with exactly one pair of parallel sides.

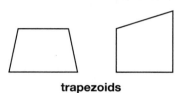

trapezoids   not **trapezoids**

tree diagram *diagrama de árbol* *(32)*	A way to organize and illustrate possible combinations.

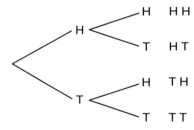

*This **tree diagram** identifies the four possible outcomes of tossing a coin twice.*

triangle *triángulo* *(20)*	A polygon with three sides.

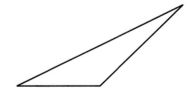

trinomial *trinomio* *(80)*	A polynomial with three terms. *The expression $x^2 + 5x + 4$ is a **trinomial.***

trigonometry *trigonometría* *(112)*	The branch of mathematics that explores relationships between the angles and sides of triangles.

GLOSSARY

union *unión* (90)	The combining of two sets, denoted by the symbol ∪. 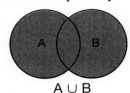 A ∪ B *The shaded Venn diagram illustrates the **union** of sets A and B.*
unit conversion *conversión de unidades* (52)	The process of changing a measure to an equivalent measure that has different units. *Through **unit conversion,** we can write 2 feet as 24 inches.*
unit multiplier *factor de conversión* (52)	A ratio equal to 1 that is composed of two equivalent measures expressed in different units. $$\frac{12\ inches}{1\ foot} = 1$$ *We can use this **unit multiplier** to convert feet to inches.*
unit price *precio unitario* (maintenance)	The price of one unit of measure of a product. *The **unit price** of bananas is $1.19 per pound.*
unit rate *tasa unitaria* (7)	A division relationship between two measures where the second measure equals 1. *65 miles per hour (65 mph) and 15 cents per ounce ($0.15/oz) are **unit rates.***
U.S. Customary System *Sistema usual de EE.UU.* (6)	A system of measurement used almost exclusively in the United States. *Pounds, quarts, and feet are units in the **U.S. Customary System.***

V

variable *variable* (2)	A letter used to represent a number that has not been designated. *In the statement x + 7 = y, the letters x and y are **variables.***
Venn diagram *diagrama de Venn* (Inv. 3)	A way to illustrate sets and subsets. 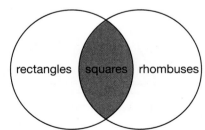 *This **Venn diagram** shows that the set of squares is a subset of both the set of rectangles and the set of rhombuses.*

vertex
vértice
(19)

(Plural: *vertices*) A point of an angle, polygon, or polyhedron where two or more lines, rays, or segments meet.

*One **vertex** of this cube is colored. A cube has eight **vertices.***

vertical angles
ángulos verticales
(54)

A pair of nonadjacent angles formed by a pair of intersecting lines. Vertical angles share the same vertex and are congruent.

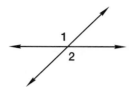

*Angles 1 and 2 are **vertical angles.***

volume
volumen
(42)

The amount of space a solid shape occupies. Volume is measured in cubic units.

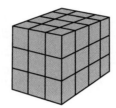

*This rectangular prism is 3 units wide, 3 units high, and 4 units deep. Its **volume** is $3 \cdot 3 \cdot 4 = 36$ cubic units.*

W

whole numbers
números enteros
(1)

The members of the set {0, 1, 2, 3, 4, …}.

*0, 25, and 134 are **whole numbers.***
*-3, 0.56, and $100\frac{3}{4}$ are not **whole numbers.***

X

x-axis
eje de las x
(Inv. 1)

The horizontal number line of a coordinate plane.

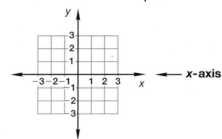

Y

y-axis
eje de las y
(Inv. 1)

The vertical number line of a coordinate plane.

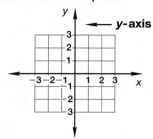

y-intercept
intercepción en y
(56)

The point on a coordinate plane where the graph of an equation intersects the *y*-axis.

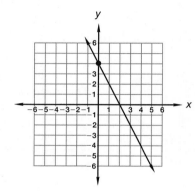

*This line crosses the y-axis at y = 4, so the **y-intercept** of this line is 4.*

Z

zero property of multiplication
propiedad del cero en la multiplicación
(1)

Zero times any number is zero. In symbolic form, $0 \times a = 0$.

*The **zero property of multiplication** tells us that $89 \times 0 = 0$.*

Formulas (cont.)
    surface area, 294–296
        of cones, 377, 666
        of cylinders, 569–570
        of prisms, 569–570
        of pyramids, 665
        of spheres, 734
    transforming, 525–530
    of variation, direct and inverse, 463–464, 658
    volume
        of cones, 575
        of cubes, 288, 508
        of cylinders, 574, 732–733
        of prisms, 288, 574
        of pyramids, 576
        of rectangular prisms, 508
        of spheres, 733
        of triangular prisms, 508–509
Four-step problem-solving process, 6, 12, 26, 47, 54, 72, 92, 97, 120, 139, 146, 153, 176, 202, 250
45°-45°-90° triangles, 447–449
Fractions, 31–35, *See also* Mixed numbers
    adding, 85–90
    comparing, 32, 63–64
    complex, 773–777
    converting
        decimal equivalents, 79–80, 192, 395
        decimal equivalents, non-terminating, 429–430, 722–726
        percent equivalents, 73–75, 82
    decimals versus, for computations, 429, 563–564
    dividing, 146–150, 156
    equal to one, 486, 774
    equivalent, 62–64
    improper, 64–65, 86
    least common denominator, 87
    multiplying, 146–150, 156
    with negative exponents, 431–432
    rates as, 435
    rationalizing denominators, 778–781
    reciprocals of, 148–149, 432
    regrouping, 88
    rewriting mixed numbers as, 154
    simplest form of, 62
    simplifying, 62–63, 148
    solving equations containing, 788–790
    subtracting, 85–90
Function notation, 816–817
Functions
    characteristics of, 652
    defined, 280
    graphing, 280–283, 319–323, 817
    illustrating, 280
    linear, 279–283
    non-linear, 727–730
    proportional, 280–281
    quadratic, 728–729
    relations and, 651–655
    step, 321

G

Geometric figures. *See also* Height of geometric figures
    apex of, 634–636, 664
    bases of. *See* Bases of geometric figures
    height of
        defined, 634
        prisms and cylinders, 507–508
        slant, 634–639
    transformations of, 169–174
Geometric measures with radicals, 640–645
Geometric probability, 675–680
Geometric sequence, 416–417, 681
Geometric solids. *See also specific solids*
    drawing, 271–276
    nets of, 375–378
    similar, 712–716
    surface area of, 294–297, 707–711
    volume of, 287–290, 707–711
Golden Rectangle, 199
Grams (g), 36
Graphing. *See also* Calculators; Graphing calculator
    calculators for, 542
    the coordinate plane, 69–71, 538–541, 831–836
    equations
        slope-intercept form, 382–386, 551, 790–792, 795–797, 851–853
        standard form of linear equations, 550–553
        with two unknowns, 593–598
        union of solutions, 631
        using intercepts, 550–553
    functions, 278–284, 319–323, 817
    on number lines, 8–9
        inequalities, 422–426, 514
        inequality pairs, 629–633
    sequences, 415, 798–800
    systems of inequalities, 840–841
    transformations, 342–345
Graphing calculator. *See also* Calculators.
    Graphing Calculator Activities, references to, 44, 69, 89, 100, 154, 182, 218, 238, 322, 344, 350, 414, 464, 478, 492, 542, 551, 594, 660, 687, 718, 764, 770
Graphs
    arrowheads on lines in, 423, 466
    bar, 412
    best-fit line, 539–544, 654
    box-and-whisker plots, 686–688
    circle, 414
    of direct-variation relationships, 465–466, 586
    histograms, 361, 413
    of inverse-variation relationships, 660
    linear, 465, 586
    line plot, 361, 686–688
    lines on. *See* Lines
    of non-linear functions, 727–730
    points on. *See* Points on a graph
    of proportional relationships, 465–466, 586
    quadratic functions, 728–730
    scatterplots, 538–544, 654, 742–747
    of sequences, 417

Multiplication (cont.)
    commutative property of, 14
    of decimal numbers, 163–167
    distributive property of, 140–141
    of expressions with exponents, 178, 348–349
    of fractions, 146–149, 156
    by fractions vs. decimals, 563–567
    identity property of, 14, 63, 86, 224, 774
    of inequalities, by negative numbers, 514
    of integers, 237–240
    as inverse of division, 251, 759–760
    of measurements, 765
    of mixed numbers, 153–156
    of numbers in scientific notation, 313–316
    and order of operations, 141
    properties of, 14, 15, 140–141
    repeated, with exponents, 98
    rule for dependent events, 558
    of signed numbers, 238–239, 620
    significant digits in, 765
    of small numbers, 389–391
    of square roots, 497–498, 519–524
    symbols for, 14
    of two variables, 658–661
    of variables, by coefficients, 789
    word problems, 26–30
    zero in, 811–813
    zero property of, 14
Multiplication Counting Principle, 458
Multiplication Rule for Probability, 457–460
multiplicative inverse, 148–149
Multiview projection, 274–275

N

Negative coordinates, 68
Negative exponents, 346–349, 431–432
Negative numbers. *See also* Signed numbers
    adding, 204–205, 401
    dividing, 238–239
    multiplying, 238–239
    on number line, 7
    subtracting, 240
Nets, 291, 375–378
Non-linear functions, 727–730
Non-proportional relationships, 585–592
*n*th root, 100
Number lines. *See also* Graphs; Lines
    absolute value and, 8
    graphing inequalities on, 422–426, 515
    graphing inequality pairs on, 629–633
    graphs on, 8–9
    integers on, 8–9, 237
    line plot graphs on, 686–688
    rational numbers on, 61
    real numbers on, 104
    zero on, 758–759
Number pairs
    inequalities, 629–633
    relations, 651
Number sentences, 4
Numerators, 31, 62

O

Obtuse angles, 116, 546
Odd numbers, 9, 41
Odds, 211
Operations
    of arithmetic, 12–18
    order of, 141–143
Opinion surveys, 413
Opposite angles, 367–368
Opposite solutions symbol (±), 624
Order of operations, 141–143
Origin, 7, 68
Outcomes, counting. *See* Probability
Outliers, 539
Output number, 278

P

Parabolic curve, 727
Parallel lines, 115
    angle relationships and, 368–370
Parallelograms, 197, 198
    area of, 406–409, 502
Parallel projection, 273–274
Patterns. *See also* Sequence; Problem-solving
        strategies
    counting, 60
Pentagons, 122
Percent change
    dilation, 479
    of dimensions, 479–483
    fractions vs. decimals for determining, 563–564
    ratio tables in calculating, 452–456
    reduction (contraction), 479
    scale factor in calculating, 479–482
Percents
    converting to
        decimal equivalents, 81–82, 395, 430
        fraction equivalents, 73–76, 82, 395, 430
    problems, solving with equations, 394–396
    of a whole, 326–328
Perfect squares, 100
Perimeter, 48–51
    of circles (circumference), 257–260
    estimating, 110
    percent change calculations, 480–481
    of rectangles, 49, 93, 140
    scaling's effect on, 480–481, 611
    of squares, 93
Perpendicular lines, 115
Perspective, 275–276
Pi (π), 258–259, 264
Place value, 108
Planes, 114
    coordinate. *See* Coordinate plane
Points
    on a coordinate plane,
    finding slope using, 793–795
    linear, 465, 586
    on a number line, 8–9
    outliers, 539, 743

Pyramids (cont.)
 drawing lateral surface area of, 274
 nets of, 375–378
 right, 664–669
 slant height of, 634–639
 surface area, 635, 664–669
 volume of, 574–579
Pythagorean puzzle, 137–138
Pythagorean Theorem, 132–138, 246, 447, 499
 calculating slant height using, 635–636
 converse of, 135
 proof, 782–784
Pythagorean triplet, 134, 135–136

Q

Quadrants, 68
Quadratic equations, 624–626
 factoring, 822–824
 quadratic function, 728
 solving
  by factoring, 829–831, 837–839
  using the quadratic formula, 846–851
Quadrilaterals, 122
 classification of, 197–201
Qualitative data, 412
Quantitative data, 412
Quotients, 13, 601

R

Radicals. *See also* Roots
 adding, 640, 805–806
 in fractions, 778–781
 geometric measures with, 640–645
Radical sign ($\sqrt{}$), 99, 239, 519, 625
Radicand, 99, 519, 805
Radius (radii), 258
Random number generators, 607–608
Random sample, 606–608
Range, 44, 319–321, 361
Rate, 41–43
 of change, 303
 converting with unit multipliers, 435–437
 as ratios, 435
 unit, 41
Rate problems, 330–333, 698
Rational numbers, 60–62, 429–432
 converting to decimals, 81, 193
 and equivalent fractions, 62–64
 as ratios, 429
 selecting appropriate, 563–567
 square roots of, 103–104
 symbol for, 601
Ratios, 186–188, 210
 of areas, 611, 612
 constant, 416, 681
 of dimensions. *See* Scale factor
 of direct-variation variables, 463, 471
 equivalent, 787–788
 horizontal/vertical in graphs, 790
 of lengths, 611, 612
 of percents, 326–328

Ratios (cont.)
 of perimeters, 611
 proportions, 223–225, 318
 in rate problems, 330–333
 rates as, 435
 rational numbers as, 429
 ratio tables, 452–454, 580–581
 reducing equal, 224
 scale factors as. *See* Scale factor
 of side lengths, 231–232
 side lengths of right triangles, 737–741, 768–772
 in similar figures. *See* Similar figures
 slope, 302, 790–793
 in trigonometry, 768
 unit multipliers, 354–356
 word problems, 226
  involving totals, 308–310
Rays, 114
Reading math, 7, 37, 99, 258, 368, 415, 420, 427, 508, 759
Reading mathematical expressions
 "and" in inequalities, 631
 angle measures, 368
 approximately equal to, 37, 258, 508
 "between" in inequalities, 422–423
 ellipsis, 7, 415, 429
 inequalities, 422–423, 629, 631
 infinity, 759
 opposite solutions, 624
 "or" in inequalities, 631
 square inch, 99
 term in a sequence, 416–417
Real numbers, 104, 601
Real-world application problems, 11, 19–23, 25, 27–29, 32–34, 37–40, 42–45, 51–53, 57, 58, 66, 75, 76, 82–84, 89–92, 94–97, 100, 101, 105–108, 110–113, 117, 118, 124, 125, 127, 129–130, 138, 139, 144, 145, 150, 152, 153, 156, 159–161, 163, 164, 166, 167, 174, 175, 178, 180, 183, 187–191, 195, 204, 207–209, 214, 215, 217–222, 224, 226–228, 233, 234, 236–238, 242, 243, 245, 248–250, 253–256, 259–261, 263, 265–267, 269, 270, 279, 281, 284–291, 295, 297–299, 303, 305–312, 315–320, 322–325, 327–329, 332–334, 336, 338, 340, 341, 350, 351, 354–358, 362–367, 370–373, 378–382, 386–389, 391, 392, 394–398, 401–404, 406, 409–411, 419–421, 424, 426–428, 431–433, 436, 437, 440–445, 448–456, 460–462, 466–468, 471–475, 479, 481, 483–491, 494, 495, 499–501, 503–506, 511, 516, 517, 522, 528, 532, 533, 535–537, 546–549, 553–555, 558, 560–562, 564–567, 569–572, 574, 575, 577, 578, 581, 582, 585, 588–592, 596, 597, 603, 604, 610, 611, 613–616, 618, 621, 622, 624, 626–628, 633, 637, 638, 642–645, 650, 654–662, 667, 675, 676, 678–684, 688, 689, 692–695, 697–700, 703–705, 707, 710, 712–714, 718–722, 724, 725, 734, 735, 737, 739, 742, 743, 745, 748, 752, 756, 761, 764–766, 771, 776, 780
Reasonable answers, 110
Reciprocals, 148–149, 347
 of fractions, 432
 negative exponents and, 431–432